T0320603

An Introduction to Continuum Mechanics, Second Edition

This best-selling textbook presents the concepts of continuum mechanics in a simple yet rigorous manner. The book introduces the invariant form as well as the component form of the basic equations and their applications to problems in elasticity, fluid mechanics, and heat transfer and offers a brief introduction to linear viscoelasticity. The book is ideal for advanced undergraduates and beginning graduate students looking to gain a strong background in the basic principles common to all major engineering fields and for those who will pursue further work in fluid dynamics, elasticity, plates and shells, viscoelasticity, plasticity, and interdisciplinary areas such as geomechanics, biomechanics, mechanobiology, and nanoscience. The book features derivations of the basic equations of mechanics in invariant (vector and tensor) form and specification of the governing equations to various coordinate systems, and numerous illustrative examples, chapter summaries, and exercise problems. This second edition includes additional explanations, examples, and problems.

J. N. Reddy is a University Distinguished Professor, Regents Professor, and Oscar S. Wyatt Endowed Chair in the Department of Mechanical Engineering at Texas A&M University. Dr. Reddy is internationally known for his contributions to theoretical and applied mechanics and computational mechanics. He is the author of more than 450 journal papers and 17 books. Dr. Reddy is the recipient of numerous awards, including the Walter L. Huber Civil Engineering Research Prize of the American Society of Civil Engineers, the Worcester Reed Warner Medal and the Charles Russ Richards Memorial Award of the American Society of Mechanical Engineers, the 1997 Archie Higdon Distinguished Educator Award from the American Society of Engineering Education, the 1998 Nathan M. Newmark Medal from the American Society of Civil Engineers, the 2000 Excellence in the Field of Composites from the American Society of Composites, the 2003 Bush Excellence Award for Faculty in International Research from Texas A&M University, and the 2003 Computational Solid Mechanics Award from the U.S. Association of Computational Mechanics. Dr. Reddy received an Honoris Causa from the Technical University of Lisbon, Portugal, in 2009 and an honorary degree from Odlar Yurdu University, Baku, Azerbaijan, in 2011. Dr. Reddy is a Fellow of AIAA, ASCE, ASME, American Academy of Mechanics, the American Society of Composites, the U.S. Association of Computational Mechanics, the International Association of Computational Mechanics, and the Aeronautical Society of India. Dr. Reddy is the Editor-in-Chief of *Mechanics of Advanced Materials and Structures*, *International Journal of Computational Methods in Engineering Science and Mechanics*, and *International Journal of Structural Stability and Dynamics*. He also serves on the editorial boards of more than two dozen other journals, including *International Journal for Numerical Methods in Engineering*, *Computer Methods in Applied Mechanics and Engineering*, and *International Journal of Non-Linear Mechanics*. Dr. Reddy is one of the selective researchers in engineering around the world who is recognized by ISI Highly Cited Researchers with 10,000-plus citations with an H-index of more than 50.

An Introduction to Continuum Mechanics, Second Edition

J. N. REDDY

Texas A & M University

CAMBRIDGE
UNIVERSITY PRESS

CAMBRIDGE
UNIVERSITY PRESS

University Printing House, Cambridge CB2 8BS, United Kingdom

One Liberty Plaza, 20th Floor, New York, NY 10006, USA

477 Williamstown Road, Port Melbourne, VIC 3207, Australia

314-321, 3rd Floor, Plot 3, Splendor Forum, Jasola District Centre, New Delhi - 110025, India

79 Anson Road, #06-04/06, Singapore 079906

Cambridge University Press is part of the University of Cambridge.

It furthers the University's mission by disseminating knowledge in the pursuit of education, learning and research at the highest international levels of excellence.

www.cambridge.org
Information on this title: www.cambridge.org/9781107025431

First edition published 2008
Second edition published 2013

A catalogue record for this publication is available from the British Library

Library of Congress Cataloging in Publication data
Reddy, J. N. (Junuthula Narasimha), 1945–
An introduction to continuum mechanics / J. N. Reddy. – Second edition.
    pages cm
Includes bibliographical references and index.
ISBN 978-1-107-02543-1 (hardback)
1. Continuum mechanics – Textbooks.   I. Title.
QA808.2.R43   2013
531–dc23   2013002793

ISBN 978-1-107-02543-1 Hardback

To
Rohan, Asha, and Mira
Who have filled my life with joy

Contents

List of Symbols

The symbols that are used throughout the book for various important quantities are defined in the following list. In some cases, the same symbol has different meaning in different parts of the book; it should be clear from the context.

$\mathbf{a}$	Acceleration vector, $\frac{D\mathbf{v}}{Dt}$
$\mathbf{A}$	Matrix of normalized eigenvectors [see Eq. (3.9.8)]
$\mathbf{B}$	Left Cauchy–Green deformation tensor (or Finger tensor), $\mathbf{B} = \mathbf{F} \cdot \mathbf{F}^{\mathrm{T}}$; magnetic flux density vector
$\tilde{\mathbf{B}}$	Cauchy strain tensor, $\tilde{\mathbf{B}} = \mathbf{F}^{-\mathrm{T}} \cdot \mathbf{F}^{-1}$; $\tilde{\mathbf{B}}^{-1} = \mathbf{B}$
$B(\ ,\ )$	Bilinear form
c	Specific heat, moisture concentration
$\mathbf{c}$	Couple vector
c_v, c_p	Specific heat at constant volume and pressure
$\mathbf{C}$	Right Cauchy–Green deformation tensor, $\mathbf{C} = \mathbf{F}^{\mathrm{T}} \cdot \mathbf{F}$; fourth-order elasticity tensor [see Eq. (6.3.4)]
C_{ij}	Elastic stiffness coefficients
$\mathbf{d}$	Third-order tensor of piezoelectric moduli
$\mathcal{D}$	Internal dissipation
$d\mathbf{a}$	Area element (vector) in spatial description
$d\mathbf{A}$	Area element (vector) in material description
ds	Surface element in current configuration
dS	Surface element in reference configuration
$d\mathbf{x}$	Line element (vector) in current configuration
$d\mathbf{X}$	Line element (vector) in reference configuration
$\mathbf{D}$	Symmetric part of the velocity gradient tensor, $\mathbf{L} = (\nabla\mathbf{v})^{\mathrm{T}}$; that is, $\mathbf{D} = \frac{1}{2}\left[(\nabla\mathbf{v})^{\mathrm{T}} + \nabla\mathbf{v}\right]$; electric flux vector; mass diffusivity tensor
D/Dt	Material time derivative
D_i	Internal diameter
e	Specific internal energy
$\mathbf{e}$	Almansi strain tensor, $\mathbf{e} = \frac{1}{2}\left(\mathbf{I} - \mathbf{F}^{-\mathrm{T}} \cdot \mathbf{F}^{-1}\right)$
$\hat{\mathbf{e}}$	A unit vector
$\hat{\mathbf{e}}_A$	A unit basis vector in the direction of vector $\mathbf{A}$
$\mathbf{e}_i$	A basis vector in the x_i–direction
e_{ijk}	Components of alternating tensor, $\mathcal{E}$
$\mathbf{E}$	Green–Lagrange strain tensor, $\mathbf{E} = \frac{1}{2}\left(\mathbf{F}^{\mathrm{T}} \cdot \mathbf{F} - \mathbf{I}\right)$;
E, E_1, E_2	Young's modulus (modulus of elasticity)
$\hat{\mathbf{E}}_i$	Unit base vector along the X_i material coordinate direction electric field intensity vector
E_{ij}	Components of the Green–Lagrange strain tensor
f	Load per unit length of a bar
$\mathbf{f}$	Body force vector
$f(\ )$	Function
f_x, f_y, f_z	Body force components in the x, y, and z directions

$\mathbf{F}$	Deformation gradient, $\mathbf{F} = \left(\mathbf{\nabla}_0 \mathbf{x}\right)^{\mathrm{T}}$; force vector
$\mathcal{F}$	Functional mapping
g	Acceleration due to gravity; function; internal heat generation
$\mathbf{g}$	Gradient of temperature, $\mathbf{g} = \mathbf{\nabla}\theta$
G	Shear modulus (modulus of rigidity)
h	Height of the beam; thickness; heat transfer coefficient
H	Total entropy (see Section 5.4.3.1); unit step function
$\mathbf{H}$	Nonlinear deformation tensor [see Eq. 6.6.25)]; magnetic field intensity vector
I	Second moment of area of a beam cross section; functional
$\mathbf{I}$	Unit second-order tensor
I_1, I_2, I_3	Principal invariants of stress tensor
J	Determinant of the matrix of deformation gradient (Jacobian); polar second moment of area of a shaft cross section
$\mathbf{J}$	Current density vector; creep compliance
J_i	Principal invariants of strain tensor $\mathbf{E}$ or rate of deformation tensor $\mathbf{D}$
k	Spring constant; thermal conductivity
$\mathbf{k}$	Thermal conductivity tensor
K	Kinetic energy
K_{ij}	Stiffness coefficients
K_s	Shear correction factor in Timoshenko beam theory
ℓ_{ij}	Direction cosines [see Eq. (2.2.71) or Eq. (4.3.4)]
L	Length; Lagrangian function
$\mathbf{L}$	Velocity gradient tensor, $\mathbf{L} = \left(\mathbf{\nabla}\mathbf{v}\right)^{\mathrm{T}}$
$L(\ )$	Linear form
$[L]$	Matrix of direction cosines, ℓ_{ij}
m	A scalar memory function (or relaxation kernel)
$\mathbf{m}$	Couple traction vector [see Eq. (5.3.33)]
M	Bending moment in beam problems
$\mathbf{M}$	Couple stress tensor; magnetization vector
$\hat{\mathbf{n}}$	Unit normal vector in the current configuration
n_i	ith component of the unit normal vector $\hat{\mathbf{n}}$
N	Axial force in beam problems
$\hat{\mathbf{N}}$	Unit normal vector in the reference configuration
N_I	Ith component of the unit normal vector $\hat{\mathbf{N}}$
p	Pressure (hydrostatic or thermodynamic)
$\mathbf{p}$	Angular momentum vector; vector of pyroelectric coefficients
P	Point load in beams; perimeter
$\mathbf{P}$	First Piola–Kirchhoff stress tensor; polarization vector
q	Distributed transverse load on a beam
q_0	Intensity of the distributed transverse load in beams
$\mathbf{q}_0$	Heat flux vector in the reference configuration

q_n	Heat flux normal to the boundary, $q_n = \nabla \cdot \hat{\mathbf{n}}$
$\mathbf{q}_f$	Moisture flux vector
$\mathbf{q}_i$	Force components
$\mathbf{q}$	Heat flux vector in the current configuration
Q	First moment of area; volume rate of flow
$\mathbf{Q}$	Rotation tensor [see Eq. (3.8.12)]
Q_h	Heat input
Q_J	Joule heating
r	Radial coordinate in the cylindrical polar system
r_0	Internal heat generation per unit mass in the reference configuration
r_h	Internal heat generation per unit mass in the current configuration
R	Radial coordinate in the spherical coordinate system; universal gas constant
$\mathbf{R}$	Position vector in the spherical coordinate system; proper orthogonal tensor
$\mathbf{S}$	A second-order tensor; second Piola–Kirchhoff stress tensor
$\mathbf{S}_e$	Electric susceptibility tensor
S_{ij}	Elastic compliance coefficients
t	Time
$\mathbf{t}$	Stress vector; traction vector
T	Torque; temperature
$\mathbf{u}$	Displacement vector
u, v, w	Displacements in the x, y, and z directions
u_1, u_2, u_3	Displacements in the x_1, x_2, and x_3 directions
U	Internal (or strain) energy
$\mathbf{U}$	Right Cauchy stretch tensor
v	Velocity, $v = |\mathbf{v}|$
$\mathbf{v}$	Velocity vector, $\mathbf{v} = \frac{D\mathbf{x}}{Dt}$
V	Shear force in beam problems; potential energy due to loads
$\mathbf{V}$	Left Cauchy stretch tensor
V_f	Scalar potential
$\mathbf{w}$	Vorticity vector, $\mathbf{w} = \frac{1}{2}\nabla \times \mathbf{v}$
W	Power input
$\mathbf{W}$	Skew symmetric part of the velocity gradient tensor, $\mathbf{L} = (\nabla\mathbf{v})^{\mathrm{T}}$; that is, $\mathbf{W} = \frac{1}{2}\left[(\nabla\mathbf{v})^{\mathrm{T}} - \nabla\mathbf{v}\right]$
$\mathbf{x}$	Spatial coordinates
x, y, z	Rectangular Cartesian coordinates
x_1, x_2, x_3	Rectangular Cartesian coordinates
$\mathbf{X}$	Material coordinates
Y	Relaxation modulus
z	Transverse coordinate in the beam problem; axial coordinate in the torsion problem

Greek symbols

α	Angle; coefficient of thermal expansion
α_{ij}	Thermal coefficients of expansion
β_{ij}	Material coefficients, $\beta_{ij} = C_{ijk\ell}\alpha_{k\ell}$
χ	Deformation mapping
δ	Variational operator used in Chapter 7; Dirac delta
δ_{ij}	Components of the unit tensor, $\mathbf{I}$ (Kronecker delta)
Δ	Change of (followed by another symbol)
ε	Infinitesimal strain tensor
$\tilde{\varepsilon}$	Symmetric part of the displacement gradient tensor, $(\boldsymbol{\nabla}\mathbf{u})^{\mathrm{T}}$; that is, $\tilde{\varepsilon} = \frac{1}{2}\left[(\boldsymbol{\nabla}\mathbf{u})^{\mathrm{T}} + \boldsymbol{\nabla}\mathbf{u}\right]$
ϵ_0	Permittivity of free space
ε_{ij}	Rectangular components of the infinitesimal strain tensor
ϕ	A typical variable; angular coordinate in the spherical coordinate system; electric potential; relaxation function
ϕ_f	Moisture source
Φ	Viscous dissipation, $\Phi = \boldsymbol{\tau} : \mathbf{D}$; Gibb's potential; Airy stress function
γ	Shear strain in one-dimensional problems
Γ	Internal entropy production; total boundary
η	Entropy density per unit mass; dashpot constant
η_0	Viscosity coefficient
κ_0, κ	Reference and current configurations
λ	Extension ratio; Lamé constant; eigenvalue
μ	Lamé constant; viscosity; principal value of strain
μ_0	Permeability of free space
ν	Poisson's ratio; ν_{ij} Poisson's ratios
Π	Total potential energy functional
θ	Angular coordinate in the cylindrical and spherical coordinate systems; angle; twist per unit length; absolute temperature
Θ	Twist
ρ	Density in the current configuration; charge density
ρ_0	Density in the reference configuration
σ	Boltzman constant
$\tilde{\sigma}$	Mean stress
$\boldsymbol{\sigma}$	Cauchy stress tensor
τ	Shear stress; retardation or relaxation time
$\boldsymbol{\tau}$	Viscous stress tensor
Ω	Domain of a problem
$\boldsymbol{\Omega}$	Skew symmetric part of the displacement gradient tensor, $(\boldsymbol{\nabla}\mathbf{u})^{\mathrm{T}}$; that is $\boldsymbol{\Omega} = \frac{1}{2}\left[(\boldsymbol{\nabla}\mathbf{u})^{\mathrm{T}} - \boldsymbol{\nabla}\mathbf{u}\right]$

ω	Angular velocity
$\boldsymbol{\omega}$	Infinitesimal rotation vector, $\boldsymbol{\omega} = \frac{1}{2}\boldsymbol{\nabla} \times \mathbf{u}$
ψ	Warping function; stream function; creep function
Ψ	Helmholtz free energy density; Prandtl stress function
$\boldsymbol{\nabla}$	Gradient operator with respect to $\mathbf{x}$
$\boldsymbol{\nabla}_0$	Gradient operator with respect to $\mathbf{X}$
$[\ ]$	Matrix associated with the enclosed quantity
$\{\ \}$	Column vector associated with the enclosed quantity
$\|\ \|$	Magnitude or determinant of the enclosed quantity
$\dot{(\ )}$	Time derivative of the enclosed quantity
$(\ )^*$	Enclosed quantity with superposed rigid-body motion
$(\ )'$	Deviatoric tensors associated with the enclosed tensor

Note:

Quotes by various people included in this book were found at different web sites; for example, visit:

http://naturalscience.com/dsqhome.html,

http://thinkexist.com/quotes/david_hilbert/, and

http://www.yalescientific.org/2010/10/from-the-editor-imagination-in-science/.

The author cannot vouch for their accuracy; this author is motivated to include the quotes at various places in his book for their wit and wisdom.

Preface to the Second Edition

Tis the good reader that makes the good book; in every book he finds passages which seem confidences or asides hidden from all else and unmistakeably meant for his ear; the profit of books is according to the sensibility of the reader; the profoundest thought or passion sleeps as in a mine, until it is discovered by an equal mind and heart.
— Ralph Waldo Emerson (1803–1882)

You cannot teach a man anything, you can only help him find it within himself.
— Galileo Galilei (1564–1642)

Engineers are problem solvers. They construct mathematical models, develop analytical and numerical approaches and methodologies, and design and manufacture various types of devices, systems, or processes. Mathematical development and engineering analysis are aids to designing systems for specific functionalities, and they involve (1) mathematical model development, (2) data acquisition by measurements, (3) numerical simulation, and (4) evaluation of the results in light of known information. Mathematical models are developed using laws of physics and assumptions concerning the behavior of the system under consideration. The most difficult step in arriving at a design that is both functional and cost-effective is the construction of a suitable mathematical model of the system's behavior. It is in this context that a course on continuum mechanics or elasticity provides engineers with the background to formulate a suitable mathematical model and evaluate it in the context of the functionality and design constraints placed on the system.

Most classical books on continuum mechanics are very rigorous in mathematical treatments of the subject but short on detailed explanations and including few examples and problem sets. Such books serve as reference books but not as textbooks. This textbook provides illustrative examples and problem sets that enable readers to test their understanding of the subject matter and utilize the tools developed in the formulation of engineering problems.

This second edition of *Introduction to Continuum Mechanics* has the same objective as the first, namely, to facilitate an easy and thorough understanding of continuum mechanics and elasticity concepts. The course also helps engineers who depend on canned programs to analyze problems to interpret the results produced by such programs. **The book offers a concise yet rigorous treatment of the subject of continuum mechanics and elasticity at the introductory level.** In all of the chapters of the second edition, additional explanations, examples, and problems have been added. No attempt has been made to enlarge the scope or increase the number of topics covered.

The book may be used as a textbook for a first course on continuum mechanics as well as elasticity (omitting Chapter 8 on fluid mechanics and heat transfer). A solutions manual has also been prepared for the book. The solution manual is available from the publisher only to instructors who adopt the book as a textbook for a course.

Since the publication of the first edition, several users of the book communicated their comments and compliments as well as errors they found, for which the author thanks them. All of the errors known to the author have been corrected in the current edition. The author is grateful, in particular, to Drs. Karan Surana (University of Kansas), Arun Srinivasa (Texas A&M University), Rebecca Brannon (University of Utah), Vinu Unnikrishnan (University of Alabama), Wenbin Yu (Utah State University), Srikanth Vedantam (Indian Institute of Technology, Madras), Shailendra Joshi (National University of Singapore), Ganesh Subbarayan (Purdue University), S. H. Khan (Indian Institute of Technology, Kanpur), and Jaehyung Ju (University of North Texas) for their constructive comments and help. The author also expresses his sincere thanks to Mr. Peter Gordon, Senior Editor (Engineering) at Cambridge University Press, for his continued encouragement, friendship, and support in producing this book. The author requests readers to send their comments and corrections to *jn_reddy@yahoo.com.*

J. N. Reddy
College Station, Texas

What is there that confers the noblest delight? What is that which swells a man's breast with pride above that which any other experience can bring to him? Discovery! To know that you are walking where none others have walked ... —— Mark Twain (1835–1910)

You can get into a habit of thought in which you enjoy making fun of all those other people who don't see things as clearly as you do. We have to guard carefully against it.
—— Carl Sagan (1934–1996)

Preface to the First Edition

If I have been able to see further, it was only because I stood on the shoulders of giants.

— Isaac Newton (1643–1727)

Many of the mathematical models of natural phenomena are based on fundamental scientific laws of physics or otherwise, extracted from centuries of research on the behavior of physical systems under the action of natural "forces." Today this subject is referred to simply as *mechanics* – a phrase that encompasses broad fields of science concerned with the behavior of fluids, solids, and complex materials. Mechanics is vitally important to virtually every area of technology and remains an intellectually rich subject taught in all major universities. It is also the focus of research in departments of aerospace, chemical, civil, and mechanical engineering, and engineering science and mechanics, as well as applied mathematics and physics. The last several decades have witnessed a great deal of research in continuum mechanics and its application to a variety of problems. As most modern technologies are no longer discipline-specific but involve multidisciplinary approaches, scientists and engineers should be trained to think and work in such environments. Therefore, it is necessary to introduce the subject of mechanics to senior undergraduate and beginning graduate students so that they have a strong background in the basic principles common to all major engineering fields. A first course on *continuum mechanics* or *elasticity* is the one that provides the basic principles of mechanics and prepares engineers and scientists for advanced courses in traditional as well as emerging fields such as biomechanics and nanomechanics.

There are many books on mechanics of continua. These books fall into two major categories: those that present the subject as a highly mathematical and abstract subject, and those that are too elementary to be of use for those who will pursue further work in fluid dynamics, elasticity, plates and shells, viscoelasticity, plasticity, and interdisciplinary areas such as geomechanics, biomechanics, mechanobiology, and nanoscience. As is the case with all other books written (solely) by the author, the objective is to facilitate an easy understanding of the topics covered. It is hoped that the book is simple in presenting the main concepts yet mathematically rigorous enough in providing the invariant form as well as component form of the governing equations for analysis of practical problems of engineering. In particular, the book contains formulations and applications to specific problems from heat transfer, fluid mechanics, and solid mechanics.

The motivation and encouragement that led to the writing of this book came from the experience of teaching a course on continuum mechanics at Virginia Polytechnic Institute and State University and Texas A&M University. A course on continuum mechanics takes different forms – from abstract to very applied – when taught by different people. The primary objective of the course taught by the author is two-fold: (1) formulation of equations that describe the motion and thermomechanical response of materials and (2) solution of these equations for specific problems from elasticity, fluid flows, and heat transfer. The present book is a formal presentation of the author's notes developed for such a course over the last two and half decades.

With a brief discussion of the concept of a continuum in Chapter 1, a review of vectors and tensors is presented in Chapter 2. Since the language of mechanics is mathematics, it is necessary for all readers to familiarize themselves with the notation and operations of vectors and tensors. The subject of kinematics is discussed in Chapter 3. Various measures of strain are introduced here. The deformation gradient, Cauchy–Green deformation, Green–Lagrange strain, Cauchy and Euler strain, rate of deformation, and vorticity tensors are introduced, and the polar decomposition theorem is discussed in this chapter. In Chapter 4, various measures of stress – Cauchy stress and Piola–Kirchhoff stress measures – are introduced, and stress equilibrium equations are presented.

Chapter 5 is dedicated to the derivation of the field equations of continuum mechanics, which forms the heart of the book. The field equations are derived using the principles of conservation of mass and balance of momenta and energy. Constitutive relations that connect the kinematic variables (e.g., density, temperature, and deformation) to the kinetic variables (e.g., internal energy, heat flux, and stresses) are discussed in Chapter 6 for elastic materials, viscous and viscoelastic fluids, and heat transfer.

Chapters 7 and 8 are devoted to the application of the field equations derived in Chapter 5 and constitutive models of Chapter 6 to problems of linearized elasticity, and fluid mechanics and heat transfer, respectively. Simple boundary-value problems, mostly linear, are formulated and their solutions are discussed. The material presented in these chapters illustrates how physical problems are analytically formulated with the aid of continuum equations. Chapter 9 deals with linear viscoelastic constitutive models and their application to simple problems of solid mechanics. Since a continuum mechanics course is mostly offered by solid mechanics programs, the coverage in this book is slightly more directed, in terms of the amount and type of material covered, to solid and structural mechanics.

The book was written keeping undergraduate seniors and first-year graduate students of engineering in mind. Therefore, it is most suitable as a text book for adoption for a first course on continuum mechanics or elasticity. The book also serves as an excellent precursor to courses on viscoelasticity, plasticity, nonlinear elasticity, and nonlinear continuum mechanics.

The book contains so many mathematical equations that it is hardly possible not to have typographical and other kinds of errors. I wish to thank in advance those readers who are willing to draw the author's attention to typos and errors, using the e-mail address: *jn_reddy@yahoo.com*.

J. N. Reddy
College Station, Texas

About the Author

J. N. Reddy is a University Distinguished Professor, Regents Professor, and the holder of the Oscar S. Wyatt Endowed Chair in the Department of Mechanical Engineering at Texas A&M University, College Station. Prior to this current position, he was the Clifton C. Garvin Professor in the Department of Engineering Science and Mechanics at Virginia Polytechnic Institute and State University (VPI&SU), Blacksburg.

Dr. Reddy is internationally known for his contributions to theoretical and applied mechanics and computational mechanics. He is the author of more than 480 journal papers and 18 books. Professor Reddy is the recipient of numerous awards including the Walter L. Huber Civil Engineering Research Prize of the American Society of Civil Engineers (ASCE), the Worcester Reed Warner Medal and the Charles Russ Richards Memorial Award of the American Society of Mechanical Engineers (ASME), the 1997 Archie Higdon Distinguished Educator Award from the American Society of Engineering Education (ASEE), the 1998 Nathan M. Newmark Medal from ASCE, the 2000 Excellence in the Field of Composites from the American Society of Composites (ASC), the 2003 Bush Excellence Award for Faculty in International Research from Texas A&M University, and the 2003 Computational Solid Mechanics Award from the U.S. Association of Computational Mechanics (USACM). Dr. Reddy received honorary degrees (Honoris Causa) from the Technical University of Lisbon, Portugal, in 2009 and Odlar Yurdu University, Baku, Azerbaijan in 2011.

Professor Reddy is a Fellow of the the American Academy of Mechanics (AAM), American Institution of Aeronautics and Astronautics (AIAA), ASC, ASCE, ASME, USACM, the International Association of Computational Mechanics (IACM), and the Aeronautical Society of India (ASI). Professor Reddy is the Editor-in-Chief of *Mechanics of Advanced Materials and Structures* and *International Journal of Computational Methods in Engineering Science and Mechanics*, and Co-Editor of *International Journal of Structural Stability and Dynamics*; he also serves on the editorial boards of more than two dozen other journals, including *International Journal for Numerical Methods in Engineering*, *Computer Methods in Applied Mechanics and Engineering*, and *International Journal of Non-Linear Mechanics*.

Dr. Reddy is one of the selective researchers in engineering around the world who is recognized by *ISI Highly Cited Researchers* with more than 13,000 citations (without self-citations more than 12,000) with an h-index of more than 54 as per Web of Science, 2013; as per Google Scholar the number of citations is more than 29,000 and the h-index is 71. A more complete resume with links to journal papers can be found at

http://isihighlycited.com/   or   http://www.tamu.edu/acml.

INTRODUCTION

I can live with doubt and uncertainty and not knowing. I think it is much more interesting to live not knowing than to have answers that might be wrong.
— Richard Feynmann (1918–1988)

What we need is not the will to believe but the will to find out.
— Bertrand Russell (1872-1970)

1.1 Continuum Mechanics

The subject of *mechanics* deals with the study of deformations and forces in matter, whether it is a solid, liquid, or gas. In such a study, we make the simplifying assumption, for analytical purposes, that the matter is distributed continuously, without gaps or empty spaces (i.e., we disregard the molecular structure of matter). Such a hypothetical continuous matter is termed a *continuum*. In essence, in a continuum all quantities such as mass density, displacements, velocities, stresses, and so on vary continuously so that their spatial derivatives exist and are continuous.[1] The continuum assumption allows us to shrink an arbitrary volume of material to a point, in much the same way as we take the limit in defining a derivative, so that we can define quantities of interest at a point. For example, mass density (mass per unit volume) of a material at a point is defined as the ratio of the mass Δm of the material to its volume ΔV surrounding the point in the limit that ΔV becomes a value ϵ^3, where ϵ is small compared with the mean distance between molecules

$$\rho = \lim_{\Delta V \to \epsilon^3} \frac{\Delta m}{\Delta V}. \tag{1.1.1}$$

In fact, we take the limit $\epsilon \to 0$. A mathematical study of the mechanics of such an idealized continuum is called *continuum mechanics*.

The primary objectives of this book are (1) to study the conservation principles in mechanics of continua and formulate the equations that describe the motion and mechanical behavior of materials, and (2) to present the applications of these equations to simple problems associated with flows of fluids, conduction of heat, and deformations of solid bodies. Although the first of these objectives is important, the reason for the formulation of the equations is to gain a quantitative understanding of the behavior of an engineering system. This quantitative understanding is useful in design and manufacture of better products. Typical examples of engineering problems, which are sufficiently simple to be

---

[1]The continuity is violated when we consider shock waves in gas dynamics (discontinuity in density and velocity) as well as dissimilar-material interfaces. In such cases, in addition to the concepts to be discussed here, certain jump conditions are employed to deal with discontinuities. We do not consider such situations in this book.

covered in this book, are described in the examples discussed next. At this stage of discussion, it is sufficient to rely on the reader's intuitive understanding of concepts from basic courses in fluid mechanics, heat transfer, and mechanics of materials about the meaning of stress and strain and what constitutes viscosity, conductivity, modulus, and so on used in the examples.

Problem 1 (solid mechanics)

We wish to design a diving board (which enables a swimmer to gain momentum before jumping into the pool) of given length L, assumed to be fixed at one end and free at the other end (see Fig. 1.1.1). The board is initially straight and horizontal and of uniform cross section. The design process consists of selecting the material (with Young's modulus E) and cross-sectional dimensions b and h such that the board carries the (moving) weight W of the swimmer. The design criteria are that the stresses developed do not exceed the allowable stress values and the deflection of the free end does not exceed a pre-specified value δ. A preliminary design of such systems is often based on mechanics of materials equations. The final design involves the use of more sophisticated equations, such as the three-dimensional (3D) elasticity equations. The equations of elementary beam theory may be used to find a relation between the deflection δ of the free end in terms of the length L, cross-sectional dimensions b and h, Young's modulus E, and weight W:

$$\delta = \frac{4WL^3}{Ebh^3} \, . \tag{1.1.2}$$

Given δ (allowable deflection) and load W (maximum possible weight of a swimmer), one can select the material (Young's modulus, E) and dimensions L, b, and h (which must be restricted to the standard sizes fabricated by a manufacturer). In addition to the deflection criterion, one must also check if the board develops stresses that exceed the allowable stresses of the material selected. Analysis of pertinent equations provides the designer with alternatives to select the material and dimensions of the board so as to have a cost-effective but functionally reliable structure.

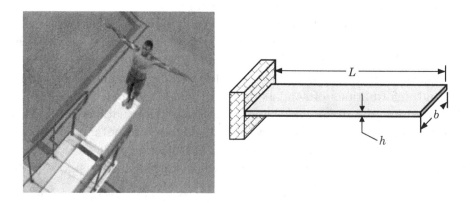

Fig. 1.1.1: A diving board fixed at the left end and free at the right end.

Problem 2 (fluid mechanics)

We wish to measure the viscosity μ of a lubricating oil used in rotating machinery to prevent the damage of the parts in contact. Viscosity, like Young's modulus of solid materials, is a material property that is useful in the calculation of shear stresses developed between a fluid and a solid body. A capillary tube is used to determine the viscosity of a fluid via the formula

$$\mu = \frac{\pi d^4}{128Q} \frac{p_1 - p_2}{L}, \qquad (1.1.3)$$

where d is the internal diameter and L is the length of the capillary tube, p_1 and p_2 are the pressures at the two ends of the tube (oil flows from one end to the other, as shown in Fig. 1.1.2), and Q is the volume rate of flow at which the oil is discharged from the tube. Equation (1.1.3) is derived, as we shall see later in this book, using the principles of continuum mechanics.

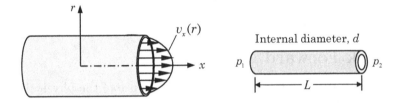

Fig. 1.1.2: Measurement of the viscosity of a fluid using a capillary tube.

Problem 3 (heat transfer)

We wish to determine the heat loss through the wall of a furnace. The wall typically consists of layers of brick, cement mortar, and cinder block (see Fig. 1.1.3). Each of these materials provides a varying degree of thermal resistance. The Fourier heat conduction law,

$$q = -k\frac{dT}{dx}, \qquad (1.1.4)$$

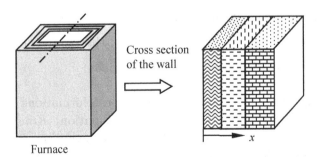

Furnace

Fig. 1.1.3: Heat transfer through the composite wall of a furnace.

provides a relation between the heat flux q (heat flow per unit area) and gradient of temperature T. Here k denotes thermal conductivity ($1/k$ is the thermal resistance) of the material. The negative sign in Eq. (1.1.4) indicates that heat flows from a high-temperature region to a low-temperature region. Using the continuum mechanics equations, one can determine the heat loss when the temperatures inside and outside of the building are known. A building designer can select the materials as well as thicknesses of various components of the wall to reduce the heat loss (while ensuring necessary structural strength – a structural analysis aspect).

The foregoing examples provide some indication of the need for studying the mechanical response of materials under the influence of external loads. The response of a material is consistent with the laws of physics and the constitutive behavior of the material. The present book aims to describe the physical principles and derive the equations governing the stress and deformation of continuous materials, and then solve some simple problems from various branches of engineering to illustrate the applications of the principles discussed and equations derived.

1.2 A Look Forward

The primary objective of this book is two fold: (1) use of the physical principles to derive the equations that govern the motion and thermomechanical response of materials, and (2) application of these equations for the solution of specific problems of linearized elasticity, heat transfer, and fluid mechanics. The governing equations for the study of deformation and stress of a continuous material are nothing but an analytical representation of the global laws of conservation of mass and balance of momenta and energy and the constitutive response of the continuum. They are applicable to all materials that are treated as a continuum. Tailoring these equations to particular problems and solving them constitutes the bulk of engineering analysis and design.

The study of motion and deformation of a continuum (or a "body" consisting of continuously distributed material) can be broadly classified into four basic categories:

(1) Kinematics (strain-displacement equations)

(2) Kinetics (balance of linear and angular momentum)

(3) Thermodynamics (first and second laws of thermodynamics)

(4) Constitutive equations (stress–strain relations)

Kinematics is the study of geometric changes or deformations in a continuum, without consideration of forces causing the deformation. *Kinetics* is the study of the equilibrium of forces and moments acting on a continuum, using the principles of balance of linear and angular momentum. This study leads to equations of motion as well as the symmetry of stress tensor in the absence of body couples. *Thermodynamic principles* are concerned with the balance of energy and relations among heat, mechanical work, and thermodynamic properties of a continuum. *Constitutive equations* describe thermomechanical behavior of the material of the continuum, and they relate the dependent variables introduced in the kinetic description to those introduced in the kinematic and thermodynamic descriptions. Table 1.2.1 provides a brief summary of the relationship between physical principles and governing equations and physical entities involved in the equations.

Table 1.2.1: The major four topics of study, physical principles used, resulting governing equations, and variables involved.

Topic of study	Physical law	Equations	Variables
1. Kinematics	None (based on geometric changes)	Strain–displacement relations	Displacements and strains
		Strain rate–velocity relations	Velocities and strain rates
2. Kinetics	Conservation of linear momentum	Equations of motion	Stresses and velocities
	Conservation of angular momentum	Symmetry of stress tensor	Stresses
3. Thermodynamics	First law	Energy equation	Temperature, heat flux, stresses, and velocities
	Second law	Clasius–Duhem inequality	Temperature, heat flux, and entropy
4. Constitutive equations*	Constitutive axioms	Hooke's law	Stresses, strains, heat flux, and temperature
		Newtonian fluids	Stresses, pressure, and velocities
		Fourier's law	heat flux and temperature
		Equations of state	Density, pressure, and temperature

*Not all relations are listed.

1.3 Summary

In this chapter, the concept of a continuous medium is discussed and the major objectives of the present study, namely, to use the physical principles to derive the equations governing a continuous medium and to present application of the equations in the solution of specific problems of linearized elasticity, heat transfer, and fluid mechanics are presented. The study of physical principles is broadly divided into four topics, as outlined in Table 1.2.1. These four topics are the subjects of Chapters 3 through 6, respectively. Mathematical formulation of the governing equations of a continuous medium necessarily requires the use of vectors and tensors, objects that facilitate invariant analytical formulation of the natural laws. Therefore, it is useful to study certain operational properties of vectors and tensors first. Chapter 2 is dedicated for this purpose.

Although the present book is self-contained for an introduction to continuum mechanics or elasticity, other books are available that may provide an advanced treatment of the subject. Many of the classical books on the subject do not contain example and/or exercise problems to test readers' understanding of the concepts. Interested readers may consult the list of references at the end of this book.

Problems

1.1 Newton's second law can be expressed as

$$\mathbf{F} = m\mathbf{a}, \tag{1}$$

where $\mathbf{F}$ is the net force acting on the body, m is the mass of the body, and $\mathbf{a}$ is the acceleration of the body in the direction of the net force. Use Eq. (1) to determine the governing equation of a free-falling body. Consider only the forces due to gravity and the air resistance, which is assumed to be proportional to the square of the velocity of the falling body.

1.2 Consider steady-state heat transfer through a cylindrical bar of nonuniform cross section. The bar is subject to a known temperature T_0 (°C) at the left end and exposed, both on the surface and at the right end, to a medium (such as cooling fluid or air) at temperature T_∞. Assume that temperature is uniform at any section of the bar, $T = T(x)$, and neglect thermal expansion of the bar (i.e., assume rigid). Use the principle of balance of energy (which requires that the rate of change (increase) of internal energy is equal to the sum of heat gained by conduction, convection, and internal heat generation) to a typical element of the bar (see Fig. P1.2) to derive the governing equations of the problem.

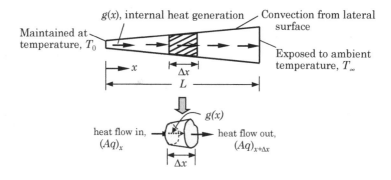

Fig. P1.2

1.3 The Euler–Bernoulli hypothesis concerning the kinematics of bending deformation of a beam assumes that straight lines perpendicular to the beam axis before deformation remain (1) straight, (2) perpendicular to the tangent line to the beam axis, and (3) inextensible during deformation. These assumptions lead to the following displacement field:

$$u_1(x, y) = -y\frac{dv}{dx}, \quad u_2 = v(x), \quad u_3 = 0, \tag{1}$$

where (u_1, u_2, u_3) are the displacements of a point (x, y, z) along the $x, y,$ and z coordinates, respectively, and v is the vertical displacement of the beam at point $(x, 0, 0)$. Suppose that the beam is subjected to a distributed transverse load $q(x)$. Determine the governing equation by summing the forces and moments on an element of the beam (see Fig. P1.3). Note that the sign conventions for the moment and shear force are based on the definitions

$$V = \int_A \sigma_{xy}\, dA, \quad M = \int_A y\,\sigma_{xx}\, dA,$$

and may not agree with the sign conventions used in some mechanics of materials books.

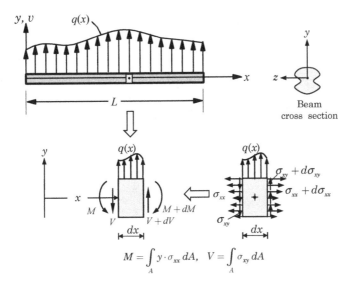

Fig. P1.3

1.4 A cylindrical storage tank of diameter D contains a liquid column of height $h(x, t)$. Liquid is supplied to the tank at a rate of q_i (m^3/day) and drained at a rate of q_o (m^3/day). Assume that the fluid is incompressible (i.e., constant mass density ρ) and use the principle of conservation of mass to obtain a differential equation governing $h(x, t)$.

1.5 (*Surface tension*). Forces develop at the interface between two immiscible liquids, causing the interface to behave as if it were a membrane stretched over the fluid mass. Molecules in the interior of the fluid mass are surrounded by molecules that are attracted to each other, whereas molecules along the surface (i.e., inside the imaginary membrane) are subjected to a net force toward the interior. This force imbalance creates a tensile force in the membrane and is called *surface tension* (measured per unit length). Let the difference between the pressure inside the drop and the external pressure be p and the surface tension, t_s. Determine the relation between p and t_s for a spherical drop of radius R.

VECTORS AND TENSORS

A mathematical theory is not to be considered complete until you have made it so clear that you can explain it to the first man whom you meet on the street.

—— David Hilbert (1862–1943)

2.1 Background and Overview

In the mathematical description of equations governing a continuous medium, we derive relations between various quantities that characterize the stress and deformation of the continuum by means of the laws of nature (such as Newton's laws, balance of energy, and so on). As a means of expressing a natural law, a coordinate system in a chosen frame of reference is often introduced. The mathematical form of the law thus depends on the chosen coordinate system and may appear different in another type of coordinate system. The laws of nature, however, should be independent of the choice of the coordinate system, and we may seek to represent the law in a manner independent of the particular coordinate system. A way of doing this is provided by vector and tensor analysis. When vector notation is used, a particular coordinate system need not be introduced. Consequently, the use of vector notation in formulating natural laws leaves them *invariant* to coordinate transformations. A study of physical phenomena by means of vector equations often leads to a deeper understanding of the problem in addition to bringing simplicity and versatility into the analysis.

In basic engineering courses, the term *vector* is used often to imply a *physical vector* that has "magnitude and direction and satisfies the parallelogram law of addition." In mathematics, vectors are more abstract objects than physical vectors. Like physical vectors, *tensors* are more general objects that possess a magnitude and multiple direction(s) and satisfy rules of tensor addition and scalar multiplication. In fact, physical vectors are often termed the first-order tensors. As will be shown shortly, the specification of a stress component (i.e., force per unit area) requires a magnitude and two directions – one normal to the plane on which the stress component is measured and the other is its direction – to specify it uniquely.

This chapter is dedicated to the study of the elements of algebra and calculus of vectors and tensors. Useful elements of the matrix theory and eigenvalue problems associated with second-order tensors are discussed. Index and summation notations, which are extensively used throughout the book, are also introduced. Those who are familiar with the material covered in any of the sections may skip them and go to the next section or to Chapter 3.

2.2 Vector Algebra

In this section, we present a review of the formal definition of a geometric (or physical) vector, discuss various products of vectors and physically interpret them, introduce index notation to simplify representations of vectors in terms of their components as well as vector operations, and develop transformation equations among the components of a vector expressed in two different coordinate systems. Many of these concepts, with the exception of the index notation, may be familiar to most students of engineering, physics, and mathematics and may be skipped.

2.2.1 Definition of a Vector

The quantities encountered in analytical descriptions of physical phenomena may be classified into two groups according to the information needed to specify them completely: scalars and nonscalars. The scalars are given by a single number. Nonscalars have not only a magnitude specified, but also additional information, such as direction. Nonscalars that obey certain rules (such as the parallelogram law of addition) are called *vectors*. Not all nonscalar quantities are vectors (e.g., a finite rotation is not a vector).

A physical vector is often shown as a directed line segment with an arrowhead at the end of the line. The length of the line represents the magnitude of the vector and the arrow indicates the direction. Thus, a physical vector, possessing magnitude, is known as a *normed vector space*. In written material, it is customary to place an arrow over the letter denoting the physical vector, such as $\vec{A}$. In printed material the vector letter is commonly denoted by a boldface letter, $\mathbf{A}$, such as is used in this book. The magnitude of the vector $\mathbf{A}$, to be formally defined shortly, is denoted by $|\mathbf{A}|$ or A. The magnitude of a vector is a scalar.

A vector of unit length is called a *unit vector*. The unit vector along $\mathbf{A}$ may be defined as follows:

$$\hat{\mathbf{e}}_A = \frac{\mathbf{A}}{|\mathbf{A}|} = \frac{\mathbf{A}}{A}. \tag{2.2.1}$$

We may now write a vector $\mathbf{A}$ as

$$\mathbf{A} = A\,\hat{\mathbf{e}}_A. \tag{2.2.2}$$

Thus, *any vector may be represented as a product of its magnitude and a unit vector along the vector*. A unit vector is used to designate direction; it does not have any physical dimensions. However, $|\mathbf{A}|$ has the physical dimensions. A "hat" (caret) above the boldface letter, $\hat{\mathbf{e}}$, is used to signify that it is a vector of unit magnitude. A vector of zero magnitude is called a *zero vector* or a *null vector*, and denoted by boldface zero, $\mathbf{0}$. Note that a lightface zero, 0, is a scalar and boldface zero, $\mathbf{0}$, is the zero vector. Also, a zero vector has no direction associated with it.

2.2.1.1 Vector addition

Let $\mathbf{A}$, $\mathbf{B}$, and $\mathbf{C}$ be any vectors. Then there exists a vector $\mathbf{A} + \mathbf{B}$, called the sum of $\mathbf{A}$ and $\mathbf{B}$, with the properties

(1) $\mathbf{A} + \mathbf{B} = \mathbf{B} + \mathbf{A}$ (commutative)

(2) $(\mathbf{A} + \mathbf{B}) + \mathbf{C} = \mathbf{A} + (\mathbf{B} + \mathbf{C})$ (associative)

(3) There exists a unique vector, $\mathbf{0}$, independent of $\mathbf{A}$

such that $\mathbf{A} + \mathbf{0} = \mathbf{A}$ (existence of the zero vector). $\qquad$ (2.2.3)

(4) To every vector $\mathbf{A}$ there exists a unique vector $-\mathbf{A}$

(that depends on $\mathbf{A}$) such that

$\mathbf{A} + (-\mathbf{A}) = \mathbf{0}$ (existence of the negative vector).

The negative vector $-\mathbf{A}$ has the same magnitude as $\mathbf{A}$, but has the opposite *sense*. Subtraction of vectors is carried out along the same lines. To form the difference $\mathbf{A} - \mathbf{B}$, we write $\mathbf{A} + (-\mathbf{B})$, and subtraction reduces to the operation of addition.

2.2.1.2 Multiplication of a vector by a scalar

Let $\mathbf{A}$ and $\mathbf{B}$ be vectors and α and β be real numbers (scalars). To every vector $\mathbf{A}$ and every real number α, there corresponds a unique vector $\alpha\mathbf{A}$ such that

(1) $\alpha(\beta\mathbf{A}) = (\alpha\beta)\mathbf{A}$ (scalar multiplication is associative)

(2) $(\alpha + \beta)\mathbf{A} = \alpha\mathbf{A} + \beta\mathbf{A}$ (scalar addition is distributive)

(3) $\alpha(\mathbf{A} + \mathbf{B}) = \alpha\mathbf{A} + \alpha\mathbf{B}$ (vector addition is distributive)

(4) $1 \cdot \mathbf{A} = \mathbf{A} \cdot 1 = \mathbf{A}, \quad 0 \cdot \mathbf{A} = \mathbf{0}$

$\qquad$ (2.2.4)

Equations (2.2.3) and (2.2.4) clearly show that the laws that govern addition, subtraction, and scalar multiplication of vectors are identical with those governing the operations of scalar algebra.

Two vectors $\mathbf{A}$ and $\mathbf{B}$ are equal if their magnitudes are equal, $|\mathbf{A}| = |\mathbf{B}|$, *and* if their directions are equal. Consequently, a vector is not changed if it is moved parallel to itself. This means that the position of a vector in space, that is, the point from which the line segment is drawn (or the end without arrowhead), may be chosen arbitrarily. In certain applications, however, the actual point of location of a vector may be important, for instance, a moment or a force acting on a body. A vector associated with a given point is known as a localized or bound vector. The fact that vectors can be represented graphically is an *incidental* rather than a fundamental feature of the vector concept.

2.2.1.3 Linear independence of vectors

The concepts of collinear and coplanar vectors can be stated in algebraic terms. A set of n vectors is said to be *linearly dependent* if a set of n numbers $\beta_1, \beta_2, \cdots, \beta_n$ can be found such that

$$\beta_1 \mathbf{A}_1 + \beta_2 \mathbf{A}_2 + \cdots + \beta_n \mathbf{A}_n = \mathbf{0}, \qquad (2.2.5)$$

where $\beta_1, \beta_2, \cdots, \beta_n$ cannot all be zero. If this expression cannot be satisfied, the vectors are said to be *linearly independent*. If two vectors are linearly dependent, then they are *collinear*. If three vectors are linearly dependent, then they are *coplanar*. Four or more vectors in three-dimensional space are always linearly dependent.

Example 2.2.1 ——————————————————————————————————

Determine whether the following set of vectors is linearly independent:

$$\mathbf{A} = \hat{\mathbf{e}}_1 + \hat{\mathbf{e}}_2, \quad \mathbf{B} = \hat{\mathbf{e}}_2 + \hat{\mathbf{e}}_4, \quad \mathbf{C} = \hat{\mathbf{e}}_3 + \hat{\mathbf{e}}_4, \quad \mathbf{D} = \hat{\mathbf{e}}_1 + \hat{\mathbf{e}}_2 + \hat{\mathbf{e}}_3 + \hat{\mathbf{e}}_4.$$

Here $\hat{\mathbf{e}}_i$ are orthonormal unit base vectors in $\Re^3$.

Solution: We set

$$\alpha \mathbf{A} + \beta \mathbf{B} + \gamma \mathbf{C} + \lambda \mathbf{D} = \mathbf{0},$$

which gives

$$\alpha + \lambda = 0, \quad \alpha + \beta + \lambda = 0, \quad \gamma + \lambda = 0, \quad \beta + \gamma + \lambda = 0.$$

The solution of these equations yields $\beta = 0$ and $\alpha = \gamma = -\lambda$. Thus, the set is not linearly independent. The reader may verify that the sets $(\mathbf{A}, \mathbf{B}, \mathbf{C})$, $(\mathbf{A}, \mathbf{B}, \mathbf{D})$, and $(\mathbf{B}, \mathbf{C}, \mathbf{D})$ are linearly independent, but the set $(\mathbf{A}, \mathbf{C}, \mathbf{D})$ is not.

2.2.2 Scalar and Vector Products

Besides addition and subtraction of vectors, and multiplication of a vector by a scalar, we also encounter the product of two vectors. There are several ways the product of two vectors can be defined. We consider first the so-called scalar product.

2.2.2.1 Scalar product

When a force $\mathbf{F}$ acts on a mass point and moves through a displacement vector $\mathbf{d}$, the work done by the force vector is defined by the *projection* of the force in the direction of the displacement, as shown in Fig. 2.2.1, times the magnitude of the displacement. Such an operation may be defined for any two vectors. Since the result of the product is a scalar, it is called the *scalar product*. We denote this product as $\mathbf{F} \cdot \mathbf{d} \equiv (\mathbf{F}, \mathbf{d})$ and it is defined as follows:

$$\mathbf{F} \cdot \mathbf{d} \equiv (\mathbf{F}, \mathbf{d}) = Fd \cos \theta, \qquad 0 \leq \theta \leq \pi. \tag{2.2.6}$$

The scalar product is also known as the *dot product* or *inner product*.

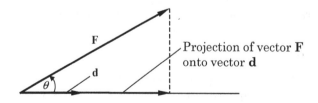

Fig. 2.2.1: Vector representation of work done.

The following simple results follow from the definition in Eq. (2.2.6):

(1) The scalar product is commutative: $\mathbf{A} \cdot \mathbf{B} = \mathbf{B} \cdot \mathbf{A}$.

(2) If the vectors $\mathbf{A}$ and $\mathbf{B}$ are perpendicular to each other, then $\mathbf{A} \cdot \mathbf{B} = AB\cos(\pi/2) = 0$. Conversely, if $\mathbf{A} \cdot \mathbf{B} = 0$, then either $\mathbf{A}$ or $\mathbf{B}$ is zero or $\mathbf{A}$ is perpendicular, or *orthogonal*, to $\mathbf{B}$.

(3) If two vectors $\mathbf{A}$ and $\mathbf{B}$ are parallel and in the same direction, then $\mathbf{A} \cdot \mathbf{B} = AB\cos 0 = AB$, because $\cos 0 = 1$. Thus the scalar product of a vector multiplied with itself is equal to the square of its magnitude ($|\mathbf{A}| = A$):

$$\mathbf{A} \cdot \mathbf{A} = AA = A^2. \tag{2.2.7}$$

(4) The orthogonal projection of a vector $\mathbf{A}$ in direction $\hat{\mathbf{e}}$ is given by $\mathbf{A} \cdot \hat{\mathbf{e}}$.

(5) The scalar product also follows the distributive law:

$$\mathbf{A}{\cdot}(\mathbf{B} + \mathbf{C}) = (\mathbf{A} \cdot \mathbf{B}) + (\mathbf{A} \cdot \mathbf{C}). \tag{2.2.8}$$

2.2.2.2 Vector product

To see the need for the *vector product*, consider the concept of the *moment* due to a force about a point. Let us describe the moment about a point O of a force $\mathbf{F}$ acting at a point P, such as shown in Fig. 2.2.2(a). By definition, the magnitude of the moment is given by

$$M = F\ell, \quad F = |\mathbf{F}|, \tag{2.2.9}$$

where ℓ is the perpendicular distance from the point O to the force $\mathbf{F}$ (called lever arm). If $\mathbf{r}$ denotes the vector $\mathbf{OP}$ and θ the angle between $\mathbf{r}$ and $\mathbf{F}$ such that $0 \leq \theta \leq \pi$, we have $\ell = r\sin\theta$ and thus

$$M = Fr\sin\theta. \tag{2.2.10}$$

A direction can now be assigned to the moment. Drawing the vectors $\mathbf{F}$ and $\mathbf{r}$ from the common origin O, we note that the rotation due to $\mathbf{F}$ tends to bring $\mathbf{r}$ into $\mathbf{F}$, as can be seen from Fig. 2.2.2(b). We now set up an axis of rotation

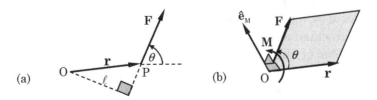

Fig. 2.2.2: (a) Representation of a moment about a point. (b) Direction of rotation.

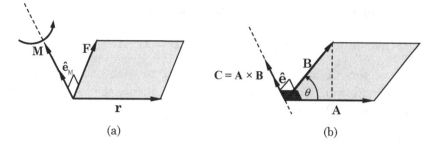

Fig. 2.2.3: (a) Axis of rotation. (b) Representation of the vector product.

perpendicular to the plane formed by $\mathbf{F}$ and $\mathbf{r}$. Along this axis of rotation we set up a preferred direction as one in which a right-handed screw would advance when turned in the direction of rotation due to the moment, as can be seen from Fig. 2.2.3(a). Along this axis of rotation we draw a unit vector $\hat{\mathbf{e}}_M$ and agree that it represents the direction of the moment $\mathbf{M}$. Thus we have

$$\mathbf{M} = Fr\ \sin\theta\ \hat{\mathbf{e}}_M = \mathbf{r} \times \mathbf{F}. \tag{2.2.11}$$

According to this expression, $\mathbf{M}$ may be looked on as resulting from a special operation between the two vectors $\mathbf{F}$ and $\mathbf{r}$. It is thus the basis for defining a product between any two vectors. Due to the fact that the result of such a product is a vector, it may be called the *vector product*.

The product of two vectors $\mathbf{A}$ and $\mathbf{B}$ is a vector $\mathbf{C}$ whose magnitude is equal to the product of the magnitude of $\mathbf{A}$ and $\mathbf{B}$ times the sine of the angle measured from $\mathbf{A}$ to $\mathbf{B}$ such that $0 \leq \theta \leq \pi$, and whose direction is specified by the condition that $\mathbf{C}$ be perpendicular to the plane of the vectors $\mathbf{A}$ and $\mathbf{B}$ and points in the direction in which a right-handed screw advances when turned so as to bring $\mathbf{A}$ into $\mathbf{B}$, as shown in Fig. 2.2.3(b). The vector product is usually denoted by

$$\mathbf{C} = \mathbf{A} \times \mathbf{B} = AB\ \sin(\mathbf{A}, \mathbf{B})\ \hat{\mathbf{e}} = AB\ \sin\theta\ \hat{\mathbf{e}}, \tag{2.2.12}$$

where $\sin(\mathbf{A}, \mathbf{B})$ denotes the sine of the angle between vectors $\mathbf{A}$ and $\mathbf{B}$. This product is called the *cross product, skew product,* and also *outer product,* as well as the vector product. When $\mathbf{A} = a\,\hat{\mathbf{e}}_A$ and $\mathbf{B} = b\,\hat{\mathbf{e}}_B$ are the vectors representing the sides of a parallelogram, with a and b denoting the lengths of the sides, then the vector product $\mathbf{A} \times \mathbf{B}$ represents the area of the parallelogram, $AB\sin\theta$. The unit vector $\hat{\mathbf{e}} = \hat{\mathbf{e}}_A \times \hat{\mathbf{e}}_B$ denotes the normal to the plane area. Thus an area can be represented as a vector (see Section 2.2.3 for additional discussion).

The description of the velocity of a point of a rotating rigid body is an important example of geometrical and physical applications of vectors. Suppose a rigid body is rotating with an angular velocity ω about an axis, and we wish to describe the velocity of some point P of the body, as shown in Fig. 2.2.4(a). Let $\mathbf{v}$ denote the velocity at the point. Each point of the body describes a circle that lies in a plane perpendicular to the axis with its center on the axis. The

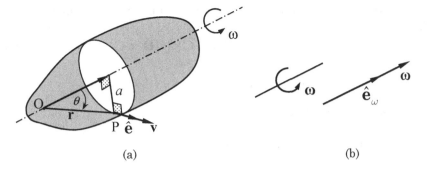

(a)   (b)

Fig. 2.2.4: (a) Velocity at a point in a rotating rigid body. (b) Angular velocity as a vector.

radius of the circle, a, is the perpendicular distance from the axis to the point of interest. The magnitude of the velocity is equal to ωa. The direction of $\mathbf{v}$ is perpendicular to a and to the axis of rotation. We denote the direction of the velocity by the unit vector $\hat{\mathbf{e}}$. Thus we can write

$$\mathbf{v} = \omega a\, \hat{\mathbf{e}}. \qquad (2.2.13)$$

Let O be a reference point on the axis of revolution, and let $\mathbf{OP} = \mathbf{r}$. We then have $a = r \sin \theta$, so that

$$\mathbf{v} = \omega r \sin \theta\, \hat{\mathbf{e}}. \qquad (2.2.14)$$

The angular velocity is a vector because it has an assigned direction, magnitude, and obeys the parallelogram law of addition. We denote it by $\boldsymbol{\omega}$ and represent its direction in the sense of a right-handed screw, as shown in Fig. 2.2.4(b). If we further let $\hat{\mathbf{e}}_r$ be a unit vector in the direction of $\mathbf{r}$, we see that

$$\hat{\mathbf{e}}_\omega \times \hat{\mathbf{e}}_r = \hat{\mathbf{e}} \sin \theta. \qquad (2.2.15)$$

With these relations we have

$$\mathbf{v} = \boldsymbol{\omega} \times \mathbf{r}. \qquad (2.2.16)$$

Thus the velocity of a point of a rigid body rotating about an axis is given by the vector product of $\boldsymbol{\omega}$ and a position vector $\mathbf{r}$ drawn from any reference point on the axis of revolution.

From the definition of vector product, a few simple results follow.

(1) The products $\mathbf{A} \times \mathbf{B}$ and $\mathbf{B} \times \mathbf{A}$ are not equal. In fact, we have

$$\mathbf{A} \times \mathbf{B} \equiv -\mathbf{B} \times \mathbf{A}. \qquad (2.2.17)$$

Thus the vector product *does not commute*.

(2) If two vectors $\mathbf{A}$ and $\mathbf{B}$ are parallel to each other, then $\theta = 0, \pi$, and $\sin \theta = 0$. Thus

$$\mathbf{A} \times \mathbf{B} = \mathbf{0}.$$

Conversely, if $\mathbf{A} \times \mathbf{B} = \mathbf{0}$, then either $\mathbf{A}$ or $\mathbf{B}$ is zero, or they are parallel vectors. It follows that the vector product of a vector with itself is zero; that is, $\mathbf{A} \times \mathbf{A} = \mathbf{0}$.

(3) The distributive law still holds, but the order of the factors must be maintained:

$$(\mathbf{A} + \mathbf{B}) \times \mathbf{C} = (\mathbf{A} \times \mathbf{C}) + (\mathbf{B} \times \mathbf{C}). \qquad (2.2.18)$$

2.2.2.3 Triple products of vectors

Now consider the various products of three vectors:

$$\mathbf{A}(\mathbf{B} \cdot \mathbf{C}), \quad \mathbf{A} \cdot (\mathbf{B} \times \mathbf{C}), \quad \mathbf{A} \times (\mathbf{B} \times \mathbf{C}). \qquad (2.2.19)$$

The product $\mathbf{A}(\mathbf{B} \cdot \mathbf{C})$ is merely a multiplication of the vector $\mathbf{A}$ by the scalar $\mathbf{B} \cdot \mathbf{C}$. The product $\mathbf{A} \cdot (\mathbf{B} \times \mathbf{C})$ is a scalar and it is termed *the scalar triple product*. It can be seen from Fig. 2.2.5 that the product $\mathbf{A} \cdot (\mathbf{B} \times \mathbf{C})$, except for the algebraic sign, is the volume of the parallelepiped formed by the vectors $\mathbf{A}$, $\mathbf{B}$, and $\mathbf{C}$.

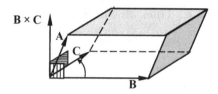

Fig. 2.2.5: Scalar triple product representation of the volume of a parallelepiped.

We note the following properties of a scalar triple product:

(1) The dot and cross can be interchanged without changing the value:

$$\mathbf{A} \cdot \mathbf{B} \times \mathbf{C} = \mathbf{A} \times \mathbf{B} \cdot \mathbf{C} \equiv [\mathbf{ABC}]. \qquad (2.2.20)$$

(2) A cyclical permutation of the order of the vectors leaves the result unchanged:

$$\mathbf{A} \cdot \mathbf{B} \times \mathbf{C} = \mathbf{C} \cdot \mathbf{A} \times \mathbf{B} = \mathbf{B} \cdot \mathbf{C} \times \mathbf{A} \equiv [\mathbf{ABC}]. \qquad (2.2.21)$$

(3) If the cyclic order is changed, the sign changes:

$$\mathbf{A} \cdot \mathbf{B} \times \mathbf{C} = -\mathbf{A} \cdot \mathbf{C} \times \mathbf{B} = -\mathbf{C} \cdot \mathbf{B} \times \mathbf{A} = -\mathbf{B} \cdot \mathbf{A} \times \mathbf{C}. \qquad (2.2.22)$$

(4) A necessary and sufficient condition for any three vectors, $\mathbf{A}$, $\mathbf{B}$, and $\mathbf{C}$ to be coplanar is that $\mathbf{A} \cdot (\mathbf{B} \times \mathbf{C}) = 0$. Note also that the scalar triple product is zero when any two vectors are the same.

The *vector triple product* $\mathbf{A} \times (\mathbf{B} \times \mathbf{C})$ is a vector normal to the plane formed by $\mathbf{A}$ and $(\mathbf{B} \times \mathbf{C})$. The vector $(\mathbf{B} \times \mathbf{C})$, however, is perpendicular to the plane formed by $\mathbf{B}$ and $\mathbf{C}$. This means that $\mathbf{A} \times (\mathbf{B} \times \mathbf{C})$ lies in the plane formed by

B and **C** and is perpendicular to **A**, as shown in Fig. 2.2.6. Thus $\mathbf{A} \times (\mathbf{B} \times \mathbf{C})$ can be expressed as a linear combination of **B** and **C**:

$$\mathbf{A} \times (\mathbf{B} \times \mathbf{C}) = m_1 \mathbf{B} + n_1 \mathbf{C}. \tag{2.2.23}$$

Likewise, we would find that

$$(\mathbf{A} \times \mathbf{B}) \times \mathbf{C} = m_2 \mathbf{A} + n_2 \mathbf{B}. \tag{2.2.24}$$

Thus, the parentheses *cannot* be interchanged or removed. It can be shown that $m_1 = \mathbf{A} \cdot \mathbf{C}$ and $n_1 = -\mathbf{A} \cdot \mathbf{B}$, and hence that

$$\begin{aligned} \mathbf{A} \times (\mathbf{B} \times \mathbf{C}) &= (\mathbf{A} \cdot \mathbf{C})\mathbf{B} - (\mathbf{A} \cdot \mathbf{B})\mathbf{C}, \\ (\mathbf{A} \times \mathbf{B}) \times \mathbf{C} &= (\mathbf{A} \cdot \mathbf{C})\mathbf{B} - (\mathbf{B} \cdot \mathbf{C})\mathbf{A}, \end{aligned} \tag{2.2.25}$$

and one can show that $\mathbf{A} \times (\mathbf{B} \times \mathbf{C}) = (\mathbf{A} \times \mathbf{B}) \times \mathbf{C}$ if and only if $\mathbf{B} \times (\mathbf{C} \times \mathbf{A}) = 0$.

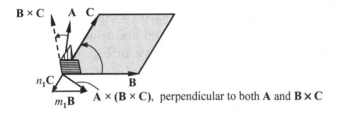

Fig. 2.2.6: The vector triple product.

Example 2.2.2 _____

Let **A** and **B** be any two vectors in space. Express vector **A** in terms of its components along (i.e., parallel) and perpendicular to vector **B**.

Solution: The component of **A** along **B** is given by $(\mathbf{A} \cdot \hat{\mathbf{e}}_B)$, where $\hat{\mathbf{e}}_B = \mathbf{B}/B$ is the unit vector in the direction of **B**. The component of **A** perpendicular to **B** and in the plane of **A** and **B** is given by the vector triple product $\hat{\mathbf{e}}_B \times (\mathbf{A} \times \hat{\mathbf{e}}_B)$. Thus,

$$\mathbf{A} = (\mathbf{A} \cdot \hat{\mathbf{e}}_B)\hat{\mathbf{e}}_B + \hat{\mathbf{e}}_B \times (\mathbf{A} \times \hat{\mathbf{e}}_B). \tag{2.2.26}$$

Alternatively, Eq. (2.2.25) gives the same result with $\mathbf{A} = \mathbf{C} = \hat{\mathbf{e}}_B$ and $\mathbf{B} = \mathbf{A}$:

$$\hat{\mathbf{e}}_B \times (\mathbf{A} \times \hat{\mathbf{e}}_B) = \mathbf{A} - (\hat{\mathbf{e}}_B \cdot \mathbf{A})\hat{\mathbf{e}}_B \quad \text{or} \quad \mathbf{A} = (\mathbf{A} \cdot \hat{\mathbf{e}}_B)\hat{\mathbf{e}}_B + \hat{\mathbf{e}}_B \times (\mathbf{A} \times \hat{\mathbf{e}}_B).$$

2.2.3 Plane Area as a Vector

The magnitude of the vector $\mathbf{C} = \mathbf{A} \times \mathbf{B}$ is equal to the area of the parallelogram formed by the vectors **A** and **B**, as shown in Fig. 2.2.7(a). In fact, the vector **C** may be considered to represent *both* the magnitude and the direction of the product of **A** and **B**. Thus, a plane area may be looked upon as possessing a direction in addition to a magnitude, the directional character arising out of the need to specify an orientation of the plane in space. Representation of an area as a vector has many uses in continuum mechanics, as will be seen in the chapters that follow.

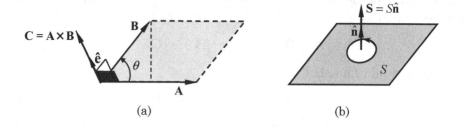

Fig. 2.2.7: (a) Plane area as a vector. (b) Unit normal vector and sense of travel.

It is customary to denote the direction of a plane area by means of a unit vector drawn normal to that plane. To fix the direction of the normal, we assign a *sense of travel* along the contour of the boundary of the plane area in question. The direction of the normal is taken by convention as that in which a right-handed screw advances as it is rotated according to the sense of travel along the boundary curve or contour, as shown in Fig. 2.2.7(b). Let the unit normal vector be given by $\hat{\mathbf{n}}$. Then the area can be denoted by $\mathbf{S} = S\hat{\mathbf{n}}$.

Representation of a plane as a vector has many uses. The vector can be used to determine the area of an inclined plane in terms of its projected area, as illustrated in Example 2.2.3.

Example 2.2.3 ———————————————————————————

(a) Determine the plane area of the surface obtained by cutting a cylinder of cross-sectional area S_0 with an inclined plane whose normal is $\hat{\mathbf{n}}$, as shown in Fig 2.2.8(a).
(b) Consider a cube (or a prism) cut by an inclined plane whose normal is $\hat{\mathbf{n}}$, as shown in Fig. 2.2.8(b). Express the areas of the sides of the resulting tetrahedron in terms of the area S of the inclined surface.

Solution: (a) Let the plane area of the inclined surface be S, as shown in Fig 2.2.8(a). First, we express the areas as vectors

$$\mathbf{S}_0 = S_0\,\hat{\mathbf{n}}_0 \quad \text{and} \quad \mathbf{S} = S\,\hat{\mathbf{n}}. \tag{2.2.27}$$

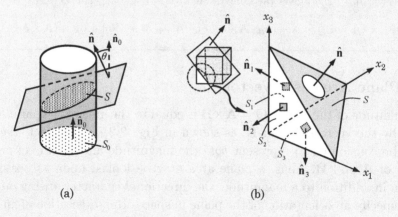

Fig. 2.2.8: Vector representation of inclined plane areas and their components.

Due to the fact that S_0 is the projection of $\mathbf{S}$ along $\hat{\mathbf{n}}_0$ (if the angle between $\hat{\mathbf{n}}$ and $\hat{\mathbf{n}}_0$ is acute; otherwise the negative of it), we have

$$S_0 = \mathbf{S} \cdot \hat{\mathbf{n}}_0 = S\hat{\mathbf{n}} \cdot \hat{\mathbf{n}}_0 = S \cos\theta. \tag{2.2.28}$$

The scalar product $\hat{\mathbf{n}} \cdot \hat{\mathbf{n}}_0$ is the cosine of the angle between the two unit normal vectors, $\hat{\mathbf{n}} \cdot \hat{\mathbf{n}}_0 = \cos\theta$.

(b) For reference purposes we label the sides of the tetrahedron by 1, 2, and 3 and the normals and surface areas by $(\hat{\mathbf{n}}_1, S_1), (\hat{\mathbf{n}}_2, S_2),$ and $(\hat{\mathbf{n}}_3, S_3)$, respectively (that is, S_i is the surface area of the plane perpendicular to the ith line or $\hat{\mathbf{n}}_i$ vector), as shown in Fig. 2.2.8(b). Then we have

$$S_1 = S\,\hat{\mathbf{n}} \cdot \hat{\mathbf{n}}_1, \quad S_2 = S\,\hat{\mathbf{n}} \cdot \hat{\mathbf{n}}_2, \quad S_3 = S\,\hat{\mathbf{n}} \cdot \hat{\mathbf{n}}_3. \tag{2.2.29}$$

2.2.4 Reciprocal Basis

2.2.4.1 Components of a vector

So far we have considered a geometrical description of a vector. We now embark on an analytical description of a vector based on the notion of its components. In the following discussion, we shall consider a three-dimensional space, and the extensions to n dimensions will be evident. In a three-dimensional space a set of no more than three linearly independent vectors can be found. Let us choose any set and denote it as $\mathbf{e}_1, \mathbf{e}_2, \mathbf{e}_3$. This set is called a *basis*. We can represent any vector in three-dimensional space as a linear combination of the basis vectors

$$\mathbf{A} = A^1\mathbf{e}_1 + A^2\mathbf{e}_2 + A^3\mathbf{e}_3. \tag{2.2.30}$$

The vectors $A^1\mathbf{e}_1$, $A^2\mathbf{e}_2$, and $A^3\mathbf{e}_3$ are called the *vector components* of $\mathbf{A}$, and A^1, A^2, and A^3 are called the *scalar components* of $\mathbf{A}$ associated with the basis $(\mathbf{e}_1, \mathbf{e}_2, \mathbf{e}_3)$, as indicated in Fig. 2.2.9.

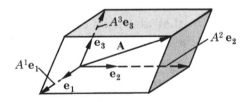

Fig. 2.2.9: Components of a vector.

2.2.4.2 General basis

For any arbitrary basis $(\mathbf{e}_1, \mathbf{e}_2, \mathbf{e}_3)$, we can derive another basis from it [see Reddy and Rasmussen (1991)]. We make use of the fact that the cross product $\mathbf{e}_i \times \mathbf{e}_j$ (for fixed i and j) when dotted with either $\mathbf{e}_i$ or $\mathbf{e}_j$ produces zero because $\mathbf{e}_i \times \mathbf{e}_j$ is perpendicular to both $\mathbf{e}_i$ and $\mathbf{e}_j$. Thus, taking the scalar product of the vector $\mathbf{A}$ in Eq. (2.2.30) with the cross product $\mathbf{e}_1 \times \mathbf{e}_2$, we obtain

$$\mathbf{A} \cdot (\mathbf{e}_1 \times \mathbf{e}_2) = A^3\mathbf{e}_3 \cdot (\mathbf{e}_1 \times \mathbf{e}_2).$$

Solving for A^3 gives

$$A^3 = \mathbf{A} \cdot \frac{\mathbf{e}_1 \times \mathbf{e}_2}{\mathbf{e}_3 \cdot (\mathbf{e}_1 \times \mathbf{e}_2)} = \mathbf{A} \cdot \frac{\mathbf{e}_1 \times \mathbf{e}_2}{[\mathbf{e}_1\mathbf{e}_2\mathbf{e}_3]}.$$

Similarly, we obtain

$$A^1 = \mathbf{A} \cdot \frac{\mathbf{e}_2 \times \mathbf{e}_3}{[\mathbf{e}_1\mathbf{e}_2\mathbf{e}_3]}, \quad A^2 = \mathbf{A} \cdot \frac{\mathbf{e}_3 \times \mathbf{e}_1}{[\mathbf{e}_1\mathbf{e}_2\mathbf{e}_3]}.$$

where, in the evaluation of the cross products, we shall always use the right-hand rule (i.e., 1-2-3, 2-3-1, or 3-1-2). Thus we can obtain the components A^1, A^2, and A^3 by taking the scalar product of the vector $\mathbf{A}$ with special vectors, which we denote as follows:

$$\mathbf{e}^1 = \frac{\mathbf{e}_2 \times \mathbf{e}_3}{[\mathbf{e}_1\mathbf{e}_2\mathbf{e}_3]}, \quad \mathbf{e}^2 = \frac{\mathbf{e}_3 \times \mathbf{e}_1}{[\mathbf{e}_1\mathbf{e}_2\mathbf{e}_3]}, \quad \mathbf{e}^3 = \frac{\mathbf{e}_1 \times \mathbf{e}_2}{[\mathbf{e}_1\mathbf{e}_2\mathbf{e}_3]}. \tag{2.2.31}$$

The set of vectors $(\mathbf{e}^1, \mathbf{e}^2, \mathbf{e}^3)$ is linearly independent because in the linear relation

$$\alpha_1 \mathbf{e}^1 + \alpha_2 \mathbf{e}^2 + \alpha_3 \mathbf{e}^3 = \mathbf{0}$$

all of the scalars α_1, α_2, and α_3 are zero. Indeed, by taking the scalar product of the above relation with $\mathbf{e}_i$ we obtain $\alpha_i = 0$ for $i = 1, 2$, and 3. The set $(\mathbf{e}^1, \mathbf{e}^2, \mathbf{e}^3)$, being a linearly independent set, constitutes a basis, called the *dual* or *reciprocal basis*. It is possible, because the dual basis is linearly independent, to express a vector $\mathbf{A}$ in terms of the dual basis:

$$\mathbf{A} = A_1\mathbf{e}^1 + A_2\mathbf{e}^2 + A_3\mathbf{e}^3 = \sum_{i=1}^{3} A_i\mathbf{e}^i. \tag{2.2.32}$$

Notice now that the components associated with the dual basis have subscripts to distinguish from the components with respect to the original basis [see Eq. (2.2.30)].

By a process analogous to that above we can show that the original basis can be expressed in terms of the dual basis as

$$\mathbf{e}_1 = \frac{\mathbf{e}^2 \times \mathbf{e}^3}{[\mathbf{e}^1\mathbf{e}^2\mathbf{e}^3]}, \quad \mathbf{e}_2 = \frac{\mathbf{e}^3 \times \mathbf{e}^1}{[\mathbf{e}^1\mathbf{e}^2\mathbf{e}^3]}, \quad \mathbf{e}_3 = \frac{\mathbf{e}^1 \times \mathbf{e}^2}{[\mathbf{e}^1\mathbf{e}^2\mathbf{e}^3]}. \tag{2.2.33}$$

From the basic definitions we have the following relations among the two sets of bases ($\mathbf{e}^1 \cdot \mathbf{e}_1 = \mathbf{e}^2 \cdot \mathbf{e}_2 = \mathbf{e}^3 \cdot \mathbf{e}_3 = 1$):

$$\mathbf{e}^i \cdot \mathbf{e}_j = \begin{cases} 1, & \text{if } i \text{ has the same value as } j \\ 0, & \text{if } i \text{ has a different value than } j. \end{cases} \tag{2.2.34}$$

$$\mathbf{e}^i \times \mathbf{e}^j \cdot \mathbf{e}^k = \begin{cases} 1, & \text{if } i, j, \text{ and } k \text{ take the values 1-2-3, 2-3-1, or 3-1-2} \\ -1, & \text{if } i, j, \text{ and } k \text{ take the values 1-3-2, 3-2-1, or 2-1-3} \\ 0, & \text{if any two of the three indices have the same value.} \end{cases} \tag{2.2.35}$$

From Eqs. (2.2.30)–(2.2.35) one can obtain

$$\begin{array}{llll} A^1 = \mathbf{A} \cdot \mathbf{e}^1, & A^2 = \mathbf{A} \cdot \mathbf{e}^2, & A^3 = \mathbf{A} \cdot \mathbf{e}^3 & (A^i = \mathbf{A} \cdot \mathbf{e}^i), \\ A_1 = \mathbf{A} \cdot \mathbf{e}_1, & A_2 = \mathbf{A} \cdot \mathbf{e}_2, & A_3 = \mathbf{A} \cdot \mathbf{e}_3 & (A_i = \mathbf{A} \cdot \mathbf{e}_i). \end{array} \tag{2.2.36}$$

In summary, we have two ways of expressing the same vector for a given basis. This gives rise to the description *cogredient* (A_1, A_2, A_3) and *contragredient* (A^1, A^2, A^3) components of a vector. This terminology is based on the way the components transform as a given coordinate system is transformed to another coordinate system. The components (A_1, A_2, A_3) transform in the same way as the basis $(\mathbf{e}_1, \mathbf{e}_2, \mathbf{e}_3)$, whereas the components (A^1, A^2, A^3) transform by another rule (contrary-wise!) in the same way as $(\mathbf{e}^1, \mathbf{e}^2, \mathbf{e}^3)$. As we shall find, a particular basis, called a *unitary basis*, can be defined in terms of a related coordinate system. For a unitary basis, the term cogredient is called *covariant* and contragredient is called *contravariant*.

2.2.4.3 Orthonormal basis

When the basis vectors are constant, that is, with fixed lengths (with the same units) and directions, the basis is called *Cartesian*. The general Cartesian system is oblique. When a basis is unit and orthogonal, it is termed *orthonormal*. When the basis vectors are orthonormal, the basis system is called *rectangular Cartesian*. Notations used for Cartesian rectangular basis are

$$(\hat{\mathbf{i}}, \hat{\mathbf{j}}, \hat{\mathbf{k}}), \quad (\hat{\mathbf{e}}_x, \hat{\mathbf{e}}_y, \hat{\mathbf{e}}_z), \quad (\hat{\mathbf{i}}_1, \hat{\mathbf{i}}_2, \hat{\mathbf{i}}_3), \quad (\hat{\mathbf{e}}_1, \hat{\mathbf{e}}_2, \hat{\mathbf{e}}_3). \tag{2.2.37}$$

For an orthonormal basis, we have

$$[\mathbf{e}_1 \mathbf{e}_2 \mathbf{e}_3] = 1 \quad \text{and} \quad \mathbf{e}^1 = \mathbf{e}_1 \equiv \hat{\mathbf{e}}_1, \quad \mathbf{e}^2 = \mathbf{e}_2 \equiv \hat{\mathbf{e}}_2, \quad \mathbf{e}^3 = \mathbf{e}_3 \equiv \hat{\mathbf{e}}_3.$$

Thus, for an orthonormal system, there is no distinction between cogredient and contragredient components. In most situations an orthonormal basis simplifies calculations.

For an orthonormal basis, a vector $\mathbf{A}$ can be expressed in terms of its components as

$$\mathbf{A} = A_1 \hat{\mathbf{e}}_1 + A_2 \hat{\mathbf{e}}_2 + A_3 \hat{\mathbf{e}}_3 = \sum_{i=1}^{3} A_i \hat{\mathbf{e}}_i, \tag{2.2.38}$$

where $\hat{\mathbf{e}}_i$ $(i = 1, 2, 3)$ is the orthonormal basis and A_i are the corresponding *physical components* (i.e., the components have the same physical dimensions as the vector). In an orthonormal basis, the scalar and vector products of vectors $\mathbf{A}$ and $\mathbf{B}$ can be expressed in terms of their components as

$$\mathbf{A} \cdot \mathbf{B} = \sum_{i,j=1}^{3} A_i B_j (\hat{\mathbf{e}}_i \cdot \hat{\mathbf{e}}_j) = \sum_{i=1}^{3} A_i B_i,$$

$$(\mathbf{A} \times \mathbf{B})_k = \sum_{i,j=1;\ i,j \neq k}^{3} A_i B_j (\hat{\mathbf{e}}_i \times \hat{\mathbf{e}}_j \cdot \hat{\mathbf{e}}_k), \tag{2.2.39}$$

where $(\cdot)_k$ indicates the kth component of the enclosed vector. Shortly, we shall introduce specific symbols to denote the scalars $\hat{\mathbf{e}}_i \cdot \hat{\mathbf{e}}_j$ and $\hat{\mathbf{e}}_i \times \hat{\mathbf{e}}_j \cdot \hat{\mathbf{e}}_k$.

Example 2.2.4

Let the vectors $(\hat{\mathbf{i}}, \hat{\mathbf{j}}, \hat{\mathbf{k}})$ constitute an orthonormal basis. In terms of this basis, define a cogredient basis by

$$\mathbf{e}_1 = \frac{\sqrt{3}}{4}\hat{\mathbf{i}} + \frac{1}{4}\hat{\mathbf{j}}, \quad \mathbf{e}_2 = \frac{1}{2}\hat{\mathbf{i}} + \frac{3}{2}\hat{\mathbf{j}}, \quad \mathbf{e}_3 = \hat{\mathbf{k}}.$$

Determine

(a) the dual or reciprocal (contragredient) basis $(\mathbf{e}^1, \mathbf{e}^2, \mathbf{e}^3)$ in terms of the orthonormal basis $(\hat{\mathbf{i}}, \hat{\mathbf{j}}, \hat{\mathbf{k}})$,

(b) the magnitudes (or norms) $|\mathbf{e}_1|, |\mathbf{e}_2|, |\mathbf{e}_3|, |\mathbf{e}^1|, |\mathbf{e}^2|$, and $|\mathbf{e}^3|$, and

(c) the cogredient components A_1, A_2, and A_3 of a vector $\mathbf{A}$ if its contragredient components are given by $A^1 = 1, A^2 = 2, A^3 = 3$.

Solution: (a) Note that

$$\mathbf{e}_1 \times \mathbf{e}_2 = \left(\frac{3\sqrt{3}-1}{8}\right)\hat{\mathbf{k}}, \quad \mathbf{e}_2 \times \mathbf{e}_3 = -\frac{1}{2}\hat{\mathbf{j}} + \frac{3}{2}\hat{\mathbf{i}}, \quad \mathbf{e}_3 \times \mathbf{e}_1 = \frac{\sqrt{3}}{4}\hat{\mathbf{j}} - \frac{1}{4}\hat{\mathbf{i}}, \quad [\mathbf{e}_1\mathbf{e}_2\mathbf{e}_3] = \frac{3\sqrt{3}-1}{8}.$$

Then

$$\mathbf{e}^1 = \frac{3}{2\Delta}\hat{\mathbf{i}} - \frac{1}{2\Delta}\hat{\mathbf{j}}, \quad \mathbf{e}^2 = -\frac{1}{4\Delta}\hat{\mathbf{i}} + \frac{\sqrt{3}}{4\Delta}\hat{\mathbf{j}}, \quad \mathbf{e}^3 = \hat{\mathbf{k}},$$

where $\Delta = [\mathbf{e}_1\mathbf{e}_2\mathbf{e}_3]$. The reader is asked to sketch both sets of vectors to an approximate scale.

(b) The magnitudes of the base vectors are

$$|\mathbf{e}_1| = \frac{1}{2}, \quad |\mathbf{e}_2| = \sqrt{\frac{5}{2}}, \quad |\mathbf{e}_3| = 1, \quad |\mathbf{e}^1| = \frac{1}{\Delta}\sqrt{\frac{5}{2}}, \quad |\mathbf{e}^2| = \frac{1}{2\Delta}, \quad |\mathbf{e}^3| = 1.$$

(c) We have $\mathbf{A} = A^1\mathbf{e}_1 + A^2\mathbf{e}_2 + A^3\mathbf{e}_3 = \mathbf{e}_1 + 2\mathbf{e}_2 + 3\mathbf{e}_3$. We obtain

$$A_1 = (\mathbf{e}_1 + 2\mathbf{e}_2 + 3\mathbf{e}_3) \cdot \mathbf{e}_1 = \frac{1}{4} + \frac{\sqrt{3}+3}{4} = 1 + \frac{\sqrt{3}}{4},$$

$$A_2 = (\mathbf{e}_1 + 2\mathbf{e}_2 + 3\mathbf{e}_3) \cdot \mathbf{e}_2 = \frac{\sqrt{3}+3}{8} + 2 \times \frac{10}{4} = \frac{\sqrt{3}+43}{8},$$

$$A_3 = (\mathbf{e}_1 + 2\mathbf{e}_2 + 3\mathbf{e}_3) \cdot \mathbf{e}_3 = 3.$$

2.2.4.4 The Gram–Schmidt orthonormalization

If $(\mathbf{e}_1, \mathbf{e}_2, \mathbf{e}_3)$ is a linearly independent set but not orthonormal, the *Gram–Schmidt orthonormalization* process can be used to convert the set to an orthonormal set, as follows. Let $(\mathbf{e}_1, \mathbf{e}_2, ...\mathbf{e}_n)$ be any linearly independent set of vectors. We construct an orthonormal set $(\hat{\mathbf{e}}_1, \hat{\mathbf{e}}_2, ..., \hat{\mathbf{e}}_n)$ using the following procedure: Due to the fact that $\mathbf{e}_1$ is an element of a linearly independent set, $\mathbf{e}_1 \neq \mathbf{0}$, and therefore $|\mathbf{e}_1| > 0$. Let $\hat{\mathbf{e}}_1 \equiv \mathbf{e}_1/|\mathbf{e}_1|$ so that $|\hat{\mathbf{e}}_1| = 1$. Next choose the second vector $\mathbf{e}_2$ from the original set and require the vector $\mathbf{e}'_2 = \mathbf{e}_2 - \alpha\hat{\mathbf{e}}_1$ to be orthogonal to $\hat{\mathbf{e}}_1$:

$$0 = \hat{\mathbf{e}}_1 \cdot (\mathbf{e}_2 - \alpha\hat{\mathbf{e}}_1) = (\hat{\mathbf{e}}_1 \cdot \mathbf{e}_2) - \alpha|\hat{\mathbf{e}}_1|^2 \implies \alpha = \hat{\mathbf{e}}_1 \cdot \mathbf{e}_2.$$

The second element of the desired set is then obtained by

$$\hat{\mathbf{e}}_2 = [\mathbf{e}_2 - (\hat{\mathbf{e}}_1 \cdot \mathbf{e}_2)\hat{\mathbf{e}}_1]\frac{1}{|\mathbf{e}'_2|}.$$

Continuing the procedure, we obtain the $(r+1)$st element as

$$\mathbf{e}'_{r+1} = \mathbf{e}_{r+1} - (\hat{\mathbf{e}}_1 \cdot \mathbf{e}_{r+1})\hat{\mathbf{e}}_1 - (\hat{\mathbf{e}}_2 \cdot \mathbf{e}_{r+1})\hat{\mathbf{e}}_2 - \cdots - (\hat{\mathbf{e}}_r \cdot \mathbf{e}_{r+1})\hat{\mathbf{e}}_r,$$

$$\hat{\mathbf{e}}_{r+1} = \frac{\mathbf{e}'_{r+1}}{|\mathbf{e}'_{r+1}|}. \tag{2.2.40}$$

Example 2.2.5 ———————————————————————————————————————

Construct an orthonormal basis from the set of cogredient vectors in Example 2.2.4:

$$\mathbf{e}_1 = \tfrac{\sqrt{3}}{4}\hat{\mathbf{i}} + \tfrac{1}{4}\hat{\mathbf{j}}, \quad \mathbf{e}_2 = \tfrac{1}{2}\hat{\mathbf{i}} + \tfrac{3}{2}\hat{\mathbf{j}}, \quad \mathbf{e}_3 = \hat{\mathbf{k}}.$$

Solution: We begin with $\mathbf{e}_1$ and determine $\hat{\mathbf{e}}_1$:

$$\hat{\mathbf{e}}_1 = \frac{\mathbf{e}_1}{|\mathbf{e}_1|} = \tfrac{\sqrt{3}}{2}\hat{\mathbf{i}} + \tfrac{1}{2}\hat{\mathbf{j}}.$$

Then $\hat{\mathbf{e}}_2$ is constructed using Eq. (2.2.40):

$$\mathbf{e}'_2 \equiv \mathbf{e}_2 - (\hat{\mathbf{e}}_1 \cdot \mathbf{e}_2)\hat{\mathbf{e}}_1 = \tfrac{1}{2}\hat{\mathbf{i}} + \tfrac{3}{2}\hat{\mathbf{j}} - \left[\left(\tfrac{\sqrt{3}}{2}\hat{\mathbf{i}} + \tfrac{1}{2}\hat{\mathbf{j}}\right) \cdot \left(\tfrac{1}{2}\hat{\mathbf{i}} + \tfrac{3}{2}\hat{\mathbf{j}}\right)\right]\left(\tfrac{\sqrt{3}}{2}\hat{\mathbf{i}} + \tfrac{1}{2}\hat{\mathbf{j}}\right)$$

$$= \tfrac{1-3\sqrt{3}}{8}\hat{\mathbf{i}} + \tfrac{9-\sqrt{3}}{8}\hat{\mathbf{j}}$$

$$\hat{\mathbf{e}}_2 = \frac{\mathbf{e}'_2}{|\mathbf{e}'_2|}, \quad |\mathbf{e}'_2|^2 = \frac{14 - 3\sqrt{3}}{8}$$

Finally, $\hat{\mathbf{e}}_3$ is given by $\mathbf{e}'_3 \equiv \mathbf{e}_3 - (\hat{\mathbf{e}}_1 \cdot \mathbf{e}_3)\hat{\mathbf{e}}_1 - (\hat{\mathbf{e}}_2 \cdot \mathbf{e}_3)\hat{\mathbf{e}}_2 = \hat{\mathbf{k}} \Rightarrow \hat{\mathbf{e}}_3 = \hat{\mathbf{k}}.$

2.2.5 Summation Convention

Equations governing a continuous medium contain, especially in three dimensions, long expressions with many additive terms. Often these terms have a similar structure so that a typical term of the expression can be identified. For example, consider the component form of vector $\mathbf{A}$ [see Eq. (2.2.30)]:

$$\mathbf{A} = A^1\mathbf{e}_1 + A^2\mathbf{e}_2 + A^3\mathbf{e}_3, \tag{2.2.41}$$

A typical term in Eq. (2.2.41) is of the form $A^i\mathbf{e}_i$, where i takes the values of 1, 2, and 3. Therefore, the expression can be abbreviated as

$$\mathbf{A} = \sum_{i=1}^{3} A^i\mathbf{e}_i \quad \text{or} \quad \mathbf{A} = \sum_{m=1}^{3} A^m\mathbf{e}_m. \tag{2.2.42}$$

The summation index i or m is arbitrary as long as the same index is used for both A and $\mathbf{e}$. The expression can be shortened further by omitting the summation sign and understanding that a repeated index means summation over all values of that index. Thus, the three-term expression $\mathbf{A}$ can be written simply as

$$\mathbf{A} = A^i\mathbf{e}_i = A^m\mathbf{e}_m \tag{2.2.43}$$

This notation is called the *summation convention*. For example, an arbitrary vector **A** can be expressed in terms of its components [see Eq. (2.2.36)] as

$$\mathbf{A} = \left(\mathbf{A} \cdot \mathbf{e}^i\right) \mathbf{e}_i = \left(\mathbf{A} \cdot \mathbf{e}_j\right) \mathbf{e}^j. \qquad (2.2.44)$$

2.2.5.1 Dummy index

The repeated index is called a *dummy index* because it can be replaced by *any other symbol that has not already been used* in that expression. Thus, the expression in Eq. (2.2.43) can also be written as

$$\mathbf{A} = A^i \mathbf{e}_i = A^j \mathbf{e}_j = A^m \mathbf{e}_m, \qquad (2.2.45)$$

and so on. As a rule, no index can appear more than twice in an expression. For example, $A_i B_i C_i$ is not a valid expression because the index i appears more than twice. Other examples of dummy indices are

$$F_i = A_i B_j C_j, \quad G_k = H_k(2 - 3A_i B_i) + P_j Q_j F_k. \qquad (2.2.46)$$

The first equation in Eq. (2.2.46), for example, expresses three equations when the range of i and j is 1 to 3. We have

$$F_1 = A_1 \left(B_1 C_1 + B_2 C_2 + B_3 C_3\right),$$
$$F_2 = A_2 \left(B_1 C_1 + B_2 C_2 + B_3 C_3\right),$$
$$F_3 = A_3 \left(B_1 C_1 + B_2 C_2 + B_3 C_3\right).$$

2.2.5.2 Free index

A *free index* is one that appears in every expression of an equation, except for expressions that contain real numbers (scalars) only. Index i in the equation $F_i = A_i B_j C_j$ and k in the equation $G_k = H_k(2 - 3A_i B_i) + P_j Q_j F_k$ are free indices. Another example is

$$A_i = 2 + B_i + C_i + D_i + (F_j G_j - H_j P_j)E_i.$$

This expression contains three equations ($i = 1, 2, 3$). The expressions $A_i = B_j C_k$, $A_i = B_j$, and $F_k = A_i B_j C_k$ do not make sense and should not arise, because the indices on both sides of the equal sign do not match.

One must be careful when substituting a quantity with an index into an expression with indices or solving for one quantity with index in terms of the others with indices in an equation. For example, consider the equations $p_i = a_i b_j c_j$ and $c_k = d_i e_i q_k$. It is correct to write $a_i = p_i/(b_j c_j)$ but it is incorrect to write $b_j c_j = p_i/a_i$, which has a totally different meaning. Before substituting for c_k from one expression into the other, one should examine the indices and determine which ones are free and which ones are dummy indices, and then make suitable changes of the indices before substitution. For example, we can rewrite the second expression as $c_j = d_k e_k q_j$ and substitute it into the first expression, $p_i = a_i b_j d_k e_k q_j = a_i(b_j q_j)(d_k e_k)$. Other correct ways to write are

$$p_i = a_i b_k c_k = a_i b_k d_j e_j q_k = a_i(b_k q_k)(d_j e_j), \quad p_k = a_k b_j d_i e_i q_j = a_k(b_j q_j)(d_i e_i).$$

2.2.5.3 Kronecker delta

It is convenient to introduce the Kronecker delta δ_{ij} because it allows simple representation of the scalar product of orthonormal vectors in a right-handed basis system. We define the scalar (or dot) product $\hat{\mathbf{e}}_i \cdot \hat{\mathbf{e}}_j$ as [also see Eq. (2.2.34)]

$$\delta_{ij} \equiv \hat{\mathbf{e}}_i \cdot \hat{\mathbf{e}}_j = \delta_{ji}, \qquad (2.2.47)$$

where

$$\delta_{ij} = \begin{cases} 1, \text{ if } i \text{ has the same value as } j \\ 0, \text{ if } i \text{ has a different value than } j. \end{cases} \qquad (2.2.48)$$

For example, we have $\delta_{11} = \delta_{22} = \delta_{33} = 1$ and $\delta_{12} = \delta_{13} = \delta_{23} = 0$. We also have $\delta_{ii} = \delta_{11} + \delta_{22} + \delta_{33} = 3$. The Kronecker delta δ_{ij} modifies (or contracts) the subscripts in the coefficients of an expression in which it appears:

$$A_i \delta_{ij} = A_j, \quad A_i B_j \delta_{ij} = A_i B_i = A_j B_j, \quad \delta_{ij} \delta_{ik} = \delta_{jk}.$$

As we shall see shortly, δ_{ij} denotes the components of a second-order unit tensor, $\mathbf{I} = \delta_{ij} \hat{\mathbf{e}}_i \hat{\mathbf{e}}_j = \hat{\mathbf{e}}_i \hat{\mathbf{e}}_i$.

2.2.5.4 Permutation symbol

We define the vector (or cross) product $\hat{\mathbf{e}}_i \times \hat{\mathbf{e}}_j$ as

$$\hat{\mathbf{e}}_i \times \hat{\mathbf{e}}_j \equiv e_{ijk} \hat{\mathbf{e}}_k, \qquad (2.2.49)$$

where e_{ijk} is called the *alternating symbol* or *permutation symbol* [see also Eq. (2.2.35) for the meaning of e_{ijk}], which has the meaning

$$e_{ijk} \equiv \hat{\mathbf{e}}_i \times \hat{\mathbf{e}}_j \cdot \hat{\mathbf{e}}_k = \begin{cases} 1, \text{ if } i, j, k \text{ are in a natural cyclic order} \\ \quad \text{and not repeated } (i \neq j \neq k) \\ -1, \text{ if } i, j, k \text{ are opposite to a natural} \\ \quad \text{cyclic order and not repeated } (i \neq j \neq k) \\ 0, \text{ if any of } i, j, k \text{ are repeated.} \end{cases} \qquad (2.2.50)$$

The *natural* order of i, j, and k is the order in which they appear alphabetically. A natural cyclic order means going from i to j and k, from j to k and i, or from k to i and j, as shown in Fig. 2.2.10(a). Going opposite to a natural cyclic order is shown in Fig. 2.2.10(b). By definition, the subscripts of the permutation symbol can be permuted in a natural cyclic order, without changing its value. An interchange of any two subscripts will change the sign (hence, interchange of two subscripts twice keeps the value unchanged):

$$e_{ijk} = e_{kij} = e_{jki}, \quad e_{ijk} = -e_{jik} = e_{jki} = -e_{kji} \text{ for any } i, j, k,$$

$$e_{123} = e_{312} = e_{231} = 1, \quad e_{213} = e_{321} = e_{132} = -1, \quad e_{113} = e_{331} = e_{322} = 0.$$

One can show that (a) $e_{ijk} e_{ijk} = 6$, (b) $A_i A_j e_{ijk} = 0$, and (c) $e_{imn} e_{jmn} = 2\delta_{ij}$. An alternative formula [to Eq. (2.2.50)] to determine the value of e_{ijk} is

$$e_{ijk} = \tfrac{1}{2}(i - j)(j - k)(k - i) \text{ for any } i, j, k = 1, 2, 3. \qquad (2.2.51)$$

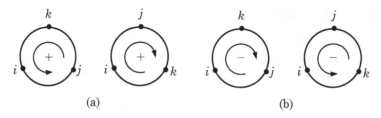

Fig. 2.2.10: (a) A natural cyclic order is going from any index to the next in the order it appears alphabetically (going from k to i makes it cyclic). (b) Opposite to a natural cyclic order is going in a direction opposite to that of a natural cyclic order.

In an orthonormal basis the scalar and vector products can be expressed in index form using the Kronecker delta and the alternating symbols [see Eq. (2.2.39)]:

$$\mathbf{A} \cdot \mathbf{B} = (A_i\hat{\mathbf{e}}_i) \cdot (B_j\hat{\mathbf{e}}_j) = A_iB_j\delta_{ij} = A_iB_i,$$
$$\mathbf{A} \times \mathbf{B} = (A_i\hat{\mathbf{e}}_i) \times (B_j\hat{\mathbf{e}}_j) = A_iB_j\,e_{ijk}\,\hat{\mathbf{e}}_k. \tag{2.2.52}$$

Note that the components of a vector in an orthonormal coordinate system can be expressed as

$$A_i = \mathbf{A} \cdot \hat{\mathbf{e}}_i, \tag{2.2.53}$$

and therefore we can express vector $\mathbf{A}$ as

$$\mathbf{A} = A_i\hat{\mathbf{e}}_i = (\mathbf{A} \cdot \hat{\mathbf{e}}_i)\hat{\mathbf{e}}_i. \tag{2.2.54}$$

Further, the Kronecker delta and the permutation symbol are related by the following identity, known as the *e-δ identity* [follows from part (d) of Problem **2.41**]:

$$e_{ijk}\, e_{imn} = \delta_{jm}\, \delta_{kn} - \delta_{jn}\, \delta_{km}. \tag{2.2.55}$$

The permutation symbol and the Kronecker delta prove to be very useful in establishing vector identities. Since a vector form of any identity is invariant (i.e., valid in any coordinate system), it suffices to prove it in one coordinate system. The following examples contain several cases of incorrect and correct uses of index notation and illustrate some of the uses of δ_{ij} and e_{ijk}.

Example 2.2.6 _____

Discuss the validity of the following expressions:

(a) $a_mb_s = c_m(d_r - f_r)$.   (b) $a_mb_s = c_m(d_s - f_s)$.   (c) $a_i = b_jc_id_i$.
(d) $x_ix_i = r^2$.   (e) $a_ib_jc_j = 3$.

Solution:

(a) Not a valid expression because the free indices r and s do not match.

(b) Valid; both m and s are free indices. There are nine equations ($m, s = 1, 2, 3$).

(c) Not a valid expression because the free index j is not matched on both sides of the equality, and index i is a dummy index in one expression and a free index in the other; i cannot be used both as a free and a dummy index in the same equation. The equation would be valid if i on the left side of the equation is replaced with j; then there will be three equations.

(d) A valid expression, containing one equation: $x_1^2 + x_2^2 + x_3^2 = r^2$. Note that $r = \sqrt{x_i x_i} \neq x_i$.

(e) A valid expression in mathematics; however, in mechanics such relations may not arise. If they do, such expressions are invalid because they violate the form-invariance under a coordinate transformation (that is, every component of a vector cannot be the same in all coordinate systems). The expression contains three equations ($i = 1, 2, 3$): $a_1(b_1c_1 + b_2c_2 + b_3c_3) = 3$, $a_2(b_1c_1 + b_2c_2 + b_3c_3) = 3$, and $a_3(b_1c_1 + b_2c_2 + b_3c_3) = 3$.

Example 2.2.7

Simplify the following expressions:

(a) $\delta_{ij}\delta_{jk}\delta_{kp}\delta_{pi}$.   (b) $e_{mjk}\,e_{njk}$.   (c) $(\mathbf{A} \times \mathbf{B}) \cdot (\mathbf{C} \times \mathbf{D})$.

Solution: (a) Successive contraction of subscripts yields the result:

$$\delta_{ij}\delta_{jk}\delta_{kp}\delta_{pi} = \delta_{ij}\delta_{jk}\delta_{ki} = \delta_{ij}\delta_{ji} = \delta_{ii} = 3.$$

(b) Expand using the e-δ identity:

$$e_{mjk}\,e_{njk} = \delta_{mn}\delta_{jj} - \delta_{mj}\delta_{nj} = 3\delta_{mn} - \delta_{mn} = 2\delta_{mn}.$$

In particular, the expression $e_{ijk}\,e_{ijk}$ is equal to $2\delta_{ii} = 6$.

(c) Expanding the expression using the index notation, we obtain

$$\begin{aligned}
(\mathbf{A} \times \mathbf{B}) \cdot (\mathbf{C} \times \mathbf{D}) &= (A_i B_j e_{ijk}\hat{\mathbf{e}}_k) \cdot (C_m D_n\,e_{mnp}\,\hat{\mathbf{e}}_p) \\
&= A_i B_j C_m D_n\,e_{ijk}\,e_{mnp}\,\delta_{kp} = A_i B_j C_m D_n\,e_{ijk}\,e_{mnk} \\
&= A_i B_j C_m D_n(\delta_{im}\delta_{jn} - \delta_{in}\delta_{jm}) \\
&= A_i B_j C_m D_n\,\delta_{im}\delta_{jn} - A_i B_j C_m D_n\,\delta_{in}\delta_{jm} \\
&= A_i B_j C_i D_j - A_i B_j C_j D_i = A_i C_i B_j D_j - A_i D_i B_j C_j \\
&= (\mathbf{A} \cdot \mathbf{C})(\mathbf{B} \cdot \mathbf{D}) - (\mathbf{A} \cdot \mathbf{D})(\mathbf{B} \cdot \mathbf{C}),
\end{aligned}$$

where we have used the e-δ identity (2.2.55). Although the vector identity is established in an orthonormal coordinate system, it holds in a general coordinate system.

Example 2.2.8

Rewrite the expression $e_{ijk}\,A_i\,B_j\,C_k$ in vector form.

Solution: Examining the indices in the permutation symbol and those of the coefficients, it is clear that there are three possibilities: (1) $\mathbf{A}$ and $\mathbf{B}$ must have a cross product between them and the resulting vector must have a dot product with $\mathbf{C}$; (2) $\mathbf{B}$ and $\mathbf{C}$ must have a cross product between them and the resulting vector must have a dot product with $\mathbf{A}$; or (3) $\mathbf{C}$ and $\mathbf{A}$ must have a cross product between them and the resulting vector must have a dot product with $\mathbf{B}$. Thus we have

$$e_{ijk}\,A_i\,B_j\,C_k = (\mathbf{A} \times \mathbf{B}) \cdot \mathbf{C} = (\mathbf{B} \times \mathbf{C}) \cdot \mathbf{A} = (\mathbf{C} \times \mathbf{A}) \cdot \mathbf{B} \equiv [\mathbf{ABC}] = [\mathbf{BCA}] = [\mathbf{CAB}].$$

The parentheses can be removed in the above expressions as the dot product has no meaning unless the expression in the parentheses is a vector. Also, the interchange of dot product and cross product $\mathbf{A} \times \mathbf{B} \cdot \mathbf{C} = \mathbf{A} \cdot \mathbf{B} \times \mathbf{C}$, or a cyclical permutation of the order of the vectors $\mathbf{A} \times \mathbf{B} \cdot \mathbf{C} = \mathbf{B} \times \mathbf{C} \cdot \mathbf{A}$, leaves the result unchanged [see Eq. (2.2.20)].

2.2.6 Transformation Law for Different Bases

2.2.6.1 General transformation laws

In addition to the basis (e_1, e_2, e_3) and its dual (e^1, e^2, e^3), consider a second (barred) basis: $(\bar{e}_1, \bar{e}_2, \bar{e}_3)$ and its dual $(\bar{e}^1, \bar{e}^2, \bar{e}^3)$. Now we can express the same vector in four ways:

$$\mathbf{A} = A^i \mathbf{e}_i = A_j \mathbf{e}^j, \quad \text{in unbarred basis,} \tag{2.2.56}$$
$$= \bar{A}^m \bar{\mathbf{e}}_m = \bar{A}_n \bar{\mathbf{e}}^n, \quad \text{in barred basis.} \tag{2.2.57}$$

Note that i, j, m, and n are all dummy indices. From Eq. (2.2.44) we have

$$A^i = \mathbf{A} \cdot \mathbf{e}^i, \quad A_j = \mathbf{A} \cdot \mathbf{e}_j, \quad \bar{A}^m = \mathbf{A} \cdot \bar{\mathbf{e}}^m, \quad \bar{A}_n = \mathbf{A} \cdot \bar{\mathbf{e}}_n, \tag{2.2.58}$$

and from Eqs. (2.2.56)–(2.2.58) it follows that

$$\bar{A}^m = (\mathbf{e}_i \cdot \bar{\mathbf{e}}^m) A^i = (\mathbf{e}^j \cdot \bar{\mathbf{e}}^m) A_j, \tag{2.2.59}$$
$$\bar{A}_n = (\mathbf{e}^j \cdot \bar{\mathbf{e}}_n) A_j = (\mathbf{e}_i \cdot \bar{\mathbf{e}}_n) A^i. \tag{2.2.60}$$

The first two terms of Eqs. (2.2.59) and (2.2.60) give the transformation rules between the contragredient and the cogredient components in the two basis systems.

By means of Eq. (2.2.44) we find that the basis systems are related by

$$\bar{\mathbf{e}}^s = (\bar{\mathbf{e}}^s \cdot \mathbf{e}_j) \mathbf{e}^j = (\bar{\mathbf{e}}^s \cdot \mathbf{e}^j) \mathbf{e}_j. \tag{2.2.61}$$
$$\bar{\mathbf{e}}_s = (\bar{\mathbf{e}}_s \cdot \mathbf{e}^j) \mathbf{e}_j = (\bar{\mathbf{e}}_s \cdot \mathbf{e}_j) \mathbf{e}^j. \tag{2.2.62}$$

If we now write

$$a^j_s \equiv \bar{\mathbf{e}}_s \cdot \mathbf{e}^j, \qquad b^s_i \equiv \bar{\mathbf{e}}^s \cdot \mathbf{e}_i, \tag{2.2.63}$$

then we have in summary

$$\bar{\mathbf{e}}_s = a^j_s \mathbf{e}_j, \quad \bar{A}_s = a^j_s A_j, \quad \text{cogredient law.} \tag{2.2.64}$$
$$\bar{\mathbf{e}}^s = b^s_i \mathbf{e}^i, \quad \bar{A}^s = b^s_i A^i, \quad \text{contragredient law.} \tag{2.2.65}$$

Thus there are two transformation laws, and the subscripts and superscripts are assigned according to which law is satisfied. The subscripted basis vectors and the subscripted components transform according to the same law, the cogredient law, and the superscripted basis vectors and superscripted components transform according to another law, the contragredient law.

There are also relations between the subscripted and superscripted basis vectors and components in the two systems. These relations are called *mixed* laws. Let

$$c_{sj} \equiv \bar{\mathbf{e}}_s \cdot \mathbf{e}_j, \quad d^{sj} \equiv \bar{\mathbf{e}}^s \cdot \mathbf{e}^j. \tag{2.2.66}$$

Then we have

$$\left. \begin{array}{ll} \bar{\mathbf{e}}_s = c_{sj} \mathbf{e}^j, & \bar{A}_s = c_{sj} A^j \\ \bar{\mathbf{e}}^s = d^{sj} \bar{\mathbf{e}}_j, & \bar{A}^s = d^{sj} A_j \end{array} \right\}, \quad \text{mixed laws.} \tag{2.2.67}$$

2.2.6.2 Transformation laws for orthonormal systems

In much of our study, we shall deal with Cartesian bases. Let us denote an orthonormal Cartesian basis by

$$\{\hat{\mathbf{e}}_x, \hat{\mathbf{e}}_y, \hat{\mathbf{e}}_z\} \quad \text{or} \quad \{\hat{\mathbf{e}}_1, \hat{\mathbf{e}}_2, \hat{\mathbf{e}}_3\}.$$

The Cartesian coordinates are denoted by (x, y, z) or (x_1, x_2, x_3). The familiar rectangular Cartesian coordinate system is shown in Fig. 2.2.11(a). We shall always use right-handed coordinate systems.

A position vector $\mathbf{x} = \mathbf{r}$ to an arbitrary point $(x, y, z) = (x_1, x_2, x_3)$ is

$$\mathbf{x} = x\hat{\mathbf{e}}_x + y\hat{\mathbf{e}}_y + z\hat{\mathbf{e}}_z = x_1\hat{\mathbf{e}}_1 + x_2\hat{\mathbf{e}}_2 + x_3\hat{\mathbf{e}}_3, \tag{2.2.68}$$

or, in summation notation,

$$\mathbf{x} = x_j\hat{\mathbf{e}}_j, \quad \mathbf{x} \cdot \mathbf{x} = r^2 = x_i x_i. \tag{2.2.69}$$

Next, we establish the relationship between the components of two different orthonormal coordinate systems, say, unbarred and barred [see Fig. 2.2.11(b)]. Consider the unbarred coordinate basis

$$(\hat{\mathbf{e}}_1, \hat{\mathbf{e}}_2, \hat{\mathbf{e}}_3)$$

and the barred coordinate basis

$$(\hat{\bar{\mathbf{e}}}_1, \hat{\bar{\mathbf{e}}}_2, \hat{\bar{\mathbf{e}}}_3).$$

Then Eq. (2.2.62) yields, as a special case, the following relations:

$$\hat{\bar{\mathbf{e}}}_i = \ell_{ij}\,\hat{\mathbf{e}}_j, \quad \bar{A}_i = \ell_{ij}\,A_j, \tag{2.2.70}$$

where

$$\ell_{ij} = \hat{\bar{\mathbf{e}}}_i \cdot \hat{\mathbf{e}}_j. \tag{2.2.71}$$

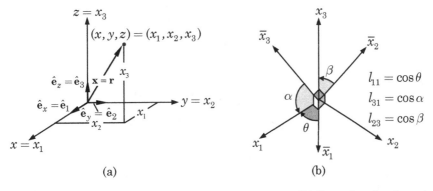

Fig. 2.2.11: (a) A rectangular Cartesian coordinate system. (b) Barred and unbarred coordinate systems.

Equation (2.2.70) gives the relationship between the base vectors as well as the components of the barred and unbarred coordinate systems. The relationship between the components $(\bar{A}_1, \bar{A}_2, \bar{A}_3)$ and (A_1, A_2, A_3) is called the *transformation rule* between the barred and unbarred components in the two orthogonal coordinate systems. The coefficients ℓ_{ij} are the *direction cosines* of the barred coordinate system with respect to the unbarred coordinate system:

$$\ell_{ij} = \text{cosine of the angle between } \hat{\bar{e}}_i \text{ and } \hat{e}_j. \tag{2.2.72}$$

Note that the first subscript of ℓ_{ij} comes from the barred coordinate system and the second subscript from the unbarred system. Obviously, ℓ_{ij} is not symmetric (i.e., $\ell_{ij} \neq \ell_{ji}$). The direction cosines allow us to relate components of a vector (or a tensor) in the unbarred coordinate system to components of the same vector (or tensor) in the barred coordinate system. Example 2.2.9 illustrates the computation of direction cosines.

Example 2.2.9 ————————————————————————————

Let $\hat{e}_i$ $(i = 1, 2, 3)$ be a set of orthonormal base vectors. Then define a new right-handed coordinate basis by (note that $\hat{\bar{e}}_1 \cdot \hat{\bar{e}}_2 = 0$):

$$\hat{\bar{e}}_1 = \tfrac{1}{3}\left(2\hat{e}_1 + 2\hat{e}_2 + \hat{e}_3\right), \quad \hat{\bar{e}}_2 = \tfrac{1}{\sqrt{2}}\left(\hat{e}_1 - \hat{e}_2\right), \quad \hat{\bar{e}}_3 = \hat{\bar{e}}_1 \times \hat{\bar{e}}_2 = \tfrac{1}{3\sqrt{2}}\left(\hat{e}_1 + \hat{e}_2 - 4\hat{e}_3\right).$$

The original and new coordinate systems are depicted in Fig. 2.2.12. Determine the direction cosines ℓ_{ij} of the transformation and display them in a rectangular array.

Solution: From Eq. (2.2.71) we have

$$\ell_{11} = \hat{\bar{e}}_1 \cdot \hat{e}_1 = \tfrac{2}{3}, \qquad \ell_{12} = \hat{\bar{e}}_1 \cdot \hat{e}_2 = \tfrac{2}{3}, \qquad \ell_{13} = \hat{\bar{e}}_1 \cdot \hat{e}_3 = \tfrac{1}{3},$$

$$\ell_{21} = \hat{\bar{e}}_2 \cdot \hat{e}_1 = \tfrac{1}{\sqrt{2}}, \qquad \ell_{22} = \hat{\bar{e}}_2 \cdot \hat{e}_2 = -\tfrac{1}{\sqrt{2}}, \qquad \ell_{23} = \hat{\bar{e}}_2 \cdot \hat{e}_3 = 0,$$

$$\ell_{31} = \hat{\bar{e}}_3 \cdot \hat{e}_1 = \tfrac{1}{3\sqrt{2}}, \qquad \ell_{32} = \hat{\bar{e}}_3 \cdot \hat{e}_2 = \tfrac{1}{3\sqrt{2}}, \qquad \ell_{33} = \hat{\bar{e}}_3 \cdot \hat{e}_3 = -\tfrac{4}{3\sqrt{2}}.$$

The rectangular array of these components is denoted by $[L]$ and has the form (see Section 2.3 for the concept of a matrix)

$$[L] = \tfrac{1}{3\sqrt{2}} \begin{bmatrix} 2\sqrt{2} & 2\sqrt{2} & \sqrt{2} \\ 3 & -3 & 0 \\ 1 & 1 & -4 \end{bmatrix}.$$

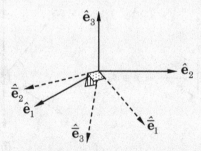

Fig. 2.2.12: Original and transformed coordinate systems defined in Example 2.2.9.

2.3 Theory of Matrices

2.3.1 Definition

In the preceding sections we studied the algebra of ordinary vectors and the transformation of vector components from one coordinate system to another. For example, the transformation in Eq. (2.2.70) relates the components of a vector in the barred coordinate system to those in the unbarred coordinate system. Writing Eq. (2.2.70) in expanded form,

$$\bar{A}_1 = \ell_{11}A_1 + \ell_{12}A_2 + \ell_{13}A_3,$$
$$\bar{A}_2 = \ell_{21}A_1 + \ell_{22}A_2 + \ell_{23}A_3, \qquad (2.3.1)$$
$$\bar{A}_3 = \ell_{31}A_1 + \ell_{32}A_2 + \ell_{33}A_3,$$

we see that there are nine coefficients relating the components A_i to $\bar{A}_i$. The form of these linear equations suggests writing down the scalars of ℓ_{ij} (jth components in the ith equation) in the rectangular array

$$[L] = \begin{bmatrix} \ell_{11} & \ell_{12} & \ell_{13} \\ \ell_{21} & \ell_{22} & \ell_{23} \\ \ell_{31} & \ell_{32} & \ell_{33} \end{bmatrix}.$$

This rectangular array $[L]$ of scalars ℓ_{ij} is called a *matrix*, and the quantities ℓ_{ij} are called the *elements* of $[L]$[1].

If a matrix has m rows and n columns, we say that is m by n ($m \times n$), the number of rows always being listed first. The element in the ith row and jth column of a matrix $[A]$ is generally denoted by a_{ij}, and we will sometimes designate a matrix by $[A] = [a_{ij}]$. A square matrix is one that has the same number of rows as columns. An $n \times n$ matrix is said to be of *order* n. The elements of a square matrix for which the row number and the column number are the same (i.e., a_{ij} for $i = j$) are called *diagonal elements* or simply the *diagonal*. A square matrix is said to be a *diagonal matrix* if all of the off-diagonal elements are zero. An *identity matrix*, denoted by $[I]$, is a diagonal matrix whose elements are all 1's. Examples of a diagonal and an identity matrix are given below:

$$\begin{bmatrix} 5 & 0 & 0 & 0 \\ 0 & -2 & 0 & 0 \\ 0 & 0 & 1 & 0 \\ 0 & 0 & 0 & 3 \end{bmatrix}, \qquad [I] = \begin{bmatrix} 1 & 0 & 0 & 0 \\ 0 & 1 & 0 & 0 \\ 0 & 0 & 1 & 0 \\ 0 & 0 & 0 & 1 \end{bmatrix}.$$

The sum of the diagonal elements is called the *trace* of the matrix.

---

[1]The word "matrix" was first used in 1850 by James Sylvester (1814–1897), an English algebraist. However, Arthur Caley (1821–1895), professor of mathematics at Cambridge, was the first one to explore properties of matrices. Significant contributions in the early years were made by Charles Hermite, Georg Frobenius, and Camille Jordan, among others.

If the matrix has only one row or one column, we normally use only a single subscript to designate its elements. For example,

$$\{X\} = \begin{Bmatrix} x_1 \\ x_2 \\ x_3 \end{Bmatrix} \quad \text{and} \quad \{Y\} = \{y_1 \ y_2 \ y_3\}$$

denote a column matrix and a row matrix, respectively. Row and column matrices can be used to represent the components of a vector.

2.3.2 Matrix Addition and Multiplication of a Matrix by a Scalar

The *sum* of two matrices of the same size is defined to be a matrix of the same size obtained by simply adding the corresponding elements. If $[A]$ is an $m \times n$ matrix and $[B]$ is an $m \times n$ matrix, their sum is an $m \times n$ matrix, $[C]$, with

$$c_{ij} = a_{ij} + b_{ij} \qquad \text{for all } i, j. \qquad (2.3.2)$$

A constant multiple of a matrix is equal to the matrix obtained by multiplying all of the elements by the constant. That is, the multiplication of a matrix $[A]$ by a scalar α, $\alpha[A]$, is the matrix obtained by multiplying each of its elements with α:

$$[A] = \begin{bmatrix} a_{11} & a_{12} & \cdots & a_{1n} \\ a_{21} & a_{22} & \cdots & a_{2n} \\ \vdots & \vdots & \cdots & \vdots \\ a_{m1} & a_{m2} & \cdots & a_{mn} \end{bmatrix}, \quad \alpha[A] = \begin{bmatrix} \alpha a_{11} & \alpha a_{12} & \cdots & \alpha a_{1n} \\ \alpha a_{21} & \alpha a_{22} & \cdots & \alpha a_{2n} \\ \cdots & \cdots & \cdots & \cdots \\ \alpha a_{m1} & \alpha a_{m2} & \cdots & \alpha a_{mn} \end{bmatrix}.$$

Matrix addition has the following properties:

(1) Addition is commutative: $[A] + [B] = [B] + [A]$.

(2) Addition is associative: $[A] + ([B] + [C]) = ([A] + [B]) + [C]$.

(3) There exists a unique matrix $[0]$, such that $[A] + [0] = [0] + [A] = [A]$. The matrix $[0]$ is called *zero matrix* when all elements are zeros.

(4) For each matrix $[A]$, there exists a unique matrix $-[A]$ such that $[A] + (-[A]) = [0]$.

(5) Addition is distributive with respect to scalar multiplication: $\alpha([A] + [B]) = \alpha[A] + \alpha[B]$.

(6) Addition is distributive with respect to matrix multiplication, which is discussed in Section 2.3.5:

$$([A] + [B])[C] = [A][C] + [B][C]; \quad [D]([A] + [B]) = [D][A] + [D][B],$$

where $[A]$ and $[B]$ are $m \times n$ matrices, $[C]$ is a $n \times p$ matrix, and $[D]$ is a $q \times m$ matrix (so that the products of matrices appearing in the above expressions are meaningful).

2.3.3 Matrix Transpose

If $[A]$ is an $m \times n$ matrix, then the $n \times m$ matrix obtained by interchanging its rows and columns is called the *transpose* of $[A]$ and is denoted by $[A]^T$. For example, the transposes of the matrices

$$[A] = \begin{bmatrix} 5 & -2 & 1 \\ 8 & 7 & 6 \\ 2 & 4 & 3 \\ -1 & 9 & 0 \end{bmatrix} \quad \text{and} \quad [B] = \begin{bmatrix} 3 & -1 & 2 & 4 \\ -6 & 3 & 5 & 7 \\ 9 & 6 & -2 & 1 \end{bmatrix}.$$

are

$$[A]^T = \begin{bmatrix} 5 & 8 & 2 & -1 \\ -2 & 7 & 4 & 9 \\ 1 & 6 & 3 & 0 \end{bmatrix} \quad \text{and} \quad [B]^T = \begin{bmatrix} 3 & -6 & 9 \\ -1 & 3 & 6 \\ 2 & 5 & -2 \\ 4 & 7 & 1 \end{bmatrix}.$$

Two basic properties are: $([A]^T)^T = [A]$; $([A] + [B])^T = [A]^T + [B]^T$.

2.3.4 Symmetric and Skew Symmetric Matrices

A square matrix $[A]$ of real numbers is said to be *symmetric* if $[A]^T = [A]$, and it is said to be *skew symmetric* if $[A]^T = -[A]$. In terms of the elements of $[A]$, $[A]$ is symmetric if and only if $a_{ij} = a_{ji}$, and it is skew symmetric if and only if $a_{ij} = -a_{ji}$. Note that the diagonal elements of a skew symmetric matrix are always zero because $a_{ij} = -a_{ij}$ implies $a_{ij} = 0$ for $i = j$. Examples of symmetric and skew symmetric matrices, respectively, are

$$\begin{bmatrix} 5 & -2 & 12 & 21 \\ -2 & 2 & 16 & -3 \\ 12 & 16 & 13 & 8 \\ 21 & -3 & 8 & 19 \end{bmatrix}, \quad \begin{bmatrix} 0 & -11 & 32 & 4 \\ 11 & 0 & 25 & 7 \\ -32 & -25 & 0 & 15 \\ -4 & -7 & -15 & 0 \end{bmatrix}.$$

Every square matrix $[A]$ can be expressed as a sum of a symmetric matrix and a skew symmetric matrix:

$$[A] = \tfrac{1}{2}\left([A] + [A]^T\right) + \tfrac{1}{2}\left([A] - [A]^T\right). \tag{2.3.3}$$

For example, consider the unsymmetric matrix

$$\begin{bmatrix} 4 & 6 & 8 \\ 2 & 10 & 12 \\ 18 & 14 & 6 \end{bmatrix},$$

which can be expressed as

$$\begin{bmatrix} 4 & 6 & 8 \\ 2 & 10 & 12 \\ 18 & 14 & 6 \end{bmatrix} = \begin{bmatrix} 2 & 3 & 4 \\ 1 & 5 & 6 \\ 9 & 7 & 3 \end{bmatrix} + \begin{bmatrix} 2 & 1 & 9 \\ 3 & 5 & 7 \\ 4 & 6 & 3 \end{bmatrix} + \begin{bmatrix} 2 & 3 & 4 \\ 1 & 5 & 6 \\ 9 & 7 & 3 \end{bmatrix} - \begin{bmatrix} 2 & 1 & 9 \\ 3 & 5 & 7 \\ 4 & 6 & 3 \end{bmatrix}$$

$$= \begin{bmatrix} 4 & 4 & 13 \\ 4 & 10 & 13 \\ 13 & 13 & 6 \end{bmatrix} + \begin{bmatrix} 0 & 2 & -5 \\ -2 & 0 & -1 \\ 5 & 1 & 0 \end{bmatrix}.$$

2.3.5 Matrix Multiplication

Consider a vector $\mathbf{A} = a_1\hat{\mathbf{e}}_1 + a_2\hat{\mathbf{e}}_2 + a_3\hat{\mathbf{e}}_3$ in a Cartesian system. We can represent $\mathbf{A}$ as a *product* of a row matrix with a column matrix,

$$\mathbf{A} = \{a_1 \ a_2 \ a_3\} \left\{ \begin{array}{c} \hat{\mathbf{e}}_1 \\ \hat{\mathbf{e}}_2 \\ \hat{\mathbf{e}}_3 \end{array} \right\} = \{\hat{\mathbf{e}}_1 \ \hat{\mathbf{e}}_2 \ \hat{\mathbf{e}}_3\} \left\{ \begin{array}{c} a_1 \\ a_2 \\ a_3 \end{array} \right\}.$$

Note that the vector $\mathbf{A}$ is obtained by multiplying the ith element in the row matrix with the ith element in the column matrix and adding them. This gives us a strong reason to define the product of two matrices.

Let $\{X\}$ and $\{Y\}$ be the vectors (matrices with one column)

$$\{X\} = \left\{ \begin{array}{c} x_1 \\ x_2 \\ \vdots \\ x_m \end{array} \right\}, \quad \{Y\} = \left\{ \begin{array}{c} y_1 \\ y_2 \\ \vdots \\ y_m \end{array} \right\}. \tag{2.3.4}$$

We define the product $\{X\}^{\mathrm{T}}\{Y\}$ to be the scalar

$$\{X\}^{\mathrm{T}}\{Y\} = \{x_1, x_2, \ldots, x_m\} \left\{ \begin{array}{c} y_1 \\ y_2 \\ \vdots \\ y_m \end{array} \right\} = x_1 y_1 + x_2 y_2 + \cdots + x_m y_m = \sum_{i=1}^{m} x_i y_i,$$

$$= \sum_{i=1}^{m} y_i x_i = \{y_1, y_2, \ldots, y_m\} \left\{ \begin{array}{c} x_1 \\ x_2 \\ \vdots \\ x_m \end{array} \right\} = \{Y\}^{\mathrm{T}}\{X\}. \tag{2.3.5}$$

It follows from Eq. (2.3.5) that $\{X\}^{\mathrm{T}}\{Y\} = \{Y\}^{\mathrm{T}}\{X\}$. More generally, let $[A] = [a_{ij}]$ be $m \times n$ and $[B] = [b_{ij}]$ be $n \times p$ matrices. The product $[A][B]$ is defined as the $m \times p$ matrix $[C] = [c_{ij}]$, with

$$c_{ij} = \{i\text{th row of } [A]\} \left\{ \begin{array}{c} j\text{th} \\ \text{column of} \\ [B] \end{array} \right\} = \{a_{i1}, a_{i2}, \ldots, a_{in}\} \left\{ \begin{array}{c} b_{1j} \\ b_{2j} \\ \vdots \\ b_{nj} \end{array} \right\}$$

$$= a_{i1}b_{1j} + a_{i2}b_{2j} + \cdots + a_{in}b_{nj} = \sum_{k=1}^{n} a_{ik}b_{kj}. \tag{2.3.6}$$

The next example illustrates the computation of the product of a square matrix with a column matrix.

The following remarks concerning matrix multiplication are in order, where $[A]$ denotes an $m \times n$ matrix and $[B]$ denotes a $p \times q$ matrix:

(1) The product $[A][B]$ is defined only if the number of columns n in $[A]$ is equal to the number of rows p in $[B]$ (i.e., $p = n$). Similarly, the product $[B][A]$ is defined only if $q = m$.

(2) If $[A][B]$ is defined (i.e., $p = n$), $[B][A]$ may (if $q = m$) or may not (if $q \neq m$) be defined. If both $[A][B]$ and $[B][A]$ are defined (i.e., $p = n$ and $q = m$), it is not necessary that they be of the same size; $[A][B]$ is $m \times m$ and $[B][A]$ is $n \times n$.

(3) The products $[A][B]$ and $[B][A]$ are of the same size if and only if both $[A]$ and $[B]$ are square matrices of the same size.

(4) The products $[A][B]$ and $[B][A]$ are, in general, not equal: $[A][B] \neq [B][A]$ (even if they are of equal size). That is, matrix multiplication is not *commutative*.

(5) For any real square matrix $[A]$, $[A]$ is said to be *normal* if $[A][A]^{\mathrm{T}} = [A]^{\mathrm{T}}[A]$. The product $[A][A]^{\mathrm{T}}$ is always symmetric: $([A][A]^{\mathrm{T}})^{\mathrm{T}} = [A][A]^{\mathrm{T}}$.

(6) If $[A]$ is a square matrix, the powers of $[A]$ are defined by $[A]^2 = [A][A]$, $[A]^3 = [A][A]^2 = [A]^2[A]$, and so on.

(7) Matrix multiplication is associative: $([A][B])[C] = [A]([B][C])$.

(8) The product of any square matrix with the identity matrix is the matrix itself, $[A][I] = [A]$ and $[I][A] = [A]$.

(9) The transpose of the product is $([A][B])^{\mathrm{T}} = [B]^{\mathrm{T}}[A]^{\mathrm{T}}$ (note the order).

Example 2.3.1

Verify property 3 in the preceding list using matrices $[A]$ and $[B]$ of Section 2.3.3.

Solution: The product of matrices $[A]$ and $[B]$ is

$$[A][B] = \begin{bmatrix} 5 & -2 & 1 \\ 8 & 7 & 6 \\ 2 & 4 & 3 \\ -1 & 9 & 0 \end{bmatrix} \begin{bmatrix} 3 & -1 & 2 & 4 \\ -6 & 3 & 5 & 7 \\ 9 & 6 & -2 & 1 \end{bmatrix} = \begin{bmatrix} 36 & -5 & -2 & 7 \\ 36 & 49 & 39 & 87 \\ 9 & 28 & 18 & 39 \\ -57 & 28 & 43 & 59 \end{bmatrix},$$

and

$$([A][B])^{\mathrm{T}} = \begin{bmatrix} 36 & 36 & 9 & -57 \\ -5 & 49 & 28 & 28 \\ -2 & 39 & 18 & 43 \\ 7 & 87 & 39 & 59 \end{bmatrix}.$$

Now compute the product

$$[B]^{\mathrm{T}}[A]^{\mathrm{T}} = \begin{bmatrix} 3 & -6 & 9 \\ -1 & 3 & 6 \\ 2 & 5 & -2 \\ 4 & 7 & 1 \end{bmatrix} \begin{bmatrix} 5 & 8 & 2 & -1 \\ -2 & 7 & 4 & 9 \\ 1 & 6 & 3 & 0 \end{bmatrix} = \begin{bmatrix} 36 & 36 & 9 & -57 \\ -5 & 49 & 28 & 28 \\ -2 & 39 & 18 & 43 \\ 7 & 87 & 39 & 59 \end{bmatrix}.$$

Thus, $([A][B])^{\mathrm{T}} = [B]^{\mathrm{T}}[A]^{\mathrm{T}}$ is verified.

2.3.6 Inverse and Determinant of a Matrix

If $[A]$ is an $n \times n$ matrix and $[B]$ is any $n \times n$ matrix such that $[A][B] = [B][A] = [I]$, then $[B]$ is called an *inverse* of $[A]$. If it exists, the inverse of a matrix is unique (a consequence of the associative law). If both $[B]$ and $[C]$ are inverses for $[A]$, then by definition,

$$[A][B] = [B][A] = [A][C] = [C][A] = [I].$$

Due to the fact that matrix multiplication is associative, we have

$$[B][A][C] = ([B][A])[C] = [I][C] = [C]$$
$$= [B]([A][C]) = [B][I] = [B].$$

This shows that $[C] = [B]$, and the inverse is unique. The inverse of $[A]$ is denoted by $[A]^{-1}$. A procedure for computing the inverse will be presented shortly.

A matrix is said to be *singular* if it does not have an inverse. If $[A]$ is nonsingular, then the transpose of the inverse is equal to the inverse of the transpose: $([A]^{-1})^{\mathrm{T}} = ([A]^{\mathrm{T}})^{-1} \equiv [A]^{-\mathrm{T}}$.

Let $[A] = [a_{ij}]$ be an $n \times n$ matrix. We wish to associate with $[A]$ a scalar that in some sense measures the "size" of $[A]$ and indicates whether or not $[A]$ is nonsingular. The *determinant* of the matrix $[A] = [a_{ij}]$ is defined to be the scalar $\det[A] = |A|$ computed according to the rule

$$\det[A] = |A| = \sum_{i=1}^{n} (-1)^{i+1} a_{i1} |A_{i1}|, \qquad (2.3.7)$$

where $|A_{ij}|$ is the determinant of the $(n-1) \times (n-1)$ matrix that remains on deleting the ith row and the jth column of $[A]$. For convenience we define the determinant of a zeroth-order matrix to be unity. For 1×1 matrices the determinant is defined according to $|a_{11}| = a_{11}$. For a 2×2 matrix $[A]$ the determinant is defined by

$$[A] = \begin{bmatrix} a_{11} & a_{12} \\ a_{21} & a_{22} \end{bmatrix}, \quad |A| = \begin{vmatrix} a_{11} & a_{12} \\ a_{21} & a_{22} \end{vmatrix} = a_{11}\, a_{22} - a_{21}\, a_{12}.$$

In the above definition special attention is given to the first column of the matrix $[A]$. We call it the expansion of $|A|$ according to the first column of $[A]$. One can expand $|A|$ according to any column or row:

$$|A| = \sum_{i \text{ or } j=1}^{n} (-1)^{i+j} a_{ij} |A_{ij}| \text{ for fixed } j \text{ or } i.$$

A matrix is said to be singular if its determinant is zero. Obviously, there is a connection between the inverse and the determinant of a matrix, as we shall see shortly. A numerical example of the calculation of a determinant is presented next.

Example 2.3.2

Compute the determinant of the matrix

$$[A] = \begin{bmatrix} 2 & 5 & -1 \\ 1 & 4 & 3 \\ 2 & -3 & 5 \end{bmatrix}.$$

Solution: Using definition in Eq. (2.3.7) and expanding by the first column, we have

$$|A| = \sum_{i=1}^{3}(-1)^{i+1}a_{i1}|A_{i1}| = (-1)^2 a_{11}|A_{11}| + (-1)^3 a_{21}|A_{21}| + (-1)^4 a_{31}|A_{31}|$$

$$= a_{11}\begin{vmatrix} 4 & 3 \\ -3 & 5 \end{vmatrix} - a_{21}\begin{vmatrix} 5 & -1 \\ -3 & 5 \end{vmatrix} + a_{31}\begin{vmatrix} 5 & -1 \\ 4 & 3 \end{vmatrix}$$

$$= 2[(4)(5) - (-3)(3)] + (-1)[(5)(5) - (-3)(-1)] + 2[(5)(3) - (4)(-1)]$$

$$= 2(20 + 9) - (25 - 3) + 2(15 + 4) = 74.$$

Let $\mathbf{A} = A_i\hat{\mathbf{e}}_i$, $\mathbf{B} = B_i\hat{\mathbf{e}}_i$, and $\mathbf{C} = C_i\hat{\mathbf{e}}_i$. Then the cross product of vectors $\mathbf{A}$ and $\mathbf{B}$ can be expressed as the determinant

$$\mathbf{A} \times \mathbf{B} \equiv \begin{vmatrix} \hat{\mathbf{e}}_1 & \hat{\mathbf{e}}_2 & \hat{\mathbf{e}}_3 \\ A_1 & A_2 & A_3 \\ B_1 & B_2 & B_3 \end{vmatrix}, \tag{2.3.8}$$

and the scalar triple product can be expressed as the determinant

$$\mathbf{A}\cdot(\mathbf{B} \times \mathbf{C}) \equiv \begin{vmatrix} A_1 & A_2 & A_3 \\ B_1 & B_2 & B_3 \\ C_1 & C_2 & C_3 \end{vmatrix}. \tag{2.3.9}$$

Consequently, the determinant of a 3×3 matrix $[A]$ can be expressed as $|A| = e_{ijk}a_{1i}a_{2j}a_{3k}$ [let $\mathbf{A} = \mathbf{A}_1$, $\mathbf{B} = \mathbf{A}_2$, and $\mathbf{C} = \mathbf{A}_3$ in Eq. (2.3.9)].

We note the following properties of determinants of square matrices:

(1) $\det([A][B]) = \det[A] \cdot \det[B]$.

(2) $\det[A]^{\mathrm{T}} = \det[A]$.

(3) $\det(\alpha[A]) = \alpha^n \det[A]$, where α is a scalar and n is the order of $[A]$.

(4) If $[A']$ is a matrix obtained from $[A]$ by multiplying a row (or column) of $[A]$ by a scalar α, then $\det[A'] = \alpha\det[A]$.

(5) If $[A']$ is the matrix obtained from $[A]$ by interchanging any two rows (or columns) of $[A]$, then $\det[A'] = -\det[A]$.

(6) If $[A]$ has two rows (or columns) one of which is a scalar multiple of another (i.e., linearly dependent), $\det[A] = 0$.

(7) If $[A']$ is the matrix obtained from $[A]$ by adding a multiple of one row (or column) to another, then $\det[A'] = \det[A]$.

A matrix is said to be *singular* if and only if its determinant is zero; hence, does not have an inverse. By property 6 in the foregoing list the determinant of a matrix is zero if it has linearly dependent rows (or columns).

For an $n \times n$ matrix $[A]$ the determinant of the $(n-1) \times (n-1)$ sub-matrix of $[A]$ obtained by deleting row i and column j of $[A]$ is called the *minor* of a_{ij} and is denoted by $M_{ij}([A])$. The quantity $\text{cof}_{ij}([A]) \equiv (-1)^{i+j} M_{ij}([A])$ is called the *cofactor* of a_{ij}. The determinant of $[A]$ can be cast in terms of the minor and cofactor of a_{ij}

$$\det[A] = \sum_{i=1}^{n} (-1)^{i+j} a_{ij} M_{ij}([A]) = \sum_{i=1}^{n} a_{ij} \, \text{cof}_{ij}([A]), \qquad (2.3.10)$$

for any fixed value of j.

The *adjunct* (also called *adjoint*) of a matrix $[A]$ is the transpose of the matrix obtained from $[A]$ by replacing each element by its cofactor. The adjunct of $[A]$ is denoted by $\text{Adj}[A]$.

Now we have the essential tools to compute the inverse of a matrix. If $[A]$ is nonsingular (i.e., $|A| \neq 0$), the inverse $[A]^{-1}$ of $[A]$ can be computed according to

$$[A]^{-1} = \frac{1}{\det[A]} \text{Adj}[A]. \qquad (2.3.11)$$

Example 2.3.3 illustrates the computation of the inverse of a matrix following the procedure just outlined.

Example 2.3.3

Determine the inverse of the matrix $[A]$ of Example 2.3.2:

$$[A] = \begin{bmatrix} 2 & 5 & -1 \\ 1 & 4 & 3 \\ 2 & -3 & 5 \end{bmatrix}.$$

Solution: The determinant is given by (expanding by the first row)

$$|A| = (2)(29) + (-)(5)(-1) + (-1)(-11) = 74.$$

Then we compute the minors M_{ij}. For example, we have

$$M_{11}([A]) = \begin{vmatrix} 4 & 3 \\ -3 & 5 \end{vmatrix}, \quad M_{12}([A]) = \begin{vmatrix} 1 & 3 \\ 2 & 5 \end{vmatrix}, \quad M_{13}([A]) = \begin{vmatrix} 1 & 4 \\ 2 & -3 \end{vmatrix},$$

$$\text{cof}_{11}([A]) = (-1)^2 M_{11}([A]) = 4 \times 5 - (-3)3 = 29,$$
$$\text{cof}_{12}([A]) = (-1)^3 M_{12}([A]) = -(1 \times 5 - 2 \times 3) = 1,$$
$$\text{cof}_{13}([A]) = (-1)^4 M_{13}([A]) = 1 \times (-3) - 2 \times 4 = -11,$$

and so on. The $\text{Adj}([A])$ is given by

$$\text{Adj}([A]) = \begin{bmatrix} \text{cof}_{11}([A]) & \text{cof}_{12}([A]) & \text{cof}_{13}([A]) \\ \text{cof}_{21}([A]) & \text{cof}_{22}([A]) & \text{cof}_{23}([A]) \\ \text{cof}_{31}([A]) & \text{cof}_{32}([A]) & \text{cof}_{33}([A]) \end{bmatrix}^{\text{T}}$$

$$= \begin{bmatrix} 29 & -22 & 19 \\ 1 & 12 & 19 \\ -11 & 16 & 3 \end{bmatrix}.$$

The inverse of **A** can be now computed using Eq. (2.3.11),

$$[A]^{-1} = \frac{1}{74} \begin{bmatrix} -29 & -22 & 19 \\ 1 & 12 & -7 \\ -11 & 16 & 3 \end{bmatrix}.$$

It can be easily verified that $[A][A]^{-1} = [I]$.

2.3.7 Positive-Definite and Orthogonal Matrices

A symmetric matrix $[A]$ is said to be *positive* or *positive-definite* if there exists a nonzero vector $\{X\}$ such that

$$\{X\}^{\mathrm{T}}[A]\{X\} > 0. \tag{2.3.12}$$

The expression $\{X\}^{\mathrm{T}}[A]\{X\}$ represents a quadratic polynomial associated with matrix $[A]$ with respect to vector $\{X\}$. A corollary to this definition is that a matrix $[A]$ is positive if and only if there exists a nonsingular matrix $[T]$ such that

$$[A] = [T]^{\mathrm{T}}[T]. \tag{2.3.13}$$

Then we have

$$\{X\}^{\mathrm{T}}[T]^{\mathrm{T}}[T]\{X\} = \{Y\}^{\mathrm{T}}\{Y\}, \quad \{Y\} \equiv [T]\{X\}, \tag{2.3.14}$$

which is always positive for all nonzero vectors $\{Y\}$.

A nonsingular matrix $[Q]$ is said to be *orthogonal* if the following condition holds:

$$[Q][Q]^{\mathrm{T}} = [Q]^{\mathrm{T}}[Q] = [I] \quad \text{or} \quad [Q]^{-1} = [Q]^{\mathrm{T}}. \tag{2.3.15}$$

From the foregoing definition it follows that the determinant of an orthogonal matrix is $|Q| = \pm 1$. If the determinant of $[Q]$ is $+1$, then $[Q]$ is called a rotation or a *proper orthogonal* matrix; otherwise, it is called an improper orthogonal matrix. A proper orthogonal matrix transforms a right-handed coordinate system into another right-handed coordinate system.

Example 2.3.4 ————————————————————————

Consider the matrices

$$[A] = \begin{bmatrix} 2 & 3 \\ 3 & -4 \end{bmatrix}, \quad [B] = \begin{bmatrix} 2 & 3 \\ 3 & 5 \end{bmatrix}.$$

Determine if the matrices are positive.

Solution: Compute the product

$$\{x_1 \ x_2\} \begin{bmatrix} 2 & 3 \\ 3 & -4 \end{bmatrix} \begin{Bmatrix} x_1 \\ x_2 \end{Bmatrix} = 2x_1^2 + 6x_1 x_2 - 4x_2^2,$$

which is *not* positive for all nonzero vectors $\{X\}^{\mathrm{T}} = \{x_1 \ x_2\}$. Clearly, for $\{X\}^{\mathrm{T}} = \{1 \ -1\}$, we have $\{X\}^{\mathrm{T}}[A]\{X\} = -8$. Thus, the matrix is not positive.

Next, consider the matrix

$$[B] = \begin{bmatrix} 2 & 3 \\ 3 & 5 \end{bmatrix}.$$

Forming the quadratic form

$$\{x_1 \ x_2\} \begin{bmatrix} 2 & 3 \\ 3 & 5 \end{bmatrix} \begin{Bmatrix} x_1 \\ x_2 \end{Bmatrix} = 2x_1^2 + 6x_1x_2 + 5x_2^2 = (x_1 + 2x_2)^2 + (x_1 + x_2)^2$$

The matrix is positive because $\{X\}^T[A]\{X\} > 0$ for all nonzero vectors $\{X\}^T = \{x_1 \ x_2\}$. In addition, for the choice of

$$[T] = \begin{bmatrix} 1 & 0 & 0 \\ 0 & 1 & 0 \\ 0 & 1 & 1 \end{bmatrix},$$

we find that

$$[A] = [T]^T[T] = [T][T]^T = \begin{bmatrix} 1 & 0 & 0 \\ 0 & 2 & 1 \\ 0 & 1 & 1 \end{bmatrix}$$

is positive because $\{X\}^T[A]\{X\} = x_1^2 + 2x_2^2 + x_3^2 + 2x_2x_3 > 0$ for all nonzero vectors $\{X\}^T = \{x_1 \ x_2 \ x_3\}$. However, $[T]$ is not an orthogonal matrix because $[T]^T[T] \neq [I]$.

2.4 Vector Calculus

2.4.1 Differentiation of a Vector with Respect to a Scalar

Suppose that a vector is given as a function of a scalar t, say $\mathbf{A} = \mathbf{A}(t)$. In general, vector $\mathbf{A}$ will have different magnitudes and different directions for different values of t, which is pictured schematically in Fig. 2.4.1(a). With the tails of the vector being at the same point for different values of the scalar t, the tip of the arrow draws out a trajectory, as shown in Fig. 2.4.1(a).

Consider now two values of t differing by an infinitesimal amount, say t and $t + \Delta t$. Then the variation in Fig. 2.4.1(a) becomes as shown in Fig. 2.4.1(b). With this picture in mind, it is easy to visualize the definition of the derivative of a vector with respect to t:

$$\frac{d\mathbf{A}}{dt} = \lim_{\Delta t \to 0} \frac{\mathbf{A}(t + \Delta t) - \mathbf{A}(t)}{\Delta t}. \tag{2.4.1}$$

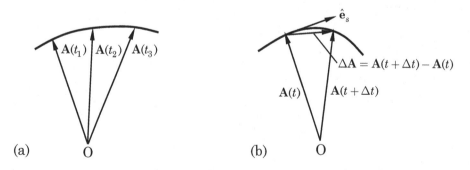

Fig. 2.4.1: (a) Variation of a vector as a function of a scalar t. (b) Differential change in a vector $\mathbf{A}(t)$.

Now let $\Delta s = |\Delta \mathbf{A}|$ so that s is the distance measured along the trajectory. Then we have

$$\frac{d\mathbf{A}}{dt} = \lim_{\Delta t \to 0} \frac{\Delta \mathbf{A}}{\Delta s} \frac{\Delta s}{\Delta t}.$$

In the limit, $\Delta \mathbf{A}/\Delta s$ is a unit vector that is tangent to the trajectory, say $\hat{\mathbf{e}}_s$. In the limit, therefore, we have

$$\frac{d\mathbf{A}}{dt} = \frac{ds}{dt}\hat{\mathbf{e}}_s. \tag{2.4.2}$$

It is clear that the derivative of the vector has a different direction and magnitude than the vector itself. The magnitude of the derivative ds/dt is the rate of change of distance s with respect to t along the trajectory. Observe that the distance s has the same dimensions as the vector $\mathbf{A}$ itself.

An important special case occurs when the vector has a constant length. In general, we note that

$$\mathbf{A} \cdot \mathbf{A} = A^2(t). \tag{2.4.3}$$

From differentiation of both sides of this equation it follows that

$$\mathbf{A} \cdot \frac{d\mathbf{A}}{dt} = A\frac{dA}{dt}. \tag{2.4.4}$$

When $\mathbf{A}$ has a constant length, then $dA/dt = 0$, and we have

$$\mathbf{A} \cdot \frac{d\mathbf{A}}{dt} = 0. \tag{2.4.5}$$

We deduce from this result that either $\mathbf{A} = \mathbf{0}$, or $d\mathbf{A}/dt = \mathbf{0}$ or $\mathbf{A}$ is perpendicular to $d\mathbf{A}/dt$. If $d\mathbf{A}/dt = \mathbf{0}$, then the vector has not only a constant length, but also a constant direction. If the direction of $\mathbf{A}$ varies, but its length is fixed, then the *derivative $d\mathbf{A}/dt$ is perpendicular to $\mathbf{A}$.*

Consider now the derivative of a vector in terms of its components. If we write

$$\mathbf{A} = A^1(t)\mathbf{e}_1(t) + A^2(t)\mathbf{e}_2(t) + A^3(t)\mathbf{e}_3(t),$$

we must remember that the basis vectors are also functions of the scalar t. Thus we have

$$\frac{d\mathbf{A}}{dt} = \frac{dA^1}{dt}\mathbf{e}_1 + \frac{dA^2}{dt}\mathbf{e}_2 + \frac{dA^3}{dt}\mathbf{e}_3 + A^1\frac{d\mathbf{e}_1}{dt} + A^2\frac{d\mathbf{e}_2}{dt} + A^3\frac{d\mathbf{e}_3}{dt}. \tag{2.4.6}$$

Basis vectors that are *constant* are associated with *Cartesian* systems. Only in these systems do the derivatives $d\mathbf{e}_i/dt$ vanish.

2.4.2 Curvilinear Coordinates

Consider a transformation from the coordinate system (x_1, x_2, x_3) to a new set of coordinates (q^1, q^2, q^3):

$$q^i = q^i(x_1, x_2, x_3) = q^i(x, y, z), \quad i = 1, 2, 3, \tag{2.4.7}$$

and the inverse transformation is

$$x_i = x_i(q^1, q^2, q^3), \quad i = 1, 2, 3. \tag{2.4.8}$$

The inverse transformation exists if and only if the Jacobian matrix $[J] = [\partial x_j / \partial q^i] \neq 0$; in fact, the determinant J of the matrix $[J]$ must be positive in order that the transformation produces a right-hand coordinate system (q^1, q^2, q^3) from a right-hand coordinate system (x_1, x_2, x_3). The functions $q^i(x, y, z) = \text{con-}$ stant, where $i = 1, 2, 3$, denote three surfaces in space. The intersection of any two of these surfaces defines a *coordinate curve*, as shown in Fig. 2.4.2. When the aforementioned transformation is nonlinear, the coordinate curves denoted by q^1, q^2, and q^3 are curved lines, and the (q^1, q^2, q^3) system is called *curvilinear*. When the transformation is linear, the coordinate lines will be straight lines, but not necessarily parallel to the original (x, y, z) system, and a new Cartesian coordinate system is defined.

In view of the transformations defined, the position vector $\mathbf{r}$ is a function of the new coordinates (q^1, q^2, q^3):

$$\mathbf{r} = \mathbf{r}(q^1, q^2, q^3). \tag{2.4.9}$$

The differential $d\mathbf{r}$ can now be written

$$d\mathbf{r} = \frac{\partial \mathbf{r}}{\partial q^1} dq^1 + \frac{\partial \mathbf{r}}{\partial q^2} dq^2 + \frac{\partial \mathbf{r}}{\partial q^3} dq^3 = \frac{\partial \mathbf{r}}{\partial q^i} dq^i. \tag{2.4.10}$$

The partial derivatives $\partial \mathbf{r}/\partial q^1$, $\partial \mathbf{r}/\partial q^2$, and $\partial \mathbf{r}/\partial q^3$ are vectors that are tangent to the coordinate curves, as shown in Fig. 2.4.2. These vectors can be taken as a basis system associated with the coordinates (q^1, q^2, q^3). This basis is referred

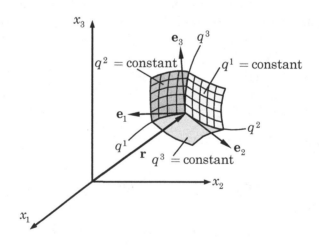

Fig. 2.4.2: Curvilinear coordinates.

to as the *unitary basis*. Note that these vectors in general are not unit nor orthogonal. We now denote the unitary basis by $(\mathbf{e}_1, \mathbf{e}_2, \mathbf{e}_3)$ as follows:

$$\mathbf{e}_1 \equiv \frac{\partial \mathbf{r}}{\partial q^1}, \qquad \mathbf{e}_2 \equiv \frac{\partial \mathbf{r}}{\partial q^2}, \qquad \mathbf{e}_3 \equiv \frac{\partial \mathbf{r}}{\partial q^3}. \tag{2.4.11}$$

A differential distance is denoted by

$$dr = dq^1\mathbf{e}_1 + dq^2\mathbf{e}_2 + dq^3\mathbf{e}_3 = dq^i\mathbf{e}_i. \tag{2.4.12}$$

Observe that the q's have superscripts whereas the unitary basis has subscripts. The dq's thus are referred to as the *contravariant components* of the differential vector dr. The unitary basis will satisfy the *covariant* transformation law and thus is a *covariant basis*.

The unitary basis can be described in terms of the Cartesian basis $(\hat{\mathbf{e}}_x, \hat{\mathbf{e}}_y, \hat{\mathbf{e}}_z)$ as

$$\begin{aligned}
\mathbf{e}_1 &= \frac{\partial \mathbf{r}}{\partial q^1} = \frac{\partial x}{\partial q^1}\hat{\mathbf{e}}_x + \frac{\partial y}{\partial q^1}\hat{\mathbf{e}}_y + \frac{\partial z}{\partial q^1}\hat{\mathbf{e}}_z, \\
\mathbf{e}_2 &= \frac{\partial \mathbf{r}}{\partial q^2} = \frac{\partial x}{\partial q^2}\hat{\mathbf{e}}_x + \frac{\partial y}{\partial q^2}\hat{\mathbf{e}}_y + \frac{\partial z}{\partial q^2}\hat{\mathbf{e}}_z, \\
\mathbf{e}_3 &= \frac{\partial \mathbf{r}}{\partial q^3} = \frac{\partial x}{\partial q^3}\hat{\mathbf{e}}_x + \frac{\partial y}{\partial q^3}\hat{\mathbf{e}}_y + \frac{\partial z}{\partial q^3}\hat{\mathbf{e}}_z.
\end{aligned} \tag{2.4.13}$$

In the summation convention we have

$$\mathbf{e}_i \equiv \frac{\partial \mathbf{r}}{\partial q^i} = \frac{\partial x_j}{\partial q^i}\hat{\mathbf{e}}_j, \qquad i = 1, 2, 3. \tag{2.4.14}$$

We can associate with the covariant base vectors $\mathbf{e}_i$, a dual or reciprocal basis, defined by Eq. (2.2.31).

2.4.3 The Fundamental Metric

The square of the infinitesimal distance between two points can now be written in the unitary basis as

$$(ds)^2 = dr \cdot dr = (\mathbf{e}_i \cdot \mathbf{e}_j)dq^i\,dq^j = g_{ij}\,dq^i\,dq^j, \tag{2.4.15}$$

$$g_{ij} \equiv \mathbf{e}_i \cdot \mathbf{e}_j, \tag{2.4.16}$$

where g_{ij} are the covariant components of the *fundamental metric tensor*, $\mathbf{g}$. According to this definition g_{ij} is symmetric, that is, $g_{ij} = g_{ji}$. In view of Eq. (2.2.44), the unitary basis and its dual basis are related by

$$\mathbf{e}_i = (\mathbf{e}_i \cdot \mathbf{e}_j)\mathbf{e}^j = g_{ij}\,\mathbf{e}^j, \quad \mathbf{e}^i = (\mathbf{e}^i \cdot \mathbf{e}^j)\mathbf{e}_j \equiv g^{ij}\,\mathbf{e}_j, \quad g^{ij} = \mathbf{e}^i \cdot \mathbf{e}^j, \tag{2.4.17}$$

where g^{ij} are the contravariant components of the metric tensor $\mathbf{g}$. Note that

$$\delta^i_j = \mathbf{e}^i \cdot \mathbf{e}_j = g^{im}g_{jn}\mathbf{e}_m \cdot \mathbf{e}^n = g^{in}g_{jn}. \tag{2.4.18}$$

Thus, $[g^{ij}]$ is the inverse of $[g_{ij}]$. Similar relations hold between the contravariant and covariant components of a vector. We can write a vector $\mathbf{A}$ either in terms of its components in the covariant basis or contravariant basis as

$$\mathbf{A} = A^i\,\mathbf{e}_i, \quad \mathbf{A} = A_i\,\mathbf{e}^i. \tag{2.4.19}$$

Then

$$A_j = g_{ji} A^i, \quad A^i = g^{ij} A_j. \tag{2.4.20}$$

Analogous to Eq. (2.2.49), we can define the permutation symbol in a general right-handed curvilinear system by

$$\mathbf{e}_i \times \mathbf{e}_j = \mathcal{E}_{ijk} \mathbf{e}^k, \tag{2.4.21}$$

where $\mathcal{E}_{ijk}$ is the permutation symbol in a general curvilinear system:

$$\mathcal{E}_{ijk} = \mathbf{e}_i \times \mathbf{e}_j \cdot \mathbf{e}_k. \tag{2.4.22}$$

The length of a unitary vector $\mathbf{e}_i$ (for fixed i) is defined as

$$|\mathbf{e}_i| = [\mathbf{e}_i \cdot \mathbf{e}_i]^{\frac{1}{2}} = \sqrt{g_{ii}} \quad (\text{no sum on } i). \tag{2.4.23}$$

The fundamental metric describes the nature of a space, or manifold. A manifold may be either "curved" or "flat." A flat space is said to be *Euclidean*. If it is possible to find a transformation to a coordinate system such that all the g_{ij}'s are constant, the space is Euclidean. If it is not possible, it is said to be non-Euclidean or *Riemannian*. For most engineering applications we will be concerned with Euclidean spaces.

2.4.4 Derivative of a Scalar Function of a Vector

The basic notions of vector and scalar calculus, especially with regard to physical applications, are closely related to the rate of change of a scalar field, such as the velocity potential or temperature, with distance. Let us denote a scalar field by $\phi = \phi(\mathbf{x})$, $\mathbf{x}$ being the position vector, as shown in Fig. 2.4.3.

In general coordinates we can write $\phi = \phi(q^1, q^2, q^3)$, and a differential of ϕ is given by

$$d\phi = \frac{\partial \phi}{\partial q^1} dq^1 + \frac{\partial \phi}{\partial q^2} dq^2 + \frac{\partial \phi}{\partial q^3} dq^3 = \frac{\partial \phi}{\partial q^i} dq^i.$$

The differentials dq^1, dq^2, dq^3 are components of $d\mathbf{r} = d\mathbf{x}$ [see Eq. (2.4.12)]. We would now like to write $d\phi$ in such a way that we elucidate *the direction* as well as the magnitude of $d\mathbf{x}$. In view of the identity $\mathbf{e}^i \cdot \mathbf{e}_j = \delta^i_j$, we can write

$$d\phi = \mathbf{e}^1 \frac{\partial \phi}{\partial q^1} \cdot \mathbf{e}_1 \, dq^1 + \mathbf{e}^2 \frac{\partial \phi}{\partial q^2} \cdot \mathbf{e}_2 \, dq^2 + \mathbf{e}^3 \frac{\partial \phi}{\partial q^3} \cdot \mathbf{e}_3 \, dq^3$$

$$= (dq^1 \mathbf{e}_1 + dq^2 \mathbf{e}_2 + dq^3 \mathbf{e}_3) \cdot \left(\mathbf{e}^1 \frac{\partial \phi}{\partial q^1} + \mathbf{e}^2 \frac{\partial \phi}{\partial q^2} + \mathbf{e}^3 \frac{\partial \phi}{\partial q^3} \right)$$

$$= d\mathbf{x} \cdot \left(\mathbf{e}^1 \frac{\partial \phi}{\partial q^1} + \mathbf{e}^2 \frac{\partial \phi}{\partial q^2} + \mathbf{e}^3 \frac{\partial \phi}{\partial q^3} \right). \tag{2.4.24}$$

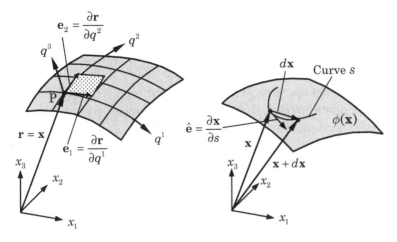

Fig. 2.4.3: Directional derivative of a scalar function.

Let us now denote the magnitude of $d\mathbf{x}$ by $ds \equiv |d\mathbf{x}|$. Then $\hat{\mathbf{e}} = d\mathbf{x}/ds$ is a unit vector in the direction of $d\mathbf{x}$, and we have

$$\left(\frac{d\phi}{ds}\right)_{\hat{\mathbf{e}}} = \hat{\mathbf{e}} \cdot \left(\mathbf{e}^1 \frac{\partial \phi}{\partial q^1} + \mathbf{e}^2 \frac{\partial \phi}{\partial q^2} + \mathbf{e}^3 \frac{\partial \phi}{\partial q^3}\right). \tag{2.4.25}$$

The derivative $(d\phi/ds)_{\hat{\mathbf{e}}}$ is called the *directional derivative*[2] of ϕ. We see that it is the *rate of change* of ϕ with respect to distance and that it depends on the direction $\hat{\mathbf{e}}$ in which the distance is taken.

The vector in Eq. (2.4.25) that is taken a scalar product with $\hat{\mathbf{e}}$ can be obtained immediately whenever the scalar field ϕ is given. Since the magnitude of this vector is equal to the maximum value of the directional derivative, it is called the *gradient vector* and is denoted by grad ϕ:

$$\text{grad}\,\phi \equiv \mathbf{e}^1 \frac{\partial \phi}{\partial q^1} + \mathbf{e}^2 \frac{\partial \phi}{\partial q^2} + \mathbf{e}^3 \frac{\partial \phi}{\partial q^3} = \mathbf{e}^i \frac{\partial \phi}{\partial q^i}. \tag{2.4.26}$$

From this representation it can be seen that

$$\frac{\partial \phi}{\partial q^1}, \qquad \frac{\partial \phi}{\partial q^2}, \qquad \frac{\partial \phi}{\partial q^3}$$

are the *covariant components* of the gradient vector.

2.4.5 The del Operator

It is convenient to write the gradient vector, grad ϕ, as

$$\boldsymbol{\nabla}\phi \equiv \left(\mathbf{e}^1 \frac{\partial}{\partial q^1} + \mathbf{e}^2 \frac{\partial}{\partial q^2} + \mathbf{e}^3 \frac{\partial}{\partial q^3}\right)\phi, \tag{2.4.27}$$

---

[2]The directional derivative is also defined as $[d\phi(\mathbf{q} + \alpha\hat{\mathbf{e}})/d\alpha]_{\alpha=0}$.

and interpret $\nabla\phi$ as some operator ∇ operating on ϕ. This operator ∇ is defined by

$$\nabla \equiv \mathbf{e}^1 \frac{\partial}{\partial q^1} + \mathbf{e}^2 \frac{\partial}{\partial q^2} + \mathbf{e}^3 \frac{\partial}{\partial q^3} = \mathbf{e}^i \frac{\partial}{\partial q^i}, \qquad (2.4.28)$$

and it is called the *del operator*. The del operator is a *vector differential* operator, and the "components" $\partial/\partial q^1$, $\partial/\partial q^2$, and $\partial/\partial q^3$ appear as covariant components.

In Cartesian systems we have the simple form

$$\nabla \equiv \hat{\mathbf{e}}_x \frac{\partial}{\partial x} + \hat{\mathbf{e}}_y \frac{\partial}{\partial y} + \hat{\mathbf{e}}_z \frac{\partial}{\partial z}, \qquad (2.4.29)$$

or, in the summation convention, we have

$$\nabla \equiv \hat{\mathbf{e}}_i \frac{\partial}{\partial x_i}. \qquad (2.4.30)$$

It is important to note that although the del operator has some of the properties of a vector, it does not have them all, because it is an operator. For instance, $\nabla \cdot \mathbf{A}$ is a scalar (called the divergence of $\mathbf{A}$) whereas $\mathbf{A} \cdot \nabla$ is a scalar *differential operator*. Thus the del operator ∇ does not commute in this sense. *Also, in general, the derivatives of the base vectors $\mathbf{e}^i$ with respect to the coordinates q^i are not zero.*

When the scalar function $\phi(\mathbf{x})$ is set equal to a constant, $\phi(\mathbf{x}) = $ constant, a family of surfaces is generated. A different surface is designated by different values of the constant, and each surface is called a *level surface*, as shown in Fig. 2.4.4(a). The unit vector $\hat{\mathbf{e}}$ is tangent to a level surface. If the direction in which the directional derivative is taken lies within a level surface, then $d\phi/ds$ is zero because ϕ is a constant on a level surface. It follows, therefore, that if $d\phi/ds$ is zero, then $\nabla\phi$ must be perpendicular to $\hat{\mathbf{e}}$, and hence *perpendicular to a level surface*. Thus, if any surface is defined by $\phi(\mathbf{x}) = $ constant, the unit normal to the surface is determined from [see Fig. 2.4.4(b)]

$$\hat{\mathbf{n}} = \pm\frac{\nabla\phi}{|\nabla\phi|}, \quad \hat{\mathbf{n}} = n_i \hat{\mathbf{e}}_i \,(= n_x\,\hat{\mathbf{e}}_x + n_y\,\hat{\mathbf{e}}_y + n_z\,\hat{\mathbf{e}}_z), \quad n_i = \frac{1}{|\nabla\phi|}\frac{\partial\phi}{\partial x_i}. \qquad (2.4.31)$$

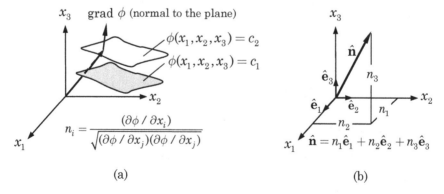

(a)   (b)

Fig. 2.4.4: (a) Level surfaces and unit normal vector. (b) Components of unit normal vector.

In general, the normal vector is a function of position $\mathbf{x}$; $\hat{\mathbf{n}}$ is independent of $\mathbf{x}$ only when ϕ is a plane (i.e., linear function of $\mathbf{x}$). The plus or minus sign appears in Eq. (2.4.31) because the direction of $\hat{\mathbf{n}}$ may point in either direction away from the surface. If the surface is closed, the usual convention is to take $\hat{\mathbf{n}}$ pointing outward from the surface.

An important note concerning the del operator ∇ is in order. Two types of gradients are used in continuum mechanics: forward and backward gradients. The forward gradient is the usual gradient and the backward gradient is the transpose of the forward gradient operator. To see the difference between the two types of gradients, consider a vector function $\mathbf{A} = A_i(\mathbf{x})\hat{\mathbf{e}}_i$. The forward and backward gradients of a vector $\mathbf{A}$ are

$$\vec{\nabla}\mathbf{A} \equiv \nabla\mathbf{A} = \hat{\mathbf{e}}_j \frac{\partial}{\partial x_j}\left(A_i\hat{\mathbf{e}}_i\right) = \frac{\partial A_i}{\partial x_j}\hat{\mathbf{e}}_j\hat{\mathbf{e}}_i = A_{i,j}\hat{\mathbf{e}}_j\hat{\mathbf{e}}_i, \qquad (2.4.32)$$

$$\mathbf{A}\overleftarrow{\nabla} \equiv (\nabla\mathbf{A})^{\mathrm{T}} = \frac{\partial A_i}{\partial x_j}\left(\hat{\mathbf{e}}_j\hat{\mathbf{e}}_i\right)^{\mathrm{T}} = A_{i,j}\hat{\mathbf{e}}_i\hat{\mathbf{e}}_j, \qquad (2.4.33)$$

where $A_{i,j} = \partial A_i/\partial x_j$. Both are second-order tensors,[3] as discussed in Section 2.5. The backward gradient $\overleftarrow{\nabla}$, a more natural one, is often used in defining the deformation gradient tensor, displacement gradient tensor, and velocity gradient tensor, which is introduced in Chapter 3. In the present book only one gradient operator (in bold), namely, the forward gradient operator ∇, a more common one, is used. To clarify, the transpose of the forward gradient is used to denote the backward gradient operator.

2.4.6 Divergence and Curl of a Vector

The dot product of a del operator with a vector is called the *divergence of a vector* and is denoted by

$$\nabla \cdot \mathbf{A} \equiv \operatorname{div}\mathbf{A}. \qquad (2.4.34)$$

If we take the divergence of the gradient of a scalar function $\phi(\mathbf{x})$, we have

$$\operatorname{div}(\operatorname{grad}\phi) \equiv \nabla \cdot \nabla\phi = (\nabla \cdot \nabla)\phi = \nabla^2\phi. \qquad (2.4.35)$$

The notation $\nabla^2 = \nabla \cdot \nabla$ is called the *Laplacian operator*. In Cartesian systems this reduces to the simple form

$$\nabla^2\phi = \frac{\partial^2\phi}{\partial x_i\partial x_i} = \frac{\partial^2\phi}{\partial x^2} + \frac{\partial^2\phi}{\partial y^2} + \frac{\partial^2\phi}{\partial z^2}. \qquad (2.4.36)$$

The Laplacian of a scalar appears frequently in the partial differential equations governing physical phenomena.

The *curl of a vector* is defined as the del operator operating on a vector by means of the cross product:

$$\operatorname{curl}\mathbf{A} = \nabla \times \mathbf{A} = e_{ijk}\hat{\mathbf{e}}_i\frac{\partial A_k}{\partial x_j}. \qquad (2.4.37)$$

---

[3]Operationally it is more appropriate to use the notation $\mathbf{A}\overleftarrow{\nabla}$ instead of $\overleftarrow{\nabla}\mathbf{A}$.

The quantity $\hat{\mathbf{n}} \cdot \operatorname{grad} \phi$ of a scalar function ϕ is called the *normal derivative* of ϕ and is denoted by

$$\frac{\partial \phi}{\partial n} \equiv \hat{\mathbf{n}} \cdot \nabla \phi. \tag{2.4.38}$$

In a Cartesian system this becomes

$$\frac{\partial \phi}{\partial n} = \frac{\partial \phi}{\partial x} n_x + \frac{\partial \phi}{\partial y} n_y + \frac{\partial \phi}{\partial z} n_z, \tag{2.4.39}$$

where n_x, n_y and n_z are the direction cosines of the unit normal [see Fig. 2.4.4(b)]

$$\hat{\mathbf{n}} = n_x \hat{\mathbf{e}}_x + n_y \hat{\mathbf{e}}_y + n_z \hat{\mathbf{e}}_z. \tag{2.4.40}$$

Next, we present several examples to illustrate the use of index notation to prove certain identities involving vector calculus.

Example 2.4.1

Establish the following identities using index notation (we use the notations $\mathbf{x} = \mathbf{r}$ and $|\mathbf{x}| = r$)

(a) $\nabla(r) = \frac{\mathbf{r}}{r}$.

(b) $\nabla(r^n) = n r^{n-2} \mathbf{r}$.

(c) $\nabla \times (\nabla F) = \mathbf{0}$.

(d) $\nabla \cdot (\nabla F \times \nabla G) = 0$.

(e) $\nabla \times (\nabla \times \mathbf{A}) = \nabla(\nabla \cdot \mathbf{A}) - \nabla^2 \mathbf{A}$.

(f) $\nabla \cdot (\mathbf{A} \times \mathbf{B}) = \nabla \times \mathbf{A} \cdot \mathbf{B} - \nabla \times \mathbf{B} \cdot \mathbf{A}$.

(g) $\mathbf{A} \times (\nabla \times \mathbf{A}) = \frac{1}{2} \nabla(\mathbf{A} \cdot \mathbf{A}) - \mathbf{A} \cdot \nabla \mathbf{A}$.

In the above expressions F and G denote scalar functions and $\mathbf{A}$ and $\mathbf{B}$ denote vector functions of position $\mathbf{x}$ with continuous first and second derivatives.

Solution: First note that $\partial x_j / \partial x_i = \delta_{ij}$ and $\partial x_i / \partial x_i = \delta_{ii} = 3$.

(a) Consider

$$\nabla(r) = \hat{\mathbf{e}}_i \frac{\partial r}{\partial x_i} = \hat{\mathbf{e}}_i \frac{\partial}{\partial x_i} (x_j x_j)^{\frac{1}{2}}$$

$$= \frac{1}{2} \hat{\mathbf{e}}_i \left(r^2\right)^{\frac{1}{2}-1} \left(\frac{\partial x_j}{\partial x_i} x_j + x_j \frac{\partial x_j}{\partial x_i}\right) = x_i \hat{\mathbf{e}}_i \left(r^2\right)^{-\frac{1}{2}} = \frac{\mathbf{x}}{r}, \tag{1}$$

from which we note the identity

$$\frac{\partial r}{\partial x_i} = \frac{x_i}{r}. \tag{2}$$

(b) Similar to (a), we have

$$\nabla(r^n) = \hat{\mathbf{e}}_i \frac{\partial}{\partial x_i} (r^n) = n r^{n-1} \hat{\mathbf{e}}_i \frac{\partial r}{\partial x_i} = n r^{n-2} x_i \hat{\mathbf{e}}_i = n r^{n-2} \mathbf{r}.$$

(c) Consider the expression

$$\nabla \times (\nabla F) = \left(\hat{\mathbf{e}}_i \frac{\partial}{\partial x_i}\right) \times \left(\hat{\mathbf{e}}_j \frac{\partial F}{\partial x_j}\right) = e_{ijk} \hat{\mathbf{e}}_k \frac{\partial^2 F}{\partial x_i \partial x_j}.$$

Note that $\frac{\partial^2 F}{\partial x_i \partial x_j}$ is symmetric in i and j. Consider the kth component of the vector $\nabla \times (\nabla F)$:

$$e_{ijk}\frac{\partial^2 F}{\partial x_i \partial x_j} = -e_{jik}\frac{\partial^2 F}{\partial x_i \partial x_j} \quad \text{(interchanged i and j)}$$

$$= -e_{ijk}\frac{\partial^2 F}{\partial x_j \partial x_i} \quad \text{(renamed i as j and j as i)}$$

$$= -e_{ijk}\frac{\partial^2 F}{\partial x_i \partial x_j} \quad \text{(used the symmetry of $\frac{\partial^2 F}{\partial x_i \partial x_j}$).}$$

Thus, the expression is equal to its own negative, implying that it is zero.

(d) We have

$$\nabla \cdot (\nabla F \times \nabla G) = \left(\hat{e}_i \frac{\partial}{\partial x_i}\right) \cdot \left(\hat{e}_j \frac{\partial F}{\partial x_j} \times \hat{e}_k \frac{\partial G}{\partial x_k}\right)$$

$$= e_{jk\ell}(\hat{e}_i \cdot \hat{e}_\ell)\left(\frac{\partial^2 F}{\partial x_i \partial x_j}\frac{\partial G}{\partial x_k} + \frac{\partial F}{\partial x_j}\frac{\partial^2 G}{\partial x_i \partial x_k}\right)$$

$$= e_{ijk}\left(\frac{\partial^2 F}{\partial x_i \partial x_j}\frac{\partial G}{\partial x_k} + \frac{\partial F}{\partial x_j}\frac{\partial^2 G}{\partial x_i \partial x_k}\right) = 0.$$

(e) Observe that

$$\nabla \times (\nabla \times \mathbf{A}) = \hat{e}_i \frac{\partial}{\partial x_i} \times \left(\hat{e}_j \frac{\partial}{\partial x_j} \times A_k\hat{e}_k\right)$$

$$= \hat{e}_i \frac{\partial}{\partial x_i} \times \left(e_{jk\ell}\frac{\partial A_k}{\partial x_j}\hat{e}_\ell\right) = e_{i\ell m}\, e_{jk\ell}\frac{\partial^2 A_k}{\partial x_i \partial x_j}\hat{e}_m.$$

$$= (\delta_{mj}\delta_{ik} - \delta_{mk}\delta_{ij})\frac{\partial^2 A_k}{\partial x_i \partial x_j}\hat{e}_m = \frac{\partial^2 A_i}{\partial x_i \partial x_j}\hat{e}_j - \frac{\partial^2 A_k}{\partial x_i \partial x_i}\hat{e}_k$$

$$= \hat{e}_j \frac{\partial}{\partial x_j}\left(\frac{\partial A_i}{\partial x_i}\right) - \frac{\partial^2}{\partial x_i \partial x_i}(A_k\hat{e}_k) = \nabla(\nabla \cdot \mathbf{A}) - \nabla^2 \mathbf{A}.$$

This result is sometimes used as the definition of the Laplacian of a vector, that is,

$$\nabla^2 \mathbf{A} = \nabla(\nabla \cdot \mathbf{A}) - \nabla \times \nabla \times \mathbf{A}.$$

(f) Expanding the vector expression:

$$\nabla \cdot (\mathbf{A} \times \mathbf{B}) = \hat{e}_i \cdot \frac{\partial}{\partial x_i}(e_{jk\ell}A_j B_k\hat{e}_\ell)$$

$$= e_{ijk}\left(\frac{\partial A_j}{\partial x_i}B_k + A_j\frac{\partial B_k}{\partial x_i}\right)$$

$$= \nabla \times \mathbf{A} \cdot \mathbf{B} - \nabla \times \mathbf{B} \cdot \mathbf{A}.$$

(g) Expanding the left side of the expression using index notation:

$$\mathbf{A} \times (\nabla \times \mathbf{A}) = A_i\hat{e}_i \times \left(\frac{\partial A_k}{\partial x_j}\hat{e}_j \times \hat{e}_k\right) = e_{jkm}A_i\frac{\partial A_k}{\partial x_j}(\hat{e}_i \times \hat{e}_m)$$

$$= e_{jkm}e_{imn}A_i\frac{\partial A_k}{\partial x_j}\hat{e}_n = A_i\frac{\partial A_k}{\partial x_j}(\delta_{jn}\delta_{ki} - \delta_{ji}\delta_{kn})\hat{e}_n$$

$$= A_i\frac{\partial A_i}{\partial x_j}\hat{e}_j - A_i\frac{\partial A_k}{\partial x_i}\hat{e}_k = \frac{1}{2}\frac{\partial(A_iA_i)}{\partial x_j}\hat{e}_j - A_i\frac{\partial A_k}{\partial x_i}\hat{e}_k$$

$$= \frac{1}{2}\hat{e}_j\frac{\partial(A_iA_i)}{\partial x_j} - A_i\frac{\partial}{\partial x_i}(A_k\hat{e}_k)$$

$$= \frac{1}{2}\nabla(\mathbf{A} \cdot \mathbf{A}) - \mathbf{A} \cdot \nabla\mathbf{A}.$$

These examples illustrate the convenience of using index notation in establishing vector identities and simplifying vector expressions. The difficult step in these proofs is recognizing vector expressions from scalar expressions involving components with indices; for example, seeing that $e_{ijk}A_jB_k$ is the ith component of the vector product $\mathbf{A} \times \mathbf{B}$ or recognizing $e_{ijk}\frac{\partial A_j}{\partial x_i}B_k$ as $\boldsymbol{\nabla} \times \mathbf{A} \cdot \mathbf{B}$ and $B_i\frac{\partial A_j}{\partial x_i}\hat{\mathbf{e}}_j$ as $\mathbf{B} \cdot \boldsymbol{\nabla}\mathbf{A}$ is not always easy but practice makes one good at it. To aid readers in this pursuit, a list of vector operations in both vector and Cartesian component forms is presented in Table 2.4.1. Some of the identities are left as an exercise to the reader (see Problems **2.30**–**2.32**). *Note that the gradient operation increases, divergence decreases, and curl operation keeps the order the same of a tensor field when operated on by $\boldsymbol{\nabla}$.* For example, the gradient of a scalar function ϕ is a vector $\boldsymbol{\nabla}\phi$, and the divergence of a vector-valued function $\mathbf{A}$ is a scalar $\boldsymbol{\nabla} \cdot \mathbf{A}$; similarly, the gradient of a vector-valued function $\mathbf{A}$ is a second-order tensor $\boldsymbol{\nabla}\mathbf{A}$, whereas the divergence of a second-order tensor-valued function $\mathbf{S}$ is a vector $\boldsymbol{\nabla} \cdot \mathbf{S}$.

Table 2.4.1: Vector expressions and their Cartesian component forms [$\mathbf{A}$, $\mathbf{B}$, and $\mathbf{C}$ are vector functions, U is a scalar function, $\mathbf{x}$ is the position vector; $(\hat{\mathbf{e}}_1, \hat{\mathbf{e}}_2, \hat{\mathbf{e}}_3)$ are the Cartesian unit vectors in a rectangular Cartesian coordinate system; see Fig. 2.2.11].

No.	Vector form and its equivalence	Component form
1.	$\mathbf{A} \cdot \mathbf{B}$	A_iB_i
2.	$\mathbf{A} \times \mathbf{B}$	$e_{ijk}A_iB_j\hat{\mathbf{e}}_k$
3.	$\mathbf{A} \cdot (\mathbf{B} \times \mathbf{C})$	$e_{ijk}A_iB_jC_k$
4.	$\mathbf{A} \times (\mathbf{B} \times \mathbf{C}) = \mathbf{B}(\mathbf{A} \cdot \mathbf{C}) - \mathbf{C}(\mathbf{A} \cdot \mathbf{B})$	$e_{ijk}e_{klm}A_jB_lC_m\hat{\mathbf{e}}_i$
5.	$\boldsymbol{\nabla}\mathbf{A}$	$\frac{\partial A_j}{\partial x_i}\hat{\mathbf{e}}_i\hat{\mathbf{e}}_j$
6.	$\boldsymbol{\nabla} \cdot \mathbf{A}$	$\frac{\partial A_i}{\partial x_i}$
7.	$\boldsymbol{\nabla} \times \mathbf{A}$	$e_{ijk}\frac{\partial A_j}{\partial x_i}\hat{\mathbf{e}}_k$
8.	$\boldsymbol{\nabla} \cdot (\boldsymbol{\nabla} \times \mathbf{A}) = 0$	$e_{ijk}\frac{\partial^2 A_j}{\partial x_i\partial x_k}$
9.	$\boldsymbol{\nabla} \times (\boldsymbol{\nabla}U) = 0$	$e_{ijk}\hat{\mathbf{e}}_k\frac{\partial^2 U}{\partial x_i\partial x_j}$
10.	$\boldsymbol{\nabla} \cdot (\mathbf{A} \times \mathbf{B}) = \mathbf{B} \cdot (\boldsymbol{\nabla} \times \mathbf{A}) - \mathbf{A} \cdot (\boldsymbol{\nabla} \times \mathbf{B})$	$e_{ijk}\frac{\partial}{\partial x_i}(A_jB_k)$
11.	$(\boldsymbol{\nabla} \times \mathbf{A}) \times \mathbf{B} = \mathbf{B} \cdot [\boldsymbol{\nabla}\mathbf{A} - (\boldsymbol{\nabla}\mathbf{A})^{\mathrm{T}}]$	$e_{ijk}e_{klm}B_l\frac{\partial A_j}{\partial x_i}\hat{\mathbf{e}}_m$
12.	$\mathbf{A} \times (\boldsymbol{\nabla} \times \mathbf{A}) = \frac{1}{2}\boldsymbol{\nabla}(\mathbf{A} \cdot \mathbf{A}) - (\mathbf{A} \cdot \boldsymbol{\nabla})\mathbf{A}$	$e_{nim}e_{jkm}A_i\frac{\partial A_k}{\partial x_j}\hat{\mathbf{e}}_n$
13.	$\boldsymbol{\nabla} \cdot (\boldsymbol{\nabla}\mathbf{A}) = \nabla^2\mathbf{A}$	$\frac{\partial^2 A_j}{\partial x_i\partial x_i}\hat{\mathbf{e}}_j$
14.	$\boldsymbol{\nabla} \times \boldsymbol{\nabla} \times \mathbf{A} = \boldsymbol{\nabla}(\boldsymbol{\nabla} \cdot \mathbf{A}) - (\boldsymbol{\nabla} \cdot \boldsymbol{\nabla})\mathbf{A}$	$e_{mil}e_{jkl}\frac{\partial^2 A_k}{\partial x_i\partial x_j}\hat{\mathbf{e}}_m$
15.	$(\mathbf{A} \cdot \boldsymbol{\nabla})\mathbf{B}$	$A_j\frac{\partial B_i}{\partial x_j}\hat{\mathbf{e}}_i$
16.	$\mathbf{A}(\boldsymbol{\nabla} \cdot \mathbf{B})$	$A_i\hat{\mathbf{e}}_i\frac{\partial B_j}{\partial x_j}$
17.	$\boldsymbol{\nabla} \cdot (U\mathbf{A}) = U\boldsymbol{\nabla} \cdot \mathbf{A} + \boldsymbol{\nabla}U \cdot \mathbf{A}$	$\frac{\partial}{\partial x_i}(UA_i)$
18.	$\boldsymbol{\nabla} \times (U\mathbf{A}) = \boldsymbol{\nabla}U \times \mathbf{A} + U\boldsymbol{\nabla} \times \mathbf{A}$	$e_{ijk}\frac{\partial}{\partial x_j}(UA_k)\hat{\mathbf{e}}_i$
19.	$\boldsymbol{\nabla}(U\mathbf{A}) = \boldsymbol{\nabla}U\mathbf{A} + U\boldsymbol{\nabla}\mathbf{A}$	$\hat{\mathbf{e}}_j\frac{\partial}{\partial x_j}(UA_k\hat{\mathbf{e}}_k)$
20.	$\boldsymbol{\nabla}(\mathbf{A} \cdot \mathbf{x}) = \mathbf{A} + \boldsymbol{\nabla}\mathbf{A} \cdot \mathbf{x}$	$\frac{\partial}{\partial x_j}(A_ix_i)\hat{\mathbf{e}}_j$

2.4.7 Cylindrical and Spherical Coordinate Systems

Two commonly used orthogonal curvilinear coordinate systems are the *cylindrical* coordinate system [see Fig. 2.4.5(a)] and the *spherical* coordinate system [see Fig. 2.4.5(b)]. Table 2.4.2 contains a summary of the basic information, such as the base vectors and their derivatives with respect to the coordinates; definition of the del operator; the gradient, divergence, and curl of a vector; and the gradient of a second-order tensor for the two coordinate systems. The matrices of direction cosines between the orthogonal rectangular Cartesian system (x, y, z) and the orthogonal curvilinear systems (r, θ, z) and (R, ϕ, θ) are given in Eqs. (2.4.41)–(2.4.44).

Cylindrical coordinates

$$\begin{Bmatrix} \hat{\mathbf{e}}_r \\ \hat{\mathbf{e}}_\theta \\ \hat{\mathbf{e}}_z \end{Bmatrix} = \begin{bmatrix} \cos\theta & \sin\theta & 0 \\ -\sin\theta & \cos\theta & 0 \\ 0 & 0 & 1 \end{bmatrix} \begin{Bmatrix} \hat{\mathbf{e}}_x \\ \hat{\mathbf{e}}_y \\ \hat{\mathbf{e}}_z \end{Bmatrix}, \tag{2.4.41}$$

$$\begin{Bmatrix} \hat{\mathbf{e}}_x \\ \hat{\mathbf{e}}_y \\ \hat{\mathbf{e}}_z \end{Bmatrix} = \begin{bmatrix} \cos\theta & -\sin\theta & 0 \\ \sin\theta & \cos\theta & 0 \\ 0 & 0 & 1 \end{bmatrix} \begin{Bmatrix} \hat{\mathbf{e}}_r \\ \hat{\mathbf{e}}_\theta \\ \hat{\mathbf{e}}_z \end{Bmatrix}. \tag{2.4.42}$$

Spherical coordinates

$$\begin{Bmatrix} \hat{\mathbf{e}}_R \\ \hat{\mathbf{e}}_\phi \\ \hat{\mathbf{e}}_\theta \end{Bmatrix} = \begin{bmatrix} \sin\phi\cos\theta & \sin\phi\sin\theta & \cos\phi \\ \cos\phi\cos\theta & \cos\phi\sin\theta & -\sin\phi \\ -\sin\theta & \cos\theta & 0 \end{bmatrix} \begin{Bmatrix} \hat{\mathbf{e}}_x \\ \hat{\mathbf{e}}_y \\ \hat{\mathbf{e}}_z \end{Bmatrix}, \tag{2.4.43}$$

$$\begin{Bmatrix} \hat{\mathbf{e}}_x \\ \hat{\mathbf{e}}_y \\ \hat{\mathbf{e}}_z \end{Bmatrix} = \begin{bmatrix} \sin\phi\cos\theta & \cos\phi\cos\theta & -\sin\theta \\ \sin\phi\sin\theta & \cos\phi\sin\theta & \cos\theta \\ \cos\phi & -\sin\phi & 0 \end{bmatrix} \begin{Bmatrix} \hat{\mathbf{e}}_R \\ \hat{\mathbf{e}}_\phi \\ \hat{\mathbf{e}}_\theta \end{Bmatrix}. \tag{2.4.44}$$

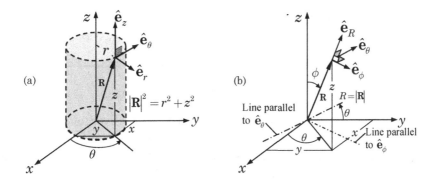

Fig. 2.4.5: (a) Cylindrical coordinate system. (b) Spherical coordinate system.

Table 2.4.2: Base vectors and operations with the del operator in cylindrical and spherical coordinate systems; see Fig. 2.4.5.

- *Cylindrical coordinate system* (r, θ, z)

$x = r\cos\theta, \; y = r\sin\theta, \; z = z, \quad \mathbf{R} = r\hat{\mathbf{e}}_r + z\hat{\mathbf{e}}_z, \quad \mathbf{A} = A_r\hat{\mathbf{e}}_r + A_\theta\hat{\mathbf{e}}_\theta + A_z\hat{\mathbf{e}}_z$ (a vector)

$\hat{\mathbf{e}}_r = \cos\theta\,\hat{\mathbf{e}}_x + \sin\theta\,\hat{\mathbf{e}}_y, \; \hat{\mathbf{e}}_\theta = -\sin\theta\,\hat{\mathbf{e}}_x + \cos\theta\,\hat{\mathbf{e}}_y, \; \hat{\mathbf{e}}_z = \hat{\mathbf{e}}_z$

$\frac{\partial\hat{\mathbf{e}}_r}{\partial\theta} = -\sin\theta\,\hat{\mathbf{e}}_x + \cos\theta\,\hat{\mathbf{e}}_y = \hat{\mathbf{e}}_\theta, \; \frac{\partial\hat{\mathbf{e}}_\theta}{\partial\theta} = -\cos\theta\,\hat{\mathbf{e}}_x - \sin\theta\,\hat{\mathbf{e}}_y = -\hat{\mathbf{e}}_r$

All other derivatives of the base vectors are zero.

$\nabla = \hat{\mathbf{e}}_r\frac{\partial}{\partial r} + \frac{1}{r}\hat{\mathbf{e}}_\theta\frac{\partial}{\partial\theta} + \hat{\mathbf{e}}_z\frac{\partial}{\partial z}, \quad \nabla^2 = \frac{1}{r}\left[\frac{\partial}{\partial r}\left(r\frac{\partial}{\partial r}\right) + \frac{1}{r}\frac{\partial^2}{\partial\theta^2} + r\frac{\partial^2}{\partial z^2}\right]$

$\nabla\cdot\mathbf{A} = \frac{1}{r}\left[\frac{\partial(rA_r)}{\partial r} + \frac{\partial A_\theta}{\partial\theta} + r\frac{\partial A_z}{\partial z}\right]$

$\nabla\times\mathbf{A} = \left(\frac{1}{r}\frac{\partial A_z}{\partial\theta} - \frac{\partial A_\theta}{\partial z}\right)\hat{\mathbf{e}}_r + \left(\frac{\partial A_r}{\partial z} - \frac{\partial A_z}{\partial r}\right)\hat{\mathbf{e}}_\theta + \frac{1}{r}\left[\frac{\partial(rA_\theta)}{\partial r} - \frac{\partial A_r}{\partial\theta}\right]\hat{\mathbf{e}}_z$

$\nabla\mathbf{A} = \frac{\partial A_r}{\partial r}\hat{\mathbf{e}}_r\hat{\mathbf{e}}_r + \frac{\partial A_\theta}{\partial r}\hat{\mathbf{e}}_r\hat{\mathbf{e}}_\theta + \frac{1}{r}\left(\frac{\partial A_r}{\partial\theta} - A_\theta\right)\hat{\mathbf{e}}_\theta\hat{\mathbf{e}}_r + \frac{\partial A_z}{\partial r}\hat{\mathbf{e}}_r\hat{\mathbf{e}}_z + \frac{\partial A_r}{\partial z}\hat{\mathbf{e}}_z\hat{\mathbf{e}}_r$

$\qquad + \frac{1}{r}\left(A_r + \frac{\partial A_\theta}{\partial\theta}\right)\hat{\mathbf{e}}_\theta\hat{\mathbf{e}}_\theta + \frac{1}{r}\frac{\partial A_z}{\partial\theta}\hat{\mathbf{e}}_\theta\hat{\mathbf{e}}_z + \frac{\partial A_\theta}{\partial z}\hat{\mathbf{e}}_z\hat{\mathbf{e}}_\theta + \frac{\partial A_z}{\partial z}\hat{\mathbf{e}}_z\hat{\mathbf{e}}_z$

- *Spherical coordinate system* (R, ϕ, θ)

$x = R\sin\phi\cos\theta, \; y = R\sin\phi\sin\theta, \; z = R\cos\phi, \quad \mathbf{R} = R\hat{\mathbf{e}}_R, \quad \mathbf{A} = A_R\hat{\mathbf{e}}_R + A_\phi\hat{\mathbf{e}}_\phi + A_\theta\hat{\mathbf{e}}_\theta$

$\hat{\mathbf{e}}_R = \sin\phi\,(\cos\theta\,\hat{\mathbf{e}}_x + \sin\theta\,\hat{\mathbf{e}}_y) + \cos\phi\,\hat{\mathbf{e}}_z, \; \hat{\mathbf{e}}_\phi = \cos\phi\,(\cos\theta\,\hat{\mathbf{e}}_x + \sin\theta\,\hat{\mathbf{e}}_y) - \sin\phi\,\hat{\mathbf{e}}_z$

$\hat{\mathbf{e}}_\theta = -\sin\theta\,\hat{\mathbf{e}}_x + \cos\theta\,\hat{\mathbf{e}}_y$

$\hat{\mathbf{e}}_x = \cos\theta\,(\sin\phi\,\hat{\mathbf{e}}_R + \cos\phi\,\hat{\mathbf{e}}_\phi) - \sin\theta\,\hat{\mathbf{e}}_\theta, \; \hat{\mathbf{e}}_y = \sin\theta\,(\sin\phi\,\hat{\mathbf{e}}_R + \cos\phi, \hat{\mathbf{e}}_\phi) + \cos\theta\,\hat{\mathbf{e}}_\theta$

$\hat{\mathbf{e}}_z = \cos\phi\,\hat{\mathbf{e}}_R - \sin\phi\,\hat{\mathbf{e}}_\phi$

$\frac{\partial\hat{\mathbf{e}}_R}{\partial\phi} = \hat{\mathbf{e}}_\phi, \; \frac{\partial\hat{\mathbf{e}}_R}{\partial\theta} = \sin\phi\,\hat{\mathbf{e}}_\theta, \; \frac{\partial\hat{\mathbf{e}}_\phi}{\partial\phi} = -\hat{\mathbf{e}}_R, \; \frac{\partial\hat{\mathbf{e}}_\phi}{\partial\theta} = \cos\phi\,\hat{\mathbf{e}}_\theta, \; \frac{\partial\hat{\mathbf{e}}_\theta}{\partial\theta} = -\sin\phi\,\hat{\mathbf{e}}_R - \cos\phi\,\hat{\mathbf{e}}_\phi$

All other derivatives of the base vectors are zero.

$\nabla = \hat{\mathbf{e}}_R\frac{\partial}{\partial R} + \frac{\hat{\mathbf{e}}_\phi}{R}\frac{\partial}{\partial\phi} + \frac{\hat{\mathbf{e}}_\theta}{R\sin\phi}\frac{\partial}{\partial\theta}, \; \nabla^2 = \frac{1}{R^2}\left[\frac{\partial}{\partial R}\left(R^2\frac{\partial}{\partial R}\right) + \frac{1}{\sin\phi}\frac{\partial}{\partial\phi}\left(\sin\phi\frac{\partial}{\partial\phi}\right) + \frac{1}{\sin^2\phi}\frac{\partial^2}{\partial\theta^2}\right]$

$\nabla\cdot\mathbf{A} = \frac{2A_R}{R} + \frac{\partial A_R}{\partial R} + \frac{1}{R\sin\phi}\frac{\partial(A_\phi\sin\phi)}{\partial\phi} + \frac{1}{R\sin\phi}\frac{\partial A_\theta}{\partial\theta}$

$\nabla\times\mathbf{A} = \frac{1}{R\sin\phi}\left[\frac{\partial(\sin\phi A_\theta)}{\partial\phi} - \frac{\partial A_\phi}{\partial\theta}\right]\hat{\mathbf{e}}_R + \left[\frac{1}{R\sin\phi}\frac{\partial A_R}{\partial\theta} - \frac{1}{R}\frac{\partial(RA_\theta)}{\partial R}\right]\hat{\mathbf{e}}_\phi$

$\qquad + \frac{1}{R}\left[\frac{\partial(RA_\phi)}{\partial R} - \frac{\partial A_R}{\partial\phi}\right]\hat{\mathbf{e}}_\theta$

$\nabla\mathbf{A} = \frac{\partial A_R}{\partial R}\hat{\mathbf{e}}_R\hat{\mathbf{e}}_R + \frac{\partial A_\phi}{\partial R}\hat{\mathbf{e}}_R\hat{\mathbf{e}}_\phi + \frac{1}{R}\left(\frac{\partial A_R}{\partial\phi} - A_\phi\right)\hat{\mathbf{e}}_\phi\hat{\mathbf{e}}_R + \frac{\partial A_\theta}{\partial R}\hat{\mathbf{e}}_R\hat{\mathbf{e}}_\theta$

$\qquad + \frac{1}{R\sin\phi}\left(\frac{\partial A_R}{\partial\theta} - A_\theta\sin\phi\right)\hat{\mathbf{e}}_\theta\hat{\mathbf{e}}_R + \frac{1}{R}\left(A_R + \frac{\partial A_\phi}{\partial\phi}\right)\hat{\mathbf{e}}_\phi\hat{\mathbf{e}}_\phi + \frac{1}{R}\frac{\partial A_\theta}{\partial\phi}\hat{\mathbf{e}}_\phi\hat{\mathbf{e}}_\theta$

$\qquad + \frac{1}{R\sin\phi}\left(\frac{\partial A_\phi}{\partial\theta} - A_\theta\cos\phi\right)\hat{\mathbf{e}}_\theta\hat{\mathbf{e}}_\phi + \frac{1}{R\sin\phi}\left(A_R\sin\phi + A_\phi\cos\phi + \frac{\partial A_\theta}{\partial\theta}\right)\hat{\mathbf{e}}_\theta\hat{\mathbf{e}}_\theta$

2.4.8 Gradient, Divergence, and Curl Theorems

Integral identities involving the gradient of a vector, divergence of a vector, and curl of a vector can be established from integral relations between volume integrals and surface integrals. These identities will be useful in later chapters when we derive the equations of a continuous medium.

Let Ω denote a region in $\Re^3$ bounded by the closed surface Γ. Let ds be a differential element of surface and $\hat{\mathbf{n}}$ the unit outward normal, and let $d\mathbf{x}$ be a differential volume element in Ω. The following relations, known from advanced calculus, hold:

$$\int_\Omega \nabla \phi \, d\mathbf{x} = \oint_\Gamma \hat{\mathbf{n}} \, \phi \, ds \quad \text{(Gradient theorem)} \tag{2.4.45}$$

$$\int_\Omega \nabla \cdot \mathbf{A} \, d\mathbf{x} = \oint_\Gamma \hat{\mathbf{n}} \cdot \mathbf{A} \, ds \quad \text{(Divergence theorem)} \tag{2.4.46}$$

$$\int_\Omega \nabla \times \mathbf{A} \, d\mathbf{x} = \oint_\Gamma \hat{\mathbf{n}} \times \mathbf{A} \, ds \quad \text{(Curl theorem)} \tag{2.4.47}$$

The combination $\hat{\mathbf{n}} \cdot \mathbf{A} \, ds$ is called the *outflow of* $\mathbf{A}$ *through the differential surface* ds. The integral is called the total or net outflow through the surrounding surface Δs. This is easier to see if one imagines that $\mathbf{A}$ is a velocity vector and the outflow is the amount of fluid flow. In the limit as the region shrinks to a point, the net outflow per unit volume is associated therefore with the *divergence of the vector field*. The integral forms presented in Eqs. (2.4.45)–(2.4.47) are known as the *invariant forms* because they do not depend on the coordinate system.

2.5 Tensors

2.5.1 Dyads and Dyadics

As we have already seen in Eqs. (2.4.32) and (2.4.33), the expression for the gradient of a vector contains two basis vectors standing next to each other (sometimes called the *tensor product*), indicating that the entity is characterized by two directions, one coming from the gradient (vector) operator and other from the vector itself. One may also recall from the elementary mechanics of materials course that the stress, which is force per unit area, depends not only on the orientation of the plane on which it is acting but also on the direction of the force.[4] Thus, specification of stress at a point requires two vectors, one to denote the plane (by a vector perpendicular to the plane) on which the force is acting and the other to denote the direction of the force. Objects that are composed of two vectors standing next to each other, without any vector operation between them, are known as *dyads*, or what we shall call here a second-order tensor (the two phrases are interchangeable). Thus, a dyad is defined as two vectors standing side by side and acting as a unit. A linear combination of dyads is called a *dyadic*.

Let $\mathbf{A}_1, \mathbf{A}_2, \cdots, \mathbf{A}_n$ and $\mathbf{B}_1, \mathbf{B}_2, \cdots, \mathbf{B}_n$ be arbitrary vectors. Then the following linear combination, denoted $\mathbf{S}$, constitutes a dyadic:

$$\mathbf{S} = \mathbf{A}_1 \mathbf{B}_1 + \mathbf{A}_2 \mathbf{B}_2 + \cdots + \mathbf{A}_n \mathbf{B}_n. \tag{2.5.1}$$

The transpose of a dyadic is defined as the result obtained by the interchange of the two vectors in each of the dyads. For example, the transpose of the dyadic in Eq. (2.5.1) is

$$\mathbf{S}^{\mathrm{T}} = \mathbf{B}_1 \mathbf{A}_1 + \mathbf{B}_2 \mathbf{A}_2 + \cdots + \mathbf{B}_n \mathbf{A}_n.$$

---

[4]The concept of stress is discussed in more detail in Chapter 4.

One of the properties of a dyadic is defined by the dot product with a vector, say $\mathbf{V}$:

$$\mathbf{S} \cdot \mathbf{V} = \mathbf{A}_1(\mathbf{B}_1 \cdot \mathbf{V}) + \mathbf{A}_2(\mathbf{B}_2 \cdot \mathbf{V}) + \cdots + \mathbf{A}_n(\mathbf{B}_n \cdot \mathbf{V}),$$
$$\mathbf{V} \cdot \mathbf{S} = (\mathbf{V} \cdot \mathbf{A}_1)\mathbf{B}_1 + (\mathbf{V} \cdot \mathbf{A}_2)\mathbf{B}_2 + \cdots + (\mathbf{V} \cdot \mathbf{A}_n)\mathbf{B}_n. \tag{2.5.2}$$

The dot operation of a dyad with a vector produces another vector. In the first case the dyad acts as a *prefactor* and in the second case as a *postfactor*. The two operations in general produce different vectors. The dot product between a dyad and a vector can be written in alternative forms using the definition of the transpose of a dyad,

$$\mathbf{V} \cdot \mathbf{S} = \mathbf{S}^{\mathrm{T}} \cdot \mathbf{V}, \quad \mathbf{S} \cdot \mathbf{V} = \mathbf{V} \cdot \mathbf{S}^{\mathrm{T}}. \tag{2.5.3}$$

In general, one can show (see Problem **2.47**) that the transpose of the dot product of tensors (of any order) follows the rule

$$(\mathbf{R} \cdot \mathbf{S})^{\mathrm{T}} = \mathbf{S}^{\mathrm{T}} \cdot \mathbf{R}^{\mathrm{T}}, \quad (\mathbf{R} \cdot \mathbf{S} \cdot \mathbf{T})^{\mathrm{T}} = \mathbf{T}^{\mathrm{T}} \cdot \mathbf{S}^{\mathrm{T}} \cdot \mathbf{R}^{\mathrm{T}}. \tag{2.5.4}$$

The dot product of a dyad (or a second-order tensor) with itself is a dyad, and it is denoted by

$$\mathbf{S} \cdot \mathbf{S} = \mathbf{S}^2, \quad \mathbf{S}^3 = \mathbf{S}^2 \cdot \mathbf{S}, \quad \cdots, \quad \mathbf{S}^n = \mathbf{S}^{n-1} \cdot \mathbf{S}. \tag{2.5.5}$$

Some authors [see, e.g., Gurtin (1981)] use the notation

$$\mathbf{A} \otimes \mathbf{B} \text{ for } \mathbf{A}\mathbf{B} \quad \text{and} \quad \mathbf{A}\mathbf{B} \text{ for } \mathbf{A} \cdot \mathbf{B}. \tag{2.5.6}$$

Thus, when there is no operation (other than the usual multiplication) between vectors, they use the symbol $\otimes$, called *tensor product*; and when there is a dot product between vectors and tensors no dot is placed between them. The notation adopted here is explicit, that is, indicates the operation (e.g., dot product or cross product) between vectors and tensors. If there is no operation between vectors, other than the usual multiplication of the underlying scalars, no symbol is placed. For example, we use $\mathbf{A}\mathbf{B}$ instead of $\mathbf{A} \otimes \mathbf{B}$ and $\mathbf{S} \cdot \mathbf{A}$ instead of $\mathbf{S}\mathbf{A}$ for the dot product between the tensor $\mathbf{S}$ and vector $\mathbf{A}$. Also, in writing the gradient of a vector or tensor in component form, we use the forward gradient operator and list the base vectors in the order they appear in the operation.

2.5.2 Nonion Form of a Second-Order Tensor

Let each of the vectors in a dyad $\mathbf{S} = \mathbf{A}\mathbf{B}$ be represented in a given basis system. In Cartesian system, we can write $\mathbf{A} = A_m \hat{\mathbf{e}}_m$ and $\mathbf{B} = B_n \hat{\mathbf{e}}_n$, with summations on m and n as implied by the repeated indices. Then $\mathbf{S} = A_m B_n \hat{\mathbf{e}}_m \hat{\mathbf{e}}_n$, or $\mathbf{S} = S_{mn} \hat{\mathbf{e}}_m \hat{\mathbf{e}}_n$, with $S_{mn} = A_m B_n$.

We can display all of the components of a dyad $\mathbf{S}$ in the rectangular Cartesian basis $\hat{\mathbf{e}}_i$ as $\mathbf{S} = S_{ij} \hat{\mathbf{e}}_i \hat{\mathbf{e}}_j$ by letting the j index run to the right and the i index run downward:

$$\mathbf{S} = S_{ij}\,\hat{\mathbf{e}}_i\hat{\mathbf{e}}_j = s_{11}\hat{\mathbf{e}}_1\hat{\mathbf{e}}_1 + s_{12}\hat{\mathbf{e}}_1\hat{\mathbf{e}}_2 + s_{13}\hat{\mathbf{e}}_1\hat{\mathbf{e}}_3$$
$$+ s_{21}\hat{\mathbf{e}}_2\hat{\mathbf{e}}_1 + s_{22}\hat{\mathbf{e}}_2\hat{\mathbf{e}}_2 + s_{23}\hat{\mathbf{e}}_2\hat{\mathbf{e}}_3$$
$$+ s_{31}\hat{\mathbf{e}}_3\hat{\mathbf{e}}_1 + s_{32}\hat{\mathbf{e}}_3\hat{\mathbf{e}}_2 + s_{33}\hat{\mathbf{e}}_3\hat{\mathbf{e}}_3. \tag{2.5.7}$$

This form is called the *nonion* form of the dyadic $\mathbf{S}$. Equation (2.5.7) illustrates that a dyadic in three-dimensional space has nine independent components in general, each component associated with a certain dyad pair. The components are thus said to be ordered. When the ordering is understood, such as suggested by the nonion form in Eq. (2.5.7), the explicit writing of the dyads can be suppressed and the dyadic is written as a matrix:

$$[S] = \begin{bmatrix} s_{11} & s_{12} & s_{13} \\ s_{21} & s_{22} & s_{23} \\ s_{31} & s_{32} & s_{33} \end{bmatrix} \quad \text{and} \quad \mathbf{S} = \begin{Bmatrix} \hat{\mathbf{e}}_1 \\ \hat{\mathbf{e}}_2 \\ \hat{\mathbf{e}}_3 \end{Bmatrix}^{\mathrm{T}} [S] \begin{Bmatrix} \hat{\mathbf{e}}_1 \\ \hat{\mathbf{e}}_2 \\ \hat{\mathbf{e}}_3 \end{Bmatrix}. \tag{2.5.8}$$

This representation is simpler than Eq. (2.5.7), but it is taken to mean the same. The determinant of a second-order tensor $\mathbf{S}$ is the determinant of the matrix of its components, $\det [S] = |\mathbf{S}|$. The determinant of $\mathbf{S}$ can be expressed in the form [see Problem **2.41(a)**]

$$|\mathbf{S}| = e_{ijk}\, s_{1i}\, s_{2j}\, s_{3k}, \tag{2.5.9}$$

Although the definition of the determinant in Eq. (2.5.9) involves components of $\mathbf{S}$ (and therefore, depends on the coordinate system in which the components are defined), the value is actually independent of the coordinate system, as can be seen from the alternative definition of the determinant of a second-order tensor $\mathbf{S}$ (see Problem **2.48**)

$$|\mathbf{S}| = \frac{[(\mathbf{S}\cdot\mathbf{A})\times(\mathbf{S}\cdot\mathbf{B})]\cdot(\mathbf{S}\cdot\mathbf{C})}{\mathbf{A}\times\mathbf{B}\cdot\mathbf{C}}, \tag{2.5.10}$$

where $\mathbf{A}$, $\mathbf{B}$, and $\mathbf{C}$ are arbitrary vectors.

The unit dyad or unit second-order tensor is defined in terms of the Cartesian components δ_{ij} as

$$\mathbf{I} = \delta_{ij}\hat{\mathbf{e}}_i\hat{\mathbf{e}}_j = \hat{\mathbf{e}}_i\hat{\mathbf{e}}_i = \hat{\mathbf{e}}_1\hat{\mathbf{e}}_1 + \hat{\mathbf{e}}_2\hat{\mathbf{e}}_2 + \hat{\mathbf{e}}_3\hat{\mathbf{e}}_3. \tag{2.5.11}$$

It is clear that the unit second-order tensor $\mathbf{I}$ is symmetric. The unit second-order tensor in an orthogonal Cartesian coordinate system can be written alternatively as

$$\mathbf{I} = \begin{Bmatrix} \hat{\mathbf{e}}_1 \\ \hat{\mathbf{e}}_2 \\ \hat{\mathbf{e}}_3 \end{Bmatrix}^{\mathrm{T}} [I] \begin{Bmatrix} \hat{\mathbf{e}}_1 \\ \hat{\mathbf{e}}_2 \\ \hat{\mathbf{e}}_3 \end{Bmatrix}, \quad [I] = \begin{bmatrix} 1 & 0 & 0 \\ 0 & 1 & 0 \\ 0 & 0 & 1 \end{bmatrix}. \tag{2.5.12}$$

Two types of "double-dot products" between two second-order tensors are useful in the sequel. The horizontal double-dot product and the vertical double-dot product between a dyad $(\mathbf{AB})$ and another dyad $(\mathbf{CD})$ (that is, $\mathbf{A}$, $\mathbf{B}$, $\mathbf{C}$, and $\mathbf{D}$ are vectors) are defined as the scalars

$$(\mathbf{AB})\cdots(\mathbf{CD}) \equiv (\mathbf{B}\cdot\mathbf{C})(\mathbf{A}\cdot\mathbf{D}),$$
$$(\mathbf{AB}):(\mathbf{CD}) \equiv (\mathbf{A}\cdot\mathbf{C})(\mathbf{B}\cdot\mathbf{D}).$$

The double-dot products, by this definition, are commutative. The two double-dot product between two dyads $\mathbf{S}$ and $\mathbf{T}$ in a rectangular Cartesian system are

$$\begin{aligned}
\mathbf{S} : \mathbf{T} &= (s_{ij}\hat{\mathbf{e}}_i\hat{\mathbf{e}}_j) : (t_{mn}\hat{\mathbf{e}}_m\hat{\mathbf{e}}_n) \\
&= s_{ij}t_{mn}(\hat{\mathbf{e}}_i \cdot \hat{\mathbf{e}}_m)(\hat{\mathbf{e}}_j \cdot \hat{\mathbf{e}}_n) \\
&= s_{ij}t_{mn}\delta_{im}\delta_{jn} = s_{ij}\,t_{ij}, \\
\mathbf{S} \cdot\cdot \mathbf{T} &= (s_{ij}\hat{\mathbf{e}}_i\hat{\mathbf{e}}_j) \cdot\cdot (t_{mn}\hat{\mathbf{e}}_m\hat{\mathbf{e}}_n) \\
&= s_{ij}t_{mn}(\hat{\mathbf{e}}_j \cdot \hat{\mathbf{e}}_m)(\hat{\mathbf{e}}_i \cdot \hat{\mathbf{e}}_n) \\
&= s_{ij}t_{mn}\delta_{jm}\delta_{in} = s_{ij}\,t_{ji}.
\end{aligned}$$

$(2.5.13)$

$(2.5.14)$

The trace of a second-order tensor (i.e., the sum of the diagonal terms of the matrix representing the tensor) is defined to be the double-dot product of the tensor with the unit tensor

$$\operatorname{tr}\mathbf{S} = \mathbf{S} : \mathbf{I} = \mathbf{S} \cdot\cdot \mathbf{I}. \tag{2.5.15}$$

Certain combinations of the components of a tensor remain the same in all coordinate systems; that is, they are invariant under coordinate transformations. Such quantities (involving sums and products of the components of a tensor) are termed *invariants*. For example, the determinant of a tensor is the same in all coordinate systems. Similarly, the trace of the matrix representing a tensor is an *invariant*. Among many invariants of a tensor, the following three invariants, called *principal invariants*, are identified because of their role in finding eigenvalues of the tensor (readers should be aware of the fact that the definitions of the second and third principal invariants may differ from those in other books):

$$I_1 = \operatorname{tr}\mathbf{S}, \quad I_2 = \tfrac{1}{2}\left[(\operatorname{tr}\mathbf{S})^2 - \operatorname{tr}\left(\mathbf{S}^2\right)\right], \quad I_3 = |\mathbf{S}|. \tag{2.5.16}$$

In terms of the rectangular Cartesian components of $\mathbf{S}$, the three principal invariants have the form

$$I_1 = s_{ii}, \quad I_2 = \tfrac{1}{2}\left(s_{ii}s_{jj} - s_{ij}s_{ji}\right), \quad I_3 = |\mathbf{S}|. \tag{2.5.17}$$

In the general scheme that is developed so far, scalars are the *zeroth-order tensors*, vectors are the *first-order tensors*, and dyads are the *second-order tensors*. The order of a tensor can be determined by counting the number of basis vectors in the representation of a tensor. However, the definition of a tensor of any order must obey certain rules, such as the physical vector discussed in Section 2.2.1.1. We list them in Section 2.5.4.

In view of the matrix representation of a second-order tensor, many of the definitions and properties introduced for matrices can be extended to second-order tensors, $\mathbf{S}$. They are summarized here.

(1) $\mathbf{S}$ is *symmetric* if and only if $\mathbf{S} = \mathbf{S}^{\mathrm{T}}$ $(s_{ij} = s_{ji})$.

(2) $\mathbf{S}$ is *skew symmetric* if and only if $\mathbf{S} = -\mathbf{S}^{\mathrm{T}}$ $(s_{ij} = -s_{ji}$ for $i \neq j$ and $s_{(i)(i)} = 0$ for any fixed i).

(3) $\mathbf{S}$ can be represented as a sum of symmetric and skew symmetric parts:

$$\mathbf{S} = \tfrac{1}{2}\left(\mathbf{S} + \mathbf{S}^{\mathrm{T}}\right) + \tfrac{1}{2}\left(\mathbf{S} - \mathbf{S}^{\mathrm{T}}\right) \equiv \mathbf{S}^{\mathrm{sym}} + \mathbf{S}^{\mathrm{skew}}.$$

(4) If **S** is symmetric, **W** is skew symmetric, and **T** is an arbitrary tensor, then

$$\mathbf{S} : \mathbf{W} = \mathbf{W} : \mathbf{S} = 0, \quad \mathbf{S} : \mathbf{T} = \mathbf{S} : \mathbf{T}^{\text{sym}},$$

$$\mathbf{W} : \mathbf{T} = -\mathbf{W} : \mathbf{T}^{\text{T}} = \mathbf{W} : \mathbf{T}^{\text{skew}}.$$

Also, if $\mathbf{S} : \mathbf{T} = 0$ for any tensor **S**, then $\mathbf{T} = \mathbf{0}$. If $\mathbf{S} : \mathbf{T} = 0$ for any symmetric tensor **S**, then **T** is skew symmetric. The converse also holds: If $\mathbf{S} : \mathbf{W} = 0$ for any skew tensor **W**, then **S** is symmetric.

(5) The inverse **T** of any second-order tensor **S**, denoted $\mathbf{T} = \mathbf{S}^{-1}$, is defined to be $\mathbf{S}^{-1} \cdot \mathbf{S} = \mathbf{S} \cdot \mathbf{S}^{-1} = \mathbf{I}$; the inverse of **S** can be represented in terms of the inverse of its (nonsingular) matrix: $\mathbf{S}^{-1} = \{\hat{\mathbf{e}}\}^{\text{T}} [S]^{-1} \{\hat{\mathbf{e}}\}$. We use the notation $(\mathbf{S}^{-1})^{\text{T}} = (\mathbf{S}^{\text{T}})^{-1} = \mathbf{S}^{-\text{T}}$ and note $(\mathbf{S} \cdot \mathbf{T})^{-1} = \mathbf{T}^{-1} \cdot \mathbf{S}^{-1}$ and $(\mathbf{S} \cdot \mathbf{T})^{-\text{T}} = \mathbf{S}^{-\text{T}} \cdot \mathbf{T}^{-\text{T}}$ (see parts (e) and (f) of Problem **2.47**).

(6) A necessary and sufficient condition for a second-order tensor **Q** to be *orthogonal* is $\mathbf{Q} \cdot \mathbf{Q}^{\text{T}} = \mathbf{I}$. It can be shown that an orthogonal tensor **Q** preserves the inner product in the sense that $(\mathbf{Q} \cdot \mathbf{u}) \cdot (\mathbf{Q} \cdot \mathbf{v}) = \mathbf{u} \cdot \mathbf{v}$. In the case of physical vectors, this amounts to preserving the lengths of the vectors **u** and **v** as well as the angle between them, as shown in Fig. 2.5.1.

(7) A second-order tensor **S** is said to be *positive-definite* if and only if $\mathbf{u} \cdot \mathbf{S} \cdot \mathbf{u} > 0$ for all nonzero vectors **u**.

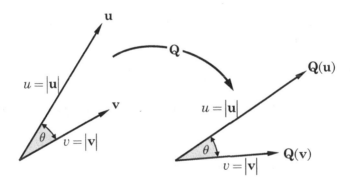

Fig. 2.5.1: An orthogonal tensor **Q** preserves the lengths $|\mathbf{u}|$ and $|\mathbf{v}|$ and angle θ between any two vectors **u** and **v**.

2.5.3 Transformation of Components of a Tensor

A *second-order Cartesian tensor* **S** (i.e., tensor with Cartesian components) may be represented in barred and unbarred Cartesian coordinate systems [see Fig. 2.2.11(b)] as

$$\mathbf{S} = s_{ij}\, \hat{\mathbf{e}}_i\, \hat{\mathbf{e}}_j = \bar{s}_{mn}\, \hat{\bar{\mathbf{e}}}_m\, \hat{\bar{\mathbf{e}}}_n. \tag{2.5.18}$$

The unit base vectors in the unbarred and barred systems are related by

$$\hat{\mathbf{e}}_j = \ell_{ij}\, \hat{\bar{\mathbf{e}}}_i \quad \text{and} \quad \hat{\bar{\mathbf{e}}}_i = \ell_{ij}\, \hat{\mathbf{e}}_j, \tag{2.5.19}$$

where ℓ_{ij} denotes the direction cosines between barred and unbarred systems, $\ell_{ij} = \hat{\bar{\mathbf{e}}}_i \cdot \hat{\mathbf{e}}_j$ [see Eq. (2.2.71)]. In orthogonal coordinate systems the determinant of the matrix of direction cosines is unity and its inverse is equal to the transpose (i.e., $[L]$ is an orthogonal matrix):

$$[L]^{-1} = [L]^{\mathrm{T}} \quad \text{or} \quad [L][L]^{\mathrm{T}} = [I]. \tag{2.5.20}$$

Using Eq. (2.5.19) to replace the unbarred base vectors in Eq. (2.5.18), we obtain

$$\left(\bar{s}_{mn} - s_{ij}\ell_{mi}\ell_{nj}\right)\hat{\bar{\mathbf{e}}}_m\,\hat{\bar{\mathbf{e}}}_n = 0 \quad \Rightarrow \quad \bar{s}_{mn} - s_{ij}\ell_{mi}\ell_{nj} = 0.$$

Thus the components of a second-order tensor transform according to

$$\bar{s}_{mn} = s_{ij}\,\ell_{mi}\,\ell_{nj} \quad \text{or} \quad [\bar{S}] = [L][S][L]^{\mathrm{T}}. \tag{2.5.21}$$

Mathematically, a tensor is defined as a quantity whose components transform according to Eq. (2.5.21).

2.5.4 Higher-Order Tensors

Third-order tensors can be viewed as those derived from *triadics*, or three vectors standing side by side. A tensor $\mathbf{T}$ of any order n is a quantity defined by 3^n components, which may be written as $a_{ijk\cdots n}$, provided the components transform according to the law

$$\bar{a}_{pqr\cdots t} = \ell_{pi}\,\ell_{qj}\,\ell_{rk}\,\cdots\,\ell_{tn}\,a_{ijk\cdots n}. \tag{2.5.22}$$

For example, the components of third- and fourth-order tensors, $\mathbf{T}$ and $\mathbf{C}$, transform according to the rules

$$\bar{t}_{ijk} = \ell_{im}\,\ell_{jn}\,\ell_{kp}\,t_{mnp}, \quad \bar{c}_{ijkl} = \ell_{im}\,\ell_{jn}\,\ell_{kp}\,\ell_{lq}\,c_{mnpq}.$$

The permutation symbol e_{ijk} can be viewed as the Cartesian components of a third-order tensor of a special kind,

$$\mathcal{E} = e_{ijk}\,\hat{\mathbf{e}}_i\,\hat{\mathbf{e}}_j\,\hat{\mathbf{e}}_k. \tag{2.5.23}$$

Tensors of various orders, especially the first-, second-, and fourth-order, appear in the study of a continuous medium. As we shall see in Chapter 6, the tensor that characterizes the material constitution is a fourth-order tensor. Tensors whose components are the same in all coordinate systems, that is, the components are invariant under coordinate transformations, are known as *isotropic* tensors. By definition, all zero-order tensors (i.e., scalars) are isotropic and the only isotropic tensor of order 1 is the zero vector $\mathbf{0}$. Every isotropic tensor $\mathbf{T}$ of order 2 can be written as $\mathbf{T} = \lambda\mathbf{I}$, and the components $C_{ijk\ell}$ of every fourth-order isotropic tensor $\mathbf{C}$ can be expressed as

$$C_{ijk\ell} = \lambda\delta_{ij}\delta_{k\ell} + \mu\left(\delta_{ik}\delta_{j\ell} + \delta_{i\ell}\delta_{jk}\right) + \kappa\left(\delta_{ik}\delta_{j\ell} - \delta_{i\ell}\delta_{jk}\right), \tag{2.5.24}$$

where λ, μ, and κ are scalars.

In general, tensors of all orders obey the following rules:

$$\alpha(\mathbf{T}) = (\alpha\mathbf{T}) = \mathbf{T}\alpha,$$
$$\alpha(\mathbf{T}_1 + \mathbf{T}_2) = \alpha\mathbf{T}_1 + \alpha\mathbf{T}_2,$$
$$\mathbf{T}_1 + \mathbf{T}_2 = \mathbf{T}_2 + \mathbf{T}_1,$$
$$\mathbf{T}_1 + (\mathbf{T}_2 + \mathbf{T}_3) = (\mathbf{T}_1 + \mathbf{T}_2) + \mathbf{T}_3,$$
$$\mathbf{T}_1 * (\mathbf{T}_2 + \mathbf{T}_3) = \mathbf{T}_1 * \mathbf{T}_2 + \mathbf{T}_1 * \mathbf{T}_3,$$

where * denotes the dot product, cross product, or no operation (other than the multiplication of the scalars involved). However, in general, we do not have the property $\mathbf{T}_1 * \mathbf{T}_2 = \mathbf{T}_2 * \mathbf{T}_1$.

2.5.5 Tensor Calculus

Here we discuss the calculus of tensors whose components are functions of position $\mathbf{x}$. We begin with the gradient of a vector, which is a second-order tensor:

$$\nabla\mathbf{A} = \begin{cases} \mathbf{e}^i \frac{\partial}{\partial q^i}(A_j\mathbf{e}^j) = \frac{\partial A_j}{\partial q^i}\mathbf{e}^i\mathbf{e}^j + A_j\mathbf{e}^i\frac{\partial \mathbf{e}^j}{\partial q^i}, \\ \mathbf{e}^i \frac{\partial}{\partial q^i}(A^j\mathbf{e}_j) = \frac{\partial A^j}{\partial q^i}\mathbf{e}^i\mathbf{e}_j + A^j\mathbf{e}^i\frac{\partial \mathbf{e}_j}{\partial q^i}. \end{cases}$$

Note that the order of the base vectors is kept intact, that is, kept in the same order in which they come into the operation. Thus, the derivatives of the unitary base vectors are, in general, not zero; they are expressed in terms of the *Christoffel symbols*, and we shall not discuss them here.

The gradient of a vector, and hence a second-order tensor, can be expressed as the sum of symmetric and skew symmetric parts by adding and subtracting $(1/2)(\nabla\mathbf{A})^{\mathrm{T}}$

$$\nabla\mathbf{A} = \tfrac{1}{2}\left[\nabla\mathbf{A} + (\nabla\mathbf{A})^{\mathrm{T}}\right] + \tfrac{1}{2}\left[\nabla\mathbf{A} - (\nabla\mathbf{A})^{\mathrm{T}}\right], \tag{2.5.25}$$

or in rectangular Cartesian component form

$$\nabla\mathbf{A} = \tfrac{1}{2}\left(\frac{\partial A_j}{\partial x_i} + \frac{\partial A_i}{\partial x_j}\right)\hat{\mathbf{e}}_i\hat{\mathbf{e}}_j + \tfrac{1}{2}\left(\frac{\partial A_j}{\partial x_i} - \frac{\partial A_i}{\partial x_j}\right)\hat{\mathbf{e}}_i\hat{\mathbf{e}}_j.$$

Analogously to the divergence of a vector, the divergence of a second-order Cartesian tensor is defined as

$$\nabla \cdot \mathbf{S} = \begin{cases} \mathbf{e}^i \frac{\partial}{\partial q^i}\left(S_{mn}\mathbf{e}^m\mathbf{e}^n\right) & \text{(covariant components)} \\ \mathbf{e}^i \frac{\partial}{\partial q^i}\left(S^{mn}\mathbf{e}_m\mathbf{e}_n\right) & \text{(contravariant components)} \\ \mathbf{e}^i \frac{\partial}{\partial q^i}\left(S^m_n\mathbf{e}_m\mathbf{e}^n\right) & \text{(mixed components)} \\ \mathbf{e}^i \frac{\partial}{\partial q^i}\left(S^n_m\mathbf{e}^m\mathbf{e}_n\right) & \text{(mixed components)} \end{cases} \tag{2.5.26}$$

In rectangular Cartesian component form, we have

$$\nabla \cdot \mathbf{S} = \hat{\mathbf{e}}_i \cdot \frac{\partial}{\partial x_i}(s_{mn}\hat{\mathbf{e}}_m\hat{\mathbf{e}}_n) = \frac{\partial s_{mn}}{\partial x_i}(\hat{\mathbf{e}}_i \cdot \hat{\mathbf{e}}_m)\hat{\mathbf{e}}_n = \frac{\partial s_{in}}{\partial x_i}\hat{\mathbf{e}}_n.$$

Thus, the divergence of a second-order tensor is a vector.

The integral theorems of vectors presented in Section 2.4.8 are also valid for tensors, but it is important that the order of the operations be observed. The gradient and divergence of a tensor can be expressed in cylindrical and spherical coordinate systems by writing the del operator and the tensor in component form (see Table 2.4.2). For example, the gradient of a vector **A** in the cylindrical coordinate system [see Fig. 2.4.5(a)] can be obtained by writing

$$\nabla = \hat{e}_r \frac{\partial}{\partial r} + \frac{1}{r} \hat{e}_\theta \frac{\partial}{\partial \theta} + \hat{e}_z \frac{\partial}{\partial z}, \quad \mathbf{A} = A_r \hat{e}_r + A_\theta \hat{e}_\theta + A_z \hat{e}_z.$$

Then we have

$$
\begin{aligned}
\nabla \mathbf{A} = {}& \hat{e}_r \hat{e}_r \frac{\partial A_r}{\partial r} + \hat{e}_r \hat{e}_\theta \frac{\partial A_\theta}{\partial r} + \hat{e}_\theta \hat{e}_r \frac{1}{r} \left(\frac{\partial A_r}{\partial \theta} - A_\theta \right) \\
& + \hat{e}_r \hat{e}_z \frac{\partial A_z}{\partial r} + \hat{e}_z \hat{e}_r \frac{\partial A_r}{\partial z} + \hat{e}_\theta \hat{e}_\theta \frac{1}{r} \left(A_r + \frac{\partial A_\theta}{\partial \theta} \right) \\
& + \frac{1}{r} \hat{e}_\theta \hat{e}_z \frac{\partial A_z}{\partial \theta} + \hat{e}_z \hat{e}_\theta \frac{\partial A_\theta}{\partial z} + \hat{e}_z \hat{e}_z \frac{\partial A_z}{\partial z},
\end{aligned}
\tag{2.5.27}
$$

where the following derivatives of the base vectors are used:

$$\frac{\partial \hat{e}_r}{\partial \theta} = \hat{e}_\theta, \quad \frac{\partial \hat{e}_\theta}{\partial \theta} = -\hat{e}_r. \tag{2.5.28}$$

Similarly, the gradient of a vector in spherical coordinates [see Fig. 2.4.5(b)] can be obtained by expressing ∇ and vector **A** in the spherical coordinates (see Problem **2.52**):

$$\nabla = \hat{e}_R \frac{\partial}{\partial R} + \frac{1}{R} \hat{e}_\phi \frac{\partial}{\partial \phi} + \frac{1}{R \sin \phi} \hat{e}_\theta \frac{\partial}{\partial \theta}, \quad \mathbf{A} = A_R \hat{e}_R + A_\phi \hat{e}_\phi + A_\theta \hat{e}_\theta.$$

$$
\begin{aligned}
\nabla \mathbf{A} = {}& \frac{\partial A_R}{\partial R} \hat{e}_R \hat{e}_R + \frac{\partial A_\phi}{\partial R} \hat{e}_R \hat{e}_\phi + \frac{1}{R} \left(\frac{\partial A_R}{\partial \phi} - A_\phi \right) \hat{e}_\phi \hat{e}_R + \frac{\partial A_\theta}{\partial R} \hat{e}_R \hat{e}_\theta \\
& + \frac{1}{R \sin \theta} \left(\frac{\partial A_R}{\partial \theta} - A_\theta \sin \phi \right) \hat{e}_\theta \hat{e}_R + \frac{1}{R} \left(A_R + \frac{\partial A_\phi}{\partial \phi} \right) \hat{e}_\phi \hat{e}_\phi \\
& + \frac{1}{R} \frac{\partial A_\theta}{\partial \phi} \hat{e}_\phi \hat{e}_\theta + \frac{1}{R \sin \phi} \left(\frac{\partial A_\phi}{\partial \theta} - A_\theta \cos \phi \right) \hat{e}_\theta \hat{e}_\phi \\
& + \frac{1}{R \sin \phi} \left(A_R \sin \phi + A_\phi \cos \phi + \frac{\partial A_\theta}{\partial \theta} \right) \hat{e}_\theta \hat{e}_\theta,
\end{aligned}
\tag{2.5.29}
$$

where the following derivatives of the base vectors are used:

$$\frac{\partial \hat{e}_R}{\partial \phi} = \hat{e}_\phi, \quad \frac{\partial \hat{e}_\phi}{\partial \phi} = -\hat{e}_R, \quad \frac{\partial \hat{e}_R}{\partial \theta} = \sin \phi \, \hat{e}_\theta,$$

$$\frac{\partial \hat{e}_\phi}{\partial \theta} = \cos \phi \, \hat{e}_\theta, \quad \frac{\partial \hat{e}_\theta}{\partial \theta} = -\sin \phi \, \hat{e}_R - \cos \phi \, \hat{e}_\phi. \tag{2.5.30}$$

In the same way, one can compute the curl and divergence of a tensor in cylindrical and spherical coordinate systems. Example 2.5.1 illustrates the procedure (see also Problems **2.49–2.52**).

Example 2.5.1

Suppose that the second-order tensor $\mathbf{E}$ referred to the cylindrical coordinate system ($\hat{\mathbf{e}}_r$, $\hat{\mathbf{e}}_\theta$, $\hat{\mathbf{e}}_z$) is of the form

$$\mathbf{E} = E_{rr}(r, z)\,\hat{\mathbf{e}}_r\,\hat{\mathbf{e}}_r + E_{\theta\theta}(r, z)\,\hat{\mathbf{e}}_\theta\,\hat{\mathbf{e}}_\theta.$$

Determine (a) $\boldsymbol{\nabla} \times \mathbf{E}$ and its transpose, and (b) $\boldsymbol{\nabla} \cdot \mathbf{E}$ and its gradient.

Solution: We note that $(\partial E_{rr}/\partial\theta) = (\partial E_{\theta\theta}/\partial\theta) = 0$ because E_{rr} and $E_{\theta\theta}$ are not functions of θ (e.g., an axisymmetric problem).

(a) Using the del operator in the cylindrical coordinate system, we can write $\boldsymbol{\nabla} \times \mathbf{E}$ as

$$
\begin{aligned}
\boldsymbol{\nabla} \times \mathbf{E} &= \left(\hat{\mathbf{e}}_r\frac{\partial}{\partial r} + \frac{\hat{\mathbf{e}}_\theta}{r}\frac{\partial}{\partial\theta} + \hat{\mathbf{e}}_z\frac{\partial}{\partial z}\right) \times \left(E_{rr}\hat{\mathbf{e}}_r\hat{\mathbf{e}}_r + E_{\theta\theta}\hat{\mathbf{e}}_\theta\hat{\mathbf{e}}_\theta\right) \\
&= \hat{\mathbf{e}}_r \times \frac{\partial}{\partial r}\left(E_{rr}\hat{\mathbf{e}}_r\hat{\mathbf{e}}_r + E_{\theta\theta}\hat{\mathbf{e}}_\theta\hat{\mathbf{e}}_\theta\right) + \frac{1}{r}\hat{\mathbf{e}}_\theta \times \frac{\partial}{\partial\theta}\left(E_{rr}\hat{\mathbf{e}}_r\hat{\mathbf{e}}_r + E_{\theta\theta}\hat{\mathbf{e}}_\theta\hat{\mathbf{e}}_\theta\right) \\
&\quad + \hat{\mathbf{e}}_z \times \frac{\partial}{\partial z}\left(E_{rr}\hat{\mathbf{e}}_r\hat{\mathbf{e}}_r + E_{\theta\theta}\hat{\mathbf{e}}_\theta\hat{\mathbf{e}}_\theta\right) \\
&= \hat{\mathbf{e}}_r \times \left(\frac{\partial E_{\theta\theta}}{\partial r}\hat{\mathbf{e}}_\theta\hat{\mathbf{e}}_\theta\right) + \frac{1}{r}\hat{\mathbf{e}}_\theta \times \left(E_{rr}\hat{\mathbf{e}}_r\frac{\partial\hat{\mathbf{e}}_r}{\partial\theta} + E_{\theta\theta}\frac{\partial\hat{\mathbf{e}}_\theta}{\partial\theta}\hat{\mathbf{e}}_\theta\right) \\
&\quad + \hat{\mathbf{e}}_z \times \left(\frac{\partial E_{rr}}{\partial z}\hat{\mathbf{e}}_r\hat{\mathbf{e}}_r + \frac{\partial E_{\theta\theta}}{\partial z}\hat{\mathbf{e}}_\theta\hat{\mathbf{e}}_\theta\right) \\
&= \frac{\partial E_{\theta\theta}}{\partial r}\left(\hat{\mathbf{e}}_r \times \hat{\mathbf{e}}_\theta\right)\hat{\mathbf{e}}_\theta + \frac{1}{r}E_{rr}\left(\hat{\mathbf{e}}_\theta \times \hat{\mathbf{e}}_r\right)\frac{\partial\hat{\mathbf{e}}_r}{\partial\theta} \\
&\quad + \frac{1}{r}E_{\theta\theta}\left(\hat{\mathbf{e}}_\theta \times \frac{\partial\hat{\mathbf{e}}_\theta}{\partial\theta}\right)\hat{\mathbf{e}}_\theta + \frac{\partial E_{rr}}{\partial z}\left(\hat{\mathbf{e}}_z \times \hat{\mathbf{e}}_r\right)\hat{\mathbf{e}}_r + \frac{\partial E_{\theta\theta}}{\partial z}\left(\hat{\mathbf{e}}_z \times \hat{\mathbf{e}}_\theta\right)\hat{\mathbf{e}}_\theta \\
&= \left[\frac{\partial E_{\theta\theta}}{\partial r} + \frac{1}{r}\left(E_{\theta\theta} - E_{rr}\right)\right]\hat{\mathbf{e}}_z\,\hat{\mathbf{e}}_\theta + \frac{\partial E_{rr}}{\partial z}\hat{\mathbf{e}}_\theta\,\hat{\mathbf{e}}_r - \frac{\partial E_{\theta\theta}}{\partial z}\hat{\mathbf{e}}_r\,\hat{\mathbf{e}}_\theta.
\end{aligned}
$$

The transpose is obtained by switching the base vectors in each expression:

$$(\boldsymbol{\nabla} \times \mathbf{E})^{\mathrm{T}} = \left[\frac{\partial E_{\theta\theta}}{\partial r} + \frac{1}{r}\left(E_{\theta\theta} - E_{rr}\right)\right]\hat{\mathbf{e}}_\theta\,\hat{\mathbf{e}}_z + \frac{\partial E_{rr}}{\partial z}\hat{\mathbf{e}}_r\,\hat{\mathbf{e}}_\theta - \frac{\partial E_{\theta\theta}}{\partial z}\hat{\mathbf{e}}_\theta\,\hat{\mathbf{e}}_r.$$

(b) The divergence of $\mathbf{E}$ is

$$
\begin{aligned}
\boldsymbol{\nabla} \cdot \mathbf{E} &= \left(\hat{\mathbf{e}}_r\frac{\partial}{\partial r} + \frac{\hat{\mathbf{e}}_\theta}{r}\frac{\partial}{\partial\theta} + \hat{\mathbf{e}}_z\frac{\partial}{\partial z}\right) \cdot \left(E_{rr}\hat{\mathbf{e}}_r\hat{\mathbf{e}}_r + E_{\theta\theta}\hat{\mathbf{e}}_\theta\hat{\mathbf{e}}_\theta\right) \\
&= \hat{\mathbf{e}}_r \cdot \frac{\partial}{\partial r}\left(E_{rr}\hat{\mathbf{e}}_r\hat{\mathbf{e}}_r + E_{\theta\theta}\hat{\mathbf{e}}_\theta\hat{\mathbf{e}}_\theta\right) + \frac{1}{r}\hat{\mathbf{e}}_\theta \cdot \frac{\partial}{\partial\theta}\left(E_{rr}\hat{\mathbf{e}}_r\hat{\mathbf{e}}_r + E_{\theta\theta}\hat{\mathbf{e}}_\theta\hat{\mathbf{e}}_\theta\right) \\
&\quad + \hat{\mathbf{e}}_z \cdot \frac{\partial}{\partial z}\left(E_{rr}\hat{\mathbf{e}}_r\hat{\mathbf{e}}_r + E_{\theta\theta}\hat{\mathbf{e}}_\theta\hat{\mathbf{e}}_\theta\right) \\
&= \hat{\mathbf{e}}_r \cdot \left(\frac{\partial E_{rr}}{\partial r}\hat{\mathbf{e}}_r\hat{\mathbf{e}}_r\right) + \frac{1}{r}\hat{\mathbf{e}}_\theta \cdot \left(E_{rr}\frac{\partial\hat{\mathbf{e}}_r}{\partial\theta}\hat{\mathbf{e}}_r + E_{\theta\theta}\hat{\mathbf{e}}_\theta\frac{\partial\hat{\mathbf{e}}_\theta}{\partial\theta}\right) \\
&= \frac{\partial E_{rr}}{\partial r}\hat{\mathbf{e}}_r + \frac{1}{r}\left(E_{rr} - E_{\theta\theta}\right)\hat{\mathbf{e}}_r = \left[\frac{\partial E_{rr}}{\partial r} + \frac{1}{r}\left(E_{rr} - E_{\theta\theta}\right)\right]\hat{\mathbf{e}}_r.
\end{aligned}
$$

and the gradient of the divergence of $\mathbf{E}$ is

$$
\begin{aligned}
\boldsymbol{\nabla}(\boldsymbol{\nabla} \cdot \mathbf{E}) &= \left[\frac{\partial^2 E_{rr}}{\partial r^2} - \frac{1}{r^2}\left(E_{rr} - E_{\theta\theta}\right) + \frac{1}{r}\left(\frac{\partial E_{rr}}{\partial r} - \frac{\partial E_{\theta\theta}}{\partial r}\right)\right]\hat{\mathbf{e}}_r\hat{\mathbf{e}}_r \\
&\quad + \frac{1}{r}\left[\frac{\partial E_{rr}}{\partial r} + \frac{1}{r}\left(E_{rr} - E_{\theta\theta}\right)\right]\hat{\mathbf{e}}_\theta\hat{\mathbf{e}}_\theta \\
&\quad + \left[\frac{\partial^2 E_{rr}}{\partial r\partial z} + \frac{1}{r}\left(\frac{\partial E_{rr}}{\partial z} - \frac{\partial E_{\theta\theta}}{\partial z}\right)\right]\hat{\mathbf{e}}_z\hat{\mathbf{e}}_r.
\end{aligned}
$$

2.5.6 Eigenvalues and Eigenvectors

2.5.6.1 Eigenvalue problem

It is conceptually useful to regard a tensor as an operator that changes a vector into another vector (by means of the dot product). In this regard it is of interest to inquire whether there are certain vectors that have only their lengths, and not their orientation, changed when operated on by a given tensor $\mathbf{S}$ (that is, seek vectors that are transformed into multiples of themselves). If such vectors $\mathbf{x}$ exist, they must satisfy the equation

$$\mathbf{S} \cdot \mathbf{x} = \lambda \mathbf{x}. \tag{2.5.31}$$

Such vectors $\mathbf{x}$ are called *characteristic vectors, eigenvectors,* or *principal planes* (in mechanics) associated with $\mathbf{S}$. The parameter λ is called a *characteristic value, eigenvalue,* or *principal value,* and it represents the change in length of the eigenvector $\mathbf{x}$ after it has been operated on by $\mathbf{S}$.

In view of the fact that $\mathbf{x}$ can be expressed as $\mathbf{x} = \mathbf{I} \cdot \mathbf{x}$, Eq. (2.5.31) can also be written as

$$(\mathbf{S} - \lambda \mathbf{I}) \cdot \mathbf{x} = \mathbf{0}, \quad \text{or in matrix form} \quad ([S] - \lambda[I])\{X\} = \{0\}. \tag{2.5.32}$$

Equation (2.5.32) represents a homogeneous set of linear equations for $\{X\}$. Therefore, a nontrivial solution, that is, vector with at least one component of $\mathbf{x}$ is nonzero, will not exist unless the determinant of the matrix $[S] - \lambda[I]$ vanishes:

$$|\mathbf{S} - \lambda \mathbf{I}| = 0. \tag{2.5.33}$$

The vanishing of this determinant yields an algebraic equation of degree n in λ, called the *characteristic equation,* when $[S]$ is an $n \times n$ matrix associated with tensor $\mathbf{S}$.

Second-order tensors, such as strain and stress tensors, are of interest in mechanics, where the eigenvalues and eigenvectors represent principal values and directions. Since the underlying matrix is 3×3, the characteristic equation resulting from Eq. (2.5.33) is cubic in λ and yields three eigenvalues, say λ_1, λ_2, and λ_3. The character of these eigenvalues depends on the nature (i.e., real-valued, symmetric, positive-definite, and so on) of the tensor $\mathbf{S}$. For a real-valued second-order tensor $\mathbf{S}$, at least one of the eigenvalues will be real. The other two may be real and distinct, real and repeated, or complex conjugates. The vanishing of a determinant implies that the columns or rows of the matrix are linearly dependent and the three eigenvectors $\mathbf{x}^{(1)}$, $\mathbf{x}^{(2)}$, and $\mathbf{x}^{(3)}$ are not unique. An infinite number of solutions exist, within a multiplicative constant, having at least $n = 3$ different orientations. Since only orientation is important, it is useful to represent the three eigenvectors by three unit eigenvectors $\hat{\mathbf{x}}^{(1)}$, $\hat{\mathbf{x}}^{(2)}$, and $\hat{\mathbf{x}}^{(3)}$, which denote three different orientations, each associated with a particular eigenvalue.

2.5.6.2 Eigenvalues and eigenvectors of a real symmetric tensor

A matrix $[S]$ associated with a real-valued symmetric tensor $\mathbf{S}$ of order n has some desirable consequences as far as the eigenvalues and eigenvectors are concerned. These properties are listed next.

(1) All eigenvalues of $[S]$ are real.

(2) If $[S]$ is positive-definite, then the eigenvalues are strictly positive.

(3) Eigenvectors[5] $\mathbf{x}^{(1)}$ and $\mathbf{x}^{(2)}$ associated with two distinct eigenvalues λ_1 and λ_2 are orthogonal: $\mathbf{x}^{(1)} \cdot \mathbf{x}^{(2)} = 0$. If all eigenvalues are distinct, then the associated eigenvectors are all orthogonal to each other.

(4) $[S]$ always has n linearly independent eigenvectors, regardless of the algebraic multiplicities of the eigenvalues.

(5) For an eigenvalue of algebraic multiplicity m, it is possible to choose m eigenvectors that are mutually orthogonal. Hence, the set of n vectors can always be chosen to be linearly independent.

We now prove some of the aforementioned properties for a real-valued tensor. When n is odd, at least one of the eigenvalues is real and the remaining even number of eigenvalues will be complex conjugate pairs (and repeated eigenvalues will be real). Suppose that λ_1 and λ_2 are two distinct eigenvalues and $\mathbf{x}^{(1)}$ and $\mathbf{x}^{(2)}$ are their corresponding eigenvectors. Then from Eq. (2.5.31) we have

$$\mathbf{S} \cdot \mathbf{x}^{(1)} = \lambda_1 \mathbf{x}^{(1)}, \qquad \mathbf{S} \cdot \mathbf{x}^{(2)} = \lambda_2 \mathbf{x}^{(2)}. \tag{2.5.34}$$

Taking the scalar product (from the left) of the first equation with $\mathbf{x}^{(2)}$ and the second equation with $\mathbf{x}^{(1)}$, and subtracting the second equation from the first equation, we obtain

$$\mathbf{x}^{(2)} \cdot \mathbf{S} \cdot \mathbf{x}^{(1)} - \mathbf{x}^{(1)} \cdot \mathbf{S} \cdot \mathbf{x}^{(2)} = (\lambda_1 - \lambda_2)\,\mathbf{x}^{(1)} \cdot \mathbf{x}^{(2)}. \tag{2.5.35}$$

Because

$$\mathbf{x}^{(2)} \cdot \mathbf{S} \cdot \mathbf{x}^{(1)} = \mathbf{x}^{(1)} \cdot \mathbf{S}^{\mathrm{T}} \cdot \mathbf{x}^{(2)},$$

and $\mathbf{S}$ is symmetric, $\mathbf{S} = \mathbf{S}^{\mathrm{T}}$, the left-hand side of Eq. (2.5.35) vanishes, giving

$$0 = (\lambda_1 - \lambda_2)\,\mathbf{x}^{(1)} \cdot \mathbf{x}^{(2)}. \tag{2.5.36}$$

In view of the fact that $\mathbf{S}$ is a real-valued tensor, either the eigenvalues λ_1 and λ_2 are real or they are complex conjugate pairs. Suppose that λ_1 and λ_2 are complex conjugate pairs

$$\lambda_1 = \lambda_R + i\,\lambda_I, \quad \lambda_2 = \lambda_R - i\,\lambda_I, \quad i = \sqrt{-1},$$

---

[5]The vectors $\mathbf{x}^{(i)}$ are constructed from $\mathbf{x}^{(i)} = X_1^{(i)}\hat{\mathbf{e}}_1 + X_2^{(i)}\hat{\mathbf{e}}_2 + \cdots + X_n^{(i)}\hat{\mathbf{e}}_n$.

and $\mathbf{x}^{(1)}$ and $\mathbf{x}^{(2)}$ are the complex conjugate vectors associated with λ_1 and λ_2

$$\mathbf{x}^{(1)} = \mathbf{x}_R + i\,\mathbf{x}_I, \quad \mathbf{x}^{(2)} = \mathbf{x}_R - i\,\mathbf{x}_I.$$

Then $\lambda_1 - \lambda_2 = 2i\lambda_I$, and $\mathbf{x}^{(1)} \cdot \mathbf{x}^{(2)}$ is always positive $\mathbf{x}^{(1)} \cdot \mathbf{x}^{(2)} = \mathbf{x}_R \cdot \mathbf{x}_R + \mathbf{x}_I \cdot \mathbf{x}_I > 0$. Then it follows from Eq. (2.5.36) that $\lambda_I = 0$ (and $\mathbf{x}_I = \mathbf{0}$) and, therefore, *that the n eigenvalues associated with a symmetric matrix are all real.*

If the tensor $\mathbf{S}$ is positive-definite, $\mathbf{x} \cdot \mathbf{S} \cdot \mathbf{x} > 0$, then from Eq. (2.5.34) it follows that

$$\mathbf{x} \cdot \mathbf{S} \cdot \mathbf{x} = \lambda \mathbf{x} \cdot \mathbf{x} > 0 \;\Rightarrow\; \lambda > 0 \text{ for all } \mathbf{x} \neq \mathbf{0}.$$

Thus, *when $\mathbf{S}$ is positive-definite the eigenvalues are strictly positive.*

Next, assume that λ_1 and λ_2 are real and distinct such that $\lambda_1 - \lambda_2$ is not zero. It then follows from Eq. (2.5.36) that $\mathbf{x}^{(1)} \cdot \mathbf{x}^{(2)} = 0$. Thus the *eigenvectors associated with distinct eigenvalues of a symmetric second-order tensor are orthogonal.* If the three eigenvalues are all distinct, then the three eigenvectors are mutually orthogonal.

If an eigenvalue is repeated, say $\lambda_3 = \lambda_2$, then $\mathbf{x}^{(3)}$ must also be perpendicular to $\mathbf{x}^{(1)}$, as deduced by an argument similar to that for $\mathbf{x}^{(2)}$ stemming from Eq. (2.5.36). Neither $\mathbf{x}^{(2)}$ nor $\mathbf{x}^{(3)}$ is preferred, and they are both arbitrary, except insofar as they are both perpendicular to $\mathbf{x}^{(1)}$. It is useful, however, to select $\mathbf{x}^{(3)}$ such that it is perpendicular to both $\mathbf{x}^{(1)}$ and $\mathbf{x}^{(2)}$. We do this by choosing $\mathbf{x}^{(3)} = \mathbf{x}^{(1)} \times \mathbf{x}^{(2)}$ and thus establishing a mutually orthogonal set of eigenvectors.

2.5.6.3 Spectral theorem

Let $\mathbf{S}$ be a real-valued, symmetric, second-order tensor defined on $\Re^3$. Then there exists an orthonormal basis $(\hat{\mathbf{e}}_1, \hat{\mathbf{e}}_2, \hat{\mathbf{e}}_3)$ consisting of the eigenvectors of $\mathbf{S}$. The eigenvalues λ_1, λ_2, and λ_3 form the entire spectrum of $\mathbf{S}$, that is, $\mathbf{S}$ can be expressed as

$$\mathbf{S} = \lambda_1 \,\hat{\mathbf{e}}_1\,\hat{\mathbf{e}}_1 + \lambda_2 \,\hat{\mathbf{e}}_2\,\hat{\mathbf{e}}_2 + \lambda_3 \,\hat{\mathbf{e}}_3\,\hat{\mathbf{e}}_3. \tag{2.5.37}$$

That is, the matrix associated with $\mathbf{S}$ with respect to the basis $(\hat{\mathbf{e}}_1, \hat{\mathbf{e}}_2, \hat{\mathbf{e}}_3)$ is a diagonal matrix. Conversely, if $\mathbf{S}$ has the form in Eq. (2.5.37), then $\{\hat{\mathbf{e}}_i\}$ are the orthonormal eigenvectors and λ_i are the corresponding eigenvalues. In addition, we have the following properties:

(1) $\mathbf{S}$ has exactly three distinct eigenvalues if and only if the characteristic vectors $(\hat{\mathbf{e}}_1, \hat{\mathbf{e}}_2, \hat{\mathbf{e}}_3)$ are all mutually orthogonal.

(2) $\mathbf{S}$ has two distinct eigenvalues, λ_1 and λ_2, if and only if $\mathbf{S}$ admits the representation

$$\mathbf{S} = \lambda_1 \,\hat{\mathbf{e}}\,\hat{\mathbf{e}} + \lambda_2 \,(\mathbf{I} - \hat{\mathbf{e}}\,\hat{\mathbf{e}}), \quad |\hat{\mathbf{e}}| = 1. \tag{2.5.38}$$

(3) $\mathbf{S}$ has only one eigenvalue if and only if

$$\mathbf{S} = \lambda \,\mathbf{I}. \tag{2.5.39}$$

2.5.6.4 Calculation of eigenvalues and eigenvectors

Returning to Eq. (2.5.32), let $[S]$ be the matrix representation of a second-order tensor $\mathbf{S}$ with respect to a rectangular Cartesian basis. Then the eigenvalue problem has the explicit form

$$\begin{bmatrix} s_{11} - \lambda & s_{12} & s_{13} \\ s_{21} & s_{22} - \lambda & s_{23} \\ s_{31} & s_{32} & s_{33} - \lambda \end{bmatrix} \begin{Bmatrix} x_1 \\ x_2 \\ x_3 \end{Bmatrix} = \begin{Bmatrix} 0 \\ 0 \\ 0 \end{Bmatrix}. \tag{2.5.40}$$

For a nontrivial solution (i.e., at least one of the components x_1, x_2, and x_3 is nonzero), we require that the determinant of the coefficient matrix be zero:

$$\begin{vmatrix} s_{11} - \lambda & s_{12} & s_{13} \\ s_{21} & s_{22} - \lambda & s_{23} \\ s_{31} & s_{32} & s_{33} - \lambda \end{vmatrix} = 0. \tag{2.5.41}$$

The characteristic equation associated with Eq. (2.5.41) can be expressed in the form

$$-\lambda^3 + I_1 \lambda^2 - I_2 \lambda + I_3 = 0, \tag{2.5.42}$$

where I_1, I_2, and I_3 are the invariants of $\mathbf{S}$ as defined in Eq. (2.5.17), which, for a second-order tensor, have the specific form

$$I_1 = s_{11} + s_{22} + s_{33}, \quad I_3 = |S|,$$
$$I_2 = \tfrac{1}{2} \left(I_1^2 - s_{11}^2 - s_{22}^2 - s_{33}^2 - s_{12}^2 - s_{13}^2 - s_{23}^2 - s_{21}^2 - s_{31}^2 - s_{32}^2 \right). \tag{2.5.43}$$

The invariants can also be expressed in terms of the eigenvalues (when known),

$$I_1 = \lambda_1 + \lambda_2 + \lambda_3, \quad I_2 = (\lambda_1 \lambda_2 + \lambda_2 \lambda_3 + \lambda_3 \lambda_1), \quad I_3 = \lambda_1 \lambda_2 \lambda_3. \tag{2.5.44}$$

The eigenvector $\mathbf{x}^{(i)}$ associated with any particular eigenvalue λ_i is calculated using Eq. (2.5.40), which gives only two independent relations among the three components $x_1^{(i)}$, $x_2^{(i)}$, and $x_3^{(i)}$. Thus, two of the three components can be written in terms of the third, whose value is arbitrary (but nonzero). In other words, we can determine the eigenvectors only within a multiplicative constant. If the eigenvector is normalized such that it is a unit vector, then we use the following additional (i.e., third) condition to determine all three components:

$$(x_1^{(i)})^2 + (x_2^{(i)})^2 + (x_3^{(i)})^2 = 1. \tag{2.5.45}$$

For example, if $x_1^{(i)}$ and $x_2^{(i)}$ are expressed in terms of $x_3^{(i)}$ [using Eq. (2.5.40)], say $x_1^{(i)} = \alpha \, x_3^{(i)}$ and $x_2^{(i)} = \beta \, x_3^{(i)}$, then Eq. (2.5.45) yields

$$(x_3^{(i)})^2 = \left(\alpha^2 + \beta^2 + 1 \right)^{-1} \quad \text{or} \quad x_3^{(i)} = \pm \frac{1}{\sqrt{\alpha^2 + \beta^2 + 1}}, \tag{2.5.46}$$

and the normalized eigenvector is

$$\{\hat{X}\}^{(i)} = \pm \frac{1}{\sqrt{\alpha^2 + \beta^2 + 1}} \begin{Bmatrix} \alpha \\ \beta \\ 1 \end{Bmatrix}. \tag{2.5.47}$$

The sign $\pm$ on the three eigenvectors should be selected such that we have a right-hand coordinate system (when the eigenvectors are orthonormal):

$$\hat{\mathbf{x}}^{(1)} = \hat{\mathbf{x}}^{(2)} \times \hat{\mathbf{x}}^{(3)}. \tag{2.5.48}$$

If the matrix under consideration is of order 2×2, the characteristic equation is of the form

$$\lambda^2 - I_1\lambda + I_3 = 0, \quad I_1 = s_{11} + s_{22}, \quad I_3 = s_{11}s_{22} - s_{12}s_{21}. \tag{2.5.49}$$

The roots of this quadratic equation are (ordered lowest to the highest)

$$\lambda_1 = \tfrac{1}{2}\left(I_1 - \sqrt{I_1^2 - 4I_3}\,\right), \quad \lambda_2 = \tfrac{1}{2}\left(I_1 + \sqrt{I_1^2 - 4I_3}\,\right). \tag{2.5.50}$$

The eigenvector $\mathbf{x}^{(i)}$ associated with λ_i $(i = 1, 2)$ is determined from

$$\begin{bmatrix} s_{11} - \lambda_i & s_{12} \\ s_{21} & s_{22} - \lambda_i \end{bmatrix} \begin{Bmatrix} x_1^{(i)} \\ x_2^{(i)} \end{Bmatrix} = \begin{Bmatrix} 0 \\ 0 \end{Bmatrix}. \tag{2.5.51}$$

This matrix equation gives only one independent relation between $x_1^{(i)}$ and $x_2^{(i)}$. One can arbitrarily select the value of one of the two components and determine the other; alternatively, one can use normalization $(x_1^{(i)})^2 + (x_2^{(i)})^2 = 1$ to determine the vector $\hat{\mathbf{x}}^{(i)}$. One may also normalize a eigenvector with respect to the largest component of the eigenvector. Example 2.5.2 illustrates the procedure.

Example 2.5.2 —————————————————————————————

Determine the eigenvalues and eigenvectors of the following matrix:

$$[S] = \begin{bmatrix} 5 & -1 \\ 3 & 1 \end{bmatrix}.$$

Solution: The eigenvalue problem associated with the matrix $[S]$ is $[S]\{X\} - \lambda\{X\} = \{0\}$:

$$\begin{bmatrix} 5 & -1 \\ 3 & 1 \end{bmatrix} \begin{Bmatrix} x_1 \\ x_2 \end{Bmatrix} - \lambda \begin{bmatrix} 1 & 0 \\ 0 & 1 \end{bmatrix} \begin{Bmatrix} x_1 \\ x_2 \end{Bmatrix} = \begin{Bmatrix} 0 \\ 0 \end{Bmatrix} \;\rightarrow\; \begin{bmatrix} 5 - \lambda & -1 \\ 3 & 1 - \lambda \end{bmatrix} \begin{Bmatrix} x_1 \\ x_2 \end{Bmatrix} = \begin{Bmatrix} 0 \\ 0 \end{Bmatrix}. \tag{1}$$

The characteristic equation is obtained either from $\lambda^2 - I_1\lambda + I_3 = 0$ or from $|S - \lambda I| = 0$. Thus, $\lambda^2 - I_1\lambda + I_3 = \lambda^2 - 6\lambda + 8 = 0$; alternatively,

$$\begin{vmatrix} 5 - \lambda & -1 \\ 3 & 1 - \lambda \end{vmatrix} = (5 - \lambda)(1 - \lambda) - (3)(-1) = 0 \;\Rightarrow\; \lambda^2 - 6\lambda + 8 = 0.$$

The two roots of the quadratic equation are

$$\lambda_1 = \tfrac{1}{2}\left(6 - \sqrt{6^2 - 4 \times 8}\,\right) = 2, \quad \lambda_2 = \tfrac{1}{2}\left(6 + \sqrt{6^2 - 4 \times 8}\,\right) = 4$$

To find the eigenvectors, we return to Eq. (1) and substitute for λ each of the eigenvalues λ_i and solve the resulting algebraic equations for $(x_1^{(i)}, x_2^{(i)})$. For $\lambda = \lambda_1 = 2$, we have

$$\begin{bmatrix} 5 - 2 & -1 \\ 3 & 1 - 2 \end{bmatrix} \begin{Bmatrix} x_1^{(1)} \\ x_2^{(1)} \end{Bmatrix} = \begin{Bmatrix} 0 \\ 0 \end{Bmatrix}. \tag{2}$$

Each row of the matrix equation in (2) yields the same condition $3x_1^{(1)} - x_2^{(1)} = 0$ or $x_2^{(1)} = 3x_1^{(1)}$. The eigenvector $\mathbf{x}^{(1)}$ s given by

$$\{X\}^{(1)} = \begin{Bmatrix} 1 \\ 3 \end{Bmatrix} x_1^{(1)}, \quad x_1^{(1)} \neq 0, \text{ arbitrary.}$$

Usually, we take $x_1^{(1)} = 1$, as we are interested in the direction of the vector $\{X\}^{(1)}$ rather than in its magnitude. One may also normalize the eigenvector by using the condition

$$(x_1^{(1)})^2 + (x_2^{(1)})^2 = 1. \tag{3}$$

Then we obtain the following normalized eigenvector:

$$\{\hat{X}\}^{(1)} = \pm \frac{1}{\sqrt{10}} \begin{Bmatrix} 1 \\ 3 \end{Bmatrix} = \pm \begin{Bmatrix} 0.3162 \\ 0.9487 \end{Bmatrix}. \tag{4}$$

Using the same procedure, we can determine the eigenvector associated with $\lambda_2 = 4$. Substituting for $\lambda = \lambda_2 = 4$ into Eq. (1), we obtain

$$\begin{bmatrix} 5-4 & -1 \\ 3 & 1-4 \end{bmatrix} \begin{Bmatrix} x_1^{(2)} \\ x_2^{(2)} \end{Bmatrix} = \begin{Bmatrix} 0 \\ 0 \end{Bmatrix}, \tag{5}$$

from which we obtain the condition $x_1^{(2)} - x_2^{(2)} = 0$ or $x_2^{(2)} = x_1^{(2)}$. The eigenvector $\mathbf{x}^{(2)}$ is

$$\{X\}^{(2)} = \begin{Bmatrix} 1 \\ 1 \end{Bmatrix} \quad \text{or} \quad \{\hat{X}\}^{(2)} = \pm \frac{1}{\sqrt{2}} \begin{Bmatrix} 1 \\ 1 \end{Bmatrix} = \pm \begin{Bmatrix} 0.7071 \\ 0.7071 \end{Bmatrix}. \tag{6}$$

Because the matrix under consideration is not symmetric, we do not expect the eigenvectors to be orthogonal.

When matrix $[S]$ is of order 3×3, finding the roots of the cubic equation in Eq. (2.5.42) is not always easy. However, if the 3×3 matrix is of the special form

$$\begin{bmatrix} s_{11} & 0 & 0 \\ 0 & s_{22} & s_{23} \\ 0 & s_{32} & s_{33} \end{bmatrix}, \tag{2.5.52}$$

then one of the roots is $\lambda_1 = s_{11}$ with eigenvector $\mathbf{x}^{(1)} = \hat{\mathbf{e}}_1$, and the remaining two roots can be found, as in the case of a 2×2 matrix, from the quadratic equation

$$\begin{vmatrix} s_{22} - \lambda & s_{23} \\ s_{32} & s_{33} - \lambda \end{vmatrix} = (s_{22} - \lambda)(s_{33} - \lambda) - s_{23}s_{32} = 0.$$

That is,

$$\lambda_{2,3} = \tfrac{1}{2}(s_{22} + s_{33}) \pm \tfrac{1}{2}\sqrt{(s_{22} + s_{33})^2 - 4(s_{22}s_{33} - s_{23}s_{32})}.$$

When the matrix $[S]$ is a full 3×3 matrix, we use a method that facilitates analytical computation of eigenvalues. In this method we seek the eigenvalues of the so-called *deviatoric tensor* $\mathbf{S}'$ associated with tensor $\mathbf{S}$:

$$\mathbf{S}' = \mathbf{S} - \tfrac{1}{3}\text{tr}(\mathbf{S})\,\mathbf{I} \quad \left(s'_{ij} \equiv s_{ij} - \tfrac{1}{3}s_{kk}\delta_{ij}\right). \tag{2.5.53}$$

Note that

$$\text{tr}(\mathbf{S}') = \mathbf{0} \quad \text{or} \quad s'_{ii} = s_{ii} - s_{kk} = 0. \tag{2.5.54}$$

That is, the first invariant I_1' of the deviatoric tensor is zero. As a result, the characteristic equation associated with the deviatoric tensor is of the form

$$-(\lambda')^3 - I_2'\lambda' + I_3' = 0, \qquad (2.5.55)$$

where λ' is the eigenvalue of the deviatoric tensor. The eigenvalues λ associated with $\mathbf{S}$ itself can be computed from

$$\lambda = \lambda' + \tfrac{1}{3}s_{kk}. \qquad (2.5.56)$$

The cubic equation in (2.5.55) is of a special form that allows a direct computation of its roots. Equation (2.5.55) can be solved explicitly by introducing the transformation

$$\lambda' = 2\left(-\tfrac{1}{3}I_2'\right)^{\frac{1}{2}}\cos\alpha, \qquad (2.5.57)$$

which transforms (2.5.55) into

$$2\left(-\tfrac{1}{3}I_2'\right)^{\frac{3}{2}}\left(4\cos^3\alpha - 3\cos\alpha\right) = I_3'.$$

The expression $4\cos^3\alpha - 3\cos\alpha$ is equal to $\cos 3\alpha$. Hence

$$\cos 3\alpha = \tfrac{1}{2}I_3'\left(-\frac{3}{I_2'}\right)^{\frac{3}{2}}. \qquad (2.5.58)$$

If α_1 is the angle satisfying $0 \le 3\alpha_1 \le \pi$ whose cosine is given by Eq. (2.5.58), then $3\alpha_1$, $3\alpha_1 + 2\pi$, and $3\alpha_1 - 2\pi$ all have the same cosine, and furnish three independent roots of Eq. (2.5.55),

$$\lambda_i' = 2\left(-\tfrac{1}{3}I_2'\right)^{\frac{1}{2}}\cos\alpha_i, \quad i = 1, 2, 3, \qquad (2.5.59)$$

where

$$\alpha_1 = \tfrac{1}{3}\left\{\cos^{-1}\left[\frac{I_3'}{2}\left(-\frac{3}{I_2'}\right)^{3/2}\right]\right\}, \quad \alpha_2 = \alpha_1 + \tfrac{2}{3}\pi, \quad \alpha_3 = \alpha_1 - \tfrac{2}{3}\pi. \qquad (2.5.60)$$

Finally, we can compute λ_i from Eq. (2.5.56). Example 2.5.3 is an application of the procedures discussed.

Example 2.5.3

Determine the eigenvalues and eigenvectors of the following matrix:

$$[S] = \begin{bmatrix} 2 & 1 & 0 \\ 1 & 4 & 1 \\ 0 & 1 & 2 \end{bmatrix}.$$

Solution: The characteristic equation is $-\lambda^3 + I_1\lambda^2 - I_2\lambda + I_3 = 0$, with

$$I_1 = 2 + 4 + 2 = 8, \quad I_2 = \tfrac{1}{2}\left(8^2 - 2^2 - 4^2 - 2^2 - 2\times 1^2 - 2\times 1^2\right) = 18, \quad |S| = 12.$$

Thus, the characteristic equation is (not always possible to factor out a root):

$$-\lambda^3 + 8\lambda^2 - 18\lambda + 12 = 0 \quad \text{or} \quad (2 - \lambda)\left(\lambda^2 - 6\lambda + 6\right).$$

The roots of this cubic equation are (ordered from the largest to the smallest value):

$$\lambda_1 = 3 + \sqrt{3} = 4.7321, \quad \lambda_2 = 2, \quad \lambda_3 = 3 - \sqrt{3} = 1.2679.$$

Alternatively, using Eqs. (2.5.53)–(2.5.60), we have

$$[S'] = \begin{bmatrix} 2 - \frac{8}{3} & 1 & 0 \\ 1 & 4 - \frac{8}{3} & 1 \\ 0 & 1 & 2 - \frac{8}{3} \end{bmatrix}, \quad I_1' = 0, \quad I_3' = |S'| = \tfrac{52}{27},$$

$$I_2' = \tfrac{1}{2}\left(s_{ii}'s_{jj}' - s_{ij}'s_{ij}'\right) = -\tfrac{1}{2}s_{ij}'s_{ij}'$$

$$= -\frac{1}{2}\left[\left(-\tfrac{2}{3}\right)^2 + \left(-\tfrac{2}{3}\right)^2 + \left(\tfrac{4}{3}\right)^2 + 2 \times 1^2 + 2 \times 1^2\right] = -\tfrac{10}{3}.$$

From Eq. (2.5.60),

$$\alpha_1 = \tfrac{1}{3}\left\{\cos^{-1}\left[\tfrac{52}{54}\left(\tfrac{9}{10}\right)^{3/2}\right]\right\} = 11.565°, \quad \alpha_2 = 131.565°, \quad \alpha_3 = -108.435°,$$

and from Eq. (2.5.59),

$$\lambda_1' = 2.065384, \quad \lambda_2' = -0.66667, \quad \lambda_3' = -1.3987.$$

Finally, using Eq. (2.5.56), we obtain the following eigenvalues:

$$\lambda_1 = \lambda_1' + \tfrac{8}{3} = 4.7321, \quad \lambda_2 = -0.66667 + \tfrac{8}{3} = 2, \quad \lambda_3 = -1.3987 + \tfrac{8}{3} = 1.2679.$$

The eigenvector corresponding to $\lambda_1 = 3 + \sqrt{3}$ is calculated Eq. (2.5.40). We have

$$\begin{bmatrix} 2 - (3 + \sqrt{3}) & 1 & 0 \\ 1 & 4 - (3 + \sqrt{3}) & 1 \\ 0 & 1 & 2 - (3 + \sqrt{3}) \end{bmatrix} \begin{Bmatrix} x_1^{(1)} \\ x_2^{(1)} \\ x_3^{(1)} \end{Bmatrix} = \begin{Bmatrix} 0 \\ 0 \\ 0 \end{Bmatrix}.$$

This gives the following independent relations:

$$-(1 + \sqrt{3})x_1^{(1)} + x_2^{(1)} = 0, \quad x_2^{(1)} - (1 + \sqrt{3})x_3^{(1)} = 0 \quad \Rightarrow \quad x_1^{(1)} = x_3^{(1)} = \frac{1}{(1 + \sqrt{3})}x_2^{(1)}.$$

Hence we have the following eigenvector:

$$\{X\}^{(1)} = \frac{1}{(1 + \sqrt{3})}\begin{Bmatrix} 1 \\ (1 + \sqrt{3}) \\ 1 \end{Bmatrix} x_2^{(1)} = \begin{Bmatrix} 0.366 \\ 1.000 \\ 0.366 \end{Bmatrix} x_2^{(1)}.$$

Normalizing the vector, we obtain

$$\{\hat{X}\}^{(1)} = \pm\frac{1}{\sqrt{(6 + 2\sqrt{3})}}\begin{Bmatrix} 1 \\ (1 + \sqrt{3}) \\ 1 \end{Bmatrix} = \pm\begin{Bmatrix} 0.325 \\ 0.888 \\ 0.325 \end{Bmatrix}.$$

The eigenvector corresponding to $\lambda_2 = 2$ is calculated from

$$\begin{bmatrix} 2 - 2 & 1 & 0 \\ 1 & 4 - 2 & 1 \\ 0 & 1 & 2 - 2 \end{bmatrix} \begin{Bmatrix} x_1^{(2)} \\ x_2^{(2)} \\ x_3^{(2)} \end{Bmatrix} = \begin{Bmatrix} 0 \\ 0 \\ 0 \end{Bmatrix},$$

which gives $x_2^{(2)} = 0$ and $x_1^{(2)} = -x_3^{(2)}$. The eigenvector (without and with normalization) associated with $\lambda_2 = 2$ is

$$\{X\}^{(2)} = \left\{\begin{matrix} -1 \\ 0 \\ 1 \end{matrix}\right\} x_3^{(2)}, \quad \{\hat{X}\}^{(2)} = \pm\frac{1}{\sqrt{2}}\left\{\begin{matrix} -1 \\ 0 \\ 1 \end{matrix}\right\} = \pm\left\{\begin{matrix} -0.707 \\ 0.000 \\ 0.707 \end{matrix}\right\}.$$

Finally, the eigenvector corresponding to $\lambda_3 = 3 - \sqrt{3}$ is calculated from

$$\begin{bmatrix} 2 - (3 - \sqrt{3}) & 1 & 0 \\ 1 & 4 - (3 + \sqrt{3}) & 1 \\ 0 & 1 & 2 - (3 - \sqrt{3}) \end{bmatrix}\left\{\begin{matrix} x_1^{(3)} \\ x_2^{(3)} \\ x_3^{(3)} \end{matrix}\right\} = \left\{\begin{matrix} 0 \\ 0 \\ 0 \end{matrix}\right\},$$

which gives

$$-(1 - \sqrt{3})x_1^{(3)} + x_2^{(3)} = 0, \quad x_2^{(3)} - (1 - \sqrt{3})x_3^{(3)} = 0 \quad \Rightarrow \quad x_1^{(3)} = x_3^{(3)} = \frac{1}{(1-\sqrt{3})}x_2^{(3)}.$$

Hence the eigenvector is

$$\{X\}^{(3)} = \frac{1}{(1-\sqrt{3})}\left\{\begin{matrix} 1 \\ (1 - \sqrt{3}) \\ 1 \end{matrix}\right\} x_2^{(3)} = \left\{\begin{matrix} -1.366 \\ 1.000 \\ -1.366 \end{matrix}\right\} x_2^{(3)}.$$

Normalizing the vector, we obtain

$$\{\hat{X}\}^{(3)} = \pm\frac{1}{\sqrt{(6-2\sqrt{3})}}\left\{\begin{matrix} 1 \\ (1 - \sqrt{3}) \\ 1 \end{matrix}\right\} = \pm\left\{\begin{matrix} 0.628 \\ -0.460 \\ 0.628 \end{matrix}\right\}.$$

We can check to see if $\hat{\mathbf{x}}^{(1)} \times \hat{\mathbf{x}}^{(2)} = \hat{\mathbf{x}}^{(3)}$ without consideration of the minus sign. We find that the vectors without $\pm$ in front of them constitute a right-hand system. We can also verify that they are also orthonormal. Therefore, we can write

$$[Q][S][Q]^{\mathrm{T}} = \begin{bmatrix} \lambda_1 & 0 & 0 \\ 0 & \lambda_2 & 0 \\ 0 & 0 & \lambda_3 \end{bmatrix},$$

where $[Q]$ is the matrix of eigenvector components (components of each vector are arranged in rows)

$$[Q] = \begin{bmatrix} 0.325 & 0.888 & 0.325 \\ -0.707 & 0.000 & 0.707 \\ 0.628 & -0.460 & 0.628 \end{bmatrix}.$$

In other words, we can represent $\mathbf{S}$ with respect to the orthonormal vectors $(\hat{\mathbf{x}}^{(1)}, \hat{\mathbf{x}}^{(2)}, \hat{\mathbf{x}}^{(3)})$, which form a basis, as

$$\mathbf{S} = \lambda_1\,\hat{\mathbf{x}}^{(1)}\hat{\mathbf{x}}^{(1)} + \lambda_2\,\hat{\mathbf{x}}^{(2)}\hat{\mathbf{x}}^{(2)} + \lambda_3\,\hat{\mathbf{x}}^{(3)}\hat{\mathbf{x}}^{(3)} = (3+\sqrt{3})\hat{\mathbf{x}}^{(1)}\hat{\mathbf{x}}^{(1)} + 2\hat{\mathbf{x}}^{(2)}\hat{\mathbf{x}}^{(2)} + (3-\sqrt{3})\hat{\mathbf{x}}^{(3)}\hat{\mathbf{x}}^{(3)}.$$

When $[S]$ in Eq. (2.5.32) is an $n \times n$ matrix, Eq. (2.5.33) is a polynomial of degree n in λ, and therefore, there are n eigenvalues $\lambda_1, \lambda_2, \cdots, \lambda_n$, some of which may be repeated. In general, if an eigenvalue appears m times as a root of Eq. (2.5.33), then that eigenvalue is said to have *algebraic multiplicity* m. An eigenvalue of algebraic multiplicity m may have r linearly independent eigenvectors. The number r is called the *geometric multiplicity* of the eigenvalue, and r lies (not shown here) in the range $1 \leq r \leq m$. Thus, a square matrix $[S]$ of order n may have fewer than n linearly independent eigenvectors when $[S]$ has one or more repeated eigenvalues. Example 2.5.4 illustrates the calculation of eigenvectors of a matrix when it has repeated eigenvalues.

Example 2.5.4 ————————————————————————————————

Determine the eigenvalues and eigenvectors of the following matrix:

$$[S] = \begin{bmatrix} 0 & 1 & 1 \\ 1 & 0 & 1 \\ 1 & 1 & 0 \end{bmatrix}.$$

Solution: The characteristic equation $-\lambda^3 + I_1\lambda^2 - I_2\lambda + I_3 = 0$ in the present case is

$$-\lambda^3 - (-3)\lambda + 2 = 0,$$

which can be expressed as $(2 - \lambda)(1 + \lambda)^2 = 0$. Thus, the three roots are

$$\lambda_1 = 2, \quad \lambda_2 = -1, \quad \lambda_3 = -1.$$

We note that $\lambda = -1$ is an eigenvalue with algebraic multiplicity of 2.
The eigenvector components associated with $\lambda = 2$ are obtained from

$$\begin{bmatrix} -2 & 1 & 1 \\ 1 & -2 & 1 \\ 1 & 1 & -2 \end{bmatrix} \begin{Bmatrix} x_1^{(1)} \\ x_2^{(1)} \\ x_3^{(1)} \end{Bmatrix} = \begin{Bmatrix} 0 \\ 0 \\ 0 \end{Bmatrix},$$

which gives

$$-2x_1^{(1)} + x_2^{(1)} + x_3^{(1)} = 0, \quad x_1^{(1)} - 2x_2^{(1)} + x_3^{(1)} = 0, \quad x_1^{(1)} + x_2^{(1)} - 2x_3^{(1)} = 0.$$

Solution of these equations gives $x_1^{(1)} = x_2^{(1)} = x_3^{(1)}$. Thus the eigenvector associated with $\lambda_1 = 2$ is the vector

$$\{X\}^{(1)} = \begin{Bmatrix} 1 \\ 1 \\ 1 \end{Bmatrix} x_1^{(1)} \quad \text{or} \quad \{\hat{X}\}^{(1)} = \pm \frac{1}{\sqrt{3}} \begin{Bmatrix} 1 \\ 1 \\ 1 \end{Bmatrix},$$

where $\{\hat{X}\}^{(1)}$ denotes the normalized (unit) vector.
The eigenvector components associated with $\lambda = -1$ are obtained from

$$\begin{bmatrix} 1 & 1 & 1 \\ 1 & 1 & 1 \\ 1 & 1 & 1 \end{bmatrix} \begin{Bmatrix} x_1^{(2)} \\ x_2^{(2)} \\ x_3^{(2)} \end{Bmatrix} = \begin{Bmatrix} 0 \\ 0 \\ 0 \end{Bmatrix}.$$

All three equations yield the same single relation:

$$x_1^{(2)} + x_2^{(2)} + x_3^{(2)} = 0.$$

Thus, values of two of the three components $(x_1^{(2)}, x_2^{(2)}, x_3^{(2)})$ can be chosen arbitrarily. For the choice of $x_2^{(2)} = 0$ and $x_3^{(2)} = 1$, we obtain the vector (or any nonzero multiples of it):

$$\{X\}^{(2)} = \begin{Bmatrix} -1 \\ 0 \\ 1 \end{Bmatrix} x_1^{(2)} \quad \text{or} \quad \{\hat{X}\}^{(2)} = \pm \frac{1}{\sqrt{2}} \begin{Bmatrix} -1 \\ 0 \\ 1 \end{Bmatrix}.$$

A second independent vector can be found by choosing $x_2^{(2)} = 1$ and $x_3^{(2)} = 0$. We obtain

$$\{X\}^{(3)} = \begin{Bmatrix} -1 \\ 1 \\ 0 \end{Bmatrix} x_1^{(3)} \quad \text{or} \quad \{\hat{X}\}^{(3)} = \pm \frac{1}{\sqrt{2}} \begin{Bmatrix} -1 \\ 1 \\ 0 \end{Bmatrix}.$$

Thus, in the present case, there exist two linearly independent eigenvectors associated with the double eigenvalue. The eigenvectors are not mutually orthogonal.

We note that the matrix $[S]$ considered here is symmetric (but not positive). Clearly, properties 1 and 3 concerning the eigenvalues and eigenvectors of a real-valued symmetric matrix are satisfied. As far as property 5 is concerned, it is possible to choose the values of the two of the three components (x_1, x_2, x_3) to have a set of linearly independent eigenvectors that are orthogonal. The second vector associated with $\lambda = -1$ could have been chosen by setting $x_1^{(2)} = x_3^{(2)} = 1$. We obtain

$$\{X\}^{(3)} = \left\{ \begin{array}{c} 1 \\ -2 \\ 1 \end{array} \right\} x_1^{(2)} \quad \text{or} \quad \{\hat{X}\}^{(3)} = \pm \frac{1}{\sqrt{6}} \left\{ \begin{array}{c} 1 \\ -2 \\ 1 \end{array} \right\}.$$

Thus the three eigenvectors (check for the right-handed system)

$$\{\hat{X}\}^{(1)} = \frac{1}{\sqrt{3}} \left\{ \begin{array}{c} 1 \\ 1 \\ 1 \end{array} \right\}, \quad \{\hat{X}\}^{(2)} = \frac{1}{\sqrt{2}} \left\{ \begin{array}{c} -1 \\ 0 \\ 1 \end{array} \right\}, \quad \{\hat{X}\}^{(3)} = \frac{1}{\sqrt{6}} \left\{ \begin{array}{c} 1 \\ -2 \\ 1 \end{array} \right\}$$

are mutually orthogonal. Hence, we can write

$$\frac{1}{6} \begin{bmatrix} \sqrt{2} & \sqrt{2} & \sqrt{2} \\ -\sqrt{3} & 0 & \sqrt{3} \\ 1 & -2 & 1 \end{bmatrix} \begin{bmatrix} 0 & 1 & 1 \\ 1 & 0 & 1 \\ 1 & 1 & 0 \end{bmatrix} \begin{bmatrix} \sqrt{2} & -\sqrt{3} & 1 \\ \sqrt{2} & 0 & -2 \\ \sqrt{2} & \sqrt{3} & 1 \end{bmatrix} = \begin{bmatrix} \lambda_1 & 0 & 0 \\ 0 & \lambda_2 & 0 \\ 0 & 0 & \lambda_3 \end{bmatrix} = \begin{bmatrix} 2 & 0 & 0 \\ 0 & -1 & 0 \\ 0 & 0 & -1 \end{bmatrix}.$$

2.6 Summary

In this chapter, mathematical preliminaries needed for this course are reviewed. In particular, the notion of geometric vector, vector algebra, vector calculus, theory of matrices, tensors, and tensor calculus are thoroughly reviewed. The index notation for writing vectors and tensors in terms of their components is introduced. Transformations of vector and tensor components are presented, and eigenvalue problems associated with second-order tensors are discussed. A number of examples are presented throughout the chapter to illustrate the concepts and definitions introduced. The material included in this chapter is indispensable for the rest of the book, and readers are urged to make themselves familiar with the concepts as well as notation introduced here. The following is a brief review of the notations used:

a—(italic and lightface letter) a *scalar*

$\mathbf{A}$—(boldface roman) a *tensor* of order 1 or higher; vectors are tensors of order 1; and scalars are tensors of order 0

$\mathbf{A} \cdot \mathbf{B}$—the *dot product* of two tensors, $\mathbf{A}$ and $\mathbf{B}$, of order 1 or higher

$\mathbf{A} \times \mathbf{B}$—the *cross product* of two tensors, $\mathbf{A}$ and $\mathbf{B}$, of order 1 or higher

$\mathbf{AB}$—a *dyad* formed by vectors $\mathbf{A}$ and $\mathbf{B}$

$\boldsymbol{\nabla}$—the *del operator*, a vector differential operator

$\boldsymbol{\nabla}\mathbf{A}$—the *gradient* of a tensor $\mathbf{A}$ of order 0 or higher

$\boldsymbol{\nabla} \cdot \mathbf{A}$—the *divergence* of a tensor $\mathbf{A}$ of order 1 or higher

$\boldsymbol{\nabla} \times \mathbf{A}$—the *curl* of a tensor $\mathbf{A}$ of order 1 or higher

e_i—*unitary base vectors* in a general coordinate system

e^i—*dual base vectors* in a dual (to unitary) coordinate system

$\hat{e}_i$—*unit base vectors* in an orthonormal coordinate system

$\hat{n}$—*unit outward normal* vector

ℓ_{ij}—*direction cosines,* $\ell_{ij} = \hat{e}_i \cdot \hat{e}_j$ between barred $(\bar{x}_1, \bar{x}_2, \bar{x}_3)$ and unbarred (x_1, x_2, x_3) coordinates

δ_{ij}—the *Kronecker delta* symbol; components of the *unit tensor,* **I**

e_{ijk}—the *permutation* symbol; components of the third-order tensor, $\mathcal{E}$

$\{X\}$—a *column* or *row vector*

$[A]$—a *matrix* whose elements are denoted as a_{ij}; a_{ij} is an element in the ith row and jth column of the matrix; a matrix underlying a tensor **A** is denoted as $[A]$

$[A]^T$—*transpose of matrix* $[A]$, obtained by interchanging the rows and columns of matrix $[A]$

$\mathbf{A}^T$—*transpose of a tensor* **A**, obtained by transposing the base vectors.

$\|[A]\|$—the *determinant* of matrix $[A]$

$|\mathbf{A}|$—the *determinant* of matrix $[A]$ associated with a tensor **A**

When the same symbol is used in different places and contexts, the reader should not assume that they have the same meaning; the meaning of the symbol will be evident within the context.

Problems

VECTOR ALGEBRA

2.1 Find the equation of a line (or a set of lines) passing through the terminal point of a vector **A** and in the direction of vector **B**.

2.2 Obtain the equation of a plane perpendicular to a vector **A** and passing through the terminal point of vector **B**, without using any coordinate system.

2.3 Find the equation of a plane connecting the terminal points of vectors **A**, **B**, and **C**. Assume that all three vectors are referred to a common origin.

2.4 Let A and B denote two points in space, and let these points be represented by two vectors **A** and **B** with a common origin O, as shown in Fig. P2.4. Show that the straight line through points A and B can be represented by the vector equation

$$(\mathbf{r} - \mathbf{A}) \times (\mathbf{B} - \mathbf{A}) = \mathbf{0}.$$

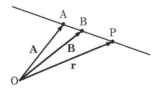

Fig. P2.4

2.5 Prove with the help of vectors that the diagonals of a parallelogram bisect each other.

2.6 Show that the position vector $\mathbf{r}$ that divides a line PQ in the ratio $k : l$ is given by

$$\mathbf{r} = \tfrac{l}{k+l}\, \mathbf{A} + \tfrac{k}{k+l}\, \mathbf{B},$$

where $\mathbf{A}$ and $\mathbf{B}$ are the vectors that designate points P and Q, respectively.

2.7 Represent a tetrahedron by the three noncoplanar vectors $\mathbf{A}$, $\mathbf{B}$, and $\mathbf{C}$, as shown in Fig. P2.7. Show that the vectorial sum of the areas of the tetrahedron sides is zero.

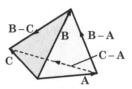

Fig. P2.7

2.8 Deduce that the vector equation for a sphere with its center located at point A and with a radius R is given by

$$(\mathbf{r} - \mathbf{A}) \cdot (\mathbf{r} - \mathbf{A}) = R^{2},$$

where $\mathbf{A}$ is the vector connecting the origin to point A and $\mathbf{r}$ is the position vector.

2.9 Verify that the following identity holds (without using index notation):

$$(\mathbf{A} \cdot \mathbf{B})^{2} + (\mathbf{A} \times \mathbf{B}) \cdot (\mathbf{A} \times \mathbf{B}) = |\mathbf{A}|^{2}\, |\mathbf{B}|^{2},$$

where $\mathbf{A}$ and $\mathbf{B}$ are arbitrary vectors. *Hint:* Use Eqs. (2.2.21) and (2.2.25).

2.10 If $\mathbf{A}, \mathbf{B}$, and $\mathbf{C}$ are noncoplanar vectors (that is, $\mathbf{A}, \mathbf{B}$, and $\mathbf{C}$ are linearly independent), determine if the following set of vectors is linearly independent:

$$\mathbf{r}_1 = \mathbf{A} - 3\mathbf{B} + 2\mathbf{C}, \quad \mathbf{r}_2 = 2\mathbf{A} - 5\mathbf{B} + 3\mathbf{C}, \quad \mathbf{r}_3 = \mathbf{A} - 5\mathbf{B} + 4\mathbf{C}.$$

2.11 Determine whether the following set of vectors is linearly independent:

$$\mathbf{A} = 2\hat{\mathbf{e}}_1 - \hat{\mathbf{e}}_2 + \hat{\mathbf{e}}_3, \quad \mathbf{B} = -\hat{\mathbf{e}}_2 - \hat{\mathbf{e}}_3, \quad \mathbf{C} = -\hat{\mathbf{e}}_1 + \hat{\mathbf{e}}_2.$$

Here $\hat{\mathbf{e}}_i$ are orthonormal unit base vectors in $\Re^3$.

2.12 Let the vectors $(\hat{\mathbf{i}}, \hat{\mathbf{j}}, \hat{\mathbf{k}})$ constitute an orthonormal basis. In terms of this basis, define a cogredient basis by

$$\mathbf{e}_1 = -\hat{\mathbf{i}} - \hat{\mathbf{j}}, \quad \mathbf{e}_2 = \hat{\mathbf{i}} + 2\hat{\mathbf{j}} - 2\hat{\mathbf{k}}, \quad \mathbf{e}_3 = 2\hat{\mathbf{i}} + \hat{\mathbf{j}} + \hat{\mathbf{k}}.$$

Determine

(a) the dual or reciprocal (contragredient) basis $(\mathbf{e}^1, \mathbf{e}^2, \mathbf{e}^3)$ in terms of the orthonormal basis $(\hat{\mathbf{i}}, \hat{\mathbf{j}}, \hat{\mathbf{k}})$,

(b) the magnitudes (or norms) $|\mathbf{e}_1|, |\mathbf{e}_2|, |\mathbf{e}_3|, |\mathbf{e}^1|, |\mathbf{e}^2|$, and $|\mathbf{e}^3|$, and

(c) the cogredient components A_1, A_2, and A_3 of a vector $\mathbf{A}$ if its contragredient components are given by $A^1 = 1, A^2 = 2, A^3 = 3$.

2.13 Using the Gram–Schmidt orthonormalization process, construct the orthonormal sets associated with the following sets of vectors:

(a) $\mathbf{e}_1 = \hat{\mathbf{i}}_1 + \hat{\mathbf{i}}_3, \quad \mathbf{e}_2 = \hat{\mathbf{i}}_1 + 2\hat{\mathbf{i}}_2 + 2\hat{\mathbf{i}}_3, \quad \mathbf{e}_3 = 2\hat{\mathbf{i}}_1 - \hat{\mathbf{i}}_2 + \hat{\mathbf{i}}_3.$

(b) $\mathbf{e}_1 = 2\hat{\mathbf{i}}_1 + \hat{\mathbf{i}}_2, \quad \mathbf{e}_2 = \hat{\mathbf{i}}_1 - 2\hat{\mathbf{i}}_2 + \hat{\mathbf{i}}_3, \quad \mathbf{e}_3 = -2\hat{\mathbf{i}}_1 + \hat{\mathbf{i}}_2 + \hat{\mathbf{i}}_3.$

Here $(\hat{\mathbf{i}}_1, \hat{\mathbf{i}}_2, \hat{\mathbf{i}}_3)$ denotes an orthonormal Cartesian basis.

INDEX NOTATION

2.14 Prove the following vector identity using index notation:
$$\mathbf{A} \times (\mathbf{B} \times \mathbf{C}) = (\mathbf{A} \cdot \mathbf{C})\mathbf{B} - (\mathbf{A} \cdot \mathbf{B})\mathbf{C}.$$

2.15 Prove the following vector identity using index notation:
$$(\mathbf{A} \cdot \mathbf{B})^2 + (\mathbf{A} \times \mathbf{B}) \cdot (\mathbf{A} \times \mathbf{B}) = |\mathbf{A}|^2 |\mathbf{B}|^2.$$

2.16 Use index notation and the e-δ identity to rewrite the vector expression as a sum (or difference) of two vector expressions:
$$(\nabla \times \mathbf{A}) \times \mathbf{B},$$

where $\mathbf{A}$ and $\mathbf{B}$ are vector functions.

2.17 Simplify the vector expression $\nabla \cdot \left(\frac{\mathbf{x}-\mathbf{y}}{\rho}\right)$, where $\rho = |\mathbf{x} - \mathbf{y}|$ and $\mathbf{y}$ is a fixed point, and $\mathbf{x}$ is the position vector of a point in a 3D space. Express the final result in terms of ρ only.

2.18 Using index notation, prove the following identities among vectors $\mathbf{A}$, $\mathbf{B}$, $\mathbf{C}$, and $\mathbf{D}$:

(a) $(\mathbf{A} \times \mathbf{B}) \cdot (\mathbf{B} \times \mathbf{C}) \times (\mathbf{C} \times \mathbf{A}) = (\mathbf{A} \cdot (\mathbf{B} \times \mathbf{C}))^2.$

(b) $(\mathbf{A} \times \mathbf{B}) \times (\mathbf{C} \times \mathbf{D}) = [\mathbf{A} \cdot (\mathbf{C} \times \mathbf{D})]\mathbf{B} - [\mathbf{B} \cdot (\mathbf{C} \times \mathbf{D}]\mathbf{A}.$

2.19 Prove that
$$[ABC][DEF] = \begin{vmatrix} \mathbf{A} \cdot \mathbf{D} & \mathbf{A} \cdot \mathbf{E} & \mathbf{A} \cdot \mathbf{F} \\ \mathbf{B} \cdot \mathbf{D} & \mathbf{B} \cdot \mathbf{E} & \mathbf{B} \cdot \mathbf{F} \\ \mathbf{C} \cdot \mathbf{D} & \mathbf{C} \cdot \mathbf{E} & \mathbf{C} \cdot \mathbf{F} \end{vmatrix},$$

and from there show that
$$e_{ijk}\, e_{rst} = \begin{vmatrix} \delta_{ir} & \delta_{is} & \delta_{it} \\ \delta_{jr} & \delta_{js} & \delta_{jt} \\ \delta_{kr} & \delta_{ks} & \delta_{kt} \end{vmatrix}.$$

2.20 Establish the following identities :

(a) $\quad e_{ijk} = \begin{vmatrix} \delta_{i1} & \delta_{i2} & \delta_{i3} \\ \delta_{j1} & \delta_{j2} & \delta_{j3} \\ \delta_{k1} & \delta_{k2} & \delta_{k3} \end{vmatrix}.$

(b) $\quad e_{ijk}e_{pqr} = \begin{vmatrix} \delta_{ip} & \delta_{iq} & \delta_{ir} \\ \delta_{jp} & \delta_{jq} & \delta_{jr} \\ \delta_{kp} & \delta_{kq} & \delta_{kr} \end{vmatrix}.$

(c) $\quad e_{ijk}\, e_{ijk} = 6.$

(d) $\quad e_{ijk}e_{mnk} = \delta_{im}\delta_{jn} - \delta_{in}\delta_{jm}.$

COORDINATE TRANSFORMATIONS

2.21 Consider two rectangular Cartesian coordinate systems that are translated and rotated with respect to each other. The transformation between the two coordinate systems is given by
$$\{\bar{X}\} = \{C\} + [L]\{X\},$$

where $\{C\}$ is a constant vector and $[L] = [\ell_{ij}]$ is the matrix of direction cosines
$$\ell_{ij} \equiv \hat{\mathbf{e}}_i \cdot \hat{\mathbf{e}}_j.$$

Deduce that the following orthogonality conditions hold:
$$[L][L]^{\mathrm{T}} = [I] \quad \text{or} \quad \ell_{ik}\ell_{kj} = \delta_{ij}.$$

That is, $[L]$ is an orthogonal matrix.

2.22 Determine the transformation matrix relating the orthonormal basis vectors $(\hat{\mathbf{e}}_1, \hat{\mathbf{e}}_2, \hat{\mathbf{e}}_3)$ and $(\hat{\mathbf{e}}_1', \hat{\mathbf{e}}_2', \hat{\mathbf{e}}_3')$, when $\hat{\mathbf{e}}_i'$ are given by

(a) $\hat{\mathbf{e}}_1'$ is along the vector $\hat{\mathbf{e}}_1 - \hat{\mathbf{e}}_2 + \hat{\mathbf{e}}_3$ and $\hat{\mathbf{e}}_2'$ is perpendicular to the plane $2x_1 + 3x_2 + x_3 - 5 = 0$.

(b) $\hat{e}_1'$ is along the line segment connecting point $(1, -1, 3)$ to $(2, -2, 4)$ and $\hat{e}_3' = (-\hat{e}_1 + \hat{e}_2 + 2\hat{e}_3)/\sqrt{6}$.

2.23 The angles between the barred and unbarred coordinate lines are given by

	$\hat{e}_1$	$\hat{e}_2$	$\hat{e}_3$
$\hat{\bar{e}}_1$	$60°$	$30°$	$90°$
$\hat{\bar{e}}_2$	$150°$	$60°$	$90°$
$\hat{\bar{e}}_3$	$90°$	$90°$	$0°$

Determine the direction cosines of the transformation.

2.24 The angles between the barred and unbarred coordinate lines are given by

	x_1	x_2	x_3
$\bar{x}_1$	$45°$	$90°$	$45°$
$\bar{x}_2$	$60°$	$45°$	$120°$
$\bar{x}_3$	$120°$	$45°$	$60°$

Determine the transformation matrix.

MATRICES

2.25 Write the following sets of equations in matrix form $[A]\{X\} = \{Y\}$:

(a) $2x_1 + x_2 - 2x_3 = 1,$ (b) $2x_1 + x_2 - x_3 = 0,$
$\quad\;\; x_1 - 2x_2 + x_3 = 5,$ $\quad\;\; 3x_1 \quad\;\; - x_3 = 2,$
$\quad\;\; 3x_1 + x_2 - x_3 = 4.$ $\quad\;\; x_1 + x_2 + x_3 = 1.$

2.26 Determine the cofactors and the determinants of the coefficient matrices in Problem **2.25**.

2.27 Find the inverses of the coefficient matrices in Problem **2.25**.

2.28 Determine if the following matrices are positive:

(a) $\begin{bmatrix} 2 & 1 & -2 \\ 1 & -2 & 1 \\ 3 & 1 & -1 \end{bmatrix}$, (b) $\begin{bmatrix} 2 & 1 & -1 \\ 3 & 2 & -1 \\ 1 & 1 & 0 \end{bmatrix}$, (c) $\begin{bmatrix} 2 & 1 & 1 \\ 1 & -1 & 2 \\ 3 & 2 & 1 \end{bmatrix}$.

2.29 Check to see if the following $[Q]$ is nonsingular, and if it is, construct the positive matrix associated with it:

$$[Q] = \begin{bmatrix} 1 & 0 & 0 \\ 0 & 1 & 2 \\ 1 & 1 & 1 \end{bmatrix}.$$

VECTOR ALGEBRA[6]

2.30 Let $\mathbf{r}$ denote a position vector $\mathbf{r} = \mathbf{x} = x_i \hat{e}_i$ $(r^2 = x_i x_i)$ and $\mathbf{A}$ be an arbitrary constant vector. Use index notation to show that:

(a) $\nabla^2(r^n) = n(n + 1)r^{n-2}.$ (b) $\nabla(\mathbf{r} \cdot \mathbf{A}) = \mathbf{A}.$
(c) $\nabla \cdot (\mathbf{r} \times \mathbf{A}) = 0.$ (d) $\nabla \times (\mathbf{r} \times \mathbf{A}) = -2\mathbf{A}.$
(e) $\nabla \cdot (r\mathbf{A}) = \dfrac{1}{r}(\mathbf{r} \cdot \mathbf{A}).$ (f) $\nabla \times (r\mathbf{A}) = \dfrac{1}{r}(\mathbf{r} \times \mathbf{A}).$

2.31 Let $\mathbf{A}$ and $\mathbf{B}$ be vector functions of position vector $\mathbf{x}$ with continuous first and second derivatives, and let F and G be scalar functions of position $\mathbf{x}$ with continuous first and second derivatives. Use index notation to show that:

(a) $\nabla \cdot (\nabla \times \mathbf{A}) = 0.$

---

[6]Many problems here have several parts, and the instructor may assign selected parts as separate problems.

(b) $\boldsymbol{\nabla} \times (\boldsymbol{\nabla} F) = 0$.

(c) $\boldsymbol{\nabla} \cdot (\boldsymbol{\nabla} F \times \boldsymbol{\nabla} G) = 0$.

(d) $\boldsymbol{\nabla} \cdot (F\mathbf{A}) = \mathbf{A} \cdot \boldsymbol{\nabla} F + F \boldsymbol{\nabla} \cdot \mathbf{A}$.

(e) $\boldsymbol{\nabla} \times (F\mathbf{A}) = F \boldsymbol{\nabla} \times \mathbf{A} - \mathbf{A} \times \boldsymbol{\nabla} F$.

(f) $\boldsymbol{\nabla}(\mathbf{A} \cdot \mathbf{B}) = \mathbf{A} \cdot \boldsymbol{\nabla} \mathbf{B} + \mathbf{B} \cdot \boldsymbol{\nabla} \mathbf{A} + \mathbf{A} \times (\boldsymbol{\nabla} \times \mathbf{B}) + \mathbf{B} \times (\boldsymbol{\nabla} \times \mathbf{A})$.

(g) $\boldsymbol{\nabla} \cdot (\mathbf{A} \times \mathbf{B}) = \boldsymbol{\nabla} \times \mathbf{A} \cdot \mathbf{B} - \boldsymbol{\nabla} \times \mathbf{B} \cdot \mathbf{A}$.

2.32 Let $\mathbf{A}$ and $\mathbf{B}$ be vector functions of position vector $\mathbf{x}$ with continuous first and second derivatives, and let F and G be scalar functions of position $\mathbf{x}$ with continuous first and second derivatives. Use index notation to show that:

(a) $\boldsymbol{\nabla} \times (\mathbf{A} \times \mathbf{B}) = \mathbf{B} \cdot \boldsymbol{\nabla} \mathbf{A} - \mathbf{A} \cdot \boldsymbol{\nabla} \mathbf{B} + \mathbf{A} \boldsymbol{\nabla} \cdot \mathbf{B} - \mathbf{B} \boldsymbol{\nabla} \cdot \mathbf{A}$.

(b) $(\boldsymbol{\nabla} \times \mathbf{A}) \times \mathbf{A} = \mathbf{A} \cdot \boldsymbol{\nabla} \mathbf{A} - \boldsymbol{\nabla} \mathbf{A} \cdot \mathbf{A}$.

(c) $\nabla^2(FG) = F\nabla^2 G + 2\boldsymbol{\nabla} F \cdot \boldsymbol{\nabla} G + G\nabla^2 F$.

(d) $\nabla^2(F\mathbf{x}) = 2\boldsymbol{\nabla} F + \mathbf{x}\nabla^2 F$.

(e) $\mathbf{A} \cdot \boldsymbol{\nabla} \mathbf{A} = \frac{1}{2}\boldsymbol{\nabla}(\mathbf{A} \cdot \mathbf{A}) - \mathbf{A} \times \boldsymbol{\nabla} \times \mathbf{A}$.

(f) $\boldsymbol{\nabla}(\mathbf{A} \cdot \mathbf{x}) = \mathbf{A} + \boldsymbol{\nabla} \mathbf{A} \cdot \mathbf{x}$.

(g) $\nabla^2(\mathbf{A} \cdot \mathbf{x}) = 2\boldsymbol{\nabla} \cdot \mathbf{A} + \mathbf{x} \cdot \nabla^2 \mathbf{A}$.

2.33 Show that
$$\boldsymbol{\nabla}(R^n\mathbf{x}) = R^n\mathbf{I} + nR^{n-2}\mathbf{x}\mathbf{x}, \quad R^2 = \mathbf{x} \cdot \mathbf{x}$$
using (a) index notation and (b) the spherical coordinate system.

2.34 Show that
$$\nabla^2(R^n\mathbf{x}) = n(n+3)R^{n-2}\mathbf{x}, \quad R^2 = \mathbf{x} \cdot \mathbf{x}$$
using (a) index notation, and (b) the spherical coordinate system.

2.35 Show that the vector area of a closed surface is zero, that is,
$$\oint_s \hat{\mathbf{n}}\, ds = \mathbf{0}.$$

2.36 Show that the volume of the region Ω enclosed by a boundary surface Γ is
$$\text{volume} = \frac{1}{6}\oint_\Gamma \boldsymbol{\nabla}(r^2) \cdot \hat{\mathbf{n}}\, ds = \frac{1}{3}\oint_\Gamma \mathbf{r} \cdot \hat{\mathbf{n}}\, ds.$$

2.37 Let $\phi(\mathbf{r})$ be a scalar field. Show that
$$\int_\Omega \nabla^2\phi\, d\mathbf{x} = \oint_\Gamma \frac{\partial\phi}{\partial n}\, ds.$$

2.38 In the divergence theorem (2.4.34), set $\mathbf{A} = \phi\boldsymbol{\nabla}\psi$ and $\mathbf{A} = \psi\boldsymbol{\nabla}\phi$ successively and obtain the integral forms

(a) $\displaystyle\int_\Omega \left[\phi\nabla^2\psi + \boldsymbol{\nabla}\phi \cdot \boldsymbol{\nabla}\psi\right] d\mathbf{x} = \oint_\Gamma \phi\frac{\partial\psi}{\partial n}\, ds,$

(b) $\displaystyle\int_\Omega \left[\phi\nabla^2\psi - \psi\nabla^2\phi\right] d\mathbf{x} = \oint_\Gamma \left[\phi\frac{\partial\psi}{\partial n} - \psi\frac{\partial\phi}{\partial n}\right] ds,$

(c) $\displaystyle\int_\Omega \left[\phi\nabla^4\psi - \nabla^2\phi\nabla^2\psi\right] d\mathbf{x} = \oint_\Gamma \left[\phi\frac{\partial}{\partial n}(\nabla^2\psi) - \nabla^2\psi\frac{\partial\phi}{\partial n}\right] ds,$

where Ω denotes a (2D or 3D) region with bounding surface Γ. The first two identities are sometimes called *Green's first and second theorems*.

2.39 Let $\mathbf{V}$ and $\mathbf{S}$ be smooth vector and second-order tensor fields defined in Ω and on Γ (the closed boundary of Ω) and let $\hat{\mathbf{n}}$ be the unit outward normal to Γ. Establish the identity
$$\oint_\Gamma \mathbf{V} \cdot \mathbf{S} \cdot \hat{\mathbf{n}}\, ds = \int_\Omega \boldsymbol{\nabla} \cdot (\mathbf{S}^\mathsf{T} \cdot \mathbf{V})\, d\mathbf{x}.$$

2.40 Let **S** be a smooth second-order tensor field defined in Ω and on Γ (the closed boundary of Ω) and let $\hat{\mathbf{n}}$ be the unit outward normal to Γ. Use index notation to establish the identity

$$\oint_\Gamma \mathbf{x} \times (\hat{\mathbf{n}} \cdot \mathbf{S}) \, ds = \int_\Omega [\mathbf{x} \times (\nabla \cdot \mathbf{S}) + \mathcal{E} : \mathbf{S}] \, d\mathbf{x},$$

where **x** is the position vector and $\mathcal{E}$ is the third-order permutation tensor [see Eq. (2.5.23)].

TENSORS AND TENSOR CALCULUS

2.41 Establish the following identities for a second-order tensor **S**:

(a) $|\mathbf{S}| = e_{ijk}\, s_{1i}\, s_{2j}\, s_{3k}.$ (b) $|\mathbf{S}| = \frac{1}{6} s_{ir}\, s_{js}\, s_{kt}\, e_{rst}\, e_{ijk}.$

(c) $e_{rst}|\mathbf{S}| = e_{ijk}\, s_{ir}\, s_{js}\, s_{kt}.$ (d) $\begin{vmatrix} s_{im} & s_{in} & s_{ip} \\ s_{jm} & s_{jn} & s_{jp} \\ s_{km} & s_{kn} & s_{kp} \end{vmatrix} = e_{ijk}\, e_{mnp}\, |\mathbf{S}|.$

2.42 Given vector **A** and second-order tensors **S** and **T** with the following components:

$$\{A\} = \begin{Bmatrix} 2 \\ -1 \\ 4 \end{Bmatrix}, \quad [S] = \begin{bmatrix} -1 & 0 & 5 \\ 3 & 7 & 4 \\ 9 & 8 & 6 \end{bmatrix}, \quad [T] = \begin{bmatrix} 8 & -1 & 6 \\ 5 & 4 & 9 \\ -7 & 8 & -2 \end{bmatrix}$$

determine

(a) $\text{tr}(\mathbf{S})$.   (b) $\mathbf{S}:\mathbf{S}$.   (c) $\mathbf{S}:\mathbf{S}^\mathrm{T}$.

(d) $\mathbf{A}\cdot\mathbf{S}$.   (e) $\mathbf{S}\cdot\mathbf{A}$.   (f) $\mathbf{S}\cdot\mathbf{T}\cdot\mathbf{A}$.

2.43 Determine the rotation transformation matrix such that the new base vector $\hat{\mathbf{e}}_1$ is along $\hat{\mathbf{e}}_1 - \hat{\mathbf{e}}_2 + \hat{\mathbf{e}}_3$, and $\hat{\mathbf{e}}_2$ is along the normal to the plane $2x_1 + 3x_2 + x_3 = 5$. If **S** is the tensor whose components in the unbarred system are given by $s_{11} = 1$, $s_{12} = s_{21} = 0$, $s_{13} = s_{31} = -1$, $s_{22} = 3$, $s_{23} = s_{32} = -2$, and $s_{33} = 0$, find the components in the barred coordinates.

2.44 Suppose that the new axes $\bar{x}_i$ are obtained by rotating x_i through $60°$ about the x_2-axis. Determine the components $\bar{A}_i$ of a vector **A** whose components with respect to the x_i coordinates are $(2, 1, 3)$.

2.45 Show that the following expressions for an arbitrary tensor **S** are invariant: (a) S_{ii}, (b) $S_{ij}S_{ij}$, and (c) $S_{ij}S_{jk}S_{ki}$.

2.46 If **A** and **B** are arbitrary vectors and **S** and **T** are arbitrary dyads, verify that:

(a) $(\mathbf{A}\cdot\mathbf{S})\cdot\mathbf{B} = \mathbf{A}\cdot(\mathbf{S}\cdot\mathbf{B})$.   (b) $(\mathbf{S}\cdot\mathbf{T})\cdot\mathbf{A} = \mathbf{S}\cdot(\mathbf{T}\cdot\mathbf{A})$.

(c) $\mathbf{A}\cdot(\mathbf{S}\cdot\mathbf{T}) = (\mathbf{A}\cdot\mathbf{S})\cdot\mathbf{T}$.   (d) $(\mathbf{S}\cdot\mathbf{A})\cdot(\mathbf{T}\cdot\mathbf{B}) = \mathbf{A}\cdot(\mathbf{S}^\mathrm{T}\cdot\mathbf{T})\cdot\mathbf{B}$.

2.47 If **A** is an arbitrary vector and **R** and **S** are arbitrary dyads, verify that:

(a) $(\mathbf{I}\times\mathbf{A})\cdot\mathbf{R} = \mathbf{A}\times\mathbf{R}$.   (b) $(\mathbf{A}\times\mathbf{I})\cdot\mathbf{R} = \mathbf{A}\times\mathbf{R}$.

(c) $(\mathbf{R}\times\mathbf{A})^\mathrm{T} = -\mathbf{A}\times\mathbf{R}^\mathrm{T}$.   (d) $(\mathbf{R}\cdot\mathbf{S})^\mathrm{T} = \mathbf{S}^\mathrm{T}\cdot\mathbf{R}^\mathrm{T}$.

(e) $(\mathbf{R}\cdot\mathbf{S})^{-1} = \mathbf{S}^{-1}\cdot\mathbf{R}^{-1}$.   (f) $(\mathbf{R}\cdot\mathbf{S})^{-\mathrm{T}} = \mathbf{R}^{-\mathrm{T}}\cdot\mathbf{S}^{-\mathrm{T}}$.

2.48 The determinant of a second-order tensor **S** is also defined by the expression

$$|\mathbf{S}| = \frac{[(\mathbf{S}\cdot\mathbf{A})\times(\mathbf{S}\cdot\mathbf{B})]\cdot(\mathbf{S}\cdot\mathbf{C})}{\mathbf{A}\times\mathbf{B}\cdot\mathbf{C}}$$

where **A**, **B**, and **C** are arbitrary vectors. Verify the identity in an orthonormal basis $\{\hat{\mathbf{e}}_i\}$.

2.49 For an arbitrary second-order tensor $\mathbf{S}$ show that $\nabla \cdot \mathbf{S}$ in the cylindrical coordinate system is given by

$$
\begin{aligned}
\nabla \cdot \mathbf{S} = & \left[\frac{\partial S_{rr}}{\partial r} + \frac{1}{r} \frac{\partial S_{\theta r}}{\partial \theta} + \frac{\partial S_{zr}}{\partial z} + \frac{1}{r} (S_{rr} - S_{\theta\theta}) \right] \hat{\mathbf{e}}_r \\
& + \left[\frac{\partial S_{r\theta}}{\partial r} + \frac{1}{r} \frac{\partial S_{\theta\theta}}{\partial \theta} + \frac{\partial S_{z\theta}}{\partial z} + \frac{1}{r} (S_{r\theta} + S_{\theta r}) \right] \hat{\mathbf{e}}_\theta \\
& + \left[\frac{\partial S_{rz}}{\partial r} + \frac{1}{r} \frac{\partial S_{\theta z}}{\partial \theta} + \frac{\partial S_{zz}}{\partial z} + \frac{1}{r} S_{rz} \right] \hat{\mathbf{e}}_z.
\end{aligned}
$$

2.50 For an arbitrary second-order tensor $\mathbf{S}$ show that $\nabla \times \mathbf{S}$ in the cylindrical coordinate system is given by

$$
\begin{aligned}
\nabla \times \mathbf{S} = & \; \hat{\mathbf{e}}_r \hat{\mathbf{e}}_r \left(\frac{1}{r} \frac{\partial S_{zr}}{\partial \theta} - \frac{\partial S_{\theta r}}{\partial z} - \frac{1}{r} S_{z\theta} \right) + \hat{\mathbf{e}}_\theta \hat{\mathbf{e}}_\theta \left(\frac{\partial S_{r\theta}}{\partial z} - \frac{\partial S_{z\theta}}{\partial r} \right) + \\
& \hat{\mathbf{e}}_z \hat{\mathbf{e}}_z \left(\frac{1}{r} S_{\theta z} - \frac{1}{r} \frac{\partial S_{rz}}{\partial \theta} + \frac{\partial S_{\theta z}}{\partial r} \right) + \hat{\mathbf{e}}_r \hat{\mathbf{e}}_\theta \left(\frac{1}{r} \frac{\partial S_{z\theta}}{\partial \theta} - \frac{\partial S_{\theta\theta}}{\partial z} + \frac{1}{r} S_{zr} \right) + \\
& \hat{\mathbf{e}}_\theta \hat{\mathbf{e}}_r \left(\frac{\partial S_{rr}}{\partial z} - \frac{\partial S_{zr}}{\partial r} \right) + \hat{\mathbf{e}}_r \hat{\mathbf{e}}_z \left(\frac{1}{r} \frac{\partial S_{zz}}{\partial \theta} - \frac{\partial S_{\theta z}}{\partial z} \right) + \\
& \hat{\mathbf{e}}_z \hat{\mathbf{e}}_r \left(\frac{\partial S_{\theta r}}{\partial r} - \frac{1}{r} \frac{\partial S_{rr}}{\partial \theta} + \frac{1}{r} S_{r\theta} + \frac{1}{r} S_{\theta r} \right) + \hat{\mathbf{e}}_\theta \hat{\mathbf{e}}_z \left(\frac{\partial S_{rz}}{\partial z} - \frac{\partial S_{zz}}{\partial r} \right) + \\
& \hat{\mathbf{e}}_z \hat{\mathbf{e}}_\theta \left(\frac{\partial S_{\theta\theta}}{\partial r} + \frac{1}{r} S_{\theta\theta} - \frac{1}{r} S_{rr} - \frac{1}{r} \frac{\partial S_{r\theta}}{\partial \theta} \right).
\end{aligned}
$$

2.51 For an arbitrary second-order tensor $\mathbf{S}$ show that $\nabla \cdot \mathbf{S}$ in the spherical coordinate system is given by

$$
\begin{aligned}
\nabla \cdot \mathbf{S} = & \left\{ \frac{\partial S_{RR}}{\partial R} + \frac{1}{R} \frac{\partial S_{\phi R}}{\partial \phi} + \frac{1}{R \sin \phi} \frac{\partial S_{\theta R}}{\partial \theta} + \frac{1}{R} [2S_{RR} - S_{\phi\phi} - S_{\theta\theta} + S_{\phi R} \cot \phi] \right\} \hat{\mathbf{e}}_R \\
& + \left\{ \frac{\partial S_{R\phi}}{\partial R} + \frac{1}{R} \frac{\partial S_{\phi\phi}}{\partial \phi} + \frac{1}{R \sin \phi} \frac{\partial S_{\theta\phi}}{\partial \theta} + \frac{1}{R} [(S_{\phi\phi} - S_{\theta\theta}) \cot \phi + S_{\phi R} + 2S_{R\phi}] \right\} \hat{\mathbf{e}}_\phi \\
& + \left\{ \frac{\partial S_{R\theta}}{\partial R} + \frac{1}{R} \frac{\partial S_{\phi\theta}}{\partial \phi} + \frac{1}{R \sin \phi} \frac{\partial S_{\theta\theta}}{\partial \theta} + \frac{1}{R} [(S_{\phi\theta} + S_{\theta\phi}) \cot \phi + 2S_{R\theta} + S_{\theta R}] \right\} \hat{\mathbf{e}}_\theta.
\end{aligned}
$$

2.52 Show that $\nabla \mathbf{u}$ in the spherical coordinate system is given by

$$
\begin{aligned}
\nabla \mathbf{u} = & \; \frac{\partial u_R}{\partial R} \hat{\mathbf{e}}_R \hat{\mathbf{e}}_R + \frac{\partial u_\phi}{\partial R} \hat{\mathbf{e}}_R \hat{\mathbf{e}}_\phi + \frac{\partial u_\theta}{\partial R} \hat{\mathbf{e}}_R \hat{\mathbf{e}}_\theta \\
& + \frac{1}{R} \left(\frac{\partial u_R}{\partial \phi} - u_\phi \right) \hat{\mathbf{e}}_\phi \hat{\mathbf{e}}_R + \frac{1}{R} \left(\frac{\partial u_\phi}{\partial \phi} + u_R \right) \hat{\mathbf{e}}_\phi \hat{\mathbf{e}}_\phi + \frac{1}{R} \frac{\partial u_\theta}{\partial \phi} \hat{\mathbf{e}}_\phi \hat{\mathbf{e}}_\theta \\
& + \frac{1}{R \sin \phi} \left[\left(\frac{\partial u_R}{\partial \theta} - u_\theta \sin \phi \right) \hat{\mathbf{e}}_\theta \hat{\mathbf{e}}_R + \left(\frac{\partial u_\phi}{\partial \theta} - u_\theta \cos \phi \right) \hat{\mathbf{e}}_\theta \hat{\mathbf{e}}_\phi \right. \\
& \left. + \left(\frac{\partial u_\theta}{\partial \theta} + u_R \sin \phi + u_\phi \cos \phi \right) \hat{\mathbf{e}}_\theta \hat{\mathbf{e}}_\theta \right].
\end{aligned}
$$

2.53 Prove the following identities when $\mathbf{A}$ and $\mathbf{B}$ are vectors and $\mathbf{S}$, $\mathbf{R}$, and $\mathbf{T}$ are second-order tensors:

(a) $\text{tr}(\mathbf{AB}) = \mathbf{A} \cdot \mathbf{B}$. (b) $\text{tr}(\mathbf{S}^{\mathrm{T}}) = \text{tr}\,\mathbf{S}$.

(c) $\text{tr}(\mathbf{R} \cdot \mathbf{S}) = \mathbf{R} \cdot \cdot \mathbf{S}$. (d) $\text{tr}(\mathbf{R}^{\mathrm{T}} \cdot \mathbf{S}) = \mathbf{R} : \mathbf{S}$.

(e) $\text{tr}(\mathbf{R} \cdot \mathbf{S}) = \text{tr}(\mathbf{S} \cdot \mathbf{R})$. (f) $\text{tr}(\mathbf{R} \cdot \mathbf{S} \cdot \mathbf{T}) = \text{tr}(\mathbf{T} \cdot \mathbf{R} \cdot \mathbf{S}) = \text{tr}(\mathbf{S} \cdot \mathbf{T} \cdot \mathbf{R})$.

EIGENVALUE PROBLEMS

2.54 Show that the characteristic equation for a symmetric second-order tensor $\boldsymbol{\Phi}$ can be expressed as

$$
-\lambda^3 + I_1 \lambda^2 - I_2 \lambda + I_3 = 0,
$$

where

$$I_1 = \phi_{kk}, \quad I_2 = \frac{1}{2}(\phi_{ii}\phi_{jj} - \phi_{ij}\phi_{ji}),$$

$$I_3 = \frac{1}{6}(2\phi_{ij}\phi_{jk}\phi_{ki} - 3\phi_{ij}\phi_{ji}\phi_{kk} + \phi_{ii}\phi_{jj}\phi_{kk}) = \det(\phi_{ij}).$$

2.55 Find the eigenvalues and eigenvectors of the following matrices:

(a) $\begin{bmatrix} 4 & -4 & 0 \\ -4 & 0 & 0 \\ 0 & 0 & 3 \end{bmatrix}$.   (b) $\begin{bmatrix} 2 & -\sqrt{3} & 0 \\ -\sqrt{3} & 4 & 0 \\ 0 & 0 & 4 \end{bmatrix}$.

(c) $\begin{bmatrix} 1 & 0 & 0 \\ 0 & 3 & -1 \\ 0 & -1 & 3 \end{bmatrix}$ (d) $\begin{bmatrix} 2 & -1 & 1 \\ -1 & 0 & 1 \\ 1 & 1 & 2 \end{bmatrix}$.

2.56 Find the eigenvalues and eigenvectors of the following matrices:

(a) $\begin{bmatrix} 3 & 5 & 8 \\ 5 & 1 & 0 \\ 8 & 0 & 2 \end{bmatrix}$.   (b) $\begin{bmatrix} 1 & -1 & 0 \\ -1 & 2 & -1 \\ 0 & -1 & 2 \end{bmatrix}$.

(c) $\begin{bmatrix} 1 & 2 & 0 \\ 1 & -1 & 1 \\ 0 & -2 & 1 \end{bmatrix}$.   (d) $\begin{bmatrix} 3 & 2 & 0 \\ 2 & 0 & 0 \\ 1 & 0 & 2 \end{bmatrix}$.

2.57 Find the eigenvalues and eigenvectors associated with the matrix

$$[S] = \begin{bmatrix} -2 & 2 & 10 \\ 2 & -11 & 8 \\ 10 & 8 & -5 \end{bmatrix}.$$

2.58 If $p(x) = a_0 + a_1 x^2 + \cdots + a_n x^n$, and $[A]$ is any square matrix, we define the polynomial in $[A]$ by

$$p(A) = a_0[I] + a_1[A] + a_2[A]^2 + \cdots + a_n[A]^n.$$

If

$$[A] = \begin{bmatrix} 1 & -1 \\ -1 & 1 \end{bmatrix},$$

and $p(x) = 1 - 2x + x^2$, compute $p(A)$.

2.59 *Cayley–Hamilton Theorem* Consider a square matrix $[S]$ of order n. Denote by $p(\lambda)$ the determinant of $|[S] - \lambda[I]|$ [that is, $p(\lambda) \equiv p(S - \lambda I)$], called the *characteristic polynomial*. Then the Cayley–Hamilton Theorem states that $p(\lambda) = 0$ (i.e., every matrix satisfies its own characteristic equation). Here $p(\lambda)$ is as defined in Problem **2.58**. Use matrix computation to verify the Cayley–Hamilton theorem for each of the following matrices:

(a) $\begin{bmatrix} 1 & -1 \\ 2 & 1 \end{bmatrix}$.   (b) $\begin{bmatrix} 2 & -1 & 1 \\ 0 & 1 & 0 \\ 1 & -2 & 1 \end{bmatrix}$.

2.60 Consider the matrix in Example 2.5.3:

$$[S] = \begin{bmatrix} 2 & 1 & 0 \\ 1 & 4 & 1 \\ 0 & 1 & 2 \end{bmatrix}.$$

Verify the Cayley–Hamilton theorem and use it to compute the inverse of $[S]$.

3

KINEMATICS OF CONTINUA

The man who cannot occasionally imagine events and conditions of existence that are contrary to the causal principle as he knows it will never enrich his science by the addition of a new idea.
— Max Planck (1858–1947)

It is through science that we prove, but through intuition that we discover.
— Henri Poincaré (1854–1912)

3.1 Introduction

Material or matter is composed of discrete molecules, which in turn are made up of atoms. An atom consists of negatively charged electrons, positively charged protons, and neutrons. Electrons form chemical bonds. The study of matter at molecular or atomistic levels is very useful for understanding a variety of phenomena, but studies at these scales are not useful to solve common engineering problems. *Continuum mechanics* is concerned with a study of various forms of matter at the macroscopic level. Central to this study is the assumption that the discrete nature of matter can be overlooked, provided the length scales of interest are large compared to the length scales of discrete molecular structures. Thus, matter at sufficiently large length scales can be treated as a *continuum*,[1] in which all physical quantities of interest, including density, are continuously differentiable almost everywhere.

Engineers and scientists undertake the study of continuous systems to understand their behavior under "working conditions," so that the systems can be designed to function properly and to be produced economically. For example, if we were to repair or replace a damaged artery in the human body, we must understand the function of the original artery and the conditions that led to its damage. An artery carries blood from the heart to different parts of the body. Conditions such as high blood pressure and increase in cholesterol levels in the blood may lead to deposition of particles in the arterial wall, as shown in Fig. 3.1.1. With time, accumulation of these particles in the arterial wall hardens and constricts the passage, leading to cardiovascular diseases. A possible remedy for such diseases is to repair or replace the damaged portion of the artery. This in turn requires an understanding of the deformations and stresses caused in the arterial wall by the blood flow. The understanding is then used to design a vascular prosthesis (that is, artificial artery).

---

[1]We mean a differentiable manifold with a boundary. Inherent in this assumption is that material particles that are neighbors will remain neighbors during the motion.

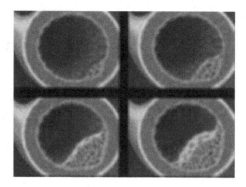

Fig. 3.1.1: Progressive damage of an artery due to deposition of particles in the arterial wall.

The present chapter is devoted to the study of geometric changes in a continuous medium (such as the artery) that is in equilibrium. The study of geometric changes in a continuum without regard to the forces causing the changes is known as *kinematics*. Sections or subsections that are considered to be too advanced for a first course can be skipped without loss of continuity (or returned to when needed).

3.2 Descriptions of Motion

3.2.1 Configurations of a Continuous Medium

Consider a body $\mathcal{B}$ of known geometry in a three-dimensional Euclidean space $\Re^3$; $\mathcal{B}$ may be viewed as a set of particles, each particle representing a large collection of molecules with a continuous distribution of matter in space and time. An example of a body $\mathcal{B}$ is a diving board. Under external stimuli, body $\mathcal{B}$ will undergo macroscopic geometric changes, which are termed *deformations*. The geometric changes are accompanied by stresses that are induced in the body. If the applied loads are time dependent, the deformation of the body will be a function of time; that is, the geometry of the body $\mathcal{B}$ will change with time. If the loads are applied slowly so that the deformation is dependent only on the loads, the body will occupy a sequence of geometrical regions. The region occupied by the continuum at a given time t is termed a *configuration* and denoted by κ. Thus, the positions occupied in space $\Re^3$ by all material points of the continuum $\mathcal{B}$ at different instants of time are called configurations.

Suppose that the continuum initially occupies a configuration κ_0, in which a particle X occupies position $\mathbf{X}$, referred to a *reference frame* of right-handed, rectangular Cartesian axes (X_1, X_2, X_3) at a fixed origin O with orthonormal basis vectors $\hat{\mathbf{E}}_i$, as shown in Fig. 3.2.1. Note that X (lightface roman letter) is the name of the particle that occupies location $\mathbf{X}$ (boldface letter) in configuration κ_0, and therefore (X_1, X_2, X_3) are called the *material coordinates*. After the application of some external stimuli (e.g., loads), the continuum changes its geometric shape and thus assumes a new configuration κ, called the *current* or *deformed configuration*. Particle X now occupies position $\mathbf{x}$ in the deformed configuration κ, as shown in Fig. 3.2.1. The mapping $\chi : \mathcal{B}_{\kappa_0} \to \mathcal{B}_{\kappa}$ is called the *deformation mapping* of the body $\mathcal{B}$ from κ_0 to κ. The deformation mapping $\chi(\mathbf{X}, t)$ takes the position vector $\mathbf{X}$ from the reference configuration and places the same point in the deformed configuration as $\mathbf{x} = \chi(\mathbf{X}, t)$. The inverse mapping $\chi^{-1} : \mathcal{B}_{\kappa} \to \mathcal{B}_{\kappa_0}$ takes the position vector $\mathbf{x}$ from the deformed configuration κ back to the reference configuration κ_0, $\mathbf{X} = \chi^{-1}(\mathbf{x}, t)$. It is not always possible to construct the inverse mapping $\chi^{-1}(\mathbf{x}, t)$ from a known deformation mapping $\chi(\mathbf{X}, t)$. In the following discussion, we shall use $\chi(\mathbf{X}, t)$ to denote the deformation mapping and $\mathbf{x}$ to denote the value of $\chi(\mathbf{X}, t)$.

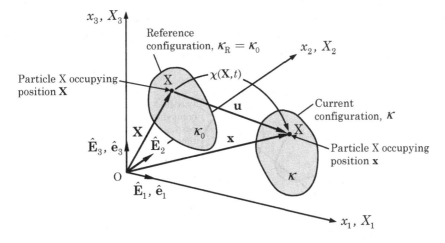

Fig. 3.2.1: Reference and deformed configurations of a body.

A nonrotating frame of reference is chosen, explicitly or implicitly, to describe the deformation. A frame of reference is a coordinate system with respect to which a configuration is described (or measured). We shall use the same coordinate system to describe reference and current configurations. The components X_i and x_i of vectors $\mathbf{X} = X_i \hat{\mathbf{E}}_i$ and $\mathbf{x} = x_i \hat{\mathbf{e}}_i$ are along the coordinates used, with the origins of the basis vectors $\hat{\mathbf{E}}_i$ and $\hat{\mathbf{e}}_i$ being the same.

The mathematical description of the deformation of a continuous body follows one of two approaches: (1) the material description or (2) the spatial description. The material description is also known as the *Lagrangian description*, and the spatial description is known as the *Eulerian description*. These descriptions are discussed next.

3.2.2 Material Description

In the material description, the motion of the body is referred to a reference configuration κ_R, which is often chosen to be the initial configuration[2], $\kappa_R = \kappa_0$, although any other known configuration can serve as a reference configuration. Thus, in the Lagrangian description, the current coordinates $\mathbf{x} \in \kappa$ are expressed in terms of the reference coordinates $\mathbf{X} \in \kappa_0$:

---

[2]Typically, the initial configuration is one without any stimuli and hence undeformed.

$$\mathbf{x} = \chi(\mathbf{X}, t), \quad \mathbf{X} = \chi(\mathbf{X}, 0), \qquad (3.2.1)$$

and the variation of a typical variable ϕ over the body is described with respect to the material coordinates $\mathbf{X}$ and time t:

$$\phi = \phi(\mathbf{x}(\mathbf{X}), t) = \phi(\mathbf{X}, t). \qquad (3.2.2)$$

For a fixed value of $\mathbf{X} \in \kappa_0$, $\phi(\mathbf{X}, t)$ gives the value of ϕ at time t associated with the fixed material particle X whose position in the reference configuration is $\mathbf{X}$, as shown in Fig. 3.2.2. Thus, a change in time t implies that the *same* material particle X, occupying position $\mathbf{X}$ in κ_0, has a different value ϕ. Figure 3.2.3 shows the deformation of a *fixed material volume* with time. Thus the attention is focused on the fixed material of the continuum.

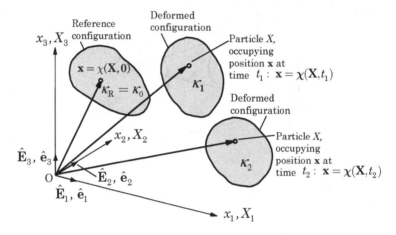

Fig. 3.2.2: Reference and deformed configurations in the material description.

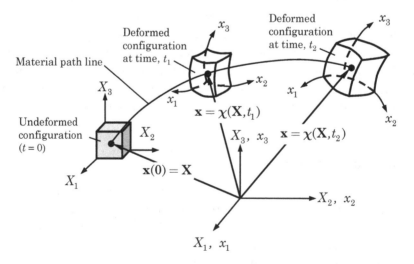

Fig. 3.2.3: Deformation of *fixed material volume* with time.

3.2.3 Spatial Description

In the spatial description, the motion is referred to the current configuration κ occupied by the body $\mathcal{B}$, and ϕ is described with respect to the current position $\mathbf{x} \in \kappa$ in space, currently occupied by material particle X:

$$\phi = \phi(\mathbf{x}, t), \quad \mathbf{X} = \mathbf{X}(\mathbf{x}, t) = \boldsymbol{\chi}^{-1}(\mathbf{x}, t). \tag{3.2.3}$$

The coordinates $\mathbf{x}$ are termed the *spatial coordinates*. For a fixed value of $\mathbf{x} \in \kappa$, $\phi(\mathbf{x}, t)$ gives the value of ϕ associated with a fixed point $\mathbf{x}$ in space, which will be the value of ϕ associated with different material points at different times, because different material points occupy position $\mathbf{x} \in \kappa$ at different times, as shown in Fig. 3.2.4. Thus, a change in time t implies that a different value ϕ is observed at the *same* spatial location $\mathbf{x} \in \kappa$, now probably occupied by a different material particle X. Hence, attention is focused on a spatial position $\mathbf{x} \in \kappa$. The notation $\mathbf{X} = \mathbf{X}(\mathbf{x}, t)$ is only symbolic because in all practical cases one is not interested (or does not know) where the material particle X comes from before it occupies the current position $\mathbf{x}$ and where it goes when it leaves the position. That is, material particles are of no interest in the spatial description.

When ϕ is known in the material description, $\phi = \phi(\mathbf{X}, t)$, its total time derivative, D/Dt, is simply the partial derivative with respect to time because the material coordinates $\mathbf{X}$ do not change with time:

$$\frac{D}{Dt}[\phi(\mathbf{X}, t)] \equiv \frac{\partial}{\partial t}[\phi(\mathbf{X}, t)]\bigg|_{\mathbf{x} \text{ fixed}} = \frac{\partial \phi}{\partial t}. \tag{3.2.4}$$

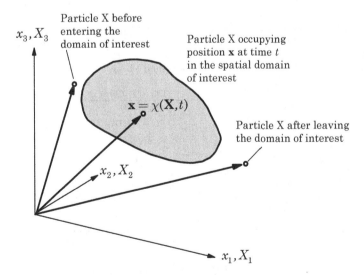

Fig. 3.2.4: Material points within and outside the spatial domain of interest in the spatial description.

However, when ϕ is known in the spatial description, $\phi = \phi(\mathbf{x}, t)$, its time derivative for a given particle, known as the *material derivative*,[3] is

$$\frac{D}{Dt}[\phi(\mathbf{x}, t)] = \frac{\partial}{\partial t}[\phi(\mathbf{x}, t)] + \frac{Dx_i}{Dt}\frac{\partial}{\partial x_i}[\phi(\mathbf{x}, t)]$$

$$= \frac{\partial \phi}{\partial t} + v_i \frac{\partial \phi}{\partial x_i} = \frac{\partial \phi}{\partial t} + \mathbf{v} \cdot \nabla \phi, \qquad (3.2.5)$$

where $\mathbf{v}$ is the velocity $\mathbf{v} = D\mathbf{x}/Dt = \dot{\mathbf{x}}$. Thus, if the velocity of a particle in the spatial description is $v(x, t)$, then the acceleration of a particle is

$$\mathbf{a} = \frac{D\mathbf{v}}{Dt} = \frac{\partial \mathbf{v}}{\partial t} + \mathbf{v} \cdot \nabla \mathbf{v} \quad \left(a_i = \frac{\partial v_i}{\partial t} + v_j \frac{\partial v_i}{\partial x_j} \right). \qquad (3.2.6)$$

Example 3.2.1 illustrates the determination of the inverse of a given mapping and computation of the material time derivative of a given function.

Example 3.2.1

Suppose that the motion of a continuous medium $\mathcal{B}$ is described by the mapping $\boldsymbol{\chi} : \kappa_0 \to \kappa$,

$$\boldsymbol{\chi}(\mathbf{X}, t) = \mathbf{x} = (X_1 + AtX_2)\hat{\mathbf{e}}_1 + (X_2 - AtX_1)\hat{\mathbf{e}}_2 + X_3\,\hat{\mathbf{e}}_3,$$

and that the temperature T in the continuum in the spatial description is given by

$$T(\mathbf{x}, t) = c_1(x_1 + c_2\,tx_2) = x_1 + tx_2,$$

where, in the interest of brevity, constants c_1 and c_2 are omitted; in SI units, $c_1 = 1$ K/m and $c_2 = 1$ /s. Determine (a) the inverse of the mapping $\boldsymbol{\chi}$, (b) the velocity components, and (c) the total time derivatives of T in the two descriptions.

Solution: A known deformation mapping $\boldsymbol{\chi}(\mathbf{X}, t)$ relates the material coordinates (X_1, X_2, X_3) to the spatial coordinates (x_1, x_2, x_3) of a particle X. In the present case, we have

$$x_1 = X_1 + At\,X_2, \quad x_2 = X_2 - At\,X_1, \quad x_3 = X_3 \quad \text{or} \quad \begin{Bmatrix} x_1 \\ x_2 \\ x_3 \end{Bmatrix} = \begin{bmatrix} 1 & At & 0 \\ -At & 1 & 0 \\ 0 & 0 & 1 \end{bmatrix} \begin{Bmatrix} X_1 \\ X_2 \\ X_3 \end{Bmatrix}. \qquad (1)$$

Clearly, the relationships between x_i and X_i are linear (that is, the mapping is linear). Therefore, polygons are mapped into polygons. In particular, a unit square is mapped into a square that is rotated in a clockwise direction, as shown in Fig. 3.2.5. This can be verified by checking where the four corner points have moved in the "deformed" body:

(X_1, X_2, X_3)		(x_1, x_2, x_3)
$(0, 0, 0)$	$\to$	$(0, 0, 0)$
$(1, 0, 0)$	$\to$	$(1, -At, 0)$
$(0, 1, 0)$	$\to$	$(At, 1, 0)$
$(1, 1, 0)$	$\to$	$(1 + At, 1 - At, 0)$

Note that, in general, the deformed square is not a unit square as the side now has a length of $1/\cos\alpha$, where $\alpha = \tan^{-1}(At)$. The reference configuration and deformed configurations at four different times, $t = 1, 2, 3$, and 4, for a value of $A = 0.25$, are shown in Fig. 3.2.6.

(a) The inverse mapping can be determined, when possible, by expressing (x_1, x_2, x_3) in terms of (X_1, X_2, X_3). In the present case, it is possible to invert the relations in Eq. (1) and obtain

$$\begin{Bmatrix} X_1 \\ X_2 \\ X_3 \end{Bmatrix} = \frac{1}{(1 + A^2t^2)} \begin{bmatrix} 1 & -At & 0 \\ At & 1 & 0 \\ 0 & 0 & 1 + A^2t^2 \end{bmatrix} \begin{Bmatrix} x_1 \\ x_2 \\ x_3 \end{Bmatrix}. \qquad (2)$$

---

[3] As opposed to d/dt, here we use Stokes's notation D/Dt for material derivative.

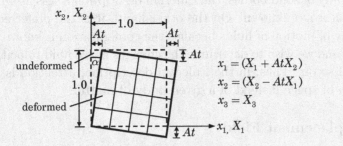

Fig. 3.2.5: A sketch of the mapping χ as applied to a unit square.

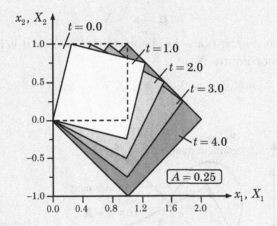

Fig. 3.2.6: Deformed configurations of the unit square at four different times ($A = 0.25$).

Therefore, we can write the inverse mapping as $\chi^{-1} : \kappa \to \kappa_0$ as

$$\chi^{-1}(\mathbf{x}, t) = \left(\frac{x_1 - Atx_2}{1 + A^2t^2} \right) \hat{\mathbf{E}}_1 + \left(\frac{x_2 + Atx_1}{1 + A^2t^2} \right) \hat{\mathbf{E}}_2 + x_3 \, \hat{\mathbf{E}}_3. \tag{3}$$

(b) The velocity vector is given by $\mathbf{v} = v_1 \hat{\mathbf{E}}_1 + v_2 \hat{\mathbf{E}}_2$, with

$$v_1 = \frac{Dx_1}{Dt} = AX_2, \quad v_2 = \frac{Dx_2}{Dt} = -AX_1. \tag{4}$$

(c) The time rate of change of temperature of a material particle in $\mathcal{B}$ is simply

$$\frac{D}{Dt}[T(\mathbf{X}, t)] = \frac{\partial}{\partial t}[T(\mathbf{X}, t)]\bigg|_{\mathbf{x} \text{ fixed}} = -2AtX_1 + (1 + A)X_2. \tag{5}$$

On the other hand, the time rate of change of temperature at point $\mathbf{x}$, which is now occupied by particle X, is

$$\frac{D}{Dt}[T(\mathbf{x}, t)] = \frac{\partial T}{\partial t} + v_i \frac{\partial T}{\partial x_i} = x_2 + v_1 \cdot 1 + v_2 \cdot t$$
$$= -2AtX_1 + (1 + A)X_2. \tag{6}$$

In the study of solid bodies, the Eulerian description is less useful because the configuration κ is unknown. On the other hand, it is the preferred description for the study of motion of fluids because the configuration is known and remains unchanged, and we wish to determine the changes in the fluid velocities, pressure, density, and so on. Thus, in the Eulerian description, attention is focused on a given region of space instead of a given body of matter.

3.2.4 Displacement Field

The phrase "deformation of a continuum" refers to relative displacements and changes in the geometry experienced by the continuum $\mathcal{B}$ under the influence of a force system. The displacement of the particle X is defined, as shown in Fig. 3.2.7, by

$$\mathbf{u} = \mathbf{x} - \mathbf{X}. \tag{3.2.7}$$

In the Lagrangian description, the displacement vector $\mathbf{u}$ is expressed in terms of the material coordinates $\mathbf{X}$:

$$\mathbf{u}(\mathbf{X}, t) = \mathbf{x}(\mathbf{X}, t) - \mathbf{X}. \tag{3.2.8}$$

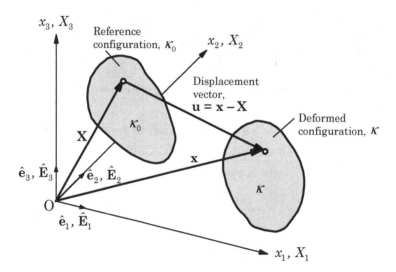

Fig. 3.2.7: Position vectors in the initial and current configurations and the displacement $\mathbf{u}$ of a particle X.

If the displacement of every particle in the body $\mathcal{B}$ is known, we can construct the current configuration κ from the reference configuration κ_0, $\chi(\mathbf{X}, t) = \mathbf{x} = \mathbf{X} + \mathbf{u}(\mathbf{X}, t)$. On the other hand, in the Eulerian description the displacements are expressed in terms of the spatial coordinates (or current position) $\mathbf{x}$:

$$\mathbf{u}(\mathbf{x}, t) = \mathbf{x} - \chi^{-1}(\mathbf{x}, t) = \mathbf{x} - \mathbf{X}(\mathbf{x}, t). \tag{3.2.9}$$

To see the difference between the two descriptions further, consider the one-dimensional mapping $\chi(X, t) = x = X(1 + 0.5t)$ defining the motion of a rod of initial length two units. The inverse mapping is $\chi^{-1}(x, t) = X = x/(1 + 0.5t)$. The rod experiences a temperature distribution T given by the material description $T = 2t^2 X$ or by the spatial description $T = 2t^2 x/(1+0.5t)$, as shown in Fig. 3.2.8 [see Bonet and Wood (2008)].

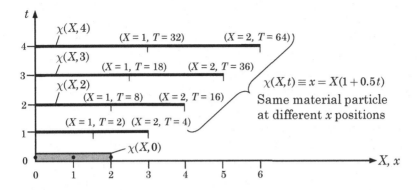

Fig. 3.2.8: Material and spatial descriptions of motion.

From Fig. 3.2.8, we see that the particle's material coordinate X remains associated with the particle while its spatial position x changes. The temperature at a given time can be found in one of the two ways: for example, at time $t = 3$ the temperature of the particle labeled 2 with material coordinate $X = 2$ is $T = 2 \times 2(3)^2 = 36$; alternatively, the temperature of the same particle, which at $t = 3$ is at a spatial position $x = 2(1 + 0.5 \times 3) = 5$, is $T = 2 \times 5(3)^2/(1 + 0.5 \times 3) = 36$. The displacement of a material point X occupying position X in κ_0 is

$$u(X, t) = x - X = X(1 + 0.5t) - X = 0.5Xt.$$

A *rigid body* is one in which the distance between any two material particles remains the same whereas a *deformable body* is one in which the material particles can move relative to each other under the action of external stimuli. A rigid-body motion is one in which *all* material particles of the body undergo the same displacement. Then the deformation of a continuum can be determined only by considering the change of distance between any two arbitrary but infinitesimally close points of the continuum.

3.3 Analysis of Deformation

3.3.1 Deformation Gradient

One of the key quantities in deformation analysis is the *deformation gradient* of κ relative to the reference configuration κ_0, denoted $\mathbf{F}_\kappa$, which provides the relationship between a material line $d\mathbf{X}$ before deformation and the line $d\mathbf{x}$, consisting of the same material as $d\mathbf{X}$ after deformation. It is defined as follows (in the interest of brevity, the subscript κ on $\mathbf{F}$ is dropped):

$$d\mathbf{x} = \mathbf{F} \cdot d\mathbf{X} = d\mathbf{X} \cdot \mathbf{F}^{\mathrm{T}}, \tag{3.3.1}$$

$$\mathbf{F} = \left(\frac{\partial \boldsymbol{\chi}}{\partial \mathbf{X}}\right)^{\mathrm{T}} = \left(\frac{\partial \mathbf{x}}{\partial \mathbf{X}}\right)^{\mathrm{T}} \equiv \mathbf{x}\overleftarrow{\nabla}_0 = (\nabla_0 \mathbf{x})^{\mathrm{T}}. \tag{3.3.2}$$

Here $\overleftarrow{\nabla}$ denotes the backward gradient operator (see the note at the end of Section 2.4.5) and ∇ is the forward gradient operator with respect to $\mathbf{x}$. The subscript 0 on the del operator denotes that the differentiation is with respect to $\mathbf{X}$ (the material coordinates). Many authors [see, e.g., Gurtin (1981); Bonet and Wood (2008); and Gurtin, Fried, and Anand (2010)] use ∇ in defining the deformation gradient, but a close look reveals that they actually mean the backward gradient operator discussed in Section 2.4.5.

By definition, $\mathbf{F}$ is a function of both position $\mathbf{X}$ and time t; $\mathbf{F}$ is sometimes referred to as a *two-point tensor*[4] (or a linear transformation of points in the small neighborhood of $\mathbf{X}$ from κ_0 into the neighborhood of $\mathbf{x}$ in κ) because it describes the local deformation of a material line element at point $\mathbf{X}$ in the reference configuration κ_0 to the point $\mathbf{x}$ in the current configuration κ; $\mathbf{F}$ involves, in general, both stretch and rotation. For example, in the case of pure stretch followed by rotation, the deformation of $d\mathbf{X}$ into $d\mathbf{X}'$ involves only pure stretch and the deformation from $d\mathbf{X}'$ into $d\mathbf{x}$ involves only pure rotation (although, in reality, stretch and rotation occur simultaneously). Thus, we can write $d\mathbf{X}' = \mathbf{U} \cdot d\mathbf{X}$ and $d\mathbf{x} = \mathbf{R} \cdot d\mathbf{X}'$, where $\mathbf{U}$ is a stretch tensor and $\mathbf{R}$ is a *proper* orthogonal tensor, $\mathbf{R}^{\mathrm{T}} \cdot \mathbf{R} = \mathbf{I}$ and $|\mathbf{R}| = 1$. The tensors $\mathbf{U}$ and $\mathbf{R}$ can be interpreted as linear transformations, $\mathbf{U} : d\mathbf{X} \rightarrow d\mathbf{X}'$ and $\mathbf{R} : d\mathbf{X}' \rightarrow d\mathbf{x}$. In other words, the linear transformation $\mathbf{F} : d\mathbf{X} \rightarrow d\mathbf{x}$ is replaced by the composite transformation $\mathbf{F} = \mathbf{R} \cdot \mathbf{U}$. The requirement that $\mathbf{R}$ be a proper orthogonal transformation follows from the fact that a pure rotation should not change the length of the line element $d\mathbf{X}'$:

$$d\mathbf{x} \cdot d\mathbf{x} = (\mathbf{R} \cdot d\mathbf{X}') \cdot (\mathbf{R} \cdot d\mathbf{X}') = d\mathbf{X}' \cdot \mathbf{R}^{\mathrm{T}} \cdot \mathbf{R} \cdot d\mathbf{X}' = d\mathbf{X}' \cdot d\mathbf{X}'.$$

The multiplicative decomposition of $\mathbf{F}$ into pure stretch $\mathbf{U}$ and pure rotation $\mathbf{R}$ is discussed further in Section 3.9 on polar decomposition.

In index notation, Eq. (3.3.2) can be written as

$$\mathbf{F} = F_{iJ}\,\hat{\mathbf{e}}_i\,\hat{\mathbf{E}}_J, \qquad F_{iJ} = \frac{\partial x_i}{\partial X_J}. \tag{3.3.3}$$

More explicitly, we have

$$[F] = \begin{bmatrix} \dfrac{\partial x_1}{\partial X_1} & \dfrac{\partial x_1}{\partial X_2} & \dfrac{\partial x_1}{\partial X_3} \\[2mm] \dfrac{\partial x_2}{\partial X_1} & \dfrac{\partial x_2}{\partial X_2} & \dfrac{\partial x_2}{\partial X_3} \\[2mm] \dfrac{\partial x_3}{\partial X_1} & \dfrac{\partial x_3}{\partial X_2} & \dfrac{\partial x_3}{\partial X_3} \end{bmatrix}, \tag{3.3.4}$$

where the lowercase indices refer to the current (spatial) Cartesian coordinates, whereas uppercase indices refer to the reference (material) Cartesian coordinates. The determinant of $[F]$ is called the *Jacobian of the motion*, and it is denoted by $J = |F|$. The equation $\mathbf{F} \cdot d\mathbf{X} = \mathbf{0}$ for $d\mathbf{X} \neq \mathbf{0}$ implies that a material line in the reference configuration is reduced to zero by the deformation. Since this is physically not realistic, we conclude that $\mathbf{F} \cdot d\mathbf{X} \neq \mathbf{0}$ for $d\mathbf{X} \neq \mathbf{0}$. That

---

[4]Strictly speaking, $\mathbf{F}$ is not a tensor because its components do not transform like those of a tensor.

is, $\mathbf{F}$ is a nonsingular tensor, $J \neq 0$. Hence, $\mathbf{F}$ has an inverse $\mathbf{F}^{-1}$, and we can write

$$dX = \mathbf{F}^{-1} \cdot dx = dx \cdot \mathbf{F}^{-T}, \quad \text{where} \quad \mathbf{F}^{-T} = \frac{\partial \mathbf{X}}{\partial \mathbf{x}} \equiv \nabla \mathbf{X}, \tag{3.3.5}$$

and in index notation

$$\mathbf{F}^{-1} = F_{Ji}^{-1} \hat{\mathbf{E}}_J \hat{\mathbf{e}}_i, \quad F_{Ji}^{-1} = \frac{\partial X_J}{\partial x_i}. \tag{3.3.6}$$

In explicit form the matrix associated with $\mathbf{F}^{-1}$ is

$$[F]^{-1} = \begin{bmatrix} \frac{\partial X_1}{\partial x_1} & \frac{\partial X_1}{\partial x_2} & \frac{\partial X_1}{\partial x_3} \\ \frac{\partial X_2}{\partial x_1} & \frac{\partial X_2}{\partial x_2} & \frac{\partial X_2}{\partial x_3} \\ \frac{\partial X_3}{\partial x_1} & \frac{\partial X_3}{\partial x_2} & \frac{\partial X_3}{\partial x_3} \end{bmatrix}. \tag{3.3.7}$$

Note that $[F]$ stretches or compresses a material vector $\{X\}$ into the current vector $\{x\}$, while χ gives the current position $\mathbf{x} \in \kappa$ of the material point X that occupied position $\mathbf{X} \in \kappa_0$. The deformation gradient and its inverse can be expressed in terms of the displacement vector as

$$\mathbf{F} = (\nabla_0 \mathbf{x})^{\mathrm{T}} = (\nabla_0 \mathbf{u} + \mathbf{I})^{\mathrm{T}} \quad \text{or} \quad \mathbf{F}^{-1} = (\nabla \mathbf{X})^{\mathrm{T}} = (\mathbf{I} - \nabla \mathbf{u})^{\mathrm{T}}. \tag{3.3.8}$$

Example 3.3.1 illustrates computation of the components of the deformation gradient and the displacement vector from known mapping of motion.

Example 3.3.1 ———————————————————————

Consider the uniform deformation of a square block of side 2 units and initially centered at $\mathbf{X} = (0,0)$. If the deformation is defined by the mapping

$$\chi(\mathbf{X}) = (3.5 + X_1 + 0.5 X_2)\,\hat{\mathbf{e}}_1 + (4 + X_2)\,\hat{\mathbf{e}}_2 + X_3\,\hat{\mathbf{e}}_3,$$

(a) sketch the deformation, (b) determine the deformation gradient $\mathbf{F}$, and (c) compute the displacements.

Solution: (a) From the given mapping, we have $x_1 = 3.5 + X_1 + 0.5 X_2$, $x_2 = 4 + X_2$, and $x_3 = X_3$; in matrix form, we have

$$\begin{Bmatrix} x_1 \\ x_2 \\ x_3 \end{Bmatrix} = \begin{bmatrix} 1.0 & 0.5 & 0.0 \\ 0.0 & 1.0 & 0.0 \\ 0.0 & 0.0 & 1.0 \end{bmatrix} \begin{Bmatrix} X_1 \\ X_2 \\ X_3 \end{Bmatrix} + \begin{Bmatrix} 3.5 \\ 4.0 \\ 0.0 \end{Bmatrix}. \tag{1}$$

These are linear relations and therefore the square is mapped, in general, into a parallelogram. To see where the corner points of the square are mapped to, apply the above equations to the corner points (no change in X_3):

(X_1, X_2)		(x_1, x_2)
$(-1, -1)$	$\rightarrow$	$(2, 3)$
$(\ 1, -1)$	$\rightarrow$	$(4, 3)$
$(\ 1,\ 1)$	$\rightarrow$	$(5, 5)$
$(-1,\ 1)$	$\rightarrow$	$(3, 5)$

Thus, under the mapping the square moved and became a parallelogram, centered at $(3.5, 4)$, as shown in Fig. 3.3.1. The base and height of the parallelogram remained 2 units each.

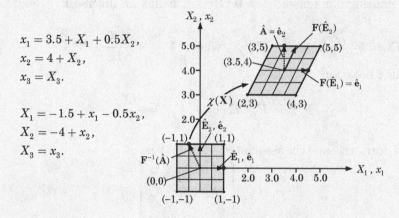

$$x_1 = 3.5 + X_1 + 0.5X_2,$$
$$x_2 = 4 + X_2,$$
$$x_3 = X_3.$$

$$X_1 = -1.5 + x_1 - 0.5x_2,$$
$$X_2 = -4 + x_2,$$
$$X_3 = x_3.$$

Fig. 3.3.1: Uniform deformation of a square block of material.

The linear relations in Eq. (1) can be inverted to obtain X_1, X_2, and X_3 in terms of x_1, x_2, and x_3:

$$\begin{Bmatrix} X_1 \\ X_2 \\ X_3 \end{Bmatrix} = \begin{bmatrix} 1.0 & -0.5 & 0.0 \\ 0.0 & 1.0 & 0.0 \\ 0.0 & 0.0 & 1.0 \end{bmatrix} \left(\begin{Bmatrix} x_1 \\ x_2 \\ x_3 \end{Bmatrix} - \begin{Bmatrix} 3.5 \\ 4.0 \\ 0.0 \end{Bmatrix} \right), \tag{2}$$

or $X_1 = -1.5 + x_1 - 0.5x_2$, $X_2 = -4 + x_2$, and $X_3 = x_3$. Thus, the inverse mapping is

$$\chi^{-1}(\mathbf{x}) = \left(-1.5 + x_1 - 0.5x_2\right)\hat{\mathbf{E}}_1 + \left(-4 + x_2\right)\hat{\mathbf{E}}_2 + x_3\,\hat{\mathbf{E}}_3, \tag{3}$$

which recovers the square shape from the parallelogram shape shown in Fig. 3.3.1. This type of deformation is known as *simple shear*, in which there exists a set of line elements, in the present case, lines parallel to the X_1-axis, whose orientation is such that they are unchanged in length and orientation by the deformation.

(b) The components of the deformation gradient and its inverse in matrix form are

$$[F] = \begin{bmatrix} \frac{\partial x_1}{\partial X_1} & \frac{\partial x_1}{\partial X_2} & \frac{\partial x_1}{\partial X_3} \\ \frac{\partial x_2}{\partial X_1} & \frac{\partial x_2}{\partial X_2} & \frac{\partial x_2}{\partial X_3} \\ \frac{\partial x_3}{\partial X_1} & \frac{\partial x_3}{\partial X_2} & \frac{\partial x_3}{\partial X_3} \end{bmatrix} = \begin{bmatrix} 1.0 & 0.5 & 0.0 \\ 0.0 & 1.0 & 0.0 \\ 0.0 & 0.0 & 1.0 \end{bmatrix}, \quad [F]^{-1} = \begin{bmatrix} \frac{\partial X_1}{\partial x_1} & \frac{\partial X_1}{\partial x_2} & \frac{\partial X_1}{\partial x_3} \\ \frac{\partial X_2}{\partial x_1} & \frac{\partial X_2}{\partial x_2} & \frac{\partial X_2}{\partial x_3} \\ \frac{\partial X_3}{\partial x_1} & \frac{\partial X_3}{\partial x_2} & \frac{\partial X_3}{\partial x_3} \end{bmatrix} = \begin{bmatrix} 1.0 & -0.5 & 0.0 \\ 0.0 & 1.0 & 0.0 \\ 0.0 & 0.0 & 1.0 \end{bmatrix}.$$

The unit vectors $\hat{\mathbf{E}}_1$ and $\hat{\mathbf{E}}_2$ in the initial configuration deform to the lengths

$$\begin{bmatrix} 1.0 & 0.5 & 0.0 \\ 0.0 & 1.0 & 0.0 \\ 0.0 & 0.0 & 1.0 \end{bmatrix} \begin{Bmatrix} 1 \\ 0 \\ 0 \end{Bmatrix} = \begin{Bmatrix} 1 \\ 0 \\ 0 \end{Bmatrix}, \quad \begin{bmatrix} 1.0 & 0.5 & 0.0 \\ 0.0 & 1.0 & 0.0 \\ 0.0 & 0.0 & 1.0 \end{bmatrix} \begin{Bmatrix} 0 \\ 1 \\ 0 \end{Bmatrix} = \begin{Bmatrix} 0.5 \\ 1.0 \\ 0.0 \end{Bmatrix}.$$

The unit vector $\hat{\mathbf{A}} = \hat{\mathbf{e}}_2$ in the current configuration is deformed from the vector

$$\begin{bmatrix} 1.0 & -0.5 & 0.0 \\ 0.0 & 1.0 & 0.0 \\ 0.0 & 0.0 & 1.0 \end{bmatrix} \begin{Bmatrix} 0 \\ 1 \\ 0 \end{Bmatrix} = \begin{Bmatrix} -0.5 \\ 1.0 \\ 0.0 \end{Bmatrix} \quad \Rightarrow \quad \mathbf{F}^{-1}(\hat{\mathbf{A}}) = -0.5\hat{\mathbf{E}}_1 + \hat{\mathbf{E}}_2.$$

(c) The displacement vector is given by

$$\mathbf{u} = \mathbf{x} - \mathbf{X} = \left(3.5 + 0.5X_2\right)\mathbf{e}_1 + 4\,\mathbf{e}_2,$$

which is independent of X_1. The displacement components are $u_1 = 3.5 + 0.5X_2$, $u_2 = 4$, and $u_3 = 0$. Thus, a $X_2 =$ constant line moved 4 units up and $3.5 + 0.5X_2$ units to the right.

3.3.2 Isochoric, Homogeneous, and Inhomogeneous Deformations

3.3.2.1 Isochoric deformation

If the Jacobian is unity $J = 1$, then the deformation is volume-preserving or the current and reference configurations coincide. If volume does not change locally during the deformation, the deformation is said to be *isochoric* at $\mathbf{X}$. If $J = 1$ everywhere in the body $\mathcal{B}$, then the deformation of the body is isochoric.

3.3.2.2 Homogeneous deformation

In general, the deformation gradient $\mathbf{F}$ is a function of $\mathbf{X}$. If $\mathbf{F} = \mathbf{I}$ everywhere in the body, then the body is not rotated but might be rigidly translated. If $\mathbf{F}$ has the same value at every material point in a body (that is, $\mathbf{F}$ is independent of $\mathbf{X}$), then the mapping $\mathbf{x} = \mathbf{x}(\mathbf{X}, t)$ is said to be a *homogeneous motion* of the body and the deformation is said to be homogeneous. In general, at any given time $t > 0$, a mapping $\mathbf{x} = \mathbf{x}(\mathbf{X}, t)$ is said to be a homogeneous motion if and only if it can be expressed as (so that $\mathbf{F}$ is a constant tensor)

$$\mathbf{x} = \mathbf{A} \cdot \mathbf{X} + \mathbf{c}, \tag{3.3.9}$$

where the second-order tensor $\mathbf{A}$ and vector $\mathbf{c}$ are functions of time t only; $\mathbf{c}$ represents a rigid-body translation. Note that for a homogeneous motion we have $\mathbf{F} = \mathbf{A}$. Clearly, the motion described by the mapping of Example 3.3.1 is homogeneous and isochoric. Next, we consider several simple forms of homogeneous deformations.

Pure dilatation. Consider a cube of material with edges of length L and ℓ in the reference and current configurations, respectively. If the deformation mapping has the form (see Fig. 3.3.2)

$$\boldsymbol{\chi}(\mathbf{X}) = \lambda X_1 \, \hat{\mathbf{e}}_1 + \lambda X_2 \, \hat{\mathbf{e}}_2 + \lambda X_3 \, \hat{\mathbf{e}}_3, \quad \lambda = \frac{L}{\ell}, \tag{3.3.10}$$

then $\mathbf{F}$ has the matrix representation

$$[F] = \begin{bmatrix} \lambda & 0 & 0 \\ 0 & \lambda & 0 \\ 0 & 0 & \lambda \end{bmatrix}. \tag{3.3.11}$$

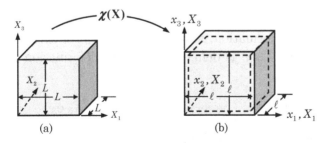

Fig. 3.3.2: A deformation mapping of pure dilatation.

This deformation is known as pure dilatation or pure stretch, and it is isochoric if and only if $\lambda = 1$ (λ is called the principal stretch), as shown in Fig. 3.3.2.

Simple extension. An example of homogeneous extension in the X_1-direction, as shown in Fig. 3.3.3, is provided by the deformation mapping

$$\boldsymbol{\chi}(\mathbf{X}) = (1 + \alpha)X_1\,\hat{\mathbf{e}}_1 + X_2\,\hat{\mathbf{e}}_2 + X_3\,\hat{\mathbf{e}}_3. \tag{3.3.12}$$

The inverse mapping is [because $x_1 = (1 + \alpha)X_1$, $x_2 = X_2$, and $x_3 = X_3$]

$$\boldsymbol{\chi}^{-1}(\mathbf{x}) = \frac{1}{(1 + \alpha)}x_1\,\hat{\mathbf{e}}_1 + x_2\,\hat{\mathbf{e}}_2 + x_3\,\hat{\mathbf{e}}_3.$$

The matrices of the deformation gradient and its inverse are

$$[F] = \begin{bmatrix} 1+\alpha & 0 & 0 \\ 0 & 1 & 0 \\ 0 & 0 & 1 \end{bmatrix}, \quad [F]^{-1} = \frac{1}{(1+\alpha)}\begin{bmatrix} 1 & 0 & 0 \\ 0 & 1+\alpha & 0 \\ 0 & 0 & 1+\alpha \end{bmatrix}. \tag{3.3.13}$$

For example, a line $X_2 = a + bX_1$ in the initial configuration κ_0 transforms under the mapping to the line

$$x_2 = X_2 = a + bX_1 = a + \frac{b}{1+\alpha}x_1$$

in the current configuration κ.

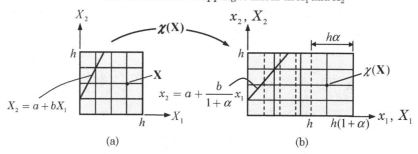

Fig. 3.3.3: A deformation mapping of simple extension. Typical material lines inside the body are also shown.

Simple shear. This deformation, also known as *uniform shear deformation*, as discussed in Example 3.3.1, is defined by a linear deformation mapping of the form (see Fig. 3.3.4)

$$\boldsymbol{\chi}(\mathbf{X}) = (X_1 + \gamma X_2)\hat{\mathbf{e}}_1 + X_2\,\hat{\mathbf{e}}_2 + X_3\,\hat{\mathbf{e}}_3, \tag{3.3.14}$$

where material planes $X_2 = $ constant slide in the X_1-direction in linear proportion to X_2, the proportionality constant being γ, which is a measure of the amount of shear. The planes $x_2 = $ constant are the shear planes and the direction along x_1 is the shear direction. The matrix representation of the deformation gradient in this case is

$$[F] = \begin{bmatrix} 1 & \gamma & 0 \\ 0 & 1 & 0 \\ 0 & 0 & 1 \end{bmatrix}. \tag{3.3.15}$$

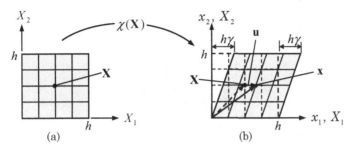

The deformation mapping is linear in X_1 and X_2

(a) (b)

Fig. 3.3.4: A deformation mapping of simple shear. Typical material lines inside the body are also shown.

3.3.2.3 Nonhomogeneous deformation

A nonhomogeneous deformation is one in which the deformation gradient **F** is a function of **X**. An example of nonhomogeneous deformation mapping is provided, as shown in Fig. 3.3.5, by

$$\chi(\mathbf{X}) = X_1(1 + \gamma_1 X_2)\hat{\mathbf{e}}_1 + X_2(1 + \gamma_2 X_1)\hat{\mathbf{e}}_2 + X_3\,\hat{\mathbf{e}}_3. \tag{3.3.16}$$

The matrix representation of the deformation gradient is

$$[F] = \begin{bmatrix} 1 + \gamma_1 X_2 & \gamma_1 X_1 & 0 \\ \gamma_2 X_2 & 1 + \gamma_2 X_1 & 0 \\ 0 & 0 & 1 \end{bmatrix}. \tag{3.3.17}$$

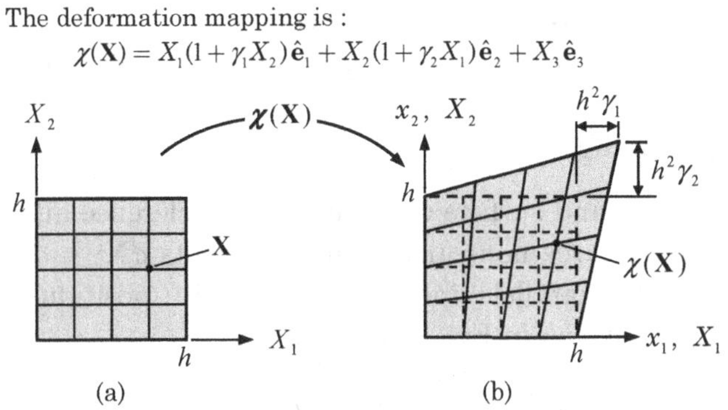

The deformation mapping is :
$$\chi(\mathbf{X}) = X_1(1 + \gamma_1 X_2)\,\hat{\mathbf{e}}_1 + X_2(1 + \gamma_2 X_1)\,\hat{\mathbf{e}}_2 + X_3\,\hat{\mathbf{e}}_3$$

(a) (b)

Fig. 3.3.5: A deformation mapping of combined shearing and extension. Typical material lines inside the body are also shown.

It is rather difficult to invert the mapping even for this simple nonhomogeneous deformation.

Figure 3.3.6 shows the deformed configuration for the values of $h = 1$ and $\gamma_2 = 3\gamma_1 = 3\gamma = 3$. Note that the straight line AC (that is, line $X_2 = X_1$) in the undeformed configuration becomes a curve in the deformed configuration, although the edges of the deformed configuration remain as straight lines.

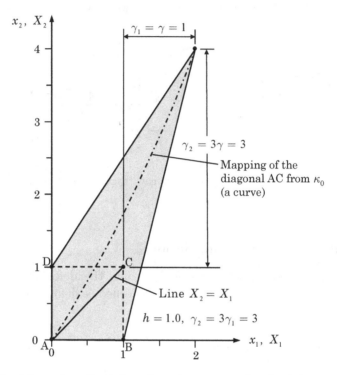

Fig. 3.3.6: The deformed configuration of a unit square under the deformation mapping in Eq. (3.3.16) for $\gamma_1 = 1$ and $\gamma_2 = 3$.

3.3.3 Change of Volume and Surface

Here we study how deformation mapping affects surface areas and volumes of a continuum. The motivation for this study comes from the need to write global equilibrium statements that involve integrals over areas and volumes.

3.3.3.1 Volume change

We can define volume and surface elements in the reference and deformed configurations. Consider three noncoplanar line elements $d\mathbf{X}^{(1)}$, $d\mathbf{X}^{(2)}$, and $d\mathbf{X}^{(3)}$ forming the edges of a parallelepiped at point P with position vector $\mathbf{X}$ in the reference body $\mathcal{B}$, as shown in Fig. 3.3.7, so that

$$d\mathbf{x}^{(i)} = \mathbf{F} \cdot d\mathbf{X}^{(i)}, \quad i = 1, 2, 3. \tag{3.3.18}$$

Note that the vectors $d\mathbf{x}^{(i)}$ are not necessarily parallel to or have the same length as the vectors $d\mathbf{X}^{(i)}$ due to shearing and stretching of the parallelepiped. We assume that the triad $(d\mathbf{X}^{(1)}, d\mathbf{X}^{(2)}, d\mathbf{X}^{(3)})$ is positively oriented in the sense that the triple scalar product $d\mathbf{X}^{(1)} \cdot d\mathbf{X}^{(2)} \times d\mathbf{X}^{(3)} > 0$. We denote the volume of the parallelepiped as

$$dV = d\mathbf{X}^{(1)} \cdot d\mathbf{X}^{(2)} \times d\mathbf{X}^{(3)} = \left(\hat{\mathbf{N}}_1 \cdot \hat{\mathbf{N}}_2 \times \hat{\mathbf{N}}_3 \right) dX^{(1)} dX^{(2)} dX^{(3)}$$
$$= dX^{(1)} dX^{(2)} dX^{(3)}, \tag{3.3.19}$$

where $\hat{\mathbf{N}}_i$ denotes the unit vector along $d\mathbf{X}^{(i)}$. The corresponding volume in the deformed configuration is given by

$$dv = d\mathbf{x}^{(1)} \cdot d\mathbf{x}^{(2)} \times d\mathbf{x}^{(3)} = \left(\mathbf{F} \cdot \hat{\mathbf{N}}_1 \right) \cdot \left(\mathbf{F} \cdot \hat{\mathbf{N}}_2 \right) \times \left(\mathbf{F} \cdot \hat{\mathbf{N}}_3 \right) dX^{(1)} dX^{(2)} dX^{(3)},$$

or

$$dv = |F| dX^{(1)} dX^{(2)} dX^{(3)} = J \, dV. \tag{3.3.20}$$

We assume that the volume elements are positive so that the relative orientation of the line elements is preserved under the deformation, that is, $J > 0$. Thus, J has the physical meaning of being the local ratio of current to reference volume of a material volume element.

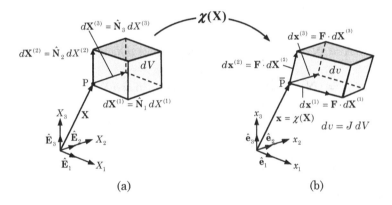

Fig. 3.3.7: Transformation of a volume element under a deformation mapping.

3.3.3.2 Area change

Next, consider an infinitesimal vector element of material surface $d\mathbf{A}$ in a neighborhood of the point $\mathbf{X}$ in the reference configuration, as shown in Fig. 3.3.8. The surface vector can be expressed as $d\mathbf{A} = dA\,\hat{\mathbf{N}}$, where $\hat{\mathbf{N}}$ is the positive unit normal to the surface in the reference configuration.

Suppose that $d\mathbf{A}$ from the reference configuration becomes $d\mathbf{a}$ in the current configuration, where $d\mathbf{a} = da\,\hat{\mathbf{n}}$, $\hat{\mathbf{n}}$ being the outward unit normal to the surface in the current configuration. The outward unit normals in the reference and current configurations can be expressed as

$$\hat{\mathbf{N}} = \frac{\hat{\mathbf{N}}_1 \times \hat{\mathbf{N}}_2}{|\hat{\mathbf{N}}_1 \times \hat{\mathbf{N}}_2|}, \quad \hat{\mathbf{n}} = \frac{\mathbf{F} \cdot \hat{\mathbf{n}}_1 \times \mathbf{F} \cdot \hat{\mathbf{n}}_2}{|\mathbf{F} \cdot \hat{\mathbf{n}}_1 \times \mathbf{F} \cdot \hat{\mathbf{n}}_2|}. \tag{3.3.21}$$

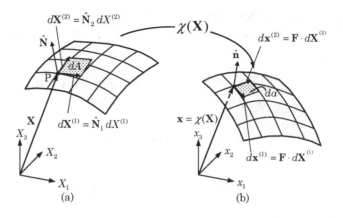

Fig. 3.3.8: Transformation of a surface element under a deformation mapping.

The areas of the parallelograms in the reference and current configurations are

$$dA \equiv |\hat{\mathbf{N}}_1 \times \hat{\mathbf{N}}_2| \, dX_1 \, dX_2, \quad da \equiv |\mathbf{F} \cdot \hat{\mathbf{n}}_1 \times \mathbf{F} \cdot \hat{\mathbf{n}}_2| \, dx_1 \, dx_2. \tag{3.3.22}$$

The area vectors are

$$dA = \hat{\mathbf{N}} \, dA = \frac{\hat{\mathbf{N}}_1 \times \hat{\mathbf{N}}_2}{|\hat{\mathbf{N}}_1 \times \hat{\mathbf{N}}_2|} \, |\hat{\mathbf{N}}_1 \times \hat{\mathbf{N}}_2| \, dX_1 \, dX_2$$

$$= \left(\hat{\mathbf{N}}_1 \times \hat{\mathbf{N}}_2\right) dX_1 \, dX_2 = \left(\hat{\mathbf{n}}_1 \times \hat{\mathbf{n}}_2\right) dX_1 \, dX_2, \tag{3.3.23}$$

$$da = \hat{\mathbf{n}} \, da = \frac{\mathbf{F} \cdot \hat{\mathbf{n}}_1 \times \mathbf{F} \cdot \hat{\mathbf{n}}_2}{|\mathbf{F} \cdot \hat{\mathbf{n}}_1 \times \mathbf{F} \cdot \hat{\mathbf{n}}_2|} \, |\mathbf{F} \cdot \hat{\mathbf{n}}_1 \times \mathbf{F} \cdot \hat{\mathbf{n}}_2| \, dx_1 \, dx_2$$

$$= \left(\mathbf{F} \cdot \hat{\mathbf{n}}_1 \times \mathbf{F} \cdot \hat{\mathbf{n}}_2\right) dx_1 \, dx_2. \tag{3.3.24}$$

Then it can be shown that (see the result of Problem **3.16**)

$$da = J \mathbf{F}^{-T} \cdot d\mathbf{A} \quad \text{or} \quad \hat{\mathbf{n}} \, da = J \mathbf{F}^{-T} \cdot \hat{\mathbf{N}} \, dA. \tag{3.3.25}$$

3.4 Strain Measures

3.4.1 Cauchy–Green Deformation Tensors

The geometric changes that a continuous medium experiences can be measured in a number of ways. Here, we discuss a general measure of deformation of a continuous medium, independent of both translation and rotation.

Consider two material particles P and Q in the neighborhood of each other, separated by $d\mathbf{X}$ in the reference configuration, as shown in Fig. 3.4.1. In the current (deformed) configuration the material points P and Q occupy positions $\bar{\mathrm{P}}$ and $\bar{\mathrm{Q}}$, and they are separated by $d\mathbf{x}$. We wish to determine the change in the distance $d\mathbf{X}$ between the material points P and Q as the body deforms and the material points move to the new locations $\bar{\mathrm{P}}$ and $\bar{\mathrm{Q}}$.

The distances between points P and Q and points $\bar{\text{P}}$ and $\bar{\text{Q}}$ are given, respectively, by

$$(dS)^2 = d\mathbf{X} \cdot d\mathbf{X}, \tag{3.4.1}$$

$$(ds)^2 = d\mathbf{x} \cdot d\mathbf{x} = d\mathbf{X} \cdot (\mathbf{F}^{\mathrm{T}} \cdot \mathbf{F}) \cdot d\mathbf{X} \equiv d\mathbf{X} \cdot \mathbf{C} \cdot d\mathbf{X}, \tag{3.4.2}$$

where $\mathbf{C}$ is called the *right Cauchy–Green deformation tensor*

$$\mathbf{C} = \mathbf{F}^{\mathrm{T}} \cdot \mathbf{F}. \tag{3.4.3}$$

By definition, $\mathbf{C}$ is a symmetric second-order tensor. The *left Cauchy–Green deformation tensor* or *Finger tensor* is defined by

$$\mathbf{B} = \mathbf{F} \cdot \mathbf{F}^{\mathrm{T}}, \tag{3.4.4}$$

which is also a symmetric tensor.

Recall from Eq. (2.4.25) that the directional (or tangential) derivative of a field $\phi(\mathbf{X})$ is given by

$$\frac{d\phi}{dS} = \hat{\mathbf{N}} \cdot \boldsymbol{\nabla}_0 \phi, \quad \hat{\mathbf{N}} = \frac{d\mathbf{X}}{|d\mathbf{X}|} = \frac{d\mathbf{X}}{dS}, \tag{3.4.5}$$

where $\hat{\mathbf{N}}$ is the unit vector in the direction of $d\mathbf{X}$ at point $\mathbf{X}$. Therefore, a parameterized curve in the deformed configuration is determined by the deformation mapping $\mathbf{x}(S) = \boldsymbol{\chi}(\mathbf{X}(S))$, and we have ($\mathbf{F} = F_{iJ}\,\hat{\mathbf{e}}_i\,\hat{\mathbf{E}}_J$ and $\hat{\mathbf{N}} = N_K\hat{\mathbf{E}}_K$)

$$\frac{d\mathbf{x}}{dS} = \frac{d\mathbf{X}}{dS} \cdot \boldsymbol{\nabla}_0\boldsymbol{\chi}(\mathbf{X}) = \mathbf{F} \cdot \frac{d\mathbf{X}}{dS} = \mathbf{F} \cdot \hat{\mathbf{N}} = F_{iJ}N_J\,\hat{\mathbf{e}}_i. \tag{3.4.6}$$

Note that $d\mathbf{x}/dS = F_{iJ}N_J\,\hat{\mathbf{e}}_i$ is a vector defined in the current (deformed) configuration because $\hat{\mathbf{e}}_i$ is a unit vector in the current configuration.

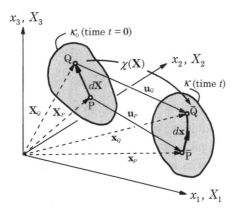

Fig. 3.4.1: Points P and Q separated by a distance $d\mathbf{X}$ in the reference configuration κ_0 take up positions $\bar{\text{P}}$ and $\bar{\text{Q}}$, respectively, in the deformed configuration κ, where they are separated by distance $d\mathbf{x}$.

The *stretch* of a curve at a point in the deformed configuration is defined as the ratio of the deformed length of the curve to its original length, $\lambda = ds/dS$. Let us consider an infinitesimal length dS of the curve in the neighborhood of the material point $\mathbf{X}$. Then the stretch λ of the curve is simply the length of the tangent vector $\mathbf{F} \cdot \hat{\mathbf{N}}$ in the deformed configuration

$$\lambda^2(S) = (\mathbf{F} \cdot \hat{\mathbf{N}}) \cdot (\mathbf{F} \cdot \hat{\mathbf{N}}) \tag{3.4.7}$$

$$= \hat{\mathbf{N}} \cdot (\mathbf{F}^{\mathrm{T}} \cdot \mathbf{F}) \cdot \hat{\mathbf{N}} = \hat{\mathbf{N}} \cdot \mathbf{C} \cdot \hat{\mathbf{N}}. \tag{3.4.8}$$

Equation (3.4.8) holds for any arbitrary curve with $d\mathbf{X} = dS\,\hat{\mathbf{N}}$, and thus allows us to compute the stretch in any direction at a given point. In particular, the square of the stretch in the direction of the unit base vector $\hat{\mathbf{E}}_I$ is given by

$$\lambda^2(\hat{\mathbf{E}}_I) = \hat{\mathbf{E}}_I \cdot \mathbf{C} \cdot \hat{\mathbf{E}}_I = C_{II} \quad \text{(no sum on } I). \tag{3.4.9}$$

That is, the diagonal terms of the right Cauchy–Green deformation tensor $\mathbf{C}$ represent the squares of the stretches in the direction of the coordinate axes (X_1, X_2, X_3). The off-diagonal elements of $\mathbf{C}$ give a measure of the angle of shearing between two base vectors $\hat{\mathbf{E}}_I$ and $\hat{\mathbf{E}}_J$, for $I \neq J$, under the deformation mapping χ. Further, the squares of the principal stretches at a point are equal to the eigenvalues of $\mathbf{C}$. We shall return to this aspect in Section 3.9 on the polar decomposition theorem.

3.4.2 Green–Lagrange Strain Tensor

The change in the squared lengths that occurs as a body deforms from the reference to the current configuration can be expressed relative to the original length as

$$(ds)^2 - (dS)^2 = 2\,d\mathbf{X} \cdot \mathbf{E} \cdot d\mathbf{X}, \tag{3.4.10}$$

where $\mathbf{E}$ is called the *Green–St. Venant (Lagrangian) strain tensor*, the *Green–Lagrange strain tensor*, or simply the *Green strain tensor*.[5] The Green–Lagrange strain tensor can be expressed, in view of Eqs. (3.4.1)–(3.4.3), as

$$\mathbf{E} = \tfrac{1}{2}\left(\mathbf{F}^{\mathrm{T}} \cdot \mathbf{F} - \mathbf{I}\right) = \tfrac{1}{2}\left(\mathbf{C} - \mathbf{I}\right)$$

$$= \tfrac{1}{2}\left[(\mathbf{I} + \boldsymbol{\nabla}_0\mathbf{u}) \cdot (\mathbf{I} + \boldsymbol{\nabla}_0\mathbf{u})^{\mathrm{T}} - \mathbf{I}\right]$$

$$= \tfrac{1}{2}\left[\boldsymbol{\nabla}_0\mathbf{u} + (\boldsymbol{\nabla}_0\mathbf{u})^{\mathrm{T}} + (\boldsymbol{\nabla}_0\mathbf{u}) \cdot (\boldsymbol{\nabla}_0\mathbf{u})^{\mathrm{T}}\right], \tag{3.4.11}$$

where Eq. (3.3.8) is used in writing $\mathbf{F}$ in terms of $\boldsymbol{\nabla}_0\mathbf{u}$. By definition, the Green–Lagrange strain tensor is a symmetric second-order tensor. Also, the change in the squared lengths is zero if and only if $\mathbf{E} = \mathbf{0}$.

The vector form of the Green–Lagrange strain tensor in Eq. (3.4.11) allows us to express it in terms of its components in any coordinate system. In particular, in the rectangular Cartesian coordinate system (X_1, X_2, X_3), the components of $\mathbf{E}$ are given by

---

[5]Readers should not confuse the symbol $\mathbf{E}$ used for the Green–Lagrange strain tensor and $\hat{\mathbf{E}}_i$ used for the basis vectors in the reference configuration. One should always pay attention to different typefaces and subscripts used.

$$E_{IJ} = \tfrac{1}{2}\left(\frac{\partial u_I}{\partial X_J} + \frac{\partial u_J}{\partial X_I} + \frac{\partial u_K}{\partial X_I}\frac{\partial u_K}{\partial X_J}\right), \tag{3.4.12}$$

where summation on repeated index (K) is implied. Clearly, the last term in Eqs. (3.4.11) and (3.4.12) is nonlinear in the displacement gradients. In expanded notation, the Green–Lagrange strain tensor components are given by

$$E_{11} = \frac{\partial u_1}{\partial X_1} + \frac{1}{2}\left[\left(\frac{\partial u_1}{\partial X_1}\right)^2 + \left(\frac{\partial u_2}{\partial X_1}\right)^2 + \left(\frac{\partial u_3}{\partial X_1}\right)^2\right],$$

$$E_{22} = \frac{\partial u_2}{\partial X_2} + \frac{1}{2}\left[\left(\frac{\partial u_1}{\partial X_2}\right)^2 + \left(\frac{\partial u_2}{\partial X_2}\right)^2 + \left(\frac{\partial u_3}{\partial X_2}\right)^2\right],$$

$$E_{33} = \frac{\partial u_3}{\partial X_3} + \frac{1}{2}\left[\left(\frac{\partial u_1}{\partial X_3}\right)^2 + \left(\frac{\partial u_2}{\partial X_3}\right)^2 + \left(\frac{\partial u_3}{\partial X_3}\right)^2\right], \tag{3.4.13}$$

$$E_{12} = \frac{1}{2}\left(\frac{\partial u_1}{\partial X_2} + \frac{\partial u_2}{\partial X_1} + \frac{\partial u_1}{\partial X_1}\frac{\partial u_1}{\partial X_2} + \frac{\partial u_2}{\partial X_1}\frac{\partial u_2}{\partial X_2} + \frac{\partial u_3}{\partial X_1}\frac{\partial u_3}{\partial X_2}\right),$$

$$E_{13} = \frac{1}{2}\left(\frac{\partial u_1}{\partial X_3} + \frac{\partial u_3}{\partial X_1} + \frac{\partial u_1}{\partial X_1}\frac{\partial u_1}{\partial X_3} + \frac{\partial u_2}{\partial X_1}\frac{\partial u_2}{\partial X_3} + \frac{\partial u_3}{\partial X_1}\frac{\partial u_3}{\partial X_3}\right),$$

$$E_{23} = \frac{1}{2}\left(\frac{\partial u_2}{\partial X_3} + \frac{\partial u_3}{\partial X_2} + \frac{\partial u_1}{\partial X_2}\frac{\partial u_1}{\partial X_3} + \frac{\partial u_2}{\partial X_2}\frac{\partial u_2}{\partial X_3} + \frac{\partial u_3}{\partial X_2}\frac{\partial u_3}{\partial X_3}\right).$$

The components E_{11}, E_{22}, and E_{33} are termed normal strains and E_{12}, E_{23}, and E_{13} are called shear strains, as shown in Fig. 3.4.2.

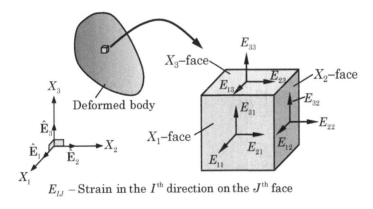

Fig. 3.4.2: Green–Lagrange strain tensor components in rectangular Cartesian coordinates.

3.4.3 Physical Interpretation of Green–Lagrange Strain Tensor Components

To see the physical meaning of the normal strain component E_{11}, consider a line element initially parallel to the X_1-axis, that is, $d\mathbf{X} = dX_1\hat{\mathbf{E}}_1$ in the reference configuration of the body, as shown in Fig. 3.4.3. Then

$$(ds)^2 - (dS)^2 = 2E_{IJ}\,dX_I\,dX_J = 2E_{11}\,dX_1\,dX_1 = 2E_{11}\,(dS)^2.$$

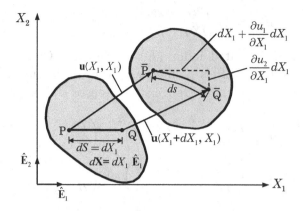

Fig. 3.4.3: Physical interpretation of normal strain component E_{11}.

Solving for E_{11}, we obtain

$$E_{11} = \tfrac{1}{2} \frac{(ds)^2 - (dS)^2}{(dS)^2} = \tfrac{1}{2} \left[\left(\frac{ds}{dS} \right)^2 - 1 \right],$$

where the initial and final lengths are given by (approximating the curve as a straight line)

$$(dS)^2 = (dX_1)^2, \quad (ds)^2 = \left(dX_1 + \frac{\partial u_1}{\partial X_1} dX_1 \right)^2 + \left(\frac{\partial u_2}{\partial X_1} dX_1 \right)^2. \tag{3.4.14}$$

Thus, we have

$$E_{11} = \tfrac{1}{2} \left[2 \frac{\partial u_1}{\partial X_1} + \left(\frac{\partial u_1}{\partial X_1} \right)^2 + \left(\frac{\partial u_2}{\partial X_1} \right)^2 \right].$$

We can also write

$$E_{11} = \tfrac{1}{2} \frac{(ds)^2 - (dS)^2}{(dS)^2} = \tfrac{1}{2} \left[\left(\frac{ds}{dS} \right)^2 - 1 \right] = \tfrac{1}{2} \left(\lambda^2 - 1 \right), \tag{3.4.15}$$

where λ is the stretch of the line element $d\mathbf{X}$:

$$\lambda = \frac{ds}{dS} = \sqrt{1 + 2E_{11}}. \tag{3.4.16}$$

The shear strain components E_{IJ}, $I \neq J$, can be interpreted as a measure of the change in the angle between line elements that were perpendicular to each other in the reference configuration. To see this, consider line elements $d\mathbf{X}^{(1)} = dX_1 \hat{\mathbf{E}}_1$ and $d\mathbf{X}^{(2)} = dX_2 \hat{\mathbf{E}}_2$ in the reference configuration of the body, which are perpendicular to each other, as shown in Fig. 3.4.4. The material line elements $d\mathbf{X}^{(1)}$ and $d\mathbf{X}^{(2)}$ occupy positions $d\mathbf{x}^{(1)}$ and $d\mathbf{x}^{(2)}$, respectively, in the deformed body. Then the cosine of the angle between the line elements $\bar{O}\bar{Q}$ and $\bar{O}\bar{P}$ in the deformed body is given by [see Eq. (3.3.1)]

$$
\begin{aligned}
\cos\theta_{12} = \hat{\mathbf{n}}_1 \cdot \hat{\mathbf{n}}_2 &= \frac{d\mathbf{x}^{(1)} \cdot d\mathbf{x}^{(2)}}{|d\mathbf{x}^{(1)}|\,|d\mathbf{x}^{(2)}|} \\
&= \frac{[d\mathbf{X}^{(1)} \cdot \mathbf{F}^{\mathrm{T}}] \cdot [\mathbf{F} \cdot d\mathbf{X}^{(2)}]}{\sqrt{d\mathbf{X}^{(1)} \cdot \mathbf{C} \cdot d\mathbf{X}^{(1)}}\sqrt{d\mathbf{X}^{(2)} \cdot \mathbf{C} \cdot d\mathbf{X}^{(2)}}}.
\end{aligned}
\tag{3.4.17}
$$

In view of the relations

$$
\mathbf{C} = \mathbf{F}^{\mathrm{T}} \cdot \mathbf{F}, \quad \hat{\mathbf{N}}_1 = \hat{\mathbf{E}}_1, \quad \hat{\mathbf{N}}_2 = \hat{\mathbf{E}}_2,
\tag{3.4.18}
$$

we have

$$
\cos\theta_{12} = \frac{\hat{\mathbf{N}}_1 \cdot \mathbf{C} \cdot \mathbf{N}_2}{\sqrt{\hat{\mathbf{N}}_1 \cdot \mathbf{C} \cdot \mathbf{N}_1}\,\sqrt{\hat{\mathbf{N}}_2 \cdot \mathbf{C} \cdot \mathbf{N}_2}} = \frac{C_{12}}{\sqrt{C_{11}}\sqrt{C_{22}}},
$$

or

$$
\cos\theta_{12} = \frac{C_{12}}{\lambda_1\lambda_2} = \frac{2E_{12}}{\sqrt{(1 + 2E_{11})}\sqrt{(1 + 2E_{22})}}.
\tag{3.4.19}
$$

Thus, $2E_{12}$ is equal to the cosine of the angle between the line elements, θ_{12}, multiplied by the product of extension ratios λ_1 and λ_2. Clearly, the finite strain E_{12} depends not only on the angle θ_{12} but also on the stretches of elements involved. When the unit extensions and the angle changes are small compared to unity, we find that $2E_{12}$ is the *decrease* from $\pi/2$:

$$
\tfrac{\pi}{2} - \theta_{12} \approx \sin(\tfrac{\pi}{2} - \theta_{12}) = \cos\theta_{12} \approx 2E_{12}.
\tag{3.4.20}
$$

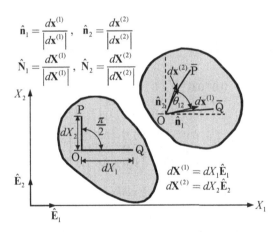

Fig. 3.4.4: Physical interpretation of shear strain component E_{12}.

3.4.4 Cauchy and Euler Strain Tensors

Returning to the strain measures, the change in the squared lengths that occurs as the body deforms from the initial to the current configuration can be expressed relative to the current length. First, we express dS in terms of $d\mathbf{x}$ as

$$
(dS)^2 = d\mathbf{X} \cdot d\mathbf{X} = d\mathbf{x} \cdot (\mathbf{F}^{-\mathrm{T}} \cdot \mathbf{F}^{-1}) \cdot d\mathbf{x} \equiv d\mathbf{x} \cdot \tilde{\mathbf{B}} \cdot d\mathbf{x},
\tag{3.4.21}
$$

where $\tilde{\mathbf{B}}$ is called the *Cauchy strain tensor*

$$\tilde{\mathbf{B}} = \mathbf{F}^{-\mathrm{T}} \cdot \mathbf{F}^{-1}, \quad \tilde{\mathbf{B}}^{-1} \equiv \mathbf{B} = \mathbf{F} \cdot \mathbf{F}^{\mathrm{T}}, \tag{3.4.22}$$

and $\mathbf{B}$ is the left Cauchy–Green tensor or Finger tensor. We can write

$$(ds)^2 - (dS)^2 = 2\,d\mathbf{x} \cdot \mathbf{e} \cdot d\mathbf{x}, \tag{3.4.23}$$

where $\mathbf{e}$, called the *Almansi–Hamel (Eulerian) strain tensor* or simply the *Euler strain tensor*, is defined as

$$\mathbf{e} = \tfrac{1}{2}\left(\mathbf{I} - \mathbf{F}^{-\mathrm{T}} \cdot \mathbf{F}^{-1}\right) = \tfrac{1}{2}\left(\mathbf{I} - \tilde{\mathbf{B}}\right) \tag{3.4.24}$$

$$= \tfrac{1}{2}\left[\mathbf{I} - (\mathbf{I} - \nabla\mathbf{u}) \cdot (\mathbf{I} - \nabla\mathbf{u})^{\mathrm{T}}\right]$$

$$= \tfrac{1}{2}\left[\nabla\mathbf{u} + (\nabla\mathbf{u})^{\mathrm{T}} - (\nabla\mathbf{u}) \cdot (\nabla\mathbf{u})^{\mathrm{T}}\right]. \tag{3.4.25}$$

The rectangular Cartesian components of $\mathbf{C}$, $\tilde{\mathbf{B}}$, and $\mathbf{e}$ are given by

$$C_{IJ} = \frac{\partial x_k}{\partial X_I}\frac{\partial x_k}{\partial X_J}, \quad \tilde{B}_{ij} = \frac{\partial X_K}{\partial x_i}\frac{\partial X_K}{\partial x_j}, \tag{3.4.26}$$

$$e_{ij} = \tfrac{1}{2}\left(\delta_{ij} - \frac{\partial X_K}{\partial x_i}\frac{\partial X_K}{\partial x_j}\right) = \tfrac{1}{2}\left(\frac{\partial u_i}{\partial x_j} + \frac{\partial u_j}{\partial x_i} - \frac{\partial u_k}{\partial x_i}\frac{\partial u_k}{\partial x_j}\right). \tag{3.4.27}$$

Examples 3.4.1 and 3.4.2 illustrate the calculation of various measures of strain.

Example 3.4.1

For the deformation given in Example 3.3.1, determine the Cartesian components of the right Cauchy–Green deformation tensor $\mathbf{C}$, the Cauchy strain tensor $\tilde{\mathbf{B}}$, and the Green–Lagrange and Almansi strain tensors, $\mathbf{E}$ and $\mathbf{e}$.

Solution: The components of the right Cauchy–Green deformation tensor and the Cauchy strain tensor are

$$[C] = [F]^{\mathrm{T}}[F] = \begin{bmatrix} 1.0 & 0.0 & 0.0 \\ 0.5 & 1.0 & 0.0 \\ 0.0 & 0.0 & 1.0 \end{bmatrix}\begin{bmatrix} 1.0 & 0.5 & 0.0 \\ 0.0 & 1.0 & 0.0 \\ 0.0 & 0.0 & 1.0 \end{bmatrix} = \begin{bmatrix} 1.0 & 0.50 & 0.0 \\ 0.5 & 1.25 & 0.0 \\ 0.0 & 0.00 & 1.0 \end{bmatrix},$$

$$[\tilde{B}] = [F]^{-\mathrm{T}}[F]^{-1} = \begin{bmatrix} 1.0 & 0.0 & 0.0 \\ -0.5 & 1.0 & 0.0 \\ 0.0 & 0.0 & 1.0 \end{bmatrix}\begin{bmatrix} 1.0 & -0.5 & 0.0 \\ 0.0 & 1.0 & 0.0 \\ 0.0 & 0.0 & 1.0 \end{bmatrix} = \begin{bmatrix} 1.0 & -0.50 & 0.0 \\ -0.5 & 1.25 & 0.0 \\ 0.0 & 0.00 & 1.0 \end{bmatrix}.$$

The Green–Lagrange and Almansi strain tensor components in matrix form are given by

$$[E] = \tfrac{1}{2}\left([C] - [I]\right) = \tfrac{1}{2}\begin{bmatrix} 0.0 & 0.50 & 0.0 \\ 0.5 & 0.25 & 0.0 \\ 0.0 & 0.00 & 0.0 \end{bmatrix}; \quad [e] = \tfrac{1}{2}\left([I] - [\tilde{B}]\right) = \tfrac{1}{2}\begin{bmatrix} 0.0 & 0.50 & 0.0 \\ 0.5 & -0.25 & 0.0 \\ 0.0 & 0.00 & 0.0 \end{bmatrix}.$$

Example 3.4.2

Consider the uniform deformation of a square block $\mathcal{B}$ of side length 2 units, initially centered at $\mathbf{X} = (0,0)$, as shown in Fig. 3.4.5. The deformation is defined by the mapping

$$\boldsymbol{\chi}(\mathbf{X}) = \tfrac{1}{4}(18 + 4X_1 + 6X_2)\hat{\mathbf{e}}_1 + \tfrac{1}{4}(14 + 6X_2)\hat{\mathbf{e}}_2 + X_3\hat{\mathbf{e}}_3.$$

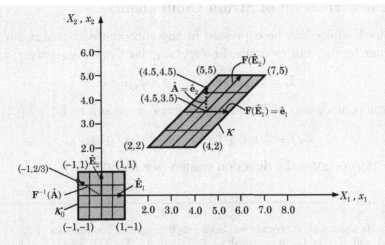

Fig. 3.4.5: Undeformed (κ_0) and deformed (κ) configurations of a rectangular block.

(a) Sketch the deformed configuration κ of the body $\mathcal{B}$.

(b) Compute the components of the deformation gradient $\mathbf{F}$ and its inverse.

(c) Compute the components of the right Cauchy–Green deformation tensor $\mathbf{C}$ and Cauchy strain tensor $\tilde{\mathbf{B}}$.

(d) Compute Green's and Almansi's strain tensor components (E_{IJ} and e_{ij}).

Solution:
(a) A sketch of the deformed configuration of the body is shown in Fig. 3.4.5.

(b) Note that the inverse transformation is given by ($X_3 = x_3$)

$$\left\{\begin{matrix} X_1 \\ X_2 \end{matrix}\right\} = \frac{1}{6}\begin{bmatrix} 6 & -6 \\ 0 & 4 \end{bmatrix}\left(\left\{\begin{matrix} x_1 \\ x_2 \end{matrix}\right\} - \frac{1}{4}\left\{\begin{matrix} 18 \\ 14 \end{matrix}\right\}\right) = -\frac{1}{3}\left\{\begin{matrix} 3 \\ 7 \end{matrix}\right\} + \frac{1}{3}\begin{bmatrix} 3 & -3 \\ 0 & 2 \end{bmatrix}\left\{\begin{matrix} x_1 \\ x_2 \end{matrix}\right\},$$

or

$$\boldsymbol{\chi}^{-1}(\mathbf{x}) = (-1 + x_1 - x_2)\hat{\mathbf{E}}_1 + \tfrac{1}{3}(-7 + 2x_2)\hat{\mathbf{E}}_2 + x_3\,\hat{\mathbf{E}}_3.$$

The matrix form of the deformation gradient and its inverse are

$$[F] = \begin{bmatrix} \frac{\partial x_1}{\partial X_1} & \frac{\partial x_1}{\partial X_2} \\ \frac{\partial x_2}{\partial X_1} & \frac{\partial x_2}{\partial X_2} \end{bmatrix} = \frac{1}{2}\begin{bmatrix} 2 & 3 \\ 0 & 3 \end{bmatrix}; \quad [F]^{-1} = \begin{bmatrix} \frac{\partial X_1}{\partial x_1} & \frac{\partial X_1}{\partial x_2} \\ \frac{\partial X_2}{\partial x_1} & \frac{\partial X_2}{\partial x_2} \end{bmatrix} = \frac{1}{3}\begin{bmatrix} 3 & -3 \\ 0 & 2 \end{bmatrix}.$$

(c) The right Cauchy–Green deformation tensor and Cauchy strain tensor are, respectively,

$$[C] = [F]^{\mathrm{T}}[F] = \tfrac{1}{2}\begin{bmatrix} 2 & 3 \\ 3 & 9 \end{bmatrix}, \quad [B] = [F][F]^{\mathrm{T}} = \tfrac{1}{4}\begin{bmatrix} 13 & 9 \\ 9 & 9 \end{bmatrix}.$$

(d) The Green and Almansi strain tensor components in matrix form are, respectively,

$$[E] = \tfrac{1}{2}\left([F]^{\mathrm{T}}[F] - [I]\right) = \tfrac{1}{4}\begin{bmatrix} 0 & 3 \\ 3 & 7 \end{bmatrix}, \quad [e] = \tfrac{1}{2}\left([I] - [F]^{-\mathrm{T}}[F]^{-1}\right) = \tfrac{1}{18}\begin{bmatrix} 0 & 9 \\ 9 & -4 \end{bmatrix}.$$

3.4.5 Transformation of Strain Components

The tensors $\mathbf{E}$ and $\mathbf{e}$ can be expressed in any coordinate system much like any second-order tensor. For example, in a rectangular Cartesian system, we have

$$\mathbf{E} = E_{IJ}\,\hat{\mathbf{E}}_I\hat{\mathbf{E}}_J, \qquad \mathbf{e} = e_{ij}\hat{\mathbf{e}}_i\hat{\mathbf{e}}_j. \tag{3.4.28}$$

Further, the components of $\mathbf{E}$ and $\mathbf{e}$ transform according to Eq. (2.5.21):

$$\bar{E}_{IJ} = \ell_{IK}\,\ell_{JL}\,E_{KL}, \qquad \bar{e}_{ij} = \ell_{ik}\,\ell_{j\ell}\,e_{k\ell}, \tag{3.4.29}$$

where ℓ_{IJ} (ℓ_{ij}) denotes the direction cosines [see Eq. (2.2.71)].

Example 3.4.3 ——

Derive the transformation equations between the strain components E_{IJ} referred to (X_1, X_2, X_3) and $\bar{E}_{IJ}$ in the new coordinate system $(\bar{X}_1, \bar{X}_2, \bar{X}_3)$, which is obtained by rotating the former about the X_3-axis counterclockwise by the angle θ.

Solution: The two coordinate systems are related by [see Eq. (2.2.70)]

$$\begin{Bmatrix} \bar{X}_1 \\ \bar{X}_2 \\ \bar{X}_3 \end{Bmatrix} = \begin{bmatrix} \cos\theta & \sin\theta & 0 \\ -\sin\theta & \cos\theta & 0 \\ 0 & 0 & 1 \end{bmatrix} \begin{Bmatrix} X_1 \\ X_2 \\ X_3 \end{Bmatrix} \equiv [L] \begin{Bmatrix} X_1 \\ X_2 \\ X_3 \end{Bmatrix}. \tag{3.4.30}$$

Transformation of strain tensor components follows those of a second-order tensor [see Eq. (2.5.21), and note that $[L]^{-1} = [L]^T$]

$$[\bar{E}] = [L][E][L]^T; \quad [E] = [L]^T[\bar{E}][L]. \tag{3.4.31}$$

Carrying out the indicated matrix multiplications and expressing the result in single-column format, we have

$$\begin{Bmatrix} E_{11} \\ E_{22} \\ E_{33} \\ 2E_{23} \\ 2E_{13} \\ 2E_{12} \end{Bmatrix} = \begin{bmatrix} \cos^2\theta & \sin^2\theta & 0 & 0 & 0 & -\frac{1}{2}\sin 2\theta \\ \sin^2\theta & \cos^2\theta & 0 & 0 & 0 & \frac{1}{2}\sin 2\theta \\ 0 & 0 & 1 & 0 & 0 & 0 \\ 0 & 0 & 0 & \cos\theta & \sin\theta & 0 \\ 0 & 0 & 0 & -\sin\theta & \cos\theta & 0 \\ \sin 2\theta & -\sin 2\theta & 0 & 0 & 0 & \cos 2\theta \end{bmatrix} \begin{Bmatrix} \bar{E}_{11} \\ \bar{E}_{22} \\ \bar{E}_{33} \\ 2\bar{E}_{23} \\ 2\bar{E}_{13} \\ 2\bar{E}_{12} \end{Bmatrix}. \tag{3.4.32}$$

The inverse relations are

$$\begin{Bmatrix} \bar{E}_{11} \\ \bar{E}_{22} \\ \bar{E}_{33} \\ 2\bar{E}_{23} \\ 2\bar{E}_{13} \\ 2\bar{E}_{12} \end{Bmatrix} = \begin{bmatrix} \cos^2\theta & \sin^2\theta & 0 & 0 & 0 & \frac{1}{2}\sin 2\theta \\ \sin^2\theta & \cos^2\theta & 0 & 0 & 0 & -\frac{1}{2}\sin 2\theta \\ 0 & 0 & 1 & 0 & 0 & 0 \\ 0 & 0 & 0 & \cos\theta & -\sin\theta & 0 \\ 0 & 0 & 0 & \sin\theta & \cos\theta & 0 \\ -\sin 2\theta & \sin 2\theta & 0 & 0 & 0 & \cos 2\theta \end{bmatrix} \begin{Bmatrix} E_{11} \\ E_{22} \\ E_{33} \\ 2E_{23} \\ 2E_{13} \\ 2E_{12} \end{Bmatrix}. \tag{3.4.33}$$

Example 3.4.4 ——

Consider a rectangular block ($\mathcal{B}$) ABCD of dimensions $a \times b \times h$, where h is the thickness (very small compared to a and b). Suppose that block $\mathcal{B}$ is deformed into the diamond shape $\bar{A}\bar{B}\bar{C}\bar{D}$ shown in Fig. 3.4.6(a), without a change in its thickness. Determine the deformation mapping, displacements, and strains in the body. Assume that the mapping is a linear polynomial in X_1 and X_2. A *complete* linear polynomial in X_1 and X_2 is of the form $p(X_1, X2) = a_0 + a_1 X_1 + a_2 X_2 + a_3 X_1 X_2$.

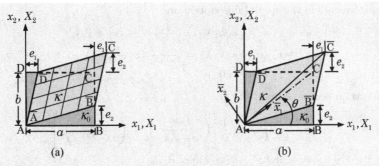

Fig. 3.4.6: Undeformed (κ_0) and deformed (κ) configurations of a rectangular block. Typical material lines inside the body are also shown.

Solution: Let (X_1, X_2, X_3) denote the coordinates of a material point in the reference configuration, κ_0. The X_3-axis is taken out of the plane of the page and not shown in the figure. By assumption, the geometry of the deformed body can be described by the mapping $\chi(\mathbf{x}) = x_1\,\hat{\mathbf{e}}_1 + x_2\,\hat{\mathbf{e}}_2 + x_3\,\hat{\mathbf{e}}_3$, where

$$x_1 = a_0 + a_1 X_1 + a_2 X_2 + a_3 X_1 X_2,$$
$$x_2 = b_0 + b_1 X_1 + b_2 X_2 + b_3 X_1 X_2,$$
$$x_3 = X_3,$$

and $a_0, a_1, a_2, a_3, b_0, b_1, b_2$, and b_3 are constants to be determined using the values of (X_1, X_2) from the undeformed configuration and the corresponding values of (x_1, x_2) from the deformed configuration shown in Fig. 3.4.6(a). The eight constants are determined using the 8 conditions provided by the coordinate values at points A, B, C, and D. Since point A is at the origin of the coordinate system, we immediately obtain $a_0 = b_0 = 0$. Next, we have

$$(X_1, X_2) = (a, 0), \quad (x_1, x_2) = (a, e_2) \;\to\; a_1 = 1, \quad b_1 = \frac{e_2}{a},$$

$$(X_1, X_2) = (0, b), \quad (x_1, x_2) = (e_1, b) \;\to\; a_2 = \frac{e_1}{b}, \quad b_2 = 1,$$

$$(X_1, X_2) = (a, b), \quad (x_1, x_2) = (a + e_1, b + e_2) \;\to\; a_3 = 0, \quad b_3 = 0.$$

Thus, the deformation is defined by the transformation

$$\chi(\mathbf{x}) = \left(X_1 + k_1 X_2\right)\hat{\mathbf{e}}_1 + \left(X_2 + k_2 X_1\right)\hat{\mathbf{e}}_2 + X_3\,\hat{\mathbf{e}}_3, \tag{1}$$

where $k_1 = e_1/b$ and $k_2 = e_2/a$. The inverse mapping is given by

$$\chi^{-1}(\mathbf{X}) = \tfrac{1}{1-k_1 k_2}\left(x_1 - k_1 x_2\right)\hat{\mathbf{E}}_1 + \tfrac{1}{1-k_1 k_2}\left(-k_2 x_1 + x_2\right)\hat{\mathbf{E}}_2 + x_3\,\hat{\mathbf{E}}_3. \tag{2}$$

Thus, the displacement vector of a material point in the Lagrangian description is

$$\mathbf{u} = k_1 X_2\,\hat{\mathbf{E}}_1 + k_2 X_1\,\hat{\mathbf{E}}_2. \tag{3}$$

The only nonzero Green strain tensor components are given by

$$E_{11} = \tfrac{1}{2}k_2^2, \quad 2E_{12} = k_1 + k_2, \quad E_{22} = \tfrac{1}{2}k_1^2. \tag{4}$$

For the infinitesimal case (that is, k_1 and k_2 are small), we only have the shear strain $2\varepsilon_{12} = k_1 + k_2$. The components of the deformation gradient are

$$[F] = \begin{bmatrix} 1 & k_1 & 0 \\ k_2 & 1 & 0 \\ 0 & 0 & 1 \end{bmatrix}.$$

The case in which $k_2 = 0$ is known as the *simple shear*. The Green's deformation tensor $\mathbf{C}$ is

$$\mathbf{C} = \mathbf{F}^{\mathrm{T}} \cdot \mathbf{F} \;\to\; [C] = [F]^{\mathrm{T}}[F] = \begin{bmatrix} 1+k_2^2 & k_1+k_2 & 0 \\ k_1+k_2 & 1+k_1^2 & 0 \\ 0 & 0 & 1 \end{bmatrix},$$

and $2\mathbf{E} = \mathbf{C} - \mathbf{I}$ yields the results given in Eq. (4).

The displacements in the spatial description are

$$u_1 = x_1 - X_1 = k_1 X_2 = \frac{k_1}{1 - k_1 k_2} \left(-k_2 x_1 + x_2 \right),$$

$$u_2 = x_2 - X_2 = k_2 X_1 = \frac{k_2}{1 - k_1 k_2} \left(x_1 - k_1 x_2 \right), \tag{5}$$

$$u_3 = x_3 - X_3 = 0.$$

The Almansi strain tensor components are

$$e_{11} = -\frac{k_1 k_2}{1 - k_1 k_2} - \frac{1}{2} \left[\left(\frac{k_1 k_2}{1 - k_1 k_2} \right)^2 + \left(\frac{k_2}{1 - k_1 k_2} \right)^2 \right],$$

$$2e_{12} = \frac{k_1 + k_2}{1 - k_1 k_2} + \frac{k_1 k_2 (k_1 + k_2)}{(1 - k_1 k_2)^2}, \tag{6}$$

$$e_{22} = -\frac{k_1 k_2}{1 - k_1 k_2} - \frac{1}{2} \left[\left(\frac{k_1 k_2}{1 - k_1 k_2} \right)^2 + \left(\frac{k_1}{1 - k_1 k_2} \right)^2 \right].$$

Alternatively, the same results can be obtained using the elementary mechanics of materials approach, where the strains are defined as the ratio of the difference between the final length and original length to the original length. A line element AB in the initial (undeformed) configuration κ_0 of the body $\mathcal{B}$ moves to position $\bar{A}\bar{B}$ (point $\bar{A}$ is the same as point A), as shown in Fig. 3.4.6(a). Then the Green strain in line AB is given by

$$E_{11} = E_{AB} = \frac{\bar{A}\bar{B} - AB}{AB} = \frac{1}{a} \sqrt{a^2 + e_2^2} - 1 = \sqrt{1 + \left(\frac{e_2}{a} \right)^2} - 1$$

$$= \left[1 + \frac{1}{2} \left(\frac{e_2}{a} \right)^2 + \cdots \right] - 1 \approx \frac{1}{2} \left(\frac{e_2}{a} \right)^2 = \frac{1}{2} k_2^2,$$

where cubic and higher powers of e_2/a are considered to be smaller than e_2/a and e_2^2/a^2 and thus neglected. Similarly,

$$E_{22} = \left[1 + \frac{1}{2} \left(\frac{e_1}{b} \right)^2 + \cdots \right] - 1 \approx \frac{1}{2} \left(\frac{e_1}{b} \right)^2 = \frac{1}{2} k_1^2.$$

The shear strain $2E_{12}$ is equal to the change in the angle between two line elements that were originally at $90°$, that is, change in the angle DAB. The change is equal to, as can be seen from Fig. 3.4.6(b)

$$2E_{12} = \angle DAB - \angle \bar{D}\bar{A}\bar{B} = \frac{e_1}{b} + \frac{e_2}{a} = k_1 + k_2.$$

Thus, the strains computed using mechanics of materials approach, when terms of order higher than e_1^2/b^2 and e_2^2/a^2 are neglected, yield the same as those in Eq. (4). On the other hand, if we define E_{11} and E_{22} as (consistent with the definition of the Green–Lagrange strain tensor),

$$2E_{11} = \frac{(\bar{A}\bar{B})^2 - (AB)^2}{(AB)^2} = \frac{a^2 + e_2^2}{a^2} - 1 = \frac{e_2^2}{a^2} = k_2^2,$$

$$2E_{22} = \frac{(\bar{A}\bar{D})^2 - (AD)^2}{(AD)^2} = \frac{b^2 + e_1^2}{b^2} - 1 = \frac{e_1^2}{b^2} = k_1^2,$$

we obtain the results in Eq. (4) directly, without making any order of magnitude assumption. Thus, in general, the engineering strains defined in mechanics of materials and the Green–Lagrange strains are not the same.

The axial strain in line element AC is [see Fig. 3.4.6(b)]

$$E_{AC} = \frac{\bar{A}\bar{C} - AC}{AC} = \frac{1}{\sqrt{a^2 + b^2}} \sqrt{(a + e_1)^2 + (b + e_2)^2} - 1$$

$$= \frac{1}{\sqrt{a^2 + b^2}} \sqrt{a^2 + b^2 + e_1^2 + e_2^2 + 2ae_1 + 2be_2} - 1$$

$$= \left[1 + \frac{e_1^2 + e_2^2 + 2ae_1 + 2be_2}{a^2 + b^2} \right]^{\frac{1}{2}} - 1 \approx \frac{1}{2} \frac{e_1^2 + e_2^2 + 2ae_1 + 2be_2}{a^2 + b^2}$$

$$= \frac{1}{2(a^2 + b^2)} \left[a^2 k_2^2 + 2ab(k_1 + k_2) + b^2 k_1^2 \right].$$

The axial strain E_{AC} can also be computed using the strain transformation equations (3.4.29). The line AC is oriented at $\theta = \tan^{-1}(b/a)$. Hence, we have

$$\beta_{11} = \cos\theta = \frac{a}{\sqrt{a^2 + b^2}}, \quad \beta_{12} = \sin\theta = \frac{b}{\sqrt{a^2 + b^2}},$$

$$\beta_{21} = -\sin\theta = -\frac{b}{\sqrt{a^2 + b^2}}, \quad \beta_{22} = \cos\theta = \frac{a}{\sqrt{a^2 + b^2}},$$

and

$$E_{AC} \equiv \bar{E}_{11} = \beta_{1i}\beta_{1j}E_{ij} = \beta_{11}\beta_{11}E_{11} + 2\beta_{11}\beta_{12}E_{12} + \beta_{12}\beta_{12}E_{22}$$
$$= \tfrac{1}{2(a^2+b^2)}\left[a^2 k_2^2 + 2ab(k_1 + k_2) + b^2 k_1^2\right],$$

which is the same as that computed previously.

3.4.6 Invariants and Principal Values of Strains

The principal invariants of the Green–Lagrange strain tensor $\mathbf{E}$ are [see Eqs. (2.5.16) and (2.5.17)]

$$J_1 = \text{tr}\,\mathbf{E}, \quad J_2 = \frac{1}{2}\left[(\text{tr}\,\mathbf{E})^2 - \text{tr}\left(\mathbf{E}^2\right)\right], \quad J_3 = |\mathbf{E}|, \tag{3.4.34}$$

where the trace of $\mathbf{E}$, $\text{tr}\,\mathbf{E}$, is defined as the double-dot product of $\mathbf{E}$ with the identity tensor [see Eq. (2.5.15)]

$$\text{tr}\,\mathbf{E} = \mathbf{E} : \mathbf{I}. \tag{3.4.35}$$

In terms of the rectangular Cartesian components, the three principal invariants of $\mathbf{E}$ have the form

$$J_1 = E_{II}, \quad J_2 = \tfrac{1}{2}\left(E_{II}E_{JJ} - E_{IJ}E_{JI}\right), \quad J_3 = |\mathbf{E}|. \tag{3.4.36}$$

It is of considerable interest (e.g., in the design of structures) to know the maximum and minimum values of the strain at a point. The eigenvalues of the matrix of the strain tensor (see Section 2.5.6), when ordered from large to small, characterize the maximum and minimum normal strains, and the eigenvectors represent the planes on which they occur. The maximum shear strains can be determined once the maximum normal strains are determined. The eigenvalues of a strain tensor are called the *principal values of strain*, and the corresponding eigenvectors are called the *principal directions of strain*.

The eigenvalue problem associated with the strain tensor $\mathbf{E}$ is to find μ and $\mathbf{X}$ such that

$$\mathbf{EX} = \mu\mathbf{X} \quad \text{for all } \mathbf{X} \neq \mathbf{0}; \quad |[E] - \mu[I]| = 0, \tag{3.4.37}$$

where μ are the principal values and $\{X\}$ are the principal directions. The characteristic equation is of the form

$$-\mu^3 + J_1\mu^2 - J_2\mu + J_3 = 0. \tag{3.4.38}$$

Three eigenvalues μ_1, μ_2, and μ_3 provide the three principal values (one of them is the maximum and one of them is the minimum) of normal strain.

The maximum shear strains E_{ns} are computed from (for additional discussion see Section 4.3.3)

$$
\begin{aligned}
E_{ns}^2 &= \tfrac{1}{4}(\mu_1 - \mu_2)^2 \quad \text{for} \quad \hat{\mathbf{n}} = \tfrac{1}{\sqrt{2}}(\hat{\mathbf{e}}_1 \pm \hat{\mathbf{e}}_2), \\
E_{ns}^2 &= \tfrac{1}{4}(\mu_1 - \mu_3)^2 \quad \text{for} \quad \hat{\mathbf{n}} = \tfrac{1}{\sqrt{2}}(\hat{\mathbf{e}}_1 \pm \hat{\mathbf{e}}_3), \\
E_{ns}^2 &= \tfrac{1}{4}(\mu_2 - \mu_3)^2 \quad \text{for} \quad \hat{\mathbf{n}} = \tfrac{1}{\sqrt{2}}(\hat{\mathbf{e}}_2 \pm \hat{\mathbf{e}}_3).
\end{aligned}
\tag{3.4.39}
$$

The largest shear strain is given by

$$
(E_{ns})_{\max} = \tfrac{1}{2}(\mu_{\max} - \mu_{\min}),
\tag{3.4.40}
$$

where $\mu_{\max}$ and $\mu_{\min}$ are the maximum and minimum principal values of strain, respectively. The plane of the maximum shear strain lies between the planes of the maximum and minimum principal strain (that is, oriented at $\pm 45°$ to both planes).

Example 3.4.5 deals with the computation of principal strains and their directions.

Example 3.4.5 ——————————————————————

The state of strain at a point in an elastic body is given by $(10^{-3} \text{ in./in.})$

$$
[E] = \begin{bmatrix} 4 & -4 & 0 \\ -4 & 0 & 0 \\ 0 & 0 & 3 \end{bmatrix}.
$$

Determine the principal strains and principal directions of the strain.

Solution: In this case, we know from the given matrix that $\mu = 3$ is a root with eigenvector $\mathbf{X}^{(3)} = \pm\hat{\mathbf{E}}_3$. The principal invariants of $[E]$ are

$$
J_1 = 4 + 0 + 3 = 7, \quad J_2 = \tfrac{1}{2}\left[7^2 - 4^2 - 3^2 - 2 \times (-4)^2\right] = -4, \quad J_3 = |E| = -48.
$$

Hence, the characteristic equation is

$$
-\mu^3 + 7\mu^2 + 4\mu - 48 = 0 \quad \rightarrow \quad (-\mu^2 + 4\mu + 16)(\mu - 3) = 0.
$$

Then the roots (the principal strains) of the characteristic equation are (10^{-3} in/in.)

$$
\mu_1 = 3, \quad \mu_2 = 2(1 + \sqrt{5}), \quad \mu_3 = 2(1 - \sqrt{5}).
$$

The eigenvector components $X_I^{(1)}$ associated with $E_1 = \mu_1 = 3$ are calculated from

$$
\begin{bmatrix} 4-3 & -4 & 0 \\ -4 & 0-3 & 0 \\ 0 & 0 & 3-3 \end{bmatrix} \begin{Bmatrix} X_1^{(1)} \\ X_2^{(1)} \\ X_3^{(1)} \end{Bmatrix} = \begin{Bmatrix} 0 \\ 0 \\ 0 \end{Bmatrix},
$$

which gives $X_1^{(1)} - 4X_2^{(1)} = 0$ and $-4X_1^{(1)} - 3X_2^{(1)} = 0$, or $X_1^{(1)} = X_2^{(1)} = 0$. Using the normalization $(X_1^{(1)})^2 + (X_2^{(1)})^2 + (X_3^{(1)})^2 = 1$, we obtain $X_3^{(1)} = 1$. Thus, the principal direction associated with the principal strain $E_1 = 3$ is $\{\hat{X}^{(1)}\}^{\mathrm{T}} = \pm\{0, 0, 1\}$ or $\hat{\mathbf{X}}^{(1)} = \pm\hat{\mathbf{E}}_3$.

The eigenvector components associated with principal strain $E_2 = \mu_2 = 2(1 + \sqrt{5})$ are calculated from

$$\begin{bmatrix} 4 - \mu_2 & -4 & 0 \\ -4 & -\mu_2 & 0 \\ 0 & 0 & 3 - \mu_2 \end{bmatrix} \begin{Bmatrix} X_1^{(2)} \\ X_2^{(2)} \\ X_3^{(2)} \end{Bmatrix} = \begin{Bmatrix} 0 \\ 0 \\ 0 \end{Bmatrix},$$

which gives

$$X_1^{(2)} = -\tfrac{2+2\sqrt{5}}{4} X_2^{(2)} = -1.618\, X_2^{(2)}, \quad X_3^{(2)} = 0, \;\Rightarrow\; \{\hat{X}^{(2)}\} = \pm \begin{Bmatrix} -0.851 \\ 0.526 \\ 0.000 \end{Bmatrix}.$$

Similarly, the eigenvector components associated with principal strain $E_3 = \mu_3 = 2(1 - \sqrt{5})$ are obtained as

$$X_1^{(3)} = \tfrac{2+2\sqrt{5}}{4} X_2^{(3)} = 1.618\, X_2^{(3)}, \quad X_3^{(3)} = 0, \;\Rightarrow\; \{\hat{X}^{(3)}\} = \pm \begin{Bmatrix} 0.526 \\ 0.851 \\ 0.000 \end{Bmatrix}.$$

Note that the eigenvectors $\hat{\mathbf{X}}^{(1)}$, $\hat{\mathbf{X}}^{(2)}$, and $\hat{\mathbf{X}}^{(3)}$ are mutually orthogonal, as expected.

3.5 Infinitesimal Strain Tensor and Rotation Tensor

3.5.1 Infinitesimal Strain Tensor

When all displacement gradients are small, that is, $|\nabla_0 \mathbf{u}| \ll 1$, we may neglect the nonlinear terms in the definition of the Green–Lagrange strain tensor $\mathbf{E}$ and obtain the linearized strain tensor ε, called the *infinitesimal strain tensor*. To derive ε from $\mathbf{E}$, we must linearize $\mathbf{E}$ by using a measure of smallness.

We introduce the nonnegative function

$$\epsilon(t) = \|\nabla_0 \mathbf{u}\|_\infty = \sup_{\mathbf{X} \in \kappa} |\nabla_0 \mathbf{u}|, \tag{3.5.1}$$

where "sup" stands for supremum or the least upper bound of the set of all absolute values of $\nabla_0 \mathbf{u}$ defined for all $\mathbf{X} \in \kappa$. If $f(\nabla_0 \mathbf{u})$ is a scalar-, vector-, or tensor-valued function in the neighborhood of $\nabla_0 \mathbf{u} = \mathbf{0}$ so that there exists a constant c such that

$$\|f(\nabla_0 \mathbf{u})\|_\infty < c\,\epsilon^n,$$

we say that f is of the order ϵ^n, as $\epsilon \to 0$, and write $f = \mathrm{O}(\epsilon^n)$.

If $\mathbf{E}$ is of the order $\mathrm{O}(\epsilon)$ in $\nabla_0 \mathbf{u}$, then we mean

$$\frac{\partial u_I}{\partial X_J} = \mathrm{O}(\epsilon) \quad \text{as } \epsilon \to 0.$$

If terms of the order $\mathrm{O}(\epsilon^2)$, as $\epsilon \to 0$, can be omitted in Eq. (3.4.12), then

$$E_{IJ} = \tfrac{1}{2} \left(\frac{\partial u_I}{\partial X_J} + \frac{\partial u_J}{\partial X_I} + \frac{\partial u_K}{\partial X_I} \frac{\partial u_K}{\partial X_J} \right)$$

can be approximated as

$$E_{IJ} \approx \tfrac{1}{2} \left(\frac{\partial u_I}{\partial X_J} + \frac{\partial u_J}{\partial X_I} \right) = \mathrm{O}(\epsilon) \quad \text{as } \epsilon \to 0. \tag{3.5.2}$$

Next consider

$$\frac{\partial u_I}{\partial x_j} = \frac{\partial u_I}{\partial X_K}\frac{\partial X_K}{\partial x_j} = \frac{\partial u_I}{\partial X_K}\left(\frac{\partial x_K}{\partial x_j} - \frac{\partial u_K}{\partial x_j}\right)$$

$$= \frac{\partial u_I}{\partial X_K}\delta_{Kj} + O(\epsilon^2) \text{ as } \epsilon \to 0,$$

$$\frac{\partial u_i}{\partial X_J} = \frac{\partial u_i}{\partial x_k}\frac{\partial x_k}{\partial X_J} = \frac{\partial u_i}{\partial x_k}\left(\frac{\partial u_k}{\partial X_J} + \frac{\partial X_k}{\partial X_J}\right)$$

$$= \frac{\partial u_i}{\partial x_k}\delta_{kJ} + O(\epsilon^2) \text{ as } \epsilon \to 0.$$

Thus, when terms of the order $O(\epsilon^2)$, as $\epsilon \to 0$, are neglected, it is immaterial whether the partial derivative of the displacement field $\mathbf{u}$ is taken with respect to x_j or X_j so that $\frac{\partial u_i}{\partial x_j} = \frac{\partial u_i}{\partial X_j}$; that is, $|\nabla\mathbf{u}| \approx |\nabla_0\mathbf{u}| = O(\epsilon)$. In other words, in the case of infinitesimal strains, no distinction is made between the material coordinates $\mathbf{X}$ and the spatial coordinates $\mathbf{x}$, and it is not necessary to distinguish between the Green–Lagrange strain tensor $\mathbf{E}$ and the Eulerian strain tensor $\mathbf{e}$. The infinitesimal strain tensor $\boldsymbol{\varepsilon}$ is defined as [see Eq. (3.5.2)]

$$\mathbf{E} \approx \boldsymbol{\varepsilon} = \tfrac{1}{2}\left[\nabla_0\mathbf{u} + (\nabla_0\mathbf{u})^{\mathrm{T}}\right]. \tag{3.5.3}$$

The rectangular Cartesian components of the infinitesimal strain tensor are

$$\varepsilon_{IJ} = \tfrac{1}{2}\left(\frac{\partial u_I}{\partial X_J} + \frac{\partial u_J}{\partial X_I}\right), \tag{3.5.4}$$

or, in expanded form,

$$\varepsilon_{11} = \frac{\partial u_1}{\partial X_1}; \qquad \varepsilon_{22} = \frac{\partial u_2}{\partial X_2};$$

$$\varepsilon_{33} = \frac{\partial u_3}{\partial X_3}; \qquad \varepsilon_{12} = \tfrac{1}{2}\left(\frac{\partial u_1}{\partial X_2} + \frac{\partial u_2}{\partial X_1}\right); \tag{3.5.5}$$

$$\varepsilon_{13} = \tfrac{1}{2}\left(\frac{\partial u_1}{\partial X_3} + \frac{\partial u_3}{\partial X_1}\right); \quad \varepsilon_{23} = \tfrac{1}{2}\left(\frac{\partial u_2}{\partial X_3} + \frac{\partial u_3}{\partial X_2}\right).$$

The strain components ε_{11}, ε_{22}, and ε_{33} are the infinitesimal normal strains and ε_{12}, ε_{13}, and ε_{23} are the infinitesimal shear strains. The shear strains $\gamma_{12} = 2\varepsilon_{12}$, $\gamma_{13} = 2\varepsilon_{13}$, and $\gamma_{23} = 2\varepsilon_{23}$ are called the *engineering shear strains*.

3.5.2 Physical Interpretation of Infinitesimal Strain Tensor Components

To gain insight into the physical meaning of the infinitesimal strain components, we write Eq. (3.4.10) in the form

$$(ds)^2 - (dS)^2 = 2\,d\mathbf{X} \cdot \boldsymbol{\varepsilon} \cdot d\mathbf{X} = 2\,\varepsilon_{ij}\,dX_i\,dX_j,$$

and dividing throughout by $(dS)^2$, we obtain

$$\frac{(ds)^2 - (dS)^2}{(dS)^2} = 2\,\varepsilon_{ij}\,\frac{dX_i}{dS}\,\frac{dX_j}{dS}.$$

Let $d\mathbf{X}/dS = \hat{\mathbf{N}}$, the unit vector in the direction of $d\mathbf{X}$. For small deformations, we have $ds + dS \approx 2dS$, and therefore we have

$$\frac{ds - dS}{dS} = \hat{\mathbf{N}} \cdot \boldsymbol{\varepsilon} \cdot \hat{\mathbf{N}} = \varepsilon_{ij} N_i N_j. \qquad (3.5.6)$$

The left side of Eq. (3.5.6) is the ratio of change in length per unit original length for a line element in the direction of $\hat{\mathbf{N}}$. For example, consider $\hat{\mathbf{N}}$ along the X_1-direction. Then we have

$$\frac{ds - dS}{dS} = \varepsilon_{11}.$$

Thus, the normal strain ε_{11} is the ratio of change in length of a line element that was parallel to the X_1-axis in the undeformed body to its original length. Similarly, for a line element along X_2 direction, $(ds - dS)/dS$ is the normal strain ε_{22}, and for a line element along the X_3 direction, $(ds - dS)/dS$ denotes the normal strain ε_{33}.

To understand the meaning of shear components of infinitesimal strain tensor, consider line elements $d\mathbf{X}^{(1)}$ and $d\mathbf{X}^{(2)}$ at a point in the body, which deform into line elements $d\mathbf{x}^{(1)}$ and $d\mathbf{x}^{(2)}$, respectively. Then we have [see Eqs. (3.3.1) and (3.4.3), and the first line of Eq. (3.4.11)]:

$$\begin{aligned}
d\mathbf{x}^{(1)} \cdot d\mathbf{x}^{(2)} = d\mathbf{X}^{(1)} \cdot \mathbf{F}^{\mathrm{T}} \cdot \mathbf{F} \cdot d\mathbf{X}^{(2)} &= d\mathbf{X}^{(1)} \cdot \mathbf{C} \cdot d\mathbf{X}^{(2)} \\
&= d\mathbf{X}^{(1)} \cdot (\mathbf{I} + 2\mathbf{E}) \cdot d\mathbf{X}^{(2)} \\
&= d\mathbf{X}^{(1)} \cdot d\mathbf{X}^{(2)} + 2d\mathbf{X}^{(1)} \cdot \mathbf{E} \cdot d\mathbf{X}^{(2)}. \qquad (3.5.7)
\end{aligned}$$

Now suppose that the line elements $d\mathbf{X}^{(1)}$ and $d\mathbf{X}^{(2)}$ are orthogonal to each other. Then

$$d\mathbf{x}^{(1)} \cdot d\mathbf{x}^{(2)} = 2d\mathbf{X}^{(1)} \cdot \mathbf{E} \cdot d\mathbf{X}^{(2)},$$

or

$$\begin{aligned}
2d\mathbf{X}^{(1)} \cdot \mathbf{E} \cdot d\mathbf{X}^{(2)} = dx^{(1)}dx^{(2)} \cos\theta &= dx^{(1)}dx^{(2)} \cos(\tfrac{\pi}{2} - \gamma_1 - \gamma_2) \\
&= dx^{(1)}dx^{(2)} \sin(\gamma_1 + \gamma_2) = dx^{(1)}dx^{(2)} \sin\gamma, \qquad (3.5.8)
\end{aligned}$$

where θ is the angle between the deformed line elements $dx^{(1)}$ and $dx^{(2)}$ and $\gamma = \gamma_1 + \gamma_2$ is the change in the angle from $90°$. For small deformations, we take $\sin\gamma \approx \gamma$, and obtain

$$\gamma = 2\frac{d\mathbf{X}^{(1)}}{dx^{(1)}} \cdot \mathbf{E} \cdot \frac{d\mathbf{X}^{(2)}}{dx^{(2)}} = 2\hat{\mathbf{N}}^{(1)} \cdot \boldsymbol{\varepsilon} \cdot \hat{\mathbf{N}}^{(2)}, \qquad (3.5.9)$$

where $\hat{\mathbf{N}}^{(1)} = d\mathbf{X}^{(1)}/dx^{(1)}$ and $\hat{\mathbf{N}}^{(2)} = d\mathbf{X}^{(2)}/dx^{(2)}$ are the unit vectors along the line elements $d\mathbf{X}^{(1)}$ and $d\mathbf{X}^{(2)}$, respectively. If the line elements $d\mathbf{X}^{(1)}$ and $d\mathbf{X}^{(2)}$ are taken along the X_1 and X_2 coordinates, respectively, then we have $2\varepsilon_{12} = \gamma$. Thus, the engineering shear strain $\gamma_{12} = 2\varepsilon_{12}$ represents the change in angle between line elements that were perpendicular to each other in the undeformed body.

3.5.3 Infinitesimal Rotation Tensor

The displacement gradient tensor can be expressed as the sum of a symmetric tensor and a skew symmetric tensor. We have

$$(\nabla \mathbf{u})^{\mathrm{T}} = \tfrac{1}{2}\left[(\nabla \mathbf{u})^{\mathrm{T}} + \nabla \mathbf{u}\right] + \tfrac{1}{2}\left[(\nabla \mathbf{u})^{\mathrm{T}} - \nabla \mathbf{u}\right] \equiv \tilde{\varepsilon} + \mathbf{\Omega}, \qquad (3.5.10)$$

where the symmetric part is similar to the infinitesimal strain tensor (and $\tilde{\varepsilon} \approx \varepsilon$ when $|\nabla \mathbf{u}| \approx |\nabla_0 \mathbf{u}| \ll 1$), and the skew symmetric part is known as the *infinitesimal rotation tensor*

$$\mathbf{\Omega} = \tfrac{1}{2}\left[(\nabla \mathbf{u})^{\mathrm{T}} - \nabla \mathbf{u}\right]. \qquad (3.5.11)$$

We note that there is no restriction placed on the magnitude of $\nabla \mathbf{u}$ in writing Eq. (3.5.10); $\tilde{\varepsilon}$ and $\mathbf{\Omega}$ do not have the meaning of infinitesimal strain and infinitesimal rotation tensors unless the deformation is infinitesimal (that is, $|\nabla_0 \mathbf{u}| \approx |\nabla \mathbf{u}|$).

From the definition, it follows that $\mathbf{\Omega}$ is a skew symmetric tensor, that is, $\mathbf{\Omega}^{\mathrm{T}} = -\mathbf{\Omega}$. In Cartesian component form we have

$$\Omega_{ij} = \tfrac{1}{2}\left(u_{i,j} - u_{j,i}\right), \qquad \Omega_{ij} = -\Omega_{ji}. \qquad (3.5.12)$$

Thus, there are only three independent components of $\mathbf{\Omega}$:

$$[\Omega] = \begin{bmatrix} 0 & \Omega_{12} & \Omega_{13} \\ -\Omega_{12} & 0 & \Omega_{23} \\ -\Omega_{13} & -\Omega_{23} & 0 \end{bmatrix}. \qquad (3.5.13)$$

The three components can be used to define the components of a vector $\boldsymbol{\omega}$,

$$\begin{aligned} \mathbf{\Omega} = -\mathcal{E} \cdot \boldsymbol{\omega} & \quad \text{or} \quad \boldsymbol{\omega} = -\tfrac{1}{2}\mathcal{E} : \mathbf{\Omega}, \\ \Omega_{ij} = -e_{ijk}\,\omega_k & \quad \text{or} \quad \omega_i = -\tfrac{1}{2}\,e_{ijk}\,\Omega_{jk}, \end{aligned} \qquad (3.5.14)$$

where $\mathcal{E}$ is the permutation (alternating) tensor, $\mathcal{E} = e_{ijk}\hat{\mathbf{e}}_i\hat{\mathbf{e}}_j\hat{\mathbf{e}}_k$. In view of Eqs. (3.5.11) and (3.5.14), it follows that

$$\omega_i = \tfrac{1}{2}\,e_{ijk}\,\frac{\partial u_k}{\partial x_j} \quad \text{or} \quad \boldsymbol{\omega} = \tfrac{1}{2}\nabla \times \mathbf{u}. \qquad (3.5.15)$$

Infinitesimal displacements of the form $d\mathbf{u} = \mathbf{\Omega} \cdot d\mathbf{x}$, where $\mathbf{\Omega}$ is independent of the position $\mathbf{x}$, are rigid-body rotations because

$$du_i = \Omega_{ij}\,dx_j = -e_{ijk}\,\omega_k\,dx_j = -(d\mathbf{x} \times \boldsymbol{\omega})_i = (\boldsymbol{\omega} \times d\mathbf{x})_i \quad \text{or} \quad d\mathbf{u} = \boldsymbol{\omega} \times d\mathbf{x}.$$

Thus, $\boldsymbol{\omega}$ represents the *infinitesimal rotation vector*; its magnitude is the angle of rotation and its direction gives the axis of rotation. We also note that $\nabla \cdot \boldsymbol{\omega} = 0$. Such vectors are called *solenoidal*. A *rigid-body motion* is one in which the relative distance between points is preserved.

Certain motions do not produce infinitesimal strains but they may produce finite strains. For example, consider the following deformation mapping:

$$\begin{aligned} \boldsymbol{\chi}(\mathbf{X}) = {} & (b_1 + X_1 + c_2 X_3 - c_3 X_2)\,\hat{\mathbf{e}}_1 + (b_2 + X_2 + c_3 X_1 - c_1 X_3)\,\hat{\mathbf{e}}_2 \\ & + (b_3 + X_3 + c_1 X_2 - c_2 X_1)\,\hat{\mathbf{e}}_3, \end{aligned} \qquad (3.5.16)$$

where b_i and c_i $(i = 1, 2, 3)$ are arbitrary constants. The displacement vector is

$$\mathbf{u}(\mathbf{X}) = (b_1 + c_2 X_3 - c_3 X_2)\,\hat{\mathbf{e}}_1 + (b_2 + c_3 X_1 - c_1 X_3)\,\hat{\mathbf{e}}_2$$
$$+ (b_3 + c_1 X_2 - c_2 X_1)\,\hat{\mathbf{e}}_3. \tag{3.5.17}$$

Therefore, we have

$$\frac{\partial u_1}{\partial X_1} = 0, \qquad \frac{\partial u_1}{\partial X_2} = -c_3, \qquad \frac{\partial u_1}{\partial X_3} = c_2,$$

$$\frac{\partial u_2}{\partial X_1} = c_3, \qquad \frac{\partial u_2}{\partial X_2} = 0, \qquad \frac{\partial u_2}{\partial X_3} = -c_1,$$

$$\frac{\partial u_3}{\partial X_1} = -c_2, \qquad \frac{\partial u_3}{\partial X_2} = c_1, \qquad \frac{\partial u_3}{\partial X_3} = 0,$$

Then the components of the deformation gradient $\mathbf{F}$ and left Cauchy–Green deformation tensor $\mathbf{C}$ associated with the mapping are

$$[F] = \begin{bmatrix} 1 & -c_3 & c_2 \\ c_3 & 1 & -c_1 \\ -c_2 & c_1 & 1 \end{bmatrix}, \quad [C] = \begin{bmatrix} 1 + c_2^2 + c_3^2 & -c_1 c_2 & -c_1 c_3 \\ -c_1 c_2 & 1 + c_1^2 + c_3^2 & -c_2 c_3 \\ -c_1 c_3 & -c_2 c_3 & 1 + c_1^2 + c_2^2 \end{bmatrix},$$

and the matrix of Green–Lagrange strain tensor components is

$$[E] = \frac{1}{2} \begin{bmatrix} c_2^2 + c_3^2 & -c_1 c_2 & -c_1 c_3 \\ -c_1 c_2 & c_1^2 + c_3^2 & -c_2 c_3 \\ -c_1 c_3 & -c_2 c_3 & c_1^2 + c_2^2 \end{bmatrix}. \tag{3.5.18}$$

Note that the linearized strains are all zero. Thus, for nonzero values of the constants c_i, the mapping produces nonzero finite strains. When all of the constants c_i are either zero or negligibly small (so that their products and squares are very small compared to unity), then $[F] = [I]$ and $[C] = [I]$, implying that the mapping $\mathbf{F} = \mathbf{R}$ represents a rigid-body rotation [that is, $\mathbf{U} = \mathbf{I}$; see Section 3.3.1). Figure 3.5.1 depicts the deformation for the two-dimensional case, with $b_1 = 2$, $b_2 = 3$, $c_3 = 1$, and all other constants zero. Thus, the finite strain tensor and deformation gradient give true measures of the deformation. The question of smallness of c_i in a given engineering application must be carefully examined before using linearized strains.

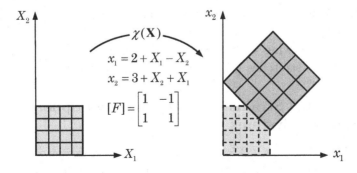

Fig. 3.5.1: A mapping that produces zero infinitesimal strains but nonzero finite strains.

Next, consider the mapping

$$\chi(\mathbf{X}) = (u_1 + X_1 \cos\theta - X_2 \sin\theta)\,\hat{\mathbf{e}}_1 + (u_2 + X_1 \sin\theta + X_2 \cos\theta)\,\hat{\mathbf{e}}_2 + X_3\,\hat{\mathbf{e}}_3,$$

$$(3.5.19)$$

where u_1 and u_2 denote the horizontal and vertical displacements of the point $(0,0,0)$, as shown in Fig. 3.5.2.

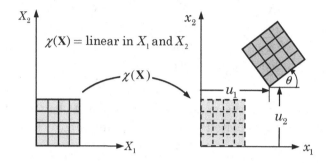

Fig. 3.5.2: A mapping that produces nonzero infinitesimal strains but zero finite strains.

The components of the deformation gradient $\mathbf{F}$ and left Cauchy–Green deformation tensor $\mathbf{C}$ are

$$[F] = \begin{bmatrix} \cos\theta & -\sin\theta & 0 \\ \sin\theta & \cos\theta & 0 \\ 0 & 0 & 1 \end{bmatrix}, \quad [C] = \begin{bmatrix} 1 & 0 & 0 \\ 0 & 1 & 0 \\ 0 & 0 & 1 \end{bmatrix}. \qquad (3.5.20)$$

Since $\mathbf{C} = \mathbf{I}$, we have $\mathbf{E} = \mathbf{0}$, indicating that the body does not experience stretching or shearing. The mapping is a rigid-body motion (both rigid-body translation and rigid-body rotation).

If we linearize the deformation mapping by making the approximations $\cos\theta \approx 1$ and $\sin\theta \approx 0$, we obtain

$$[F] = \begin{bmatrix} 1 & -\theta & 0 \\ \theta & 1 & 0 \\ 0 & 0 & 1 \end{bmatrix}, \quad [C] = \begin{bmatrix} 1+\theta^2 & 0 & 0 \\ 0 & 1+\theta^2 & 0 \\ 0 & 0 & 1 \end{bmatrix}.$$

Thus, the Green strain tensor components are no longer zero. The principal stretches $\lambda_1 = \lambda_2 = 1 + \theta^2$ are not equal to 1, as required by the definition of rigid-body motion. Owing to the artificial stretch induced by the linearization of the mapping, the stretches get larger and larger as the block rotates.

3.5.4 Infinitesimal Strains in Cylindrical and Spherical Coordinate Systems

The strains defined by Eq. (3.5.3) are valid in any coordinate system. Hence, they can be expressed in component form in any given coordinate system by expanding the strain tensors in the dyadic form and the operator $\nabla_0 = \nabla$ in that coordinate system, as given in Table 2.4.2 (see also Fig. 2.4.5).

3.5.4.1 Cylindrical coordinate system

In the cylindrical coordinate system we have

$$\mathbf{u} = u_r \hat{\mathbf{e}}_r + u_\theta \hat{\mathbf{e}}_\theta + u_z \hat{\mathbf{e}}_z, \tag{3.5.21}$$

$$\boldsymbol{\nabla}_0 = \hat{\mathbf{e}}_r \frac{\partial}{\partial r} + \frac{1}{r} \hat{\mathbf{e}}_\theta \frac{\partial}{\partial \theta} + \hat{\mathbf{e}}_z \frac{\partial}{\partial z}, \tag{3.5.22}$$

$$\frac{\partial \hat{\mathbf{e}}_r}{\partial \theta} = \hat{\mathbf{e}}_\theta, \qquad \frac{\partial \hat{\mathbf{e}}_\theta}{\partial \theta} = -\hat{\mathbf{e}}_r. \tag{3.5.23}$$

Using Eqs. (3.5.21)–(3.5.23), we obtain [see Eq. (2.5.27)]

$$\begin{aligned}
\boldsymbol{\nabla}_0 \mathbf{u} &= \hat{\mathbf{e}}_r \hat{\mathbf{e}}_r \frac{\partial u_r}{\partial r} + \hat{\mathbf{e}}_r \hat{\mathbf{e}}_\theta \frac{\partial u_\theta}{\partial r} + \frac{1}{r} \hat{\mathbf{e}}_\theta \hat{\mathbf{e}}_r \left(\frac{\partial u_r}{\partial \theta} - u_\theta \right) \\
&\quad + \hat{\mathbf{e}}_r \hat{\mathbf{e}}_z \frac{\partial u_z}{\partial r} + \hat{\mathbf{e}}_z \hat{\mathbf{e}}_r \frac{\partial u_r}{\partial z} + \frac{1}{r} \left(u_r + \frac{\partial u_\theta}{\partial \theta} \right) \hat{\mathbf{e}}_\theta \hat{\mathbf{e}}_\theta \\
&\quad + \frac{1}{r} \hat{\mathbf{e}}_\theta \hat{\mathbf{e}}_z \frac{\partial u_z}{\partial \theta} + \hat{\mathbf{e}}_z \hat{\mathbf{e}}_\theta \frac{\partial u_\theta}{\partial z} + \hat{\mathbf{e}}_z \hat{\mathbf{e}}_z \frac{\partial u_z}{\partial z},
\end{aligned} \tag{3.5.24}$$

$$\begin{aligned}
(\boldsymbol{\nabla}_0 \mathbf{u})^{\mathrm{T}} &= \hat{\mathbf{e}}_r \hat{\mathbf{e}}_r \frac{\partial u_r}{\partial r} + \hat{\mathbf{e}}_\theta \hat{\mathbf{e}}_r \frac{\partial u_\theta}{\partial r} + \frac{1}{r} \hat{\mathbf{e}}_r \hat{\mathbf{e}}_\theta \left(\frac{\partial u_r}{\partial \theta} - u_\theta \right) \\
&\quad + \hat{\mathbf{e}}_z \hat{\mathbf{e}}_r \frac{\partial u_z}{\partial r} + \hat{\mathbf{e}}_r \hat{\mathbf{e}}_z \frac{\partial u_r}{\partial z} + \frac{1}{r} \hat{\mathbf{e}}_\theta \hat{\mathbf{e}}_\theta \left(u_r + \frac{\partial u_\theta}{\partial \theta} \right) \\
&\quad + \frac{1}{r} \hat{\mathbf{e}}_z \hat{\mathbf{e}}_\theta \frac{\partial u_z}{\partial \theta} + \hat{\mathbf{e}}_\theta \hat{\mathbf{e}}_z \frac{\partial u_\theta}{\partial z} + \hat{\mathbf{e}}_z \hat{\mathbf{e}}_z \frac{\partial u_z}{\partial z}.
\end{aligned} \tag{3.5.25}$$

Substituting the above expressions into Eq. (3.5.3) and collecting the coefficients of various dyadics (that is, coefficients of $\hat{\mathbf{e}}_r \hat{\mathbf{e}}_r$, $\hat{\mathbf{e}}_r \hat{\mathbf{e}}_\theta$, and so on) we obtain the infinitesimal strain tensor components

$$\begin{aligned}
\varepsilon_{rr} &= \frac{\partial u_r}{\partial r}, & \varepsilon_{r\theta} &= \tfrac{1}{2} \left(\frac{1}{r} \frac{\partial u_r}{\partial \theta} + \frac{\partial u_\theta}{\partial r} - \frac{u_\theta}{r} \right), \\
\varepsilon_{rz} &= \tfrac{1}{2} \left(\frac{\partial u_r}{\partial z} + \frac{\partial u_z}{\partial r} \right), & \varepsilon_{\theta\theta} &= \frac{u_r}{r} + \frac{1}{r} \frac{\partial u_\theta}{\partial \theta}, \\
\varepsilon_{z\theta} &= \tfrac{1}{2} \left(\frac{\partial u_\theta}{\partial z} + \frac{1}{r} \frac{\partial u_z}{\partial \theta} \right), & \varepsilon_{zz} &= \frac{\partial u_z}{\partial z}.
\end{aligned} \tag{3.5.26}$$

3.5.4.2 Spherical coordinate system

In the spherical coordinate system, we have

$$\mathbf{u} = u_R \hat{\mathbf{e}}_R + u_\phi \hat{\mathbf{e}}_\phi + u_\theta \hat{\mathbf{e}}_\theta, \tag{3.5.27}$$

$$\boldsymbol{\nabla}_0 = \hat{\mathbf{e}}_R \frac{\partial}{\partial R} + \frac{1}{R} \hat{\mathbf{e}}_\phi \frac{\partial}{\partial \phi} + \frac{1}{R \sin \phi} \hat{\mathbf{e}}_\theta \frac{\partial}{\partial \theta}, \tag{3.5.28}$$

$$\begin{aligned}
\frac{\partial \hat{\mathbf{e}}_R}{\partial \phi} &= \hat{\mathbf{e}}_\phi, & \frac{\partial \hat{\mathbf{e}}_R}{\partial \theta} &= \sin \phi \, \hat{\mathbf{e}}_\theta, & \frac{\partial \hat{\mathbf{e}}_\phi}{\partial \phi} &= -\hat{\mathbf{e}}_R, \\
\frac{\partial \hat{\mathbf{e}}_\phi}{\partial \theta} &= \cos \phi \, \hat{\mathbf{e}}_\theta, & \frac{\partial \hat{\mathbf{e}}_\theta}{\partial \theta} &= -\sin \phi \, \hat{\mathbf{e}}_R - \cos \phi \, \hat{\mathbf{e}}_\phi.
\end{aligned} \tag{3.5.29}$$

Using Eqs. (3.5.27)–(3.5.29), we obtain [see Eq. (2.5.29)]

$$
\begin{aligned}
\boldsymbol{\nabla}_0 \mathbf{u} =\ & \frac{\partial u_R}{\partial R}\hat{\mathbf{e}}_R\hat{\mathbf{e}}_R + \frac{\partial u_\phi}{\partial R}\hat{\mathbf{e}}_R\hat{\mathbf{e}}_\phi + \frac{\partial u_\theta}{\partial R}\hat{\mathbf{e}}_R\hat{\mathbf{e}}_\theta \\
& + \frac{1}{R}\left(\frac{\partial u_R}{\partial \phi} - u_\phi\right)\hat{\mathbf{e}}_\phi\hat{\mathbf{e}}_R + \frac{1}{R}\left(\frac{\partial u_\phi}{\partial \phi} + u_R\right)\hat{\mathbf{e}}_\phi\hat{\mathbf{e}}_\phi + \frac{1}{R}\frac{\partial u_\theta}{\partial \phi}\hat{\mathbf{e}}_\phi\hat{\mathbf{e}}_\theta \\
& + \frac{1}{R\sin\phi}\left(\frac{\partial u_R}{\partial \theta} - u_\theta\sin\phi\right)\hat{\mathbf{e}}_\theta\hat{\mathbf{e}}_R + \frac{1}{R\sin\phi}\left(\frac{\partial u_\phi}{\partial \theta} - u_\theta\cos\phi\right)\hat{\mathbf{e}}_\theta\hat{\mathbf{e}}_\phi \\
& + \frac{1}{R\sin\phi}\left(\frac{\partial u_\theta}{\partial \theta} + u_R\sin\phi + u_\phi\cos\phi\right)\hat{\mathbf{e}}_\theta\hat{\mathbf{e}}_\theta,
\end{aligned}
\tag{3.5.30}
$$

$$
\begin{aligned}
(\boldsymbol{\nabla}_0 \mathbf{u})^{\mathrm{T}} =\ & \frac{\partial u_R}{\partial R}\hat{\mathbf{e}}_R\hat{\mathbf{e}}_R + \frac{\partial u_\phi}{\partial R}\hat{\mathbf{e}}_\phi\hat{\mathbf{e}}_R + \frac{\partial u_\theta}{\partial R}\hat{\mathbf{e}}_\theta\hat{\mathbf{e}}_R \\
& + \frac{1}{R}\left(\frac{\partial u_R}{\partial \phi} - u_\phi\right)\hat{\mathbf{e}}_R\hat{\mathbf{e}}_\phi + \frac{1}{R}\left(\frac{\partial u_\phi}{\partial \phi} + u_R\right)\hat{\mathbf{e}}_\phi\hat{\mathbf{e}}_\phi + \frac{1}{R}\frac{\partial u_\theta}{\partial \phi}\hat{\mathbf{e}}_\theta\hat{\mathbf{e}}_\phi \\
& + \frac{1}{R\sin\phi}\left(\frac{\partial u_R}{\partial \theta} - u_\theta\sin\phi\right)\hat{\mathbf{e}}_R\hat{\mathbf{e}}_\theta + \frac{1}{R\sin\phi}\left(\frac{\partial u_\phi}{\partial \theta} - u_\theta\cos\phi\right)\hat{\mathbf{e}}_\phi\hat{\mathbf{e}}_\theta \\
& + \frac{1}{R\sin\phi}\left(\frac{\partial u_\theta}{\partial \theta} + u_R\sin\phi + u_\phi\cos\phi\right)\hat{\mathbf{e}}_\theta\hat{\mathbf{e}}_\theta.
\end{aligned}
\tag{3.5.31}
$$

Substituting the above expressions into Eq. (3.5.3) and collecting the coefficients of various dyadics, we obtain the following infinitesimal strain tensor components in the spherical coordinate system:

$$
\begin{aligned}
& \varepsilon_{RR} = \frac{\partial u_R}{\partial R}, \quad \varepsilon_{\phi\phi} = \frac{1}{R}\left(\frac{\partial u_\phi}{\partial \phi} + u_R\right), \\
& \varepsilon_{R\phi} = \tfrac{1}{2}\left(\frac{1}{R}\frac{\partial u_R}{\partial \phi} + \frac{\partial u_\phi}{\partial R} - \frac{u_\phi}{R}\right), \\
& \varepsilon_{R\theta} = \tfrac{1}{2}\left(\frac{1}{R\sin\phi}\frac{\partial u_R}{\partial \theta} + \frac{\partial u_\theta}{\partial R} - \frac{u_\theta}{R}\right), \\
& \varepsilon_{\phi\theta} = \tfrac{1}{2}\frac{1}{R}\left(\frac{1}{\sin\phi}\frac{\partial u_\phi}{\partial \theta} + \frac{\partial u_\theta}{\partial \phi} - u_\theta\cot\phi\right), \\
& \varepsilon_{\theta\theta} = \frac{1}{R\sin\phi}\left(\frac{\partial u_\theta}{\partial \theta} + u_R\sin\phi + u_\phi\cos\phi\right).
\end{aligned}
\tag{3.5.32}
$$

3.6 Velocity Gradient and Vorticity Tensors

3.6.1 Definitions

In fluid mechanics, the velocity vector $\mathbf{v}(\mathbf{x}, t)$ is the variable of interest. Similar to the displacement gradient tensor [see Eq. (3.5.10)], we can write the *velocity gradient tensor* $\mathbf{L}$ as the sum of symmetric $\mathbf{D}$ and skew-symmetric $\mathbf{W}$ tensors:

$$
\mathbf{L} \equiv (\boldsymbol{\nabla}\mathbf{v})^{\mathrm{T}} = \tfrac{1}{2}\left[(\boldsymbol{\nabla}\mathbf{v})^{\mathrm{T}} + \boldsymbol{\nabla}\mathbf{v}\right] + \tfrac{1}{2}\left[(\boldsymbol{\nabla}\mathbf{v})^{\mathrm{T}} - \boldsymbol{\nabla}\mathbf{v}\right] \equiv \mathbf{D} + \mathbf{W},
\tag{3.6.1}
$$

where $\mathbf{D}$ is called the *rate of deformation* tensor (or *rate of strain tensor*) and $\mathbf{W}$ is called the *vorticity tensor* or *spin tensor*:

$$
\mathbf{D} = \tfrac{1}{2}\left[(\boldsymbol{\nabla}\mathbf{v})^{\mathrm{T}} + \boldsymbol{\nabla}\mathbf{v}\right], \quad \mathbf{W} = \tfrac{1}{2}\left[(\boldsymbol{\nabla}\mathbf{v})^{\mathrm{T}} - \boldsymbol{\nabla}\mathbf{v}\right].
\tag{3.6.2}
$$

It follows that

$$\mathbf{D} = \tfrac{1}{2}\left(\mathbf{L} + \mathbf{L}^{\mathrm{T}}\right), \quad \mathbf{W} = \tfrac{1}{2}\left(\mathbf{L} - \mathbf{L}^{\mathrm{T}}\right). \tag{3.6.3}$$

The skew-symmetric tensor $\mathbf{W}$ (i.e., $\mathbf{W}^{\mathrm{T}} = -\mathbf{W}$), has only three independent scalar components; they can be used to define the scalar components of a vector $\mathbf{w}$, called the *axial vector of* $\mathbf{W}$, as follows:

$$W_{ij} = -e_{ijk}w_k, \quad [W] = \begin{bmatrix} 0 & -w_3 & w_2 \\ w_3 & 0 & -w_1 \\ -w_2 & w_1 & 0 \end{bmatrix}. \tag{3.6.4}$$

The scalar components of $\mathbf{w}$ can be expressed in terms of the scalar components of $\mathbf{W}$ as

$$w_i = -\tfrac{1}{2}e_{ijk}W_{jk} = \tfrac{1}{2}e_{ijk}\frac{\partial v_k}{\partial x_j} \quad \text{or} \quad \mathbf{w} = \tfrac{1}{2}\boldsymbol{\nabla} \times \mathbf{v}. \tag{3.6.5}$$

Thus, $\mathbf{w}$ is also known as the *vorticity vector*. Note that $\boldsymbol{\nabla} \cdot \mathbf{w} = 0$ by virtue of the vector identity (that is, divergence of the curl of a vector is zero). Thus, the vorticity vector is divergence-free. As discussed in Section 3.5.3, if a velocity vector $\mathbf{v}$ is of the form $\mathbf{v} = \mathbf{W} \cdot \mathbf{x}$ for some skew symmetric tensor $\mathbf{W}$ that is independent of position $\mathbf{x}$, then the motion is a uniform rigid-body rotation about the origin with angular velocity $\mathbf{w}$. Also note that the first and third principal invariants of $\mathbf{W}$ are zero, and the second principal invariant is equal to $w_1^2 + w_2^2 + w_3^2$.

3.6.2 Relationship Between D and Ė

Note that the rate of deformation tensor $\mathbf{D}$ is not the same as the time rate of change of the infinitesimal strain tensor $\boldsymbol{\varepsilon}$ [see Eq. (3.5.1)], that is, the strain rate $\dot{\boldsymbol{\varepsilon}}$, where the superposed dot signifies the material time derivative. However, $\mathbf{D}$ is related to $\dot{\mathbf{E}}$, the time rate of change of Green–Lagrange strain tensor $\mathbf{E}$, as shown in the following paragraphs.

Taking the total time derivative of the expression in Eq. (3.4.10), we obtain

$$\frac{d}{dt}[(ds)^2 - (dS)^2] = \frac{d}{dt}[(ds)^2] = 2\,d\mathbf{X} \cdot \frac{d\mathbf{E}}{dt} \cdot d\mathbf{X} \equiv 2\,d\mathbf{X} \cdot \dot{\mathbf{E}} \cdot d\mathbf{X}, \tag{3.6.6}$$

where we used the fact that $d\mathbf{X}$ and dS are constants. On the other hand, the instantaneous rate of change of the squared length $(ds)^2$ is

$$\frac{d}{dt}[(ds)^2] = \frac{d}{dt}[d\mathbf{x} \cdot d\mathbf{x}] = 2\,d\mathbf{x} \cdot \frac{d}{dt}(d\mathbf{x}) = 2\,d\mathbf{x} \cdot d\mathbf{v}. \tag{3.6.7}$$

Because

$$\mathbf{L} = (\boldsymbol{\nabla}\mathbf{v})^{\mathrm{T}} \;\Rightarrow\; d\mathbf{v} = \mathbf{L} \cdot d\mathbf{x}, \tag{3.6.8}$$

Eq. (3.6.7) takes the form [we make use of Eq. (3.6.1)]

$$\frac{d}{dt}[(ds)^2] = 2\,d\mathbf{x} \cdot \mathbf{L} \cdot d\mathbf{x} = 2\,d\mathbf{x} \cdot (\mathbf{D} + \mathbf{W}) \cdot d\mathbf{x} = 2\,d\mathbf{x} \cdot \mathbf{D} \cdot d\mathbf{x}. \tag{3.6.9}$$

The second term is zero because of the skew symmetry of $\mathbf{W}$. Now equating the right-hand sides of Eqs. (3.6.6) and (3.6.9) and noting $d\mathbf{x} = \mathbf{F} \cdot d\mathbf{X} = d\mathbf{X} \cdot \mathbf{F}^{\mathrm{T}}$

$$dX \cdot \dot{E} \cdot dX = dx \cdot D \cdot dx = dX \cdot F^T \cdot D \cdot F \cdot dX,$$

we arrive at the result

$$\dot{E} = F^T \cdot D \cdot F. \tag{3.6.10}$$

Next consider $(dx = F \cdot dX)$

$$dv = \frac{d}{dt}(dx) = \left[\frac{dF}{dt}\right] \cdot dX + F \cdot \frac{d(dX)}{dt} = \dot{F} \cdot dX + 0, \tag{3.6.11}$$

because $d\dot{X} = 0$. Then from Eqs. (3.6.8) and (3.6.11) we have

$$\dot{F} \cdot dX = L \cdot dx. \tag{3.6.12}$$

We also have

$$\dot{F} = \frac{d}{dt}(\nabla_0 x)^T = (\nabla_0 v)^T. \tag{3.6.13}$$

Note that $\nabla_0 v$ is the gradient of the velocity vector v with respect to the material coordinates X, and it is not the same as $L = (\nabla v)^T$. From Eq. (3.6.12) or from Eq. (3.6.13) and the identity [see Problem **3.36**]

$$(\nabla_0 v)^T = L \cdot F, \tag{3.6.14}$$

we obtain

$$\dot{F} = L \cdot F \quad \text{or} \quad L = \dot{F} \cdot F^{-1}. \tag{3.6.15}$$

3.7 Compatibility Equations

3.7.1 Preliminary Comments

The task of computing strains (infinitesimal or finite) from a given displacement field is a straightforward exercise. However, sometimes we face the problem of finding the displacements from a given strain field. This is not as straightforward because there are *six* independent partial differential equations (that is, strain-displacement relations) for only *three* unknown displacements, which would in general over-determine the solution. We will find some conditions, known as *Saint-Venant's compatibility equations*, that will ensure the computation of a unique displacement field from a given strain field. The derivation is presented for infinitesimal strains. For finite strains the same steps may be followed but the process is so difficult that it is never attempted (although some general compatibility conditions may be stated to ensure integrability of the six nonlinear partial differential equations).

To understand the meaning of strain compatibility, imagine that a material body is cut up into pieces before it is strained, and then each piece is given a certain strain. The strained pieces cannot be fitted back into a single continuous body without further deformation. On the other hand, if the strain in each piece is related to or compatible with the strains in the neighboring pieces, then they can be fitted together to form a continuous body. Mathematically, the six strain–displacement relations that connect six strain components to the three displacement components should be consistent.

3.7.2 Infinitesimal Strains

The infinitesimal strain tensor $\mathbf{E} \approx \boldsymbol{\varepsilon}$ is defined in terms of the displacement vector $\mathbf{u}$ as [see Eq. (3.5.3)]

$$\boldsymbol{\varepsilon} = \tfrac{1}{2}\left[\boldsymbol{\nabla}_0\mathbf{u} + (\boldsymbol{\nabla}_0\mathbf{u})^{\mathrm{T}}\right], \quad \varepsilon_{ij} = \tfrac{1}{2}\left(\frac{\partial u_i}{\partial X_j} + \frac{\partial u_j}{\partial X_i}\right). \tag{3.7.1}$$

We begin with infinitesimal strains in two dimensions. We have the following three strain-displacement relations:

$$\frac{\partial u_1}{\partial X_1} = \varepsilon_{11},$$

$$\frac{\partial u_2}{\partial X_2} = \varepsilon_{22}, \tag{3.7.2}$$

$$\frac{\partial u_1}{\partial X_2} + \frac{\partial u_2}{\partial X_1} = 2\varepsilon_{12}.$$

If the given data $(\varepsilon_{11}, \varepsilon_{22}, \varepsilon_{12})$ are compatible (or consistent), any two of the three equations should yield the same displacement components. For example, consider the following infinitesimal strain field:

$$\varepsilon_{11} = \varepsilon_{22} = 0, \quad \varepsilon_{12} = X_1 X_2.$$

In terms of the displacement components u_1 and u_2, we have

$$\frac{\partial u_1}{\partial X_1} = 0, \quad \frac{\partial u_2}{\partial X_2} = 0, \quad \frac{\partial u_1}{\partial X_2} + \frac{\partial u_2}{\partial X_1} = 2X_1X_2.$$

Integration of the first two equations gives

$$u_1 = f(X_2), \quad u_2 = g(X_1).$$

On substitution into the shear strain, we obtain

$$\frac{df}{dX_2} + \frac{dg}{dX_1} = 2X_1X_2,$$

which cannot be satisfied; if ε_{12} is specified as $\varepsilon_{12} = c_1X_1 + c_2X_2$, it would be possible to determine f and g, and then u_1 and u_2. Thus, not all arbitrarily specified strain fields are compatible.

The compatibility of a given strain field can be established as follows. Differentiate the first equation with respect to X_2 twice, the second equation with respect to X_1 twice, and the third equation with respect to X_1 and X_2 each, and obtain

$$\frac{\partial^3 u_1}{\partial X_1 \partial X_2^2} = \frac{\partial^2 \varepsilon_{11}}{\partial X_2^2},$$

$$\frac{\partial^3 u_2}{\partial X_2 \partial X_1^2} = \frac{\partial^2 \varepsilon_{22}}{\partial X_1^2}, \tag{3.7.3}$$

$$\frac{\partial^3 u_1}{\partial X_2^2 \partial X_1} + \frac{\partial^3 u_2}{\partial X_1^2 \partial X_2} = 2\frac{\partial^2 \varepsilon_{12}}{\partial X_1 \partial X_2}.$$

Using the first two equations in the third equation, we arrive at the following relation among the three strains:

$$\frac{\partial^2 \varepsilon_{11}}{\partial X_2^2} + \frac{\partial^2 \varepsilon_{22}}{\partial X_1^2} = 2\frac{\partial^2 \varepsilon_{12}}{\partial X_1 \partial X_2}. \tag{3.7.4}$$

Equation (3.7.4) is called the strain compatibility condition among the three strains ε_{11}, ε_{22}, and ε_{12} that ensures the integrability of Eqs. (3.7.2) to determine the displacement components (u_1, u_2).

A similar procedure can be followed to obtain the strain compatibility equations for the three-dimensional case. In addition to Eq. (3.7.4), five more such conditions can be derived, as given below:

$$\frac{\partial^2 \varepsilon_{11}}{\partial X_3^2} + \frac{\partial^2 \varepsilon_{33}}{\partial X_1^2} = 2\frac{\partial^2 \varepsilon_{13}}{\partial X_1 \partial X_3}, \tag{3.7.5}$$

$$\frac{\partial^2 \varepsilon_{22}}{\partial X_3^2} + \frac{\partial^2 \varepsilon_{33}}{\partial X_2^2} = 2\frac{\partial^2 \varepsilon_{23}}{\partial X_2 \partial X_3}, \tag{3.7.6}$$

$$\frac{\partial^2 \varepsilon_{11}}{\partial X_2 \partial X_3} + \frac{\partial^2 \varepsilon_{23}}{\partial X_1^2} = \frac{\partial^2 \varepsilon_{13}}{\partial X_1 \partial X_2} + \frac{\partial^2 \varepsilon_{12}}{\partial X_1 \partial X_3}, \tag{3.7.7}$$

$$\frac{\partial^2 \varepsilon_{22}}{\partial X_1 \partial X_3} + \frac{\partial^2 \varepsilon_{13}}{\partial X_2^2} = \frac{\partial^2 \varepsilon_{23}}{\partial X_1 \partial X_2} + \frac{\partial^2 \varepsilon_{12}}{\partial X_2 \partial X_3}, \tag{3.7.8}$$

$$\frac{\partial^2 \varepsilon_{33}}{\partial X_1 \partial X_2} + \frac{\partial^2 \varepsilon_{12}}{\partial X_3^2} = \frac{\partial^2 \varepsilon_{13}}{\partial X_2 \partial X_3} + \frac{\partial^2 \varepsilon_{23}}{\partial X_1 \partial X_3}. \tag{3.7.9}$$

Equations (3.7.4)–(3.7.9) can be written as a single relation using the index notation

$$\frac{\partial^2 \varepsilon_{mn}}{\partial X_i \partial X_j} + \frac{\partial^2 \varepsilon_{ij}}{\partial X_m \partial X_n} = \frac{\partial^2 \varepsilon_{im}}{\partial X_j \partial X_n} + \frac{\partial^2 \varepsilon_{jn}}{\partial X_i \partial X_m}, \tag{3.7.10}$$

which yields $(3)^4 = 81$ equations but only 6, shown in Eqs. (3.7.4)–(3.7.9), are distinctly different. These conditions are both necessary and sufficient to determine a single-valued displacement field. Similar compatibility conditions hold for the rate of deformation tensor $\mathbf{D}$. The vector form of Eqs. (3.7.4)–(3.7.9) is given by [see Problem 3.39]

$$\nabla_0 \times (\nabla_0 \times \boldsymbol{\varepsilon})^{\mathrm{T}} = \mathbf{0} \quad \text{or} \quad e_{ikr}\, e_{jls}\, \varepsilon_{ij,kl} = 0. \tag{3.7.11}$$

Example 3.7.1 illustrates how to check the compatibility of a given strain field.

Example 3.7.1

Given the following two-dimensional, infinitesimal strain field:

$$\varepsilon_{11} = c_1 X_1 \left(X_1^2 + X_2^2 \right), \quad \varepsilon_{22} = \tfrac{1}{3} c_2 X_1^3, \quad \varepsilon_{12} = c_3 X_1^2 X_2,$$

where c_1, c_2, and c_3 are constants, determine if the strain field is compatible.

Solution: Using Eq. (3.7.4) we obtain

$$\frac{\partial^2 \varepsilon_{11}}{\partial X_2^2} + \frac{\partial^2 \varepsilon_{22}}{\partial X_1^2} - 2\frac{\partial^2 \varepsilon_{12}}{\partial X_1 \partial X_2} = 2c_1 X_1 + 2c_2 X_1 - 4c_3 X_1.$$

Thus the strain field is not compatible, unless $c_1 + c_2 - 2c_3 = 0$.

Example 3.7.2 illustrates how to determine the displacement field from a compatible strain field.

Example 3.7.2

Consider the problem of an isotropic cantilever beam bent by a load P at the free end, as shown in Fig. 3.7.1. To study the beam problem as a two-dimensional elasticity problem, consider the strain field [strains ε_{11} and ε_{12} are known from a book on mechanics of solids; see, e.g., Fenner and Reddy (2012)]:

$$\varepsilon_{11} = -\frac{PX_1X_2}{EI}, \quad \varepsilon_{12} = -\frac{(1+\nu)P}{2EI}(h^2 - X_2^2), \quad \varepsilon_{22} = -\nu\varepsilon_{11} = \frac{\nu PX_1X_2}{EI}, \quad (3.7.12)$$

where I is the moment of inertia about the X_3-axis ($I = 2bh^3/3$), ν is the Poisson ratio, E is Young's modulus, $2h$ is the height of the beam, and b is the width of the beam. Determine if the strain field is compatible and, if it is compatible, find the two-dimensional displacement field (u_1, u_2) that satisfies the kinematic boundary conditions and, therefore, is free of rigid-body translation and rotation.

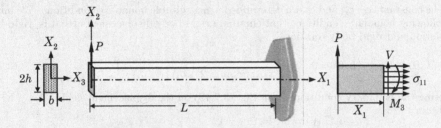

Fig. 3.7.1: Cantilever beam bent by a point load.

Solution: **(a)** Substituting ε_{ij} into the compatibility equation (3.7.4), we obtain $0 + 0 = 0$. Thus the strains in Eq. (3.7.12) satisfy the compatibility conditions for a two-dimensional state of deformation. One can verify that the three-dimensional strains are not compatible; one can show that all of the compatibility equations except Eq. (3.7.9) are satisfied.

(b) Integrating the strain-displacement equations, we obtain

$$\frac{\partial u_1}{\partial X_1} = \varepsilon_{11} = -\frac{PX_1X_2}{EI} \quad \text{or} \quad u_1 = -\frac{PX_1^2X_2}{2EI} + f(X_2), \quad (3.7.13)$$

$$\frac{\partial u_2}{\partial X_2} = \varepsilon_{22} = \frac{\nu PX_1X_2}{EI} \quad \text{or} \quad u_2 = \frac{\nu PX_1X_2^2}{2EI} + g(X_1), \quad (3.7.14)$$

where $f(X_2)$ and $g(X_1)$ are functions of integration. Substituting u_1 and u_2 into the definition of $2\varepsilon_{12}$, we obtain

$$2\varepsilon_{12} = \frac{\partial u_1}{\partial X_2} + \frac{\partial u_2}{\partial X_1} = -\frac{PX_1^2}{2EI} + \frac{df}{dX_2} + \frac{\nu PX_2^2}{2EI} + \frac{dg}{dX_1}. \quad (3.7.15)$$

But this must be equal to the shear strain known from Eq. (3.7.12):

$$-\frac{P}{2EI}X_1^2 + \frac{df}{dX_2} + \frac{\nu P}{2EI}X_2^2 + \frac{dg}{dX_1} = -\frac{(1+\nu)}{EI}P(h^2 - X_2^2).$$

Separating the terms that depend only on X_1 and those depend only on X_2 (the constant term can go with either one), we obtain

$$-\frac{dg}{dX_1} + \frac{P}{2EI}X_1^2 - \frac{(1+\nu)Ph^2}{EI} = \frac{df}{dX_2} - \frac{(2+\nu)P}{2EI}X_2^2. \quad (3.7.16)$$

Since the left side depends only on X_1 and the right side depends only on X_2, and yet the equality must hold, it follows that both sides should be equal to a constant, say c_0:

$$\frac{df}{dX_2} - \frac{(2+\nu)P}{2EI}X_2^2 = c_0, \quad -\frac{dg}{dX_1} + \frac{P}{2EI}X_1^2 - \frac{(1+\nu)Ph^2}{EI} = c_0.$$

Integrating the expressions for f and g, we obtain

$$f(X_2) = \frac{(2+\nu)P}{6EI}X_2^3 + c_0 X_2 + c_1,$$

$$g(X_1) = \frac{P}{6EI}X_1^3 - \frac{(1+\nu)Ph^2}{EI}X_1 - c_0 X_1 + c_2, \tag{3.7.17}$$

where c_1 and c_2 are constants of integration. Thus, the most general form of displacement field (u_1, u_2) that corresponds to the strains in Eq. (3.7.12) is given by

$$u_1(X_1, X_2) = -\frac{P}{2EI}X_1^2 X_2 + \frac{(2+\nu)P}{6EI}X_2^3 + c_0 X_2 + c_1,$$

$$u_2(X_1, X_2) = -\frac{(1+\nu)Ph^2}{EI}X_1 + \frac{\nu P}{2EI}X_1 X_2^2 + \frac{P}{6EI}X_1^3 - c_0 X_1 + c_2. \tag{3.7.18}$$

(c) The constants c_0, c_1, and c_2 are determined using suitable boundary conditions. We impose the following boundary conditions that eliminate rigid-body displacements (that is, rigid-body translation and rigid-body rotation):

$$u_1(L, 0) = 0, \quad u_2(L, 0) = 0, \quad \Omega_{12}\Big|_{X_1=L, X_2=0} = \frac{1}{2}\left(\frac{\partial u_2}{\partial X_1} - \frac{\partial u_1}{\partial X_2}\right)_{X_1=L, X_2=0} = 0. \tag{3.7.19}$$

Imposing the boundary conditions from Eq. (3.7.19) on the displacement field in Eq. (3.7.18), we obtain

$$u_1(L, 0) = 0 \;\Rightarrow\; c_1 = 0,$$

$$u_2(L, 0) = 0 \;\Rightarrow\; c_0 L - c_2 = -\frac{(1+\nu)Ph^2 L}{EI} + \frac{PL^3}{6EI},$$

$$\left(\frac{\partial u_2}{\partial X_1} - \frac{\partial u_1}{\partial X_2}\right)_{X_1=L, X_2=0} = 0, \;\Rightarrow\; c_0 = \frac{PL^2}{2EI} - \frac{(1+\nu)Ph^2}{2EI} \tag{3.7.20}$$

Thus, we have

$$c_0 = \frac{PL^2}{2EI} - \frac{(1+\nu)Ph^2}{2EI}, \quad c_1 = 0, \quad c_2 = \frac{PL^3}{3EI} + \frac{(1+\nu)Ph^2 L}{2EI}. \tag{3.7.21}$$

Then the final displacement field in Eq. (3.7.18) becomes

$$u_1(X_1, X_2) = \frac{PL^2 X_2}{6EI}\left[3\left(1 - \frac{X_1^2}{L^2}\right) + (2+\nu)\frac{X_2^2}{L^2} - 3(1+\nu)\frac{h^2}{L^2}\right],$$

$$u_2(X_1, X_2) = \frac{PL^3}{6EI}\left[2 - 3\frac{X_1}{L}\left(1 - \nu\frac{X_2^2}{L^2}\right) + \frac{X_1^3}{L^3} + 3(1+\nu)\frac{h^2}{L^2}\left(1 - \frac{X_1}{L}\right)\right]. \tag{3.7.22}$$

In the Euler–Bernoulli beam theory (EBT), where one assumes that $L >> 2h$ and $\nu = 0$, we have $u_1 = 0$, and u_2 is given by

$$u_2^{\text{EBT}}(X_1, X_2) = \frac{PL^3}{6EI}\left(2 - 3\frac{X_1}{L} + \frac{X_1^3}{L^3}\right),$$

while in the Timoshenko beam theory (TBT) we have $u_1 = 0$ [$E = 2(1+\nu)G$, $I = Ah^2/3$, and $A = 2bh$], and u_2 is given by

$$u_2^{\text{TBT}}(X_1, X_2) = \frac{PL^3}{6EI}\left(2 - 3\frac{X_1}{L} + \frac{X_1^3}{L^3}\right) + \frac{PL}{K_s GA}\left(1 - \frac{X_1}{L}\right).$$

Here K_s denotes the shear correction factor. Thus, the Timoshenko beam theory with shear correction factor of $K_s = 4/3$ predicts the same maximum deflection, $u_2(0, 0)$, as the two-dimensional elasticity theory [see Reddy (2002) for more details on the Timoshenko beam theory]. Both beam theory solutions, in general, are in error compared to the elasticity solution (primarily because of the Poisson effect).

3.7.3 Finite Strains

In the case of finite strains, the compatibility conditions in terms of the deformation tensor[6] are derived from the mathematical requirement that the curl of a gradient be zero. Since $\mathbf{F}$ is the gradient of $\mathbf{x}$ with respect to $\mathbf{X}$, we require that

$$\nabla_0 \times \mathbf{F}^{\mathrm{T}} = \mathbf{0}, \qquad (3.7.23)$$

or, equivalently,

$$F_{iJ,K} = F_{iK,J} \quad \text{or} \quad \frac{\partial^2 x_i}{\partial X_J \partial X_K} = \frac{\partial^2 x_i}{\partial X_K \partial X_J}. \qquad (3.7.24)$$

We close this section with a note that the compatibility conditions arise only when the strains (or stresses) are used to formulate the problem and displacements are to be determined. An example of such a situation arises in plane elasticity where stresses are expressed in terms of a single function, called *stress function*. When boundary value problems in mechanics are formulated in terms of the displacements or velocities, the question of strain compatibility never arises.

3.8 Rigid-Body Motions and Material Objectivity

3.8.1 Superposed Rigid-Body Motions

3.8.1.1 Introduction and rigid-body transformation

A rigid-body motion is one that preserves the relative distance between points. Intuitively, rigid-body motions of a whole body should have no effect on the values of computed strains, as they are based on the change of length and orientation of line elements in a small neighborhood of a point. However, mathematically, various measures of strains with superposed rigid-body motions may be expressed in different ways, although the computed values are independent of the rigid-body motion of the body[7].

Consider the motion mapping from Eq. (3.2.1), $\chi : \kappa_0 \rightarrow \kappa$,

$$\mathbf{x} = \chi(\mathbf{X}, t). \qquad (3.8.1)$$

Then a material particle X of a body, occupying position $\mathbf{X}$ in the reference configuration κ_0, now occupies a position $\mathbf{x}$ in κ at time t, as specified by the motion (3.8.1). Under a superposed rigid-body motion, the particle X that is at $\mathbf{x}$ at time t moves to a place $\mathbf{x}^*$ at time $t^* = t + a$, where a is a constant. In the following discussion, we shall use an asterisk (*) on all quantities with the superposed motion. Thus,

$$\mathbf{x}^* = \chi^*(\mathbf{X}, t^*) = \chi^*(\mathbf{X}, t). \qquad (3.8.2)$$

---

[6]The derivation of compatibility conditions in terms of the right Cauchy–Green deformation tensor $\mathbf{C}$ or the Green–Lagrange strain tensor $\mathbf{E}$ is quite involved and not attempted here.

[7]The discussion presented here closely follows that by Naghdi (2001).

Next, consider another material particle Y of the body in the reference configuration that occupies a position $\mathbf{y}$ in κ at time t, as specified by the motion (3.8.1)

$$\mathbf{y} = \chi(\mathbf{Y}, t). \tag{3.8.3}$$

Under the superposed motion the material particle that is at $\mathbf{y}$ at time t moves to a place $\mathbf{y}^*$ at time t^*. Then we have

$$\mathbf{y}^* = \chi^*(\mathbf{Y}, t^*) = \chi^*(\mathbf{Y}, t). \tag{3.8.4}$$

We can use the inverse mapping χ^{-1} to write $\mathbf{X}$ and $\mathbf{Y}$ in terms of $\mathbf{x}$ and $\mathbf{y}$, respectively. Hence, we have

$$\mathbf{x}^* = \chi^*(\mathbf{X}(\mathbf{x}, t), t) = \overline{\chi}^*(\mathbf{x}, t), \quad \mathbf{y}^* = \chi^*(\mathbf{Y}(\mathbf{y}, t), t) = \overline{\chi}^*(\mathbf{y}, t). \tag{3.8.5}$$

The superposed rigid-body motions of a whole body should preserve the distance between all pairs of material particles of the body for all times $0 \le t \le T$, where T is a finite final time; therefore, we have

$$[\overline{\chi}^*(\mathbf{x}, t) - \overline{\chi}^*(\mathbf{y}, t)]^{\mathrm{T}} \cdot [\overline{\chi}^*(\mathbf{x}, t) - \overline{\chi}^*(\mathbf{y}, t)] = (\mathbf{x} - \mathbf{y})^{\mathrm{T}} \cdot (\mathbf{x} - \mathbf{y}), \tag{3.8.6}$$

or

$$(\mathbf{x}^* - \mathbf{y}^*)^{\mathrm{T}} \cdot (\mathbf{x}^* - \mathbf{y}^*) = (\mathbf{x} - \mathbf{y})^{\mathrm{T}} \cdot (\mathbf{x} - \mathbf{y}) \quad \text{for all } \mathbf{x}, \mathbf{y} \text{ in } \kappa \text{ at time } t. \tag{3.8.7}$$

Noting that $\mathbf{x}$ and $\mathbf{y}$ are independent of each other, we can differentiate with respect to $\mathbf{x}$ and $\mathbf{y}$ successively and obtain

$$2 \left[\frac{\partial \overline{\chi}^*(\mathbf{x}, t)}{\partial \mathbf{x}} \right]^{\mathrm{T}} \cdot \left[\frac{\partial \overline{\chi}^*(\mathbf{y}, t)}{\partial \mathbf{y}} \right] = 2\mathbf{I}, \tag{3.8.8}$$

or

$$\left[\frac{\partial \overline{\chi}^*(\mathbf{x}, t)}{\partial \mathbf{x}} \right]^{\mathrm{T}} = \left[\frac{\partial \overline{\chi}^*(\mathbf{y}, t)}{\partial \mathbf{y}} \right]^{-1}.$$

Because the left side of the equality depends only on $(\mathbf{x}, t)$ and the right side depends only on $(\mathbf{y}, t)$, both must be equal to a function of time only, say $\mathbf{Q}^{\mathrm{T}}(t)$:

$$\left[\frac{\partial \overline{\chi}^*(\mathbf{x}, t)}{\partial \mathbf{x}} \right]^{\mathrm{T}} = \left[\frac{\partial \overline{\chi}^*(\mathbf{y}, t)}{\partial \mathbf{y}} \right]^{-1} = \mathbf{Q}^{\mathrm{T}}(t), \tag{3.8.9}$$

for all $\mathbf{x}$ and $\mathbf{y}$ in κ at time t. Let us set

$$\frac{\partial \overline{\chi}^*(\mathbf{x}, t)}{\partial \mathbf{x}} = \mathbf{Q}(t) \quad \text{for all } \mathbf{x} \text{ in } \kappa \text{ at time } t. \tag{3.8.10}$$

Then we must also have (because $\mathbf{y}$ is also in κ at time t)

$$\frac{\partial \overline{\chi}^*(\mathbf{y}, t)}{\partial \mathbf{y}} = \mathbf{Q}(t).$$

Therefore, we have

$$\mathbf{Q}^{\mathrm{T}} \cdot \mathbf{Q} = \mathbf{I}. \tag{3.8.11}$$

Thus, $\mathbf{Q}$ is an orthogonal tensor with $|Q| = \pm 1$. Since the motion under consideration must include the special case $\overline{\chi}^* = \mathbf{x}$, for which case $\mathbf{Q} = \mathbf{I}$, we have $|Q| = \pm 1$. Therefore, $\mathbf{Q}$ is a *proper* orthogonal tensor.

Integrating Eq. (3.8.10) with respect to $\mathbf{x}$, we obtain

$$\mathbf{x}^* = \overline{\chi}^*(\mathbf{x}, t) = \mathbf{Q} \cdot \mathbf{x} + \mathbf{c}(t), \qquad (3.8.12)$$

where $\mathbf{c}(t)$ is a vector-valued function of time t. Equation (3.8.12) represents a rigid transformation that includes translation $\mathbf{c}$ and rotation $\mathbf{Q}$. Thus, at each instant of time a rigid-body motion is a composition of rigid-body translation $\mathbf{c}$ and a rigid-body rotation $\mathbf{Q}$ about an axis of rotation, as well as a time shift $a = t^* - t$. Figure 3.8.1 shows a sequence of deformation followed by rigid-body transformation. For pure rigid-body rotation, Eq. (3.8.12) reduces to $\mathbf{x}^* = \mathbf{Q} \cdot \mathbf{x}$. The transformation in Eq. (3.8.12) preserves the distance between any two material particles as well as the angle between material lines in the small neighborhood of a material particle, as established next.

Consider the distance between two material particles occupying positions $\mathbf{x}$ and $\mathbf{y}$ in the deformed configuration,

$$|\mathbf{x}^* - \mathbf{y}^*|^2 = (\mathbf{x}^* - \mathbf{y}^*) \cdot (\mathbf{x}^* - \mathbf{y}^*) = \mathbf{Q} \cdot (\mathbf{x} - \mathbf{y}) \cdot \mathbf{Q} \cdot (\mathbf{x} - \mathbf{y})$$
$$= (\mathbf{x} - \mathbf{y}) \cdot (\mathbf{Q}^T \cdot \mathbf{Q}) \cdot (\mathbf{x} - \mathbf{y}) = (\mathbf{x} - \mathbf{y}) \cdot (\mathbf{x} - \mathbf{y}) = |\mathbf{x} - \mathbf{y}|^2.$$

Thus, the distance between any two points is preserved.

Next, we consider two material line segments in the neighborhood of point $\mathbf{x}$, one connecting $\mathbf{x}$ to $\mathbf{y}$ and the other connecting $\mathbf{x}$ to $\mathbf{z}$. The angle between the two line segments is

$$\cos \theta = \frac{(\mathbf{x} - \mathbf{y}) \cdot (\mathbf{x} - \mathbf{z})}{|\mathbf{x} - \mathbf{y}| \, |\mathbf{x} - \mathbf{z}|}.$$

Now consider the angle between the lines after superposed rigid-body motion:

$$\cos \theta^* = \frac{(\mathbf{x}^* - \mathbf{y}^*)}{|\mathbf{x}^* - \mathbf{y}^*|} \cdot \frac{(\mathbf{x}^* - \mathbf{z}^*)}{|\mathbf{x}^* - \mathbf{z}^*|}.$$

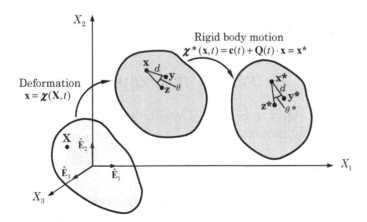

Fig. 3.8.1: Deformation followed by superposed rigid-body motion.

Since the distances are preserved, we have

$$
\begin{aligned}
\cos \theta^* &= \frac{[\mathbf{Q} \cdot (\mathbf{x} - \mathbf{y})] \cdot [\mathbf{Q} \cdot (\mathbf{x} - \mathbf{z})]}{|\mathbf{x}^* - \mathbf{y}^*| \, |\mathbf{x}^* - \mathbf{z}^*|} \\
&= \frac{(\mathbf{x} - \mathbf{y}) \cdot (\mathbf{Q}^\mathrm{T} \cdot \mathbf{Q}) \cdot (\mathbf{x} - \mathbf{z})}{|\mathbf{x} - \mathbf{y}| \, |\mathbf{x} - \mathbf{z}|} \\
&= \frac{(\mathbf{x} - \mathbf{y}) \cdot (\mathbf{x} - \mathbf{z})}{|\mathbf{x} - \mathbf{y}| \, |\mathbf{x} - \mathbf{z}|} = \cos \theta.
\end{aligned}
$$

The transformation in Eq. (3.8.12) preserves the distance between two material points and the angle between material lines in the neighborhood of every point in the body; hence, the transformation also preserves areas and volumes under the superposed rigid-body motion. Thus, when two frames of references are involved in measuring deformations (and forces) with one frame of reference moving rigidly with respect to the other, the measures will be unaffected.

3.8.1.2 Effect on F

To see the effect of superposed rigid-body motion on the deformation gradient, consider the most general rigid-body mapping in Eq. (3.8.12). Taking the derivative of Eq. (3.8.12) with respect to $\mathbf{X}$, we obtain

$$
\frac{\partial \mathbf{x}^*}{\partial \mathbf{X}} = \mathbf{Q}(t) \cdot \frac{\partial \mathbf{x}}{\partial \mathbf{X}},
$$

and therefore we have

$$
\mathbf{F}^* = \left(\mathbf{Q}(t) \cdot \frac{\partial \mathbf{x}}{\partial \mathbf{X}} \right)^\mathrm{T} = \mathbf{F} \cdot \mathbf{Q}^\mathrm{T} = \mathbf{Q} \cdot \mathbf{F}. \tag{3.8.13}
$$

Thus, the deformation gradients before and after superposed rigid-body motions are related by

$$
\mathbf{F}^*(\mathbf{X}, t) = \mathbf{Q}(t) \cdot \mathbf{F}(\mathbf{X}, t). \tag{3.8.14}
$$

Because $\mathbf{F}$ is a two-point tensor from a reference configuration, which is independent of the observer, to the current configuration, it transforms like a vector. The respective Jacobians are given by

$$
J = |F|, \quad J^* = |F^*| = |Q| \, |F| = J, \tag{3.8.15}
$$

where the fact that $|\mathbf{Q}| = 1$ is used. Thus the volume change is unaffected by superposed rigid-body motion.

3.8.1.3 Effect on C and E

To see how the right Cauchy–Green deformation tensor $\mathbf{C}$ and the Green–Lagrange strain tensor $\mathbf{E}$ change due to superposed rigid-body motion, consider

$$
\mathbf{C}^* = (\mathbf{F}^*)^\mathrm{T} \cdot \mathbf{F}^* = \left(\mathbf{F}^\mathrm{T} \cdot \mathbf{Q}^\mathrm{T} \right) \cdot (\mathbf{Q} \cdot \mathbf{F}) = \mathbf{F}^\mathrm{T} \cdot \mathbf{F} = \mathbf{C}, \tag{3.8.16}
$$

where Eq. (3.8.14) and the property $\mathbf{Q}^T \cdot \mathbf{Q} = \mathbf{I}$ of an orthogonal matrix $\mathbf{Q}$ is used. Hence, by definition [see Eq. (3.4.11)], the Green–Lagrange strain tensor $\mathbf{E}$ and the right Cauchy–Green deformation tensor $\mathbf{C}$, being defined with respect to the reference configuration, are unaffected by the superposed rigid-body motion:

$$\mathbf{E} = \mathbf{E}^*, \qquad \mathbf{C} = \mathbf{C}^*. \tag{3.8.17}$$

However, the velocities and accelerations of a material point are affected by the superposed rigid-body motion. For example, consider velocity after imposing the rigid-body motion (note that $dt/dt^* = 1$)

$$\mathbf{v}^*(\mathbf{x}^*, t^*) = \frac{d\mathbf{x}^*}{dt^*} = \frac{d}{dt^*}\left(\mathbf{c}(t) + \mathbf{Q}(t) \cdot \mathbf{x}\right) = \dot{\mathbf{c}}(t) + \dot{\mathbf{Q}}(t) \cdot \mathbf{x} + \mathbf{Q}(t) \cdot \mathbf{v}, \tag{3.8.18}$$

which shows that $\mathbf{v}^*$ and $\mathbf{v}$ are not the same, but one can be calculated from the other when $\mathbf{c}$ and $\mathbf{Q}$ are known for the superposed rigid-body motion.

3.8.1.4 Effect on L and D

Here we examine the effect of a superposed rigid-body motion on the velocity gradient tensor $\mathbf{L}$. We begin with Eq. (3.6.15)

$$\mathbf{L}^* = \dot{\mathbf{F}}^* \cdot (\mathbf{F}^*)^{-1} = \left(\dot{\mathbf{Q}} \cdot \mathbf{F} + \mathbf{Q} \cdot \dot{\mathbf{F}}\right)(\mathbf{Q} \cdot \mathbf{F})^{-1}$$

$$= \dot{\mathbf{Q}} \cdot \mathbf{Q}^T + \mathbf{Q} \cdot \dot{\mathbf{F}} \cdot \mathbf{F}^{-1} \cdot \mathbf{Q}^T = \dot{\mathbf{Q}} \cdot \mathbf{Q}^T + \mathbf{Q} \cdot \mathbf{L} \cdot \mathbf{Q}^T, \tag{3.8.19}$$

where we have used the following identities:

$$(\mathbf{Q} \cdot \mathbf{F})^{-1} = \mathbf{F}^{-1} \cdot \mathbf{Q}^{-1}, \qquad \mathbf{Q}^{-1} = \mathbf{Q}^T, \qquad \mathbf{Q} \cdot \mathbf{Q}^T = \mathbf{I}.$$

From Problem **3.48**, it is follows that $\dot{\mathbf{Q}} \cdot \mathbf{Q}^T$ is skew symmetric.

Next consider the symmetric part of $\mathbf{L}$, namely, $\mathbf{D}$. We have

$$\mathbf{D}^* = \tfrac{1}{2}\left[\mathbf{L}^* + (\mathbf{L}^*)^T\right] = \tfrac{1}{2}\left[\dot{\mathbf{Q}} \cdot \mathbf{Q}^T + \mathbf{Q} \cdot \mathbf{L} \cdot \mathbf{Q}^T + \dot{\mathbf{Q}}^T \cdot \mathbf{Q} + \mathbf{Q} \cdot \mathbf{L}^T \cdot \mathbf{Q}^T\right]$$

$$= \tfrac{1}{2}\left[\mathbf{Q} \cdot \mathbf{L} \cdot \mathbf{Q}^T + \mathbf{Q} \cdot \mathbf{L}^T \cdot \mathbf{Q}^T\right] = \mathbf{Q} \cdot \mathbf{D} \cdot \mathbf{Q}^T. \tag{3.8.20}$$

3.8.2 Material Objectivity

3.8.2.1 Observer transformation

In continuum mechanics each frame of reference represents an observer and, therefore, transformations between moving frames are termed *observer transformations*. The concept of frames of reference should not be confused with that of coordinate systems, as they are not the same at all. An observer is free to choose any coordinate system as may be convenient to observe or analyze a system's response. The equations of mechanics are used in different problems and places and, therefore, they must be independent of frames of reference, that is, invariant with respect to an observer transformation. A change of observer may be viewed as certain rigid-body motion superposed on the current configuration, as illustrated in Fig. 3.8.1.

In the analytical description of physical events, the following two requirements must be followed:

(1) Invariance of the equations with respect to stationary coordinate frames of reference

(2) Invariance of the equations with respect to frames of reference that move in arbitrary relative motion

The first requirement is readily met by expressing the equations in vector/tensor form, which is invariant. The assertion that an equation is in "invariant form" refers to the vector form that is independent of the choice of a coordinate system. The coordinate systems used in the present study were assumed to be relatively at rest. The second requirement is that the invariance property holds for reference frames (or observers) moving arbitrarily with respect to each other. This requirement is dictated by the need for forces and deformations to be the same as measured by all observers irrespective of their relative motions. Invariance with respect to changes of observer is termed *material frame indifference* or *material objectivity*.

3.8.2.2 Objectivity of various kinematic measures

Let $\mathcal{F}$ denote a reference frame with origin at O in which the $\mathbf{x}$ is the current position of a particular particle at time t. Let $\mathcal{F}^*$ be another reference frame with origin at O* with time denoted with t^*. Let ϕ be a scalar field when described in the frame $\mathcal{F}$ and ϕ^* is the same scalar field described with respect to the frame $\mathcal{F}^*$, and let $(\mathbf{u}, \mathbf{u}^*)$ and $(\mathbf{S}, \mathbf{S}^*)$ be the vector and tensor fields, respectively, in the two frames. Scalar, vector, and tensor fields are called *frame indifferent* or *objective* if they transform according to the following equations:

1. Events	$\mathbf{x}^* = \mathbf{c}(t) + \mathbf{Q}(t) \cdot \mathbf{x}, \quad t^* = t - a$
2. Scalar field	$\phi^*(\mathbf{x}^*, t^*) = \phi(\mathbf{x}, t)$
3. Displacement vector	$\mathbf{u}^*(\mathbf{x}^*, t^*) = \mathbf{Q}(t) \cdot \mathbf{u}(\mathbf{x}, t)$ (3.8.21)
4. General second-order tensors	$\mathbf{S}^*(\mathbf{x}^*, t^*) = \mathbf{Q}(t) \cdot \mathbf{S}(\mathbf{x}, t) \cdot \mathbf{Q}^{\mathrm{T}}(t)$
5. Two-point second-order tensors	$\mathbf{F}^*(\mathbf{x}^*, t^*) = \mathbf{Q}(t) \cdot \mathbf{F}(\mathbf{x}, t)$

where $\mathbf{Q}(t)$ is a proper orthogonal tensor that rotates frame $\mathcal{F}^*$ into frame $\mathcal{F}$, $\mathbf{c}(t)$ is a vector from O to O* that depends only on time t, and a is a constant. For example, $\mathbf{x}$ and $\mathbf{x}^*$ refer to the same motion, but mathematically $\mathbf{x}^*$ is the motion obtained from $\mathbf{x}$ by superposition of a rigid rotation and translation. The mapping $\mathbf{x}^* = \mathbf{c}(t) + \mathbf{Q}(t) \cdot \mathbf{x}$, derived in Eq. (3.8.12), may be interpreted as one that takes $(\mathbf{x}, t)$ to $(\mathbf{x}^*, t^*)$ as a change of observer from O to O*, so that the event that is observed at place $\mathbf{x}$ at time t by observer O is the *same* event as that observed at place $\mathbf{x}^*$ at time t^* by observer O*, where $t^* = t - a$, and a is a constant. Thus, a change of observer merely changes the *description* of an event. In short, the objectivity ensures that the direction(s) and magnitude are independent of the coordinate frame used to describe them.

We have already established in Section 3.8.1, under the following general rigid-body mapping

$$\mathbf{x}^*(\mathbf{X}, t^*) = \mathbf{c}(t) + \mathbf{Q}(t) \cdot \mathbf{x}, t), \quad t^* = t - a, \tag{3.8.22}$$

between observer O and observer O*, that (if the reference configuration is independent of the observer) the right Cauchy–Green deformation tensor $\mathbf{C}$ and the Green–Lagrange strain tensor $\mathbf{E}$ do not change under the observer transformation, that is, they are objective:

$$\mathbf{E} = \mathbf{E}^*, \qquad \mathbf{C} = \mathbf{C}^*. \tag{3.8.23}$$

In addition, the symmetric part of $\mathbf{L}$, namely, $\mathbf{D}$ is also objective in the sense

$$\mathbf{D}^* = \mathbf{Q} \cdot \mathbf{D} \cdot \mathbf{Q}^{\mathrm{T}}. \tag{3.8.24}$$

We have noted that the two observers' views of the velocity and acceleration of a given motion are different, even though the rate of change at fixed $\mathbf{X}$ is the same in each case. Thus, velocity and acceleration vectors are *not* objective.

3.8.2.3 Time rate of change in a rotating frame of reference

Next, consider two frames of reference with both having the same origin, but one is nonrotating and the other is rotating with respect to the other with an angular velocity $\boldsymbol{\omega}$. Let us use no bars on quantities in the nonrotating system and bars on quantities in the rotating system. Then the time derivatives of a vector-valued function $\mathbf{A}(t)$ in the two coordinate frames are

$$\mathbf{A}(t) = A^i \mathbf{e}_i, \quad \frac{D\mathbf{A}}{Dt} = \frac{dA^i}{dt}\mathbf{e}_i, \quad \text{nonrotating system}, \tag{3.8.25}$$

$$\mathbf{A}(t) = \bar{A}^i \bar{\mathbf{e}}_i, \quad \frac{D\mathbf{A}}{Dt} = \frac{d\bar{A}^i}{dt}\bar{\mathbf{e}}_i + \bar{A}^i \frac{d\bar{\mathbf{e}}_i}{dt}, \quad \text{rotating system}. \tag{3.8.26}$$

The rate of change $d\bar{\mathbf{e}}_i/dt$ is given by

$$\frac{d\bar{\mathbf{e}}_i}{dt} = \boldsymbol{\omega} \times \bar{\mathbf{e}}_i, \tag{3.8.27}$$

because the change is brought about by a rigid-body rotation. To an observer in the rotating frame, however, the basis vectors appear to be constant:

$$\frac{d\bar{A}^i}{dt}\bar{\mathbf{e}}_i \equiv \left(\frac{d\mathbf{A}}{dt}\right)_{\text{rot}}. \tag{3.8.28}$$

The relationship of the time derivatives in the two frames is thus given by

$$\begin{aligned} \left(\frac{d\mathbf{A}}{dt}\right)_{\text{nonrot}} &= \left(\frac{d\mathbf{A}}{dt}\right)_{\text{rot}} + \bar{A}^i(\boldsymbol{\omega} \times \bar{\mathbf{e}}_i) \\ &= \left(\frac{d\mathbf{A}}{dt}\right)_{\text{rot}} + \boldsymbol{\omega} \times \mathbf{A}. \end{aligned} \tag{3.8.29}$$

Thus, in general, the time rates of change of vectors and tensors in the two frames are related by

$$\left(\frac{d}{dt}\right)_{\text{nonrot}} = \left(\frac{d}{dt}\right)_{\text{rot}} + \boldsymbol{\omega} \times . \tag{3.8.30}$$

3.9 Polar Decomposition Theorem

3.9.1 Preliminary Comments

Recall that the deformation gradient $\mathbf{F}$ transforms a material vector $d\mathbf{X}$ at $\mathbf{X}$ into the corresponding spatial vector $d\mathbf{x}$, and it characterizes all of the deformation, stretch (elongation) as well as rotation, at $\mathbf{X}$. Therefore, it forms an essential part of the definition of any strain measure. Another role of $\mathbf{F}$ in connection with the strain measures is discussed here with the help of the polar decomposition theorem of Cauchy. The polar decomposition theorem enables one to decompose $\mathbf{F}$ uniquely into the product of a proper orthogonal tensor and a symmetric positive-definite tensor and thereby decompose the general deformation into pure stretch and pure rotation.

3.9.2 Rotation and Stretch Tensors

Suppose that $\mathbf{F}$ is nonsingular so that each line element $d\mathbf{X}$ from the reference configuration is transformed into a unique line element $d\mathbf{x}$ in the current configuration, and conversely. Then the polar decomposition theorem states that $\mathbf{F}$ has a *unique* right and left (multiplicative) decompositions of the form (see Fig. 3.9.1)

$$\mathbf{F} = \mathbf{R} \cdot \mathbf{U} = \mathbf{V} \cdot \mathbf{R} \quad (F_{iI} = R_{iK}U_{KI} = V_{ij}R_{jI}), \tag{3.9.1}$$

so that

$$d\mathbf{x} = \mathbf{F} \cdot d\mathbf{X} = (\mathbf{R} \cdot \mathbf{U}) \cdot d\mathbf{X} = (\mathbf{V} \cdot \mathbf{R}) \cdot d\mathbf{X}, \tag{3.9.2}$$

where $\mathbf{U}$ is the symmetric and positive-definite *right Cauchy stretch tensor* (stretch is the ratio of the final length to the original length), $\mathbf{V}$ is the symmetric and positive-definite *left Cauchy stretch tensor*, and $\mathbf{R}$ is the *orthogonal rotation tensor*,

$$\mathbf{R}^{\mathrm{T}} \cdot \mathbf{R} = \mathbf{I}, \quad \mathbf{U} = \mathbf{U}^{\mathrm{T}}, \quad \mathbf{V} = \mathbf{V}^{\mathrm{T}}. \tag{3.9.3}$$

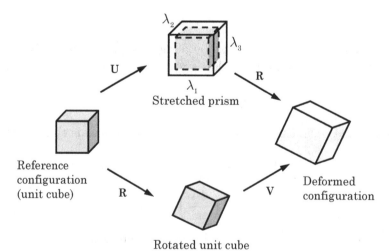

Fig. 3.9.1: The roles of $\mathbf{U}$, $\mathbf{V}$, and $\mathbf{R}$ in stretching and rotating a unit volume of material in the neighborhood of $\mathbf{X}$; λ_I ($I = 1, 2, 3$) denote the principal stretches.

In Eq. (3.9.2), $\mathbf{U} \cdot d\mathbf{X}$ describes a pure stretch deformation in which there are three mutually perpendicular directions along which the material element $d\mathbf{X}$ stretches (that is, elongates or compresses) but does not rotate. The three directions are provided by the eigenvectors of $\mathbf{U}$. The role of the rotation tensor $\mathbf{R}$ is to rotate the stretched element, $\mathbf{R} \cdot \mathbf{U} \cdot d\mathbf{X}$. These ideas are illustrated in Fig. 3.9.2, which shows[8] the material occupying the spherical volume of radius $|d\mathbf{X}|$ in the undeformed configuration being mapped by the operator $\mathbf{U}$ into an ellipsoid in the deformed configuration at $\mathbf{x}$. Then $\mathbf{R}$ rotates the ellipsoid through a rigid-body rotation.

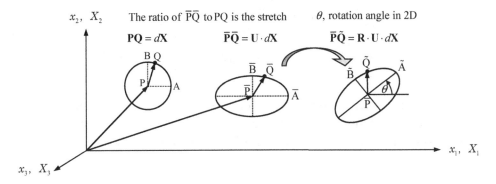

Fig. 3.9.2: The roles of $\mathbf{U}$ and $\mathbf{R}$ in transforming an ellipsoidal volume of material in the neighborhood of $\mathbf{X}$.

From Eqs. (3.9.1) and (3.9.3) it follows that

$$\mathbf{U} = \mathbf{R}^{-1} \cdot \mathbf{F} = \mathbf{R}^{\mathrm{T}} \cdot \mathbf{F}, \quad \mathbf{V} = \mathbf{F} \cdot \mathbf{R}^{-1} = \mathbf{F} \cdot \mathbf{R}^{\mathrm{T}}, \tag{3.9.4}$$

$$\mathbf{U}^2 = \mathbf{U} \cdot \mathbf{U} = \mathbf{U}^{\mathrm{T}} \cdot \mathbf{U} = \mathbf{F}^{\mathrm{T}} \cdot (\mathbf{R} \cdot \mathbf{R}^{-1}) \cdot \mathbf{F} = \mathbf{F}^{\mathrm{T}} \cdot \mathbf{F} = \mathbf{C},$$
$$\mathbf{V}^2 = \mathbf{V} \cdot \mathbf{V} = \mathbf{V} \cdot \mathbf{V}^{\mathrm{T}} = \mathbf{F} \cdot (\mathbf{R}^{-1} \cdot \mathbf{R}) \cdot \mathbf{F}^{\mathrm{T}} = \mathbf{F} \cdot \mathbf{F}^{\mathrm{T}} = \mathbf{B}, \tag{3.9.5}$$

where $\mathbf{C}$ and $\mathbf{B}$ denote the right and left Cauchy–Green deformation tensors, respectively. We also note that

$$\mathbf{F} = \mathbf{R} \cdot \mathbf{U} = (\mathbf{R} \cdot \mathbf{U}) \cdot (\mathbf{R}^{\mathrm{T}} \cdot \mathbf{R}) = (\mathbf{R} \cdot \mathbf{U} \cdot \mathbf{R}^{\mathrm{T}}) \cdot \mathbf{R} = \mathbf{V} \cdot \mathbf{R}$$
$$= \mathbf{V} \cdot \mathbf{R} = (\mathbf{R} \cdot \mathbf{R}^{\mathrm{T}}) \cdot (\mathbf{V} \cdot \mathbf{R}) = \mathbf{R} \cdot (\mathbf{R}^{\mathrm{T}} \cdot \mathbf{V} \cdot \mathbf{R}) = \mathbf{R} \cdot \mathbf{U}, \tag{3.9.6}$$

which show that

$$\mathbf{U} = \mathbf{R}^{\mathrm{T}} \cdot \mathbf{V} \cdot \mathbf{R}, \quad \mathbf{V} = \mathbf{R} \cdot \mathbf{U} \cdot \mathbf{R}^{\mathrm{T}}. \tag{3.9.7}$$

Since $\mathbf{C} = \mathbf{F}^{\mathrm{T}} \cdot \mathbf{F}$ and $\mathbf{B} = \mathbf{F} \cdot \mathbf{F}^{\mathrm{T}}$, we have $|\mathbf{C}| = |\mathbf{B}| = |\mathbf{F}|^2 = J^2$ and therefore $|\mathbf{U}| = |\mathbf{V}| = \sqrt{|\mathbf{C}|} = +J$ (positive because $\mathbf{U}$ and $\mathbf{V}$ are positive-definite matrices). In view of Eq. (3.9.1), it follows that $|\mathbf{R}| = +1$, implying that $\mathbf{R}$ is a *proper* orthogonal tensor. Because $\mathbf{F}^{\mathrm{T}} \cdot \mathbf{F}$ is real and symmetric,

---

[8]For clarity, the figure shows stretch and rotation only in the $X_1 - X_2$ plane.

there exists an orthogonal matrix $\mathbf{A}$ that transforms $\mathbf{C} = \mathbf{F}^{\mathrm{T}} \cdot \mathbf{F}$ into a diagonal matrix $\bar{\mathbf{C}}$:

$$\bar{\mathbf{C}} = \mathbf{A}^{\mathrm{T}}\mathbf{C}\mathbf{A} = \begin{bmatrix} \lambda_1^2 & 0 & 0 \\ 0 & \lambda_2^2 & 0 \\ 0 & 0 & \lambda_3^2 \end{bmatrix} = \sum_{I=1}^{3} \lambda_I^2 \hat{\mathbf{N}}^{(I)}\hat{\mathbf{N}}^{(I)}, \qquad (3.9.8)$$

where λ_I^2 are the eigenvalues of $\mathbf{C} = \mathbf{F}^{\mathrm{T}} \cdot \mathbf{F} = \mathbf{U}^2$ and $\mathbf{A}$ is the matrix of normalized eigenvectors $\hat{\mathbf{N}}^{(I)}$ (spectral theorem; see Section 2.5.6.3). The eigenvalues λ_I are called the *principal stretches* and the corresponding mutually orthogonal eigenvectors are called the *principal directions*. The tensors $\mathbf{U}$ and $\mathbf{V}$ have the same eigenvalues [see Eq. (3.9.7) and note $|\mathbf{R}| = 1$], and their eigenvectors differ only by the rotation $\mathbf{R}$; Problem **3.56** for a proof. Thus $(\bar{\mathbf{U}} = \sqrt{\bar{\mathbf{C}}})$

$$\bar{\mathbf{U}} = \begin{bmatrix} \lambda_1 & 0 & 0 \\ 0 & \lambda_2 & 0 \\ 0 & 0 & \lambda_3 \end{bmatrix}, \quad \mathbf{U} = \mathbf{A}\bar{\mathbf{U}}\mathbf{A}^{\mathrm{T}} = \sum_{I=1}^{3} \lambda_I \hat{\mathbf{N}}^{(I)}\hat{\mathbf{N}}^{(I)}, \qquad (3.9.9)$$

where $\hat{\mathbf{N}}^{(I)}$ is the normalized eigenvector associated with eigenvalue λ_I in the reference configuration. Once the stretch tensor $\mathbf{U}$ is known, the rotation tensor $\mathbf{R}$ can be obtained from Eq. (3.9.1) and left stretch tensor $\mathbf{V}$ from Eq. (3.9.4) as

$$\mathbf{R} = \mathbf{F} \cdot \mathbf{U}^{-1}, \quad \mathbf{V} = \mathbf{F} \cdot \mathbf{R}^{\mathrm{T}}. \qquad (3.9.10)$$

In view of Eq. (3.9.8), the Lagrangian and Eulerian strain tensors can be expressed in terms of $\mathbf{U}$ and $\mathbf{V}$ as

$$\mathbf{E} = \tfrac{1}{2}\left(\mathbf{U}^2 - \mathbf{I}\right) = \tfrac{1}{2}\sum_{I=1}^{3}\left(\lambda_I^2 - 1\right)\hat{\mathbf{N}}^{(I)}\hat{\mathbf{N}}^{(I)}, \qquad (3.9.11)$$

$$\mathbf{e} = \tfrac{1}{2}\left(\mathbf{I} - \mathbf{V}^{-2}\right) = \tfrac{1}{2}\sum_{i=1}^{3}\left(1 - \lambda_i^{-2}\right)\hat{\mathbf{n}}^{(i)}\hat{\mathbf{n}}^{(i)}, \qquad (3.9.12)$$

where $\hat{\mathbf{n}}^{(i)}$ is the normalized eigenvector in the current configuration. Next, we consider two examples of the use of the polar decomposition theorem.

Example 3.9.1

Consider the deformation mapping of Example 3.2.1,

$$x_1 = X_1 + At\,X_2, \quad x_2 = X_2 - At\,X_1, \quad x_3 = X_3.$$

It was shown in Example 3.2.1 that this mapping stretches a unit cube in the X_1 and X_2 directions and rotates about the X_3-axis, as shown in Fig. 3.2.5. Use the polar decomposition to determine the components of the symmetric right Cauchy stretch tensor $\mathbf{U}$, the rotation tensor $\mathbf{R}$, and the symmetric left Cauchy stretch tensor $\mathbf{V}$ associated with the deformation for $A = 0.25$ and $t = 2$.

Solution: The matrices associated with the deformation gradient $\mathbf{F}$ and the right Cauchy–Green deformation tensor $\mathbf{C}$ (depend only on A and t, as can be seen from Example 3.2.1) for $A = 0.25$ and $t = 2$ are

$$[F] = \begin{bmatrix} 1.0 & 0.5 & 0.0 \\ -0.5 & 1.0 & 0.0 \\ 0.0 & 0.0 & 1.0 \end{bmatrix}, \quad [C] = [F]^{\mathrm{T}}[F] = \begin{bmatrix} 1.25 & 0.00 & 0.00 \\ 0.00 & 1.25 & 0.00 \\ 0.00 & 0.00 & 1.00 \end{bmatrix}.$$

Thus, $[F]$ is independent of the position $\mathbf{X}$ and, therefore, the deformation is homogeneous. The eigenvalues associated with $[C]$ are $\lambda_1^2 = 1.25$, $\lambda_2^2 = 1.25$, and $\lambda_3^2 = 1.0$ for any point in the body. The matrix of normalized eigenvectors associated with these stretches is the identity matrix (the jth column is the eigenvector corresponding to the jth eigenvalue)

$$[A] = \begin{bmatrix} 1.0 & 0.0 & 0.0 \\ 0.0 & 1.0 & 0.0 \\ 0.0 & 0.0 & 1.0 \end{bmatrix}.$$

Then the matrix of the symmetric right stretch tensor $\mathbf{U}$ is determined from ($[C] = [\bar{C}]$ and $[U] = [\bar{U}]$):

$$[U]^2 = [C] = \begin{bmatrix} \lambda_1^2 & 0 & 0 \\ 0 & \lambda_2^2 & 0 \\ 0 & 0 & \lambda_3^2 \end{bmatrix} \quad \text{or} \quad [U] = \begin{bmatrix} \lambda_1 & 0 & 0 \\ 0 & \lambda_2 & 0 \\ 0 & 0 & \lambda_3 \end{bmatrix},$$

where $\lambda_1 = \lambda_2 = 1.1180$ and $\lambda_3 = 1$. The matrix of eigenvectors remains the same, and we have

$$\mathbf{U} = \lambda_1 \hat{\mathbf{e}}_1 \hat{\mathbf{e}}_1 + \lambda_2 \hat{\mathbf{e}}_2 \hat{\mathbf{e}}_2 + \lambda_3 \hat{\mathbf{e}}_3 \hat{\mathbf{e}}_3.$$

We note that $\lambda_1 = 1.1180$ is the stretch of a line parallel to the X_1-axis, $\lambda_2 = 1.1180$ is the stretch of a line parallel to the X_2-axis, and $\lambda_3 = 1$ is the stretch of a line parallel to the X_3-axis (that is, the body did not undergo deformation in the thickness direction) in the undeformed body. The stretches can be verified independently by considering, as an example, the line $X_1 = 0$ (of unit length) in the undeformed body. In the deformed body the line has a length of $l = 1/\cos\alpha = 1.1180$, where $\tan\alpha = At = 0.5$ (or $\alpha = 26.565° = 0.46365$ rad.), as shown in Fig. 3.9.3.

The matrix associated with the rotation tensor $\mathbf{R}$ is determined from Eq. (3.9.10) ($1/\lambda_1 = 0.894427$) as

$$[R] = [F][U]^{-1} = \begin{bmatrix} 1.0 & 0.5 & 0.0 \\ -0.5 & 1.0 & 0.0 \\ 0.0 & 0.0 & 1.0 \end{bmatrix} \begin{bmatrix} \frac{1}{\lambda_1} & 0 & 0 \\ 0 & \frac{1}{\lambda_2} & 0 \\ 0 & 0 & \frac{1}{\lambda_3} \end{bmatrix} = \begin{bmatrix} 0.8944 & 0.4472 & 0.0 \\ -0.4472 & 0.8944 & 0.0 \\ 0.0 & 0.0 & 1.0 \end{bmatrix}.$$

We note that the rotation tensor is of the form ($\theta = -\alpha = -26.565°$)

$$[R] = \begin{bmatrix} \cos\theta & -\sin\theta & 0 \\ \sin\theta & \cos\theta & 0 \\ 0 & 0 & 1 \end{bmatrix} = \begin{bmatrix} 0.8944 & 0.4472 & 0.0 \\ -0.4472 & 0.8944 & 0.0 \\ 0.0 & 0.0 & 1.0 \end{bmatrix},$$

which agrees with the rotation shown in Fig. 3.9.3.

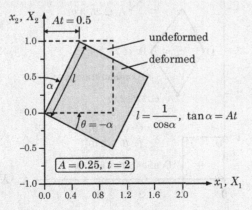

Fig. 3.9.3: Stretch and rotation of a unit square under the mapping, $x_1 = X_1 + At\,X_2$, $x_2 = X_2 - At\,X_1$, $x_3 = X_3$.

The left Cauchy stretch tensor components are determined using Eq. (3.9.10):

$$[V] = [F][R]^{\mathrm{T}} = \begin{bmatrix} 1.0 & 0.5 & 0.0 \\ -0.5 & 1.0 & 0.0 \\ 0.0 & 0.0 & 1.0 \end{bmatrix} \begin{bmatrix} 0.8944 & -0.4472 & 0.0 \\ 0.4472 & 0.8944 & 0.0 \\ 0.0000 & 0.0000 & 1.0 \end{bmatrix} = \begin{bmatrix} \lambda_1 & 0 & 0 \\ 0 & \lambda_2 & 0 \\ 0 & 0 & \lambda_3 \end{bmatrix}.$$

Example 3.9.2

Consider the deformation mapping

$$x_1 = \tfrac{1}{4}\left[4X_1 + (9 - 3X_1 - 5X_2 - X_1X_2)\,t\right], \quad x_2 = \tfrac{1}{4}\left[4X_2 + (16 + 8X_1)\,t\right], \quad x_3 = X_3.$$

For $(X_1, X_2, X_3) = (0, 0, 0)$ and time $t = 1$,

(a) determine the deformation gradient $\mathbf{F}$ and right Cauchy–Green deformation tensor $\mathbf{C}$,

(b) find the stretches λ_1 and λ_2 and the associated eigenvectors $\hat{\mathbf{N}}^{(1)}$ and $\hat{\mathbf{N}}^{(2)}$,

(c) use the polar decomposition to determine the components of the symmetric right Cauchy stretch tensor $\mathbf{U}$, the rotation tensor $\mathbf{R}$, and the symmetric left Cauchy stretch tensor $\mathbf{V}$, and

(d) use the polar decomposition to determine the components of Green–Lagrange strain tensor $\mathbf{E}$.

Solution: We note that the deformation is nonhomogeneous because of the term X_1X_2 in the mapping. The material point $(X_1, X_2, X_3) = (0, 0, 0)$ in the initial (that is, at $t = 0$) configuration occupies the location $(x_1, x_2, x_3) = (2.25, 4, 0)$ at $t = 1.0$, as shown in Fig. 3.9.4.

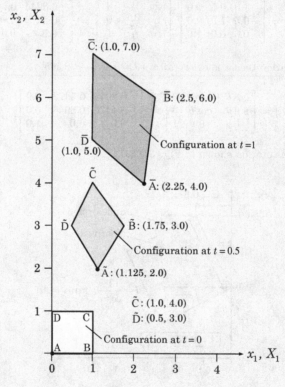

Fig. 3.9.4: Nonhomogeneous deformation of a unit square under the mapping, $x_1 = 0.25\,(9 + X_1 - 5X_2 - X_1X_2)$, $x_2 = 0.25\,(16 + 8X_1 + 4X_2)$, $x_3 = X_3$.

(a) The components of the deformation gradient $\mathbf{F}$ and right Cauchy–Green strain tensor $\mathbf{C}$ are

$$[F] = \tfrac{1}{4}\begin{bmatrix} 1 & -5 & 0 \\ 8 & 4 & 0 \\ 0 & 0 & 4 \end{bmatrix}, \quad [C] = [F]^{\mathrm{T}}[F] = \tfrac{1}{16}\begin{bmatrix} 65 & 27 & 0 \\ 27 & 41 & 0 \\ 0 & 0 & 16 \end{bmatrix}.$$

(b) The eigenvalues λ_1^2, λ_2^2, and λ_3^2 of matrix $[C]$ are determined by setting

$$||[C] - \lambda^2[I]|| = 0 \quad \rightarrow \quad \lambda_1^2 = 5.15916, \quad \lambda_2^2 = 1.46584, \quad \lambda_3^2 = 1,$$

and $\lambda_1 = 2.27138$, $\lambda_2 = 1.21072$, and $\lambda_3 = 1$. The eigenvectors are (in component form)

$$\{N^{(1)}\} = \left\{\begin{array}{c} 0.83849 \\ 0.54491 \\ 0.0 \end{array}\right\}, \quad \{N^{(2)}\} = \left\{\begin{array}{c} 0.54491 \\ -0.83849 \\ 0.0 \end{array}\right\}, \quad \{N^{(3)}\} = \left\{\begin{array}{c} 0.0 \\ 0.0 \\ 1.0 \end{array}\right\}.$$

(c) The matrix of the right Cauchy stretch tensor is computed using Eq. (3.9.9):

$$[U] = \begin{bmatrix} 0.83849 & 0.54491 & 0.0 \\ 0.54491 & -0.83849 & 0.0 \\ 0.0 & 0.0 & 1.0 \end{bmatrix}\begin{bmatrix} \lambda_1 & 0 & 0 \\ 0 & \lambda_2 & 0 \\ 0 & 0 & \lambda_3 \end{bmatrix}\begin{bmatrix} 0.83849 & 0.54491 & 0.0 \\ 0.54491 & -0.83849 & 0.0 \\ 0.0 & 0.0 & 1.0 \end{bmatrix}$$

$$= \begin{bmatrix} 1.95644 & 0.48462 & 0.0 \\ 0.48462 & 1.52566 & 0.0 \\ 0.0 & 0.0 & 1.0 \end{bmatrix},$$

and the principal stretches are $\lambda_1 = 2.27138$, $\lambda_2 = 1.21072$, and $\lambda_3 = 1$.

The tensor form of $\mathbf{U}$ is

$$\begin{aligned} \mathbf{U} &= \lambda_1\hat{\mathbf{N}}^{(1)}\hat{\mathbf{N}}^{(1)} + \lambda_2\hat{\mathbf{N}}^{(2)}\hat{\mathbf{N}}^{(2)} + \lambda_3\hat{\mathbf{N}}^{(3)}\hat{\mathbf{N}}^{(3)} \\ &= \lambda_1\left(N_1^{(1)}\hat{\mathbf{e}}_1 + N_2^{(1)}\hat{\mathbf{e}}_2 + N_3^{(1)}\hat{\mathbf{e}}_3\right)\left(N_1^{(1)}\hat{\mathbf{e}}_1 + N_2^{(1)}\hat{\mathbf{e}}_2 + N_3^{(1)}\hat{\mathbf{e}}_3\right) \\ &\quad + \lambda_2\left(N_1^{(2)}\hat{\mathbf{e}}_1 + N_2^{(2)}\hat{\mathbf{e}}_2 + N_3^{(2)}\hat{\mathbf{e}}_3\right)\left(N_1^{(2)}\hat{\mathbf{e}}_1 + N_2^{(2)}\hat{\mathbf{e}}_2 + N_3^{(2)}\hat{\mathbf{e}}_3\right) \\ &\quad + \lambda_3\left(N_1^{(3)}\hat{\mathbf{e}}_1 + N_2^{(3)}\hat{\mathbf{e}}_2 + N_3^{(3)}\hat{\mathbf{e}}_3\right)\left(N_1^{(3)}\hat{\mathbf{e}}_1 + N_2^{(3)}\hat{\mathbf{e}}_2 + N_3^{(3)}\hat{\mathbf{e}}_3\right), \end{aligned}$$

or

$$\begin{aligned} \mathbf{U} &= \left(\lambda_1[N_1^{(1)}]^2 + \lambda_2[N_1^{(2)}]^2 + \lambda_3[N_1^{(3)}]^2\right)\hat{\mathbf{e}}_1\hat{\mathbf{e}}_1 + \left(\lambda_1[N_2^{(1)}]^2 + \lambda_2[N_2^{(2)}]^2 + \lambda_3[N_2^{(3)}]^2\right)\hat{\mathbf{e}}_2\hat{\mathbf{e}}_2 \\ &\quad + \left(\lambda_1[N_3^{(1)}]^2 + \lambda_2[N_3^{(2)}]^2 + \lambda_3[N_3^{(3)}]^2\right)\hat{\mathbf{e}}_3\hat{\mathbf{e}}_3 + \left(\lambda_1 N_1^{(1)}N_2^{(1)} + \lambda_2 N_1^{(2)}N_2^{(2)}\right)(\hat{\mathbf{e}}_1\hat{\mathbf{e}}_2 + \hat{\mathbf{e}}_2\hat{\mathbf{e}}_1) \\ &\quad + \left(\lambda_1 N_1^{(1)}N_3^{(1)} + \lambda_3 N_1^{(3)}N_3^{(3)}\right)(\hat{\mathbf{e}}_1\hat{\mathbf{e}}_3 + \hat{\mathbf{e}}_3\hat{\mathbf{e}}_1) + \left(\lambda_2 N_2^{(2)}N_3^{(2)} + \lambda_3 N_2^{(3)}N_3^{(3)}\right)(\hat{\mathbf{e}}_2\hat{\mathbf{e}}_3 + \hat{\mathbf{e}}_3\hat{\mathbf{e}}_2) \\ &= 1.9564\hat{\mathbf{e}}_1\hat{\mathbf{e}}_1 + 0.4846(\hat{\mathbf{e}}_1\hat{\mathbf{e}}_2 + \hat{\mathbf{e}}_2\hat{\mathbf{e}}_1) + 1.5257\hat{\mathbf{e}}_2\hat{\mathbf{e}}_2 + \hat{\mathbf{e}}_3\hat{\mathbf{e}}_3. \end{aligned}$$

The matrix of the rotation tensor $\mathbf{R}$ is determined from Eq. (3.9.10) as

$$[R] = [F][U]^{-1} = \begin{bmatrix} 0.3590 & -0.9333 & 0 \\ 0.9333 & 0.3590 & 0 \\ 0.0 & 0.0 & 1 \end{bmatrix}.$$

It follows that the rotation angle is $\theta = 68.96°$. The left Cauchy stretch tensor components are determined using Eq. (3.9.10):

$$[V] = [F][R]^{\mathrm{T}} = \begin{bmatrix} 0.25 & -1.25 & 0.0 \\ 2.00 & 1.00 & 0.0 \\ 0.00 & 0.00 & 1.0 \end{bmatrix}\begin{bmatrix} 0.3590 & 0.9333 & 0.0 \\ -0.9333 & 0.3590 & 0.0 \\ 0.0000 & 0.0000 & 1.0 \end{bmatrix} = \begin{bmatrix} 1.2564 & -0.2153 & 0.0 \\ -0.2153 & 2.2256 & 0.0 \\ 0.0000 & 0.0000 & 1.0 \end{bmatrix}.$$

(d) The Green–Lagrange strain components at point $(X_1, X_2) = (0,0)$ are computed using Eq. (3.9.11). We have [which can be verified using $\mathbf{E} = 0.5(\mathbf{C} - \mathbf{I})$]

$$E_{11} = \tfrac{1}{2}\left[(\lambda_1^2 - 1)\, N_1^{(1)}\, N_1^{(1)} + (\lambda_2^2 - 1)\, N_1^{(2)}\, N_1^{(2)}\right]$$
$$= \tfrac{1}{2}\,(4.15916 \times 0.83849 \times 0.83849 + 0.46584 \times 0.54491 \times 0.54491) = 1.5312,$$
$$E_{22} = \tfrac{1}{2}\left[(\lambda_1^2 - 1)\, N_2^{(1)}\, N_2^{(1)} + (\lambda_2^2 - 1)\, N_2^{(2)}\, N_2^{(2)}\right]$$
$$= \tfrac{1}{2}\,(4.15916 \times 0.54491 \times 0.54491 + 0.46584 \times 0.83849 \times 0.83849) = 0.7812,$$
$$E_{12} = \tfrac{1}{2}\left[(\lambda_1^2 - 1)\, N_1^{(1)}\, N_2^{(1)} + (\lambda_2^2 - 1)\, N_1^{(2)}\, N_2^{(2)}\right]$$
$$= \tfrac{1}{2}\,(4.15916 - 0.46584) \times 0.83849 \times 0.54491 = 0.8437.$$

3.9.3 Objectivity of Stretch Tensors

Using the unique polar decomposition in Eq. (3.9.1), we can write

$$\mathbf{F}^* = \mathbf{R}^* \cdot \mathbf{U}^* = \mathbf{R}^* \cdot \mathbf{V}^*, \tag{3.9.13}$$

and objectivity of $\mathbf{F}$ gives

$$\mathbf{F}^* = \mathbf{Q} \cdot \mathbf{F} = \mathbf{Q} \cdot (\mathbf{R} \cdot \mathbf{U}) = \mathbf{Q} \cdot (\mathbf{V} \cdot \mathbf{R}). \tag{3.9.14}$$

Since $\mathbf{Q}$ and $\mathbf{R}$ are (proper) orthogonal tensors, we have

$$(\mathbf{Q} \cdot \mathbf{R}) \cdot (\mathbf{Q} \cdot \mathbf{R})^{\mathrm{T}} = \mathbf{Q} \cdot (\mathbf{R} \cdot \mathbf{R}^{\mathrm{T}}) \cdot \mathbf{Q}^{\mathrm{T}} = \mathbf{I}, \quad |\mathbf{Q} \cdot \mathbf{R}| = |\mathbf{Q}|\,|\mathbf{R}| = 1. \tag{3.9.15}$$

Thus, $\mathbf{Q} \cdot \mathbf{R}$ is also orthogonal. Therefore, in analogy to the two-point tensor $\mathbf{F}$, we can define $\mathbf{R}$ to be objective when [recall from the definition in Eq. (3.9.1) that $\mathbf{R}$ is a two-point tensor]

$$\mathbf{R}^* = \mathbf{Q} \cdot \mathbf{R}. \tag{3.9.16}$$

Then it follows from Eq. (3.9.14) that the right Cauchy stress tensor $\mathbf{U}$, being defined with respect to the reference configuration, remains unaltered by a superposed rigid-body motion and hence objective

$$\mathbf{U}^* = \mathbf{U}. \tag{3.9.17}$$

As far as the left Cauchy stretch tensor is concerned, the transformation law for it to be objective is derived as follows:

$$\mathbf{F}^* = \mathbf{V}^* \cdot \mathbf{R}^* = \mathbf{V}^* \cdot \mathbf{Q} \cdot \mathbf{R} = \mathbf{Q} \cdot \mathbf{F} = \mathbf{Q} \cdot \mathbf{V} \cdot \mathbf{R},$$

from which we arrive at the result

$$\mathbf{V}^* = \mathbf{Q} \cdot \mathbf{V} \cdot \mathbf{Q}^{\mathrm{T}}. \tag{3.9.18}$$

Thus, $\mathbf{V}$ transforms like a second-order tensor defined in the current configuration and is therefore objective.

We shall make use of the ideas presented in this section to develop the constitutive relations among the stress and strain measures (see Chapter 6).

3.10 Summary

In this chapter, deformation mapping χ of a material point occupying position $\mathbf{X}$ in a reference configuration to position $\mathbf{x}$ in the current configuration is introduced; two descriptions of motion, namely, the spatial (Eulerian) and material (Lagrange) descriptions of motion are described; and displacement $\mathbf{u}$ of a material point is defined as $\mathbf{u} = \mathbf{x} - \mathbf{X}$. The deformation gradient $\mathbf{F}$ is introduced as a two-point tensor between the reference configuration and the current configuration, $\mathbf{F} = (\boldsymbol{\nabla}_0\mathbf{x})^{\mathrm{T}}$. The deformation tensor is nonsingular and hence invertible. Isochoric, homogeneous, and inhomogeneous deformations are discussed in terms of the deformation mapping and the deformation gradient. Changes of volume and surface in going from the reference configuration to the current configuration are derived.

Several strain measures, including the Green–Lagrange strain tensor $\mathbf{E}$, Cauchy strain tensor $\tilde{\mathbf{B}}$, Euler or Almansi strain tensor $\mathbf{e}$, right Cauchy–Green deformation tensor $\mathbf{C}$, the left Cauchy–Green deformation (or Finger) tensor $\mathbf{B}$, and the Cauchy strain tensor $\tilde{\mathbf{B}} = \mathbf{B}^{-1}$ are introduced. A physical interpretation of the normal and shear strain components is also presented. Determinations of the principal strains and principal directions of strain are discussed with the help of the eigenvalue problem of Section 2.5.6. The infinitesimal strain tensor $\boldsymbol{\varepsilon}$ is obtained from $\mathbf{E}$ by retaining terms $|\boldsymbol{\nabla}_0\mathbf{u}|$ of order $\epsilon = \|\boldsymbol{\nabla}_0\mathbf{u}\|_\infty$ and omitting terms of order ϵ^2. Thus, in the infinitesimal case, the distinction between the Green–Lagrange strain tensor $\mathbf{E}$ and the Euler strain tensor $\mathbf{e}$ disappears. The displacement gradient tensor $(\boldsymbol{\nabla}\mathbf{u})^{\mathrm{T}}$ is expressed as a sum of the symmetric strain tensor $\tilde{\boldsymbol{\varepsilon}}$ and the skew symmetric rotation tensor $\boldsymbol{\Omega}$. Similarly, the rate of deformation tensor $\mathbf{L} = (\boldsymbol{\nabla}\mathbf{v})^{\mathrm{T}}$, where $\mathbf{v}$ is the velocity vector, is expressed as the sum of the symmetric part, namely, the rate of deformation tensor $\mathbf{D}$, and the skew symmetric part, the vorticity tensor $\mathbf{W}$.

The effect of superposed rigid-body motion and the concept of frame indifference that ensures nondependency on the frame of reference in measuring displacements, velocities, accelerations, and various strain measures is briefly discussed. It is shown that the measures of displacements and various measures of strains obey the frame indifference principle (i.e., they are independent of the coordinate frame of reference), while the velocities and accelerations are dependent on the coordinate of frame of reference. It is also found that the time rate of change of the displacement as well as strain measures are not objective, unless the rigid-body rotation $\mathbf{Q}$ is independent of time.

Finally, the polar decomposition theorem is presented that allows the unique (multiplicative) left and right decompositions of the deformation gradient $\mathbf{F}$ into pure stretch and pure rotation, $\mathbf{F} = \mathbf{R}\cdot\mathbf{U}$ and $\mathbf{F} = \mathbf{V}\cdot\mathbf{R}$, where $\mathbf{U}$ and $\mathbf{V}$ are the symmetric, positive-definite right and left Cauchy stretch tensors, respectively, and $\mathbf{R}$ is the proper orthogonal rotation tensor. Compatibility conditions on infinitesimal strain tensor $\boldsymbol{\varepsilon}$ and deformation tensor $\mathbf{F}$ that ensure a unique determination of displacements from a given strain field are also presented.

Numerous examples are presented throughout the chapter to illustrate the concepts introduced.

Problems

Descriptions of Motion

3.1 Given the motion

$$\chi(\mathbf{x}, t) = \mathbf{x} = (1+t)X_1\,\hat{\mathbf{e}}_1 + (1+t)X_2\,\hat{\mathbf{e}}_2 + X_3\,\hat{\mathbf{e}}_3, \quad 0 \leq t < \infty,$$

(a) determine the velocity and acceleration fields of the motion, and

(b) sketch deformations of the line $X_2 = 2X_1$, for fixed $X_3 = 1$ at $t = 1, 2$, and 3.

3.2 Determine the deformation mapping that maps a unit square into the quadrilateral shape shown in Fig. P3.2. Assume that the mapping is a complete polynomial in X_1 and X_2 up to the term X_1X_2 (note that the constant term is zero for this case).

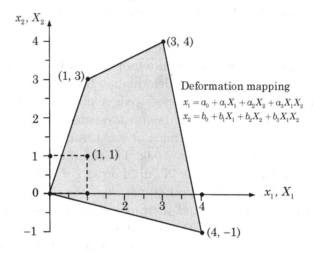

Fig. P3.2

3.3 Show that in the spatial description the acceleration components in the cylindrical coordinates are

$$a_r = \frac{\partial v_r}{\partial t} + v_r\frac{\partial v_r}{\partial r} + \frac{v_\theta}{r}\frac{\partial v_r}{\partial \theta} + v_z\frac{\partial v_r}{\partial z} - \frac{v_\theta^2}{r},$$

$$a_\theta = \frac{\partial v_\theta}{\partial t} + v_r\frac{\partial v_\theta}{\partial r} + \frac{v_\theta}{r}\frac{\partial v_\theta}{\partial \theta} + v_z\frac{\partial v_\theta}{\partial z} + \frac{v_r v_\theta}{r},$$

$$a_z = \frac{\partial v_z}{\partial t} + v_r\frac{\partial v_z}{\partial r} + \frac{v_\theta}{r}\frac{\partial v_z}{\partial \theta} + v_z\frac{\partial v_z}{\partial z}.$$

3.4 Show that in the spatial description the acceleration components in the spherical coordinates are

$$a_R = \frac{\partial v_R}{\partial t} + v_R\frac{\partial v_R}{\partial R} + \frac{v_\phi}{R}\left(\frac{\partial v_R}{\partial \phi} - v_\phi\right) + \frac{v_\theta}{R\sin\phi}\left(\frac{\partial v_R}{\partial \theta} - v_\theta\sin\phi\right),$$

$$a_\phi = \frac{\partial v_\phi}{\partial t} + v_R\frac{\partial v_\phi}{\partial R} + \frac{v_\phi}{R}\left(\frac{\partial v_\phi}{\partial \phi} + v_R\right) + \frac{v_\theta}{R\sin\phi}\left(\frac{\partial v_\phi}{\partial \theta} - v_\theta\cos\phi\right),$$

$$a_\theta = \frac{\partial v_\theta}{\partial t} + v_R\frac{\partial v_\theta}{\partial R} + \frac{v_\phi}{R}\frac{\partial v_\theta}{\partial \phi} + \frac{v_\theta}{R\sin\phi}\left(\frac{\partial v_\theta}{\partial \theta} + v_R\sin\phi + v_\phi\cos\phi\right).$$

Analysis of Deformation and Strain Measures

3.5 The motion of a continuous medium is given by

$$x_1 = \left(1 + e^{at}\right)X_1, \quad x_2 = \left(1 + e^{-2at}\right)X_2, \quad x_3 = X_3, \quad 0 \leq t < \infty,$$

where a is a positive constant. Determine

(a) the components of the deformation gradient $\mathbf{F}$ and the inverse mapping,

(b) the velocity components in the spatial description,

(c) the velocity components in the material description, and

(d) the acceleration components in the spatial description.

(e) Then verify the results of (d) by calculating first the acceleration components in the material coordinates and then using the inverse transformation in (a) to obtain the components in the spatial description.

3.6 For the deformation shown in Problem **3.2** (see Fig. P3.2), determine

(a) the components of the deformation gradient $\mathbf{F}$ and its inverse, and

(b) the components of the displacement vector.

3.7 The motion of a body is described by the mapping

$$\boldsymbol{\chi}(\mathbf{X}) = (X_1 + t^2 X_2)\,\hat{\mathbf{e}}_1 + (X_2 + t^2 X_1)\,\hat{\mathbf{e}}_2 + X_3\,\hat{\mathbf{e}}_3, \quad 0 \le t < \infty.$$

Determine

(a) the components of the deformation gradient $\mathbf{F}$ and its inverse,

(b) the components of the displacement, velocity, and acceleration vectors,

(c) the position (X_1, X_2, X_3) of the particle in undeformed configuration that occupies the position $(x_1, x_2, x_3) = (9, 6, 1)$ at time $t = 2$ in the deformed configuration, and

(d) the location at time $t = 2$ of the particle that later will be located at $\mathbf{x} = (2, 3, 1)$ at time $t = 3$.

(e) Then plot the deformed shape of a body at times $t = 0, 1, 2$, and 3, assuming that it is initially a unit cube.

3.8 *Homogeneous stretch.* Consider a body with deformation mapping of the form

$$\boldsymbol{\chi}(\mathbf{X}) = k_1 X_1\,\hat{\mathbf{e}}_1 + k_2 X_2\,\hat{\mathbf{e}}_2 + k_3 X_3\,\hat{\mathbf{e}}_3,$$

where $k_i \neq 0$ are constants. Determine the components of

(a) the deformation gradient $\mathbf{F}$, and

(b) the right and left Cauchy–Green deformation tensors $\mathbf{C}$ and $\mathbf{B}$.

3.9 *Homogeneous stretch followed by simple shear.* Consider a body with deformation mapping of the form

$$\boldsymbol{\chi}(\mathbf{X}) = (k_1 X_1 + e_0 k_2 X_2)\,\hat{\mathbf{e}}_1 + k_2 X_2\,\hat{\mathbf{e}}_2 + k_3 X_3\,\hat{\mathbf{e}}_3,$$

where $k_i \neq 0$ and e_0 are constants. Determine the components of

(a) the deformation gradient $\mathbf{F}$, and

(b) the right and left Cauchy–Green deformation tensors $\mathbf{C}$ and $\mathbf{B}$.

(c) Then plot representative shapes of a deformed unit square (let $k_1 = k_3 = 1$) that are achievable with this mapping; the suggested cases are (i) $k_2/k_1 = 1.5$, $e_0 = 0.1$; (ii) $k_2/k_1 = 1.5$, $e_0 = 0.25$; (iii) $k_2/k_1 = 1.25$, $e_0 = 0.5$; and (iv) $k_2/k_1 = 1.25$, $e_0 = 1.0$.

3.10 Suppose that the motion of a continuous medium is given by

$$x_1 = X_1 \cos At + X_2 \sin At,$$
$$x_2 = -X_1 \sin At + X_2 \cos At,$$
$$x_3 = (1 + Bt)X_3, \quad 0 \le t < \infty,$$

where A and B are constants. Determine the components of

(a) the displacement vector in the material description,

(b) the displacement vector in the spatial description,

(c) displacement vector components in the spatial description with respect to a cylindrical basis, and

(d) the Green–Lagrange and Eulerian strain tensors in the Cartesian coordinate system.

3.11 If the deformation mapping of a body is given by

$$\chi(\mathbf{X}) = (X_1 + AX_2)\,\hat{\mathbf{e}}_1 + (X_2 + BX_1)\,\hat{\mathbf{e}}_2 + X_3\,\hat{\mathbf{e}}_3,$$

where A and B are constants, determine

(a) the displacement components in the material description,

(b) the displacement components in the spatial description, and

(c) the components of the Green–Lagrange and Eulerian strain tensors.

3.12 For the deformation mapping in Problem **3.2**, determine the components of the Green–Lagrange strain tensor.

3.13 For the deformation field given in Problem **3.7**, determine the Green–Lagrange strain tensor components.

3.14 For the deformation mapping given in Problem **3.9**, determine the current positions (x_1, x_2) of material particles that were on the circle $X_1^2 + X_2^2 = R^2$ with radius R in the undeformed body.

3.15 The motion of a continuous medium is given by

$$x_1 = \tfrac{1}{2}(X_1 + X_2)e^t + \tfrac{1}{2}(X_1 - X_2)e^{-t},$$
$$x_2 = \tfrac{1}{2}(X_1 + X_2)e^t - \tfrac{1}{2}(X_1 - X_2)e^{-t},$$
$$x_3 = X_3,$$

for $0 \le t < \infty$. Determine

(a) the velocity components in the material description,

(b) the velocity components in the spatial description, and

(c) the components of the rate of deformation and vorticity tensors.

3.16 *Nanson's formula* Let the differential area in the reference configuration be dA. Then

$$\hat{\mathbf{N}}\,dA = d\mathbf{X}^{(1)} \times d\mathbf{X}^{(2)} \quad \text{or} \quad N_I\,dA = e_{IJK}\,dX_J^{(1)}dX_K^{(2)},$$

where $d\mathbf{X}^{(1)}$ and $d\mathbf{X}^{(2)}$ are two nonparallel differential vectors in the reference configuration. The mapping from the undeformed configuration to the deformed configuration maps $d\mathbf{X}^{(1)}$ and $d\mathbf{X}^{(2)}$ into $d\mathbf{x}^{(1)}$ and $d\mathbf{x}^{(2)}$, respectively. Then $\hat{\mathbf{n}}\,da = d\mathbf{x}^{(1)} \times d\mathbf{x}^{(2)}$. Show that

$$\hat{\mathbf{n}}\,da = J\mathbf{F}^{-\mathrm{T}} \cdot \hat{\mathbf{N}}\,dA.$$

3.17 Consider a rectangular block of material of thickness h and sides $3b$ and $4b$, and having a triangular hole as shown in Fig. P3.17. If the block is subjected to the deformation mapping given in Eq. (3.3.14),

$$\chi(\mathbf{X}) = (X_1 + \gamma X_2)\hat{\mathbf{e}}_1 + X_2\,\hat{\mathbf{e}}_2 + X_3\,\hat{\mathbf{e}}_3,$$

determine

(a) the equation of the line BC in the undeformed and deformed configurations,

(b) the angle ABC in the undeformed and deformed configurations, and

(c) the area of the triangle ABC in the undeformed and deformed configurations.

$$\chi(\mathbf{X}) = (X_1 + \gamma X_2)\hat{\mathbf{e}}_1 + X_2\hat{\mathbf{e}}_2 + X_3\hat{\mathbf{e}}_3$$

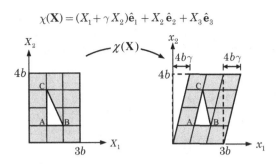

Fig. P3.17

NOTE: In Problems 3.18–3.22, undeformed and deformed configurations of bodies in equilibrium are given. In all cases, the deformation mapping can be determined uniquely with the suggested form of the mapping and the boundary data. Therefore, details of material constitution, material homogeneity, and loads causing deformation are not required to determine the kinematics of deformation.

3.18 Consider a unit square block of material of thickness h (into the plane of the paper), as shown in Fig. P3.18. If the block is subjected to a loading that deforms the square block into the shape shown (with no change in the thickness), (a) determine the deformation mapping, assuming that it is a complete polynomial in X_1 and X_2 up to the term $X_1 X_2$, (b) compute the components of the right Cauchy–Green deformation tensor $\mathbf{C}$ and Green–Lagrange strain tensor $\mathbf{E}$ at the point $\mathbf{X} = (1, 1, 0)$, and (c) compute the principal strains and directions at $\mathbf{X} = (1, 1, 0)$ for $\gamma = 1$.

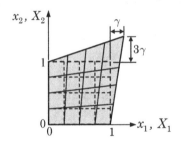

Fig. P3.18

3.19 Determine the displacements and Green–Lagrange strain tensor components for the deformed configuration shown in Fig. P3.19. The undeformed configuration is shown in dashed lines. Assume that the deformation mapping is a linear polynomial of X_1 and X_2 (note that for this case the constant terms are zero).

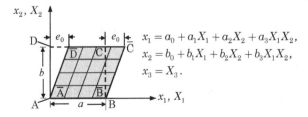

$$x_1 = a_0 + a_1 X_1 + a_2 X_2 + a_3 X_1 X_2,$$
$$x_2 = b_0 + b_1 X_1 + b_2 X_2 + b_3 X_1 X_2,$$
$$x_3 = X_3.$$

Fig. P3.19

3.20 Determine the displacements and Green–Lagrange strain components for the deformed configuration shown in Fig. P3.20. The undeformed configuration is shown in dashed lines. Use the suggested form of the deformation mapping, as implied by the deformed configuration.

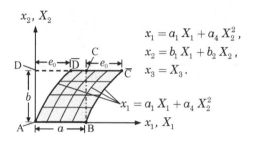

$$x_1 = a_1 X_1 + a_4 X_2^2,$$
$$x_2 = b_1 X_1 + b_2 X_2,$$
$$x_3 = X_3.$$

Fig. P3.20

3.21 Determine the displacements and Green–Lagrange strains in the (x_1, x_2, x_3) system for the deformed configuration shown in Fig. P3.21. The undeformed configuration is shown in dashed lines. Use the suggested form of the deformation mapping (for this case the constant terms are zero).

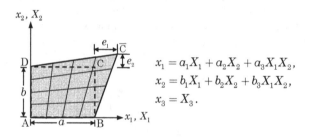

$$x_1 = a_1 X_1 + a_2 X_2 + a_3 X_1 X_2,$$
$$x_2 = b_1 X_1 + b_2 X_2 + b_3 X_1 X_2,$$
$$x_3 = X_3.$$

Fig. P3.21

3.22 Determine the displacements and Green–Lagrange strains for the deformed configuration shown in Fig. P3.22. The undeformed configuration is shown in dashed lines. Use the suggested form of the deformation mapping (note that constant terms are zero for this case).

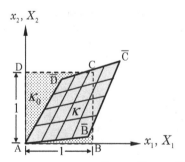

K_0	K
A: $(X_1, X_2) = (0,0)$	$\bar{A}$: $(x_1, x_2) = (0,0)$
B: $(X_1, X_2) = (1,0)$	$\bar{B}$: $(x_1, x_2) = (0.8, 0.2)$
C: $(X_1, X_2) = (1,1)$	$\bar{C}$: $(x_1, x_2) = (1.3, 1.2)$
D: $(X_1, X_2) = (0,1)$	$\bar{D}$: $(x_1, x_2) = (0.5, 0.9)$

$$x_1 = a_0 + a_1 X_1 + a_2 X_2 + a_3 X_1 X_2; \quad x_2 = b_0 + b_1 X_1 + b_2 X_2 + b_4 X_1 X_2; \quad x_3 = X_3$$

Fig. P3.22

3.23 Given the following displacement vector in a material description using a cylindrical coordinate system

$$\mathbf{u} = Ar\hat{\mathbf{e}}_r + Brz\hat{\mathbf{e}}_\theta + C\sin\theta\hat{\mathbf{e}}_z,$$

where A, B, and C are constants, determine the infinitesimal strains. Here (r,θ,z) denote the material coordinates.

3.24 Show that the components of the Green–Lagrange strain tensor in cylindrical coordinate system are given by

$$E_{rr} = \frac{\partial u_r}{\partial r} + \frac{1}{2}\left[\left(\frac{\partial u_r}{\partial r}\right)^2 + \left(\frac{\partial u_\theta}{\partial r}\right)^2 + \left(\frac{\partial u_z}{\partial r}\right)^2\right],$$

$$E_{r\theta} = \frac{1}{2}\left(\frac{1}{r}\frac{\partial u_r}{\partial\theta} + \frac{\partial u_\theta}{\partial r} - \frac{u_\theta}{r} + \frac{1}{r}\frac{\partial u_r}{\partial r}\frac{\partial u_r}{\partial\theta} + \frac{1}{r}\frac{\partial u_\theta}{\partial r}\frac{\partial u_\theta}{\partial\theta}\right.$$

$$\left. + \frac{1}{r}\frac{\partial u_z}{\partial r}\frac{\partial u_z}{\partial\theta} + \frac{u_r}{r}\frac{\partial u_\theta}{\partial r} - \frac{u_\theta}{r}\frac{\partial u_r}{\partial r}\right),$$

$$E_{rz} = \frac{1}{2}\left(\frac{\partial u_r}{\partial z} + \frac{\partial u_z}{\partial r} + \frac{\partial u_r}{\partial r}\frac{\partial u_r}{\partial z} + \frac{\partial u_\theta}{\partial r}\frac{\partial u_\theta}{\partial z} + \frac{\partial u_z}{\partial r}\frac{\partial u_z}{\partial z}\right),$$

$$E_{\theta\theta} = \frac{u_r}{r} + \frac{1}{r}\frac{\partial u_\theta}{\partial\theta} + \frac{1}{2}\left[\left(\frac{1}{r}\frac{\partial u_r}{\partial\theta}\right)^2 + \left(\frac{1}{r}\frac{\partial u_\theta}{\partial\theta}\right)^2 + \left(\frac{1}{r}\frac{\partial u_z}{\partial\theta}\right)^2\right.$$

$$\left. - \frac{2}{r^2}u_\theta\frac{\partial u_r}{\partial\theta} + \frac{2}{r^2}u_r\frac{\partial u_\theta}{\partial\theta} + \left(\frac{u_\theta}{r}\right)^2 + \left(\frac{u_r}{r}\right)^2\right],$$

$$E_{\theta z} = \frac{1}{2}\left(\frac{\partial u_\theta}{\partial z} + \frac{1}{r}\frac{\partial u_z}{\partial\theta} + \frac{1}{r}\frac{\partial u_r}{\partial\theta}\frac{\partial u_r}{\partial z} + \frac{1}{r}\frac{\partial u_\theta}{\partial\theta}\frac{\partial u_\theta}{\partial z}\right.$$

$$\left. + \frac{1}{r}\frac{\partial u_z}{\partial\theta}\frac{\partial u_z}{\partial z} - \frac{u_\theta}{r}\frac{\partial u_r}{\partial z} + \frac{u_r}{r}\frac{\partial u_\theta}{\partial z}\right),$$

$$E_{zz} = \frac{\partial u_z}{\partial z} + \frac{1}{2}\left[\left(\frac{\partial u_r}{\partial z}\right)^2 + \left(\frac{\partial u_\theta}{\partial z}\right)^2 + \left(\frac{\partial u_z}{\partial z}\right)^2\right].$$

Here (r,θ,z) denote the material coordinates (see Fig. P3.24).

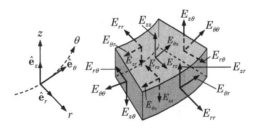

$E_{\xi\eta}$ is the component of strain in the ξ-direction on the η-plane

Fig. P3.24

3.25 The two-dimensional displacement field in a body is given by

$$u_1(\mathbf{X}) = X_1\left[X_1^2 X_2 + c_1\left(2c_2^3 + 3c_2^2 X_2 - X_2^3\right)\right],$$

$$u_2(\mathbf{X}) = -X_2\left(2c_2^3 + \frac{3}{2}c_2^2 X_2 - \frac{1}{4}X_2^3 + \frac{3}{2}c_1 X_1^2 X_2\right),$$

where c_1 and c_2 are constants. Find the linear and nonlinear Green–Lagrange strains.

3.26 Find the axial strain in the diagonal element, $\bar{A}\bar{C}$, of Problem **3.19**, using

(a) the basic definition of normal strain, and

(b) the strain transformation equations.

3.27 The biaxial state of strain at a point is given by $\varepsilon_{11} = 800 \times 10^{-6}$ in./in., $\varepsilon_{22} = 200 \times 10^{-6}$ in./in., $\varepsilon_{12} = 400 \times 10^{-6}$ in./in. Find the principal strains and their directions.

3.28 Show that the invariants J_1, J_2, and J_3 of the Green–Lagrange strain tensor $\mathbf{E}$ can be expressed in terms of the principal values λ_i of $\mathbf{E}$ as

$$J_1 = \lambda_1 + \lambda_2 + \lambda_3, \quad J_2 = \lambda_1\lambda_2 + \lambda_2\lambda_3 + \lambda_3\lambda_1, \quad J_3 = \lambda_1\lambda_2\lambda_3.$$

Of course, the above result holds for any second-order tensor.

3.29 Given the displacement field in the cylindrical coordinate system

$$u_r = U(r), \quad u_\theta = 0, \quad u_z = 0,$$

where $U(r)$ is a function of only r, determine the Green–Lagrange strain components.

3.30 Given the displacement field in the spherical coordinate system

$$u_R = U(R), \quad u_\phi = 0, \quad u_\theta = 0,$$

where $U(r)$ is a function of only r, determine the Green–Lagrange strain components.

VELOCITY GRADIENT, RATE OF DEFORMATION, AND VORTICITY TENSORS

3.31 Show that the components of the spin tensor $\mathbf{W}$ in the cylindrical coordinate system are

$$W_{r\theta} = \frac{1}{2}\left(\frac{1}{r}\frac{\partial v_r}{\partial \theta} - \frac{v_\theta}{r} - \frac{\partial v_\theta}{\partial r}\right) = -W_{\theta r},$$

$$W_{rz} = \frac{1}{2}\left(\frac{\partial v_r}{\partial z} - \frac{\partial v_z}{\partial r}\right) = -W_{zr},$$

$$W_{z\theta} = \frac{1}{2}\left(\frac{1}{r}\frac{\partial v_z}{\partial \theta} - \frac{\partial v_\theta}{\partial z}\right) = -W_{\theta z}.$$

3.32 If $\mathbf{D} = \mathbf{0}$, show that

$$\mathbf{v} = \mathbf{w} \times \mathbf{x} + \mathbf{c} \quad (v_i = e_{ijk}\, w_j\, x_k + c_i),$$

where both $\mathbf{w}$ (vorticity vector) and $\mathbf{c}$ are constant vectors.

3.33 Show that

$$\dot{\mathbf{E}} = \frac{1}{2}\dot{\mathbf{C}} = \frac{1}{2}\left(\dot{\mathbf{F}}^{\mathrm{T}} \cdot \mathbf{F} + \mathbf{F}^{\mathrm{T}} \cdot \dot{\mathbf{F}}\right),$$

and

$$\mathbf{W} = \frac{1}{2}\left(\dot{\mathbf{F}} \cdot \mathbf{F}^{-1} - \mathbf{F}^{-\mathrm{T}} \cdot \dot{\mathbf{F}}^{\mathrm{T}}\right).$$

3.34 Verify that

$$\dot{\mathbf{v}} = \frac{\partial \mathbf{v}}{\partial t} + \frac{1}{2}\boldsymbol{\nabla}(\mathbf{v} \cdot \mathbf{v}) + 2\mathbf{W} \cdot \mathbf{v}$$

$$= \frac{\partial \mathbf{v}}{\partial t} + \frac{1}{2}\boldsymbol{\nabla}(\mathbf{v} \cdot \mathbf{v}) + 2\mathbf{w} \times \mathbf{v},$$

where $\mathbf{W}$ is the spin tensor and $\mathbf{w}$ is the vorticity vector [see Eq. (3.6.5)].

3.35 Show that

$$\frac{DJ}{Dt} = (\boldsymbol{\nabla} \cdot \mathbf{v})\, J.$$

Hints: $\frac{Dx_i}{Dt} = v_i$ and $\frac{\partial v_i}{\partial X_j} = \frac{\partial v_i}{\partial x_k}\frac{\partial x_k}{\partial X_j}$. See also the list of properties of determinants highlighted in Section 2.3.6.

3.36 Establish the identities

$$d\mathbf{v} = \mathbf{L} \cdot d\mathbf{x}, \quad \text{and} \quad (\boldsymbol{\nabla}_0 \mathbf{v})^{\mathrm{T}} = \mathbf{L} \cdot \mathbf{F}.$$

3.37 Show that $\dot{\mathbf{C}} = 2\mathbf{F}^{\mathrm{T}} \cdot \mathbf{D} \cdot \mathbf{F}$, where $\mathbf{C}$, $\mathbf{D}$, and $\mathbf{F}$ are the right Cauchy–Green deformation tensor, rate of deformation tensor, and deformation gradient, respectively.

3.38 Show that the Eulerian strain rate is given by

$$\dot{\mathbf{e}} = \mathbf{D} - \left(\mathbf{e} \cdot \mathbf{L} + \mathbf{L}^{\mathrm{T}} \cdot \mathbf{e}\right),$$

and [see Eq. (3.5.10) for the definition of $\tilde{\varepsilon}$]

$$\dot{\tilde{\varepsilon}} = \mathbf{D}.$$

COMPATIBILITY CONDITIONS

3.39 Use the index notation to establish the compatibility conditions in Eq. (3.7.11)

$$\boldsymbol{\nabla}_0 \times (\boldsymbol{\nabla}_0 \times \boldsymbol{\varepsilon})^{\mathrm{T}} = \mathbf{0}$$

for the infinitesimal strains. *Hint:* Begin with $\boldsymbol{\nabla}_0 \times \boldsymbol{\varepsilon}$ and use Eq. (3.5.15).

3.40 Show that the following second-order tensor is symmetric:

$$\mathbf{S} = \boldsymbol{\nabla}_0 \times (\boldsymbol{\nabla}_0 \times \boldsymbol{\varepsilon})^{\mathrm{T}}.$$

3.41 Let [see the compatibility conditions in Eqs. (3.7.4)–(3.7.9)]

$$-S_{33} = R_3 = \frac{\partial^2 \varepsilon_{11}}{\partial X_2^2} + \frac{\partial^2 \varepsilon_{22}}{\partial X_1^2} - 2\frac{\partial^2 \varepsilon_{12}}{\partial X_1 \partial X_2}, \tag{1}$$

$$-S_{22} = R_2 = \frac{\partial^2 \varepsilon_{11}}{\partial X_3^2} + \frac{\partial^2 \varepsilon_{33}}{\partial X_1^2} - 2\frac{\partial^2 \varepsilon_{13}}{\partial X_1 \partial X_3}, \tag{2}$$

$$-S_{11} = R_1 = \frac{\partial^2 \varepsilon_{22}}{\partial X_3^2} + \frac{\partial^2 \varepsilon_{33}}{\partial X_2^2} - 2\frac{\partial^2 \varepsilon_{23}}{\partial X_2 \partial X_3}, \tag{3}$$

$$-S_{23} = U_1 = -\frac{\partial^2 \varepsilon_{11}}{\partial X_2 \partial X_3} + \frac{\partial}{\partial X_1}\left(-\frac{\partial \varepsilon_{23}}{\partial X_1} + \frac{\partial \varepsilon_{13}}{\partial X_2} + \frac{\partial \varepsilon_{12}}{\partial X_3}\right), \tag{4}$$

$$-S_{31} = U_2 = -\frac{\partial^2 \varepsilon_{22}}{\partial X_1 \partial X_3} + \frac{\partial}{\partial X_2}\left(\frac{\partial \varepsilon_{23}}{\partial X_1} - \frac{\partial \varepsilon_{13}}{\partial X_2} + \frac{\partial \varepsilon_{12}}{\partial X_3}\right), \tag{5}$$

$$-S_{12} = U_3 = -\frac{\partial^2 \varepsilon_{33}}{\partial X_1 \partial X_2} + \frac{\partial}{\partial X_3}\left(\frac{\partial \varepsilon_{23}}{\partial X_1} + \frac{\partial \varepsilon_{13}}{\partial X_2} - \frac{\partial \varepsilon_{12}}{\partial X_3}\right). \tag{6}$$

Show that

$$\frac{\partial R_1}{\partial X_1} + \frac{\partial U_3}{\partial X_2} + \frac{\partial U_2}{\partial X_3} = 0,$$

$$\frac{\partial U_3}{\partial X_1} + \frac{\partial R_2}{\partial X_2} + \frac{\partial U_1}{\partial X_3} = 0, \tag{7}$$

$$\frac{\partial U_2}{\partial X_1} + \frac{\partial U_1}{\partial X_2} + \frac{\partial R_3}{\partial X_3} = 0.$$

These relations are known as the *Bianchi formulas*.

3.42 Consider the following infinitesimal strain field:

$$\varepsilon_{11} = c_1 X_2^2, \quad \varepsilon_{22} = c_1 X_1^2, \quad 2\varepsilon_{12} = c_2 X_1 X_2,$$

$$\varepsilon_{31} = \varepsilon_{32} = \varepsilon_{33} = 0,$$

where c_1 and c_2 are constants. Determine

(a) c_1 and c_2 such that there exists a continuous, single-valued displacement field that corresponds to this strain field,

(b) the most general form of the corresponding displacement field using c_1 and c_2 obtained in (a), and

(c) the constants of integration introduced in (b) for the boundary conditions $\mathbf{u} = \mathbf{0}$ and $\boldsymbol{\Omega} = \mathbf{0}$ at $\mathbf{X} = \mathbf{0}$ (i.e., $u_1 = u_2 = 0$ and $\frac{\partial u_1}{\partial X_2} - \frac{\partial u_2}{\partial X_1} = 0$ at $X_1 = X_2 = 0$).

3.43 Determine whether the following strain fields, under the assumption of infinitesimal strains, are possible in a continuous body:

(a) $[\varepsilon] = \begin{bmatrix} (X_1^2 + X_2^2) & X_1 X_2 \\ X_1 X_2 & X_2^2 \end{bmatrix}$. (b) $[\varepsilon] = \begin{bmatrix} X_3(X_1^2 + X_2^2) & 2X_1 X_2 X_3 & X_3 \\ 2X_1 X_2 X_3 & X_2^2 & X_1 \\ X_3 & X_1 & X_3^2 \end{bmatrix}$.

3.44 Evaluate the compatibility conditions $\nabla_0 \times (\nabla_0 \times \mathbf{E})^{\mathrm{T}} = \mathbf{0}$ in cylindrical coordinates.

3.45 Given the infinitesimal strain components

$$\varepsilon_{11} = f(X_2, X_3), \quad \varepsilon_{22} = \varepsilon_{33} = -\nu f(X_2, X_3), \quad \varepsilon_{12} = \varepsilon_{13} = \varepsilon_{23} = 0,$$

determine the form of $f(X_2, X_3)$ in order that the strain field is compatible.

3.46 Given the strain tensor $\mathbf{E} = E_{rr} \hat{\mathbf{e}}_r \hat{\mathbf{e}}_r + E_{\theta\theta} \hat{\mathbf{e}}_\theta \hat{\mathbf{e}}_\theta$ in an axisymmetric body (i.e., E_{rr} and $E_{\theta\theta}$ are functions of r and z only), determine the compatibility conditions on E_{rr} and $E_{\theta\theta}$. *Hint:* See Example 2.5.1.

RIGID-BODY MOTION AND OBJECTIVITY

3.47 Determine the effect of the superposed rigid-body motion on the left Cauchy–Green deformation tensor $\mathbf{B} = \mathbf{F} \cdot \mathbf{F}^{\mathrm{T}}$.

3.48 If $\mathbf{Q}(t)$ is an orthogonal tensor-valued function of a scalar t [i.e., $\mathbf{Q}^{-1} = \mathbf{Q}^{\mathrm{T}}$], show that $\dot{\mathbf{Q}} \cdot \mathbf{Q}^{\mathrm{T}} = -(\dot{\mathbf{Q}} \cdot \mathbf{Q}^{\mathrm{T}})^{\mathrm{T}}$. That is, show that $\dot{\mathbf{Q}} \cdot \mathbf{Q}^{\mathrm{T}}$ is skew symmetric.

3.49 Show that the spin tensor $\mathbf{W}$ under superposed rigid-body motion becomes

$$\mathbf{W}^* = \mathbf{Q} \cdot \mathbf{W} \cdot \mathbf{Q}^{\mathrm{T}} + \boldsymbol{\Omega},$$

where $\boldsymbol{\Omega}$ is the skew symmetric rotation tensor, $\boldsymbol{\Omega} = \dot{\mathbf{Q}} \cdot \mathbf{Q}^{\mathrm{T}}$ [see also Eq. (3.8.19)].

3.50 Suppose that the second-order tensor $\mathbf{T}$ is objective in the sense that it satisfies the condition $\mathbf{T}^* = \mathbf{Q} \cdot \mathbf{T} \cdot \mathbf{Q}$, where quantities with and without an asterisk belong to two different frames of reference. Then show that the following second-order tensor $\mathbf{S}$ is objective (i.e., show that $\mathbf{S}^* = \mathbf{S}$):

$$\mathbf{S} = \mathbf{F}^{-1} \cdot \mathbf{T} \cdot \mathbf{F}^{-\mathrm{T}},$$

where $\mathbf{Q}$ is a proper orthogonal tensor.

3.51 Prove or disprove if the following second tensor satisfies objectivity:

$$\mathbf{T} = \mathbf{S} \cdot \mathbf{U} + \mathbf{U} \cdot \mathbf{S},$$

where $\mathbf{S}^* = \mathbf{R} \cdot \mathbf{S} \cdot \mathbf{R}^{\mathrm{T}}$, $\mathbf{U} = \mathbf{R}^{\mathrm{T}} \cdot \mathbf{F}$ is the right Cauchy stretch tensor, and $\mathbf{F}$ is the deformation gradient.

3.52 Using the transformation rule $\mathbf{F}^* = \mathbf{Q} \cdot \mathbf{F}$, show that the Euler stain tensor $\mathbf{e}$ transforms according to the rule under superposed rigid-body motion

$$\mathbf{e}^* = \mathbf{Q} \cdot \mathbf{e} \cdot \mathbf{Q}^{\mathrm{T}}.$$

3.53 Show that the spatial gradient of a vector $\mathbf{u}(\mathbf{x}, t)$ is objective, that is, prove

$$\nabla^* \mathbf{u}^*(\mathbf{x}^*, t^*) = \mathbf{Q}(t) \cdot \nabla \mathbf{u}(\mathbf{x}, t) \cdot \mathbf{Q}^{\mathrm{T}}(t).$$

3.54 Show that the material time derivatives of objective vector and tensor fields, $\mathbf{u}$ and $\mathbf{S}$, are not objective.

POLAR DECOMPOSITION

3.55 Establish the uniqueness of the decomposition $\mathbf{F} = \mathbf{R} \cdot \mathbf{U} = \mathbf{V} \cdot \mathbf{R}$. For example, if $\mathbf{F} = \mathbf{R}_1 \cdot \mathbf{U}_1 = \mathbf{R}_2 \cdot \mathbf{U}_2$, then show that $\mathbf{R}_1 = \mathbf{R}_2$ and $\mathbf{U}_1 = \mathbf{U}_2$.

3.56 Show that the eigenvalues of the left and right Cauchy stretch tensors $\mathbf{U}$ and $\mathbf{V}$ are the same and that the eigenvector of $\mathbf{V}$ is given by $\mathbf{R} \cdot \mathbf{n}$, where $\mathbf{n}$ is the eigenvector of $\mathbf{U}$.

3.57 (a) If λ is the eigenvalue and $\mathbf{n}$ is the eigenvector of $\mathbf{U}$, show that the eigenvalue of $\mathbf{C}$ is λ^2 and the eigenvector is the same as that of $\mathbf{U}$. (b) Show that a line element in the principal direction $\mathbf{n}$ of $\mathbf{C}$ becomes an element in the direction of $\mathbf{R} \cdot \mathbf{n}$ in the deformed configuration.

3.58 Show that the spin tensor $\mathbf{W}$ can be written as

$$2\mathbf{W} = 2\dot{\mathbf{R}} \cdot \mathbf{R}^{\mathrm{T}} + \mathbf{R} \cdot \left(\dot{\mathbf{U}} \cdot \mathbf{U}^{-1} - \mathbf{U}^{-1} \cdot \dot{\mathbf{U}} \right) \cdot \mathbf{R}^{\mathrm{T}},$$

where $\mathbf{R}$ is the (proper) orthogonal rotation tensor $\mathbf{R}^{-1} = \mathbf{R}^{\mathrm{T}}$ and $\mathbf{U}$ is the symmetric positive-definite right Cauchy stretch tensor. Also show that for rigid-body motion, one has $\mathbf{W} = \dot{\mathbf{R}} \cdot \mathbf{R}^{\mathrm{T}}$.

3.59 Prove the symmetry and positive-definiteness of the right Cauchy–Green deformation tensor $\mathbf{C} = \mathbf{F}^{\mathrm{T}} \cdot \mathbf{F}$.

3.60 Calculate $\sqrt{\mathbf{C}}$ when

$$[C] = \begin{bmatrix} 3 & 2 & 0 \\ 2 & 3 & 0 \\ 0 & 0 & 9 \end{bmatrix}.$$

3.61 Given that

$$[F] = \frac{1}{5} \begin{bmatrix} 2 & -5 \\ 11 & 2 \end{bmatrix},$$

determine the right and left stretch tensors.

3.62 Given that

$$[F] = \begin{bmatrix} \sqrt{3} & 1 & 0 \\ 0 & 2 & 0 \\ 0 & 0 & 1 \end{bmatrix},$$

determine the right and left stretch tensors.

3.63 Calculate the left and right Cauchy stretch tensors $\mathbf{U}$ and $\mathbf{V}$ associated with $\mathbf{F}$ of Problem **3.11** for the choice of $A = 2$ and $B = 0$.

STRESS MEASURES

Most of the fundamental ideas of science are essentially simple, and may, as a rule, be expressed in a language comprehensible to everyone. — Albert Einstein (1879–1955)

4.1 Introduction

In the beginning of Chapter 3, we briefly discussed the need for studying deformations and stresses in material systems that we may design for engineering applications. All materials have certain thresholds to withstand forces, beyond which they "fail" to perform their intended function. The force per unit area, called *stress*, is a measure of the capacity of the material to carry loads, and all designs are based on the criterion that the materials used have the capacity to carry the working loads of the system. Thus, it is necessary to determine the state of stress in a material.

In this chapter we study the concept of stress and its various measures. For instance, stress can be measured per unit deformed area or undeformed area. As we shall see shortly, stress at a point in a three-dimensional continuum can be measured in terms of nine quantities, three per plane, on three mutually perpendicular planes at the point. These nine quantities may be viewed as the components of a second-order tensor, called a *stress tensor*. Coordinate transformations and principal values associated with the stress tensor and stress equilibrium equations are also discussed.

4.2 Cauchy Stress Tensor and Cauchy's Formula

4.2.1 Stress Vector

First we introduce the true stress, that is, the stress in the deformed configuration κ that is measured per unit area of the deformed configuration κ. The surface force acting on a small element of (surface) area in a continuous medium depends not only on the magnitude of the area but also on the orientation of the area. It is customary to denote the direction of a plane area by means of a unit vector drawn normal to that plane. The direction of the normal is taken by convention as that in which a right-handed screw advances as it is rotated according to the direction of travel along the boundary curve or contour. Let the unit normal vector be denoted by $\hat{\mathbf{n}}$. Then the area is expressed as $\mathbf{A} = A\hat{\mathbf{n}}$.

If we denote by $d\mathbf{f}(\hat{\mathbf{n}})$ the force on a small area $\hat{\mathbf{n}}\,da$ located at position $\mathbf{x}$, the *stress vector* can be defined, shown graphically in Fig. 4.2.1, as

$$\mathbf{t}(\hat{\mathbf{n}}) = \lim_{\Delta a \to 0} \frac{\Delta \mathbf{f}(\hat{\mathbf{n}})}{\Delta a}.$$

(4.2.1)

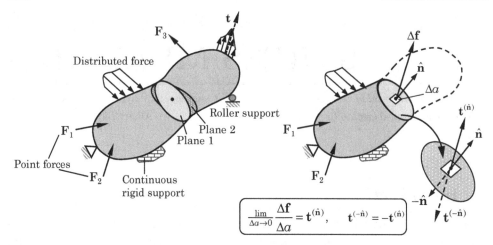

Fig. 4.2.1: Cuts through a point of a material body by planes of different orientations. The figure also shows a stress vector on a plane whose normal is $\hat{\mathbf{n}}$.

We see that the stress vector $\mathbf{t}$ is a point function of the unit normal $\hat{\mathbf{n}}$, which denotes the orientation of the plane on which $\mathbf{t}$ acts. Because of Newton's third law for action and reaction, we see that $\mathbf{t}(-\hat{\mathbf{n}}) = -\mathbf{t}(\hat{\mathbf{n}})$. It is fruitful to establish a relationship between $\mathbf{t}$ and $\hat{\mathbf{n}}$.

4.2.2 Cauchy's Formula

To establish the relationship between $\mathbf{t}$ and $\hat{\mathbf{n}}$ for the infinitesimal deformation[1], we set up an infinitesimal tetrahedron in Cartesian coordinates. The tetrahedron can come either from an interior point or from a boundary point, as indicated in Fig. 4.2.2(a). If $-\mathbf{t}_1, -\mathbf{t}_2, -\mathbf{t}_3$, and $\mathbf{t}$ denotes the stress vectors in the outward directions on the faces of the infinitesimal tetrahedron whose areas are Δa_1, Δa_2, Δa_3, and Δa, respectively, as shown in Fig. 4.2.2(b) (i.e., $-\mathbf{t}_j$ acts on the plane perpendicular to the negative x_j-axis), we have by Newton's second law for the mass inside the tetrahedron,

$$\mathbf{t}\,\Delta a - \mathbf{t}_1\,\Delta a_1 - \mathbf{t}_2\,\Delta a_2 - \mathbf{t}_3\,\Delta a_3 + \rho\,\Delta v\,\mathbf{f} = \rho\,\Delta v\,\mathbf{a}, \qquad (4.2.2)$$

where Δv is the volume of the tetrahedron, ρ is the density, $\mathbf{f}$ is the body force per unit mass, and $\mathbf{a}$ is the acceleration. Because the total vector area of a closed surface is zero (by the gradient theorem), we have

$$\Delta a\,\hat{\mathbf{n}} - \Delta a_1\,\hat{\mathbf{e}}_1 - \Delta a_2\,\hat{\mathbf{e}}_2 - \Delta a_3\,\hat{\mathbf{e}}_3 = \mathbf{0}. \qquad (4.2.3)$$

It follows that

$$\Delta a_1 = (\hat{\mathbf{e}}_1 \cdot \hat{\mathbf{n}})\Delta a, \quad \Delta a_2 = (\hat{\mathbf{e}}_2 \cdot \hat{\mathbf{n}})\Delta a, \quad \Delta a_3 = (\hat{\mathbf{e}}_3 \cdot \hat{\mathbf{n}})\Delta a. \qquad (4.2.4)$$

---

[1] The Cauchy formula can be established for the finite deformation case by considering a tetrahedron with curved faces.

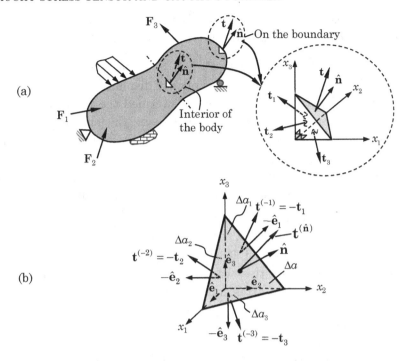

Fig. 4.2.2: A tetrahedral element with stress vectors on all its faces.

The volume of the element Δv can be expressed as

$$\Delta v = \frac{\Delta h}{3}\Delta a, \tag{4.2.5}$$

where Δh is the perpendicular distance from the origin to the slant face.

Substitution of Eqs. (4.2.4) and (4.2.5) into Eq. (4.2.2) and dividing throughout by Δa yields

$$\mathbf{t} = \mathbf{t}_1(\hat{\mathbf{e}}_1 \cdot \hat{\mathbf{n}}) + \mathbf{t}_2(\hat{\mathbf{e}}_2 \cdot \hat{\mathbf{n}}) + \mathbf{t}_3(\hat{\mathbf{e}}_3 \cdot \hat{\mathbf{n}}) + \rho\frac{\Delta h}{3}(\mathbf{a} - \mathbf{f}). \tag{4.2.6}$$

In the limit as the tetrahedron is shrunk to a point, $\Delta h \to 0$, we obtain

$$\mathbf{t} = \mathbf{t}_1(\hat{\mathbf{e}}_1 \cdot \hat{\mathbf{n}}) + \mathbf{t}_2(\hat{\mathbf{e}}_2 \cdot \hat{\mathbf{n}}) + \mathbf{t}_3(\hat{\mathbf{e}}_3 \cdot \hat{\mathbf{n}}) = \mathbf{t}_i(\hat{\mathbf{e}}_i \cdot \hat{\mathbf{n}}). \tag{4.2.7}$$

4.2.3 Cauchy Stress Tensor

It is convenient to display Eq. (4.2.7) as

$$\mathbf{t} = (\mathbf{t}_1\,\hat{\mathbf{e}}_1 + \mathbf{t}_2\,\hat{\mathbf{e}}_2 + \mathbf{t}_3\,\hat{\mathbf{e}}_3) \cdot \hat{\mathbf{n}}. \tag{4.2.8}$$

The terms in the parentheses should be treated as a dyadic, called *stress dyadic* or *stress tensor* (because its components transform like a second-order tensor), denoted $\boldsymbol{\sigma}$:

$$\boldsymbol{\sigma} \equiv \mathbf{t}_1\,\hat{\mathbf{e}}_1 + \mathbf{t}_2\,\hat{\mathbf{e}}_2 + \mathbf{t}_3\,\hat{\mathbf{e}}_3 = \mathbf{t}_j\,\hat{\mathbf{e}}_j. \tag{4.2.9}$$

The stress tensor is a property of the medium that is independent of the unit outward normal vector $\hat{\mathbf{n}}$. Thus, from Eqs. (4.2.8) and (4.2.9), we have

$$\mathbf{t}(\hat{\mathbf{n}}) = \boldsymbol{\sigma} \cdot \hat{\mathbf{n}} = \hat{\mathbf{n}} \cdot \boldsymbol{\sigma}^{\mathrm{T}} \quad (t_i = \sigma_{ij} n_j), \tag{4.2.10}$$

and the dependence of $\mathbf{t}$ on $\hat{\mathbf{n}}$ has been explicitly displayed. Equation (4.2.10) is known as the *Cauchy stress formula*, and $\boldsymbol{\sigma}$ is termed the *Cauchy stress tensor*. Thus, the Cauchy stress tensor $\boldsymbol{\sigma}$ is defined to be the *current force* per unit *deformed area*, $d\mathbf{f} = \mathbf{t}\,da = \boldsymbol{\sigma} \cdot d\mathbf{a}$, where Cauchy's formula, $\mathbf{t} = \boldsymbol{\sigma} \cdot \hat{\mathbf{n}}$, and $d\mathbf{a} = \hat{\mathbf{n}}\,da$ are used.

In Cartesian component form, the Cauchy formula in Eq. (4.2.10) can be written as $t_i = \sigma_{ij} n_j$, and it can be expressed in matrix form as

$$\begin{Bmatrix} t_1 \\ t_2 \\ t_3 \end{Bmatrix} = \begin{bmatrix} \sigma_{11} & \sigma_{12} & \sigma_{13} \\ \sigma_{21} & \sigma_{22} & \sigma_{23} \\ \sigma_{31} & \sigma_{32} & \sigma_{33} \end{bmatrix} \begin{Bmatrix} n_1 \\ n_2 \\ n_3 \end{Bmatrix}. \tag{4.2.11}$$

It is useful to resolve the stress vectors $\mathbf{t}_1, \mathbf{t}_2$, and $\mathbf{t}_3$ into their orthogonal components in a rectangular Cartesian system:

$$\mathbf{t}_j = \hat{\mathbf{e}}_1 \sigma_{1j} + \hat{\mathbf{e}}_2 \sigma_{2j} + \hat{\mathbf{e}}_3 \sigma_{3j} = \hat{\mathbf{e}}_i \sigma_{ij}, \quad j = 1, 2, 3. \tag{4.2.12}$$

Hence, the stress tensor can be expressed in the Cartesian basis as

$$\boldsymbol{\sigma} = \mathbf{t}_j \hat{\mathbf{e}}_j = \sigma_{ij} \hat{\mathbf{e}}_i \hat{\mathbf{e}}_j. \tag{4.2.13}$$

The component σ_{ij} represents the stress *in the x_i-coordinate direction and on a plane perpendicular to the x_j coordinate,* as shown in Fig. 4.2.3 on the faces of a point cube (i.e., the faces of the cube can be imagined as the planes passing through a point). We note that the symmetry of $\boldsymbol{\sigma}$ is *not* assumed.

The stress tensor can be expressed in any coordinate system. For example, in the cylindrical coordinate system, the nonion form of $\boldsymbol{\sigma}$ is (see Fig. 4.2.4)

$$\boldsymbol{\sigma} = \sigma_{rr} \hat{\mathbf{e}}_r \hat{\mathbf{e}}_r + \sigma_{r\theta} \hat{\mathbf{e}}_r \hat{\mathbf{e}}_\theta + \sigma_{\theta r} \hat{\mathbf{e}}_\theta \hat{\mathbf{e}}_r + \sigma_{rz} \hat{\mathbf{e}}_r \hat{\mathbf{e}}_z + \sigma_{zr} \hat{\mathbf{e}}_z \hat{\mathbf{e}}_r$$
$$+ \sigma_{\theta\theta} \hat{\mathbf{e}}_\theta \hat{\mathbf{e}}_\theta + \sigma_{\theta z} \hat{\mathbf{e}}_\theta \hat{\mathbf{e}}_z + \sigma_{z\theta} \hat{\mathbf{e}}_z \hat{\mathbf{e}}_\theta + \sigma_{zz} \hat{\mathbf{e}}_z \hat{\mathbf{e}}_z. \tag{4.2.14}$$

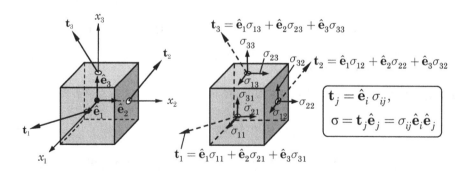

Fig. 4.2.3: Display of stress components in Cartesian rectangular coordinates.

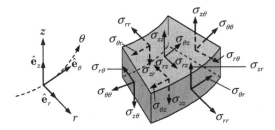

$\sigma_{\xi\eta}$ denotes the component of stress in the ξ-direction on the η-plane

Fig. 4.2.4: Display of stress components in cylindrical coordinates.

Note that $\mathbf{t}(\hat{\mathbf{n}})$, in general, is not in the direction of $\hat{\mathbf{n}}$. The component of $\mathbf{t}$ that is in the direction of $\hat{\mathbf{n}}$ is called the *normal stress*. The component of $\mathbf{t}$ that is normal to $\hat{\mathbf{n}}$ (i.e., the component lies in the surface) is termed the (projected) *shear* stress. According to the vector identity in Eq. (2.2.26), the stress vector $\mathbf{t}$ can be represented as the sum of vectors along and perpendicular to the unit normal vector $\hat{\mathbf{n}}$, as shown in Fig. 4.2.5:

$$\mathbf{t} = (\mathbf{t} \cdot \hat{\mathbf{n}})\hat{\mathbf{n}} + \hat{\mathbf{n}} \times (\mathbf{t} \times \hat{\mathbf{n}}) \equiv \mathbf{t}_{nn} + \mathbf{t}_{ns}. \tag{4.2.16}$$

Stress vectors on a plane are called *traction vectors*. The traction vector $\mathbf{t}_{nn}$ normal to the plane and its magnitude t_{nn} are

$$\mathbf{t}_{nn} = (\mathbf{t} \cdot \hat{\mathbf{n}})\hat{\mathbf{n}}; \quad t_{nn} = \mathbf{t} \cdot \hat{\mathbf{n}} = t_i n_i = n_j \sigma_{ji} n_i = \sigma_{ij} n_i n_j, \tag{4.2.17}$$

and the shear traction vector $\mathbf{t}_{ns}$ (i.e., projection of $\mathbf{t}$ along the plane) and its magnitude t_{ns} are

$$\mathbf{t}_{ns} = \mathbf{t} - \mathbf{t}_{nn}; \quad |\mathbf{t}_{ns}| = t_{ns} = \sqrt{|\mathbf{t}|^2 - t_{nn}^2}. \tag{4.2.18}$$

The tangential component lies in the $\hat{\mathbf{n}} - \mathbf{t}$ plane but perpendicular to $\hat{\mathbf{n}}$, as shown in Fig. 4.2.5. Example 4.2.1 illustrates the ideas presented here.

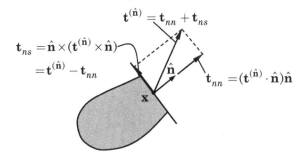

Fig. 4.2.5: The normal and shear stress vectors at a point $\mathbf{x}$ on a plane.

Example 4.2.1

With reference to a rectangular Cartesian system (x_1, x_2, x_3), the components of the stress dyadic at a certain point of a continuous medium $\mathcal{B}$ are given by (see Fig. 4.2.6)

$$[\sigma] = \begin{bmatrix} 200 & 400 & 300 \\ 400 & 0 & 0 \\ 300 & 0 & -100 \end{bmatrix} \text{ psi.}$$

Determine the stress vector $\mathbf{t}$ and its normal and tangential components at the point on the plane, $\phi(x_1, x_2, x_3) \equiv x_1 + 2x_2 + 2x_3 = $ constant, which is passing through the point.

Solution: First, we should find the unit normal to the plane on which we are required to find the stress vector. The unit normal to the plane defined by $\phi(x_1, x_2, x_3) = $ constant is

$$\hat{\mathbf{n}} = \frac{\boldsymbol{\nabla}\phi}{|\boldsymbol{\nabla}\phi|} = \tfrac{1}{3}(\hat{\mathbf{e}}_1 + 2\hat{\mathbf{e}}_2 + 2\hat{\mathbf{e}}_3).$$

The components of the stress vector are

$$\begin{Bmatrix} t_1 \\ t_2 \\ t_3 \end{Bmatrix} = \begin{bmatrix} 200 & 400 & 300 \\ 400 & 0 & 0 \\ 300 & 0 & -100 \end{bmatrix} \frac{1}{3} \begin{Bmatrix} 1 \\ 2 \\ 2 \end{Bmatrix} = \frac{1}{3} \begin{Bmatrix} 1600 \\ 400 \\ 100 \end{Bmatrix} \text{ psi,}$$

or

$$\mathbf{t}(\hat{\mathbf{n}}) = \tfrac{100}{3}(16\,\hat{\mathbf{e}}_1 + 4\,\hat{\mathbf{e}}_2 + \hat{\mathbf{e}}_3) \text{ psi.}$$

The traction vector normal to the plane is given by

$$\mathbf{t}_{nn} = (\mathbf{t}(\hat{\mathbf{n}}) \cdot \hat{\mathbf{n}})\hat{\mathbf{n}} = \tfrac{2600}{9}\hat{\mathbf{n}} = \tfrac{2600}{27}(\hat{\mathbf{e}}_1 + 2\hat{\mathbf{e}}_2 + 2\hat{\mathbf{e}}_3) \text{ psi,}$$

and the traction vector projected onto the plane (i.e., shear traction) is given by

$$\mathbf{t}_{ns} = \mathbf{t}(\hat{\mathbf{n}}) - \mathbf{t}_{nn} = \tfrac{100}{27}(118\hat{\mathbf{e}}_1 - 16\hat{\mathbf{e}}_2 - 43\hat{\mathbf{e}}_3) \text{ psi.}$$

The magnitudes are

$$|\mathbf{t}_{nn}| = t_{nn} = \tfrac{2600}{9} = 288.89 \text{ psi,} \quad |\mathbf{t}_{ns}| = t_{ns} = 468.91 \text{ psi.}$$

One can also determine t_{ns} from

$$t_{ns} = \sqrt{|\mathbf{t}|^2 - t_{nn}^2} = \tfrac{100}{9}\sqrt{(256 + 16 + 1)9 - 26 \times 26} \text{ psi} = 468.91 \text{ psi.}$$

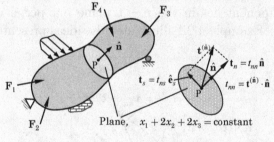

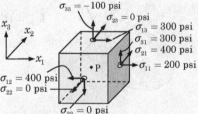

Dimensionless point cube at P

Fig. 4.2.6: Stress vector and its normal and shear components.

4.3 Transformations of Stress Components and Principal Stresses

4.3.1 Transformation of Stress Components

Since the Cauchy stress tensor $\boldsymbol{\sigma}$ is a second-order tensor, all of the properties of a second-order tensor that were discussed in Chapter 2 apply. In particular, we can define the principal invariants I_1, I_2, and I_3; transformation laws for the components of $\boldsymbol{\sigma}$; and eigenvalues (principal values) and eigenvectors (principal planes) of the Cauchy stress tensor.

4.3.1.1 Invariants

The invariants of stress tensor $\boldsymbol{\sigma}$ are defined by [see Eqs. (2.5.16) and (2.5.17)]

$$I_1 = \mathrm{tr}\,[\boldsymbol{\sigma}], \quad I_2 = \tfrac{1}{2}\left[(\mathrm{tr}\,[\boldsymbol{\sigma}])^2 - \mathrm{tr}\,([\boldsymbol{\sigma}]^2)\right], \quad I_3 = |\sigma|, \tag{4.3.1}$$

and in terms of the rectangular Cartesian components

$$I_1 = \sigma_{ii}, \quad I_2 = \tfrac{1}{2}\left(\sigma_{ii}\,\sigma_{jj} - \sigma_{ij}\,\sigma_{ji}\right), \quad I_3 = |\sigma|. \tag{4.3.2}$$

4.3.1.2 Transformation equations

The components of the Cauchy stress tensor $\boldsymbol{\sigma}$ in one rectangular Cartesian coordinate system $(\bar{x}_1, \bar{x}_2, \bar{x}_3)$ are related to the components in another rectangular Cartesian system (x_1, x_2, x_3) according to the transformation law in Eq. (2.5.21):

$$\bar{\sigma}_{ij} = \ell_{ik}\,\ell_{j\ell}\,\sigma_{k\ell} \quad \text{or} \quad [\bar{\sigma}] = [L][\sigma][L]^{\mathrm{T}}, \tag{4.3.3}$$

where ℓ_{ij} are the direction cosines

$$\ell_{ij} = \hat{\bar{\mathbf{e}}}_i \cdot \hat{\mathbf{e}}_j. \tag{4.3.4}$$

In Examples 4.3.1 and 4.3.2, symmetry of the stress tensor, which will be established in Chapter 5, is used in deriving the stress transformation equations (4.3.7) and (4.3.8) for a special coordinate transformation, whereas Eq. (4.3.3) is valid for the rectangular components of any second-order tensor and for a general coordinate transformation.

Example 4.3.1

Consider a rectangular, unidirectional fiber-reinforced composite layer shown in Fig. 4.3.1, where the fibers are symbolically shown as black lines. The rectangular coordinates (x, y, z) are taken such that the z-coordinate is normal to the plane of the layer, and the x and y coordinates are in the plane of the layer but parallel to the edges of the layer. Now suppose we define a new rectangular coordinate system (x_1, x_2, x_3) such that the x_3-coordinate coincides with the z-coordinate and the x_1-axis is taken along the fiber direction; that is, the $x_1 x_2$-plane is obtained by rotating the xy-plane about the z-axis in a counterclockwise direction by an angle θ. Determine the relations between the stress components referred to the (x, y, z) system and those referred to the coordinates system (x_1, x_2, x_3).

Solution: The coordinates of a material point in the two coordinate systems are related as follows ($z = x_3$):

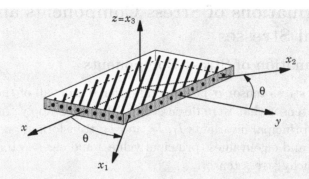

Fig. 4.3.1: Stress components in a fiber-reinforced layer referred to different rectangular Cartesian coordinate systems: (x, y, z) are parallel to the sides of the rectangular lamina, while (x_1, x_2, x_3) are taken along and transverse to the fiber direction.

$$\begin{Bmatrix} x_1 \\ x_2 \\ x_3 \end{Bmatrix} = \begin{bmatrix} \cos\theta & \sin\theta & 0 \\ -\sin\theta & \cos\theta & 0 \\ 0 & 0 & 1 \end{bmatrix} \begin{Bmatrix} x \\ y \\ z \end{Bmatrix} = [L] \begin{Bmatrix} x \\ y \\ z \end{Bmatrix}. \tag{4.3.5}$$

Next, we establish the relationship between the components of stress in the (x, y, z) and (x_1, x_2, x_3) coordinate systems. Let σ_{ij} be the components of the stress tensor $\boldsymbol{\sigma}$ in the (x_1, x_2, x_3) coordinate system, and $\sigma_{xx}, \sigma_{yy}, \sigma_{xy}$, etc. be the stress components in the (x, y, z) coordinate system. If we view (x_1, x_2, x_3) as the barred coordinate system, then ℓ_{ij} are the direction cosines defined by

$$\ell_{ij} = \hat{\bar{\mathbf{e}}}_i \cdot \hat{\mathbf{e}}_j,$$

where $\hat{\bar{\mathbf{e}}}_i$ and $\hat{\mathbf{e}}_i$ are the orthonormal basis vectors in coordinate systems (x_1, x_2, x_3) and (x, y, z), respectively. Then using Eq. (4.3.3), we can write

$$[\bar{\sigma}] = [L][\sigma][L]^\mathrm{T}, \quad [\sigma] = [L]^\mathrm{T}[\bar{\sigma}][L], \tag{4.3.6}$$

where $[L]$ is the 3×3 matrix of direction cosines defined in Eq. (4.3.5). Carrying out the indicated matrix multiplications in Eq. (4.3.6) and rearranging the equations in terms of the column vectors of stress components ($\sigma_{xy} = \sigma_{yx}$, $\sigma_{xz} = \sigma_{zx}$, $\sigma_{yz} = \sigma_{zy}$, $\sigma_{12} = \sigma_{21}$, $\sigma_{13} = \sigma_{31}$, and $\sigma_{23} = \sigma_{32}$), we obtain

$$\begin{Bmatrix} \sigma_{11} \\ \sigma_{22} \\ \sigma_{33} \\ \sigma_{23} \\ \sigma_{13} \\ \sigma_{12} \end{Bmatrix} = \begin{bmatrix} \cos^2\theta & \sin^2\theta & 0 & 0 & 0 & \sin 2\theta \\ \sin^2\theta & \cos^2\theta & 0 & 0 & 0 & -\sin 2\theta \\ 0 & 0 & 1 & 0 & 0 & 0 \\ 0 & 0 & 0 & \cos\theta & -\sin\theta & 0 \\ 0 & 0 & 0 & \sin\theta & \cos\theta & 0 \\ -\frac{1}{2}\sin 2\theta & \frac{1}{2}\sin 2\theta & 0 & 0 & 0 & \cos 2\theta \end{bmatrix} \begin{Bmatrix} \sigma_{xx} \\ \sigma_{yy} \\ \sigma_{zz} \\ \sigma_{yz} \\ \sigma_{xz} \\ \sigma_{xy} \end{Bmatrix}. \tag{4.3.7}$$

and

$$\begin{Bmatrix} \sigma_{xx} \\ \sigma_{yy} \\ \sigma_{zz} \\ \sigma_{yz} \\ \sigma_{xz} \\ \sigma_{xy} \end{Bmatrix} = \begin{bmatrix} \cos^2\theta & \sin^2\theta & 0 & 0 & 0 & -\sin 2\theta \\ \sin^2\theta & \cos^2\theta & 0 & 0 & 0 & \sin 2\theta \\ 0 & 0 & 1 & 0 & 0 & 0 \\ 0 & 0 & 0 & \cos\theta & \sin\theta & 0 \\ 0 & 0 & 0 & -\sin\theta & \cos\theta & 0 \\ \frac{1}{2}\sin 2\theta & -\frac{1}{2}\sin 2\theta & 0 & 0 & 0 & \cos 2\theta \end{bmatrix} \begin{Bmatrix} \sigma_{11} \\ \sigma_{22} \\ \sigma_{33} \\ \sigma_{23} \\ \sigma_{13} \\ \sigma_{12} \end{Bmatrix}, \tag{4.3.8}$$

The result in Eq. (4.3.8) can also be obtained from Eq. (4.3.7) by replacing θ with $-\theta$.

Example 4.3.2 ───

Consider a thin, closed, filament-wound circular cylindrical pressure vessel shown in Fig. 4.3.2. The vessel has an internal diameter $D_i = 63.5$ cm (25 in.) and thickness $h = 2$ cm, and is pressurized to $p = 1.379$ MPa (200 psi). If the filament winding angle is $\theta = 53.13°$ from the longitudinal axis of the pressure vessel, determine the shear and normal forces per unit length of the filament winding. Assume that the material used is graphite–epoxy with the following material properties [material properties are not needed to solve the problem; see Reddy (2004)]:

$$E_1 = 140 \text{ MPa } (20.3 \times 10^6 \text{ psi}), \quad E_2 = 10 \text{ MPa } (1.45 \times 10^6 \text{ psi}),$$
$$G_{12} = 7 \text{ MPa } (1.02 \times 10^6 \text{ psi}), \quad \nu_{12} = 0.3, \tag{4.3.9}$$

where MPa denotes mega (10^6) Pascal (Pa) and Pa $= \text{N/m}^2$ (1 psi $= 6,894.76$ Pa).

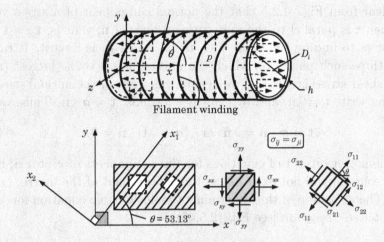

Fig. 4.3.2: A filament-wound cylindrical pressure vessel.

Solution: First, we compute the stresses in the pressure vessel using the formulas from a book on mechanics of materials [see, e.g., Fenner and Reddy (2012)]. The longitudinal (σ_{xx}) and circumferential (σ_{yy}) stresses are given by (the shear stress σ_{xy} is zero)

$$\sigma_{xx} = \frac{pD_i}{4h}, \quad \sigma_{yy} = \frac{pD_i}{2h}, \tag{4.3.10}$$

where p is internal pressure, D_i is the internal diameter, and h is the thickness of the pressure vessel. Note that the stresses are independent of material properties and depend only on the geometry and loads. We calculate the longitudinal and circumferential stresses to be

$$\sigma_{xx} = \frac{1.379 \times 0.635}{4 \times 0.02} = 10.946 \text{ MPa}, \quad \sigma_{yy} = \frac{1.379 \times 0.635}{2 \times 0.02} = 21.892 \text{ MPa}.$$

Next, we determine the shear stress σ_{12} along the fiber–matrix interface and the normal stress σ_{11} in the fiber direction using the transformation equations in Eq. (4.3.7). Noting that $\sin\theta = 0.8$, $\cos\theta = 0.6$, and $\sin 2\theta = 0.96$ for $\theta = 53.13°$, we obtain

$$\sigma_{11} = 10.946 \times (0.6)^2 + 21.892 \times (0.8)^2 = 17.951 \text{ MPa},$$
$$\sigma_{22} = 10.946 \times (0.8)^2 + 21.892 \times (0.6)^2 = 14.886 \text{ MPa},$$
$$\sigma_{12} = \tfrac{1}{2}(21.892 - 10.946) \times 0.96 = 5.254 \text{ MPa}.$$

Thus, the normal and shear forces per unit length along the fiber–matrix interface are $F_{22} = 14.886\,h$ MN and $F_{12} = 5.254\,h$ MN, whereas the force per unit length in the fiber direction is $F_{11} = 17.951\,h$ MN (MN $= 10^6$N).

4.3.2 Principal Stresses and Principal Planes

For a given state of stress, the determination of maximum normal stresses and shear stresses at a point is of considerable interest in the design of structures because failures occur when the magnitudes of stresses exceed the allowable (normal or shear) stress values, called strengths, of the material. In this regard it is of interest to determine the values and the planes on which the stresses are the maximum. Thus, we must determine the eigenvalues and eigenvectors associated with the stress tensor (see Section 2.5.6 for details).

It is clear from Fig. 4.2.5 that the normal component of a stress vector is largest when $\mathbf{t}$ is parallel to the unit outward normal $\hat{\mathbf{n}}$; that is, $\mathbf{t}_n = \mathbf{t} = |\mathbf{t}|\hat{\mathbf{n}}$. This amounts to finding the plane (i.e., $\hat{\mathbf{n}}$) on which $\mathbf{t}_n$ is largest. It turns out there are three such planes on which the normal stress is the largest (and the projected shear stress is zero). If we denote this value of the normal stress by λ, then we can write $\mathbf{t} = \lambda\hat{\mathbf{n}}$, and by Cauchy's formula, $\mathbf{t} = \boldsymbol{\sigma} \cdot \hat{\mathbf{n}}$. Thus, we have

$$\mathbf{t} = \boldsymbol{\sigma} \cdot \hat{\mathbf{n}} = \lambda\hat{\mathbf{n}} \quad \text{or} \quad (\boldsymbol{\sigma} - \lambda\mathbf{I}) \cdot \hat{\mathbf{n}} = \mathbf{0}. \tag{4.3.11}$$

This is a homogeneous set of equations for the components of vector $\hat{\mathbf{n}}$; hence, a nontrivial solution will not exist unless the determinant of the matrix $[\sigma] - \lambda[I]$ vanishes. The vanishing of this determinant yields a cubic equation for λ, called the *characteristic equation* [see Eq. (2.5.42)]:

$$-\lambda^3 + I_1\lambda^2 - I_2\lambda + I_3 = 0. \tag{4.3.12}$$

The solution of this cubic equation yields three values of λ, which are called the *principal stresses*, and the associated eigenvectors are called the *principal planes*. That is, for a given state of stress at a given point in the body $\mathcal{B}$, there exists a set of planes $\hat{\mathbf{n}}$ on which the stress vector is normal to the planes (i.e., there is no shear component on the planes).

The computation of the eigenvalues of the stress tensor is made easy by seeking the eigenvalues of the deviatoric stress tensor [see Eq. (2.5.53)]:

$$\boldsymbol{\sigma}' = \boldsymbol{\sigma} - \tfrac{1}{3}\text{tr}(\boldsymbol{\sigma})\,\mathbf{I} \quad (\sigma'_{ij} \equiv \sigma_{ij} - \tfrac{1}{3}\sigma_{kk}\,\delta_{ij}). \tag{4.3.13}$$

Let σ_m denote the mean normal stress

$$\sigma_m = \tfrac{1}{3}\text{tr}\,[\boldsymbol{\sigma}] = \tfrac{1}{3}I_1 \quad (\sigma_m = \tfrac{1}{3}\sigma_{kk}). \tag{4.3.14}$$

Then the stress tensor can be expressed as the sum of the *spherical* or the *hydrostatic* part and the *deviatoric* part of the stress tensor:

$$\boldsymbol{\sigma} = \sigma_m\mathbf{I} + \boldsymbol{\sigma}'. \tag{4.3.15}$$

Thus, the deviatoric stress tensor is defined by

$$\boldsymbol{\sigma}' = \boldsymbol{\sigma} - \tfrac{1}{3}I_1\mathbf{I} \quad (\sigma'_{ij} = \sigma_{ij} - \tfrac{1}{3}\delta_{ij}\sigma_{kk}). \tag{4.3.16}$$

The invariants I'_1, I'_2, and I'_3 of the deviatoric stress tensor are

$$I'_1 = 0, \quad I'_2 = \tfrac{1}{2}\sigma'_{ij}\sigma'_{ij}, \quad I'_3 = \tfrac{1}{3}\sigma'_{ij}\,\sigma'_{jk}\,\sigma'_{ki}. \tag{4.3.17}$$

The deviatoric stress invariants are particularly important in the determination of the principal stresses, as discussed in Section 2.5.6. Example 4.3.3 illustrates the computation of principal stresses and principal planes.

Example 4.3.3

The components of a stress dyadic at a point, referred to the (x_1, x_2, x_3) system, are:

$$[\sigma] = \begin{bmatrix} 12 & 9 & 0 \\ 9 & -12 & 0 \\ 0 & 0 & 6 \end{bmatrix} \text{ MPa.}$$

Find the principal stresses and the principal plane associated with the maximum stress.

Solution: Clearly, $\lambda = 6$ is an eigenvalue. Expanding the determinant $|\sigma - \lambda I|$ with the last row or column, we obtain

$$(6 - \lambda)[(12 - \lambda)(-12 - \lambda) - 81] = 0 \quad \Rightarrow \quad (\lambda^2 - 225)(6 - \lambda) = 0.$$

The remaining two eigenvalues are obtained from $\lambda^2 - 225 = 0 \rightarrow \lambda = \pm 15$; thus, the three principal stresses are

$$\sigma_1 = \lambda_1 = 15 \text{ MPa}, \quad \sigma_2 = \lambda_2 = 6 \text{ MPa}, \quad \sigma_3 = \lambda_3 = -15 \text{ MPa.}$$

The plane associated with the maximum principal stress $\lambda_1 = 15$ MPa can be calculated from

$$\begin{bmatrix} 12 - 15 & 9 & 0 \\ 9 & -12 - 15 & 0 \\ 0 & 0 & 6 - 15 \end{bmatrix} \begin{Bmatrix} n_1 \\ n_2 \\ n_3 \end{Bmatrix} = \begin{Bmatrix} 0 \\ 0 \\ 0 \end{Bmatrix},$$

which gives

$$-3n_1 + 9n_2 = 0, \quad 9n_1 - 27n_2 = 0, \quad -9n_3 = 0 \quad \rightarrow \quad n_3 = 0, \quad n_1 = 3n_2$$

$$\mathbf{n}^{(1)} = 3\hat{\mathbf{e}}_1 + \hat{\mathbf{e}}_2 \quad \text{or} \quad \hat{\mathbf{n}}^{(1)} = \tfrac{1}{\sqrt{10}}(3\hat{\mathbf{e}}_1 + \hat{\mathbf{e}}_2).$$

The eigenvector associated with $\lambda_2 = 6$ MPa is $\mathbf{n}^{(2)} = \hat{\mathbf{e}}_3$. Finally, the eigenvector associated with $\lambda_3 = -15$ MPa is

$$\mathbf{n}^{(3)} = \pm(\hat{\mathbf{e}}_1 - 3\hat{\mathbf{e}}_2) \quad \text{or} \quad \hat{\mathbf{n}}^{(3)} = \pm\tfrac{1}{\sqrt{10}}(\hat{\mathbf{e}}_1 - 3\hat{\mathbf{e}}_2).$$

The principal plane 1 is depicted in Fig. 4.3.3.

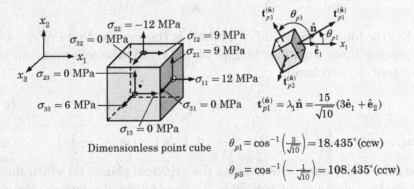

Fig. 4.3.3: Stresses on a point cube at the point of interest and orientation of the first principal plane.

4.3.3 Maximum Shear Stress

In the previous section, we studied the procedure to determine the maximum normal stresses at a point. The eigenvalues of the stress tensor at the point are the maximum normal stresses on three perpendicular planes (whose normals are the eigenvectors), and the largest of these three stresses is the true maximum normal stress. Recall that the shear stresses are zero on the principal planes. In this section, we wish to determine the maximum shear stresses and their planes.

Let λ_1, λ_2, and λ_3 denote the principal (normal) stresses and $\hat{\mathbf{n}}$ be an arbitrary unit normal vector. Then the stress vector is $\mathbf{t} = \lambda_1 n_1 \hat{\mathbf{e}}_1 + \lambda_2 n_2 \hat{\mathbf{e}}_2 + \lambda_3 n_3 \hat{\mathbf{e}}_3$ and $t_{nn} = t_i n_i = \lambda_1 n_1^2 + \lambda_2 n_2^2 + \lambda_3 n_3^2$. The square of the magnitude of the shear stress on the plane with unit normal $\hat{\mathbf{n}}$ is given by

$$t_{ns}^2(\hat{\mathbf{n}}) = |\mathbf{t}|^2 - t_{nn}^2 = \lambda_1^2 n_1^2 + \lambda_2^2 n_2^2 + \lambda_3^2 n_3^2 - \left(\lambda_1 n_1^2 + \lambda_2 n_2^2 + \lambda_3 n_3^2\right)^2. \quad (4.3.18)$$

We wish to determine the plane $\hat{\mathbf{n}}$ on which t_{ns} is the maximum. Thus, we seek the maximum of the function $F(n_1, n_2, n_3) = t_{ns}^2(n_1, n_2, n_3)$ subject to the constraint

$$n_1^2 + n_2^2 + n_3^2 - 1 = 0. \quad (4.3.19)$$

One way to determine the extremum of a function subjected to a constraint is to use the Lagrange multiplier method, in which we seek the stationary value of the modified function

$$F_L(n_1, n_2, n_3) = t_{ns}^2(n_1, n_2, n_3) + \lambda_L\left(n_1^2 + n_2^2 + n_3^2 - 1\right), \quad (4.3.20)$$

where λ_L is the Lagrange multiplier, which is to be determined along with n_1, n_2, and n_3. The necessary condition for the stationarity of F_L is

$$0 = dF_L = \frac{\partial F_L}{\partial n_1}dn_1 + \frac{\partial F_L}{\partial n_2}dn_2 + \frac{\partial F_L}{\partial n_3}dn_3 + \frac{\partial F_L}{\partial \lambda_L}d\lambda_L,$$

or, because the increments dn_1, dn_2, dn_3, and $d\lambda_L$ are linearly independent of each other, we have

$$\frac{\partial F_L}{\partial n_1} = 0, \quad \frac{\partial F_L}{\partial n_2} = 0, \quad \frac{\partial F_L}{\partial n_3} = 0, \quad \frac{\partial F_L}{\partial \lambda_L} = 0. \quad (4.3.21)$$

The last of the four relations in Eq. (4.3.21) is the same as that in Eq. (4.3.19). The remaining three equations in Eq. (4.3.21) yield the following two sets of solutions (not derived here):

$$(n_1, n_2, n_3) = (1, 0, 0), \quad (0, 1, 0), \quad (0, 0, 1), \quad (4.3.22)$$

$$(n_1, n_2, n_3) = \left(\tfrac{1}{\sqrt{2}}, \pm\tfrac{1}{\sqrt{2}}, 0\right), \quad \left(\tfrac{1}{\sqrt{2}}, 0, \pm\tfrac{1}{\sqrt{2}}\right), \quad \left(0, \tfrac{1}{\sqrt{2}}, \pm\tfrac{1}{\sqrt{2}}\right). \quad (4.3.23)$$

The first set of solutions corresponds to the principal planes, on which the shear stresses are the minimum, namely zero. The second set of solutions corresponds to the maximum shear stress planes. The maximum shear stresses on the planes are given by

$$t_{ns}^2 = \tfrac{1}{4}(\lambda_1 - \lambda_2)^2 \ \text{ for } \ \hat{n} = \tfrac{1}{\sqrt{2}}(\hat{e}_1 \pm \hat{e}_2),$$

$$t_{ns}^2 = \tfrac{1}{4}(\lambda_1 - \lambda_3)^2 \ \text{ for } \ \hat{n} = \tfrac{1}{\sqrt{2}}(\hat{e}_1 \pm \hat{e}_3), \qquad (4.3.24)$$

$$t_{ns}^2 = \tfrac{1}{4}(\lambda_2 - \lambda_3)^2 \ \text{ for } \ \hat{n} = \tfrac{1}{\sqrt{2}}(\hat{e}_2 \pm \hat{e}_3).$$

The largest shear stress is given by the largest of the three values given above. Thus, we have

$$(t_{ns})_{\max} = \tfrac{1}{2}(\lambda_{\max} - \lambda_{\min}), \qquad (4.3.25)$$

where $\lambda_{\max}$ and $\lambda_{\min}$ are the maximum and minimum principal values of stress, respectively. The plane of the maximum shear stress lies between the planes of the maximum and minimum principal stresses (i.e., oriented at $\pm 45°$ to both planes).

Example 4.3.4

For the state of stress given in Example 4.3.3, determine the maximum shear stress.

Solution: From Example 4.3.3, the principal stresses are (ordered from the minimum to the maximum)

$$\lambda_1 = -15\,\text{MPa}, \quad \lambda_2 = 6\,\text{MPa}, \quad \lambda_3 = 15\,\text{MPa}.$$

Hence, the maximum shear stress is given by

$$(t_{ns})_{\max} = \tfrac{1}{2}(\lambda_3 - \lambda_1) = \tfrac{1}{2}[15 - (-15)] = 15\,\text{MPa}.$$

The planes of the maximum principal stress ($\lambda_1 = 15$ MPa) and the minimum principal stress ($\lambda_3 = -15$ MPa) are given by their normal vectors (not unit vectors):

$$\mathbf{n}^{(1)} = 3\hat{e}_1 + \hat{e}_2, \quad \mathbf{n}^{(3)} = \hat{e}_1 - 3\hat{e}_2.$$

Then the plane of the maximum shear stress is given by the vector

$$\mathbf{n}_s = (\mathbf{n}^{(1)} - \mathbf{n}^{(3)}) = 2\hat{e}_1 + 4\hat{e}_2 \ \text{ or } \ \hat{n}_s = \tfrac{1}{\sqrt{5}}(\hat{e}_1 + 2\hat{e}_2).$$

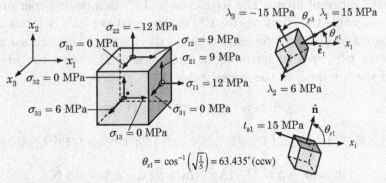

Fig. 4.3.4: Stresses on a point cube at the point of interest and orientation of the maximum shear stress plane.

4.4 Other Stress Measures

4.4.1 Preliminary Comments

The Cauchy stress tensor is the most natural and physical measure of the state of stress at a point in the deformed configuration and, measured per unit area of the deformed configuration. It is the quantity most commonly used in spatial descriptions of problems in fluid mechanics. In order to use the Lagrangian description, which is common in solid mechanics, the equations of motion or equilibrium of a material body that are derived in the deformed configuration must be expressed in terms of the known reference configuration. In doing so we introduce various other measures of stress. These measures emerge in a natural way as we transform volumes and areas from the deformed configuration to the reference configuration. These measures are purely mathematical in nature but facilitate the analysis.

4.4.2 First Piola–Kirchhoff Stress Tensor

Consider a continuum $\mathcal{B}$ subjected to a deformation mapping χ that results in the deformed configuration κ, as shown in Fig. 4.4.1. Let the force vector on an elemental area da with normal $\hat{n}$ in the deformed configuration be $d\mathbf{f}$. The force $d\mathbf{f}$ can be expressed in terms of a stress vector $\mathbf{t}$ times the deformed area da as

$$d\mathbf{f} = \mathbf{t}^{(n)}\, da = \boldsymbol{\sigma} \cdot \hat{n}\, da = \boldsymbol{\sigma} \cdot d\mathbf{a}, \qquad (4.4.1)$$

where $\boldsymbol{\sigma}$ is the Cauchy stress tensor, and Cauchy's formula (4.2.10) is invoked in arriving at the last result. Now suppose that the area element in the undeformed configuration that corresponds to da is dA. We define a stress vector $\mathbf{T}^{(N)}$ over the area element dA with normal $\mathbf{N}$ in the undeformed configuration such that it results in the *same* total force

$$d\mathbf{f} = \mathbf{t}^{(n)}\, da = \mathbf{T}^{(N)} dA. \qquad (4.4.2)$$

Clearly, both stress vectors have the same direction but different magnitudes owing to the different areas. The stress vector $\mathbf{T}^{(N)}$ is measured per unit undeformed area, while the stress vector $\mathbf{t}^{(n)}$ is measured per unit deformed area.

Analogous to Cauchy's formula relating the Cauchy stress tensor $\boldsymbol{\sigma}$ to the stress vector $\mathbf{t}^{(n)}$, we can introduce a stress tensor $\mathbf{P}$, called the *first Piola–Kirchhoff stress tensor*, to the stress vector $\mathbf{T}^{(N)}$; that is

$$\mathbf{t}^{(n)} = \boldsymbol{\sigma} \cdot \hat{n}; \qquad \mathbf{T}^{(N)} = \mathbf{P} \cdot \hat{\mathbf{N}}. \qquad (4.4.3)$$

Then using Eq. (4.4.2) and Cauchy's formulas for $\mathbf{t}^{(n)}$ and $\mathbf{T}^{(N)}$, we can write

$$d\mathbf{f} = \boldsymbol{\sigma} \cdot d\mathbf{a} = \mathbf{P} \cdot d\mathbf{A}; \quad d\mathbf{a} = da\,\hat{n}, \quad d\mathbf{A} = dA\,\hat{\mathbf{N}}. \qquad (4.4.4)$$

The first Piola–Kirchhoff stress tensor, also referred to as the *nominal stress tensor* or *Lagrangian stress tensor*, gives the *current force* per unit *undeformed area*.

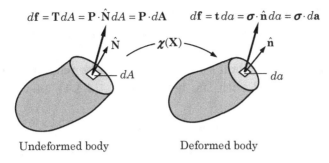

$$df = \mathbf{T}\,dA = \mathbf{P}\cdot\hat{\mathbf{N}}\,dA = \mathbf{P}\cdot d\mathbf{A} \qquad df = \mathbf{t}\,da = \boldsymbol{\sigma}\cdot\hat{\mathbf{n}}\,da = \boldsymbol{\sigma}\cdot d\mathbf{a}$$

Undeformed body   Deformed body

Fig. 4.4.1: Definition of the first Piola–Kirchhoff stress tensor.

The stress vector $\mathbf{T}^{(\mathbf{N})}$ is known as the *pseudo stress vector* associated with the first Piola–Kirchhoff stress tensor. The Cartesian component representation of tensor $\mathbf{P}$ is given by

$$\mathbf{P} = P_{iI}\,\hat{\mathbf{e}}_i\,\hat{\mathbf{E}}_I. \tag{4.4.5}$$

Clearly, the first Piola–Kirchhoff stress tensor is a *two-point tensor* (like $\mathbf{F}$) in the sense that it connects a point in the undeformed body to the corresponding point in the deformed body.

To express the first Piola–Kirchhoff stress tensor $\mathbf{P}$ in terms of the Cauchy stress tensor $\boldsymbol{\sigma}$, we must write $d\mathbf{a}$ in terms of $d\mathbf{A}$. From Nanson's formula in Eq. (3.3.25), we recall such a relation between $d\mathbf{a}$ in terms of $d\mathbf{A}$:

$$d\mathbf{a} = J\,\mathbf{F}^{-\mathrm{T}}\cdot d\mathbf{A} = J\,d\mathbf{A}\cdot\mathbf{F}^{-1}, \tag{4.4.6}$$

where J is the Jacobian, $J = |\mathbf{F}|$. Substituting the relation into Eq. (4.4.4), we obtain

$$\mathbf{P}\cdot d\mathbf{A} = \boldsymbol{\sigma}\cdot d\mathbf{a} = J\,\boldsymbol{\sigma}\cdot\mathbf{F}^{-\mathrm{T}}\cdot d\mathbf{A}. \tag{4.4.7}$$

Thus, we arrive at the relation

$$\mathbf{P} = J\boldsymbol{\sigma}\cdot\mathbf{F}^{-\mathrm{T}} \quad \text{or} \quad \boldsymbol{\sigma} = \frac{1}{J}\mathbf{P}\cdot\mathbf{F}^{\mathrm{T}}. \tag{4.4.8}$$

In general, the first Piola–Kirchhoff stress tensor $\mathbf{P}$ is unsymmetric even when the Cauchy stress tensor $\boldsymbol{\sigma}$ is symmetric (which is not yet established).

4.4.3 Second Piola–Kirchhoff Stress Tensor

Similar to the relationship between $d\mathbf{x}$ and $d\mathbf{X}$, $d\mathbf{X} = \mathbf{F}^{-1}\cdot d\mathbf{x}$, the force $d\mathbf{f}$ on the deformed elemental area $d\mathbf{a}$ can be related, by analogy to the relation between $d\mathbf{x}$ and $d\mathbf{X}$, to a force $d\mathcal{F}$ on the undeformed elemental area $d\mathbf{A}$ by

$$d\mathcal{F} = \mathbf{F}^{-1}\cdot d\mathbf{f}. \tag{4.4.9}$$

Then we can think of the existence of a stress tensor $\mathbf{S}$, in the same way as in Eq. (4.4.1), such that

$$d\mathcal{F} = \mathbf{S}\cdot d\mathbf{A}, \tag{4.4.10}$$

where **S** is called the *second Piola–Kirchhoff stress tensor* **S**. Thus, the second Piola–Kirchhoff stress tensor **S** gives the *transformed current force* per unit *undeformed area*.

The second Piola–Kirchhoff stress tensor **S** can be related to the first Piola–Kirchhoff stress tensor **P** with the help of Eqs. (4.4.4), (4.4.9), and (4.4.10) as

$$\mathbf{S} = \mathbf{F}^{-1} \cdot \mathbf{P}. \tag{4.4.11}$$

The relation between **S** and $\boldsymbol{\sigma}$ can also be established using Eqs. (4.4.4), (4.4.6), (4.4.9), and (4.4.10) as

$$\mathbf{S} \cdot d\mathbf{A} = \mathbf{F}^{-1} \cdot \boldsymbol{\sigma} \cdot d\mathbf{a} = J\,\mathbf{F}^{-1} \cdot \boldsymbol{\sigma} \cdot \mathbf{F}^{-\mathrm{T}} \cdot d\mathbf{A},$$

or

$$\mathbf{S} = J\,\mathbf{F}^{-1} \cdot \boldsymbol{\sigma} \cdot \mathbf{F}^{-\mathrm{T}} \quad \text{or} \quad \boldsymbol{\sigma} = \frac{1}{J}\mathbf{F} \cdot \mathbf{S} \cdot \mathbf{F}^{\mathrm{T}}. \tag{4.4.12}$$

Clearly, **S** is symmetric (i.e., $\mathbf{S} = \mathbf{S}^{\mathrm{T}}$) whenever $\boldsymbol{\sigma}$ is symmetric. Cartesian component representation of the tensor **S** is

$$\mathbf{S} = S_{IJ}\,\hat{\mathbf{E}}_I\,\hat{\mathbf{E}}_J. \tag{4.4.13}$$

All of the discussion in Section 4.3 concerning the transformation of components and determination of eigenvalues and eigenvectors is valid for **S**.

We can introduce the pseudo stress vector $\tilde{\mathbf{T}}$ associated with the second Piola–Kirchhoff stress tensor by

$$d\mathcal{F} = \tilde{\mathbf{T}}\,dA = \mathbf{S} \cdot \hat{\mathbf{N}}\,dA = \mathbf{S} \cdot d\mathbf{A} \quad \text{or} \quad \tilde{\mathbf{T}} = \mathbf{S} \cdot \hat{\mathbf{N}}. \tag{4.4.14}$$

An interpretation of the second Piola–Kirchhoff stress tensor is possible in the case of rigid-body motion, for which the polar decomposition theorem gives $\mathbf{F} = \mathbf{R}$ and $J = 1$. Hence, we have

$$\mathbf{S} = J\,\mathbf{F}^{-1} \cdot \boldsymbol{\sigma} \cdot \mathbf{F}^{-\mathrm{T}} = \mathbf{R}^{\mathrm{T}} \cdot \boldsymbol{\sigma} \cdot \mathbf{R}, \tag{4.4.15}$$

which resembles the stress transformation equation (4.3.3). That is, the second Piola–Kirchhoff stress components are the same as the components of the Cauchy stress tensor expressed in the local set of orthogonal axes that are obtained from rotating the global Cartesian coordinates by the rotation matrix $[L] = [R]^{\mathrm{T}}$.

We close this section with an example that illustrates the meaning of the first and second Piola–Kirchhoff stress tensors and the computation of the first and second Piola–Kirchhoff stress tensor components from the Cauchy stress tensor components [see Hjelmstad (1997)].

Example 4.4.1 ———————————————————————————

Consider a bar of cross-sectional area $A = bH$ and length L. The initial configuration of the bar is such that its longitudinal axis is along the X_1 axis, as shown in Fig. 4.4.2(a). Suppose that the bar is subjected to uniaxial tensile stress that produces a pure stretch λ along the length and a pure stretch μ along the height of the bar and then rotates it, without bending, by an angle θ, as shown in Fig. 4.4.2(a). Assume that the width b of the bar does not change during the deformation. Therefore, μ denotes the ratio of deformed to undeformed height (or cross-sectional area) of the bar. Determine (a) the deformation mapping and the components of the deformation gradient and (b) the components of the Cauchy stress tensor as well as the first and second Piola–Kirchhoff stress tensors.

Solution: (a) We use the deformed geometry to determine the deformation mapping. We have from Fig. 4.4.2(b),

$$\boldsymbol{\chi}(\mathbf{X}) = (\lambda X_1 \cos\theta - \mu X_2 \sin\theta)\hat{\mathbf{e}}_1 + (\lambda X_1 \sin\theta + \mu X_2 \cos\theta)\hat{\mathbf{e}}_2 + X_3\,\hat{\mathbf{e}}_3.$$

The ratio of volume in the deformed to undeformed configuration is $(\lambda L\,\mu H\,b)/(LHb) = \mu\,\lambda$. The components of the deformation gradient and its inverse are

$$[F] = \begin{bmatrix} \lambda\cos\theta & -\mu\sin\theta & 0 \\ \lambda\sin\theta & \mu\cos\theta & 0 \\ 0 & 0 & 1 \end{bmatrix}, \quad [F]^{-1} = \frac{1}{J}\begin{bmatrix} \mu\cos\theta & \mu\sin\theta & 0 \\ -\lambda\sin\theta & \lambda\cos\theta & 0 \\ 0 & 0 & \lambda\mu \end{bmatrix},$$

and the Jacobian is equal to $J = \mu\lambda$, which is the ratio of volumes in the deformed and undeformed configurations (i.e., $v = JV$).

(b) The unit vector normal to the undeformed cross-sectional area is $\hat{\mathbf{N}} = \hat{\mathbf{E}}_1$, and the unit vector normal to the cross-sectional area of the deformed configuration is

$$\hat{\mathbf{n}} = \cos\theta\,\hat{\mathbf{e}}_1 + \sin\theta\,\hat{\mathbf{e}}_2.$$

The Cauchy stress tensor is $\boldsymbol{\sigma} = \sigma_0\,\hat{\mathbf{n}}\hat{\mathbf{n}}$ and associated stress vector is $\mathbf{t} = \sigma_0\,\hat{\mathbf{n}}$, as shown in Fig. 4.4.3(a). The components of the Cauchy stress tensor are

$$[\sigma] = \begin{Bmatrix} \cos\theta \\ \sin\theta \\ 0 \end{Bmatrix}\sigma_0\begin{Bmatrix} \cos\theta & \sin\theta & 0 \end{Bmatrix} = \sigma_0\begin{bmatrix} \cos^2\theta & \cos\theta\sin\theta & 0 \\ \cos\theta\sin\theta & \sin^2\theta & 0 \\ 0 & 0 & 0 \end{bmatrix}.$$

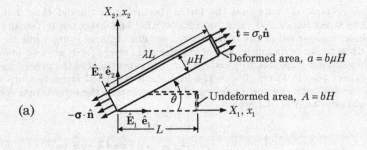

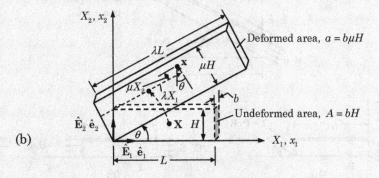

Fig. 4.4.2: (a) Undeformed and (b) deformed configurations of the bar of Example 4.4.1.

The components of the first Piola–Kirchhoff stress tensor are computed using Eq. (4.4.8)

$$[P] = J[\sigma][F]^{-T} = \sigma_0 \begin{bmatrix} \cos^2\theta & \cos\theta\sin\theta & 0 \\ \cos\theta\sin\theta & \sin^2\theta & 0 \\ 0 & 0 & 0 \end{bmatrix} \begin{bmatrix} \mu\cos\theta & -\lambda\sin\theta & 0 \\ \mu\sin\theta & \lambda\cos\theta & 0 \\ 0 & 0 & \lambda\mu \end{bmatrix}$$

$$= \mu\sigma_0 \begin{bmatrix} \cos\theta & 0 & 0 \\ \sin\theta & 0 & 0 \\ 0 & 0 & 0 \end{bmatrix}.$$

Clearly, the matrix representing $\mathbf{P}$ is not symmetric. The first Piola–Kirchhoff stress tensor is

$$\mathbf{P} = \mu\sigma_0 \begin{Bmatrix} \hat{\mathbf{e}}_1 \\ \hat{\mathbf{e}}_2 \\ \hat{\mathbf{e}}_3 \end{Bmatrix}^T \begin{bmatrix} \cos\theta & 0 & 0 \\ \sin\theta & 0 & 0 \\ 0 & 0 & 0 \end{bmatrix} \begin{Bmatrix} \hat{\mathbf{E}}_1 \\ \hat{\mathbf{E}}_2 \\ \hat{\mathbf{E}}_3 \end{Bmatrix}$$

$$= \mu\sigma_0\,(\cos\theta\,\hat{\mathbf{e}}_1 + \sin\theta\,\hat{\mathbf{e}}_2)\hat{\mathbf{E}}_1.$$

The associated stress vector is $(\hat{\mathbf{N}} = \hat{\mathbf{E}}_1)$

$$\mathbf{T} = \mathbf{P}\cdot\hat{\mathbf{N}} = \mu\sigma_0(\cos\theta\,\hat{\mathbf{e}}_1 + \sin\theta\,\hat{\mathbf{e}}_2) = \mu\sigma_0\,\hat{\mathbf{n}},$$

as shown in Fig. 4.4.3(b).

The second Piola–Kirchhoff stress tensor components can be computed either using Eq. (4.4.12) or (4.4.13). Using Eq. (4.4.12), we obtain

$$[S] = [F]^{-1}[P] = \frac{\mu\sigma_0}{J} \begin{bmatrix} \mu\cos\theta & \mu\sin\theta & 0 \\ -\lambda\sin\theta & \lambda\cos\theta & 0 \\ 0 & 0 & \lambda\mu \end{bmatrix} \begin{bmatrix} \cos\theta & 0 & 0 \\ \sin\theta & 0 & 0 \\ 0 & 0 & 0 \end{bmatrix} = \frac{\mu\sigma_0}{\lambda} \begin{bmatrix} 1 & 0 & 0 \\ 0 & 0 & 0 \\ 0 & 0 & 0 \end{bmatrix}.$$

The second Piola–Kirchhoff stress tensor and the associated pseudo stress vector are [see Fig. 4.4.3(c)]

$$\mathbf{S} = \frac{\mu\sigma_0}{\lambda}\,\hat{\mathbf{E}}_1\hat{\mathbf{E}}_1, \quad \tilde{\mathbf{T}} = \mathbf{S}\cdot\hat{\mathbf{E}} = \frac{\mu\sigma_0}{\lambda}\,\hat{\mathbf{E}}_1.$$

In closing this example we note that the forces (occurring in the deformed body) that produce the Cauchy stress tensor and the second Piola–Kirchhoff stress tensor (recall that the Cauchy stress tensor is measured as the *current force per unit deformed area* while the second Piola–Kirchhoff stress tensor is measured as the *transformed current force per unit undeformed area*) are in equilibrium [see Figs. 4.4.3(a) and 4.4.3(c)], as expected. On the other hand, there is no reason to expect pseudo forces due to the first Piola–Kirchhoff stress tensor, which is measured as the *current force per unit undeformed area*, to satisfy the equilibrium conditions in the undeformed body [see Fig. 4.4.3(b)].

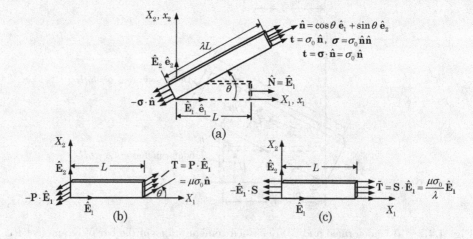

Fig. 4.4.3: Various stresses in the bar of Example 4.4.1.

4.5 Equilibrium Equations for Small Deformations

The principle of conservation of linear momentum, which is commonly known as Newton's second law of motion, is discussed along with other principles of mechanics in Chapter 5. To make the present chapter on stresses complete, we derive the equations of equilibrium of a continuous medium undergoing *small deformations* (that is, strains are infinitesimal $\mathbf{E} \approx \varepsilon$, and the difference between σ and $\mathbf{S}$ and between $\mathbf{X}$ and $\mathbf{x}$ vanishes) using Newton's second law of motion.

We isolate from the continuum an infinitesimal parallelepiped element with dimensions dx_1, dx_2, and dx_3 along coordinate x_1, x_2, and x_3, respectively, centered at point $\mathbf{x}$. The stresses acting on various faces of the parallelepiped element are shown in Fig. 4.5.1. The element is also subjected to body force $\rho_0 \mathbf{f}$ (measured per unit mass), where ρ_0 denotes the mass density. The body force components are $\rho_0 f_1$, $\rho_0 f_2$, and $\rho_0 f_3$ along the x_1-, x_2-, and x_3-coordinates, respectively. Setting the sum of all forces in the x_1-direction to zero, we obtain

$$
0 = \left(\sigma_{11} + \frac{\partial \sigma_{11}}{\partial x_1} dx_1\right) dx_2\, dx_3 - \sigma_{11}\, dx_2\, dx_3 + \left(\sigma_{12} + \frac{\partial \sigma_{12}}{\partial x_2} dx_2\right) dx_1\, dx_3
$$

$$
- \sigma_{12}\, dx_1\, dx_3 + \left(\sigma_{13} + \frac{\partial \sigma_{13}}{\partial x_3} dx_3\right) dx_1\, dx_2 - \sigma_{13}\, dx_1\, dx_2 + \rho_0 f_1\, dx_1\, dx_2\, dx_3
$$

$$
= \left(\frac{\partial \sigma_{11}}{\partial x_1} + \frac{\partial \sigma_{12}}{\partial x_2} + \frac{\partial \sigma_{13}}{\partial x_3} + \rho_0 f_1\right) dx_1\, dx_2\, dx_3. \tag{4.5.1}
$$

On dividing throughout by $dx_1\, dx_2\, dx_3$, we obtain

$$
\frac{\partial \sigma_{11}}{\partial x_1} + \frac{\partial \sigma_{12}}{\partial x_2} + \frac{\partial \sigma_{13}}{\partial x_3} + \rho_0 f_1 = 0 \quad \text{or} \quad \frac{\partial \sigma_{1j}}{\partial x_j} + \rho_0 f_1 = 0, \tag{4.5.2}
$$

for $j = 1, 2$, and 3. Similarly, by setting the sum of forces in the x_2- and x_3-directions to zero separately, we obtain

$$
\frac{\partial \sigma_{21}}{\partial x_1} + \frac{\partial \sigma_{22}}{\partial x_2} + \frac{\partial \sigma_{23}}{\partial x_3} + \rho_0 f_2 = 0 \quad \text{or} \quad \frac{\partial \sigma_{2j}}{\partial x_j} + \rho_0 f_2 = 0, \tag{4.5.3}
$$

$$
\frac{\partial \sigma_{31}}{\partial x_1} + \frac{\partial \sigma_{32}}{\partial x_2} + \frac{\partial \sigma_{33}}{\partial x_3} + \rho_0 f_3 = 0 \quad \text{or} \quad \frac{\partial \sigma_{3j}}{\partial x_j} + \rho_0 f_3 = 0. \tag{4.5.4}
$$

Equations (4.5.2)–(4.5.4) can be expressed in a single equation as

$$
\frac{\partial \sigma_{ij}}{\partial x_j} + \rho_0 f_i = 0, \quad i, j = 1, 2, 3. \tag{4.5.5}
$$

Noting that $\frac{\partial \sigma_{ij}}{\partial x_j} = (\boldsymbol{\sigma} \cdot \overleftarrow{\nabla})_i = (\boldsymbol{\nabla} \cdot \boldsymbol{\sigma}^{\mathrm{T}})_i$, we can express Eq. (4.5.5) in vector form (another example of the use of the backward gradient operator):

$$
\boldsymbol{\nabla} \cdot \boldsymbol{\sigma}^{\mathrm{T}} + \rho_0 \mathbf{f} = \mathbf{0}. \tag{4.5.6}
$$

The principle of conservation of angular momentum (i.e., Newton's second law for moments) can be used to establish the symmetry of the stress tensor when no body couples exist in the continuum. Consider the moment of all forces acting on the parallelepiped about the x_3-axis (see Fig. 4.5.1). Using the right-handed screw rule for positive moment, we obtain

$$\left[\left(\sigma_{21} + \frac{\partial \sigma_{21}}{\partial x_1} dx_1\right) dx_2\, dx_3\right] \frac{dx_1}{2} + (\sigma_{21} dx_2\, dx_3) \frac{dx_1}{2}$$

$$-\left[\left(\sigma_{12} + \frac{\partial \sigma_{12}}{\partial x_2} dx_2\right) dx_1\, dx_3\right] \frac{dx_2}{2} - (\sigma_{12} dx_1\, dx_3) \frac{dx_2}{2} = 0.$$

Note that the body force components do not have a moment because they pass through the origin of the coordinate system. Dividing throughout by $\frac{1}{2} dx_1\, dx_2\, dx_3$ and taking the limit $dx_1 \to 0$ and $dx_2 \to 0$, we obtain

$$\sigma_{21} - \sigma_{12} = 0. \tag{4.5.7}$$

Similar considerations of moments about the x_1-axis and x_2-axis give, respectively, the relations

$$\sigma_{32} - \sigma_{23} = 0, \quad \sigma_{31} - \sigma_{13} = 0. \tag{4.5.8}$$

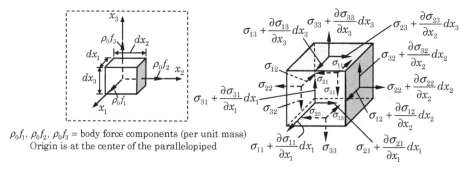

$\rho_0 f_1,\ \rho_0 f_2,\ \rho_0 f_3$ = body force components (per unit mass)
Origin is at the center of the parallelopiped

Fig. 4.5.1: Stress components on the faces of a parallelepiped element of dimensions dx_1, dx_2, and dx_3.

Thus, the stress tensor is symmetric ($\sigma_{ij} = \sigma_{ji}$). Equations (4.5.7) and (4.5.8) can be expressed in a single equation using the index notation as

$$\sigma_{ji}\, e_{ijk} = 0 \;\Rightarrow\; \sigma_{ij} = \sigma_{ji} \;\text{ or }\; \boldsymbol{\sigma}^{\mathrm{T}} = \boldsymbol{\sigma}, \tag{4.5.9}$$

where e_{ijk} are the components of the third-order permutation tensor defined in Eqs. (2.2.49)–(2.2.51). The symmetry of stress tensor with real-valued components has real principal values, and the principal directions associated with distinct principal stresses are orthogonal (see Section 2.5.6). Next, we consider two examples of application of the stress equilibrium equations.

Example 4.5.1 ──

Given the following state of stress ($\sigma_{ij} = \sigma_{ji}$) in a kinematically infinitesimal deformation,

$$\sigma_{11} = -2x_1^2, \quad \sigma_{12} = -7 + 4x_1 x_2 + x_3, \quad \sigma_{13} = 1 + x_1 - 3x_2,$$

$$\sigma_{22} = 3x_1^2 - 2x_2^2 + 5x_3, \quad \sigma_{23} = 0, \quad \sigma_{33} = -5 + x_1 + 3x_2 + 3x_3,$$

determine the body force components for which the stress field describes a state of equilibrium.

Solution: The body force components are

$$\rho_0 f_1 = -\left(\frac{\partial \sigma_{11}}{\partial x_1} + \frac{\partial \sigma_{12}}{\partial x_2} + \frac{\partial \sigma_{13}}{\partial x_3}\right) = -[(-4x_1) + (4x_1) + 0] = 0,$$

$$\rho_0 f_2 = -\left(\frac{\partial \sigma_{12}}{\partial x_1} + \frac{\partial \sigma_{22}}{\partial x_2} + \frac{\partial \sigma_{23}}{\partial x_3}\right) = -[(4x_2) + (-4x_2) + 0] = 0,$$

$$\rho_0 f_3 = -\left(\frac{\partial \sigma_{31}}{\partial x_1} + \frac{\partial \sigma_{32}}{\partial x_2} + \frac{\partial \sigma_{33}}{\partial x_3}\right) = -[1 + 0 + 3] = -4.$$

Thus, the body is in equilibrium for the body force components $\rho_0 f_1 = 0$, $\rho_0 f_2 = 0$, and $\rho_0 f_3 = -4$.

Example 4.5.2 ──

Determine if the following stress field ($\sigma_{ij} = \sigma_{ji}$) in a kinematically infinitesimal deformation satisfies the equations of equilibrium:

$$\sigma_{11} = x_2^2 + k(x_1^2 - x_2^2), \quad \sigma_{12} = -2kx_1x_2, \quad \sigma_{13} = 0,$$
$$\sigma_{22} = x_1^2 + k(x_2^2 - x_1^2), \quad \sigma_{23} = 0, \quad \sigma_{33} = k(x_1^2 + x_2^2).$$

Solution: We have

$$\frac{\partial \sigma_{11}}{\partial x_1} + \frac{\partial \sigma_{12}}{\partial x_2} + \frac{\partial \sigma_{13}}{\partial x_3} + \rho_0 f_1 = (2kx_1) + (-2kx_1) + 0 + \rho_0 f_1 = 0,$$

$$\frac{\partial \sigma_{21}}{\partial x_1} + \frac{\partial \sigma_{22}}{\partial x_2} + \frac{\partial \sigma_{23}}{\partial x_3} + \rho_0 f_2 = (-2kx_2) + (2kx_2) + 0 + \rho_0 f_2 = 0,$$

$$\frac{\partial \sigma_{31}}{\partial x_1} + \frac{\partial \sigma_{32}}{\partial x_2} + \frac{\partial \sigma_{33}}{\partial x_3} + \rho_0 f_3 = 0 + 0 + 0 + \rho_0 f_3 = 0.$$

Thus the given stress field is in equilibrium in the absence of any body forces; that is, $\rho_0 \mathbf{f} = \mathbf{0}$.

4.6 Objectivity of Stress Tensors

4.6.1 Cauchy Stress Tensor

The Cauchy stress tensor is objective if we can show that $\boldsymbol{\sigma}^* = \mathbf{Q} \cdot \boldsymbol{\sigma} \cdot \mathbf{Q}^{\mathrm{T}}$ [see Eq. (3.8.21) for the definition of the objectivity of various order tensors]. We begin with the relations

$$\mathbf{t} = \boldsymbol{\sigma} \cdot \mathbf{n}, \quad \mathbf{t}^* = \boldsymbol{\sigma}^* \cdot \mathbf{n}^*; \qquad \mathbf{t}^* = \mathbf{Q} \cdot \mathbf{t}, \quad \mathbf{n}^* = \mathbf{Q} \cdot \mathbf{n}. \qquad (4.6.1)$$

Then

$$\mathbf{t}^* = \boldsymbol{\sigma}^* \cdot \mathbf{n}^* = \boldsymbol{\sigma}^* \cdot (\mathbf{Q} \cdot \mathbf{n}),$$
$$\mathbf{t}^* = \mathbf{Q} \cdot \mathbf{t} = \mathbf{Q} \cdot \boldsymbol{\sigma} \cdot \mathbf{n}.$$

Then, we have

$$\boldsymbol{\sigma}^* \cdot \mathbf{Q} = \mathbf{Q} \cdot \boldsymbol{\sigma},$$

from which it follows that

$$\boldsymbol{\sigma}^* = \mathbf{Q} \cdot \boldsymbol{\sigma} \cdot \mathbf{Q}^{\mathrm{T}}. \qquad (4.6.2)$$

Thus, the Cauchy stress tensor is objective.

4.6.2 First Piola–Kirchhoff Stress Tensor

The first Piola–Kirchhoff stress tensor $\mathbf{P}$ is a two-point tensor, and it transforms like the other two-point tensor $\mathbf{F}$. To establish this, we begin with the relation between $\mathbf{P}$ and $\boldsymbol{\sigma}$ after superposed rigid-body motion and make use of the relations $\mathbf{F}^* = \mathbf{Q} \cdot \mathbf{F}$ and $J^* = J$,

$$\mathbf{P}^* = J^* \boldsymbol{\sigma}^* \cdot (\mathbf{F}^*)^{-\mathrm{T}} = J(\mathbf{Q} \cdot \boldsymbol{\sigma} \cdot \mathbf{Q}^{\mathrm{T}}) \cdot (\mathbf{Q} \cdot \mathbf{F})^{-\mathrm{T}}$$
$$= J\mathbf{Q} \cdot \boldsymbol{\sigma} \cdot (\mathbf{Q}^{\mathrm{T}} \cdot \mathbf{Q}^{-\mathrm{T}}) \cdot \mathbf{F}^{-\mathrm{T}} = J\mathbf{Q} \cdot \boldsymbol{\sigma} \cdot \mathbf{F}^{-\mathrm{T}} = \mathbf{Q} \cdot \mathbf{P}. \qquad (4.6.3)$$

Thus $\mathbf{P}$, being a two-point tensor, transforms like a vector under superposed rigid-body motion, and hence is objective.

4.6.3 Second Piola–Kirchhoff Stress Tensor

The second Piola–Kirchhoff stress tensor $\mathbf{S}$ is the stress tensor of choice in the study of solid mechanics. Because it is defined with respect to the reference configuration, rigid-body motion should not alter it. Using the relations $\mathbf{F}^* = \mathbf{Q} \cdot \mathbf{F}$ and $\boldsymbol{\sigma}^* = \mathbf{Q} \cdot \boldsymbol{\sigma} \cdot \mathbf{Q}^{\mathrm{T}}$, we obtain

$$\mathbf{S}^* = J^* (\mathbf{F}^*)^{-1} \cdot \boldsymbol{\sigma}^* \cdot (\mathbf{F}^*)^{-\mathrm{T}} = J (\mathbf{F}^{-1} \cdot \mathbf{Q}^{-1}) \cdot (\mathbf{Q} \cdot \boldsymbol{\sigma} \cdot \mathbf{Q}^{\mathrm{T}}) \cdot (\mathbf{Q}^{-\mathrm{T}} \cdot \mathbf{F}^{-\mathrm{T}})$$
$$= J\mathbf{F}^{-1} \cdot \boldsymbol{\sigma} \cdot \mathbf{F}^{-\mathrm{T}} = \mathbf{S}. \qquad (4.6.4)$$

Thus, $\mathbf{S}$ is not affected by the superposed rigid-body motion and, therefore, it is objective.

4.7 Summary

In this chapter, the concept of stress in a continuum is introduced and stress vector at a point is defined. It is shown that the stress vector $\mathbf{t}$ at a point depends on the orientation of the plane $(\hat{\mathbf{n}})$ on which it acts. Then a relation between the stress vector $\mathbf{t}$ acting on a plane with unit normal $\hat{\mathbf{n}}$ and stress vectors $(\mathbf{t}_1, \mathbf{t}_2, \mathbf{t}_3)$ acting on three mutually perpendicular planes whose normals are $\hat{\mathbf{e}}_1, \hat{\mathbf{e}}_2, \hat{\mathbf{e}}_3$ is established. It is here the Cauchy stress tensor $\boldsymbol{\sigma}$ is introduced as a dyadic with respect to the Cartesian basis $(\hat{\mathbf{e}}_1, \hat{\mathbf{e}}_2, \hat{\mathbf{e}}_3)$:

$$\boldsymbol{\sigma} \equiv \mathbf{t}_1 \hat{\mathbf{e}}_1 + \mathbf{t}_2 \hat{\mathbf{e}}_2 + \mathbf{t}_3 \hat{\mathbf{e}}_3 = \mathbf{t}_j \hat{\mathbf{e}}_j, \quad \mathbf{t}_j = \sigma_{ij} \hat{\mathbf{e}}_i \quad \rightarrow \quad \boldsymbol{\sigma} = \sigma_{ij} \hat{\mathbf{e}}_i \hat{\mathbf{e}}_j.$$

The stress tensor $\boldsymbol{\sigma}$ at a point $\mathbf{x}$ is shown to be related to the stress vector $\mathbf{t}$ on a plane $\hat{\mathbf{n}}$ by $\mathbf{t} = \boldsymbol{\sigma} \cdot \hat{\mathbf{n}}$, which is known as Cauchy's formula.

We encounter stress vectors $\mathbf{t}$ in two instances: (1) stress vector at a point $\mathbf{x}$ in the interior of the body on a plane whose outward normal is $\hat{\mathbf{n}}$; and (2) stress vector at a point on the surface of the body, which is either specified or to be determined. Cauchy's formula is useful not only in relating the surface traction to the state of stress inside the body at a boundary point, $\bar{\mathbf{t}} = \boldsymbol{\sigma} \cdot \hat{\mathbf{n}}$, but it also implies that when a material volume Ω_0 is removed from a body, then it is possible to maintain Ω_0 in its equilibrium state by merely applying a suitable distribution of $\mathbf{t}$ on the boundary Γ_0 of the volume Ω_0, as shown in Fig. 4.7.1.

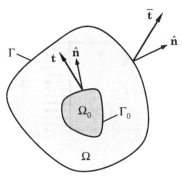

Fig. 4.7.1: Surface traction vector $\bar{\mathbf{t}}$ and internal stress vector $\mathbf{t}(\hat{\mathbf{n}})$.

Transformation relations for the components of the stress tensor in one coordinate system to its components in another coordinate system are established, and the determination of the principal values and principal planes of a stress tensor is detailed. Then two other measures of stress, namely, the first and second Piola–Kirchhoff stress tensors, $\mathbf{P}$ and $\mathbf{S}$, are introduced. Whereas the Cauchy stress tensor $\boldsymbol{\sigma}$ is measured as the current force per unit deformed area, the first Piola–Kirchhoff stress tensor $\mathbf{P}$ is measured as the current force per unit undeformed area and the second Piola–Kirchhoff stress tensor $\mathbf{S}$ is measured as the transformed current force per unit undeformed area. The first and second Piola–Kichhoff stress tensors are related to the Cauchy stress tensor by

$$\mathbf{P} = J\boldsymbol{\sigma} \cdot \mathbf{F}^{-\mathrm{T}} = \mathbf{F} \cdot \mathbf{S}, \quad \mathbf{S} = \mathbf{F}^{-1} \cdot \mathbf{P} = J\mathbf{F}^{-1} \cdot \boldsymbol{\sigma} \cdot \mathbf{F}^{-\mathrm{T}}. \tag{4.7.1}$$

The stress equilibrium equations $\nabla \cdot \boldsymbol{\sigma}^{\mathrm{T}} + \rho_0 \mathbf{f} = \mathbf{0}$ in the case of infinitesimal deformations are derived, and symmetry of the stress tensor, $\boldsymbol{\sigma}^{\mathrm{T}} = \boldsymbol{\sigma}$, in the absence of body couples, is established. Several examples are presented to illustrate the concepts and ideas introduced.

It is also shown that under superposed rigid-body transformation $\mathbf{x}^* = \mathbf{c}(t) + \mathbf{Q}(t) \cdot \mathbf{x}$, where $\mathbf{c}(t)$ is a constant vector characterizing the rigid-body translation and $\mathbf{Q}(t)$ is a proper orthogonal tensor characterizing the rigid-body rotation, the three stress tensors introduced in this chapter transform according to the following relations and are objective:

$$\boldsymbol{\sigma}^* = \mathbf{Q} \cdot \boldsymbol{\sigma} \cdot \mathbf{Q}^{\mathrm{T}}, \quad \mathbf{P}^* = \mathbf{Q} \cdot \mathbf{P}, \quad \mathbf{S} = \mathbf{S}^*. \tag{4.7.2}$$

Problems

Cauchy Stress Tensor and Cauchy's Formula

4.1 Suppose that $\mathbf{t}^{\hat{n}_1}$ and $\mathbf{t}^{\hat{n}_2}$ are stress vectors acting on planes with unit normals $\hat{\mathbf{n}}_1$ and $\hat{\mathbf{n}}_2$, respectively, and passing through a point with the stress state $\boldsymbol{\sigma}$. Show that the component of $\mathbf{t}^{\hat{n}_1}$ along $\hat{\mathbf{n}}_2$ is equal to the component of $\mathbf{t}^{\hat{n}_2}$ along the normal $\hat{\mathbf{n}}_1$ *if and only if $\boldsymbol{\sigma}$ is symmetric.*

4.2 Write the stress vectors on each boundary surface in terms of the given values and base vectors $\hat{\mathbf{i}}$ and $\hat{\mathbf{j}}$ for the system shown in Fig. P4.2.

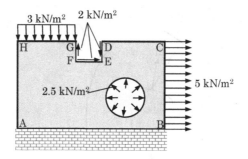

Fig. P4.2

4.3 The components of a stress tensor at a point, with respect to the (x_1, x_2, x_3) system, are (in MPa):

$$\text{(i)} \begin{bmatrix} 12 & 9 & 0 \\ 9 & -12 & 0 \\ 0 & 0 & 6 \end{bmatrix}, \quad \text{(ii)} \begin{bmatrix} 9 & 0 & 12 \\ 0 & -25 & 0 \\ 12 & 0 & 16 \end{bmatrix}, \quad \text{(iii)} \begin{bmatrix} 1 & -3 & \sqrt{2} \\ -3 & 1 & -\sqrt{2} \\ \sqrt{2} & -\sqrt{2} & 4 \end{bmatrix}.$$

Find the following:

(a) The stress vector acting on a plane perpendicular to the vector $2\hat{e}_1 - 2\hat{e}_2 + \hat{e}_3$

(b) The magnitude of the stress vector and the angle between the stress vector and the normal to the plane

(c) The magnitudes of the normal and tangential components of the stress vector

4.4 Consider a (kinematically infinitesimal) stress field whose matrix of scalar components in the vector basis $\{\hat{e}_i\}$ is

$$\begin{bmatrix} 1 & 0 & 2x_2 \\ 0 & 1 & 4x_1 \\ 2x_2 & 4x_1 & 1 \end{bmatrix} \text{ (MPa)},$$

where the Cartesian coordinate variables X_i are in meters (m) and the units of stress are MPa (10^6 Pa $= 10^6$ N/m^2).

(a) Determine the traction vector acting at point $\mathbf{X} = \hat{e}_1 + 2\hat{e}_2 + 3\hat{e}_3$ on the plane $x_1 + x_2 + x_3 = 6$.

(b) Determine the normal and projected shear tractions acting at this point on this plane.

4.5 The three-dimensional state of stress at a point $(1, 1, -2)$ within a body relative to the coordinate system (x_1, x_2, x_3) is

$$\begin{bmatrix} 2.0 & 3.5 & 2.5 \\ 3.5 & 0.0 & -1.5 \\ 2.5 & -1.5 & 1.0 \end{bmatrix} \text{ MPa.}$$

Determine the normal and shear stresses at the point and on the surface of an internal sphere whose equation is $x_1^2 + (x_2 - 2)^2 + x_3^2 = 6$.

4.6 The components of a stress tensor at a point, with respect to the (x_1, x_2, x_3) system, are

$$\begin{bmatrix} 25 & 0 & 0 \\ 0 & -30 & -60 \\ 0 & -60 & 5 \end{bmatrix} \text{ MPa.}$$

Determine

(a) the stress vector acting on a plane perpendicular to the vector $2\hat{e}_1 + \hat{e}_2 + 2\hat{e}_3$, and

(b) the magnitude of the normal and tangential components of the stress vector.

4.7 For the state of stress given in Problem **4.5**, determine the normal and shear stresses on a plane intersecting the point where the plane is defined by the points $(0, 0, 0)$, $(2, -1, 3)$, and $(-2, 0, 1)$.

4.8 The Cauchy stress tensor components at a point P in the deformed body with respect to the coordinate system (x_1, x_2, x_3) are given by

$$[\sigma] = \begin{bmatrix} 1 & 4 & -2 \\ 4 & 0 & 0 \\ -2 & 0 & 3 \end{bmatrix} \text{ MPa.}$$

(a) Determine the Cauchy stress vector $\mathbf{t}^{\hat{n}}$ at the point P on a plane passing through the point and parallel to the plane $2x_1 + 3x_2 + x_3 = 4$.

(b) Find the length of $\mathbf{t}^{\hat{n}}$ and the angle between $\mathbf{t}^{\hat{n}}$ and the vector normal to the plane.

(c) Determine the components of the Cauchy stress tensor in a rectangular coordinate system $(\bar{x}_1, \bar{x}_2, \bar{x}_3)$ whose orthonormal base vectors $\hat{\mathbf{e}}_i$ are given in terms of the base vectors $\hat{\mathbf{e}}_i$ of the coordinate system (x_1, x_2, x_3)

$$\hat{\mathbf{e}}_2 = \tfrac{1}{\sqrt{2}} (\hat{\mathbf{e}}_1 - \hat{\mathbf{e}}_3), \quad \hat{\mathbf{e}}_3 = \tfrac{1}{3} (2\hat{\mathbf{e}}_1 - \hat{\mathbf{e}}_2 + 2\hat{\mathbf{e}}_3).$$

4.9 The Cauchy stress tensor components at a point P in the deformed body with respect to the coordinate system (x_1, x_2, x_3) are given by

$$[\sigma] = \begin{bmatrix} 2 & 5 & 3 \\ 5 & 1 & 4 \\ 3 & 4 & 3 \end{bmatrix} \text{ MPa.}$$

(a) Determine the Cauchy stress vector $\mathbf{t}^{(\hat{n})}$ at the point P on a plane passing through the point whose normal is $\mathbf{n} = 3\hat{\mathbf{e}}_1 + \hat{\mathbf{e}}_2 - 2\hat{\mathbf{e}}_3$.

(b) Find the length of $\mathbf{t}^{(\hat{n})}$ and the angle between $\mathbf{t}^{(\hat{n})}$ and the vector normal to the plane.

(c) Find the normal and shear components of $\mathbf{t}^{\hat{n}}$ on the plane.

4.10 Suppose that at a point on the surface of a body the unit outward normal is $\hat{n} = (\hat{\mathbf{e}}_1 + \hat{\mathbf{e}}_2 - \hat{\mathbf{e}}_3)/\sqrt{3}$ and the traction vector is $P(\hat{\mathbf{e}}_1 + 2\hat{\mathbf{e}}_2)$, where P is a constant. Determine

(a) the normal traction vector $\mathbf{t}_n$ and the shear traction vector $\mathbf{t}_{ns}$ at this point on the surface of the body, and

(b) the conditions between the stress tensor components and the traction vector components.

4.11 Determine the traction free planes (defined by their unit normal vectors) passing through a point in the body where the stress state with respect to the rectangular Cartesian basis is

$$[\sigma] = \begin{bmatrix} 1 & 2 & 1 \\ 2 & \sigma_0 & 0 \\ 1 & 0 & -3 \end{bmatrix} \text{ MPa.}$$

What is the value of σ_0?

TRANSFORMATION EQUATIONS

4.12 Use equilibrium of forces to derive the relations between the normal and shear stresses σ_n and σ_s on a plane whose normal is $\hat{n} = \cos\theta\hat{\mathbf{e}}_1 + \sin\theta\hat{\mathbf{e}}_2$ to the stress components σ_{11}, σ_{22}, and $\sigma_{12} = \sigma_{21}$ on the $\hat{\mathbf{e}}_1$ and $\hat{\mathbf{e}}_2$ planes, as shown in Fig. P4.12:

$$\sigma_n = \sigma_{11} \cos^2\theta + \sigma_{22} \sin^2\theta + \sigma_{12} \sin 2\theta,$$
$$\sigma_s = -\tfrac{1}{2}(\sigma_{11} - \sigma_{22}) \sin 2\theta + \sigma_{12} \cos 2\theta. \tag{1}$$

Note that θ is the angle measured from the positive x_1-axis to the normal to the inclined plane (the same as that shown in Fig. 4.3.2). Then show that (a) the principal stresses at a point in a body with two-dimensional state of stress are given by

$$\sigma_{p1} = \sigma_{\max} = \frac{\sigma_{11} + \sigma_{22}}{2} + \sqrt{\left(\frac{\sigma_{11} - \sigma_{22}}{2}\right)^2 + \sigma_{12}^2} \,,$$

$$\sigma_{p2} = \sigma_{\min} = \frac{\sigma_{11} + \sigma_{22}}{2} - \sqrt{\left(\frac{\sigma_{11} - \sigma_{22}}{2}\right)^2 + \sigma_{12}^2} \,,$$

(2)

and that the orientation of the principal planes is given by

$$\theta_p = \pm\frac{1}{2} \tan^{-1}\left[\frac{2\sigma_{12}}{\sigma_{11} - \sigma_{22}}\right] ,$$

(3)

and (b) the maximum shear stress is given by

$$(\sigma_s)_{\max} = \pm\frac{\sigma_{p1} - \sigma_{p2}}{2} .$$

(4)

Also, determine the plane on which the maximum shear stress occurs.

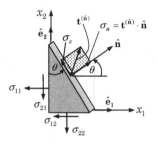

Fig. P4.12

4.13 through **4.16** Determine the normal and shear stress components on the plane indicated in Figs. P4.13–4.16.

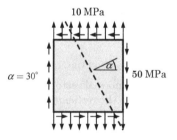

Fig. P4.13

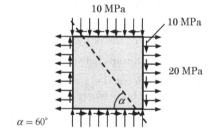

Fig. P4.14

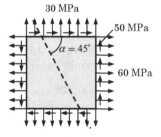

Fig. P4.15

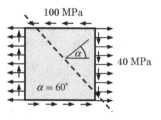

Fig. P4.16

4.17 Find the values of σ_s and σ_{22} for the state of stress shown in Fig. P4.17.

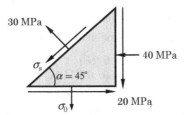

30 MPa

σ_s

$\alpha = 45°$

40 MPa

20 MPa

σ_0

Fig. P4.17

Principal Stresses and Principal Directions

4.18 For the stress state given in Problem **4.4**, determine

(a) the principal stresses and principal directions of stress at this point, and

(b) the maximum shear stress at the point.

4.19 Find the maximum and minimum normal stresses and the orientations of the principal planes for the state of stress shown in Fig. P4.15. What is the maximum shear stress at the point?

4.20 Find the maximum and minimum normal stresses and the orientations of the principal planes for the state of stress shown in Fig. P4.16. What is the maximum shear stress at the point?

4.21 Find the maximum principal stress, maximum shear stress and their orientations for the state of stress given.

$$\text{(a)} \quad [\sigma] = \begin{bmatrix} 12 & 9 & 0 \\ 9 & -12 & 0 \\ 0 & 0 & 6 \end{bmatrix} \text{MPa.} \qquad \text{(b)} \quad [\sigma] = \begin{bmatrix} 3 & 5 & 8 \\ 5 & 1 & 0 \\ 8 & 0 & 2 \end{bmatrix} \text{MPa.}$$

4.22 (*Spherical and deviatoric stress tensors*) The stress tensor can be expressed as the sum of *spherical* or *hydrostatic* stress tensor $\tilde{\sigma}$ and *deviatoric* stress tensor σ'

$$\sigma = \tilde{\sigma}\mathbf{I} + \sigma', \quad \tilde{\sigma} = \frac{1}{3}\operatorname{tr}\sigma = \frac{1}{3}I_1, \quad \sigma' = \sigma - \frac{1}{3}I_1\mathbf{I}.$$

For the state of stress shown in Fig. P4.16, compute the spherical and deviatoric components of the stress tensor.

4.23 Determine the invariants I_i' and the principal deviator stresses for the following state of stress

$$[\sigma] = \begin{bmatrix} 2 & -1 & 1 \\ -1 & 0 & 1 \\ 1 & 1 & 2 \end{bmatrix} \text{MPa.}$$

4.24 Given the following state of stress at a point in a continuum,

$$[\sigma] = \begin{bmatrix} 7 & 0 & 14 \\ 0 & 8 & 0 \\ 14 & 0 & -4 \end{bmatrix} \text{MPa,}$$

determine the principal stresses and principal directions.

4.25 Given the following state of stress ($\sigma_{ij} = \sigma_{ji}$),

$$\sigma_{11} = -2x_1^2, \quad \sigma_{12} = -7 + 4x_1 x_2 + x_3, \quad \sigma_{13} = 1 + x_1 - 3x_2,$$

$$\sigma_{22} = 3x_1^2 - 2x_2^2 + 5x_3, \quad \sigma_{23} = 0, \quad \sigma_{33} = -5 + x_1 + 3x_2 + 3x_3,$$

determine

(a) the stress vector at point (x_1, x_2, x_3) on the plane $x_1 + x_2 + x_3 = $ constant,

(b) the normal and shearing components of the stress vector at point $(1, 1, 3)$, and

(c) the principal stresses and their orientation at point $(1, 2, 1)$.

4.26 The components of a stress tensor at a point P, referred to the (x_1, x_2, x_3) system, are

$$\begin{bmatrix} 57 & 0 & 24 \\ 0 & 50 & 0 \\ 24 & 0 & 43 \end{bmatrix} \text{ MPa.}$$

Determine the principal stresses and principal directions at point P. What is the maximum shear stress at the point?

4.27 Given the following state of stress at a point in a continuum,

$$[\sigma] = \begin{bmatrix} 0 & 0 & Ax_2 \\ 0 & 0 & -Bx_3 \\ Ax_2 & -Bx_3 & 0 \end{bmatrix} \text{ MPa,}$$

where A and B are constants. Determine

(a) the body force vector such that the stress tensor corresponds to an equilibrium state,

(b) the three principal invariants of σ at the point $\mathbf{x} = B\hat{\mathbf{e}}_2 + A\hat{\mathbf{e}}_3$,

(c) the principal stress components and the associated planes at the point $\mathbf{x} = B\hat{\mathbf{e}}_2 + A\hat{\mathbf{e}}_3$, and

(d) the maximum shear stress and associated plane at the point $\mathbf{x} = B\hat{\mathbf{e}}_2 + A\hat{\mathbf{e}}_3$.

EQUILIBRIUM EQUATIONS (associated with infinitesimal deformations)

4.28 Derive the stress equilibrium equations in cylindrical coordinates by considering the equilibrium of a typical volume element shown in Fig. P4.28. Assume that the body force components are (not shown in the figure) $\rho_0 f_r$, $\rho_0 f_\theta$, and $\rho_0 f_z$ along the r, θ, and z coordinates, respectively.

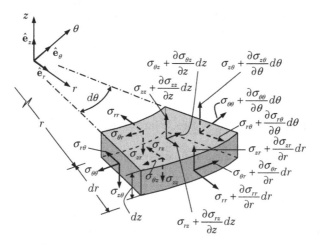

Fig. P4.28

4.29 Given the following state of stress at a point in a continuum,

$$[\sigma] = \begin{bmatrix} 1 & 0 & 2x_2 \\ 0 & 1 & 4x_1 \\ 2x_2 & 4x_1 & 1 \end{bmatrix} \text{ MPa,}$$

determine the body force vector such that the stress tensor corresponds to an equilibrium state.

4.30 Given the following state of stress at a point in a continuum,

$$[\sigma] = \begin{bmatrix} 5x_2x_3 & 3x_2^2 & 0 \\ 3x_2^2 & 0 & -x_1 \\ 0 & -x_1 & 0 \end{bmatrix} \text{ MPa},$$

determine the body force vector such that the stress tensor corresponds to an equilibrium state.

4.31 Given the following state of stress at a point in a continuum,

$$[\sigma] = \begin{bmatrix} A(x_1 - x_2) & Bx_1^2x_2 & 0 \\ Bx_1^2x_2 & -A(x_1 - x_2) & 0 \\ 0 & 0 & 0 \end{bmatrix} \text{ MPa},$$

determine the constants A and B such that the stress tensor corresponds to an equilibrium state in the absence of body forces.

4.32 Given the following state of stress at a point in a continuum,

$$[\sigma] = \begin{bmatrix} Ax_1^2x_2 & A(B^2 - x_2^2)x_1 & 0 \\ A(B^2 - x_2^2)x_1 & C(x_2^2 - 3B^2)x_2 & 0 \\ 0 & 0 & 2Bx_3^2 \end{bmatrix} \text{ MPa},$$

where A, B, and $C = A/3$ are constants, determine the body force components necessary for the body to be in equilibrium.

4.33 Given the following Cauchy stress components ($\sigma_{ij} = \sigma_{ji}$),

$$\sigma_{11} = -2x_1^2, \quad \sigma_{12} = -7 + 4x_1x_2 + x_3, \quad \sigma_{13} = 1 + x_1 - 3x_2,$$
$$\sigma_{22} = 3x_1^2 - 2x_2^2 + 5x_3, \quad \sigma_{23} = 0, \quad \sigma_{33} = -5 + x_1 + 3x_2 + 3x_3,$$

determine the body force components for which the stress field describes a state of equilibrium.

4.34 Given the following stress field, expressed in terms of its components referred to a rectangular Cartesian basis,

$$\sigma_{11} = x_1^2x_2, \quad \sigma_{12} = (c^2 - x_2^2)x_1, \quad \sigma_{13} = 0,$$
$$\sigma_{22} = \frac{1}{3}\left(x_2^3 - 3c^2x_2\right), \quad \sigma_{23} = 0, \quad \sigma_{33} = 2cx_3^2,$$

where c is a constant, find the body-force field necessary for the stress field to be in equilibrium.

4.35 The equilibrium configuration of a deformed body is described by the mapping

$$\chi(\mathbf{X}) = AX_1\,\hat{\mathbf{e}}_1 - BX_3\,\hat{\mathbf{e}}_2 + CX_2\,\hat{\mathbf{e}}_3,$$

where A, B, and C are constants. If the Cauchy stress tensor for this body is

$$[\sigma] = \begin{bmatrix} 0 & 0 & 0 \\ 0 & 0 & 0 \\ 0 & 0 & \sigma_0 \end{bmatrix} \text{ MPa},$$

where σ_0 is a constant, determine

(a) the deformation tensor and its inverse in matrix form,

(b) the matrices of the first and second Piola–Kirchhoff stress tensors, and

(c) the pseudo stress vectors associated with the first and second Piola–Kirchhoff stress tensors on the $\hat{\mathbf{e}}_3$-plane in the deformed configuration.

4.36 A body experiences deformation characterized by the mapping

$$\chi(\mathbf{X}, t) = \mathbf{x} = AX_2\,\hat{\mathbf{e}}_1 + BX_1\,\hat{\mathbf{e}}_2 + CX_3\,\hat{\mathbf{e}}_3,$$

where A, B, and C are constants. The Cauchy stress tensor components at certain point of the body are given by

$$[\sigma] = \begin{bmatrix} 0 & 0 & 0 \\ 0 & \sigma_0 & 0 \\ 0 & 0 & 0 \end{bmatrix} \text{ MPa,}$$

where σ_0 is a constant. Determine the Cauchy stress vector $\mathbf{t}$ and the first Piola–Kirchhoff stress vector $\mathbf{T}$ on a plane whose normal in the current configuration is $\hat{\mathbf{n}} = \hat{\mathbf{e}}_2$.

4.37 Express the stress equilibrium equations in Eq. (4.5.6) in terms of the stress components and body force components in the (a) cylindrical and (b) spherical coordinate systems.

4.38 Equation (4.2.7) can also be written

$$\mathbf{t} = (\hat{\mathbf{n}} \cdot \hat{\mathbf{e}}_i)\mathbf{t}_i = \hat{\mathbf{n}} \cdot (\hat{\mathbf{e}}_1\, \mathbf{t}_1 + \hat{\mathbf{e}}_2\, \mathbf{t}_2 + \hat{\mathbf{e}}_3\, \mathbf{t}_3). \tag{1}$$

The terms in the parenthesis can be defined as the *stress dyadic* or *stress tensor* $\mathbf{T}$:

$$\mathbf{T} \equiv \hat{\mathbf{e}}_1\, \mathbf{t}_1 + \hat{\mathbf{e}}_2\, \mathbf{t}_2 + \hat{\mathbf{e}}_3\, \mathbf{t}_3 = \hat{\mathbf{e}}_i\, \mathbf{t}_i. \tag{2}$$

Show that $\mathbf{T}$ is the transpose of $\boldsymbol{\sigma}$ defined in Eq. (4.2.13).

4.39 Show that the material time derivative of the Cauchy stress tensor is not objective, unless the superposed rigid-body rotation is time-independent (that is, $\mathbf{Q}$ is not a function of time); that is, show

$$\dot{\boldsymbol{\sigma}}^* \neq \mathbf{Q} \cdot \dot{\boldsymbol{\sigma}} \cdot \mathbf{Q}^{\mathrm{T}},$$

unless $\mathbf{Q}$ is independent of time.

4.40 Prove that if the stress tensor is real and symmetric, $\sigma_{ij} = \sigma_{ji}$, then its eigenvalues are real. Also, prove that the eigenvectors of a real and symmetric σ_{ij} are orthogonal.

CONSERVATION AND BALANCE LAWS

Although to penetrate into the intimate mysteries of nature and thence to learn the true causes of phenomena is not allowed to us, nevertheless it can happen that a certain fictive hypothesis may suffice for explaining many phenomena. —— Leonard Euler (1707–1783)

Nothing is too wonderful to be true if it be consistent with the laws of nature.

—— Michael Faraday (1791–1867)

5.1 Introduction

Virtually every phenomenon in nature can be described in terms of mathematical relations among certain quantities that are responsible for the phenomenon. Most mathematical models of physical phenomena are based on fundamental scientific laws of physics that are extracted from centuries of observations and research on the behavior of mechanical systems subjected to the action of natural forces. The most exciting thing about the laws of physics, which are also termed *principles of mechanics*, is that they govern biological systems as well (because of mass and energy transports). However, biological systems may require additional laws, yet to be discovered, from biology and chemistry to reasonably complete their descriptions.

This chapter is devoted to the study of fundamental laws of physics and resulting mathematical models as applied to mechanical systems. The laws of physics are expressed in analytical form with the aid of the concepts and quantities introduced in the previous chapters. The principles of mechanics to be studied are (1) the principle of conservation of mass, (2) the principle of balance of linear momentum, (3) the principle of balance of angular momentum, and (4) the principle of balance of energy. These principles allow us to write mathematical relationships – algebraic, differential, or integral type – between quantities such as displacements, velocities, temperature, stresses, and strains that arise in mechanical systems. The solution of these equations, in conjunction with the constitutive relations and boundary and initial conditions, represents the response of the system. The equations developed here not only are used not only in the later chapters of this book, but they are also useful in other engineering and applied science courses. In addition, the equations developed herein form the basis of most mathematical models employed in the study of a variety of phenomena. Thus, *the present chapter is the heart and soul of a course on continuum mechanics and elasticity.*

5.2 Conservation of Mass

5.2.1 Preliminary Discussion

It is common knowledge that the mass of a given system cannot be created or destroyed. For example, the mass flow of the blood entering a section of an artery is equal to the mass flow leaving the artery, provided that no blood is added or lost through the artery walls. Thus, mass of the blood is conserved even when the artery cross section changes along the length.

The principle of conservation of mass states that *the total mass of any part ∂B of the body B does not change in any motion.* The mathematical form of this principle is different in spatial and material descriptions of motion. Before we derive the mathematical forms of the principle, certain other identities need to be established.

5.2.2 Material Time Derivative

As discussed in Chapter 3 [see Eqs. (3.2.4) and (3.2.5)], the partial time derivative with the material coordinates $\mathbf{X}$ held constant should be distinguished from the partial time derivative with spatial coordinates $\mathbf{x}$ held constant due to the difference in the descriptions of motion. The *material time derivative*, denoted here by D/Dt, is the time derivative with the material coordinates held constant. Thus, the time derivative of a function ϕ in material description (i.e., $\phi = \phi(\mathbf{X}, t)$) with $\mathbf{X}$ held constant is nothing but the partial derivative with respect to time [see Eq. (3.2.4)]:

$$\frac{D\phi}{Dt} \equiv \left(\frac{\partial \phi}{\partial t}\right)_{\mathbf{X}=\text{const}} = \frac{\partial \phi}{\partial t} . \tag{5.2.1}$$

In particular, we have

$$\frac{D\mathbf{x}}{Dt} = \left(\frac{\partial \mathbf{x}}{\partial t}\right)_{\mathbf{X}=\text{const}} = \left(\frac{\partial \mathbf{x}}{\partial t}\right) \equiv \mathbf{v}, \tag{5.2.2}$$

where $\mathbf{v}$ is the velocity vector. Similarly, the time derivative of $\mathbf{v}$ is

$$\frac{D\mathbf{v}}{Dt} = \left(\frac{\partial \mathbf{v}}{\partial t}\right)_{\mathbf{X}=\text{const}} = \left(\frac{\partial \mathbf{v}}{\partial t}\right) \equiv \mathbf{a}, \tag{5.2.3}$$

where $\mathbf{a}$ is the acceleration vector.

In the spatial description, we have $\phi = \phi(\mathbf{x}, t)$ and the partial time derivative

$$\left(\frac{\partial \phi}{\partial t}\right)_{\mathbf{X}=\text{const}} \quad \text{is different from} \quad \left(\frac{\partial \phi}{\partial t}\right)_{\mathbf{x}=\text{const}} .$$

The time derivative

$$\left(\frac{\partial \phi}{\partial t}\right)_{\mathbf{x}=const}$$

denotes the *local rate of change* of ϕ. If $\phi = \mathbf{v}$, then it is the rate of change of $\mathbf{v}$ read by a velocity meter located at the fixed spatial location $\mathbf{x}$, which is not the same as the acceleration of the particle just passing the place $\mathbf{x}$.

To calculate the material time derivative of a function ϕ of spatial coordinates $\mathbf{x}$, $\phi = \phi(\mathbf{x}, t)$, we assume that there exists differentiable mapping $\chi(\mathbf{X}, t) = \mathbf{x}(\mathbf{X}, t)$ so that we can write $\phi(\mathbf{x}, t) = \phi[\mathbf{x}(\mathbf{X}, t), t]$ and compute the derivative using the chain rule of differentiation:

$$
\frac{D\phi}{Dt} \equiv \left(\frac{\partial\phi}{\partial t}\right)_{\mathbf{X}=\text{const}} = \left(\frac{\partial\phi}{\partial t}\right)_{\mathbf{x}=\text{const}} + \left(\frac{\partial x_i}{\partial t}\right)_{\mathbf{X}=\text{const}} \frac{\partial\phi}{\partial x_i}
$$

$$
= \left(\frac{\partial\phi}{\partial t}\right)_{\mathbf{x}=\text{const}} + v_i\frac{\partial\phi}{\partial x_i}
$$

$$
= \left(\frac{\partial\phi}{\partial t}\right)_{\mathbf{x}=\text{const}} + \mathbf{v}\cdot\nabla\phi, \tag{5.2.4}
$$

where Eq. (5.2.2) is used in the second line. Thus, the material derivative operator is given by

$$
\frac{D}{Dt} = \left(\frac{\partial}{\partial t}\right)_{\mathbf{x}=\text{const}} + \mathbf{v}\cdot\nabla. \tag{5.2.5}
$$

Example 5.2.1 illustrates the calculation of the material time derivative based on the material and spatial descriptions.

Example 5.2.1

Suppose that a motion is described by the one-dimensional mapping, $x = (1+t)X$, for $t \geq 0$. Determine (a) the velocities and accelerations in the spatial and material descriptions, and (b) the time derivative of a function $\phi(X, t) = Xt^2$ in the spatial and material descriptions.

Solution: The velocity $v \equiv Dx/Dt$ can be expressed in the material and spatial coordinates as

$$
v(X, t) = \frac{Dx}{Dt} = \frac{\partial x}{\partial t} = X, \quad v(x, t) = X(x, t) = \frac{x}{1+t}.
$$

The acceleration $a \equiv Dv/Dt$ in the two descriptions is

$$
a \equiv \frac{Dv(X, t)}{Dt} = \frac{\partial v}{\partial t} = 0,
$$

$$
a \equiv \frac{Dv(x, t)}{Dt} = \frac{\partial v}{\partial t} + v\frac{\partial v}{\partial x}
$$

$$
= -\frac{x}{(1+t)^2} + \frac{x}{1+t}\frac{1}{1+t} = 0.
$$

The material time derivative of $\phi = \phi(X, t)$ in the material description is simply

$$
\frac{D\phi(X, t)}{Dt} = \frac{\partial\phi(X, t)}{\partial t} = 2Xt.
$$

The material time derivative of $\phi = \phi(x, t) = X(x, t)t^2 = xt^2/(1+t)$ in the spatial description is

$$
\frac{D\phi}{Dt} = \frac{\partial\phi}{\partial t} + v\frac{\partial\phi}{\partial x} = \frac{2xt}{1+t} - \frac{xt^2}{(1+t)^2} + \left(\frac{x}{1+t}\right)\left(\frac{t^2}{1+t}\right) = \frac{2xt}{1+t},
$$

which is the same as that calculated before, except that it is expressed in terms of the current coordinate, x.

5.2.3 Vector and Integral Identities

In the sections of the chapter that follow, we will make use of several vector identities and the gradient and divergence theorems [see Tables 2.4.1 and 2.4.2 and Eqs. (2.4.45)–(2.4.47)]. For a ready reference, some key results are provided here.

5.2.3.1 Vector identities

For any scalar function $F(\mathbf{x})$, vector-valued function $\mathbf{A}(\mathbf{x})$, and tensor-valued functions $\mathbf{S}(\mathbf{x})$, the following identities hold:

$$\boldsymbol{\nabla} \cdot (F\mathbf{A}) = F\boldsymbol{\nabla} \cdot \mathbf{A} + \boldsymbol{\nabla}F \cdot \mathbf{A}, \tag{5.2.6a}$$

$$\mathbf{A} \cdot \boldsymbol{\nabla}\mathbf{A} = \tfrac{1}{2}\boldsymbol{\nabla}(\mathbf{A} \cdot \mathbf{A}) - \mathbf{A} \times (\boldsymbol{\nabla} \times \mathbf{A}), \tag{5.2.6b}$$

$$\boldsymbol{\nabla} \cdot (\mathbf{S} \cdot \mathbf{A}) = (\boldsymbol{\nabla} \cdot \mathbf{S}) \cdot \mathbf{A} + \mathbf{S} : \boldsymbol{\nabla}\mathbf{A}, \tag{5.2.7}$$

$$\mathbf{S} : (\boldsymbol{\nabla}\mathbf{A}) = \mathbf{S}^{\text{sym}} : (\boldsymbol{\nabla}\mathbf{A})^{\text{sym}} + \mathbf{S}^{\text{skew}} : (\boldsymbol{\nabla}\mathbf{A})^{\text{skew}}, \tag{5.2.8}$$

where : denotes the double-dot product defined in Eq. (2.5.13), and the superscripts sym and skew denote the symmetric and skew symmetric parts of the enclosed quantity [see Eq. (2.5.25)]. In addition, the del operator $\boldsymbol{\nabla}$ and the divergence of $\mathbf{A}$ and $\mathbf{S}$ in the rectangular Cartesian, cylindrical, and spherical coordinates [see Figs. 2.4.4(b) and 2.4.5 for the coordinate systems] have the forms given here (see Chapter 2 for details).

Cartesian coordinates $[\mathbf{x} = (x_1, x_2, x_3)]$

$$\boldsymbol{\nabla} = \hat{\mathbf{e}}_i\frac{\partial}{\partial x_i}, \quad \boldsymbol{\nabla} \cdot \mathbf{A} = \frac{\partial A_i}{\partial x_i}, \quad \boldsymbol{\nabla} \cdot \mathbf{S} = \frac{\partial S_{ij}}{\partial x_i}\hat{\mathbf{e}}_j. \tag{5.2.9}$$

Cylindrical coordinates $[\mathbf{x} = (r, \theta, z)]$

$$\boldsymbol{\nabla} = \hat{\mathbf{e}}_r\frac{\partial}{\partial r} + \frac{1}{r}\hat{\mathbf{e}}_\theta\frac{\partial}{\partial \theta} + \hat{\mathbf{e}}_z\frac{\partial}{\partial z}, \tag{5.2.10}$$

$$\boldsymbol{\nabla} \cdot \mathbf{A} = \frac{1}{r}\left[\frac{\partial(rA_r)}{\partial r} + \frac{\partial A_\theta}{\partial \theta} + r\frac{\partial A_z}{\partial z}\right], \tag{5.2.11}$$

$$\boldsymbol{\nabla} \cdot \mathbf{S} = \left[\frac{\partial S_{rr}}{\partial r} + \frac{1}{r}\frac{\partial S_{\theta r}}{\partial \theta} + \frac{\partial S_{zr}}{\partial z} + \frac{1}{r}(S_{rr} - S_{\theta\theta})\right]\hat{\mathbf{e}}_r$$

$$+ \left[\frac{\partial S_{r\theta}}{\partial r} + \frac{1}{r}\frac{\partial S_{\theta\theta}}{\partial \theta} + \frac{\partial S_{z\theta}}{\partial z} + \frac{1}{r}(S_{r\theta} + S_{\theta r})\right]\hat{\mathbf{e}}_\theta$$

$$+ \left[\frac{\partial S_{rz}}{\partial r} + \frac{1}{r}\frac{\partial S_{\theta z}}{\partial \theta} + \frac{\partial S_{zz}}{\partial z} + \frac{1}{r}S_{rz}\right]\hat{\mathbf{e}}_z. \tag{5.2.12}$$

Spherical coordinates $[\mathbf{x} = (R, \phi, \theta)]$

$$\boldsymbol{\nabla} = \hat{\mathbf{e}}_R\frac{\partial}{\partial R} + \frac{1}{R}\hat{\mathbf{e}}_\phi\frac{\partial}{\partial \phi} + \frac{1}{R\sin\phi}\hat{\mathbf{e}}_\theta\frac{\partial}{\partial \theta}, \tag{5.2.13}$$

$$\boldsymbol{\nabla} \cdot \mathbf{A} = 2\frac{A_R}{R} + \frac{\partial A_R}{\partial R} + \frac{1}{R\sin\phi}\frac{\partial(A_\phi\sin\phi)}{\partial \phi} + \frac{1}{R\sin\phi}\frac{\partial A_\theta}{\partial \theta}, \tag{5.2.14}$$

$$
\begin{aligned}
\nabla \cdot \mathbf{S} = \Bigg\{ & \frac{\partial S_{RR}}{\partial R} + \frac{1}{R}\frac{\partial S_{\phi R}}{\partial \phi} + \frac{1}{R\sin\phi}\frac{\partial S_{\theta R}}{\partial \theta} \\
& + \frac{1}{R}\left[2S_{RR} - S_{\phi\phi} - S_{\theta\theta} + S_{\phi R}\cot\phi\right] \Bigg\}\hat{\mathbf{e}}_R \\
& + \Bigg\{ \frac{\partial S_{R\phi}}{\partial R} + \frac{1}{R}\frac{\partial S_{\phi\phi}}{\partial \phi} + \frac{1}{R\sin\phi}\frac{\partial S_{\theta\phi}}{\partial \theta} \\
& + \frac{1}{R}\left[(S_{\phi\phi} - S_{\theta\theta})\cot\phi + S_{\phi R} + 2S_{R\phi}\right] \Bigg\}\hat{\mathbf{e}}_\phi \\
& + \Bigg\{ \frac{\partial S_{R\theta}}{\partial R} + \frac{1}{R}\frac{\partial S_{\phi\theta}}{\partial \phi} + \frac{1}{R\sin\phi}\frac{\partial S_{\theta\theta}}{\partial \theta} \\
& + \frac{1}{R}\left[(S_{\phi\theta} + S_{\theta\phi})\cot\phi + 2S_{R\theta} + S_{\theta R}\right] \Bigg\}\hat{\mathbf{e}}_\theta .
\end{aligned}
\tag{5.2.15}
$$

5.2.3.2 Integral identities

The following relations hold for a closed region Ω bounded by surface Γ with outward unit normal vector $\hat{\mathbf{n}}$:

$$
\oint_\Gamma \hat{\mathbf{n}}\,\mathbf{S}\,ds = \int_\Omega \nabla\mathbf{S}\,d\mathbf{x} \quad \text{(Gradient theorem)}
\tag{5.2.16}
$$

$$
\oint_\Gamma \phi\,\hat{\mathbf{n}}\cdot\mathbf{S}\,ds = \int_\Omega \nabla\cdot(\phi\mathbf{S})\,d\mathbf{x} \quad \text{(Divergence theorem)}
\tag{5.2.17}
$$

$$
\oint_\Gamma \mathbf{x}\times(\hat{\mathbf{n}}\cdot\mathbf{S})\,ds = \int_\Omega \left[\mathbf{x}\times(\nabla\cdot\mathbf{S}) + \mathcal{E} : \mathbf{S}\right]d\mathbf{x},
\tag{5.2.18}
$$

where $\mathbf{x}$ is the position vector, $\mathbf{S}$ is a tensor-valued function of position, $\mathcal{E}$ is the third-order permutation tensor [see Eq. (2.5.23)], ϕ is a scalar-valued function, $d\mathbf{x}$ denotes a volume element, and ds is an area element on the surface.

5.2.4 Continuity Equation in the Spatial Description

Let an arbitrary region in a continuous medium $\mathcal{B}$ be denoted by Ω, and the bounding closed surface of this region be continuous and denoted by Γ. Let each point on the bounding surface move with velocity $\mathbf{v}_s$. It can be shown that the time derivative of the volume integral of some continuous function $\phi(\mathbf{x}, t)$ is given by

$$
\begin{aligned}
\frac{d}{dt}\int_\Omega \phi(\mathbf{x}, t)\,d\mathbf{x} &\equiv \frac{\partial}{\partial t}\int_\Omega \phi\,d\mathbf{x} + \oint_\Gamma \phi\,\hat{\mathbf{n}}\cdot\mathbf{v}_s\,ds, \\
&= \int_\Omega \frac{\partial\phi}{\partial t}\,d\mathbf{x} + \oint_\Gamma \phi\,\hat{\mathbf{n}}\cdot\mathbf{v}_s\,ds.
\end{aligned}
\tag{5.2.19}
$$

This expression for the differentiation of a volume integral with variable limits is sometimes known as the three-dimensional *Leibnitz rule*.

Let each element of mass in the medium move with the velocity $\mathbf{v}(\mathbf{x}, t)$ and consider a special region Ω such that the bounding surface Γ is attached to a fixed set of material elements. Then each point of this surface moves with the material velocity, that is, $\mathbf{v}_s = \mathbf{v}$, and the region Ω thus contains a fixed total amount of mass because no mass crosses the boundary surface Γ. To distinguish the time rate of change of an integral over this material region, we replace d/dt by D/Dt and write

$$\frac{D}{Dt} \int_\Omega \phi(\mathbf{x}, t)\, d\mathbf{x} \equiv \int_\Omega \frac{\partial \phi}{\partial t}\, d\mathbf{x} + \oint_\Gamma \phi \hat{\mathbf{n}} \cdot \mathbf{v}\, ds, \qquad (5.2.20)$$

which holds for a material region, that is, a region of fixed total mass. In some books, Eq. (5.2.20) is referred to as the *Reynolds transport theorem*. The relation between the time derivative following an arbitrary region and the time derivative following a material region (fixed total mass) is

$$\frac{d}{dt} \int_\Omega \phi(\mathbf{x}, t)\, d\mathbf{x} \equiv \frac{D}{Dt} \int_\Omega \phi(\mathbf{x}, t)\, d\mathbf{x} + \oint_\Gamma \phi \hat{\mathbf{n}} \cdot (\mathbf{v}_s - \mathbf{v})\, ds. \qquad (5.2.21)$$

The velocity difference $\mathbf{v} - \mathbf{v}_s$ is the velocity of the material measured relative to the velocity of the surface. The surface integral

$$\oint_\Gamma \phi \hat{\mathbf{n}} \cdot (\mathbf{v} - \mathbf{v}_s)\, ds,$$

thus measures the total *outflow* of the property ϕ from the region Ω.

Let $\rho(\mathbf{x}, t)$ denote the mass density of a continuous region. Then the principle of conservation of mass for a fixed *material* region Ω requires that

$$\frac{D}{Dt} \int_\Omega \rho\, d\mathbf{x} = 0. \qquad (5.2.22)$$

Then from Eq. (5.2.21), with $\phi = \rho$, it follows that for a fixed *spatial region* Ω (i.e., $\mathbf{v}_s = 0$) the principle of conservation of mass can also be stated as

$$\frac{d}{dt} \int_\Omega \rho\, d\mathbf{x} = -\oint_\Gamma \rho \hat{\mathbf{n}} \cdot \mathbf{v}\, ds. \qquad (5.2.23)$$

Thus, the time rate of change of mass inside a region Ω is equal to the mass inflow (because of the negative sign) through the surface into the region. Equation (5.2.23) is known as the control volume formulation of the conservation of mass principle. In Eq. (5.2.23), Ω denotes the *control volume* (cv) and Γ the *control surface* (cs) enclosing Ω.

Using Eq. (5.2.19) with $\phi = \rho$, Eq. (5.2.23) can be expressed as

$$\int_\Omega \frac{\partial \rho}{\partial t}\, d\mathbf{x} = -\oint_\Gamma \rho \hat{\mathbf{n}} \cdot \mathbf{v}\, ds.$$

Converting the surface integral to a volume integral by means of the divergence theorem (5.2.17), we obtain

$$\int_\Omega \left[\frac{\partial \rho}{\partial t} + \nabla \cdot (\rho \mathbf{v}) \right] d\mathbf{x} = 0.$$

Since this integral vanishes for any continuous medium occupying an arbitrary region Ω, we deduce that this is true only if the integrand itself vanishes identically, giving the following local form of the principle of conservation of mass:

$$\frac{\partial \rho}{\partial t} + \nabla \cdot (\rho\, \mathbf{v}) = 0. \tag{5.2.24}$$

This equation, also known as the *continuity equation*, expresses local conservation of mass at any point in a continuous medium.

An alternative derivation of Eq. (5.2.24) that is found in fluid mechanics books is presented next. Consider an arbitrary control volume Ω in space where flow occurs into and out of the control volume. Conservation of mass in this case means that the time rate of change of mass in Ω is equal to the mass inflow through the control surface Γ into the control volume Ω. Consider an elemental area ds with unit normal $\hat{\mathbf{n}}$ around a point P on the control surface, as shown in Fig. 5.2.1. Let $\mathbf{v}$ and ρ be the velocity and mass density, respectively, at point P. The mass *outflow* (slug/s or kg/s) through the elemental surface is $\rho\, \mathbf{v} \cdot d\mathbf{s}$, where $d\mathbf{s} = \hat{\mathbf{n}}\, ds$. The total mass *inflow* through the entire surface of the control volume is

$$\oint_\Gamma (-\rho\, v_n)\, ds = -\oint_\Gamma \rho\, \hat{\mathbf{n}} \cdot \mathbf{v}\, ds = -\int_\Omega \nabla \cdot (\rho\, \mathbf{v})\, d\mathbf{x}, \tag{5.2.25}$$

where the divergence theorem (5.2.17) is used in arriving at the last expression in Eq. (5.2.25). If a continuous medium of mass density ρ fills the region Ω at time t, the total mass in Ω is $M = \int_\Omega \rho(\mathbf{x}, t)\, d\mathbf{x}$. The rate of increase of mass in the fixed region Ω is

$$\frac{\partial M}{\partial t} = \int_\Omega \frac{\partial \rho}{\partial t}\, d\mathbf{x}. \tag{5.2.26}$$

Equating Eqs. (5.2.25) and (5.2.26), we obtain

$$\int_\Omega \left[\frac{\partial \rho}{\partial t} + \nabla \cdot (\rho\, \mathbf{v}) \right] d\mathbf{x} = 0,$$

which results in the same equation as the one in Eq. (5.2.24).

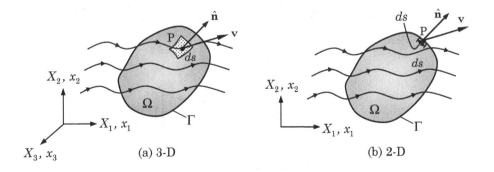

Fig. 5.2.1: A control volume for the derivation of the continuity equation.

Equation (5.2.24) can be written in an alternative form as follows [see Eq. (5.2.6a)]:

$$0 = \frac{\partial \rho}{\partial t} + \boldsymbol{\nabla} \cdot (\rho \mathbf{v}) = \frac{\partial \rho}{\partial t} + \mathbf{v} \cdot \boldsymbol{\nabla} \rho + \rho \boldsymbol{\nabla} \cdot \mathbf{v} = \frac{D\rho}{Dt} + \rho \boldsymbol{\nabla} \cdot \mathbf{v}, \qquad (5.2.27)$$

where the definition of *material time derivative*, Eq. (5.2.5), is used in arriving at the final result.

The one-dimensional version of the local form of the continuity equation (5.2.24) can be obtained by considering flow along the x-axis (see Fig. 5.2.2). The amount of mass entering (i.e., mass flow) per unit time at the left section of the elemental volume is:

$$\text{density} \times \text{cross-sectional area} \times \text{velocity of the flow} = (\rho \, A v_x)_x.$$

The mass leaving at the right section of the elemental volume is $(\rho \, A v_x)_{x+\Delta x}$, where v_x is the velocity along the x-direction. The subscript denotes the distance at which the enclosed quantity is evaluated. It is assumed that the cross-sectional area A is a function of position x but not of time t. The net mass flow *into* the elemental volume is

$$(A\rho \, v_x)_x - (A\rho \, v_x)_{x+\Delta x}.$$

On the other hand, the time rate of *increase* of the total mass inside the elemental volume is

$$\bar{A}\Delta x \frac{(\bar{\rho})_{t+\Delta t} - (\bar{\rho})_t}{\Delta t},$$

where $\bar{\rho}$ and $\bar{A}$ are the average values of the density and cross-sectional area, respectively, inside the elemental volume.

If no mass is created or destroyed inside the elemental volume, the rate of increase of mass should be equal to the mass inflow:

$$\bar{A}\Delta x \frac{(\bar{\rho})_{t+\Delta t} - (\bar{\rho})_t}{\Delta t} = (A\rho \, v_x)_x - (A\rho \, v_x)_{x+\Delta x}.$$

Dividing throughout by Δx and taking the limits $\Delta t \to 0$ and $\Delta x \to 0$, we obtain

$$\lim_{\Delta t, \, \Delta x \to 0} \bar{A} \frac{(\rho)_{t+\Delta t} - (\rho)_t}{\Delta t} + \frac{(A\rho \, v_x)_{x+\Delta x} - (A\rho \, v_x)_x}{\Delta x} = 0,$$

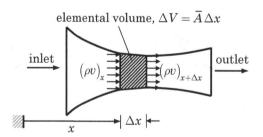

Fig. 5.2.2: Derivation of the local form of the continuity equation in one dimension.

or ($\bar{\rho} \to \rho$ and $\bar{A} \to A$ as $\Delta x \to 0$)

$$A\frac{\partial \rho}{\partial t} + \frac{\partial (A\rho v_x)}{\partial x} = 0. \tag{5.2.28}$$

Equation (5.2.28) is the same as Eq. (5.2.24) when $\mathbf{v}$ is replaced with $\mathbf{v} = v_x \hat{\mathbf{e}}_x$ and A is a constant. Note that for the steady-state case, Eq. (5.2.28) reduces to

$$\frac{\partial (A\rho v_x)}{\partial x} = 0 \rightarrow A\rho v_x = \text{constant} \Rightarrow (A\rho v_x)_1 = (A\rho v_x)_2 = \cdots = (A\rho v_x)_i, \tag{5.2.29}$$

where the subscript i refers to ith section along the direction of the (one-dimensional) flow. The quantity $Q = Av_x$ is called the *flow*, whereas $\rho A v_x$ is called the *mass flow*.

The continuity equation in Eq. (5.2.24) can also be expressed in orthogonal curvilinear coordinate systems as [see Eqs. (5.2.10)–(5.2.15); Problems **5.4–5.6** are designed to obtain these results]

Cylindrical coordinate system (r, θ, z)

$$0 = \frac{\partial \rho}{\partial t} + \frac{1}{r}\left[\frac{\partial (r\rho v_r)}{\partial r} + \frac{\partial (\rho v_\theta)}{\partial \theta} + r\frac{\partial (\rho v_z)}{\partial z}\right]. \tag{5.2.30}$$

Spherical coordinate system (R, ϕ, θ)

$$0 = \frac{\partial \rho}{\partial t} + \frac{1}{R^2}\frac{\partial (\rho R^2 v_R)}{\partial R} + \frac{1}{R\sin\phi}\frac{\partial (\rho v_\phi \sin\phi)}{\partial \theta} + \frac{1}{R\sin\phi}\frac{\partial (\rho v_\theta)}{\partial \theta}. \tag{5.2.31}$$

For steady state, we set the time derivative terms in Eqs. (5.2.24), (5.2.30), and (5.2.31) to zero. The invariant form of continuity equation for steady-state flows is (so-called divergence-free velocity field)

$$\nabla \cdot (\rho \mathbf{v}) = 0. \tag{5.2.32}$$

For materials with constant density, we set $D\rho/Dt = 0$ and obtain (the so-called divergence-free condition on the velocity field)

$$\rho \nabla \cdot \mathbf{v} = 0 \quad \text{or} \quad \nabla \cdot \mathbf{v} = 0. \tag{5.2.33}$$

Thus the motion is isochoric, and the velocity field is said to be *solenoidal*.

Next, we consider two examples of application of the principle of conservation of mass in spatial description.

Example 5.2.2

Consider a water hose with a conical-shaped nozzle at its end, as shown in Fig. 5.2.3(a). (a) Determine the pumping capacity required for the velocity of the water (assuming incompressible for the present case) exiting the nozzle to be 25 m/s. (b) If the hose is connected to a rotating sprinkler through its base, as shown in Fig. 5.2.3(b), determine the average speed of the water leaving the sprinkler nozzle.

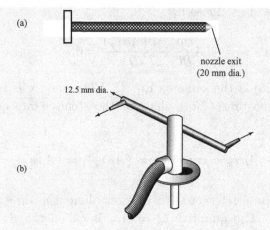

Fig. 5.2.3: (a) Water hose with a conical head. (b) Water hose connected to a sprinkler.

Solution: (**a**) The principle of conservation of mass for steady one-dimensional flow requires

$$\rho_1 A_1 v_1 = \rho_2 A_2 v_2.$$

If the exit of the nozzle is taken as the section 2 and the inlet is taken as the section 1 [see Fig. 5.2.3(a)], we can write (for an incompressible fluid, $\rho_1 = \rho_2$)

$$Q_1 = A_1 v_1 = A_2 v_2 = \frac{\pi (20 \times 10^{-3})^2}{4} 25 = 0.0025\pi \ \mathrm{m^3/s}.$$

(**b**) The average speed of the water leaving the sprinkler nozzle can be calculated using the principle of conservation of mass for steady one-dimensional flow. We obtain

$$Q_1 = 2A_2 v_2 \quad \rightarrow \quad v_2 = \frac{2Q_1}{\pi d^2} = \frac{0.005}{(12.5 \times 10^{-3})^2} = 32 \ \mathrm{m/s}.$$

Example 5.2.3

A syringe used to inoculate large animals has a cylinder, plunger, and needle combination, as shown in Fig. 5.2.4. Let the internal diameter of the cylinder be d and the plunger face area be A_p. If the liquid in the syringe is to be injected at a steady rate of Q_0, determine the speed of the plunger. Assume that the leakage rate past the plunger is 10% of the volume flow rate out of the needle.

Solution: In this problem, the control volume (shown in dotted lines in Fig. 5.2.4) is not constant. Even though there is a leakage, the plunger surface area can be taken as equal to the open cross-sectional area of the cylinder, $A_p = \pi d^2/4$. Let us consider Section 1 to be the plunger face and Section 2 to be the needle exit to apply the continuity equation.

Assuming that the flow through the needle and leakage are steady, application of the global form of the continuity equation, Eq. (5.2.23), to the control volume gives

$$0 = \frac{d}{dt} \int_{\Omega} \rho \, d\mathbf{x} + \oint_{\Gamma} \rho \, \hat{\mathbf{n}} \cdot \mathbf{v} \, ds$$

$$= \frac{d}{dt} \int_{\Omega} \rho \, d\mathbf{x} + \rho Q_0 + \rho Q_{\mathrm{leak}}. \tag{1}$$

The integral in the above equation can be evaluated as follows:

$$\frac{d}{dt} \int_{\Omega} \rho \, d\mathbf{x} = \frac{d}{dt} \left(\rho \, x A_p + \rho V_n \right) = \rho A_p \frac{dx}{dt} = -\rho A_p v_p, \tag{2}$$

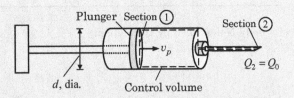

Fig. 5.2.4: The syringe discussed in Example 5.2.3.

where x is the distance between the plunger face and the end of the cylinder, V_n is the volume of the needle opening, and $v_p = -dx/dt$ is the speed of the plunger that we are after. Noting that $Q_{\text{leak}} = 0.1Q_0$, we can write the continuity equation (1) as

$$-\rho A_p v_p + 1.1\rho Q_0 = 0,$$

from which we obtain

$$v_p = 1.1\frac{Q_0}{A_p} = \frac{4.4Q_0}{\pi d^2}. \qquad (3)$$

For $Q_0 = 250\,\text{cm}^3/\text{min}$ and $d = 25\,\text{mm}$, we obtain

$$v_p = \frac{4.4 \times (250 \times 10^3)}{\pi(25 \times 25)} = 560\,\text{mm/min}.$$

5.2.5 Continuity Equation in the Material Description

Under the assumption that mass is neither created nor destroyed during motion, we require that the total mass of any material volume be the same at any instant during the motion. To express this in analytical terms, we consider a material body $\mathcal{B}$ that occupies configuration κ_0 with density ρ_0 and volume Ω_0 at time $t = 0$. The same material body occupies the configuration κ with volume Ω at time $t > 0$, and it has a density ρ. As per the principle of conservation of mass, we have

$$\int_{\Omega_0} \rho_0 \, d\mathbf{X} = \int_{\Omega} \rho \, d\mathbf{x}. \qquad (5.2.34)$$

Using the relation between $d\mathbf{X}$ and $d\mathbf{x}$, $d\mathbf{x} = J\, d\mathbf{X}$, where J is the determinant of the deformation gradient tensor $\mathbf{F}$, we arrive at

$$\int_{\Omega_0} (\rho_0 - J\rho) \, d\mathbf{X} = 0. \qquad (5.2.35)$$

This is the *global form* of the continuity equation. Since the material volume Ω_0 we selected is arbitrarily small, we can shrink the volume to a point and obtain the *local form* of the continuity equation

$$\rho_0 = J\rho. \qquad (5.2.36)$$

Example 5.2.4 illustrates the use of the material time derivative in computing velocities and use of the continuity equation to compute the density in the current configuration.

Example 5.2.4

Consider the motion of a body $\mathcal{B}$ described by the mapping

$$x_1 = \frac{X_1}{1 + tX_1}, \quad x_2 = X_2, \quad x_3 = X_3.$$

Determine the material density as a function of position $\mathbf{x}$ and time t.

Solution: The inverse mapping is given by

$$X_1 = \frac{x_1}{1 - tx_1}, \quad X_2 = x_2, \quad X_3 = x_3. \tag{1}$$

We then compute the velocity components

$$\mathbf{v} = \frac{D\mathbf{x}}{Dt} = \left(\frac{\partial \mathbf{x}}{\partial t}\right)_{\mathbf{X}=\text{fixed}} ; \quad v_i = \frac{Dx_i}{Dt} = \left(\frac{\partial x_i}{\partial t}\right)_{\mathbf{X}=\text{fixed}}. \tag{2}$$

Therefore, we have

$$v_1 = -\frac{X_1^2}{(1 + tX_1)^2} = -x_1^2, \quad v_2 = 0, \quad v_3 = 0. \tag{3}$$

Next, we compute $D\rho/Dt$ from the continuity equation (5.2.27)

$$\frac{D\rho}{Dt} = -\rho\,\boldsymbol{\nabla}\cdot\mathbf{v} = -\rho\left(\frac{\partial v_1}{\partial x_1} + \frac{\partial v_2}{\partial x_2} + \frac{\partial v_3}{\partial x_3}\right) = 2\rho\,x_1, \tag{4}$$

and in the material coordinates

$$\frac{D\rho}{Dt} = 2\rho\,\frac{X_1}{1 + tX_1}. \tag{5}$$

Integrating the above equation (for fixed X_1), we obtain

$$\int \frac{1}{\rho}\,D\rho = 2\int \frac{X_1}{1 + tX_1}\,Dt \;\Rightarrow\; \ln\rho = 2\ln(1 + tX_1) + \ln c,$$

where c is the constant of integration. If $\rho = \rho_0$ at time $t = 0$, we have $\ln c = \ln\rho_0$. Thus, the material density in the current configuration is

$$\rho = \rho_0\left(1 + tX_1\right)^2 = \frac{\rho_0}{(1 - tx_1)^2}. \tag{6}$$

It can be verified that the material time derivative of ρ gives the same result as in Eq. (4),

$$\frac{D\rho}{Dt} = \frac{\partial\rho}{\partial t} + v_1\frac{\partial\rho}{\partial x_1}$$

$$= \frac{2\rho_0\,x_1}{(1 - tx_1)^3} + (-x_1^2)\frac{2\rho_0 t}{(1 - tx_1)^3} = \frac{2\rho_0 x_1}{(1 - tx_1)^2} = 2\rho x_1.$$

The mass density in the current configuration can also be computed using the continuity equation in the material description, $\rho_0 = \rho J$. Noting that

$$dx_1 = \frac{1}{(1 + tX_1)}\,dX_1 - \frac{tX_1}{(1 + tX_1)^2}\,dX_1 = \frac{1}{(1 + tX_1)^2}\,dX_1, \quad J = \frac{dx_1}{dX_1} = \frac{1}{(1 + tX_1)^2},$$

we obtain

$$\rho = \frac{1}{J}\rho_0 = \rho_0(1 + tX_1)^2.$$

5.2.6 Reynolds Transport Theorem

The material derivative operator D/Dt corresponds to changes with respect to a fixed mass, that is, $\rho\,d\mathbf{x}$ is constant with respect to this operator. Therefore, from Eq. (5.2.20) by substituting for $\phi = \rho\,F(\mathbf{x}, t)$, where F is an arbitrary function, we obtain the result

$$\frac{D}{Dt}\int_{\Omega}\rho\,F(\mathbf{x}, t)\,d\mathbf{x} = \frac{\partial}{\partial t}\int_{\Omega}\rho\,F\,d\mathbf{x} + \oint_{\Gamma}\rho\,F\,\hat{\mathbf{n}}\cdot\mathbf{v}\,ds, \tag{5.2.38}$$

or

$$\frac{D}{Dt}\int_{\Omega}\rho\,F(\mathbf{x}, t)\,d\mathbf{x} = \int_{\Omega}\left[\rho\frac{\partial F}{\partial t} + F\frac{\partial\rho}{\partial t} + \boldsymbol{\nabla}\cdot(\rho\,F\mathbf{v})\right]d\mathbf{x}$$

$$= \int_{\Omega}\left[\rho\left(\frac{\partial F}{\partial t} + \mathbf{v}\cdot\boldsymbol{\nabla}F\right) + F\left(\frac{\partial\rho}{\partial t} + \boldsymbol{\nabla}\cdot(\rho\,\mathbf{v})\right)\right]d\mathbf{x}. \tag{5.2.39}$$

Now using the continuity equation (5.2.24) and the definition of the material time derivative, we arrive at the result

$$\frac{D}{Dt}\int_{\Omega}\rho\,F\,d\mathbf{x} = \int_{\Omega}\rho\,\frac{DF}{Dt}\,d\mathbf{x}. \tag{5.2.40}$$

Equation (5.2.40) is known as the *Reynolds transport theorem*. Equation (5.2.40) also holds when F is a vector- on tensor-valued function.

5.3 Balance of Linear and Angular Momentum

5.3.1 Principle of Balance of Linear Momentum

The principle of balance of linear momentum, also known as Newton's second law of motion, applied to a set of particles (or rigid body) can be stated as follows: *The time rate of change of (linear) momentum of a collection of particles equals the net force exerted on the collection.* Written in vector form, the principle implies

$$\frac{d}{dt}(m\mathbf{v}) = \mathbf{F}, \tag{5.3.1}$$

where m is the total mass, $\mathbf{v}$ is the velocity, and $\mathbf{F}$ is the resultant force on the collection of particles. For constant mass, Eq. (5.3.1) becomes

$$\mathbf{F} = m\frac{d\mathbf{v}}{dt} = m\mathbf{a}, \tag{5.3.2}$$

which is the familiar form of Newton's second law.

Newton's second law for a control volume Ω can be expressed as

$$\mathbf{F} = \frac{\partial}{\partial t}\int_{\Omega}\mathbf{v}(\rho\,d\mathbf{x}) + \oint_{\Gamma}\mathbf{v}(\rho\,\mathbf{v}\cdot ds), \tag{5.3.3}$$

where $\mathbf{F}$ is the resultant force and ds denotes the vector representing a surface area element of the outflow. Several simple examples that illustrate the use of Eq. (5.3.3) are presented next.

Example 5.3.1

Suppose that a jet of fluid with cross-sectional area A and mass density ρ issues from a nozzle with a velocity v and impinges against a smooth inclined flat plate, as shown in Fig. 5.3.1. Assuming that there is no frictional resistance between the jet and the plate, determine the distribution of the flow and the force required to keep the plate in position.

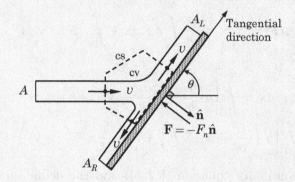

Fig. 5.3.1: Jet of fluid impinging on an inclined plate.

Solution: Since there is no change in pressure or elevation before and after impact, the velocity of the fluid remains the same before and after impact, but the flow to the left and right would be different. Let the amount of flow to the left be Q_L and to the right be Q_R. Then the total flow $Q = vA$ of the jet is equal to the sum (by the continuity equation):

$$Q = Q_L + Q_R. \tag{1}$$

Next, we use the principle of balance of linear momentum to relate Q_L and Q_R. Applying Eq. (5.3.3) to the positive tangential direction to the plate, and noting that the resultant force is zero and the first term on the right-hand side of Eq. (5.3.3) is zero by virtue of the steady-state condition, we obtain (note that the control surface has three segments that have nonzero flow across the boundary)

$$0 = \oint_{cs} v_t\, \rho \mathbf{v} \cdot d\mathbf{s} = v\cos\theta(-\rho vA) + v(\rho vA_L) + (-v)(\rho vA_R), \tag{2}$$

where the minus sign in the first term on the right side of the equality is due to the fact that the mass flow is into the control volume, and the minus sign in the third term is due to the fact that the velocity is in the opposite direction to the tangent direction (but the mass flow is out of the control volume, that is, positive). With $A_L v = Q_L$, $A_R v = Q_R$, and $Av = Q$, we obtain

$$Q_L - Q_R = Q\cos\theta.$$

Solving the two equations for Q_L and Q_R, we obtain

$$Q_L = \tfrac{1}{2}\left(1+\cos\theta\right)Q, \quad Q_R = \tfrac{1}{2}\left(1-\cos\theta\right)Q. \tag{3}$$

Thus, the total flow Q is divided into the left flow of Q_L and right flow of Q_R, as given above.

The force exerted on the plate is normal to the plate. By applying the balance of linear momentum in the normal direction (hence, the flow along the plate has zero component normal to the plate), we obtain

$$\mathbf{F} \cdot \hat{\mathbf{n}} = \oint_{cs} (\mathbf{v} \cdot \hat{\mathbf{n}})(\rho \mathbf{v} \cdot d\mathbf{s}) = (v\sin\theta)(-\rho vA),$$

or $(v_n = v\sin\theta)$

$$-F_n = \oint_{cs} v_n(\rho\mathbf{v} \cdot d\mathbf{s}) = (v\sin\theta)(-\rho vA) \;\rightarrow\; F_n = \rho Q v\sin\theta = \rho A v^2 \sin\theta. \tag{4}$$

Example 5.3.2

When a free jet of fluid impinges on a smooth (frictionless) curved vane with a velocity v, the jet is deflected in a tangential direction as shown in Fig. 5.3.2, changing the momentum and exerting a force on the vane. Assuming that the velocity is uniform throughout the jet and there is no change in the pressure, determine the force exerted on a fixed vane.

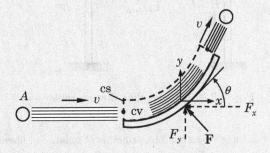

Fig. 5.3.2: Jet of fluid deflected by a curved vane.

Solution: For the steady-state condition, applying Eq. (5.3.3), we obtain

$$\mathbf{F} = \oint_{cs} \mathbf{v}(\rho \mathbf{v} \cdot d\mathbf{s}),$$
$$-F_x \,\hat{\mathbf{e}}_x + F_y \,\hat{\mathbf{e}}_y = v\hat{\mathbf{e}}_x \,(-\rho v A) + v \,(\cos\theta \,\hat{\mathbf{e}}_x + \sin\theta \,\hat{\mathbf{e}}_y)\,(\rho v A), \tag{1}$$

or

$$F_x = \rho v^2 A \,(1 - \cos\theta), \quad F_y = \rho v^2 A \sin\theta. \tag{2}$$

When a jet of water ($\rho = 10^3$ kg/m^3) discharging 80 L/s at a velocity of 60 m/s is deflected through an angle of $\theta = 60°$, we obtain ($Q = vA$)

$$F_x = 10^3 \times 0.08 \times 60 \,(1 - \cos 60°) = 2.4 \text{ kN},$$
$$F_y = 10^3 \times 0.08 \times 60 \sin 60° = 4.157 \text{ kN}.$$

When the vane moves with a horizontal velocity of $v_0 < v$, Eq. (5.3.3) becomes

$$\mathbf{F} = \oint_{cs} (\mathbf{v} - \mathbf{v_0})[\rho(\mathbf{v} - \mathbf{v_0}) \cdot d\mathbf{s}],$$
$$-F_x \,\hat{\mathbf{e}}_x + F_y \,\hat{\mathbf{e}}_y = (v - v_0) \,[-\rho(v - v_0)A\,\hat{\mathbf{e}}_x] + (v - v_0) \,(\cos\theta \,\hat{\mathbf{e}}_x + \sin\theta \,\hat{\mathbf{e}}_y)\,\rho(v - v_0)A,$$

from which we obtain

$$F_x = \rho(v - v_0)^2 A \,(1 - \cos\theta), \quad F_y = \rho(v - v_0)^2 A \sin\theta. \tag{3}$$

Example 5.3.3

A chain of total length L and mass ρ per unit length slides down from the edge of a smooth table with an initial overhang x_0 to initiate motion, as shown in Fig. 5.3.3. Assuming that the chain is rigid, find the equation of motion governing the chain and the tension in the chain.

Solution: Let x be the amount of chain sliding down the table at any instant t. By considering the entire chain as the control volume, the linear momentum of the chain is

$$\rho(L - x) \cdot \dot{x} \,\hat{\mathbf{e}}_x - \rho x \cdot \dot{x} \,\hat{\mathbf{e}}_y.$$

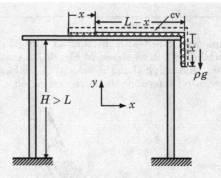

Fig. 5.3.3: Chain sliding down a table.

The resultant force in the chain is $-\rho g x \, \hat{\mathbf{e}}_y$. The principle of balance of linear momentum gives

$$-\rho\, g x \, \hat{\mathbf{e}}_y = \frac{d}{dt}\left[\rho(L-x)\,\dot{x}\,\hat{\mathbf{e}}_x - \rho\, x\, \dot{x}\,\hat{\mathbf{e}}_y\right], \qquad (1)$$

or

$$0 = (L-x)\ddot{x} - \dot{x}^2, \quad -gx = -x\ddot{x} - \dot{x}^2.$$

Eliminating $\dot{x}^2$ from the two equations, we arrive at the equation of motion:

$$\ddot{x} - \frac{g}{L}x = 0. \qquad (2)$$

The solution of this second-order differential equation is

$$x(t) = A\cosh\lambda t + B\sinh\lambda t, \quad \text{where} \quad \lambda = \sqrt{\frac{g}{L}}.$$

The constants of integration A and B are determined from the initial conditions

$$x(0) = x_0, \quad \dot{x}(0) = 0,$$

where x_0 denotes the initial overhang of the chain. We obtain

$$A = x_0, \quad B = 0,$$

and the solution becomes

$$x(t) = x_0 \cosh\lambda t, \quad \lambda = \sqrt{\frac{g}{L}}. \qquad (3)$$

The tension T in the chain can be computed by using the principle of balance of linear momentum applied to the control volume of the chain on the table as well as hanging

$$T = \frac{\partial}{\partial t}\int_\Omega \mathbf{v}(\rho\,d\mathbf{x}) + \oint_\Gamma \mathbf{v}(\rho\mathbf{v}\cdot d\mathbf{s})$$

$$= \frac{d}{dt}\left[\rho(L-x)\dot{x}\right] + \rho\dot{x}\dot{x}$$

$$= \rho(L-x)\ddot{x} = \frac{\rho g}{L}(L-x)x, \qquad (4)$$

where Eq. (2) is used in arriving at the last step.

Example 5.3.4

Consider a chain of length L and mass density ρ per unit length that is piled on a stationary table, as shown in Fig. 5.3.4. Determine the force F required to lift the chain with a constant velocity v.

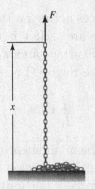

Fig. 5.3.4: Lifting of a chain piled on a table.

Solution: Let x be the height of the chain lifted off the table. Taking the control volume to be that enclosing the lifted chain and using Eq. (5.3.3) at a point, we obtain

$$F - \rho\,gx = \frac{\partial}{\partial t}\int_\Omega \mathbf{v}(\rho\,dx) + \oint_\Gamma \mathbf{v}(\rho\mathbf{v}\cdot d\mathbf{s})$$

$$= \frac{\partial}{\partial t}(\rho v) + \rho vv = 0 + \rho v^2, \tag{1}$$

or

$$F = \rho\left(gx + v^2\right).$$

The same result can be obtained using Newton's second law of motion:

$$F - \rho\,gx = \frac{d}{dt}(mv) = m\dot{v} + \dot{m}v = 0 + \dot{m}v, \tag{2}$$

where the rate of increase of mass $m = \rho x$ is $\dot{m} = \rho\dot{x} = \rho v$.

5.3.1.1 Equations of motion in the spatial description

To derive the equation of motion applied to an arbitrarily fixed region in space through which material flows (i.e., control volume), we must identify the forces acting on it. Forces acting on a volume element can be classified as *internal* and *external*. The internal forces resist the tendency of one part of the region/body to be separated from another part. The internal force per unit area is termed stress, as defined in Eq. (4.2.1). The external forces are those transmitted by the body. The external forces can be further classified as *body (or volume) forces* and *surface forces*.

Body forces act on the distribution of mass inside the body. Examples of body forces are provided by the gravitational and electromagnetic forces. Body forces are usually measured per unit mass or unit volume of the body. Let $\mathbf{f}$ denote the body force per unit mass. Consider an elemental volume $d\mathbf{x}$ inside

Ω. The body force of the elemental volume is equal to $\rho \, d\mathbf{x} \, \mathbf{f}$. Hence, the total body force of the control volume is

$$\int_{\Omega} \rho \mathbf{f} \, d\mathbf{x}. \tag{5.3.4}$$

Surface forces are contact forces acting on the boundary surface of the body, and they are measured per unit area. If $\mathbf{t}$ is the surface force per unit area, the surface force on an elemental surface area ds is $\mathbf{t} \, ds$. The total surface force acting on the closed surface of the region Ω is

$$\oint_{\Gamma} \mathbf{t} \, ds. \tag{5.3.5}$$

The principle of balance of linear momentum applied to a given mass of a medium $\mathcal{B}$, instantaneously occupying a region Ω with bounding surface Γ, and acted upon by external surface force $\mathbf{t}$ per unit area and body force $\mathbf{f}$ per unit mass, can be expressed as

$$\frac{D}{Dt} \int_{\Omega} \rho \mathbf{v} \, d\mathbf{x} = \oint_{\Gamma} \mathbf{t} \, ds + \int_{\Omega} \rho \mathbf{f} \, d\mathbf{x}, \tag{5.3.6}$$

where $\rho \mathbf{v} \, d\mathbf{x}$ denotes the linear momentum associated with elemental volume $d\mathbf{x}$, $\mathbf{v}$ being the velocity vector.

Since the stress vector $\mathbf{t}$ on the surface is related to the (internal) stress tensor $\boldsymbol{\sigma}$ by Cauchy's formula $\mathbf{t} = \boldsymbol{\sigma} \cdot \hat{\mathbf{n}}$, [see Eq. (4.2.10)], where $\hat{\mathbf{n}}$ denotes the unit normal to the surface, we can express the surface force as

$$\oint_{\Gamma} \boldsymbol{\sigma} \cdot \hat{\mathbf{n}} \, ds = \oint_{\Gamma} \hat{\mathbf{n}} \cdot \boldsymbol{\sigma}^{\mathrm{T}} \, ds = \int_{\Omega} \boldsymbol{\nabla} \cdot \boldsymbol{\sigma}^{\mathrm{T}} \, d\mathbf{x}, \tag{5.3.7}$$

where the divergence theorem (5.2.17) is used to convert the surface integral into volume integral. Thus Eq. (5.3.6) takes the form

$$\frac{D}{Dt} \int_{\Omega} \rho \mathbf{v} \, d\mathbf{x} = \int_{\Omega} \left(\boldsymbol{\nabla} \cdot \boldsymbol{\sigma}^{\mathrm{T}} + \rho \mathbf{f} \right) d\mathbf{x}. \tag{5.3.8}$$

Using the Reynolds transport theorem, Eq. (5.2.40), we arrive at

$$0 = \int_{\Omega} \left[\boldsymbol{\nabla} \cdot \boldsymbol{\sigma}^{\mathrm{T}} + \rho \mathbf{f} - \rho \frac{D\mathbf{v}}{Dt} \right] d\mathbf{x}, \tag{5.3.9}$$

which is the global form of the equation of motion. The local form is given by

$$\boldsymbol{\nabla} \cdot \boldsymbol{\sigma}^{\mathrm{T}} + \rho \mathbf{f} = \rho \frac{D\mathbf{v}}{Dt}; \quad \sigma_{ij,j} + \rho f_i = \rho \frac{Dv_i}{Dt}, \tag{5.3.10}$$

or

$$\boldsymbol{\nabla} \cdot \boldsymbol{\sigma}^{\mathrm{T}} + \rho \mathbf{f} = \rho \left(\frac{\partial \mathbf{v}}{\partial t} + \mathbf{v} \cdot \boldsymbol{\nabla} \mathbf{v} \right). \tag{5.3.11}$$

Equation (5.3.11) is known as *Cauchy's equation of motion*. In Cartesian rectangular system, we have

$$\frac{\partial \sigma_{ij}}{\partial x_j} + \rho \, f_i = \rho \left(\frac{\partial v_i}{\partial t} + v_j \frac{\partial v_i}{\partial x_j} \right). \tag{5.3.12}$$

In steady–state conditions, Eq. (5.3.11) reduces to

$$\boldsymbol{\nabla} \cdot \boldsymbol{\sigma}^{\mathrm{T}} + \rho \mathbf{f} = \rho \mathbf{v} \cdot \boldsymbol{\nabla} \mathbf{v}; \qquad \frac{\partial \sigma_{ij}}{\partial x_j} + \rho \, f_i = \rho \, v_j \frac{\partial v_i}{\partial x_j}. \tag{5.3.13}$$

When the state of stress in the medium is of the form $\boldsymbol{\sigma} = -p\mathbf{I}$ (i.e., hydrostatic state of stress), the equation of motion (5.3.10) reduces to

$$-\boldsymbol{\nabla} p + \rho \mathbf{f} = \rho \frac{D\mathbf{v}}{Dt}. \tag{5.3.14}$$

5.3.1.2 Equations of motion in the material description

To derive the equation of motion applied to an arbitrarily fixed material of density ρ_0, occupying region Ω_0 in the reference configuration, we express Eqs. (5.3.4), (5.3.7), and (5.3.8) in terms of quantities referred to the reference configuration. We have ($d\mathbf{x} = J \, d\mathbf{X}$, and $\rho_0 = \rho \, J$)

$$\int_{\Omega} \rho \, \mathbf{f}(\mathbf{x}) \, d\mathbf{x} = \int_{\Omega_0} \rho_0 \, \mathbf{f}(\mathbf{X}) \, d\mathbf{X},$$

$$\oint_{\Gamma} \boldsymbol{\sigma} \cdot \hat{\mathbf{n}} \, ds = \oint_{\Gamma_0} \mathbf{P} \cdot \hat{\mathbf{N}} \, dS = \int_{\Omega_0} \boldsymbol{\nabla}_0 \cdot \mathbf{P}^{\mathrm{T}} \, d\mathbf{X}, \tag{5.3.15}$$

$$\frac{D}{Dt} \int_{\Omega} \rho \, \mathbf{v} \, d\mathbf{x} = \frac{\partial}{\partial t} \int_{\Omega_0} \rho_0 \frac{\partial \mathbf{u}}{\partial t} \, d\mathbf{X} = \int_{\Omega_0} \rho_0 \frac{\partial^2 \mathbf{u}}{\partial t^2} \, d\mathbf{X}.$$

where $\mathbf{P}$ is the first Piola–Kirchhoff stress tensor. In arriving at the above results we have made use of Eqs. (4.4.7) and (3.3.20):

$$\boldsymbol{\sigma} \cdot d\mathbf{a} = \mathbf{P} \cdot d\mathbf{A}, \quad dv = J \, dV \quad \text{(or } d\mathbf{x} = J \, d\mathbf{X}). \tag{5.3.16}$$

Then the principle of balance of linear momentum yields

$$\boldsymbol{\nabla}_0 \cdot \mathbf{P}^{\mathrm{T}} + \rho_0 \, \mathbf{f} = \rho_0 \frac{\partial^2 \mathbf{u}}{\partial t^2}. \tag{5.3.17}$$

Then using Eq. (4.4.11), namely, $\mathbf{S} = \mathbf{F}^{-1} \cdot \mathbf{P}$ or $\mathbf{P} = \mathbf{F} \cdot \mathbf{S}$, we can express the equation of motion in terms of the second Piola–Kirchhoff stress tensor $\mathbf{S}$

$$\boldsymbol{\nabla}_0 \cdot \left(\mathbf{S}^{\mathrm{T}} \cdot \mathbf{F}^{\mathrm{T}} \right) + \rho_0 \, \mathbf{f} = \rho_0 \frac{\partial^2 \mathbf{u}}{\partial t^2}. \tag{5.3.18}$$

Expressing the deformation tensor $\mathbf{F}$ in terms of the displacement vector $\mathbf{u}$ [see Eq. (3.3.8)], we obtain

$$\boldsymbol{\nabla}_0 \cdot \left[\mathbf{S}^{\mathrm{T}} \cdot (\mathbf{I} + \boldsymbol{\nabla}_0 \mathbf{u}) \right] + \rho_0 \, \mathbf{f} = \rho_0 \frac{\partial^2 \mathbf{u}}{\partial t^2}. \tag{5.3.19}$$

In rectangular Cartesian component form, we have

$$\frac{\partial}{\partial X_J}\left[S_{KJ}\left(\delta_{KI}+\frac{\partial u_I}{\partial X_K}\right)\right]+\rho_0\,\mathbf{f}_I=\rho_0\frac{\partial^2 u_I}{\partial t^2}, \quad I=1,2,3, \tag{5.3.20}$$

where $v_I=(\partial u_I/\partial t)$. Clearly, the equations of motion expressed in terms of the second Piola–Kirchhoff stress tensor are nonlinear, because of the term $S_{KJ}(\partial u_I/\partial X_K)$, and this nonlinearity is in addition to the nonlinearity in the strain-displacement relations (see Chapter 3) and constitutive relations (to be discussed in Chapter 6).

For kinematically infinitesimal deformation, no distinction is made between $\mathbf{X}$ and $\mathbf{x}$ and between the second Piola–Kirchhoff stress tensor $\mathbf{S}$ and the Cauchy stress tensor $\boldsymbol{\sigma}$, that is, $\mathbf{X}\approx\mathbf{x}$ and $\mathbf{S}\approx\boldsymbol{\sigma}$. In this case, Eq. (5.3.20) reduces to

$$\nabla\cdot\boldsymbol{\sigma}^{\mathrm{T}}+\rho_0\,\mathbf{f}=\rho_0\frac{\partial^2\mathbf{u}}{\partial t^2}; \quad \frac{\partial\sigma_{ij}}{\partial x_j}+\rho_0\,f_i=\rho_0\frac{\partial^2 u_i}{\partial t^2}. \tag{5.3.21}$$

For bodies in static equilibrium, Eq. (5.3.21) reduces to [see Eq. (4.5.6)]

$$\nabla\cdot\boldsymbol{\sigma}^{\mathrm{T}}+\rho_0\,\mathbf{f}=\mathbf{0}; \quad \frac{\partial\sigma_{ij}}{\partial x_j}+\rho_0\,f_i=0. \tag{5.3.22}$$

Applications of the stress equilibrium equation, Eq. (5.3.22), for kinematically infinitesimal deformation were presented in Examples 4.5.1 and 4.5.2. Here we reconsider an example of application of Eq. (5.3.20).

Example 5.3.5 ————————————————————

Given the following state of stress ($S_{IJ}=S_{JI}$),

$$S_{11}=-2X_1^2, \quad S_{12}=-7+4X_1X_2+X_3, \quad S_{13}=1+X_1-3X_2,$$
$$S_{22}=3X_1^2-2X_2^2+5X_3, \quad S_{23}=0, \quad S_{33}=-5+X_1+3X_2+3X_3,$$

and displacement field,

$$u_1=AX_2, \quad u_2=BX_1, \quad u_3=0,$$

where A and B are arbitrary constants, determine the body force components for which the stress field describes a state of equilibrium.

Solution: Using Eq. (5.3.20), the body force components are

$$\rho_0\,f_I=-\frac{\partial S_{IJ}}{\partial X_J}-\frac{\partial}{\partial X_J}\left(S_{1J}\frac{\partial u_I}{\partial X_1}+S_{2J}\frac{\partial u_I}{\partial X_2}+S_{3J}\frac{\partial u_I}{\partial X_3}\right), \quad I=1,2,3.$$

We have

$$\rho_0\,f_1=-\left(\frac{\partial S_{11}}{\partial X_1}+\frac{\partial S_{12}}{\partial X_2}+\frac{\partial S_{13}}{\partial X_3}\right)$$
$$-\frac{\partial}{\partial X_1}\left(S_{11}\frac{\partial u_1}{\partial X_1}+S_{21}\frac{\partial u_1}{\partial X_2}+S_{31}\frac{\partial u_1}{\partial X_3}\right)$$
$$-\frac{\partial}{\partial X_2}\left(S_{12}\frac{\partial u_1}{\partial X_1}+S_{22}\frac{\partial u_1}{\partial X_2}+S_{32}\frac{\partial u_1}{\partial X_3}\right)$$
$$-\frac{\partial}{\partial X_3}\left(S_{13}\frac{\partial u_1}{\partial X_1}+S_{23}\frac{\partial u_1}{\partial X_2}+S_{33}\frac{\partial u_1}{\partial X_3}\right)$$
$$=-[(-4X_1)+(4X_1)+0]-A[(4X_2)+(-4X_2)+0]=0,$$

$$\rho_0\, f_2 = -\left(\frac{\partial S_{21}}{\partial X_1} + \frac{\partial S_{22}}{\partial X_2} + \frac{\partial S_{23}}{\partial X_3}\right)$$

$$- \frac{\partial}{\partial X_1}\left(S_{11}\frac{\partial u_2}{\partial X_1} + S_{21}\frac{\partial u_2}{\partial X_2} + S_{31}\frac{\partial u_2}{\partial X_3}\right)$$

$$- \frac{\partial}{\partial X_2}\left(S_{12}\frac{\partial u_2}{\partial X_1} + S_{22}\frac{\partial u_2}{\partial X_2} + S_{32}\frac{\partial u_2}{\partial X_3}\right)$$

$$- \frac{\partial}{\partial X_3}\left(S_{13}\frac{\partial u_2}{\partial X_1} + S_{23}\frac{\partial u_2}{\partial X_2} + S_{33}\frac{\partial u_2}{\partial X_3}\right)$$

$$= -[(4X_2) + (-4X_2) + 0] - B[(-4X_1) + (4X_1) + 0] = 0,$$

$$\rho_0\, f_3 = -\left(\frac{\partial S_{31}}{\partial X_1} + \frac{\partial S_{32}}{\partial X_2} + \frac{\partial S_{33}}{\partial X_3}\right)$$

$$- \frac{\partial}{\partial X_1}\left(S_{11}\frac{\partial u_3}{\partial X_1} + S_{21}\frac{\partial u_3}{\partial X_2} + S_{31}\frac{\partial u_3}{\partial X_3}\right)$$

$$- \frac{\partial}{\partial X_2}\left(S_{12}\frac{\partial u_3}{\partial X_1} + S_{22}\frac{\partial u_3}{\partial X_2} + S_{32}\frac{\partial u_3}{\partial X_3}\right)$$

$$- \frac{\partial}{\partial X_3}\left(S_{13}\frac{\partial u_3}{\partial X_1} + S_{23}\frac{\partial u_3}{\partial X_2} + S_{33}\frac{\partial u_3}{\partial X_3}\right)$$

$$= -[1 + 0 + 3] + 0 = -4.$$

Thus, the body is in equilibrium for the body force components $\rho_0\, f_1 = 0$, $\rho_0\, f_2 = 0$, and $\rho_0\, f_3 = -4$.

5.3.2 Spatial Equations of Motion in Cylindrical and Spherical Coordinates

Here we express the equations of motion in the spatial description, Eq. (5.3.11), in terms of the components in the cylindrical and spherical coordinate systems (see Figure 5.3.5). The equations are also valid for kinematically infinitesimal deformations in the material description, with the density ρ replaced with ρ_0 (also, contributions from the term $\mathbf{v}\cdot\nabla\mathbf{v}$ should be omitted).

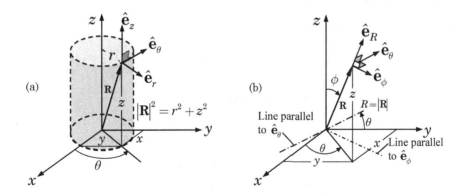

Fig. 5.3.5: (a) Cylindrical and (b) spherical coordinate systems.

5.3.2.1 Cylindrical coordinates

To express the equations of motion (5.3.11) in terms of the components in the cylindrical coordinate system, the operator ∇, velocity vector $\mathbf{v}$, body force vector $\mathbf{f}$, and stress tensor $\boldsymbol{\sigma}$ are written in the cylindrical coordinates (r, θ, z) as

$$\nabla = \hat{\mathbf{e}}_r \frac{\partial}{\partial r} + \frac{1}{r} \hat{\mathbf{e}}_\theta \frac{\partial}{\partial \theta} + \hat{\mathbf{e}}_z \frac{\partial}{\partial z},$$

$$\mathbf{v} = \hat{\mathbf{e}}_r v_r + \hat{\mathbf{e}}_\theta v_\theta + \hat{\mathbf{e}}_z v_z,$$

$$\mathbf{f} = \hat{\mathbf{e}}_r f_r + \hat{\mathbf{e}}_\theta f_\theta + \hat{\mathbf{e}}_z f_z, \qquad (5.3.23)$$

$$\boldsymbol{\sigma} = \sigma_{rr}\, \hat{\mathbf{e}}_r \hat{\mathbf{e}}_r + \sigma_{r\theta}\, \hat{\mathbf{e}}_r \hat{\mathbf{e}}_\theta + \sigma_{rz}\, \hat{\mathbf{e}}_r \hat{\mathbf{e}}_z$$
$$+ \sigma_{\theta r}\, \hat{\mathbf{e}}_\theta \hat{\mathbf{e}}_r + \sigma_{\theta\theta}\, \hat{\mathbf{e}}_\theta \hat{\mathbf{e}}_\theta + \sigma_{\theta z}\, \hat{\mathbf{e}}_\theta \hat{\mathbf{e}}_z$$
$$+ \sigma_{zr}\, \hat{\mathbf{e}}_z \hat{\mathbf{e}}_r + \sigma_{z\theta}\, \hat{\mathbf{e}}_z \hat{\mathbf{e}}_\theta + \sigma_{zz}\, \hat{\mathbf{e}}_z \hat{\mathbf{e}}_z.$$

Substituting these expressions into Eq. (5.3.11), we arrive at the following equations of motion in the cylindrical coordinate system (from the solutions of Problems **2.49** and **4.28**; see also Table 2.4.2 for the gradient of a vector):

$$\frac{\partial \sigma_{rr}}{\partial r} + \frac{1}{r}\frac{\partial \sigma_{r\theta}}{\partial \theta} + \frac{\partial \sigma_{rz}}{\partial z} + \frac{1}{r}(\sigma_{rr} - \sigma_{\theta\theta}) + \rho f_r$$
$$= \rho \left(\frac{\partial v_r}{\partial t} + v_r \frac{\partial v_r}{\partial r} + \frac{v_\theta}{r}\frac{\partial v_r}{\partial \theta} + v_z \frac{\partial v_r}{\partial z} - \frac{v_\theta^2}{r} \right),$$

$$\frac{\partial \sigma_{\theta r}}{\partial r} + \frac{1}{r}\frac{\partial \sigma_{\theta\theta}}{\partial \theta} + \frac{\partial \sigma_{\theta z}}{\partial z} + \frac{\sigma_{\theta r} + \sigma_{r\theta}}{r} + \rho f_\theta$$
$$= \rho \left(\frac{\partial v_\theta}{\partial t} + v_r \frac{\partial v_\theta}{\partial r} + \frac{v_\theta v_r}{r} + \frac{v_\theta}{r}\frac{\partial v_\theta}{\partial \theta} + v_z \frac{\partial v_\theta}{\partial z} \right), \qquad (5.3.24)$$

$$\frac{\partial \sigma_{zr}}{\partial r} + \frac{1}{r}\frac{\partial \sigma_{z\theta}}{\partial \theta} + \frac{\partial \sigma_{zz}}{\partial z} + \frac{\sigma_{zr}}{r} + \rho f_z$$
$$= \rho \left(\frac{\partial v_z}{\partial t} + v_r \frac{\partial v_z}{\partial r} + \frac{v_\theta}{r}\frac{\partial v_z}{\partial \theta} + v_z \frac{\partial v_z}{\partial z} \right). \quad .$$

5.3.2.2 Spherical coordinates

In the spherical coordinate system (R, ϕ, θ), we write

$$\nabla = \hat{\mathbf{e}}_R \frac{\partial}{\partial R} + \frac{1}{R} \hat{\mathbf{e}}_\phi \frac{\partial}{\partial \phi} + \frac{1}{R \sin \phi} \hat{\mathbf{e}}_\theta \frac{\partial}{\partial \theta},$$

$$\mathbf{v} = \hat{\mathbf{e}}_R v_R + \hat{\mathbf{e}}_\phi v_\phi + \hat{\mathbf{e}}_\theta v_\theta,$$

$$\mathbf{f} = \hat{\mathbf{e}}_R f_R + \hat{\mathbf{e}}_\phi f_\phi + \hat{\mathbf{e}}_\theta f_\theta, \qquad (5.3.25)$$

$$\boldsymbol{\sigma} = \sigma_{RR}\, \hat{\mathbf{e}}_R \hat{\mathbf{e}}_R + \sigma_{R\phi}\, \hat{\mathbf{e}}_R \hat{\mathbf{e}}_\phi + \sigma_{R\theta}\, \hat{\mathbf{e}}_R \hat{\mathbf{e}}_\theta$$
$$+ \sigma_{\phi R}\, \hat{\mathbf{e}}_\phi \hat{\mathbf{e}}_R + \sigma_{\phi\phi}\, \hat{\mathbf{e}}_\phi \hat{\mathbf{e}}_\phi + \sigma_{\phi\theta}\, \hat{\mathbf{e}}_\phi \hat{\mathbf{e}}_\theta$$
$$+ \sigma_{\theta R}\, \hat{\mathbf{e}}_\theta \hat{\mathbf{e}}_\theta + \sigma_{\theta\phi}\, \hat{\mathbf{e}}_\theta \hat{\mathbf{e}}_\phi + \sigma_{\theta\theta} \hat{\mathbf{e}}_\theta \hat{\mathbf{e}}_\theta.$$

Substituting these expressions into Eq. (5.3.11), we arrive at the following equations of motion in the spherical coordinate system (from the solution to Problem **2.51** and Table 2.4.2):

$$\frac{\partial \sigma_{RR}}{\partial R} + \frac{1}{R}\frac{\partial \sigma_{R\phi}}{\partial \phi} + \frac{1}{R\sin\phi}\frac{\partial \sigma_{R\theta}}{\partial \theta} + \frac{1}{R}\left(2\sigma_{RR} - \sigma_{\phi\phi} - \sigma_{\theta\theta} + \sigma_{R\phi}\cot\phi\right) + \rho f_R$$

$$= \rho\left(\frac{\partial v_R}{\partial t} + v_R\frac{\partial v_R}{\partial R} + \frac{v_\phi}{R}\frac{\partial v_R}{\partial \phi} + \frac{v_\theta}{R\sin\phi}\frac{\partial v_R}{\partial \theta} - \frac{v_\phi^2 + v_\theta^2}{R}\right),$$

$$\frac{\partial \sigma_{\phi R}}{\partial R} + \frac{1}{R}\frac{\partial \sigma_{\phi\phi}}{\partial \phi} + \frac{1}{R\sin\phi}\frac{\partial \sigma_{\phi\theta}}{\partial \theta} + \frac{1}{R}\left[(\sigma_{\phi\phi} - \sigma_{\theta\theta})\cot\phi + \sigma_{R\phi} + 2\sigma_{\phi R}\right] + \rho f_\phi$$

$$= \rho\left[\frac{\partial v_\phi}{\partial t} + v_R\frac{\partial v_\phi}{\partial R} + \frac{v_\phi}{R}\left(\frac{\partial v_\phi}{\partial \phi} + v_R\right) + \frac{v_\theta}{R\sin\phi}\left(\frac{\partial v_\phi}{\partial \theta} - v_\theta\cos\phi\right)\right],$$

$$\frac{\partial \sigma_{\theta R}}{\partial R} + \frac{1}{R}\frac{\partial \sigma_{\theta\phi}}{\partial \phi} + \frac{1}{R\sin\phi}\frac{\partial \sigma_{\theta\theta}}{\partial \theta} + \frac{1}{R}\left[(\sigma_{\phi\theta} + \sigma_{\theta\phi})\cot\phi + \sigma_{R\theta}\right] + \rho f_\theta$$

$$= \rho\left[\frac{\partial v_\theta}{\partial t} + v_R\frac{\partial v_\theta}{\partial R} + \frac{v_\phi}{R}\frac{\partial v_\theta}{\partial \phi} + \frac{v_\theta}{R\sin\phi}\left(\frac{\partial v_\theta}{\partial \theta} + v_\phi\cos\phi\right) + \frac{v_\theta v_R}{R}\right].$$

$$(5.3.26)$$

5.3.3 Principle of Balance of Angular Momentum

5.3.3.1 Monopolar case

This book is concerned primarily with *monopolar continuum mechanics*, where the topological features of the arrangement of matter at a micro scale, such as the distributed couples and couple stresses present at the molecular level, are overlooked. The monopolar continuum mechanics describes only macroscopic features of motion, which is sufficient in a vast majority of problems of mechanics.

The principle of balance of angular momentum for the monopolar case can be stated as follows: *The time rate of change of the total moment of momentum for a continuum is equal to the vector sum of the moments of external forces acting on the continuum.* The principle as applied to a control volume Ω with a control surface Γ can be expressed as

$$\text{moment of external forces} = \frac{\partial}{\partial t}\int_{cv}\rho\mathbf{r}\times\mathbf{v}\,dx + \int_{cs}\rho\mathbf{r}\times\mathbf{v}\,(\mathbf{v}\cdot d\mathbf{s}), \quad (5.3.27)$$

where cv and cs denote the control volume and control surface, respectively. An application of the principle is presented in Example 5.3.6.

Example 5.3.6 _____

Consider the top view of a sprinkler as shown in Fig. 5.3.6. The sprinkler discharges water outward in a horizontal plane (which is in the plane of the paper). The sprinkler exits are oriented at an angle of θ from the tangent line to the circle formed by rotating the sprinkler about its vertical centerline. The sprinkler has a constant cross-sectional flow area of A and discharges a flow rate of Q when $\omega = 0$ at time $t = 0$. Hence, the radial velocity is equal to $v_r = Q/2A$. Determine ω (counterclockwise) as a function of time.

Solution: Suppose that the moment of inertia of the empty sprinkler head is I_z and the resisting torque due to friction (from bearings and seals) is T (clockwise). The control volume is taken to be the cylinder of unit height (into the plane of the page) and radius R, formed by the rotating sprinkler head. The inflow, being along the axis, has no moment of momentum. Thus the time rate of change of the moment of momentum of the sprinkler head plus the net efflux of the moment of momentum from the control surface is equal to the torque T:

$$-T\hat{e}_z = \left[2\frac{d}{dt} \int_0^R A\rho\omega r^2 \, dr + I_z \frac{d\omega}{dt} + 2R\left(\rho\frac{Q}{2}\right)(\omega R - v_r \cos\theta)\right]\hat{e}_z,$$

where the first term represents the time rate of change of the moment of momentum [moment arm times mass of a differential length dr times the velocity: $r \times (\rho A \, dr)(\omega r)$], the second term is the time rate of change of the angular momentum, and the last term represents the efflux of the moment of momentum at the control surface (i.e., exit of the sprinkler nozzles). Simplifying the equation, we arrive at

$$\left(I_z + \tfrac{2}{3}\rho A R^3\right)\frac{d\omega}{dt} + \rho Q R^2 \omega = \rho Q R v_r \cos\theta - T.$$

The above equation indicates that for rotation to start $\rho Q R v_r \cos\theta - T > 0$. The final value of ω is obtained when the sprinkler motion reaches the steady state, that is, $d\omega/dt = 0$. Thus, at steady state, we have

$$\omega_f = \frac{v_r}{R}\cos\theta - \frac{T}{\rho Q R^2}.$$

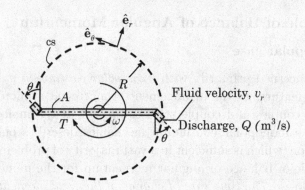

Fig. 5.3.6: A rotating sprinkler system.

The mathematical statement of the principle of balance of angular momentum as applied to a continuum is

$$\oint_\Gamma (\mathbf{x} \times \mathbf{t})ds + \int_\Omega (\mathbf{x} \times \rho\mathbf{f})d\mathbf{x} = \frac{D}{Dt}\int_\Omega (\mathbf{x} \times \rho\mathbf{v}) \, d\mathbf{x}, \tag{5.3.28}$$

where $\mathbf{t}$ denotes the stress vector and $\mathbf{f}$ denotes the body force vector (measured per unit mass). Equation (5.3.28) can be simplified with the help of the index notation in rectangular Cartesian coordinates (but the result will hold in any coordinate system). In index notation (kth component) Eq. (5.3.28) takes the form

$$\oint_\Gamma e_{ijk}\, x_i\, t_j\, ds + \int_\Omega (\rho\, e_{ijk}\, x_i\, f_j)\, d\mathbf{x} = \frac{D}{Dt}\int_\Omega \rho\, e_{ijk}\, x_i\, v_j\, d\mathbf{x}. \tag{5.3.29}$$

We use several steps to simplify the expression. First replace t_j with $t_j = n_p \sigma_{pj}$ (Cauchy's formula). Then transform the surface integral to a volume integral and use the Reynolds transport theorem, Eq. (5.2.40), for the material time derivative of a volume integral to obtain

$$\int_\Omega e_{ijk} \left(x_i\, \sigma_{jp} \right)_{,p} d\mathbf{x} + \int_\Omega \left(\rho e_{ijk}\, x_i\, f_j \right) d\mathbf{x} = \int_\Omega \rho\, e_{ijk}\, \frac{D}{Dt} \left(x_i\, v_j \right) d\mathbf{x},$$

where $(\cdot)_{,p} = \partial(\cdot)/\partial x_p$. Carrying out the indicated differentiations and noting $Dx_i/Dt = v_i$, we obtain

$$\int_\Omega e_{ijk} \left(x_i\, \sigma_{jp,p} + \delta_{ip}\, \sigma_{jp} + \rho\, x_i\, f_j \right) d\mathbf{x} = \int_\Omega \rho\, e_{ijk} \left(v_i\, v_j + x_i\, \frac{Dv_j}{Dt} \right) d\mathbf{x},$$

or (noting that $e_{ijk}\, v_i\, v_j = 0$):

$$\int_\Omega \left\{ e_{ijk} \left[x_i \left(\sigma_{jp,p} + \rho\, f_j - \rho\frac{Dv_j}{Dt} \right) \right] + e_{ijk}\, \sigma_{ji} \right\} d\mathbf{x} = 0.$$

which, in view of the equations of motion (5.3.10), yields

$$e_{ijk}\, \sigma_{ji} = 0 \quad \text{or} \quad \boldsymbol{\mathcal{E}} : \boldsymbol{\sigma}^{\mathrm{T}} = \mathbf{0}. \tag{5.3.30}$$

Thus, in the absence of body couples, we have

$$\boldsymbol{\mathcal{E}} : \boldsymbol{\sigma}^{\mathrm{T}} = \mathbf{0} \;\Rightarrow\; \boldsymbol{\sigma} = \boldsymbol{\sigma}^{\mathrm{T}} \quad (\text{or } \sigma_{ij} = \sigma_{ji}). \tag{5.3.31}$$

From Eq. (4.4.12) it follows that the second Piola–Kirchhoff stress tensor is also symmetric, $\mathbf{S} = \mathbf{S}^{\mathrm{T}}$, whenever $\boldsymbol{\sigma}$ is symmetric. Also, when $\boldsymbol{\sigma} = \boldsymbol{\sigma}^{\mathrm{T}}$, it follows from Eq. (4.4.8) that

$$\mathbf{P} \cdot \mathbf{F}^{\mathrm{T}} = \mathbf{F} \cdot \mathbf{P}^{\mathrm{T}} \quad (\text{i.e., } \mathbf{P} \cdot \mathbf{F}^{\mathrm{T}} \text{ is symmetric}). \tag{5.3.32}$$

5.3.3.2 Multipolar case

In *multipolar continuum mechanics*, a molecule may be represented by a "deformable" particle, undergoing its own internal strains (so-called "microstrains"), which represent certain types of collective behavior of the sub-particles of a molecule[1]. A multipolar continuum also contains an angular momentum vector $\rho\mathbf{p}$ and a body couple vector $\rho\mathbf{c}$ (both measured per unit mass) inside the body, in the same way as the linear momentum vector $\rho\mathbf{v}$ and the body force vector $\rho\mathbf{f}$, and a couple traction vector $\mathbf{m}$ on the boundary, in the same way as the traction vector $\mathbf{t}$. Analogous to Cauchy's formula, the couple traction vector $\mathbf{m}$ is related to the *couple stress tensor* $\mathbf{M}$ according to

$$\mathbf{m} = \mathbf{M} \cdot \hat{\mathbf{n}} \quad (\text{or } m_i = M_{ij}\, n_j). \tag{5.3.33}$$

By the principle of balance of linear momentum, body and surface couples do not enter the calculation and therefore the equation of motion (5.3.10) is unaffected. However, for a multipolar case the principle leads to additional equations. For the sake of completeness, we present these additional equations, although their use is not illustrated in this book.

---

[1]See Jaunzemis (1967) and Eringen and Hanson (2002) for further discussion.

The principle of balance of angular momentum for the multipolar case can be stated as follows: *The time rate of change of the total moment of momentum for a continuum is equal to vector sum of couples and the moments of external forces acting on the continuum.* The principle as applied to a control volume Ω with a control surface Γ can be expressed as [see Eq. (5.3.28)]

$$\oint_\Gamma (\mathbf{x} \times \mathbf{t} + \mathbf{m})\, ds + \int_\Omega (\mathbf{x} \times \rho \mathbf{f} + \rho \mathbf{c})\, d\mathbf{x} = \frac{D}{Dt}\left[\int_v (\mathbf{x} \times \rho \mathbf{v} + \rho \mathbf{p})\, d\mathbf{x}\right]. \quad (5.3.34)$$

In index notation (kth component of), Eq. (5.3.34) takes the form

$$\oint_\Gamma (e_{ijk}\, x_i\, t_j + m_k)\, ds + \int_\Omega (\rho\, e_{ijk}\, x_i\, f_j + \rho\, c_k)\, d\mathbf{x}$$

$$= \frac{D}{Dt}\int_\Omega (\rho\, e_{ijk}\, x_i\, v_j + \rho\, p_k)\, d\mathbf{x}. \quad (5.3.35)$$

Using Cauchy's formulas, $t_j = \sigma_{jp} n_p$ and $m_k = M_{kp} n_p$, and the divergence theorem, we arrive at the result

$$\int_\Omega \left[e_{ijk}\, (x_i\, \sigma_{jp})_{,p} + M_{kp,p} + \rho\, e_{ijk}\, x_i\, f_j + \rho\, c_k \right] d\mathbf{x}$$

$$= \int_\Omega \rho \left[e_{ijk}\, \frac{D}{Dt}(x_i\, v_j) + \frac{Dp_k}{Dt} \right] d\mathbf{x}. \quad (5.3.36)$$

After simplification using $Dx_i/Dt = v_i$ and $e_{ijk}\, v_i\, v_j = 0$, we obtain

$$\int_\Omega \left[e_{ijk}\, x_i \left(\sigma_{jm,m} + \rho\, f_j - \rho\, \frac{Dv_j}{Dt} \right) + M_{ki,i} + e_{ijk}\, \sigma_{ji} + \rho\, c_k - \rho\, \frac{Dp_k}{Dt} \right] d\mathbf{x} = 0,$$

or in vector form

$$\int_\Omega \left[\mathbf{x} \times \left(\boldsymbol{\nabla} \cdot \boldsymbol{\sigma}^{\mathrm{T}} + \rho \mathbf{f} - \rho\, \frac{D\mathbf{v}}{Dt} \right) + \boldsymbol{\nabla} \cdot \mathbf{M}^{\mathrm{T}} + \boldsymbol{\mathcal{E}} : \boldsymbol{\sigma}^{\mathrm{T}} + \rho \mathbf{c} - \rho\, \frac{D\mathbf{p}}{Dt} \right] d\mathbf{x} = 0,$$

$$(5.3.37)$$

where $\boldsymbol{\mathcal{E}}$ is the third-order permutation tensor. Using the equation of motion (5.3.10), we deduce the following local form, known as *Cosserat's equation*:

$$\boldsymbol{\nabla} \cdot \mathbf{M}^{\mathrm{T}} + \boldsymbol{\mathcal{E}} : \boldsymbol{\sigma}^{\mathrm{T}} + \rho \mathbf{c} = \rho\, \frac{D\mathbf{p}}{Dt}. \quad (5.3.38)$$

Thus, for a multipolar continuum, the stress tensor is no longer symmetric.

5.4 Thermodynamic Principles

5.4.1 Introduction

The first law of thermodynamics is commonly known as the principle of balance of energy, and it can be regarded as a statement of the interconvertibility of heat and work, while the total energy remains constant. The law does not place any restrictions on the direction of the process. For instance, in the study of mechanics of particles and rigid bodies, the kinetic energy and potential energy can be fully transformed from one to the other in the absence of friction and other dissipative mechanisms. From our experience, we know that mechanical energy that is converted into heat cannot all be converted back into mechanical energy. For example, the motion (kinetic energy) of a flywheel can all be converted into heat (internal energy) by means of a friction brake; if the whole system is insulated, the internal energy causes the temperature of the system to rise. Although the first law does not restrict the reversal process, namely the conversion of heat to internal energy and internal energy to motion (of the flywheel), such a reversal cannot occur because the frictional dissipation is an *irreversible process*. The second law of thermodynamics provides the restriction on the interconvertibility of energies.

5.4.2 Balance of Energy

The first law of thermodynamics states that *the time rate of change of the total energy is equal to the sum of the rate of work done by the external forces and the change of heat content per unit time.* The total energy is the sum of the kinetic energy and the internal energy. The principle of the balance of energy can be expressed as

$$\frac{D}{Dt}(K + U) = W + Q_h. \tag{5.4.1}$$

Here, K denotes the kinetic energy, U is the internal energy, W is the power input, and Q_h is the heat input to the system.

5.4.2.1 Energy equation in the spatial description

The kinetic energy of the system is given by

$$K = \tfrac{1}{2} \int_\Omega \rho \mathbf{v} \cdot \mathbf{v} \, d\mathbf{x}, \tag{5.4.2}$$

where $\mathbf{v}$ is the velocity vector. If e is the energy per unit mass (or *specific internal energy*), the total internal energy of the system is given by

$$U = \int_\Omega \rho e \, d\mathbf{x}. \tag{5.4.3}$$

The kinetic energy (K) is the energy associated with the macroscopically observable velocity of the continuum. The kinetic energy associated with the (microscopic) motions of molecules of the continuum is a part of the internal energy; the elastic strain energy and other forms of energy are also parts of the internal energy, U.

Here we consider only the nonpolar case, that is, body couples are zero and the stress tensor is symmetric. The power input consists of the rate of work done by external surface tractions $\mathbf{t}$ per unit area and body forces $\mathbf{f}$ per unit volume of the region Ω bounded by Γ:

$$W = \oint_\Gamma \mathbf{t} \cdot \mathbf{v} \, ds + \int_\Omega \rho \mathbf{f} \cdot \mathbf{v} \, d\mathbf{x}. \tag{5.4.4}$$

The rate of heat input consists of conduction through the boundary Γ and heat generation inside the region Ω (possibly from a radiation field or transmission of electric current). Let $\mathbf{q}$ be the heat flux vector and r_h be the internal heat generation per unit mass. Then the heat inflow across the surface element ds is $-\mathbf{q} \cdot \hat{\mathbf{n}}\, ds$, and internal heat generation in volume element $d\mathbf{x}$ is $\rho r_h d\mathbf{x}$. Hence, the total heat input is

$$Q_h = -\oint_\Gamma \hat{\mathbf{n}} \cdot \mathbf{q}\, ds + \int_\Omega \rho r_h\, d\mathbf{x}. \tag{5.4.5}$$

Substituting expressions for K, U, W, and Q_h from Eqs. (5.4.2)–(5.4.5) into Eq. (5.4.1), we obtain

$$\frac{D}{Dt} \int_\Omega \rho \left(\frac{1}{2} \mathbf{v} \cdot \mathbf{v} + e \right) d\mathbf{x} = \oint_\Gamma \mathbf{t} \cdot \mathbf{v}\, ds + \int_\Omega \rho \mathbf{f} \cdot \mathbf{v}\, d\mathbf{x}$$
$$- \oint_\Gamma \hat{\mathbf{n}} \cdot \mathbf{q}\, ds + \int_\Omega \rho r_h\, d\mathbf{x}. \tag{5.4.6}$$

Equation (5.4.6) can be simplified using a number of previously derived equations and identities, as explained next.

We begin with the expression for W (symmetry of $\boldsymbol{\sigma}$ allows us to write $\boldsymbol{\sigma} \cdot \hat{\mathbf{n}} = \hat{\mathbf{n}} \cdot \boldsymbol{\sigma}$)

$$W = \oint_\Gamma \mathbf{t} \cdot \mathbf{v}\, ds + \int_\Omega \rho \mathbf{f} \cdot \mathbf{v}\, d\mathbf{x} = \oint_\Gamma (\hat{\mathbf{n}} \cdot \boldsymbol{\sigma}) \cdot \mathbf{v}\, ds + \int_\Omega \rho \mathbf{f} \cdot \mathbf{v}\, d\mathbf{x}$$
$$= \int_\Omega [\boldsymbol{\nabla} \cdot (\boldsymbol{\sigma} \cdot \mathbf{v}) + \rho \mathbf{f} \cdot \mathbf{v}]\, d\mathbf{x} = \int_\Omega [(\boldsymbol{\nabla} \cdot \boldsymbol{\sigma} + \rho \mathbf{f}) \cdot \mathbf{v} + \boldsymbol{\sigma} : \boldsymbol{\nabla}\mathbf{v}]\, d\mathbf{x}$$
$$= \int_\Omega \left(\rho \frac{D\mathbf{v}}{Dt} \cdot \mathbf{v} + \boldsymbol{\sigma} : \boldsymbol{\nabla}\mathbf{v} \right) d\mathbf{x}, \tag{5.4.7}$$

where : denotes the "double-dot product" $\mathbf{S} : \mathbf{T} = S_{ij} T_{ij}$ [see Eq. (2.5.13)]. The Cauchy formula, vector identity in Eq. (5.2.7), and the equation of motion (5.3.10) are used in arriving at the last step. We note that [see Eq. (5.2.8)]

$$\boldsymbol{\sigma} : \boldsymbol{\nabla}\mathbf{v} = \boldsymbol{\sigma}^{\text{sym}} : \mathbf{D} - \boldsymbol{\sigma}^{\text{skew}} : \mathbf{W},$$

where $\mathbf{D}$ is the symmetric part, called the rate of deformation tensor, and $\mathbf{W}$ is the skew symmetric part, called the vorticity (or spin) tensor, of $(\boldsymbol{\nabla}\mathbf{v})^{\text{T}}$ [see Eq. (3.6.2)],

$$\mathbf{D} = \tfrac{1}{2} \left[\boldsymbol{\nabla}\mathbf{v} + (\boldsymbol{\nabla}\mathbf{v})^{\text{T}} \right], \quad \mathbf{W} = \tfrac{1}{2} \left[(\boldsymbol{\nabla}\mathbf{v})^{\text{T}} - \boldsymbol{\nabla}\mathbf{v} \right], \tag{5.4.8}$$

and $\boldsymbol{\sigma}^{\text{sym}}$ and $\boldsymbol{\sigma}^{\text{skew}}$ are the symmetric and skew symmetric parts of $\boldsymbol{\sigma}$. When $\boldsymbol{\sigma}$ is symmetric, we have $\boldsymbol{\sigma}^{\text{sym}} = \boldsymbol{\sigma}$ and $\boldsymbol{\sigma}^{\text{skew}} = \mathbf{0}$. Hence, Eq. (5.4.7) becomes

$$W = \tfrac{1}{2} \int_\Omega \rho \frac{D}{Dt} (\mathbf{v} \cdot \mathbf{v})\, d\mathbf{x} + \int_\Omega \boldsymbol{\sigma} : \mathbf{D}\, d\mathbf{x} = \tfrac{1}{2} \frac{D}{Dt} \int_\Omega \rho \mathbf{v} \cdot \mathbf{v}\, d\mathbf{x} + \int_\Omega \boldsymbol{\sigma} : \mathbf{D}\, d\mathbf{x},$$

where the Reynolds transport theorem, Eq. (5.2.40), is used to write the final expression. Next, Q_h can be expressed as

$$Q_h = -\oint_\Gamma \hat{\mathbf{n}} \cdot \mathbf{q}\, ds + \int_\Omega \rho r_h\, d\mathbf{x} = \int_\Omega (-\boldsymbol{\nabla} \cdot \mathbf{q} + \rho r_h)\, d\mathbf{x}. \tag{5.4.9}$$

With the new expressions for W and Q_h, Eq. (5.4.6) can be written as

$$0 = \int_\Omega \left(\rho \frac{De}{Dt} - \boldsymbol{\sigma} : \mathbf{D} + \boldsymbol{\nabla} \cdot \mathbf{q} - \rho r_h \right) dx, \qquad (5.4.10)$$

which is the global form of the energy equation. The local form of the energy equation is given by

$$\rho \frac{De}{Dt} = \boldsymbol{\sigma} : \mathbf{D} - \boldsymbol{\nabla} \cdot \mathbf{q} + \rho r_h, \qquad (5.4.11)$$

which is known as the *thermodynamic form* of the energy equation for a continuum. The term $\boldsymbol{\sigma} : \mathbf{D}$ is known as the *stress power*, which can be regarded as the internal work production.

5.4.2.2 Energy equation in the material description

To derive the energy equation in the material description, we write K, U, W, and Q_h in the material description:

$$K = \tfrac{1}{2} \int_\Omega \rho \mathbf{v} \cdot \mathbf{v} \, dx = \tfrac{1}{2} \int_{\Omega_0} \rho_0 \mathbf{v} \cdot \mathbf{v} \, d\mathbf{X}, \qquad (5.4.12)$$

$$U = \int_\Omega \rho e \, dx = \int_{\Omega_0} \rho_0 e \, d\mathbf{X}, \qquad (5.4.13)$$

$$W = \oint_\Gamma \mathbf{t} \cdot \mathbf{v} \, ds + \int_\Omega \rho \mathbf{f} \cdot \mathbf{v} \, dx = \oint_{\Gamma_0} \mathbf{T} \cdot \mathbf{v} \, dS + \int_{\Omega_0} \rho_0 \mathbf{f} \cdot \mathbf{v} \, d\mathbf{X}$$

$$= \oint_{\Gamma_0} (\hat{\mathbf{N}} \cdot \mathbf{P}^{\mathrm{T}}) \cdot \mathbf{v} \, dS + \int_{\Omega_0} \rho_0 \mathbf{f} \cdot \mathbf{v} \, d\mathbf{X} = \int_{\Omega_0} \left[\boldsymbol{\nabla}_0 \cdot (\mathbf{P}^{\mathrm{T}} \cdot \mathbf{v}) + \rho_0 \mathbf{f} \cdot \mathbf{v} \right] d\mathbf{X}$$

$$= \int_{\Omega_0} \left[(\boldsymbol{\nabla}_0 \cdot \mathbf{P}^{\mathrm{T}} + \rho_0 \mathbf{f}) \cdot \mathbf{v} + \mathbf{P}^{\mathrm{T}} : \boldsymbol{\nabla}_0 \mathbf{v} \right] d\mathbf{X}$$

$$= \int_\Omega \left[\tfrac{1}{2} \rho_0 \frac{\partial}{\partial t} (\mathbf{v} \cdot \mathbf{v}) + \mathbf{P}^{\mathrm{T}} : \boldsymbol{\nabla}_0 \mathbf{v} \right] d\mathbf{X}, \qquad (5.4.14)$$

$$Q_h = -\oint_\Gamma \hat{\mathbf{n}} \cdot \mathbf{q} \, ds + \int_\Omega \rho r_h \, dx = -\oint_{\Gamma_0} \hat{\mathbf{N}} \cdot \mathbf{q}_0 \, dS + \int_{\Omega_0} \rho_0 r_h \, d\mathbf{X}$$

$$= \int_{\Omega_0} \left[-\boldsymbol{\nabla}_0 \cdot \mathbf{q}_0 + \rho_0 r_h \right] d\mathbf{X}, \qquad (5.4.15)$$

where Eqs. (5.3.15) and (5.3.17), and the following relations are used to write the final expressions for K, W, and Q_h:

$$\hat{\mathbf{n}} \, ds = J \mathbf{F}^{-\mathrm{T}} \cdot \hat{\mathbf{N}} \, dS = J \hat{\mathbf{N}} \cdot \mathbf{F}^{-1} \, dS, \quad \rho \, dx = \rho_0 \, d\mathbf{X}, \qquad (5.4.16)$$

and all variables are now function of $\mathbf{X}$, the material coordinates.

Substitution of expressions from Eqs. (5.4.12)–(5.4.15) into Eq. (5.4.1), we obtain the following local form of the energy equation in the material description:

$$\rho_0 \frac{\partial e}{\partial t} = \mathbf{P}^{\mathrm{T}} : \boldsymbol{\nabla}_0 \mathbf{v} - \boldsymbol{\nabla}_0 \cdot \mathbf{q}_0 + \rho_0 r_h. \qquad (5.4.17)$$

In terms of the second Piola–Krichhoff stress tensor, we have

$$\rho_0 \frac{\partial e}{\partial t} = (\mathbf{S} \cdot \mathbf{F}^{\mathrm{T}}) : \boldsymbol{\nabla}_0 \mathbf{v} - \boldsymbol{\nabla}_0 \cdot \mathbf{q}_0 + \rho_0 r_h. \qquad (5.4.18)$$

5.4.3 Entropy Inequality

The concept of entropy is a difficult one to explain in simple terms; it has its roots in statistical physics and thermodynamics and is generally considered as a measure of the tendency of the atoms toward a disorder. For example, carbon has a lower entropy in the form of diamond, a hard crystal with atoms closely bound in a highly ordered array.

Temperature cannot be decreased below a certain absolute minimum. We introduce θ as the absolute temperature whose greatest lower bound is zero. We also recall that in an admissible deformation, the deformation gradient tensor $\mathbf{F}$ should be nonsingular, that is, $J = |\mathbf{F}| \neq 0$. Thus each thermodynamic process should satisfy the conditions

$$\theta \geq 0, \quad |\mathbf{F}| \neq 0.$$

5.4.3.1 Homogeneous processes

Let us denote total entropy by the symbol H, and define *internal dissipation*, $\mathcal{D}$, as

$$\mathcal{D} = \theta \dot{H} - Q_h, \tag{5.4.19}$$

where Q_h denotes the rate of heat supply to the body; $\theta \dot{H}$ is interpreted as the time rate of change of the heat content of the body. The ratio of $\mathcal{D}$ to θ is called the *internal entropy production*,

$$\Gamma \equiv \frac{\mathcal{D}}{\theta} = \dot{H} - \frac{Q_h}{\theta}. \tag{5.4.20}$$

The second law of thermodynamics states that the internal entropy production is always positive, which is known as the *Clausius–Duhem inequality*, and it is expressed, for homogeneous processes, as

$$\mathcal{D} = \theta \dot{H} - Q_h \geq 0. \tag{5.4.21}$$

If $\mathcal{D} = 0$, then the process is said to be *reversible*, and we have $\dot{H} = Q_h/\theta$; otherwise, the process is said to be *irreversible*. The processes in which $Q_h = 0$, hence $\dot{H} \geq 0$, are said to be *adiabatic*. Processes in which $\dot{H} = 0$ (i.e., $Q_h \leq 0$) are called *isentropic*. The second law of thermodynamics essentially states that the time rate of change of the heat content $\theta \dot{H}$ of a body can never be less than the rate of heat supply Q_h.

5.4.3.2 Nonhomogeneous processes

To derive the Clausius–Duhem inequality for nonhomogeneous processes (that is, processes that depend not only on time but also on position), let us introduce the entropy density per unit mass, η, so that

$$H = \int_{\Omega} \rho \eta \, d\mathbf{x}. \tag{5.4.22}$$

We define the entropy production as [in the same form as Eq. (5.4.20) for the homogeneous case]

$$\Gamma = \frac{D}{Dt} \int_\Omega \rho\eta \, d\mathbf{x} - \left[-\oint_\Gamma \frac{1}{\theta}\mathbf{q}\cdot\hat{\mathbf{n}} \, ds + \int_\Omega \frac{\rho r_h}{\theta} \, d\mathbf{x} \right]$$

$$= \int_\Omega \rho\frac{D\eta}{Dt} \, d\mathbf{x} + \int_\Omega \left[\mathbf{\nabla}\cdot\left(\frac{\mathbf{q}}{\theta}\right) - \frac{\rho r_h}{\theta} \right] d\mathbf{x}. \tag{5.4.23}$$

Then, the second law of thermodynamics requires $\Gamma \geq 0$, giving

$$\int_\Omega \rho\frac{D\eta}{Dt} \, d\mathbf{x} \geq \int_\Omega \left[\left(\frac{\rho r_h}{\theta}\right) - \mathbf{\nabla}\cdot\left(\frac{\mathbf{q}}{\theta}\right) \right] d\mathbf{x}. \tag{5.4.24}$$

The local form of the Clausius–Duhem inequality, or *entropy inequality* is

$$\frac{D\eta}{Dt} \geq \frac{r_h}{\theta} - \frac{1}{\rho}\mathbf{\nabla}\cdot\left(\frac{\mathbf{q}}{\theta}\right) \quad \text{or} \quad \rho\theta\frac{D\eta}{Dt} - \rho r_h + \mathbf{\nabla}\cdot\mathbf{q} - \frac{1}{\theta}\mathbf{q}\cdot\mathbf{\nabla}\theta \geq 0. \tag{5.4.25}$$

The quantity $\mathbf{q}/\theta$ is known as the *entropy flux* and r_h/θ is the *entropy supply density*.

The sum of internal energy (e) and irreversible heat energy ($-\theta\eta$) is known as the *Helmhotz free energy density*:

$$\Psi = e - \theta\eta. \tag{5.4.26}$$

Substituting Eq. (5.4.26) into Eq. (5.4.11), we obtain

$$\rho\frac{D\Psi}{Dt} = \boldsymbol{\sigma}:\mathbf{D} - \rho\frac{D\theta}{Dt}\eta - \mathcal{D}, \tag{5.4.27}$$

where $\mathbf{D}$ is the symmetric part of the velocity gradient tensor [see Eq. (5.4.8)], and $\mathcal{D}$ is the *internal dissipation*

$$\mathcal{D} = \rho\theta\frac{D\eta}{Dt} + \mathbf{\nabla}\cdot\mathbf{q} - \rho r_h. \tag{5.4.28}$$

In view of Eq. (5.4.25) we can write

$$\mathcal{D} - \frac{1}{\theta}\mathbf{q}\cdot\mathbf{\nabla}\theta \geq 0. \tag{5.4.29}$$

We have $\mathcal{D} > 0$ for an irreversible process, and $\mathcal{D} = 0$ for a reversible process. Expressing the entropy inequality (5.4.25) in terms of the Helmholtz free energy density, we obtain

$$\boldsymbol{\sigma}:\mathbf{L} - \rho\dot{\theta}\eta - \rho\dot{\Psi} - \frac{1}{\theta}\mathbf{q}\cdot\mathbf{\nabla}\theta \geq 0, \tag{5.4.30}$$

where the superposed dot denotes the material time derivative, and $\mathbf{L}$ is the velocity gradient tensor, $\mathbf{L} = (\mathbf{\nabla}\mathbf{v})^{\mathrm{T}} = \mathbf{D} + \mathbf{W}$, $\mathbf{W}$ being the skew symmetric spin tensor in Eq. (5.4.8). Note that when $\boldsymbol{\sigma}$ is symmetric, we have $\boldsymbol{\sigma}:\mathbf{L} = \boldsymbol{\sigma}:\mathbf{D}$.

5.5 Summary

5.5.1 Preliminary Comments

This chapter was devoted to the derivation of the field equations governing a continuous medium using the principle of conservation of mass and balance of momenta and energy, and therefore constitutes the heart of the book. The equations are derived in invariant (i.e., vector and tensor) form so that they can be expressed in any chosen coordinate system (e.g., rectangular, cylindrical, spherical, or curvilinear system). The principle of conservation of mass results in the continuity equation; the principle of balance of linear momentum, which is equivalent to Newton's second law of motion, leads to the equations of motion in terms of the Cauchy stress tensor; the principle of balance of angular momentum yields, in the absence of body couples, the symmetry of the Cauchy stress tensor; and the principles of thermodynamics – the first and second laws of thermodynamics – give rise to the energy equation and Clausius–Duhem inequality.

In closing this chapter, we summarize the invariant form of the equations resulting from the application of conservation principles to a continuum. The variables appearing in the equations were already defined and are not repeated here. In this study, we will be concerned only with monopolar media, where no body couples and body moments are accounted for. This amounts to assuming the symmetry of the stress tensors:

$$\boldsymbol{\sigma} = \boldsymbol{\sigma}^{\mathrm{T}}, \quad \mathbf{S} = \mathbf{S}^{\mathrm{T}}, \quad \mathbf{P} \cdot \mathbf{F}^{\mathrm{T}} = \mathbf{F} \cdot \mathbf{P}^{\mathrm{T}}.$$

5.5.2 Conservation and Balance Equations in the Spatial Description

Conservation of mass

$$\frac{\partial \rho}{\partial t} + \boldsymbol{\nabla} \cdot (\rho \mathbf{v}) = 0 \tag{5.5.1}$$

Balance of linear momentum

$$\boldsymbol{\nabla} \cdot \boldsymbol{\sigma} + \rho \mathbf{f} = \rho \left(\frac{\partial \mathbf{v}}{\partial t} + \mathbf{v} \cdot \boldsymbol{\nabla} \mathbf{v} \right) \tag{5.5.2}$$

Balance of angular momentum

$$\boldsymbol{\sigma}^{\mathrm{T}} = \boldsymbol{\sigma} \tag{5.5.3}$$

Balance of energy

$$\rho \frac{De}{Dt} = \boldsymbol{\sigma} : \mathbf{D} - \boldsymbol{\nabla} \cdot \mathbf{q} + \rho r_h \tag{5.5.4}$$

Entropy inequality

$$\rho \theta \frac{D\eta}{Dt} - \rho r_h + \boldsymbol{\nabla} \cdot \mathbf{q} - \frac{1}{\theta} \mathbf{q} \cdot \boldsymbol{\nabla} \theta \geq 0 \tag{5.5.5}$$

5.5.3 Conservation and Balance Equations in the Material Description

Conservation of mass

$$\rho_0 = \rho \, J \tag{5.5.6}$$

Balance of linear momentum

$$\nabla_0 \cdot (\mathbf{S} \cdot \mathbf{F}^{\mathrm{T}}) + \rho_0 \, \mathbf{f} = \rho_0 \, \frac{\partial^2 \mathbf{u}}{\partial t^2} \tag{5.5.7}$$

Balance of angular momentum

$$\mathbf{S}^{\mathrm{T}} = \mathbf{S} \tag{5.5.8}$$

Balance of energy

$$\rho_0 \frac{\partial e}{\partial t} = (\mathbf{S} \cdot \mathbf{F}^{\mathrm{T}}) : \nabla_0 \mathbf{v} - \nabla_0 \cdot \mathbf{q}_0 + \rho_0 \, r_0 \tag{5.5.9}$$

Entropy inequality

$$\rho_0 \theta \frac{D\eta}{Dt} - \rho_0 \, r_0 + \nabla_0 \cdot \mathbf{q}_0 - \frac{1}{\theta} \mathbf{q}_0 \cdot \nabla_0 \theta \geq 0 \tag{5.5.10}$$

We shall return to these equations in the subsequent chapters as needed. These equations may be supplemented by other field equations, such as Maxwell's equations governing electromagnetics, depending on the field of study.

5.5.4 Closing Comments

The subject of continuum mechanics is concerned primarily with the determination of the behavior (e.g., $\mathbf{F}$, θ, $\nabla\theta = \mathbf{g}$, etc.) of a body under externally applied causes (e.g., $\mathbf{f}$, r_h, and so on). After introducing suitable constitutive relations for $\boldsymbol{\sigma}$, e, η, and $\mathbf{q}$ (to be discussed in Chapter 6), this task involves solving the initial-boundary-value problem described by partial differential equations (5.5.1)–(5.5.4) under specified initial and boundary conditions. The role of the entropy inequality in formulating the problem is to make sure that the behavior of a body is consistent with the inequality (5.5.5). Often, the constitutive relations developed are required to be consistent with the second law of thermodynamics (i.e., satisfy the entropy inequality). The *entropy principle* states that constitutive relations be such that the entropy inequality is satisfied identically for any thermodynamic process.

An examination of the conservation principles presented in this chapter shows that all of the mathematical statements resulting from the principles share a common mathematical structure. These all can be expressed in terms of a general balance (or conservation) equation in the spatial description as

$$\frac{D}{Dt} \int_\Omega \phi(\mathbf{x}, t) \, d\mathbf{x} = \oint_\Gamma \psi(\mathbf{x}, t, \hat{\mathbf{n}}) \, ds + \int_\Omega f(\mathbf{x}, t) \, d\mathbf{x}, \tag{5.5.11}$$

where ϕ is a tensor field of order n measured per unit volume; $\psi(\mathbf{x}, t, \hat{\mathbf{n}})$ is a surface tensor of order $n-1$, measured per unit current area and depends on the surface orientation $\hat{\mathbf{n}}$; and $f(\mathbf{x}, t)$ is a source tensor of order n, also measured per unit volume $(n = 0, 1)$. The variables ϕ, ψ, and f associated with the balance equations resulting from the principles of conservation of mass, balance of linear and angular momentum, and the first and second laws of thermodynamics are presented in Table 5.5.1.

Table 5.5.1: Expressions for variables ϕ, ψ, and f in Eq. (5.5.11) for the four conservation principles.

No.	$\phi(\mathbf{x}, t)$	$\psi(\mathbf{x}, t, \hat{\mathbf{n}})$	$f(\mathbf{x}, t)$	number
1.	ρ	0	0	(5.2.22)
2.	$\rho\,\mathbf{v}$	$\mathbf{t}$	$\rho\,\mathbf{f}$	(5.3.6)
3.	$\mathbf{x} \times \rho\,\mathbf{v}$	$\mathbf{x} \times \mathbf{t}$	$\rho\,\mathbf{x} \times \mathbf{f}$	(5.3.28)
4.	$\rho\,(v^2/2 + e)$	$\mathbf{t} \cdot \mathbf{v} - \hat{\mathbf{n}} \cdot \mathbf{q}$	$\rho\,\mathbf{f} \cdot \mathbf{v} + \rho r$	(5.4.6)
5.	$\rho\,\eta$	$-\frac{\mathbf{q} \cdot \hat{\mathbf{n}}}{\theta}$	$\frac{\rho r}{\theta}$	(5.4.23)

To complete the mathematical description of the behavior of a continuous medium, the conservation equations derived in this chapter must be supplemented with the constitutive equations that relate $\boldsymbol{\sigma}$, e, η, and $\mathbf{q}$ to $\mathbf{F}$, θ, and $\mathbf{g} \equiv \boldsymbol{\nabla}\theta$. The strain (or strain rate) measures ($\mathbf{e}$, $\mathbf{D}$, $\mathbf{E}$, $\mathbf{F}$, $\mathbf{C}$) introduced in Chapter 3 and the stress measures ($\boldsymbol{\sigma}$, $\mathbf{P}$, $\mathbf{S}$) introduced in Chapter 4 are objective and, therefore, they are suitable candidates for the description of material response, which should be independent of the observer. Chapter 6 is dedicated to the discussion of the material constitutive relations. Applications of the governing equations to linearized elasticity problems and fluid mechanics and heat transfer problems are discussed in Chapters 7 and 8, respectively.

Problems

CONSERVATION OF MASS

5.1 The acceleration of a material element in a continuum is described by

$$\frac{D\mathbf{v}}{Dt} \equiv \frac{\partial \mathbf{v}}{\partial t} + \mathbf{v} \cdot \boldsymbol{\nabla}\mathbf{v}, \tag{1}$$

where $\mathbf{v}$ is the velocity vector. Show by means of vector identities that the acceleration can also be written as

$$\frac{D\mathbf{v}}{Dt} \equiv \frac{\partial \mathbf{v}}{\partial t} + \boldsymbol{\nabla}\left(\frac{v^2}{2}\right) - \mathbf{v} \times \boldsymbol{\nabla} \times \mathbf{v}, \quad v^2 = \mathbf{v} \cdot \mathbf{v}. \tag{2}$$

5.2 Show that the local form of the principle of conservation of mass, Eq. (5.2.22), can be expressed as

$$\frac{D}{Dt}(\rho J) = 0.$$

5.3 Use the equation

$$\frac{D}{Dt}(\rho J) = 0,$$

to derive the continuity equation

$$\frac{D\rho}{Dt} + \rho\,\boldsymbol{\nabla} \cdot \mathbf{v} = 0.$$

5.4 Derive the continuity equation in the cylindrical coordinate system by considering a differential volume element shown in Fig. P5.4.

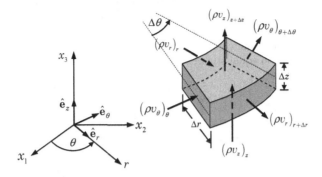

Fig. P5.4

5.5 Express the continuity equation (5.2.24) in the cylindrical coordinate system (see Table 2.4.2 for various operators). The result should match the one in Eq. (5.2.30).

5.6 Express the continuity equation (5.2.24) in the spherical coordinate system (see Table 2.4.2 for various operators). The result should match the one in Eq. (5.2.31).

5.7 Determine if the following velocity fields for an incompressible flow satisfy the continuity equation:

(a) $\quad v_1(x_1, x_2) = -\dfrac{x_1}{r^2}, \quad v_2(x_1, x_2) = -\dfrac{x_2}{r^2} \quad$ where $\quad r^2 = x_1^2 + x_2^2.$

(b) $\quad v_r = 0, \ v_\theta = 0, \ v_z = c\left(1 - \dfrac{r^2}{R^2}\right)$

where c and R are constants.

5.8 The velocity distribution between two parallel plates separated by distance b is

$$v_x(y) = \frac{y}{b}v_0 - c\frac{y}{b}\left(1 - \frac{y}{b}\right), \quad v_y = 0, \quad v_z = 0, \quad 0 < y < b,$$

where y is measured from and normal to the bottom plate, x is taken along the plates, v_x is the velocity component parallel to the plates, v_0 is the velocity of the top plate in the x direction, and c is a constant. Determine if the velocity field satisfies the continuity equation and find the volume rate of flow and the average velocity.

BALANCE OF LINEAR MOMENTUM

5.9 Calculate the force exerted by a water ($\rho = 10^3 \ \text{kg/m}^3$) jet of diameter $d = 8$ mm and velocity $v = 12$ m/s that impinges against a smooth inclined flat plate at an angle of $45°$ to the axis of the jet, as shown in Fig. P5.9.

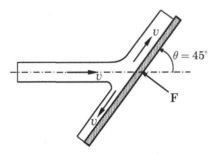

Fig. P5.9

5.10 Calculate the force exerted by a water ($\rho = 10^3$ kg/m^3) jet of diameter $d = 60$ mm and velocity $v = 6$ m/s that impinges against a smooth inclined flat plate at an angle of $60°$ to the axis of the jet. Also calculate the volume flow rates Q_L and Q_R.

5.11 A jet of air ($\rho = 1.206$ kg/m^3) impinges on a smooth vane with a velocity $v = 50$ m/s at the rate of $Q = 0.4$ m^3/s. Determine the force required to hold the plate in position for the two different vane configurations shown in Fig. P5.11. Assume that the vane splits the jet into two equal streams, and neglect any energy loss in the streams.

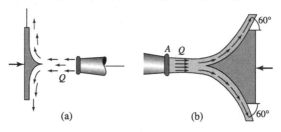

(a)　　　　　　　　　　　　　　(b)

Fig. P5.11

5.12 In Example 5.3.3, determine (a) the velocity and accelerations as functions of x, and (b) the velocity as the chain leaves the table.

5.13 Using the definition of ∇, vector forms of the velocity vector, body force vector, and the dyadic form of σ [see Eq. (5.3.23)], express the equation of motion (5.3.11) in the cylindrical coordinate system as given in Eq. (5.3.24).

5.14 Using the definition of ∇, vector forms of the velocity vector, body force vector, and the dyadic form of σ [see Eq. (5.3.25)], express the equation of motion (5.3.11) in the spherical coordinate system as given in Eq. (5.3.26).

5.15 Use the continuity equation and the equation of motion to obtain the so-called *conservation form* of the linear momentum equation

$$\frac{\partial}{\partial t}(\rho\mathbf{v}) + \text{div}\left(\rho\mathbf{v}\mathbf{v} - \boldsymbol{\sigma}^{\text{T}}\right) = \rho\mathbf{f}.$$

5.16 Show that

$$\rho\frac{D}{Dt}\left(\frac{v^2}{2}\right) = \mathbf{v}\cdot\nabla\cdot\boldsymbol{\sigma}^{\text{T}} + \rho\mathbf{v}\cdot\mathbf{f} \qquad (v = |\mathbf{v}|).$$

5.17 Deduce that

$$\nabla\times\left(\frac{D\mathbf{v}}{Dt}\right) \equiv \frac{D\mathbf{w}}{Dt} + \mathbf{w}\nabla\cdot\mathbf{v} - \mathbf{w}\cdot\nabla\mathbf{v}, \qquad (a)$$

where $\mathbf{w} \equiv \frac{1}{2}\nabla\times\mathbf{v}$ is the vorticity vector. *Hint:* Use the result of Problem **5.1** and the identity (you need to prove it)

$$\nabla\times(\mathbf{A}\times\mathbf{B}) = \mathbf{B}\cdot\nabla\mathbf{A} - \mathbf{A}\cdot\nabla\mathbf{B} + \mathbf{A}\nabla\cdot\mathbf{B} - \mathbf{B}\nabla\cdot\mathbf{A}. \qquad (2)$$

5.18 If the stress field σ in a continuum has the following components in a rectangular Cartesian coordinate system

$$[\sigma] = a\begin{bmatrix} x_1^2 x_2 & (b^2 - x_2^2)x_1 & 0 \\ (b^2 - x_2^2)x_1 & \frac{1}{3}(x_2^2 - 3b^2)x_2 & 0 \\ 0 & 0 & 2bx_3^2 \end{bmatrix},$$

where a and b are constants, determine the body force components necessary for the body to be in equilibrium.

5.19 If the stress field σ in a continuum has the following components in a rectangular Cartesian coordinate system

$$[\sigma] = \begin{bmatrix} x_1 x_2 & x_1^2 & -x_2 \\ x_1^2 & 0 & 0 \\ -x_2 & 0 & x_1^2 + x_2^2 \end{bmatrix},$$

determine the body force components necessary for the body to be in equilibrium.

5.20 A two-dimensional state of stress σ exists in a continuum with no body forces. The following components of stress tensor are given ($\sigma_{21} = \sigma_{12}$):

$$\sigma_{11} = c_1 x_2^3 + c_2 x_1^2 x_2 - c_3 x_1, \quad \sigma_{22} = c_4 x_2^3 - c_5, \quad \sigma_{12} = c_6 x_1 x_2^2 + c_7 x_1^2 x_2 - c_8,$$

where c_i are constants. Determine the conditions on the constants so that the stress field is in equilibrium.

5.21 Given the following stress field with respect to the cylindrical coordinate system in a body that is in equilibrium ($\sigma_{\theta r} = \sigma_{r\theta}$):

$$\sigma_{rr} = 2A \left(r + \frac{B}{r^3} - \frac{C}{r} \right) \sin \theta,$$

$$\sigma_{\theta\theta} = 2A \left(3r - \frac{B}{r^3} - \frac{C}{r} \right) \sin \theta,$$

$$\sigma_{r\theta} = -2A \left(r + \frac{B}{r^3} - \frac{C}{r} \right) \cos \theta,$$

where A, B, and C are constants, determine if the stress field satisfies the equilibrium equations when the body forces are zero. Assume that all other stress components are zero.

5.22 Given the following stress field with respect to the spherical coordinate system in a body that is in equilibrium:

$$\sigma_{RR} = -\left(A + \frac{B}{R^3} \right), \quad \sigma_{\phi\phi} = \sigma_{\theta\theta} = -\left(A + \frac{C}{R^3} \right),$$

where A, B, and C are constants, determine if the stress field satisfies the equilibrium equations when the body forces are zero and all other stress components are zero.

5.23 For a cantilevered beam bent by a point load at the free end, for kinematically infinitesimal deformations, the bending moment M_3 about the x_3-axis is given by $M_3 = -Px_1$ (see Fig. P5.23). The bending stress σ_{11} is given by

$$\sigma_{11} = \frac{M_3 x_2}{I_3} = -\frac{P x_1 x_2}{I_3},$$

where I_3 is the moment of inertia of the cross section about the x_3-axis. Starting with this equation, use the two-dimensional equilibrium equations to determine stresses σ_{22} and σ_{12} as functions of x_1 and x_2.

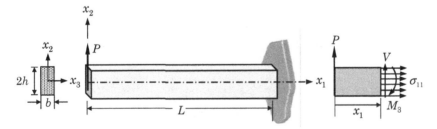

Fig. P5.23

5.24 For a cantilevered beam bent by a uniformly distributed load (see Fig. P5.24), for kinematically infinitesimal deformations, the bending stress σ_{11} is given by [because $M_3 = -q_0 x_1^2/2$]

$$\sigma_{11} = \frac{M_3 x_2}{I_3} = -\frac{q_0 x_1^2 x_2}{2I_3},$$

where I_3 is the moment of inertia of the cross section about the x_3-axis. Starting with this equation, use the two-dimensional equilibrium equations to determine the stresses σ_{22} and σ_{12} as functions of x_1 and x_2.

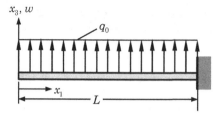

Fig. P5.24

5.25 Given the following components of the second Piola–Kirchhoff stress tensor $\mathbf{S}$ and displacement vector $\mathbf{u}$ in a body without body forces:

$$S_{11} = c_1 X_2^3 + c_2 X_1^2 X_2 - c_3 X_1, \quad S_{22} = c_4 X_2^3 - c_5, \quad S_{12} = c_6 X_1 X_2^2 + c_7 X_1^2 X_2 - c_8,$$

$$S_{13} = S_{23} = S_{33} = 0, \quad u_1 = c_9 X_2, \quad u_2 = c_{10} X_1, \quad u_3 = 0,$$

where c_i are constants, determine the conditions on the constants so that the stress field is in equilibrium.

5.26 Given the following components of the second Piola–Kirchhoff stress tensor $\mathbf{S}$ and displacement vector $\mathbf{u}$ in a body without body forces (expressed in the cylindrical coordinate system):

$$S_{rr} = -c_1 \frac{\cos\theta}{r}, \quad S_{r\theta} = S_{\theta\theta} = 0,$$

$$u_r = c_2 \log\left(\frac{r}{a}\right)\cos\theta + c_3\,\theta\sin\theta, \quad u_\theta = -c_2 \log\left(\frac{r}{a}\right)\sin\theta + c_3\,\theta\cos\theta - c_4\sin\theta,$$

where c_i are constants, determine the conditions on the constants so that the stress field is in equilibrium for (a) the linear (i.e., infinitesimal deformations) case and (b) the finite deformation case. Assume a two-dimensional state of stress and deformation in r and θ coordinates.

CONSERVATION OF ANGULAR MOMENTUM

5.27 A sprinkler with four nozzles, each nozzle having an exit area of $A = 0.25$ cm^2, rotates at a constant angular velocity of $w = 20$ rad/s and distributes water ($\rho = 10^3$ kg/m^3) at the rate of $Q = 0.5$ L/s (see Fig. P5.27). Determine

  (a) the torque T required on the shaft of the sprinkler to maintain the given motion and

  (b) the angular velocity w_0 at which the sprinkler rotates when no external torque is applied. Take $r = 0.1$ m.

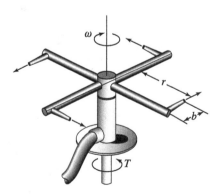

Fig. P5.27

5.28 Consider an unsymmetric sprinkler head shown in Fig. P5.28. If the discharge is $Q = 0.5$ L/s through each nozzle, determine the angular velocity of the sprinkler. Assume that no external torque is exerted on the system. Take $A = 10^{-4}$ m^2.

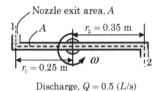

Discharge, $Q = 0.5$ (L/s)

Fig. P5.28

BALANCE OF ENERGY

5.29 Show that for a multipolar continuum the Clausius–Duhem inequality (5.4.24) remains unchanged.

5.30 Establish the following alternative form of the energy equation ($\boldsymbol{\sigma}^{\mathrm{T}} = \boldsymbol{\sigma}$):

$$\rho \frac{D}{Dt}\left(e + \frac{v^2}{2}\right) = \boldsymbol{\nabla} \cdot (\boldsymbol{\sigma} \cdot \mathbf{v}) + \rho \mathbf{f} \cdot \mathbf{v} + \rho r_h - \boldsymbol{\nabla} \cdot \mathbf{q}.$$

5.31 Establish the following *thermodynamic form* of the energy equation ($\boldsymbol{\sigma}^{\mathrm{T}} = \boldsymbol{\sigma}$):

$$\rho \frac{De}{Dt} = \boldsymbol{\nabla} \cdot (\boldsymbol{\sigma} \cdot \mathbf{v}) - \mathbf{v} \cdot \boldsymbol{\nabla} \cdot \boldsymbol{\sigma} + \rho r_h - \boldsymbol{\nabla} \cdot \mathbf{q}.$$

5.32 The total rate of work done by the surface stresses per unit volume is given by $\boldsymbol{\nabla} \cdot (\boldsymbol{\sigma} \cdot \mathbf{v})$. The rate of work done by the resultant of the surface stresses per unit volume is given by $\mathbf{v} \cdot \boldsymbol{\nabla} \cdot \boldsymbol{\sigma}$. The difference between these two terms yields the rate of work done by the surface stresses in deforming the material particle, per unit volume. Show that this difference can be written as $\boldsymbol{\sigma} : \mathbf{D}$, where $\mathbf{D}$ is the strain rate tensor defined in Eq. (5.4.8).

5.33 The rate of internal work done (power) in a continuous medium in the current configuration can be expressed as

$$W = \tfrac{1}{2} \int_\Omega \boldsymbol{\sigma} : \mathbf{D}\, d\mathbf{x}, \tag{1}$$

where $\boldsymbol{\sigma}$ is the Cauchy stress tensor and $\mathbf{D}$ is the strain rate tensor (i.e., symmetric part of the velocity gradient tensor)

$$\mathbf{D} = \tfrac{1}{2}\left[(\boldsymbol{\nabla}\mathbf{v})^{\mathrm{T}} + \boldsymbol{\nabla}\mathbf{v}\right], \quad \mathbf{v} = \frac{d\mathbf{x}}{dt}. \tag{2}$$

The pair $(\boldsymbol{\sigma}, \mathbf{D})$ is said to be *energetically conjugate* because it produces the (strain) energy stored in a deformable medium. Show that

(a) the first Piola–Kirchhoff stress tensor $\mathbf{P}$ is energetically conjugate to the rate of deformation gradient $\dot{\mathbf{F}}$, and

(b) the second Piola–Kirchhoff stress tensor $\mathbf{S}$ is energetically conjugate to the rate of Green strain tensor $\dot{\mathbf{E}}$.

Hints: Note the following identities:

$$d\mathbf{x} = J\, d\mathbf{X}, \quad \mathbf{L} \equiv (\boldsymbol{\nabla}\mathbf{v})^{\mathrm{T}} = \dot{\mathbf{F}} \cdot \mathbf{F}^{-1}, \quad \mathbf{P} = J\mathbf{F}^{-1} \cdot \boldsymbol{\sigma}, \quad \boldsymbol{\sigma} = \frac{1}{J}\mathbf{F} \cdot \mathbf{S} \cdot \mathbf{F}^{\mathrm{T}}.$$

CONSTITUTIVE EQUATIONS

The truth is, the science of Nature has been already too long made only a work of the brain and the fancy. It is now high time that it should return to the plainness and soundness of observations on material and obvious things. —— Robert Hooke (1635–1703)

There are two possible outcomes: If the result confirms the hypothesis, then you've made a measurement. If the result is contrary to the hypothesis, then you've made a discovery. —— Enrico Fermi (1901–1954)

6.1 Introduction

6.1.1 General Comments

The kinematic relations developed in Chapter 3, and the principles of conservation of mass, balance of momenta, and thermodynamic principles discussed in Chapter 5, are applicable to any continuum irrespective of its physical constitution. The kinematic variables such as strains and temperature gradient, and kinetic variables such as stresses and heat flux were introduced independently of each other. *Constitutive equations* arc thosc rclations that connect the *primary* field variables (e.g., ρ, θ, $\nabla\theta$, $\mathbf{u}$, $\nabla\mathbf{u}$, $\mathbf{v}$, and $\nabla\mathbf{v}$) to the *secondary* field variables (e.g., e, η, $\mathbf{q}$, and $\boldsymbol{\sigma}$), and they involve the intrinsic physical properties of a continuum. Constitutive equations are *not* derived from any physical principles, although they are subject to obeying certain rules and the entropy inequality. In essence, constitutive equations are mathematical models of the real behavior of materials that are validated against experimental results. The differences between theoretical predictions and experimental findings are often attributed to an inaccurate mathematical representation of the constitutive behavior. *Fluid mechanics*, which deals with liquids and gases, and *solid mechanics*, which deals with metals, fiber-reinforced composites, rubber, ceramics, and so on, share the same field equations (developed in Chapter 5), but their constitutive equations differ considerably.

The main objective of this chapter is to study the most commonly known phenomenological constitutive equations that describe the macroscopic nature of the material response of idealized continua. Constitutive equations from solid mechanics, fluid mechanics, and heat transfer are discussed. We begin with certain terminologies that can be found in introductory texts on mechanics of materials and fluid mechanics.

- A continuum is said to be *homogeneous* if the material properties are the same throughout the continuum (i.e., material properties are independent of position). In a *heterogeneous* continuum, the material properties are a function of position.

- An *anisotropic* continuum is one that has different values of a material property in different directions at a point, that is, material properties are direction dependent.

- An *isotropic* material is one for which a material property is the same in all directions at a point.

An isotropic or anisotropic material can be nonhomogeneous or homogeneous.

6.1.2 General Principles of Constitutive Theory

Constitutive equations are often postulated based on experimental observations. Although experiments are necessary in the determination of various parameters (e.g., elastic constants, thermal conductivity, thermal coefficient of expansion, and coefficients of viscosity) appearing in the constitutive equations, the formulation of the constitutive equations for a given material is guided by certain rules. The approach typically involves assuming the form of the constitutive equation and then restricting the form to a specific one by appealing to certain physical requirements, which are summarized here.

(1) *Consistency (or physical admissibility).* All constitutive equations should be consistent with the conservation of mass, balance of momenta and energy, and the entropy inequality.

(2) *Coordinate frame invariance.* The constitutive equations should not depend on any particular choice of coordinate frame. Although they may have different forms in different coordinate systems, the actual response should be independent of the chosen coordinate system.

(3) *Material frame indifference.* The constitutive equations must be invariant with respect to observer transformations (see Section 3.8). That is, the form of the constitutive function should not change if the material is studied in a different frame of reference. The consequences of this are more apparent in the three-dimensional setting.

(4) *Material symmetry.* The constitutive equations must be form-invariant with respect to a group of unimodular transformations of the material frame of reference. That is, the constitutive equations should reflect material symmetries such as isotropy (infinite number of planes of symmetry) and orthotropy (three mutually perpendicular planes of symmetry).

(5) *Equipresence.* A quantity appearing as an independent variable in one constitutive equation should appear in all constitutive equations, unless

the appearance contradicts the balance laws or the second law of thermo-dynamics. That is, all dependent variables should be functions of the same list of independent variables; one should not *a priori* omit any independent variable.

(6) *Determinism.* The values of the constitutive variables (e.g., stress, heat flux, entropy, and internal energy) at a material point at any time are determined by the histories of motion and temperature of all points of the continuum.

(7) *Local action.* The constitutive variables at a point $\mathbf{x}$ are not apprecia-bly affected by the values of the dependent variables (e.g., displacements, strains, temperature, pressure, etc.) at points distant from $\mathbf{x}$.

(8) *Dimensionality.* The constitutive functionals should be dimensionally con-sistent in the sense that all terms appearing on either side of the constitu-tive equations should be the same.

(9) *Fading memory.* The current values of the constitutive variables are not appreciably affected by their values at past times. This is the time domain counterpart of the axiom of local action.

(10) *Causality.* The variables entering the description of motion of a continuum and temperature are considered as the self-evident observable effects in ev-ery thermomechanical behavior of a continuum. The remaining quantities (i.e., those derivable from the motion and temperature) that enter the expression of entropy production are "causes" or dependent constitutive variables.

These principles/axioms ensure that the initial value problems resulting from the conservation principles, constitutive equations, and physically meaningful boundary and initial conditions are well-posed in the sense that the solution exists and it is unique.

In a continuum theory of constitutive equations, one begins with a general form of functional constitutive equations; seeks to determine if certain variables should be included in the constitutive equation based on some general rules, such as those listed above; and then specializes the equations to certain type, as dictated by the material response being studied. This kind of formal ap-proach enables one to account properly for all possible coupling effects (e.g., thermomechanical, electromechanical, electromagnetics, and so on). An exten-sive discussion of this formal approach is beyond the scope of this introductory book on continuum mechanics and elasticity [see Truesdell and Knoll (1965) for a comprehensive treatment].

6.1.3 Material Frame Indifference

The effect of superposed rigid-body motion $\mathbf{x}^* = \mathbf{c}(t) + \mathbf{Q} \cdot \mathbf{x}$, where $\mathbf{c}$ de-notes rigid-body translation and $\mathbf{Q}$ is a proper orthogonal tensor that represents rigid-body rotation, and the importance of frame indifference in calculating/measuring various quantities introduced in the kinematic and kinetic descriptions of a continuum were discussed in Chapters 3 and 4 [see Eq. (3.8.21)]. In summary, the following relations were established to show that the displacement vector $\mathbf{u}$, deformation gradient $\mathbf{F}$, the right Cauchy–Green deformation tensor $\mathbf{C}$, the Green–Lagrange strain tensor $\mathbf{E}$, the rate of deformation tensor $\mathbf{D}$, the Cauchy stress tensor $\boldsymbol{\sigma}$, the first Piola–Kirchhoff stress tensor $\mathbf{P}$, and the second Piola–Kirchhoff stress tensor $\mathbf{S}$ are objective:

$$\mathbf{u}^* = \mathbf{Q} \cdot \mathbf{u}, \quad \mathbf{F}^* = \mathbf{Q} \cdot \mathbf{F}, \quad \mathbf{C}^* = \mathbf{C}, \quad \mathbf{E}^* = \mathbf{E}, \quad \mathbf{D}^* = \mathbf{Q} \cdot \mathbf{D} \cdot \mathbf{Q}^{\mathrm{T}}$$
$$\boldsymbol{\sigma}^* = \mathbf{Q} \cdot \boldsymbol{\sigma} \cdot \mathbf{Q}^{\mathrm{T}}, \quad \mathbf{P}^* = \mathbf{Q} \cdot \mathbf{P}, \quad \mathbf{S}^* = \mathbf{S},$$

where quantities with an asterisk (*) are those with superposed rigid-body motion; that is, they are the quantities observed in a frame of reference that is undergoing a rigid-body motion with respect to a stationary frame of reference in which quantities without an asterisk are observed. The central idea of the (principle of) material frame indifference is that the constitutive equations relating the quantities introduced in the kinematic description to those appearing in the kinetic description must be independent of the frame of reference (i.e., invariant under observer transformations). *One must make sure that the quantities entering any constitutive equation must be the same type – objective or not – on both sides of the equation.*

6.1.4 Restrictions Placed by the Entropy Inequality

To ensure thermodynamic equilibrium of the processes under consideration, its constitutive equations must be derived using the entropy inequality; if derived in other ways, the constitutive equations must satisfy the conditions resulting from the entropy inequality. Following the axioms discussed in Section 6.1.2 and by examining the momentum and energy equations and the entropy inequality, it can be concluded that the stress tensor $\boldsymbol{\sigma}$, Helmholtz free energy density Ψ, specific entropy η, and heat flux vector $\mathbf{q}$ must be the dependent variables in the constitutive models for a homogeneous and isotropic material. The arguments of these variables depend on the physics of the process. For isotropic and homogeneous materials, $\mathbf{F}$ is a measure of the deformation, the temperature gradient vector $\mathbf{g}$ is needed because of $\mathbf{q}$, and the temperature θ is an obvious choice as an argument for thermoelastic solids. Hence, we have the following functional forms based on the principle of equipresence[1]:

$$\begin{aligned} \boldsymbol{\sigma}(\mathbf{x}, t) &= \mathcal{F}_\sigma[\mathbf{F}, \theta, \mathbf{g}], \\ \Psi(\mathbf{x}, t) &= \mathcal{F}_\Psi[\mathbf{F}, \theta, \mathbf{g}], \\ \eta(\mathbf{x}, t) &= \mathcal{F}_\eta[\mathbf{F}, \theta, \mathbf{g}], \\ \mathbf{q}(\mathbf{x}, t) &= \mathcal{F}_q[\mathbf{F}, \theta, \mathbf{g}], \end{aligned} \qquad (6.1.1)$$

where $\mathcal{F}$ denotes the functional mapping, referred to as the *response function*, whose actual form will become apparent in the following discussion. Any of

---

[1]One can use either Ψ or e, as they are interdependent.

the three arguments $(\mathbf{F}, \theta, \mathbf{g})$ that contradict the constitutive axioms, especially material symmetry, objectivity, or the entropy inequality, will be removed from the argument list in any of the response functions listed in Eq. (6.1.1). The entropy inequality provides guidelines for the form of the constitutive relations.

For example, suppose that Ψ is a function of $\mathbf{F}$, θ, and $\mathbf{g}$, $\Psi = \Psi(\mathbf{F}, \theta, \mathbf{g})$. The entropy inequality from Eq. (5.4.30) is

$$-\rho\dot{\Psi} + \boldsymbol{\sigma} : \mathbf{L} - \rho\dot{\theta}\eta - \frac{1}{\theta}\mathbf{q}\cdot\mathbf{g} \geq 0, \tag{6.1.2}$$

where $\mathbf{g} = \nabla\theta$; $\mathbf{L}$ is the velocity gradient tensor, $\mathbf{L} = (\nabla\mathbf{v})^\mathrm{T} = \mathbf{D} + \mathbf{W}$; $\mathbf{D}$ is the symmetric part; and $\mathbf{W}$ is the skew symmetric part of the velocity gradient tensor $\mathbf{L}$. We can write

$$\dot{\Psi} = \frac{\partial\Psi}{\partial\mathbf{F}} : \dot{\mathbf{F}}^\mathrm{T} + \frac{\partial\Psi}{\partial\theta}\dot{\theta} + \frac{\partial\Psi}{\partial\mathbf{g}}\cdot\dot{\mathbf{g}}. \tag{6.1.3}$$

Substituting for $\dot{\mathbf{F}}$ from Eq. (3.6.15) into Eq. (6.1.3) and the result into Eq. (6.1.2), we obtain (note that $\boldsymbol{\sigma} : \mathbf{L} = \boldsymbol{\sigma} : \mathbf{L}^\mathrm{T}$ when $\boldsymbol{\sigma}$ is symmetric)

$$\left(\boldsymbol{\sigma} - \rho\frac{\partial\Psi}{\partial\mathbf{F}}\cdot\mathbf{F}^\mathrm{T}\right) : \mathbf{L}^\mathrm{T} - \rho\left(\eta + \frac{\partial\Psi}{\partial\theta}\right)\dot{\theta} - \frac{\partial\Psi}{\partial\mathbf{g}}\cdot\dot{\mathbf{g}} - \frac{1}{\theta}\mathbf{q}\cdot\mathbf{g} \geq 0. \tag{6.1.4}$$

In index notation, we have

$$\left(\sigma_{ij} - \rho\frac{\partial\Psi}{\partial F_{iK}}F_{jK}\right)L_{ij} - \rho\left(\eta + \frac{\partial\Psi}{\partial\theta}\right)\dot{\theta} - \frac{\partial\Psi}{\partial g_i}\dot{g}_i - \frac{1}{\theta}q_i g_i \geq 0. \tag{6.1.5}$$

Since $\mathbf{L}$, $\dot{\theta}$, and $\dot{\mathbf{g}}$ are linearly independent of each other, it follows that

$$\boldsymbol{\sigma} - \rho\frac{\partial\Psi}{\partial\mathbf{F}}\cdot\mathbf{F}^\mathrm{T} = \mathbf{0}, \tag{6.1.6}$$

$$\eta + \frac{\partial\Psi}{\partial\theta} = 0, \tag{6.1.7}$$

$$-\frac{\partial\Psi}{\partial\mathbf{g}} = 0, \tag{6.1.8}$$

$$-\mathbf{q}\cdot\mathbf{g} \geq 0. \tag{6.1.9}$$

Equation (6.1.8) implies that Ψ is not a function of the temperature gradient $\mathbf{g}$. Also, Eq. (6.1.9) implies that

$$\mathbf{q}\cdot\mathbf{g} \leq 0. \tag{6.1.10}$$

Therefore, $\mathbf{q}$ is proportional to the negative of the gradient of the temperature $\mathbf{g} = \nabla\theta$, as we will see shortly from the Fourier heat conduction law. Equation (6.1.7) implies that η can be determined from Ψ and, hence, cannot be a dependent variable in the constitutive model. Furthermore, we conclude from Eq. (6.1.6), because Ψ is a function of only $\mathbf{F}$ and θ, that $\boldsymbol{\sigma}$ can depend only on $\mathbf{F}$ and θ. Thus, Eq. (6.1.1) is modified to read

$$\begin{aligned} \Psi(\mathbf{x}, t) &= \mathcal{F}_\Psi[\mathbf{F}, \theta], \\ \eta(\mathbf{x}, t) &= \mathcal{F}_\eta[\mathbf{F}, \theta], \\ \mathbf{q}(\mathbf{x}, t) &= \mathcal{F}_q[\mathbf{F}, \theta, \mathbf{g}]. \end{aligned} \tag{6.1.11}$$

6.2 Elastic Materials

6.2.1 Cauchy-Elastic Materials

A material is called *Cauchy-elastic* or *elastic* if the stress field at time t depends only on the state of deformation and temperature at that time, and not on the history of these variables. The constitutive relation for an elastic body under isothermal conditions (i.e., no change in the temperature from the reference configuration) relates the Cauchy stress tensor $\sigma(\mathbf{x}, t)$ at a point $\mathbf{x} = \chi(\mathbf{X}, t)$ and time t to the deformation gradient $\mathbf{F}(\mathbf{X}, t)$ [see Eq. (6.1.1)]:

$$\sigma(\mathbf{x}, t) = \mathcal{F}[\mathbf{F}(\mathbf{X}, t), \mathbf{X}], \tag{6.2.1}$$

where $\mathcal{F}$ is the response function, and σ denotes the value of $\mathcal{F}$, which characterizes the material properties of an isothermal Cauchy-elastic material. The requirement that the response function $\mathcal{F}$ be unaffected by superposed rigid-body motions places a restriction on $\mathcal{F}$.

Consider the Cauchy stress tensor after superposed rigid-body motion [see Eq. (6.1.1) for the transformation equations of objective quantities]:

$$\sigma^* = \mathcal{F}(\mathbf{F}^*) = \mathcal{F}(\mathbf{Q} \cdot \mathbf{F}), \tag{6.2.2}$$

but

$$\sigma^* = \mathbf{Q} \cdot \sigma \cdot \mathbf{Q}^{\mathrm{T}} = \mathbf{Q} \cdot \mathcal{F}(\mathbf{F}) \cdot \mathbf{Q}^{\mathrm{T}}. \tag{6.2.3}$$

These two relations place the following restriction on $\mathcal{F}$:

$$\mathcal{F}(\mathbf{Q} \cdot \mathbf{F}) = \mathbf{Q} \cdot \mathcal{F}(\mathbf{F}) \cdot \mathbf{Q}^{\mathrm{T}}. \tag{6.2.4}$$

Using the right-hand polar decomposition of $\mathbf{F}$, $\mathbf{F} = \mathbf{R} \cdot \mathbf{U}$, in Eq. (6.2.4), we obtain

$$\mathcal{F}(\mathbf{Q} \cdot \mathbf{R} \cdot \mathbf{U}) = \mathbf{Q} \cdot \mathcal{F}(\mathbf{F}) \cdot \mathbf{Q}^{\mathrm{T}}. \tag{6.2.5}$$

Since $\mathbf{R}$ is a proper orthogonal rotation matrix, we can take $\mathbf{Q} = \mathbf{R}^{\mathrm{T}}$ and obtain $\mathcal{F}(\mathbf{Q} \cdot \mathbf{Q}^{\mathrm{T}} \cdot \mathbf{U}) = \mathcal{F}(\mathbf{U})$. Thus, from Eq. (6.2.5), we have

$$\mathcal{F}(\mathbf{F}) = \mathbf{R} \cdot \mathcal{F}(\mathbf{U}) \cdot \mathbf{R}^{\mathrm{T}}, \tag{6.2.6}$$

which constitutes the restriction on the response function in order that it is objective.

6.2.2 Green-Elastic or Hyperelastic Materials

A hyperelastic material, also known as the *Green-elastic material*, is one for which there exists a *Helmholtz free-energy potential* Ψ (measured per unit volume) whose derivative with respect to a strain gives the corresponding stress and whose derivative with respect to temperature gives the heat flux vector. When Ψ is solely a function of $\mathbf{F}$, $\mathbf{C}$, or some strain tensor, it is called the *strain energy density function* and denoted by U_0 (measured per unit mass). For example, if $U_0 = U_0(\mathbf{F})$, we have [see Eq. (6.1.6)]

$$\mathbf{P} = \rho_0 \frac{\partial U_0(\mathbf{F})}{\partial \mathbf{F}} \quad \left(P_{iJ} = \rho_0 \frac{\partial U_0}{\partial F_{iJ}} \right), \tag{6.2.7}$$

$$\boldsymbol{\sigma} = \frac{1}{J} \mathbf{P} \cdot \mathbf{F}^{\mathrm{T}} = \rho \frac{\partial U_0(\mathbf{F})}{\partial \mathbf{F}} \cdot \mathbf{F}^{\mathrm{T}}, \tag{6.2.8}$$

$$\mathbf{S} = \mathbf{F}^{-1} \cdot \mathbf{P} = \rho_0 \mathbf{F}^{-1} \cdot \frac{\partial U_0(\mathbf{F})}{\partial \mathbf{F}}. \tag{6.2.9}$$

For an incompressible elastic material (i.e., material for which the volume is preserved and hence $J = 1$), we postulate the existence of a strain energy density function in the form

$$\rho_0 \hat{U}_0 = \rho_0 U_0(\mathbf{F}) - p(J - 1), \tag{6.2.10}$$

where p denotes a *hydrostatic pressure*, and $\hat{U}_0(\mathbf{F})$ is the strain energy density for the case $J = 1$. Equation (6.2.10) can be viewed as one in which the strain energy density for the case $J = 1$ is constructed from U_0 by treating $J - 1 = 0$ as a constraint, and using the *Lagrange multiplier method* to include the constraint; it turns out that the *Lagrange multiplier is* $\lambda = -p$. Then the constitutive equation for incompressible hyperelastic material is ($J = 1$):

$$\mathbf{P} = \rho_0 \frac{\partial \hat{U}_0(\mathbf{F})}{\partial \mathbf{F}} = -p \mathbf{F}^{-\mathrm{T}} + \rho_0 \frac{\partial U_0(\mathbf{F})}{\partial \mathbf{F}}, \tag{6.2.11}$$

$$\boldsymbol{\sigma} = \mathbf{P} \cdot \mathbf{F}^{\mathrm{T}} = -p\mathbf{I} + \rho_0 \frac{\partial U_0(\mathbf{F})}{\partial \mathbf{F}} \cdot \mathbf{F}^{\mathrm{T}}, \tag{6.2.12}$$

$$\mathbf{S} = \mathbf{F}^{-1} \cdot \mathbf{P} = -p \mathbf{F}^{-1} \cdot \mathbf{F}^{-\mathrm{T}} + \rho_0 \mathbf{F}^{-1} \cdot \frac{\partial U_0(\mathbf{F})}{\partial \mathbf{F}}, \tag{6.2.13}$$

where the derivative of J with respect to $\mathbf{F}$ is (the reader is asked to verify this)

$$\frac{\partial J}{\partial \mathbf{F}} = J \mathbf{F}^{-\mathrm{T}}. \tag{6.2.14}$$

6.2.3 Linearized Hyperelastic Materials: Infinitesimal Strains

Here we present the constitutive equations for the case of infinitesimal deformation (ie., $|\nabla \mathbf{u}| = O(\epsilon) \ll 1$). Hence, we do not distinguish between various measures of stress and strain, and use $\mathbf{S} \approx \boldsymbol{\sigma}$ for the stress tensor and $\mathbf{E} \approx \boldsymbol{\varepsilon}$ for the strain tensor in the material description used in solid mechanics. For such materials, the Helmholtz free energy density Ψ is the same as the strain energy density U_0, and it is more meaningful to assume that the strain energy density is a function of the strain, $\boldsymbol{\varepsilon}$, rather than the deformation gradient, although one may also assume that U_0 is a function of the strain invariants.

The constitutive equation to be developed here for stress tensor $\boldsymbol{\sigma}$ does not include creep at constant stress and stress relaxation at constant strain. Thus, the material coefficients that specify the constitutive relationship between the stress and strain components are assumed to be constant during the deformation. This does not automatically imply that we neglect temperature effects on deformation. We account for the thermal expansion of the material, which can produce strains or stresses as large as those produced by the applied mechanical forces.

The constitutive equation for linearized hyperelastic materials can be derived using

$$\boldsymbol{\sigma} = \rho_0 \frac{\partial U_0(\boldsymbol{\varepsilon})}{\partial \boldsymbol{\varepsilon}} \quad \left(\sigma_{ij} = \rho_0 \frac{\partial U_0}{\partial \varepsilon_{ij}} \right). \tag{6.2.15}$$

As indicated earlier, here we assume that U_0 is a function of $\boldsymbol{\varepsilon}$ and expand it in Taylor's series about the strain $\boldsymbol{\varepsilon} = 0$ in the reference configuration,

$$\rho_0 U_0 = C_0 + C_{ij}\, \varepsilon_{ij} + \frac{1}{2!} C_{ijk\ell}\, \varepsilon_{ij}\, \varepsilon_{k\ell} + \frac{1}{3!} C_{ijk\ell mn}\, \varepsilon_{ij}\, \varepsilon_{k\ell}\, \varepsilon_{mn} + \dots, \tag{6.2.16}$$

where C_0, C_{ij}, $C_{ijk\ell}$, and so on are material stiffness coefficients that are independent of the deformation. For linear elastic materials U_0 is a quadratic function of the strain tensor, and for nonlinear elastic materials, U_0 is a cubic function of the strain tensor $\boldsymbol{\varepsilon}$. For linear elastic materials, the mechanical pressure is the same as the negative of the mean normal stress.

This chapter is focused primarily on constitutive relations for Hookean solids (linear elastic solids), Newtonian fluids (fluids with linear relations between stress and strain rate), and Fourier heat conduction law (a linear relation between the heat flux vector and the temperature gradient vector). The constitutive equations presented in Section 6.3 for elastic solids are based on an assumption of small strain. Nonlinear constitutive relations for elastic solids are briefly discussed in Section 6.4. In Section 6.5, constitutive relations for Newtonian fluids are presented, and in Section 6.6 differential and integral generalized Newtonian constitutive relations are reviewed. The Fourier heat conduction law is presented in Section 6.7. Finally, constitutive relations for coupled problems, for example, electromagnetics, electroelasticity, and thermoelasticity , are presented in Section 6.8.

6.3 Hookean Solids

6.3.1 Generalized Hooke's Law

To develop the stress–strain relations for a linear elastic solid, we set up a coordinate system in which the material parameters are measured. This coordinate system is termed the *material coordinate system*, not to be confused with the *material description* of Chapter 5. The coordinate system used to write the equations of motion and strain-displacement equations is called the *problem coordinates* to distinguish it from the material coordinate system. In the remaining discussion of this section, we use the Lagrangian description with coordinates (x, y, z) to describe the kinematics, stress state, and the field equations, and use the material coordinate system (x_1, x_2, x_3) to describe the constitutive response. The material coordinate system is one that is aligned with the *planes of material symmetry* (to be defined shortly), so that measurement of material parameters becomes simple. Of course, the constitutive relations have to be transformed to the problem coordinates in order to solve the final boundary-value problem. When no preferred planes of material symmetry exist, the material is called isotropic, and the material coordinates are taken to be the same as the problem coordinates. All tensor quantities measured in (x_1, x_2, x_3) will have integer subscripts, for example, σ_{ij}, ε_{ij}, and so on, whereas those measured in (x, y, z) will have letter subscripts, for example, $\sigma_{xx}, \sigma_{xy}, \cdots$, and $\varepsilon_{xx}, \varepsilon_{xy}, \cdots$, and so on.

We begin with the quadratic form of U_0 in the material coordinate system:

$$\rho_0 U_0 = C_0 + C_{ij}\,\varepsilon_{ij} + \tfrac{1}{2}C_{ijk\ell}\,\varepsilon_{ij}\,\varepsilon_{k\ell}, \tag{6.3.1}$$

where C_0 is a reference value of U_0 from which the strain energy density function is measured. From Eq. (6.2.15), we have

$$\sigma_{mn} = \rho_0 \frac{\partial U_0}{\partial \varepsilon_{mn}} = C_{ij}\,\delta_{mi}\,\delta_{nj} + \tfrac{1}{2}C_{ijk\ell}\,(\varepsilon_{k\ell}\,\delta_{im}\,\delta_{jn} + \varepsilon_{ij}\delta_{km}\delta_{\ell n})$$

$$= C_{mn} + \tfrac{1}{2}C_{mnk\ell}\,\varepsilon_{k\ell} + \tfrac{1}{2}C_{ijmn}\,\varepsilon_{ij} = C_{mn} + \tfrac{1}{2}\left(C_{mnij} + C_{ijmn}\right)\varepsilon_{ij}$$

$$= C_{mn} + C_{mnij}\,\varepsilon_{ij}, \tag{6.3.2}$$

where

$$C_{mnij} = \tfrac{1}{2}\left(C_{mnij} + C_{ijmn}\right) = \rho_0 \frac{\partial^2 U_0}{\partial \varepsilon_{ij}\partial \varepsilon_{mn}} = \rho_0 \frac{\partial^2 U_0}{\partial \varepsilon_{mn}\partial \varepsilon_{ij}} = C_{ijmn}. \tag{6.3.3}$$

Clearly, C_{mn} have the same units as σ_{mn}, and they represent the *residual stress* components of a solid. We shall assume, without loss of generality, that the body is free of stress prior to the load application so that we may write

$$\boldsymbol{\sigma} = \mathbf{C} : \boldsymbol{\varepsilon} \qquad (\sigma_{ij} = C_{ijk\ell}\,\varepsilon_{k\ell}). \tag{6.3.4}$$

Equation (6.3.4) is known as the generalized Hooke's law. The coefficients $C_{ijk\ell}$ are called elastic *stiffness* coefficients. In general, there are $3^4 = 81$ scalar components of the fourth-order tensor[2] $\mathbf{C}$. The number of coefficients is significantly reduced because (a) the components $C_{ijk\ell}$ satisfy the symmetry conditions implied by Eq. (6.3.3), and (b) the stress and strain tensors are symmetric, requiring Eq. (6.3.4) to be valid when subscripts i and j are interchanged as well as k and ℓ are interchanged. Thus, we have

$$C_{ijk\ell} = C_{k\ell ij}, \quad C_{ijk\ell} = C_{jik\ell}, \quad C_{ij\ell k} = C_{ijk\ell}, \quad C_{ijk\ell} = C_{ji\ell k}, \tag{6.3.5}$$

and the stress–strain relations (6.3.4) take the form

$$\begin{Bmatrix} \sigma_{11} \\ \sigma_{22} \\ \sigma_{33} \\ \sigma_{23} \\ \sigma_{13} \\ \sigma_{12} \end{Bmatrix} = \begin{bmatrix} C_{1111} & C_{1122} & C_{1133} & C_{1123} & C_{1113} & C_{1112} \\ & C_{2222} & C_{2233} & C_{2223} & C_{2213} & C_{2212} \\ & & C_{3333} & C_{3323} & C_{3313} & C_{3312} \\ & & & C_{2323} & C_{2313} & C_{2312} \\ & \text{symmetric} & & & C_{1313} & C_{1312} \\ & & & & & C_{1212} \end{bmatrix} \begin{Bmatrix} \varepsilon_{11} \\ \varepsilon_{22} \\ \varepsilon_{33} \\ 2\varepsilon_{23} \\ 2\varepsilon_{13} \\ 2\varepsilon_{12} \end{Bmatrix}. \tag{6.3.6}$$

---

[2]In this chapter $\mathbf{C}$ denotes the fourth-order elasticity tensor $\mathbf{C}$, not the right Cauchy–Green deformation tensor $\mathbf{C}$.

Thus the number of independent coefficients C_{ijmn} is only $6+5+4+3+2+1 = 21$. Materials that obey Eq. (6.3.6) are called *triclinic* materials.

We can express Eq. (6.3.4) in an alternative form using a single subscript notation for stresses and strains and a two subscript notation for the material stiffness coefficients:

$$\sigma_1 = \sigma_{11}, \ \sigma_2 = \sigma_{22}, \ \sigma_3 = \sigma_{33}, \ \sigma_4 = \sigma_{23}, \ \sigma_5 = \sigma_{13}, \ \sigma_6 = \sigma_{12},$$
$$\varepsilon_1 = \varepsilon_{11}, \ \varepsilon_2 = \varepsilon_{22}, \ \varepsilon_3 = \varepsilon_{33}, \ \varepsilon_4 = 2\varepsilon_{23}, \ \varepsilon_5 = 2\varepsilon_{13}, \ \varepsilon_6 = 2\varepsilon_{12}.$$
$$(6.3.7)$$

$$11 \to 1 \quad 22 \to 2 \quad 33 \to 3 \quad 23 \to 4 \quad 13 \to 5 \quad 12 \to 6. \qquad (6.3.8)$$

It should be cautioned that the single subscript notation used for stresses and strains and the two-subscript components C_{ij} render them non-tensor components; that is, σ_i, ε_i, and C_{ij} do not transform like the components of a tensor, $\sigma_i \neq \ell_{ij}\sigma_j$. The single subscript notation for stresses and strains is called the *engineering notation* or the *Voigt–Kelvin notation*. Equation (6.3.6) now takes the form

$$\begin{Bmatrix} \sigma_1 \\ \sigma_2 \\ \sigma_3 \\ \sigma_4 \\ \sigma_5 \\ \sigma_6 \end{Bmatrix} = \begin{bmatrix} C_{11} & C_{12} & C_{13} & C_{14} & C_{15} & C_{16} \\ C_{21} & C_{22} & C_{23} & C_{24} & C_{25} & C_{26} \\ C_{31} & C_{32} & C_{33} & C_{34} & C_{35} & C_{36} \\ C_{41} & C_{42} & C_{43} & C_{44} & C_{45} & C_{46} \\ C_{51} & C_{52} & C_{53} & C_{54} & C_{55} & C_{56} \\ C_{61} & C_{62} & C_{63} & C_{64} & C_{65} & C_{66} \end{bmatrix} \begin{Bmatrix} \varepsilon_1 \\ \varepsilon_2 \\ \varepsilon_3 \\ \varepsilon_4 \\ \varepsilon_5 \\ \varepsilon_6 \end{Bmatrix}, \qquad (6.3.9)$$

or simply

$$\sigma_i = C_{ij}\,\varepsilon_j, \qquad (6.3.10)$$

where summation on repeated subscripts is implied (now i and j take values from 1 to 6). Note that the coefficients C_{ij} are symmetric, $C_{ij} = C_{ji}$, a property inherited from Eq. (6.3.6).

We assume that the stress–strain relations (6.3.10) are invertible. Thus, the components of strain are related to the components of stress by

$$\varepsilon_i = S_{ij}\,\sigma_j, \qquad (6.3.11)$$

where $S_{ij} = S_{ji}$ are the material *compliance* coefficients with $[S] = [C]^{-1}$ [i.e., the compliance tensor is the inverse of the stiffness tensor: $\mathbf{S} = \mathbf{C}^{-1}$]. In matrix form, Eq. (6.3.11) becomes

$$\begin{Bmatrix} \varepsilon_1 \\ \varepsilon_2 \\ \varepsilon_3 \\ \varepsilon_4 \\ \varepsilon_5 \\ \varepsilon_6 \end{Bmatrix} = \begin{bmatrix} S_{11} & S_{12} & S_{13} & S_{14} & S_{15} & S_{16} \\ S_{21} & S_{22} & S_{23} & S_{24} & S_{25} & S_{26} \\ S_{31} & S_{32} & S_{33} & S_{34} & S_{35} & S_{36} \\ S_{41} & S_{42} & S_{43} & S_{44} & S_{45} & S_{46} \\ S_{51} & S_{52} & S_{53} & S_{54} & S_{55} & S_{56} \\ S_{61} & S_{62} & S_{63} & S_{64} & S_{65} & S_{66} \end{bmatrix} \begin{Bmatrix} \sigma_1 \\ \sigma_2 \\ \sigma_3 \\ \sigma_4 \\ \sigma_5 \\ \sigma_6 \end{Bmatrix}. \qquad (6.3.12)$$

The strain–stress relations are more suitable in determining the material constants in a laboratory because experiments involve the application of loads and measurement of changes in the geometry (i.e., determine strains from an applied stress state).

6.3.2 Material Symmetry Planes

Further reduction in the number of independent stiffness (or compliance) parameters comes from the so-called material symmetry. When elastic material parameters at a point have *the same values* for every pair of coordinate systems that are mirror images of each other in a certain plane, that plane is called a *material plane of symmetry* (for example, symmetry of internal structure due to crystallographic form, *regular* arrangement of fibers or molecules, and so on). We note that the symmetry under discussion is a directional property and not a positional property. Thus, a material may have a certain elastic symmetry at every point of a material body and the properties may vary from point to point. Positional dependence of material properties is what we called inhomogeneity of the material.

In the following, we discuss various planes of symmetry and forms of associated stress–strain relations. Note that the use of components of stress and strain tensors is necessary in the following discussion because transformation equations are valid only for components of tensors from two different coordinate systems. The components σ_{ij} and ε_{ij} of second-order tensors $\boldsymbol{\sigma}$ and $\boldsymbol{\varepsilon}$ and the components C_{ijkl} of a fourth-order elasticity tensor $\mathbf{C}$ transform according to the relations

$$\bar{\sigma}_{ij} = \ell_{ip}\,\ell_{jq}\,\sigma_{pq}, \quad \bar{\varepsilon}_{ij} = \ell_{ip}\,\ell_{jq}\,\varepsilon_{pq}, \quad \bar{C}_{ijkl} = \ell_{ip}\,\ell_{jq}\,\ell_{kr}\,\ell_{ls}\,C_{pqrs}, \qquad (6.3.13)$$

where ℓ_{ij} are the direction cosines associated with the coordinate systems $(\bar{x}_1, \bar{x}_2, \bar{x}_3)$ and (x_1, x_2, x_3), and $\bar{C}_{ijkl}$ and C_{pqrs}, for example, are the components of the fourth-order tensor $\mathbf{C}$ in the barred and unbarred coordinates systems, respectively [see Eqs. (2.2.70), (2.2.71), (3.4.29), and (4.3.3)].

A trivial material symmetry transformation is one in which the barred coordinate system is obtained from the unbarred coordinate system by simply reversing their directions (i.e., mirror reflection): $\bar{x}_1 = -x_1$, $\bar{x}_2 = -x_2$, and $\bar{x}_3 = -x_3$; that is, $\hat{\mathbf{e}}_i = -\hat{\mathbf{e}}_i$ and $\ell_{ij} = -\delta_{ij}$ (it does not matter that it is a left-handed coordinate system as it does not affect the discussion), as shown in Fig. 6.3.1(a).

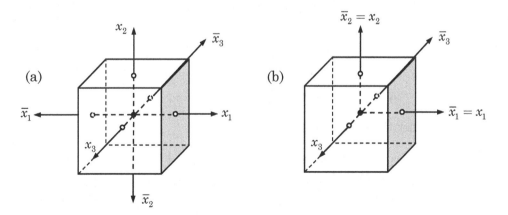

Fig. 6.3.1: (a) Transformation defined by $\hat{\bar{\mathbf{e}}}_i = -\hat{\mathbf{e}}_i$, $i = 1, 2, 3$. (b) Transformation defined by $\hat{\bar{\mathbf{e}}}_\alpha = \hat{\mathbf{e}}_\alpha$, $\alpha = 1, 2$, and $\hat{\bar{\mathbf{e}}}_3 = -\hat{\mathbf{e}}_3$.

Thus, the transformation matrix is

$$[L] = \begin{bmatrix} -1 & 0 & 0 \\ 0 & -1 & 0 \\ 0 & 0 & -1 \end{bmatrix}. \tag{6.3.14}$$

For this transformation, it follows that

$$\bar{\sigma}_{ij} = (-\delta_{ip})(-\delta_{jq})\sigma_{pq} = \sigma_{ij}; \quad \bar{\varepsilon}_{ij} = (-\delta_{ip})(-\delta_{jq})\varepsilon_{pq} = \varepsilon_{ij},$$
$$\bar{C}_{ijk\ell} = (-1)^4 \delta_{ip}\,\delta_{jq}\,\delta_{kr}\,\delta_{\ell s}\,C_{pqrs} = C_{ijk\ell}.$$

Thus, the transformation $\ell_{ij} = -\delta_{ij}$ does not alter the constitutive relation (6.3.6) of triclinic materials.

6.3.3 Monoclinic Materials

When the elastic coefficients at a point have the same value for every pair of coordinate systems that are the mirror images of each other with respect to a plane, the material is called *monoclinic* at the point. For example, let (x_1, x_2, x_3) and $(\bar{x}_1, \bar{x}_2, \bar{x}_3)$ be two coordinates systems, with the x_1, x_2-plane parallel to the plane of symmetry. Choose the $\bar{x}_3$-axis such that $\bar{x}_3 = -x_3$ so that one system is the mirror image of the other, as shown in Fig. 6.3.1(b). This symmetry transformation can be expressed by the transformation matrix ($\bar{x}_1 = x_1$, $\bar{x}_2 = x_2$, $\bar{x}_3 = -x_3$)

$$[L] = \begin{bmatrix} 1 & 0 & 0 \\ 0 & 1 & 0 \\ 0 & 0 & -1 \end{bmatrix}, \tag{6.3.15}$$

or

$$\ell_{\alpha\beta} = \delta_{\alpha\beta}, \quad \ell_{3\alpha} = 0, \quad \ell_{\alpha 3} = 0, \quad \ell_{33} = -1, \quad \text{for } \alpha, \beta = 1, 2.$$

Then the stress and strain transformation equations

$$[\bar{\sigma}] = [L][\sigma][L]^{\mathrm{T}}, \quad [\bar{\varepsilon}] = [L][\varepsilon][L]^{\mathrm{T}}$$

give the relations

$$\bar{\sigma}_{ij} = \sigma_{ij}, \quad \text{except for} \quad \bar{\sigma}_{13} = -\sigma_{13}, \quad \bar{\sigma}_{23} = -\sigma_{23},$$
$$\bar{\varepsilon}_{ij} = \varepsilon_{ij}, \quad \text{except for} \quad \bar{\varepsilon}_{13} = -\varepsilon_{13}, \quad \bar{\varepsilon}_{23} = -\varepsilon_{23}.$$

Now consider the stress–strain relations

$$\sigma_{11} = C_{1111}\,\varepsilon_{11} + C_{1122}\,\varepsilon_{22} + C_{1133}\,\varepsilon_{33} + C_{1123}\,\varepsilon_{23} + C_{1113}\,\varepsilon_{13} + C_{1112}\,\varepsilon_{12},$$
$$\bar{\sigma}_{11} = C_{1111}\,\bar{\varepsilon}_{11} + C_{1122}\,\bar{\varepsilon}_{22} + C_{1133}\,\bar{\varepsilon}_{33} + C_{1123}\,\bar{\varepsilon}_{23} + C_{1113}\,\bar{\varepsilon}_{13} + C_{1112}\,\bar{\varepsilon}_{12}$$
$$= C_{1111}\,\varepsilon_{11} + C_{1122}\,\varepsilon_{22} + C_{1133}\,\varepsilon_{33} - C_{1123}\,\varepsilon_{23} - C_{1113}\,\varepsilon_{13} + C_{1112}\,\varepsilon_{12}.$$

Since $\sigma_{11} = \bar{\sigma}_{11}$, from the preceding two relations it follows that

$$C_{1123}\,\varepsilon_{23} + C_{1113}\,\varepsilon_{13} = -C_{1123}\,\varepsilon_{23} - C_{1113}\,\varepsilon_{13},$$

which must hold for any independent set of strain components, ε_{23} and ε_{13}. This implies that $C_{1123} = 0$ and $C_{1113} = 0$. Similarly, from the constitutive relations for σ_{22} and $\bar{\sigma}_{22}$, σ_{33} and $\bar{\sigma}_{33}$, and σ_{12} and $\bar{\sigma}_{12}$ we obtain $C_{2223} = C_{2213} = 0$, $C_{3323} = C_{3313} = 0$, and $C_{1223} = C_{1213} = 0$.

Next consider the constitutive relations for σ_{23} and $\bar{\sigma}_{23}$ (note $C_{ijk\ell} = C_{k\ell ij}$)

$$\sigma_{23} = C_{2311}\,\varepsilon_{11} + C_{2322}\,\varepsilon_{22} + C_{2333}\,\varepsilon_{33} + C_{2323}\,\varepsilon_{23} + C_{2313}\,\varepsilon_{13} + C_{2312}\,\varepsilon_{12}$$
$$= C_{2323}\,\varepsilon_{23} + C_{2313}\,\varepsilon_{13}$$
$$\bar{\sigma}_{23} = C_{2311}\,\bar{\varepsilon}_{11} + C_{2322}\,\bar{\varepsilon}_{22} + C_{2333}\,\bar{\varepsilon}_{33} + C_{2323}\,\bar{\varepsilon}_{23} + C_{2313}\,\bar{\varepsilon}_{13} + C_{2312}\,\bar{\varepsilon}_{12}$$
$$= -C_{2323}\,\varepsilon_{23} - C_{2313}\,\varepsilon_{13}.$$

Since $\sigma_{23} = -\bar{\sigma}_{23}$, these two relations are consistent. In the same way, no new conditions are obtained by considering the constitutive relations for σ_{13} and $\bar{\sigma}_{13}$.

In summary, for monoclinic materials, 8 of the 21 coefficients are zero:

$$C_{1123} = C_{1113} = C_{2223} = C_{2213} = C_{3323} = C_{3313} = C_{1223} = C_{1213} = 0.$$

Therefore, the stress–strain relations of Eq. (6.3.9) become

$$\begin{Bmatrix} \sigma_1 \\ \sigma_2 \\ \sigma_3 \\ \sigma_4 \\ \sigma_5 \\ \sigma_6 \end{Bmatrix} = \begin{bmatrix} C_{11} & C_{12} & C_{13} & 0 & 0 & C_{16} \\ C_{12} & C_{22} & C_{23} & 0 & 0 & C_{26} \\ C_{13} & C_{23} & C_{33} & 0 & 0 & C_{36} \\ 0 & 0 & 0 & C_{44} & C_{45} & 0 \\ 0 & 0 & 0 & C_{45} & C_{55} & 0 \\ C_{16} & C_{26} & C_{36} & 0 & 0 & C_{66} \end{bmatrix} \begin{Bmatrix} \varepsilon_1 \\ \varepsilon_2 \\ \varepsilon_3 \\ \varepsilon_4 \\ \varepsilon_5 \\ \varepsilon_6 \end{Bmatrix}, \qquad (6.3.16)$$

which has only 13 independent parameters. We note that monoclinic materials exhibit shear-extensional coupling, that is, a shear strain can produce a normal stress and a normal stress can produce a shear strain.

6.3.4 Orthotropic Materials

When three mutually orthogonal planes of material symmetry exist at a point, the number of elastic coefficients is reduced to nine using arguments similar to those given for a single material symmetry plane. Such materials are called *orthotropic* at the point. The transformation matrices associated with the three planes of symmetry are

$$[L^{(1)}] = \begin{bmatrix} 1 & 0 & 0 \\ 0 & 1 & 0 \\ 0 & 0 & -1 \end{bmatrix}, \quad [L^{(2)}] = \begin{bmatrix} -1 & 0 & 0 \\ 0 & 1 & 0 \\ 0 & 0 & 1 \end{bmatrix}, \quad [L^{(3)}] = \begin{bmatrix} 1 & 0 & 0 \\ 0 & -1 & 0 \\ 0 & 0 & 1 \end{bmatrix}. \quad (6.3.17)$$

Under these transformations, we obtain $C_{1112} = C_{16} = 0$, $C_{2212} = C_{26} = 0$, $C_{3312} = C_{36} = 0$, and $C_{2313} = C_{45} = 0$. In view of the aforementioned result, the stress–strain relations for an orthotropic material take the form

$$\begin{Bmatrix} \sigma_1 \\ \sigma_2 \\ \sigma_3 \\ \sigma_4 \\ \sigma_5 \\ \sigma_6 \end{Bmatrix} = \begin{bmatrix} C_{11} & C_{12} & C_{13} & 0 & 0 & 0 \\ C_{12} & C_{22} & C_{23} & 0 & 0 & 0 \\ C_{13} & C_{23} & C_{33} & 0 & 0 & 0 \\ 0 & 0 & 0 & C_{44} & 0 & 0 \\ 0 & 0 & 0 & 0 & C_{55} & 0 \\ 0 & 0 & 0 & 0 & 0 & C_{66} \end{bmatrix} \begin{Bmatrix} \varepsilon_1 \\ \varepsilon_2 \\ \varepsilon_3 \\ \varepsilon_4 \\ \varepsilon_5 \\ \varepsilon_6 \end{Bmatrix}. \tag{6.3.18}$$

As stated earlier, in practice we apply stresses and determine the strains. Hence we must write the inverse of the relations in Eq. (6.3.18):

$$\begin{Bmatrix} \varepsilon_1 \\ \varepsilon_2 \\ \varepsilon_3 \\ \varepsilon_4 \\ \varepsilon_5 \\ \varepsilon_6 \end{Bmatrix} = \begin{bmatrix} S_{11} & S_{12} & S_{13} & 0 & 0 & 0 \\ S_{12} & S_{22} & S_{23} & 0 & 0 & 0 \\ S_{13} & S_{23} & S_{33} & 0 & 0 & 0 \\ 0 & 0 & 0 & S_{44} & 0 & 0 \\ 0 & 0 & 0 & 0 & S_{55} & 0 \\ 0 & 0 & 0 & 0 & 0 & S_{66} \end{bmatrix} \begin{Bmatrix} \sigma_1 \\ \sigma_2 \\ \sigma_3 \\ \sigma_4 \\ \sigma_5 \\ \sigma_6 \end{Bmatrix}. \tag{6.3.19}$$

The material compliance coefficients S_{ij} are often determined in a laboratory in terms of engineering material parameters such as Young's modulus, shear modulus, and so on. These constants are measured using simple tests such as a uniaxial tension test or a pure shear test. Because of their direct and obvious physical meaning, engineering constants are used in place of the abstract compliance coefficients S_{ij}. Next, we discuss how the compliance coefficients S_{ij} are determined in terms of the engineering parameters.

One of the consequences of linearity (both kinematic and material linearizations) is that the principle of superposition applies. That is, if the applied loads and geometric constraints are independent of deformation, the sum of the displacements (and hence strains) produced by two sets of loads is equal to the displacements (and strains) produced by the sum of the two sets of loads. In particular, the strains of the same kind as produced by the application of individual stress components can be superposed. For example, the extensional strain $\varepsilon_{11}^{(1)}$ in the material coordinate direction x_1 due to the stress σ_{11} in the same direction is σ_{11}/E_1, as shown in Fig. 6.3.2; here E_1 denotes Young's modulus of the material in the x_1 direction. The extensional strain $\varepsilon_{11}^{(2)}$, experienced as a result of the Poisson effect, due to the stress σ_{22} applied in the x_2 direction, is $-\nu_{21}(\sigma_{22}/E_2)$, where ν_{21} is Poisson's ratio (note that the first subscript in ν_{ij}, $i \neq j$, corresponds to the load direction and the second subscript refers to the direction of the strain)

$$\varepsilon_{11} = -\nu_{21}\varepsilon_{22} \quad \text{or} \quad \nu_{21} = -\frac{\varepsilon_{11}}{\varepsilon_{22}},$$

and E_2 is Young's modulus of the material in the x_2 direction. Similarly, σ_{33} produces a strain $\varepsilon_{11}^{(3)}$ equal to $-\nu_{31}(\sigma_{33}/E_3)$. Therefore, the total strain ε_{11} due to the simultaneous application of all three normal stress components is

$$\varepsilon_{11} = \varepsilon_{11}^{(1)} + \varepsilon_{11}^{(2)} + \varepsilon_{11}^{(3)} = \frac{\sigma_{11}}{E_1} - \nu_{21}\frac{\sigma_{22}}{E_2} - \nu_{31}\frac{\sigma_{33}}{E_3}$$

$$= S_{11}\sigma_{11} + S_{12}\sigma_{22} + S_{13}\sigma_{33} \tag{6.3.20}$$

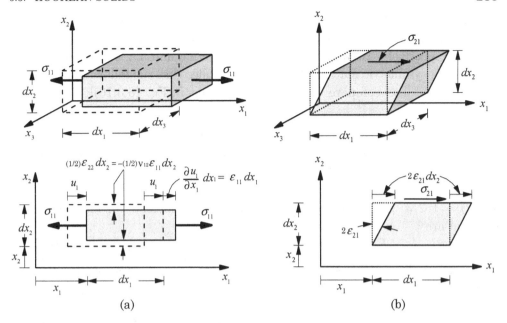

Fig. 6.3.2: Strains produced by applied (a) normal stress σ_{11} and (b) shear stress $\sigma_{21} = \sigma_{12}$ in a cube of material.

where the direction of loading is denoted by the superscript. Similarly, we can write

$$
\begin{aligned}
\varepsilon_{22} &= -\nu_{12}\frac{\sigma_{11}}{E_1} + \frac{\sigma_{22}}{E_2} - \nu_{32}\frac{\sigma_{33}}{E_3} = S_{21}\sigma_{11} + S_{22}\sigma_{22} + S_{23}\sigma_{33}, \\
\varepsilon_{33} &= -\nu_{13}\frac{\sigma_{11}}{E_1} - \nu_{23}\frac{\sigma_{22}}{E_2} + \frac{\sigma_{33}}{E_3} = S_{31}\sigma_{11} + S_{32}\sigma_{22} + S_{33}\sigma_{33}.
\end{aligned}
\tag{6.3.21}
$$

The simple shear tests with an orthotropic material give the results

$$
2\varepsilon_{12} = \frac{\sigma_{12}}{G_{12}} = S_{66}\sigma_{12}, \quad 2\varepsilon_{13} = \frac{\sigma_{13}}{G_{13}} = S_{55}\sigma_{13}, \quad 2\varepsilon_{23} = \frac{\sigma_{23}}{G_{23}} = S_{44}\sigma_{23}. \tag{6.3.22}
$$

Recall from Section 3.5.2 that $2\varepsilon_{ij}$ $(i \neq j)$ is the reduction in the right angle between two material lines parallel to the x_1 and x_2 directions at a point, σ_{ij} $(i \neq j)$ denotes the corresponding shear stress in the x_i–x_j plane, and G_{ij} $(i \neq j)$ is the shear moduli in the x_i–x_j plane.

Writing Eqs. (6.3.20)–(6.3.22) in matrix form, we obtain

$$
\begin{Bmatrix} \varepsilon_1 \\ \varepsilon_2 \\ \varepsilon_3 \\ \varepsilon_4 \\ \varepsilon_5 \\ \varepsilon_6 \end{Bmatrix} =
\begin{bmatrix}
\frac{1}{E_1} & -\frac{\nu_{21}}{E_2} & -\frac{\nu_{31}}{E_3} & 0 & 0 & 0 \\
-\frac{\nu_{12}}{E_1} & \frac{1}{E_2} & -\frac{\nu_{32}}{E_3} & 0 & 0 & 0 \\
-\frac{\nu_{13}}{E_1} & -\frac{\nu_{23}}{E_2} & \frac{1}{E_3} & 0 & 0 & 0 \\
0 & 0 & 0 & \frac{1}{G_{23}} & 0 & 0 \\
0 & 0 & 0 & 0 & \frac{1}{G_{13}} & 0 \\
0 & 0 & 0 & 0 & 0 & \frac{1}{G_{12}}
\end{bmatrix}
\begin{Bmatrix} \sigma_1 \\ \sigma_2 \\ \sigma_3 \\ \sigma_4 \\ \sigma_5 \\ \sigma_6 \end{Bmatrix}
\quad (\{\varepsilon\} = [S]\{\sigma\}), \tag{6.3.23}
$$

where E_1, E_2, and E_3 are Young's moduli in 1, 2, and 3 material directions, respectively; ν_{ij} is Poisson's ratio, defined as the ratio of transverse strain in the jth direction to the axial strain in the ith direction when stressed in the i-direction; and G_{23}, G_{13}, G_{12} are the shear moduli in the 2–3, 1–3, and 1–2 planes, respectively. Because $[S]$ is the inverse of $[C]$ and $[C]$ is symmetric, then $[S]$ is also a symmetric matrix. This in turn implies that the following reciprocal relations hold [i.e., compare the off-diagonal terms in Eq. (6.3.10)]:

$$\frac{\nu_{21}}{E_2} = \frac{\nu_{12}}{E_1}; \quad \frac{\nu_{31}}{E_3} = \frac{\nu_{13}}{E_1}; \quad \frac{\nu_{32}}{E_3} = \frac{\nu_{23}}{E_2},$$

or

$$\frac{\nu_{ij}}{E_i} = \frac{\nu_{ji}}{E_j}, \tag{6.3.24}$$

for $i, j = 1, 2, 3$. The nine independent material coefficients for an orthotropic material are

$$E_1, \ E_2, \ E_3, \ G_{23}, \ G_{13}, \ G_{12}, \ \nu_{12}, \ \nu_{13}, \ \nu_{23}. \tag{6.3.25}$$

Inversion of the strain-stress relations (6.3.23) give the stress–strain relations in Eq. (6.3.18) with

$$C_{11} = \frac{E_1}{C_0} \left(1 - \nu_{23}\nu_{32}\right), \quad C_{12} = \frac{E_1}{C_0} \left(\nu_{21} + \nu_{23}\nu_{31}\right) = \frac{E_2}{C_0} \left(\nu_{12} + \nu_{13}\nu_{32}\right),$$

$$C_{13} = \frac{E_1}{C_0} \left(\nu_{31} + \nu_{21}\nu_{32}\right) = \frac{E_3}{C_0} \left(\nu_{13} + \nu_{12}\nu_{23}\right), \quad C_{22} = \frac{E_2}{C_0} \left(1 - \nu_{13}\nu_{31}\right),$$

$$C_{23} = \frac{E_2}{C_0} \left(\nu_{32} + \nu_{31}\nu_{12}\right) = \frac{E_3}{C_0} \left(\nu_{23} + \nu_{21}\nu_{13}\right), \quad C_{33} = \frac{E_3}{C_0} \left(1 - \nu_{12}\nu_{21}\right),$$

$$C_{44} = G_{23} \quad C_{55} = G_{31}, \quad C_{66} = G_{12},$$

$$C_0 = 1 - \nu_{12}\nu_{21} - \nu_{23}\nu_{32} - \nu_{31}\nu_{13} - 2\nu_{21}\nu_{32}\nu_{13}. \tag{6.3.26}$$

The difference between ν_{12} and ν_{21} for an orthotropic material is illustrated in Fig. 6.3.3 with two cases of uniaxial stress for a square element of length a. First a stress σ_{11} is applied in the x_1-direction as shown in Fig. 6.3.3(a). The resulting strains are

$$\varepsilon_{11}^{(1)} = \frac{\sigma_{11}}{E_1}, \quad \varepsilon_{22}^{(1)} = -\frac{\nu_{12}}{E_1}\sigma_{11}, \tag{6.3.27}$$

where the direction of loading is denoted by the superscript and the negative sign indicates compression. Next, a stress σ_{22} is applied in the x_2-direction as shown in Fig. 6.3.3(b). The strains are

$$\varepsilon_{11}^{(2)} = -\frac{\nu_{21}}{E_2}\sigma_{22}, \quad \varepsilon_{22}^{(2)} = \frac{\sigma_{22}}{E_2}. \tag{6.3.28}$$

The displacements associated with each of the loads are

$$\Delta_1^{(1)} = a\frac{\sigma_{11}}{E_1}, \qquad \Delta_2^{(1)} = -a\frac{\nu_{12}}{E_1}\sigma_{11},$$

$$\Delta_1^{(2)} = -a\frac{\nu_{21}}{E_2}\sigma_{22}, \qquad \Delta_2^{(2)} = a\frac{\sigma_{22}}{E_2}, \tag{6.3.29}$$

and the reciprocal relation (6.3.24) gives, when $\sigma_{11} = \sigma_{22}$, the equality $\Delta_2^{(1)} = \Delta_1^{(2)}$, which is the statement of *Maxwell's reciprocity relation*, which is discussed in Section 7.4.3.

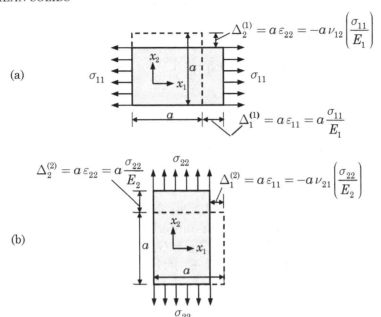

Fig. 6.3.3: Distinction between ν_{12} and ν_{21}. (a) Application of σ_{11}. (b) Application of σ_{22}.

6.3.5 Isotropic Materials

Isotropic materials are those for which the material properties are independent of the direction; that is, there exists an infinite number of material symmetry planes. An isotropic fourth-order tensor can be expressed as [see Eq. (2.5.24)]

$$C_{ijk\ell} = \lambda \delta_{ij}\delta_{k\ell} + \mu\left(\delta_{ik}\delta_{j\ell} + \delta_{i\ell}\delta_{jk}\right) + \kappa\left(\delta_{ik}\delta_{j\ell} - \delta_{i\ell}\delta_{jk}\right), \tag{6.3.30}$$

where μ, λ, and κ are called the *Lamé constants*, and summation on repeated indices is implied. In view of the symmetry of $C_{ijk\ell}$ with respect to the first two and the last two indices, the coefficient of κ is zero, giving

$$C_{ijk\ell} = \lambda \delta_{ij}\delta_{k\ell} + 2\mu\delta_{ik}\delta_{j\ell}. \tag{6.3.31}$$

Therefore, Eq. (6.3.4) takes the simple form

$$\boldsymbol{\sigma} = 2\mu\,\boldsymbol{\varepsilon} + \lambda\,\mathrm{tr}(\boldsymbol{\varepsilon})\mathbf{I}, \quad \sigma_{ij} = 2\mu\,\varepsilon_{ij} + \lambda\,\varepsilon_{kk}\,\delta_{ij}, \tag{6.3.32}$$

where $\mathrm{tr}(\cdot)$ denotes the *trace* (sum of the diagonal elements) of the enclosed tensor. Thus, only two material parameters, μ and λ, are needed to characterize the mechanical response of an isotropic material. The Lamé constants μ and λ are related to E and ν by

$$\mu = \frac{E}{2(1+\nu)}, \quad \lambda = \frac{\nu E}{(1+\nu)(1-2\nu)}, \quad 2\mu + \lambda = \frac{(1-\nu)E}{(1+\nu)(1-2\nu)}. \tag{6.3.33}$$

The stress–strain relations (6.3.32) can be expressed in terms of E and ν as

$$\boldsymbol{\sigma} = \frac{E}{1+\nu}\boldsymbol{\varepsilon} + \frac{\nu E}{(1+\nu)(1-2\nu)}\,\mathrm{tr}(\boldsymbol{\varepsilon})\,\mathbf{I}, \quad \sigma_{ij} = \frac{E}{1+\nu}\varepsilon_{ij} + \frac{\nu E}{(1+\nu)(1-2\nu)}\,\varepsilon_{kk}\,\delta_{ij}, \tag{6.3.34}$$

and the inverse relations are

$$\varepsilon = \left(\tfrac{1+\nu}{E}\right)\boldsymbol{\sigma} - \tfrac{\nu}{E}\operatorname{tr}(\boldsymbol{\sigma})\,\mathbf{I}, \quad \varepsilon_{ij} = \left(\tfrac{1+\nu}{E}\right)\sigma_{ij} - \tfrac{\nu}{E}\sigma_{kk}\,\delta_{ij}, \tag{6.3.35}$$

The strain energy density $\rho_0 U_0$ in Eq. (6.3.1) for an isotropic material takes the form

$$\rho_0 U_0(\boldsymbol{\varepsilon}) = \frac{\lambda}{2}(\operatorname{tr}\boldsymbol{\varepsilon})^2 + \mu\operatorname{tr}(\boldsymbol{\varepsilon}\cdot\boldsymbol{\varepsilon}), \quad \rho_0 U_0(\varepsilon_{ij}) = \mu\varepsilon_{ij}\varepsilon_{ij} + \tfrac{1}{2}\lambda(\varepsilon_{kk})^2. \tag{6.3.36}$$

Note that the strain energy density U_0 is positive-definite, that is,

$$U_0(\boldsymbol{\varepsilon}) > 0 \text{ whenever } \boldsymbol{\varepsilon} \neq \mathbf{0}, \text{ and } U_0(\boldsymbol{\varepsilon}) = 0 \text{ only when } \boldsymbol{\varepsilon} = \mathbf{0}. \tag{6.3.37}$$

The coefficients C_{ij} [see Eqs. (6.3.26) and (6.3.9) for the correspondence between the two-subscripted and four-subscripted C's] of Eq. (6.3.26) simplify to [with $E_1 = E_2 = E_3 = E$, $G_{12} = G_{13} = G_{23} = G = E/2(1+\nu)$, $\nu_{12} = \nu_{23} = \nu_{13} = \nu$, and $C_0 = 1 - 3\nu^2 - 2\nu^3 = (1+\nu)^2(1-2\nu)$]

$$C_{11} = C_{22} = C_{33} = 2\mu + \lambda = \frac{(1-\nu)E}{(1+\nu)(1-2\nu)},$$

$$C_{12} = C_{13} = C_{23} = \lambda = \frac{\nu E}{(1+\nu)(1-2\nu)}, \tag{6.3.38}$$

$$C_{44} = C_{55} = C_{66} = \mu = G = \frac{E}{2(1+\nu)}.$$

The stress–strain relations (6.3.34) for an isotropic material can be expressed in matrix form as

$$\begin{Bmatrix} \sigma_1 \\ \sigma_2 \\ \sigma_3 \\ \sigma_4 \\ \sigma_5 \\ \sigma_6 \end{Bmatrix} = \frac{E}{(1+\nu)(1-2\nu)} \begin{bmatrix} 1-\nu & \nu & \nu & 0 & 0 & 0 \\ \nu & 1-\nu & \nu & 0 & 0 & 0 \\ \nu & \nu & 1-\nu & 0 & 0 & 0 \\ 0 & 0 & 0 & \frac{1-2\nu}{2} & 0 & 0 \\ 0 & 0 & 0 & 0 & \frac{1-2\nu}{2} & 0 \\ 0 & 0 & 0 & 0 & 0 & \frac{1-2\nu}{2} \end{bmatrix} \begin{Bmatrix} \varepsilon_1 \\ \varepsilon_2 \\ \varepsilon_3 \\ \varepsilon_4 \\ \varepsilon_5 \\ \varepsilon_6 \end{Bmatrix}. \tag{6.3.39}$$

If the only nonzero normal stress component is $\sigma_{11} = \sigma$ and the only nonzero shear component is $\sigma_{12} = \tau$, and if we denote $\varepsilon_{11} = \varepsilon$ and $2\varepsilon_{12} = \gamma$, then Eq. (6.3.35) gives the uniaxial strain–stress relations,

$$\varepsilon = \frac{1}{E}\sigma \;\rightarrow\; \sigma = E\,\varepsilon; \quad \gamma = \frac{2(1+\nu)}{E}\tau \;\rightarrow\; \tau = G\gamma. \tag{6.3.40}$$

In summary, application of a normal stress to a rectangular block of isotropic or orthotropic material results in only extension in the direction of the applied stress and contraction perpendicular to it, whereas a monoclinic (or anisotropic) material experiences extension in the direction of the applied normal stress, contraction perpendicular to it, and shearing strain, as shown in Fig. 6.3.4. Conversely, the application of a shearing stress to a monoclinic material causes shearing strain as well as normal strains. Also, a normal stress applied to an orthotropic material at an angle to its principal material directions causes it to behave like a monoclinic material.

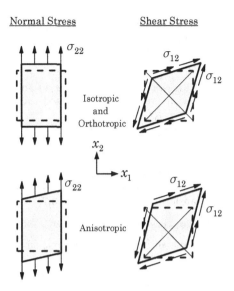

Fig. 6.3.4: Deformation of orthotropic and anisotropic rectangular blocks under uniaxial tension and pure shear.

Example 6.3.1

Consider the thin, filament-wound, closed circular cylindrical pressure vessel in Example 4.3.2, as shown in Fig. 6.3.5. The vessel has an internal diameter of $D_i = 63.5$ cm (25 in.), thickness $h = 2$ cm (0.7874 in.), and pressurized to $p = 1.379$ MPa (200 psi). Assuming a two-dimensional state of stress, determine

(a) stresses σ_{xx}, σ_{yy}, and σ_{xy} in the problem coordinates (x, y, z);

(b) stresses σ_{11}, σ_{22}, and σ_{12} in the material coordinates (x_1, x_2, x_3), with x_1 being tangent to the filament direction;

(c) strains ε_{11}, ε_{22}, and $2\varepsilon_{12}$ in the material coordinates; and

(d) strains ε_{xx}, ε_{yy}, and γ_{xy} in the problem coordinates.

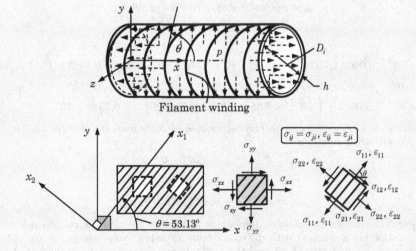

Fig. 6.3.5: A filament-wound cylindrical pressure vessel.

Use a filament winding angle of $\theta = 53.125°$ from the longitudinal axis (x) of the pressure vessel and the following material properties with respect to the material coordinates (typical of graphite-epoxy material): $E_1 = 140$ GPa (20.3 Msi), $E_2 = 10$ GPa (1.45 Msi), $G_{12} = 7$ GPa (1.02 Msi), and $\nu_{12} = 0.3$.

Solution: (a) In Example 4.3.2, the longitudinal stress (σ_{xx}) and circumferential stress (σ_{yy}) in the thin-walled cylindrical pressure vessel were calculated using the formulas

$$\sigma_{xx} = \frac{pD_i}{4h}, \quad \sigma_{yy} = \frac{pD_i}{2h}, \quad \sigma_{xy} = 0. \tag{1}$$

to be $\sigma_{xx} = 10.946$ MPa and $\sigma_{yy} = 21.892$ MPa.

(b) Next, we determine the stresses in the material coordinates (so that we have the shear stress σ_{12} at the fiber–matrix interface, tensile stress σ_{11} in the fiber, and the stress σ_{22} normal to the fiber) using the stress transformation equations (4.3.7)

$$\sigma_{11} = \sigma_{xx} \cos^2 \theta + \sigma_{yy} \sin^2 \theta + \sigma_{xy} \sin 2\theta,$$
$$\sigma_{22} = \sigma_{xx} \sin^2 \theta + \sigma_{yy} \cos^2 \theta - \sigma_{xy} \sin 2\theta, \tag{2}$$
$$\sigma_{12} = \tfrac{1}{2} (\sigma_{yy} - \sigma_{xx}) \sin 2\theta + \sigma_{xy} \cos 2\theta.$$

We obtain ($\sin \theta = 0.8$, $\cos \theta = 0.6$, $\sin 2\theta = 0.96$, and $\cos 2\theta = -0.28$ for $\theta = 53.13°$)

$$\sigma_{11} = 10.946 \times (0.6)^2 + 21.892 \times (0.8)^2 = 17.951 \text{ MPa},$$
$$\sigma_{22} = 10.946 \times (0.8)^2 + 21.892 \times (0.6)^2 = 14.886 \text{ MPa},$$
$$\sigma_{12} = \tfrac{1}{2} (21.892 - 10.946) \times 0.96 = 5.254 \text{ MPa}.$$

(c) The strains in the material coordinates can be calculated using the strain–stress relations (6.3.23). We have $(\nu_{21}/E_2 = \nu_{12}/E_1)$

$$\varepsilon_{11} = \frac{\sigma_{11}}{E_1} - \nu_{12}\frac{\sigma_{22}}{E_1} = \frac{17.95 \times 10^6}{140 \times 10^9} - 0.3\,\frac{14.885 \times 10^6}{140 \times 10^9} = 0.0963 \times 10^{-3} \text{ m/m},$$

$$\varepsilon_{22} = -\nu_{12}\frac{\sigma_{11}}{E_1} + \frac{\sigma_{22}}{E_2} = -0.3\,\frac{17.95 \times 10^6}{140 \times 10^9} + \frac{14.885 \times 10^6}{10 \times 10^9} = 1.4502 \times 10^{-3} \text{ m/m},$$

$$\varepsilon_{12} = \frac{\sigma_{12}}{2G_{12}} = \frac{5.254 \times 10^6}{2 \times 7 \times 10^9} = 0.3753 \times 10^{-3}.$$

(d) The strains in the (x, y) coordinates can be computed using the transformation equations [see Eq. (3.4.32)]

$$\varepsilon_{xx} = \varepsilon_{11} \cos^2 \theta + \varepsilon_{22} \sin^2 \theta - \varepsilon_{12} \sin 2\theta,$$
$$\varepsilon_{yy} = \varepsilon_{11} \sin^2 \theta + \varepsilon_{22} \cos^2 \theta + \varepsilon_{12} \sin 2\theta, \tag{3}$$
$$\varepsilon_{xy} = \tfrac{1}{2} (\varepsilon_{11} - \varepsilon_{22}) \sin 2\theta + \varepsilon_{12} \cos 2\theta.$$

We obtain

$$\varepsilon_{xx} = 10^{-3} \left[0.0963 \times (0.6)^2 + 1.4502 \times (0.8)^2 - 0.3753 \times 0.96\right] = 0.6023 \times 10^{-3} \text{ m/m},$$
$$\varepsilon_{yy} = 10^{-3} \left[0.0963 \times (0.8)^2 + 1.4502 \times (0.6)^2 + 0.3753 \times 0.96\right] = 0.9440 \times 10^{-3} \text{ m/m},$$
$$\varepsilon_{xy} = 10^{-3} \left[(0.0963 - 1.4502) \times 0.48 + 0.3753 \times (-0.28)\right] = -0.7549 \times 10^{-3}.$$

The strains $(\varepsilon_{xx}, \varepsilon_{yy}, \varepsilon_{xy})$ can also be determined directly from the stresses $(\sigma_{xx}, \sigma_{yy}, \sigma_{xy})$ using the strain–stress relations

$$\varepsilon_{xx} = \bar{S}_{11}\,\sigma_{xx} + \bar{S}_{12}\,\sigma_{yy} + \bar{S}_{16}\,\sigma_{xy},$$
$$\varepsilon_{yy} = \bar{S}_{12}\,\sigma_{xx} + \bar{S}_{22}\,\sigma_{yy} + \bar{S}_{26}\,\sigma_{xy}, \tag{4}$$
$$\varepsilon_{xy} = \bar{S}_{16}\,\sigma_{xx} + \bar{S}_{26}\,\sigma_{yy} + \bar{S}_{66}\,\sigma_{xy},$$

where $\bar{S}_{ij}$ are the transformed elastic compliances referred to the problem coordinates (x, y, z). A transformation law consistent with the tensor transformation equations in Eq. (6.3.13) must be used to write $\bar{S}_{ij}$ in terms of S_{ij} and the angle θ. See the answer to Problem **6.2** for the transformation relations between $\bar{S}_{ij}$ and S_{ij} and between $\bar{C}_{ij}$ and C_{ij}.

6.4 Nonlinear Elastic Constitutive Relations

Most materials exhibit nonlinear elastic behavior for certain strain thresholds, that is, the stress–strain relation is no longer linear, but recovers all its deformation upon the removal of the loads, and Hooke's law is no longer valid. Beyond certain nonlinear elastic range, permanent deformation ensues and the material is said to be inelastic or plastic, as shown in Fig. 6.4.1. Here we briefly review constitutive relations for two well-known nonlinear elastic materials, namely the Mooney–Rivlin and neo-Hookean materials. Further discussion can be found in Truesdell and Noll (1965).

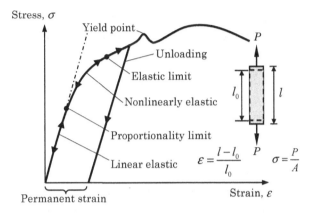

Fig. 6.4.1: A typical stress–strain curve.

Recall from Eq. (6.2.8) that for a hyperelastic material under isothermal conditions there exists a strain energy potential $\Psi = \Psi(\mathbf{F})$ such that

$$\boldsymbol{\sigma}(\mathbf{F}) = \rho \frac{\partial \Psi}{\partial \mathbf{F}} \cdot \mathbf{F}^{\mathrm{T}}, \tag{6.4.1}$$

where ρ is the material density. Some materials (e.g., rubber-like materials) undergo large deformations without appreciable change in volume (i.e., $J \approx 1$). Such materials are called *incompressible* materials. For incompressible elastic materials, the stress tensor is not completely determined by deformation. The hydrostatic pressure p affects the stress. For incompressible elastic materials, we have [see Eq. (6.2.12)]

$$\boldsymbol{\sigma}(\mathbf{F}) = -p\mathbf{I} + \rho \frac{\partial \Psi}{\partial \mathbf{F}} \cdot \mathbf{F}^{\mathrm{T}}, \tag{6.4.2}$$

where p is the hydrostatic pressure.

For a hyperelastic elastic material, Eq. (6.4.1) can also be expressed as

$$\boldsymbol{\sigma}(\mathbf{B}) = 2\rho \frac{\partial \Psi}{\partial \mathbf{B}} \cdot \mathbf{B}, \tag{6.4.3}$$

where the free energy potential Ψ is written as $\Psi = \Psi(\mathbf{B})$ and $\mathbf{B}$ is the left Cauchy–Green (or Finger) tensor $\mathbf{B} = \mathbf{F} \cdot \mathbf{F}^{\mathrm{T}}$ [see Eq. (3.4.4)]. Equations

(6.4.1)–(6.4.3), in general, are nonlinear. The free energy potential Ψ takes different forms for different materials. It is often expressed as a linear combination of unknown parameters and principal invariants of Green strain tensor $\mathbf{E}$, deformation gradient $\mathbf{F}$, or left Cauchy–Green deformation tensor $\mathbf{B}$. The parameters characterize the material and they are determined through suitable experiments.

For incompressible materials, the free energy potential Ψ is taken as a linear function of the principal invariants of $\mathbf{B}$:

$$\Psi = C_1(I_B - 3) + C_2(II_B - 3), \tag{6.4.4}$$

where C_1 and C_2 are constants and I_B and II_B are the two principal invariants of $\mathbf{B}$ (the third invariant III_B is equal to unity for incompressible materials). Materials for which the free energy potential is given by Eq. (6.4.4) are known as the *Mooney–Rivlin material*. The stress tensor in this case has the form

$$\boldsymbol{\sigma} = -p\mathbf{I} + \alpha\mathbf{B} + \beta\mathbf{B}^{-1}, \tag{6.4.5}$$

where α and β are given by

$$\alpha = 2\rho\,\frac{\partial\Psi}{\partial I_B} = 2\rho\,C_1, \quad \beta = -2\rho\,\frac{\partial\Psi}{\partial II_B} = -2\rho\,C_2. \tag{6.4.6}$$

The Mooney–Rivlin incompressible material model is most commonly used to represent the stress–strain behavior of rubber-like solid materials.

If the free energy potential is of the form $\Psi = C_1(I_B - 3)$, that is, $C_2 = 0$, the constitutive equation in Eq. (6.4.5) takes the form

$$\boldsymbol{\sigma} = -p\mathbf{I} + 2\rho\,C_1\,\mathbf{B}. \tag{6.4.7}$$

Materials whose constitutive behavior is described by Eq. (6.4.7) are called the *neo-Hookean materials*. The neo-Hookean model provides a reasonable prediction of the constitutive behavior of natural rubber for moderate strains.

6.5 Newtonian Fluids

6.5.1 Introduction

All bulk matter in nature exists in one of two forms (even before they are subjected to forces): solid or fluid. A solid body is characterized by relative immobility of its molecules, whereas a fluid state is characterized by their relative mobility. Fluids can exist either as gases or liquids. In this section we present the constitutive relations for fluids that exhibit the property that stress is proportional to velocity gradients, that is, strain rates. The proportionality parameter is known as the *viscosity* of the fluid, and the relationship is known *Newton's law of viscosity*. Fluids that behave according to Newton's law of viscosity are called *Newtonian fluids*. For such fluids, the constitutive equations for $\boldsymbol{\sigma}$ cannot be derived using the condition (6.1.6) resulting from the entropy inequality. The physics of such fluids requires the symmetric part of the velocity gradient, $\mathbf{D}$, to be an argument, in place of $\mathbf{F}$ used for solids, in the list of dependent variables.

It is assumed that the Eulerian description is used to derive all equations of mechanics.

For viscous fluids the total stress $\boldsymbol{\sigma}$ is decomposed into equilibrium and deviatoric parts. Then the conditions resulting from the entropy inequality permit derivation of constitutive relations for compressible fluids with the equilibrium stress as thermodynamic pressure $p(\rho, \theta)$, which for an incompressible fluid becomes the mechanical pressure $p(\theta)$; the entropy inequality only places the restriction that work done by the deviatoric stress be positive but provides no mechanism for its constitutive relation. Thus, $\boldsymbol{\sigma}$ is assumed to be of the form[3]

$$\boldsymbol{\sigma} = -p(\rho, \theta)\,\mathbf{I} + \mathcal{F}(\mathbf{D}), \text{ for compressible fluids,}$$

$$\boldsymbol{\sigma} = -p(\theta)\,\mathbf{I} + \mathcal{F}(\mathbf{D}), \text{ for incompressible fluids,} \tag{6.5.1}$$

where ρ is the spatial density and θ is the absolute temperature. A fluid is said to be *incompressible* if the volume change is zero:

$$\nabla \cdot \mathbf{v} = 0, \tag{6.5.2}$$

where $\mathbf{v}$ is the velocity vector. A fluid is termed *inviscid* if the viscosity is zero.

6.5.2 Ideal Fluids

An *ideal fluid* is one that is incompressible and has zero viscosity. The most general constitutive equation for an ideal fluid is of the form

$$\boldsymbol{\sigma} = -p(\rho, \theta)\mathbf{I}. \tag{6.5.3}$$

The dependence of p on ρ and θ has been experimentally verified many times over several centuries. The thermomechanical properties of an ideal fluid are the same in all directions, that is, the fluid is isotropic. It can be verified that Eq. (6.5.3) satisfies the frame indifference requirement because

$$\boldsymbol{\sigma}^* = \mathbf{Q} \cdot \boldsymbol{\sigma} \cdot \mathbf{Q}^{\mathrm{T}} = -p\,\mathbf{Q} \cdot \mathbf{I} \cdot \mathbf{Q}^{\mathrm{T}} = -p\,\mathbf{I}.$$

An explicit functional form of $p(\rho, \theta)$, valid for gases over a wide range of temperature and density, is

$$p = R\rho\theta/m, \tag{6.5.4}$$

where R is the universal gas constant, m is the mean molecular mass of the gas, and θ is the absolute temperature. Equation (6.5.4) is known to define a *perfect gas*. When p is only a function of density, the fluid is said to be "barotropic," and the barotropic constitutive model is applicable under isothermal conditions. If p is independent of both ρ and θ ($\rho = $ constant), p is determined from the equations of motion [see Eq. (5.3.14)].

---

[3]The dependence of $\mathcal{F}$ on the vorticity tensor $\mathbf{W}$ is eliminated to satisfy the frame indifference requirement.

6.5.3 Viscous Incompressible Fluids

The general constitutive equation for stress tensor in a fluid motion is assumed to be of the general form in Eq. (6.5.1). Analogous to isotropic materials, isotropic fluids are those for which the shear stress–strain rate relations are of the form [compare with Eq. (6.3.4)]

$$\mathcal{F}(\mathbf{D}) \equiv \boldsymbol{\tau} = \mathbf{C} : \mathbf{D} \quad (\tau_{ij} = C_{ijk\ell} D_{ij}). \tag{6.5.5}$$

Here $\boldsymbol{\tau}$ is the viscous stress tensor, $\mathbf{C}$ denotes the fourth-order tensor of viscosities, and $\mathbf{D}$ is the strain rate tensor [symmetric part of the velocity gradient tensor $\mathbf{L}$; see Eqs. (3.6.1) and (3.6.2)]:

$$\mathbf{D} = \tfrac{1}{2}\left[\boldsymbol{\nabla}\mathbf{v} + (\boldsymbol{\nabla}\mathbf{v})^{\mathrm{T}}\right] \quad \left[D_{ij} = \tfrac{1}{2}\left(\frac{\partial v_i}{\partial x_j} + \frac{\partial v_j}{\partial x_i}\right)\right], \tag{6.5.6}$$

where $\mathbf{v}$ is the velocity vector.

In a majority of cases a viscous fluid is characterized as an isotropic fluid. For an isotropic fluid, we have the constitutive relation [compare with Eq. (6.3.35)]

$$\boldsymbol{\sigma} = -p(\rho, \theta)\,\mathbf{I} + \boldsymbol{\tau}, \quad \boldsymbol{\tau} = 2\mu(\rho, \theta)\,\mathbf{D} + \lambda(\rho, \theta)\,\mathrm{tr}(\mathbf{D})\mathbf{I}, \tag{6.5.7}$$

or in rectangular Cartesian component form

$$\sigma_{ij} = -p\,\delta_{ij} + \tau_{ij}, \quad \tau_{ij} = 2\mu\,D_{ij} + \lambda\,D_{kk}\,\delta_{ij}, \tag{6.5.8}$$

where ρ is the spatial density, θ is the absolute temperature, and λ and μ are the Lamé parameters that have the meaning *bulk viscosity* and *shear viscosity*, respectively.

Equations (6.5.7) and (6.5.8) can be expressed in terms of the deviatoric components of stress and rate of deformation tensors,

$$\boldsymbol{\sigma}' = \boldsymbol{\sigma} - \tilde{\sigma}\mathbf{I}, \quad \mathbf{D}' = \mathbf{D} - \tfrac{1}{3}\,\mathrm{tr}(\mathbf{D})\,\mathbf{I}, \quad \tilde{\sigma} = \tfrac{1}{3}\,\mathrm{tr}(\boldsymbol{\sigma}). \tag{6.5.9}$$

We note that $\sigma'_{ii} = 0$ and $D'_{ii} = 0$. Then the Newtonian constitutive equation (6.5.8) takes the form

$$\begin{aligned} \boldsymbol{\sigma}' &= 2\mu\mathbf{D}' + \left(\tfrac{2}{3}\mu + \lambda\right)\mathrm{tr}(\mathbf{D})\,\mathbf{I} - (\tilde{\sigma} + p)\,\mathbf{I}, \\ \sigma'_{ij} &= 2\mu D'_{ij} + \left(\tfrac{2}{3}\mu + \lambda\right)D_{kk}\delta_{ij} - (\tilde{\sigma} + p)\,\delta_{ij}, \end{aligned} \tag{6.5.10}$$

from which it follows (because $\sigma'_{ii} = 0$ and $D'_{ii} = 0$) that

$$(2\mu + 3\lambda)\,D_{kk} - 3\,(\tilde{\sigma} + p) = 0. \tag{6.5.11}$$

Hence, the last two terms in Eq. (6.5.10) vanish together and we obtain

$$\boldsymbol{\sigma}' = 2\mu\mathbf{D}', \quad \sigma'_{ij} = 2\mu D'_{ij}. \tag{6.5.12}$$

Note that the mean stress $\tilde{\sigma}$ is equal to the thermodynamic pressure $-p$ if and only if one of the following two conditions is satisfied ($D_{kk} = \boldsymbol{\nabla} \cdot \mathbf{v}$):

Fluid is incompressible: $\quad \boldsymbol{\nabla} \cdot \mathbf{v} = 0,$ $\qquad\qquad$ (6.5.13)

Stokes condition: $\quad K = 2\mu + 3\lambda = 0.$ $\qquad\qquad$ (6.5.14)

In general, the Stokes condition does not hold. Thus, the constitutive equation for viscous, isotropic, incompressible fluids reduces to

$$\boldsymbol{\sigma} = -p\,\mathbf{I} + \boldsymbol{\tau}, \quad \boldsymbol{\tau} = 2\mu\,\mathbf{D}, \quad (\sigma_{ij} = -p\,\delta_{ij} + \tau_{ij}, \quad \tau_{ij} = 2\mu\,D_{ij}), \qquad (6.5.15)$$

and p represents the mean normal stress or *hydrostatic pressure*. For inviscid fluids, the constitutive equation for the stress tensor has the form

$$\boldsymbol{\sigma} = -p\,\mathbf{I} \quad (\sigma_{ij} = -p\,\delta_{ij}). \qquad (6.5.16)$$

We note that Eq. (6.5.15) does not hold for compressible fluids, unless the Stokes condition (6.5.14) is satisfied. Equation (6.5.7) is valid for compressible fluids, with p being the thermodynamic pressure.

6.6 Generalized Newtonian Fluids

6.6.1 Introduction

Fluids for which the viscosity of the fluid may be a function of the strain rate tensor (or its invariants) but the form of the constitutive equations is similar to those of the Newtonian fluid are called *generalized Newtonian fluids*. Generalized Newtonian fluids include motor oils and high molecular weight liquids such as polymers, slurries, pastes, and other complex mixtures. The processing and transporting of such fluids are central problems in the chemical, food, plastics, petroleum, and polymer industries. We note that the generalized Newtonian constitutive models presented in this section for viscous fluids are only a few of the many available in literature [see Reddy and Gartling (2001)].

Most generalized Newtonian fluids exhibit a shear rate dependent viscosity, with "shear thinning" characteristic (i.e., decreasing viscosity with increasing shear rate). Other characteristics associated with generalized Newtonian fluids are elasticity, memory effects, the Weissenberg effect, and the curvature of the free surface in an open channel flow. A discussion of these and other non-Newtonian effects is presented in the book by Bird, Armstrong, and Hassager (1971).

Generalized Newtonian fluids can be classified into two groups: (1) inelastic fluids or fluids without memory and (2) viscoelastic fluids, in which memory effects are significant. For inelastic fluids the viscosity depends on the rate of deformation of the fluid, much like nonlinear elastic solids. Viscoelastic fluids exhibit time-dependent "memory"; that is, the motion of a material point depends not only on the present stress state, but also on the deformation history of the material element. This history dependence leads to very complex constitutive equations.

The constitutive equation for the stress tensor for a generalized Newtonian fluid can be expressed as

$$\boldsymbol{\sigma} = -p\,\mathbf{I} + \boldsymbol{\tau} \quad (\sigma_{ij} = -p\,\delta_{ij} + \tau_{ij}), \qquad (6.6.1)$$

where $\boldsymbol{\tau}$ is known as the viscous or extra stress tensor.

6.6.2 Inelastic fluids

The viscosity for inelastic fluids is found to depend on the rate of deformation tensor $\mathbf{D}$. Often the viscosity is expressed as a function of the principal invariants of the rate of deformation tensor $\mathbf{D}$

$$\mu = \mu\left(J_1, J_2, J_3\right), \tag{6.6.2}$$

where the J_1, J_2, and J_3 are the principal invariants of $\mathbf{D}$,

$$J_1 = \operatorname{tr}\left(\mathbf{D}\right) = D_{ii},$$
$$J_2 = \frac{1}{2}\operatorname{tr}\left(\mathbf{D}^2\right) = \frac{1}{2}D_{ij}D_{ij}, \tag{6.6.3}$$
$$J_3 = \frac{1}{3}\operatorname{tr}\left(\mathbf{D}^3\right) = \frac{1}{3}D_{ij}D_{jk}D_{ki},$$

where $\operatorname{tr}(\cdot)$ denotes the trace of the enclosed tensor. Note that J_2 and J_3 defined above are different from $J_2 = \frac{1}{2}\left(J_1^2 - \mathbf{D} : \mathbf{D}\right)$ and $J_3 = |\mathbf{D}|$ defined in Eq. (3.4.36).

For an incompressible fluid, $J_1 = \nabla \cdot \mathbf{v} = 0$. Also, there is no theoretical or experimental evidence to suggest that the viscosity depends on J_3; thus, the dependence on the third invariant is eliminated. Equation (6.6.2) reduces to

$$\mu = \mu(J_2). \tag{6.6.4}$$

The viscosity can also depend on the thermodynamic state of the fluid, which for incompressible fluids usually implies a dependence only on the temperature. Equation (6.6.4) gives the general functional form for the viscosity function, and experimental observations and a limited theoretical base are used to provide specific forms of Eq. (6.6.4) for non-Newtonian viscosities. A variety of inelastic models have been proposed and correlated with experimental data, as discussed by Bird et al. (1971). Several of the most useful and popular models are presented next; see Reddy and Gartling (2001).

6.6.2.1 Power-law model

The simplest and most familiar non-Newtonian viscosity model is the power-law model, which has the form

$$\mu = K J_2^{(n-1)/2}, \tag{6.6.5}$$

where n and K are parameters, which are, in general, functions of temperature; n is termed the *power-law index* and K is called *consistency*. Fluids with an index $n < 1$ are termed *shear thinning* or *pseudoplastic*. A few materials are *shear thickening* or *dilatant* and have an index $n > 1$. The Newtonian viscosity is obtained with $n = 1$. The admissible range of the index n is bounded below by zero because of stability considerations.

When considering nonisothermal flows, the following empirical relations for n and K are used:

$$n = n_0 + B\left(\frac{\theta - \theta_0}{\theta_0}\right), \tag{6.6.6}$$
$$K = K_0 \exp\left(-A[\theta - \theta_0]/\theta_0\right), \tag{6.6.7}$$

where θ denotes the temperature and the subscript 0 indicates a reference value; A and B are material constants.

6.6.2.2 Carreau model

A major deficiency in the power-law model is that it fails to predict upper and lower limiting viscosities for extreme values of the deformation rate. This problem is alleviated in the Carreau model:

$$\mu = \mu_\infty + (\mu_0 - \mu_\infty)\left(1 + [\lambda\, J_2]^2\right)^{(n-1)/2}, \tag{6.6.8}$$

wherein μ_0 and μ_∞ are the initial and infinite shear rate viscosities, respectively, and λ is a time constant.

6.6.2.3 Bingham model

The *Bingham fluid* differs from most other fluids in that it can sustain an applied stress without fluid motion occurring. The fluid possesses a yield stress, τ_0, such that when the applied stresses are below τ_0 no motion occurs; when the applied stresses exceed τ_0 the material flows, with the viscous stresses being proportional to the excess of the stress over the yield condition. Typically, the constitutive equation after yield is taken to be Newtonian (Bingham model), though other forms such as a power-law equation are possible. In a general form, the Bingham model can be expressed as

$$\boldsymbol{\tau} = \left(\frac{\tau_0}{\sqrt{J_2}} + 2\mu\right)\mathbf{D} \quad \text{when } \frac{1}{2}\mathrm{tr}(\boldsymbol{\tau}^2) \geq \tau_0^2, \tag{6.6.9}$$

$$\boldsymbol{\tau} = 0 \quad \text{when } \frac{1}{2}\mathrm{tr}(\boldsymbol{\tau}^2) < \tau_0^2. \tag{6.6.10}$$

From Eq. (6.6.9) the apparent viscosity of the material beyond the yield point is $(\tau_0/\sqrt{J_2} + 2\mu)$. For a Herschel–Buckley fluid the μ in Eq. (6.6.9) is given by Eq. (6.6.5). The inequalities in Eqs. (6.6.9) and (6.6.10) describe a von Mises yield criterion.

6.6.3 Viscoelastic Constitutive Models

For a viscoelastic fluid, the constitutive equation for the extra-stress $\boldsymbol{\tau}$ in Eq. (6.6.1) is time dependent. Such a relationship is often expressed in abstract form where the current extra-stress is related to the history of deformation in the fluid as

$$\boldsymbol{\tau} = \mathcal{F}[\mathbf{G}(s), \ 0 < s < \infty], \tag{6.6.11}$$

where $\mathcal{F}$ is a tensor-valued functional, $\mathbf{G}$ is a finite deformation tensor (related to the Cauchy–Green tensor), and $s = t - t'$ is the time lapse from time t' to the present time, t. Fluids that obey constitutive equations of the form in Eq. (6.6.11) are called *simple fluids*. The functional form in Eq. (6.6.11) is not useful for general flow problems, and therefore numerous approximations of

Eq. (6.6.11) have been proposed in several different forms. Several of them are reviewed here.

The two major categories of approximate constitutive relations include the differential and integral models. For a differential model the extra-stress is determined from a differential equation that relates the stress and stress rate to the flow kinematics. The integral model represents the extra-stress in terms of an integral over past time of the fluid deformation history. In general, the specific choice is dictated by the ability of a given model to predict the non-Newtonian effects expected in a particular application.

6.6.3.1 Differential models

Constitutive models for viscoelastic fluids in differential equation form are preferable due to the ease with which they can be incorporated into the conservation and balance equations, and the resulting equations are simple to handle in a computational framework. Due to rheology of the fluid (fading memory) the deviatoric part of the stress tensor, called *extra-stress tensor*, is time dependent, and thus the constitutive models are differential equations in time between deviatoric stress tensor and the strain rate tensor.

The constitutive models for viscoelastic fluids in differential form can also be constructed using a purely phenomenological approach based on our understanding of the physics. Such models, for example, one-dimensional spring and dash-pot models discussed for viscoelastic solids in Chapter 9, serve to describe the observed physical response. However, such models do not have a thermodynamic basis, and their extension to two and three dimensions is based on an analogy with elastic constitutive relations.

In this section, we consider differential constitutive theories using the deviatoric stress tensor (derived using the theory of generators and invariants) for fluids. The well-known differential constitutive equations are generally associated with Oldroyd, Maxwell, and Jeffrey. First we define various types of material time derivatives used in these models. In the spatial description the material time derivative of a symmetric second-order tensor can be defined in several ways, all of which are frame invariant. Let $\mathbf{S}$ denote a second-order tensor. Then the *upper-convected* (or co-deformational or contravariant) derivative is defined by

$$\overset{\nabla}{\mathbf{S}} = \frac{\partial \mathbf{S}}{\partial t} + \mathbf{v} \cdot \nabla \mathbf{S} - \mathbf{L} \cdot \mathbf{S} - (\mathbf{L} \cdot \mathbf{S})^{\mathrm{T}}, \qquad (6.6.12)$$

and the *lower-convected* derivative is defined as

$$\overset{\triangle}{\mathbf{S}} = \frac{\partial \mathbf{S}}{\partial t} + \mathbf{v} \cdot \nabla \mathbf{S} + \mathbf{L}^{\mathrm{T}} \cdot \mathbf{S} + \mathbf{S}^{\mathrm{T}} \cdot \mathbf{L}, \qquad (6.6.13)$$

where $\mathbf{v}$ is the velocity vector and $\mathbf{L}$ is the velocity gradient tensor

$$\mathbf{L} = (\nabla \mathbf{v})^{\mathrm{T}} \quad \left(L_{ij} = \frac{\partial v_i}{\partial x_j} \right). \qquad (6.6.14)$$

Since both Eqs. (6.6.12) and (6.6.13) are objective (not shown here) convected derivatives, their linear combination is also objective:

$$\overset{\circ}{\mathbf{S}} = (1 - \alpha)\,\overset{\triangledown}{\mathbf{S}} + \alpha\overset{\triangle}{\mathbf{S}}, \quad 0 \leq \alpha \leq 1. \tag{6.6.15}$$

Equation (6.6.15) can be viewed as the definition of a general convected derivative, which reduces to Eq. (6.6.12) for $\alpha = 0$ and to Eq. (6.6.13) for $\alpha = 1$. When $\alpha = 0.5$ [average of Eq. (6.6.12) and Eq. (6.6.13)] the convected derivative in Eq. (6.6.15) is termed a *corotational* or the *Jaumann derivative*. The selection of one type of derivative over other is usually based on the physical plausibility of the constitutive equation, that is, matching experimental data.

The simplest differential constitutive models are the upper- and lower-convected Maxwell fluids, which are defined by the following equations:

$$\text{Upper-convected Maxwell fluid:} \quad \boldsymbol{\tau} + \lambda\overset{\triangledown}{\boldsymbol{\tau}} = 2\mu^p\,\mathbf{D}, \tag{6.6.16}$$

$$\text{Lower-convected Maxwell fluid:} \quad \boldsymbol{\tau} + \lambda\overset{\triangle}{\boldsymbol{\tau}} = 2\mu^p\,\mathbf{D} \tag{6.6.17}$$

where λ is the relaxation time for the fluid, μ^p is its viscosity, and $\mathbf{D}$ is the rate of deformation tensor. The upper-convected Maxwell model in Eq. (6.6.16) has been used extensively in testing numerical algorithms; the lower-convected and corotational forms of the Maxwell fluid predict physically unrealistic behavior and are not generally used.

Johnson–Segalman model. By employing the general convected derivative (6.6.15) in a Maxwell-like model, the Johnson–Segalman model is produced:

$$\boldsymbol{\tau} + \lambda\overset{\circ}{\boldsymbol{\tau}} = 2\mu^p\,\mathbf{D}. \tag{6.6.18}$$

Phan Thien–Tanner model. By slightly modifying Eq. (6.6.18) to include a variable coefficient for τ, the Phan Thien–Tanner model is obtained:

$$Y(\boldsymbol{\tau})\boldsymbol{\tau} + \lambda\overset{\circ}{\boldsymbol{\tau}} = 2\mu^p\,\mathbf{D}, \tag{6.6.19}$$

where

$$Y(\boldsymbol{\tau}) = 1 + (\epsilon\lambda/\mu^p)\,\text{tr}(\boldsymbol{\tau}) \tag{6.6.20}$$

and ϵ is a constant. This equation is somewhat better than (6.6.18) in representing actual material behavior.

Oldroyd model. The Johnson–Segalman and Phan Thien–Tanner models suffer from a common defect. For a monotonically increasing shear rate, there is a region where the shear stress decreases, which is a physically unrealistic behavior. To correct this anomaly, the constitutive equations are altered using the following procedure. First, the extra-stress is decomposed into two partial stresses, $\boldsymbol{\tau}^s$ and $\boldsymbol{\tau}^p$, such that

$$\boldsymbol{\tau} = \boldsymbol{\tau}^s + \boldsymbol{\tau}^p, \tag{6.6.21}$$

where $\boldsymbol{\tau}^s$ is a purely viscous and $\boldsymbol{\tau}^p$ is a viscoelastic stress component. Then $\boldsymbol{\tau}^s$ and $\boldsymbol{\tau}^p$ are expressed in terms of the rate of deformation tensor $\mathbf{D}$, using the Johnson–Segalman fluid as an example, as

$$\boldsymbol{\tau}^s = 2\mu^s\,\mathbf{D}, \quad \boldsymbol{\tau}^p + \lambda\overset{\circ}{\boldsymbol{\tau}}{}^p = 2\mu^p\,\mathbf{D}. \tag{6.6.22}$$

Finally, the partial stresses in Eqs. (6.6.21) and (6.6.22) are eliminated to produce a new constitutive relation

$$\boldsymbol{\tau} + \lambda\overset{\circ}{\boldsymbol{\tau}} = 2\bar{\mu}\left(\mathbf{D} + \lambda'\overset{\circ}{\mathbf{D}}\right), \tag{6.6.23}$$

where $\bar{\mu} = (\mu^s + \mu^p)$ and $\lambda' = \lambda\mu^s/\bar{\mu}$; and λ' is a retardation time. The constitutive equation in Eq. (6.6.23) is known as a type of Oldroyd fluid. For particular choices of the convected derivative in Eq. (6.6.23), specific models can be generated. When $\alpha = 0$ ($\overset{\circ}{\boldsymbol{\tau}} \rightarrow \overset{\triangledown}{\boldsymbol{\tau}}$), then Eq. (6.6.23) becomes the Oldroyd B fluid, and $\alpha = 1$ ($\overset{\circ}{\boldsymbol{\tau}} \rightarrow \overset{\vartriangle}{\boldsymbol{\tau}}$) produces the Oldroyd A fluid. In order to ensure a monotonically increasing shear stress, the inequality $\mu^s \geq \mu^p/8$ must be satisfied. The stress decomposition in Eq. (6.6.21) can also be used with the Phan Thien–Tanner model to produce a correct shear stress behavior.

White–Metzner model. In all of the above constitutive equations the material parameters, λ and μ^p, were assumed to be constants. For some constitutive equations the constancy of these parameters leads to material (or viscometric) functions that do not accurately represent the behavior of real elastic fluids. For example, the shear viscosity predicted by a Maxwell fluid is a constant, when in fact viscoelastic fluids normally exhibit a shear thinning behavior. This situation can be remedied to some degree by allowing the parameters λ and μ^p to be functions of the invariants of the rate of deformation tensor $\mathbf{D}$. Using the upper-convected Maxwell fluid as an example, then

$$\boldsymbol{\tau} + \lambda(J_2)\overset{\triangledown}{\boldsymbol{\tau}} = 2\mu^p\,(J_2)\mathbf{D}, \tag{6.6.24}$$

where J_2 is the second invariant of the rate of deformation tensor $\mathbf{D}$ [see Eq. (6.3.3)]. The constitutive equation in Eq. (6.6.24) is termed a White–Metzner model. White–Metzner forms of other differential models, such as the Oldroyd fluids, have also been developed and used in various situations.

6.6.3.2 Integral models

An approximate integral model for a viscoelastic fluid represents the extra-stress in terms of an integral over the past history of the fluid deformation. A general form for a single integral model can be expressed as

$$\boldsymbol{\tau} = \int_{-\infty}^{t} 2m(t - t')\mathbf{H}(t, t')\,dt', \tag{6.6.25}$$

where t is the current time, m is a scalar memory function (or relaxation kernel), and $\mathbf{H}$ is a nonlinear deformation tensor between the past time t' and current time t.

There are many possible forms for both the memory function m and the deformation tensor $\mathbf{H}$. Normally the memory function is a decreasing function of the time lapse $s = t - t'$. Typical of such a function is the exponential given by

$$m(t - t') = m(s) = \frac{\mu_0}{\lambda^2} e^{-s/\lambda}, \qquad (6.6.26)$$

where the parameters μ_0, λ, and s were defined previously. Like the choice of a convected derivative in a differential model, the selection of a deformation measure for use in Eq. (6.6.25) is somewhat arbitrary. One particular form that has received some attention is given by

$$\mathbf{H} = \phi_1(J_B, \tilde{J}_B)\mathbf{B} + \phi_2(J_B, \tilde{J}_B)\tilde{\mathbf{B}}. \qquad (6.6.27)$$

In Eq. (6.6.27) $\tilde{\mathbf{B}}$ is the Cauchy strain tensor, $\mathbf{B}$ is its inverse, called the Finger tensor [see Eq. (3.4.22)], and ϕ_1 and ϕ_2 are scalar functions of the invariants of the deformation tensors, $J_B = \mathrm{tr}(\mathbf{B})$ and $\tilde{J}_B = \mathrm{tr}(\tilde{\mathbf{B}})$. The form of the deformation measure in Eq. (6.6.27) is still quite general, though specific choices for the functions ϕ_i and the memory function m lead to several well-known constitutive models. Among these are the Kaye–BKZ fluid and the Lodge rubber-like liquid.

As a specific example of an integral model, we consider the Maxwell fluid. Setting $\phi_1 = 1$ and $\phi_2 = 0$ in Eq. (6.6.27) and using the memory function of Eq. (6.6.26), we obtain a constitutive equation of the form

$$\tau = \frac{\mu_0}{\lambda^2} \int_{-\infty}^{t} \exp\left[-(t - t')/\lambda\right] \left[\mathbf{B}(t') - \mathbf{I}\right] dt'. \qquad (6.6.28)$$

The constitutive equation (6.6.28) is an integral equivalent to the upper-convected Maxwell model shown in differential form in Eq. (6.6.16). Note that in this case, the extra-stress is given in an explicit form but its evaluation requires that the strain history be known for each fluid particle. Although the Maxwell fluid has both differential and integral forms, this is generally not true for other constitutive equations. A discussion of additional integral models can be found in the book by Bird, Armstrong, and Hassager (1971).

6.7 Heat Transfer

6.7.1 Introduction

Heat transfer is a branch of engineering that deals with the transfer of thermal energy within a medium or from one medium to another due to a temperature difference. Heat transfer may take place in one or more of the three basic forms: *conduction*, *convection*, and *radiation*. The transfer of heat within a medium due to diffusion process is called conduction heat transfer. *Fourier's law* states that the heat flow is proportional to the temperature gradient. The proportionality parameter is known as the *thermal conductivity*. Note that for heat conduction to occur there must be temperature differences between neighboring points.

Convection heat transfer is the energy transport effected by the motion of a fluid. The convection heat transfer between two dissimilar media is governed by *Newton's law of cooling*. It states that the heat flow is proportional to the difference of the temperatures between the two media. The proportionality parameter is called the *convection heat transfer coefficient* or *film conductance*. For heat convection to occur there must be a fluid or another medium that can transport energy to and from the primary medium.

Radiation is a mechanism that is different from three transport processes we discussed so far, namely, (1) *momentum transport* in Newtonian fluids that is proportional to the velocity gradient, (2) *energy transport by conduction* that is proportional to the negative of the temperature gradient, and (3) *energy transport by convection* that is proportional to the difference in temperatures of the body and the moving fluid in contact with the body. Thermal radiation is an electromagnetic mechanism, which allows energy transport with the speed of light through regions of space that are devoid of any matter. Radiant energy exchange between surfaces or between a region and its surroundings is described by the *Stefan–Boltzmann law*, which states that the radiant energy transmitted is proportional to the difference of the fourth power of the temperatures of the surfaces. The proportionality parameter is known as the *Stefan–Boltzmann parameter*.

6.7.2 Fourier's Heat Conduction Law

The Fourier heat conduction law states that the heat flow $\mathbf{q}$ is related to the temperature gradient by the relation

$$\mathbf{q} = -\mathbf{k} \cdot \nabla \theta \quad (q_i = -k_{ij}\frac{\partial \theta}{\partial x_j}), \tag{6.7.1}$$

where $\mathbf{k}$ is the thermal conductivity tensor of order two. The negative sign in (6.7.1) indicates that heat flows downhill on the temperature scale. The balance of energy (5.4.11) requires that $(e = c\,\theta)$

$$\rho c \frac{D\theta}{Dt} = \Phi - \nabla \cdot \mathbf{q} + \rho r_h, \quad \Phi = \boldsymbol{\tau} : \mathbf{D}, \tag{6.7.2}$$

which, in view of Eq. (6.7.1), becomes

$$\rho c \frac{D\theta}{Dt} = \Phi + \nabla \cdot (\mathbf{k} \cdot \nabla \theta) + \rho r_h, \tag{6.7.3}$$

where ρr_h is the internal heat generation per unit volume, ρ is the density, and c is the specific heat of the material (assumed to be independent of time t).

For heat transfer in a solid medium $(\mathbf{v} = 0)$, Eq. (6.7.3) reduces to

$$\rho c \frac{\partial \theta}{\partial t} = \nabla \cdot (\mathbf{k} \cdot \nabla \theta) + \rho r_h, \tag{6.7.4}$$

which forms the subject of the field of conduction heat transfer. For a fluid medium, Eq. (6.7.3) becomes

$$\rho c \left(\frac{\partial \theta}{\partial t} + \mathbf{v} \cdot \nabla \theta \right) = \Phi + \nabla \cdot (\mathbf{k} \cdot \nabla \theta) + \rho r_h, \tag{6.7.5}$$

where $\mathbf{v}$ is the velocity field, and Φ is the viscous dissipation function.

6.7.3 Newton's Law of Cooling

At a solid–fluid interface the heat flux is related to the difference between the temperature θ at the interface and that in the fluid

$$q_n \equiv \hat{\mathbf{n}} \cdot \mathbf{q} = h\,(\theta - \theta_{\text{fluid}}), \tag{6.7.6}$$

where $\hat{\mathbf{n}}$ is the unit normal to the surface of the body and h is known as the *heat transfer coefficient* or *film conductance*. This relation is known as Newton's law of cooling, which also defines h. Clearly, Eq. (6.7.6) defines a boundary condition on the bounding surface of a conducting medium.

6.7.4 Stefan–Boltzmann Law

The heat flow from surface 1 to surface 2 by radiation is governed by the Stefan–Boltzman law:

$$q_n = \sigma\left(\theta_1^4 - \theta_2^4\right), \tag{6.7.7}$$

where θ_1 and θ_2 are the temperatures of surfaces 1 and 2, respectively, and σ is the Stefan–Boltzman constant. Again, Eq. (6.7.7) defines a boundary condition on the surface 1 of a body.

6.8 Constitutive Relations for Coupled Problems

6.8.1 Electromagnetics

Problems involving the coupling of electromagnetic fields with fluid and thermal transport have a broad spectrum of applications ranging from astrophysics to manufacturing and to electromechanical devices and sensors. A good introduction to electromagnetic field theory is available in the textbook by Jackson (1975). Here we present a brief discussion of pertinent equations for the sake of completeness[4].

6.8.1.1 Maxwell's equations

The appropriate mathematical description of electromagnetic phenomena in a conducting material region, Ω_C, is given by the following Maxwell's equations [see Reddy and Gartling (2001) and Jackson (1975), and references therein]:

$$\nabla \times \mathbf{E} = -\frac{\partial \mathbf{B}}{\partial t}, \tag{6.8.2}$$

$$\nabla \times \mathbf{H} = \mathbf{J} + \frac{\partial \mathbf{D}}{\partial t}, \tag{6.8.3}$$

$$\nabla \cdot \mathbf{B} = 0, \tag{6.8.4}$$

$$\nabla \cdot \mathbf{D} = \rho, \tag{6.8.5}$$

where $\mathbf{E}$ is the electric field intensity, $\mathbf{H}$ is the magnetic field intensity, $\mathbf{B}$ is the magnetic flux density, $\mathbf{D}$ is the electric flux (displacement) density, $\mathbf{J}$ is the conduction current density, and ρ is the source charge density. Equation (6.8.1)

---

[4]Note that the notation used here for various fields is standard in the literature; unfortunately, some of the symbols used here were already used previously for other variables.

is referred to as *Faraday's law*, Eq. (6.8.2) as *Ampere's law* (as modified by Maxwell), and Eq. (6.8.4) as *Gauss' law*. A continuity condition on the current density is also defined by

$$\nabla \cdot \mathbf{J} = \frac{\partial \rho}{\partial t}. \tag{6.8.5}$$

Note that only three of the preceding five equations are independent; either Eqs. (6.8.1), (6.8.2), and (6.8.4) or Eqs. (6.8.1), (6.8.2), and (6.8.5) form valid sets of equations for the field variables.

6.8.1.2 Constitutive relations

To complete the formulation, the constitutive relations for the material are required. The fluxes are functionally related to the field variables by

$$\mathbf{D} = \mathcal{F}_D(\mathbf{E}, \mathbf{B}), \tag{6.8.6}$$
$$\mathbf{H} = \mathcal{F}_H(\mathbf{E}, \mathbf{B}), \tag{6.8.7}$$
$$\mathbf{J} = \mathcal{F}_J(\mathbf{E}, \mathbf{B}), \tag{6.8.8}$$

where the response functions $\mathcal{F}_D$, $\mathcal{F}_H$, and $\mathcal{F}_J$ may also depend on external variables such as temperature θ and mechanical stress σ. The form of the material response due to applied $\mathbf{E}$ or $\mathbf{B}$ fields can vary strongly depending on the microstructure and the strength of the material and on the magnitude and time-dependent nature of the applied field.

Conductive and Dielectric Materials. For conducting materials, the standard response function $\mathcal{F}_J$ gives Ohm's law, which relates the current density $\mathbf{J}$ to the electric field intensity $\mathbf{E}$

$$\mathbf{J} = \mathbf{k}_\sigma \cdot \mathbf{E}, \tag{6.8.9}$$

where $\mathbf{k}_\sigma$ is the conductivity tensor. For isotropic materials, we have $\mathbf{k}_\sigma = k_\sigma \mathbf{I}$, where k_σ is a scalar and $\mathbf{I}$ is the unit tensor. In general, the conductivity may be a function of $\mathbf{E}$ or an external variable such as temperature. This form of Ohm's law applies to stationary conductors. If the conductive material is moving in a magnetic field, then Eq. (6.8.9) is modified to read

$$\mathbf{J} = \mathbf{k}_\sigma \cdot \mathbf{E} + \mathbf{k}_\sigma \cdot (\mathbf{v} \times \mathbf{B}), \tag{6.8.10}$$

where $\mathbf{v}$ is the velocity vector describing the motion of the conductor and $\mathbf{B}$ is the magnetic flux vector.

For dielectric materials, the standard response function $\mathcal{F}_D$ relates the electric flux density $\mathbf{D}$ to the electric field $\mathbf{E}$ and polarization vector $\mathbf{P}$:

$$\mathbf{D} = \epsilon_0 \cdot \mathbf{E} + \mathbf{P}, \tag{6.8.11}$$

where ϵ_0 is the permittivity of free space. The polarization is generally related to the electric field through

$$\mathbf{P} = \epsilon_0 \mathbf{S}_e \cdot \mathbf{E} + \mathbf{P}_0, \tag{6.8.12}$$

where $\mathbf{S}_e$ is the electric susceptibility tensor that accounts for the different types of polarization, and $\mathbf{P}_0$ is the remnant polarization that may be present in some materials.

Magnetic Materials. For magnetic materials, the standard response function $\mathcal{F}_H$ relates the magnetic field intensity $\mathbf{H}$ to the magnetic flux $\mathbf{B}$

$$\mathbf{H} = \frac{1}{\mu_0}\mathbf{B} - \mathbf{M}, \tag{6.8.13}$$

where μ_0 is the permeability of free space and $\mathbf{M}$ is the magnetization vector. The magnetization vector $\mathbf{M}$ can be related to either the magnetic flux $\mathbf{B}$ or magnetic field intensity $\mathbf{H}$ by

$$\mathbf{M} = \frac{1}{\mu_0}\frac{\mathbf{S}_m}{(\mathbf{I}+\mathbf{S}_m)}\cdot\mathbf{B} + \mathbf{M}_0, \tag{6.8.14}$$

$$\mathbf{M} = \mathbf{S}_m\cdot\mathbf{H} + (\mathbf{I}+\mathbf{S}_m)\cdot\mathbf{M}_0, \tag{6.8.15}$$

where $\mathbf{S}_m$ is the magnetic susceptibility for the material, $\mathbf{M}_0$ is the remnant magnetization, and $\mathbf{I}$ is the unit tensor. If the susceptibility is negative, the material is diamagnetic, whereas a positive susceptibility defines a paramagnetic material. Generally, these susceptibilities are quite small and are often neglected. Ferromagnetic materials have large positive susceptibilities and produce a nonlinear (hysteretic) relationship between $\mathbf{B}$ and $\mathbf{H}$. These materials may also exhibit spontaneous and remnant magnetization.

Electromagnetic Forces and Volume Heating. The coupling of electromagnetic fields with a fluid or thermal problem occurs through the dependence of material properties on electromagnetic field quantities and the production of electromagnetic-induced body forces and volumetric energy production. The Lorentz body force per unit volume in a conductor due to the presence of electric currents and magnetic fields is given by

$$\mathbf{F}_B = \rho\mathbf{E} + \mathbf{J}\times\mathbf{B}, \tag{6.8.16}$$

where, in the general case, the current is defined by Eq. (6.8.10). The first term on the right-hand side of Eq. (6.8.16) is the electric field contribution to the Lorentz force; the magnetic term $\mathbf{J}\times\mathbf{B}$ is usually of more interest in applied mechanics problems. The energy generation or Joule heating in a conductor is described by

$$Q_J = \mathbf{J}\cdot\mathbf{E}, \tag{6.8.17}$$

which takes on a more familiar form if the simplified ($\mathbf{v} = \mathbf{0}$) form of Eq. (6.8.10) is used to produce

$$Q_J = \sigma^{-1}(\mathbf{J}\cdot\mathbf{J}), \tag{6.8.18}$$

where σ is the conductivity. The aforementioned forces and heat source occur in the fluid momentum and energy equations, respectively.

6.8.2 Thermoelasticity

The use of the entropy density η as an independent variable is not convenient. A more convenient thermal variable is the temperature θ, as it is fairly easy to measure and control. The constitutive equations of thermoelasticity are derived by assuming the existence of the *Helmholtz free-energy potential* $\Psi = U_0(\theta, \varepsilon) - \eta(\varepsilon_{ij})\,\theta = \Psi(\theta, \varepsilon)$:

$$\rho_0\,\Psi(\varepsilon_{ij}, \theta) = \rho_0\,U_0 - \rho_0\,(\theta - \theta_0)\eta = \frac{1}{2}C_{ijk\ell}\,\varepsilon_{ij}\,\varepsilon_{k\ell} - \beta_{ij}\,\varepsilon_{ij}\,(\theta - \theta_0) - \frac{\rho_0\,c_v}{2\theta_0}(\theta - \theta_0)^2$$

$$(6.8.19)$$

such that

$$\sigma_{ij} = \rho_0\frac{\partial\Psi}{\partial\varepsilon_{ij}} = C_{ijk\ell}\,\varepsilon_{k\ell} - \beta_{ij}\,(\theta - \theta_0), \quad \rho_0\eta = -\rho_0\frac{\partial\Psi}{\partial\theta} = \beta_{ij}\,\varepsilon_{ij} + \frac{\rho_0\,c_v}{\theta_0}(\theta - \theta_0),$$

$$(6.8.20)$$

where θ is the temperature measured from a reference value θ_0, η is the entropy density, and β_{ij} are material coefficients. In arriving at Eq. (6.8.20), it is assumed that η and σ_{ij} are initially zero (see the answer to Problem **6.35**), and c_v, β_{ij}, and $C_{ijk\ell}$ are values at the reference state. Inverting the stress–strain relations in Eq. (6.8.20), we obtain

$$\varepsilon_{ij} = S_{ijk\ell}\,\sigma_{k\ell} + \alpha_{ij}(\theta - \theta_0), \tag{6.8.21}$$

where $S_{ijk\ell}$ are the elastic compliances, and α_{ij} are the thermal coefficients of expansion, and they are related to β_{ij} by

$$\beta_{ij} = C_{ijk\ell}\,\alpha_{k\ell}. \tag{6.8.22}$$

6.8.3 Hygrothermal Elasticity

The moisture adsorption problem is mathematically similar to the heat conduction problem. The moisture concentration c in a solid is described by Fick's law (analogous to Fourier's heat conduction law):

$$\mathbf{q}_f = -\mathbf{D} \cdot \nabla c \tag{6.8.23}$$

and the diffusion process is governed by

$$\frac{\partial c}{\partial t} = -\nabla \cdot \mathbf{q}_f + \phi_f, \tag{6.8.24}$$

where $\mathbf{D}$ denotes the *mass diffusivity tensor* of order two, $\mathbf{q}_f$ is the moisture flux vector, and ϕ_f is the moisture source in the domain. The negative sign in Eq. (6.8.24) indicates that moisture seeps from a higher concentration to a lower concentration. The boundary conditions involve specifying the moisture concentration or the flux normal to the boundary:

$$c = \hat{c}(s, t) \quad \text{on } \Gamma_1, \tag{6.8.25}$$

$$\mathbf{n} \cdot \mathbf{q}_f = \hat{q}_f(s, t) \quad \text{on } \Gamma_2, \tag{6.8.26}$$

where $\Gamma = \Gamma_1 \cup \Gamma_2$, and $\Gamma_1 \cap \Gamma_2 = \emptyset$, and quantities with a hat are specified functions on the respective boundaries.

The moisture-induced strains $\{\varepsilon\}^M$ are given by

$$\{\varepsilon\}^M = \{\alpha_M\}c, \tag{6.8.27}$$

where $\{\alpha_M\}$ is the vector of *coefficients of hygroscopic expansion*. Thus, the hygrothermal strains have the same form as the thermal strains. The total strains are given by

$$\{\varepsilon\} = [S]\{\sigma\} + \{\alpha_T\}(\theta - \theta_0) + \{\alpha_M\}(c - c_0), \tag{6.8.28}$$

where $\{\alpha_T\}$ is the vector of *coefficients of thermal expansion*, and θ_0 and c_0 are reference values of temperature and concentration, respectively, from which the strains and stresses are measured. In view of the similarity between the thermal and moisture strains, thermoelasticity and hygroelasticity problems share the same solution approach.

6.8.4 Electroelasticity

Electroelasticity deals with the phenomena caused by interactions between electric and mechanical fields. The *piezoelectric effect* is one such phenomenon, and it is concerned with the effect of the electric charge on the deformation. A structure with piezoelectric layers receives actuation through an applied electric field, and the piezoelectric layers send electric signals that are used to measure the motion or deformation of the laminate. In these problems, the electric charge that is applied to actuate a structure provides an additional body force to the stress analysis problem, much the same way a temperature field induces a body force through thermal strains.

The piezoelectric effect is described by the *polarization vector* $\mathbf{P}$, which represents the electric moment per unit volume or polarization charge per unit area. It is related to the stress tensor by the relation

$$\mathbf{P} = \mathbf{d} \cdot \boldsymbol{\sigma} \quad \text{or} \quad P_i = d_{ijk}\sigma_{jk}, \tag{6.8.29}$$

where $\mathbf{d}$ is the third-order tensor of piezoelectric moduli. The inverse effect relates the electric field vector $\mathbf{E}$ to the linear strain tensor ε by

$$\varepsilon = \mathbf{E} \cdot \mathbf{d} \quad \text{or} \quad \varepsilon_{ij} = d_{kij}E_k. \tag{6.8.30}$$

Note that d_{kij} is symmetric with respect to indices i and j because of the symmetry of ε_{ij} (note that $i, j, k = 1, 2, 3$).

The *pyroelectic effect* is another phenomenon that relates temperature changes to polarization of a material. For a temperature change from a reference temperature θ_0, the change in polarization vector $\Delta\mathbf{P}$ is given by

$$\Delta\mathbf{P} = \mathbf{p}(\theta - \theta_0), \tag{6.8.31}$$

where $\mathbf{p}$ is the vector of pyroelectric coefficients.

The coupling between the mechanical, thermal, and electrical fields can be established using thermodynamical principles and Maxwell's relations. Analogous to the strain energy potential U_0 for elasticity and the Helmholtz free-energy potential Ψ for thermoelasticity, we assume the existence of a function Φ, called the electric *Gibb's free-energy potential* or *enthalpy function*,

$$\rho_0 \Phi(\varepsilon_{ij}, E_i, \theta) = \rho_0 U_0 - \rho_0 \mathbf{E} \cdot \mathbf{D} - \rho_0 \eta(\theta - \theta_0)$$

$$= \frac{1}{2} C_{ijk\ell} \varepsilon_{ij} \varepsilon_{k\ell} - e_{ijk} \varepsilon_{ij} E_k - \beta_{ij} \varepsilon_{ij}(\theta - \theta_0)$$

$$- \frac{1}{2} \epsilon_{k\ell} E_k E_\ell - p_k E_k(\theta - \theta_0) - \frac{\rho_0 c_v}{2\theta_0}(\theta - \theta_0)^2, \quad (6.8.32)$$

such that

$$\sigma_{ij} = \rho_0 \frac{\partial \Phi}{\partial \varepsilon_{ij}}, \quad \rho_0 D_i = -\rho_0 \frac{\partial \Phi}{\partial E_i}, \quad \rho_0 \eta = -\rho_0 \frac{\partial \Phi}{\partial \theta}, \qquad (6.8.33)$$

where σ_{ij} are the components of the stress tensor $\boldsymbol{\sigma}$, D_i are the components of the electric displacement vector $\mathbf{D}$, and η is the entropy. Use of Eq. (6.8.32) in Eq. (6.8.33) gives the constitutive equations of a deformable piezoelectric medium:

$$\sigma_{ij} = C_{ijk\ell} \varepsilon_{k\ell} - e_{ijk} E_k - \beta_{ij}(\theta - \theta_0), \qquad (6.8.34)$$

$$\rho_0 D_k = e_{ijk} \varepsilon_{ij} + \epsilon_{k\ell} E_\ell + p_k(\theta - \theta_0), \qquad (6.8.35)$$

$$\rho_0 \eta = \beta_{ij} \varepsilon_{ij} + p_k E_k + \frac{\rho_0 c_v}{\theta_0}(\theta - \theta_0), \qquad (6.8.36)$$

where $C_{ijk\ell}$ are the elastic moduli, e_{ijk} are the piezoelectric moduli, ϵ_{ij} are the dielectric constants, p_k are the pyroelectric constants, β_{ij} are the stress-temperature expansion coefficients, c_v is the specific heat (at constant strain or volume) per unit mass, and θ_0 is the reference temperature. In single-subscript notation for stresses and strains, Eqs. (6.8.34)–(6.8.36) can be expressed as

$$\sigma_i = C_{ij} \varepsilon_j - e_{ik} E_k - \beta_i(\theta - \theta_0), \qquad (6.8.37)$$

$$\rho_0 D_k = e_{ik} \varepsilon_i + \epsilon_{k\ell} E_\ell + p_k(\theta - \theta_0), \qquad (6.8.38)$$

$$\rho_0 \eta = \beta_i \varepsilon_i + p_k E_k + \frac{\rho_0 c_v}{\theta_0}(\theta - \theta_0). \qquad (6.8.39)$$

Note that the range of summation in Eqs. (6.8.37)–(6.8.39) is different for different terms: $i, j = 1, 2, \cdots, 6; k, \ell = 1, 2, 3$. For the general anisotropic material, there are 21 independent elastic constants, 18 piezoelectric constants, 6 dielectric constants, 3 pyroelectric constants, and 6 thermal expansion coefficients.

Maxwell's equation governing the electric displacement vector $\mathbf{D}$ is given by

$$\nabla \cdot \mathbf{D} = 0. \qquad (6.8.40)$$

It is often assumed that the electric field $\mathbf{E}$ is derivable from an electric scalar potential function ϕ:

$$\mathbf{E} = -\nabla \phi. \qquad (6.8.41)$$

This assumption allows us to write Eq. (6.8.40), in view of Eq. (6.8.38), as

$$\frac{\partial}{\partial x_1}\left(\epsilon_{11} \frac{\partial \phi}{\partial x_1}\right) + \frac{\partial}{\partial x_2}\left(\epsilon_{22} \frac{\partial \phi}{\partial x_2}\right) + \frac{\partial}{\partial x_3}\left(\epsilon_{33} \frac{\partial \phi}{\partial x_3}\right) + f_e = 0, \qquad (6.8.42)$$

where

$$f_e = -\frac{\partial}{\partial x_k}\left[e_{k\ell}\varepsilon_\ell + p_k(\theta - \theta_0)\right]. \qquad (6.8.43)$$

6.9 Summary

This chapter was dedicated to a discussion of the constitutive equations for Hookean solids, Newtonian fluids, and heat transfer in solids. Constitutive models of solids and fluids are derived using the entropy inequality or conditions resulting from the entropy inequality. Beginning with a discussion of the constitutive rules or axioms, frame indifference, and restrictions placed by the entropy inequality, general constitutive relations for the stress tensor, entropy, and heat flux were derived. Then the generalized Hooke's law governing linear elastic solids, Newtonian relations for viscous fluids, and the Fourier heat conduction equation for heat transfer in solids are presented. The generalized Hooke's law is specialized to monoclinic materials, orthotropic materials, and isotropic materials using material symmetries. Constitutive relations for nonlinear elastic solids, generalized Newtonian fluids, and coupled problems (e.g., electromagnetics, thermoelasticity, hygrothermal elasticity, and electroelasticity) are also presented for the sake of completeness.

The constitutive relations presented in this chapter along with the field equations developed in Chapter 5 will be used in Chapters 7 and 8 to analyze some typical boundary-value problems of solid mechanics, fluid mechanics, and heat transfer. The main results of this chapter that are of importance in the coming chapters are summarized here.

Hookean deformable solids (infinitesimal strains)

$$\boldsymbol{\sigma} = 2\mu\boldsymbol{\varepsilon} + \lambda\operatorname{tr}(\boldsymbol{\varepsilon})\mathbf{I}, \quad \sigma_{ij} = 2\mu\,\varepsilon_{ij} + \lambda\,\varepsilon_{kk}\,\delta_{ij} \tag{6.9.1}$$

$$\boldsymbol{\varepsilon} = \frac{1}{2}\left[\boldsymbol{\nabla}\mathbf{u} + (\boldsymbol{\nabla}\mathbf{u})^{\mathsf{T}}\right], \quad \varepsilon_{ij} = \frac{1}{2}\left(\frac{\partial u_i}{\partial x_j} + \frac{\partial u_j}{\partial x_i}\right) \tag{6.9.2}$$

Newtonian fluids (compressible)

$$\boldsymbol{\sigma} = -p(\rho, \theta)\,\mathbf{I} + 2\mu(\rho, \theta)\,\mathbf{D} + \lambda(\rho, \theta)\operatorname{tr}(\mathbf{D})\mathbf{I}$$

$$\sigma_{ij} = -p\,\delta_{ij} + 2\mu\,\varepsilon_{ij} + \lambda\,\varepsilon_{kk}\,\delta_{ij} \tag{6.9.3}$$

Newtonian Fluids (incompressible)

$$\boldsymbol{\sigma} = -p(\theta)\,\mathbf{I} + 2\mu\,\mathbf{D}, \quad \sigma_{ij} = -p(\theta)\,\delta_{ij} + 2\mu\,D_{ij} \tag{6.9.4}$$

$$\mathbf{D} = \frac{1}{2}\left[\boldsymbol{\nabla}\mathbf{v} + (\boldsymbol{\nabla}\mathbf{v})^{\mathsf{T}}\right], \quad D_{ij} = \frac{1}{2}\left(\frac{\partial v_i}{\partial x_j} + \frac{\partial v_j}{\partial x_i}\right) \tag{6.9.5}$$

Heat transfer

$$\mathbf{q} = -\mathbf{k}\cdot\boldsymbol{\nabla}\theta, \quad q_i = -k_{ij}\frac{\partial\theta}{\partial x_j} \tag{6.9.6}$$

In general, the derivation of constitutive equations of a fluid or solid matter is quite involved. The presentation here is made simple keeping in mind the introductory nature of the present course. For a detailed and advanced study of the subject, the reader may consult the books by Truesdell and Toupin (1965) and Truesdell and Noll (1965).

Problems

HOOKEAN SOLIDS

6.1 Recall from Examples 3.4.3 and 4.3.1 that under the coordinate transformation

$$\hat{e}_1 = \cos\theta\,\hat{e}_x + \sin\theta\,\hat{e}_y,$$
$$\hat{e}_2 = -\sin\theta\,\hat{e}_x + \cos\theta\,\hat{e}_y, \qquad (1)$$
$$\hat{e}_3 = \hat{e}_z,$$

the stress components and strain components ε_i and σ_i are given in terms of the components $\sigma_{xx}, \sigma_{yy}, \cdots$ and $\varepsilon_{xx}, \varepsilon_{yy}, \cdots$ by [see Eqs. (3.4.33) and (4.3.7)]

$$
\begin{Bmatrix} \varepsilon_1 \\ \varepsilon_2 \\ \varepsilon_3 \\ \varepsilon_4 \\ \varepsilon_5 \\ \varepsilon_6 \end{Bmatrix}
=
\begin{bmatrix}
\cos^2\theta & \sin^2\theta & 0 & 0 & 0 & \frac{1}{2}\sin 2\theta \\
\sin^2\theta & \cos^2\theta & 0 & 0 & 0 & -\frac{1}{2}\sin 2\theta \\
0 & 0 & 1 & 0 & 0 & 0 \\
0 & 0 & 0 & \cos\theta & -\sin\theta & 0 \\
0 & 0 & 0 & \sin\theta & \cos\theta & 0 \\
-\sin 2\theta & \sin 2\theta & 0 & 0 & 0 & \cos 2\theta
\end{bmatrix}
\begin{Bmatrix} \varepsilon_{xx} \\ \varepsilon_{yy} \\ \varepsilon_{zz} \\ \varepsilon_{yz} \\ \varepsilon_{xz} \\ \varepsilon_{xy} \end{Bmatrix},
\quad \{\bar{\varepsilon}\} = [T^\theta]\{\varepsilon\}, \qquad (2)
$$

$$
\begin{Bmatrix} \sigma_1 \\ \sigma_2 \\ \sigma_3 \\ \sigma_4 \\ \sigma_5 \\ \sigma_6 \end{Bmatrix}
=
\begin{bmatrix}
\cos^2\theta & \sin^2\theta & 0 & 0 & 0 & \sin 2\theta \\
\sin^2\theta & \cos^2\theta & 0 & 0 & 0 & -\sin 2\theta \\
0 & 0 & 1 & 0 & 0 & 0 \\
0 & 0 & 0 & \cos\theta & -\sin\theta & 0 \\
0 & 0 & 0 & \sin\theta & \cos\theta & 0 \\
-\frac{1}{2}\sin 2\theta & \frac{1}{2}\sin 2\theta & 0 & 0 & 0 & \cos 2\theta
\end{bmatrix}
\begin{Bmatrix} \sigma_{xx} \\ \sigma_{yy} \\ \sigma_{zz} \\ \sigma_{yz} \\ \sigma_{xz} \\ \sigma_{xy} \end{Bmatrix},
\quad \{\bar{\sigma}\} = [R^\theta]\{\sigma\}.
$$
$$(3)$$

Show that
$$[\bar{S}] = [T^\theta][S][T^\theta]^{\mathrm{T}}, \qquad [\bar{C}] = [R^\theta][C][R^\theta]^{\mathrm{T}}, \qquad (4)$$

where $[\bar{S}]$ is the matrix of compliance coefficients and $[\bar{C}]$ is the matrix of stiffness coefficients with respect to the (x_1, x_2, x_3) coordinates and $[S]$ is the matrix of compliance coefficients and $[C]$ is the matrix of stiffness coefficients with respect to the (x, y, z) coordinates.

6.2 Under the coordinate transformation

$$\hat{\bar{e}}_1 = \cos\theta\,\hat{e}_1 + \sin\theta\,\hat{e}_2,$$
$$\hat{\bar{e}}_2 = -\sin\theta\,\hat{e}_1 + \cos\theta\,\hat{e}_2,$$
$$\hat{\bar{e}}_3 = \hat{e}_3,$$

determine $\bar{S}_{ij}$ in terms of S_{ij} and $\bar{C}_{ij}$ in terms of C_{ij}.

6.3 Given the transformation

$$\hat{\bar{e}}_1 = \hat{e}_1, \quad \hat{\bar{e}}_2 = \hat{e}_2, \quad \hat{\bar{e}}_3 = -\hat{e}_3, \qquad (1)$$

determine the stress components $\bar{\sigma}_{ij}$ in terms of σ_{ij}, strain components $\bar{\varepsilon}_{ij}$ in terms of ε_{ij}, and the elasticity coefficients $\bar{C}_{ij}$ in terms of C_{ij}.

6.4 Establish the following relations between the Lame' constants μ and λ and engineering constants E, ν, and K:

$$\lambda = \frac{\nu E}{(1+\nu)(1-2\nu)}, \quad \mu = G = \frac{E}{2(1+\nu)}, \quad K = \frac{E}{3(1-2\nu)}.$$

6.5 Determine the longitudinal stress σ_{xx} and the hoop stress σ_{yy} in a thin-walled circular cylindrical pressure vessel with closed ends; that is, establish Eq. (1) of Example 6.3.1. Assume an internal pressure of p, internal diameter D_i, and thickness h.

6.6 Determine the stress tensor components at a point in 7075-T6 aluminum alloy body ($E = 72$ GPa, and $G = 27$ GPa) if the strain tensor at the point has the following components with respect to the Cartesian basis vectors $\hat{e}_i$:

$$
[\varepsilon] =
\begin{bmatrix}
200 & 100 & 0 \\
100 & 300 & 400 \\
0 & 400 & 0
\end{bmatrix}
\times 10^{-6} \text{ m/m }.
$$

6.7 For the state of stress and strain given in Problem **6.6**, determine the principal invariants of the stress and strain tensors.

6.8 The components of strain tensor at a point in a body made of structural steel are

$$[\varepsilon] = \begin{bmatrix} 36 & 12 & 30 \\ 12 & 40 & 0 \\ 30 & 0 & 25 \end{bmatrix} \times 10^{-6} \text{ m/m} .$$

Assuming that the Lamé constants for the structural steel are $\lambda = 207$ GPa (30×10^6 psi) and $\mu = 79.6$ GPa (11.54×10^6 psi), determine the principal invariants of the stress and strain tensors.

6.9 The components of a stress tensor at a point in a body made of structural steel are

$$[\sigma] = \begin{bmatrix} 42 & 12 & 30 \\ 12 & 15 & 0 \\ 30 & 0 & -5 \end{bmatrix} \text{ MPa.}$$

Assuming that the Lamé constants for structural steel are $\lambda = 207$ GPa (30×10^6 psi) and $\mu = 79.6$ GPa (11.54×10^6 psi), determine the principal invariants of the strain tensor.

6.10 *Plane stress-reduced constitutive relations.* Beginning with the strain-stress relations in Eq. (6.3.23) for an orthotropic material in a two-dimensional case (i.e., $\sigma_{33} = \sigma_{13} = \sigma_{23} = 0$), determine the two-dimensional stress–strain relations.

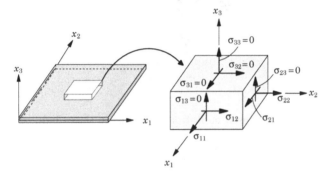

Fig. P6.10

6.11 Given the strain energy potential

$$\Psi(\mathbf{E}) = \frac{\lambda}{2}(\text{tr } \mathbf{E})^2 + \mu \, \text{tr}(\mathbf{E} \cdot \mathbf{E}),$$

determine the second Piola–Kirchhoff stress tensor $\mathbf{S}$ in terms of the Green strain tensor $\mathbf{E}$.

6.12 Given the strain energy potential for the case of infinitesimal deformations

$$\Psi(\varepsilon) = \frac{\lambda}{2}(\text{tr } \varepsilon)^2 + \mu \, \text{tr}(\varepsilon \cdot \varepsilon),$$

determine the strain energy function $\Psi(\sigma)$ in terms of the stress tensor σ.

6.13 Assuming that the strain energy density $\Psi = U_0(\sigma)$ is positive-definite, that is, $U_0 \geq 0$, with $U_0 = 0$ if and only if $\sigma = 0$, determine the restrictions placed on the elastic parameters E, K, and ν by considering the following stress states: (a) uniaxial stress state with $\sigma_{11} = \sigma$; (b) pure shear stress state, $\sigma_{12} = \tau$; and (c) hydrostatic stress state, $\sigma_{11} = \sigma_{22} = \sigma_{33} = p$.

6.14 A material is *transversely isotropic* at a point if it is symmetric with respect to an arbitrary rotation about a given axis. Aligned fiber-reinforced composites provide examples of transversely isotropic materials (see Fig. P6.14). Take the x_3-axis as the axis of symmetry with the transformation matrix

$$[L] = \begin{bmatrix} \cos\theta & \sin\theta & 0 \\ -\sin\theta & \cos\theta & 0 \\ 0 & 0 & 1 \end{bmatrix},$$

where θ is arbitrary. Show that the stress–strain relations of a transversely isotropic material are of the form

$$\begin{Bmatrix} \sigma_1 \\ \sigma_2 \\ \sigma_3 \\ \sigma_4 \\ \sigma_5 \\ \sigma_6 \end{Bmatrix} = \begin{bmatrix} C_{11} & C_{12} & C_{13} & 0 & 0 & 0 \\ C_{12} & C_{11} & C_{13} & 0 & 0 & 0 \\ C_{13} & C_{13} & C_{33} & 0 & 0 & 0 \\ 0 & 0 & 0 & C_{44} & 0 & 0 \\ 0 & 0 & 0 & 0 & C_{44} & 0 \\ 0 & 0 & 0 & 0 & 0 & \frac{1}{2}(C_{11}-C_{12}) \end{bmatrix} \begin{Bmatrix} \varepsilon_1 \\ \varepsilon_2 \\ \varepsilon_3 \\ \varepsilon_4 \\ \varepsilon_5 \\ \varepsilon_6 \end{Bmatrix}.$$

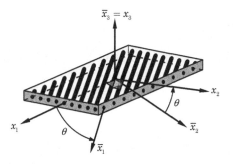

Fig. P6.14

6.15 The stress–strain relations of an isotropic material in the cylindrical coordinate system are

$$\sigma_{rr} = 2\mu\,\varepsilon_{rr} + \lambda\,(\varepsilon_{rr} + \varepsilon_{\theta\theta} + \varepsilon_{zz}),$$
$$\sigma_{\theta\theta} = 2\mu\,\varepsilon_{\theta\theta} + \lambda\,(\varepsilon_{rr} + \varepsilon_{\theta\theta} + \varepsilon_{zz}),$$
$$\sigma_{zz} = 2\mu\,\varepsilon_{zz} + \lambda\,(\varepsilon_{rr} + \varepsilon_{\theta\theta} + \varepsilon_{zz}),$$
$$\sigma_{r\theta} = 2\mu\,\varepsilon_{r\theta}, \quad \sigma_{rz} = 2\mu\,\varepsilon_{rz}, \quad \sigma_{\theta z} = 2\mu\,\varepsilon_{\theta z}.$$

Express the relations in terms of the displacements (u_r, u_θ, u_z).

6.16 Express the stress–strain relations of an isotropic material in the spherical coordinate system and express the result in terms of the displacements (u_R, u_ϕ, u_θ).

6.17 Given the displacement field in an isotropic body

$$u_r = U(r), \quad u_\theta = 0, \quad u_z = 0, \tag{1}$$

where $U(r)$ is a function of only r, determine the stress components in the cylindrical coordinate system.

6.18 Given the displacement field in an isotropic body

$$u_R = U(R), \quad u_\phi = 0, \quad u_\theta = 0, \tag{1}$$

where $U(R)$ is a function of only R, determine the stress components in the spherical coordinate system.

6.19 *The Navier equations.* Show that for an isotropic, incompressible solid with infinitesimal deformations (i.e., $\boldsymbol{\sigma} \approx \mathbf{S}$ and $\mathbf{F} \cdot \mathbf{S} \approx \mathbf{S}$), the equation of motion (5.3.11), $\nabla \cdot \boldsymbol{\sigma} + \rho_0 \mathbf{f} = \rho_0\ddot{\mathbf{u}}$, can be expressed as

$$\rho_0 \frac{\partial^2 \mathbf{u}}{\partial t^2} = \rho_0 \mathbf{f} - \nabla p + (\lambda + \mu)\,\nabla(\nabla \cdot \mathbf{u}) + \mu\nabla^2 \mathbf{u}.$$

Newtonian Fluids

6.20 Given the following motion of an isotropic continuum,

$$\chi(\mathbf{X}) = (X_1 + kt^2 X_2^2)\,\hat{\mathbf{e}}_1 + (X_2 + ktX_2)\,\hat{\mathbf{e}}_2 + X_3\,\hat{\mathbf{e}}_3,$$

determine the components of the viscous stress tensor as a function of position and time.

6.21 Express the upper and lower convective derivatives of Eqs. (6.6.12) and (6.6.13) in Cartesian component form.

6.22 Interpret the Lamé constant μ by considering the flow field

$$v_1 = f(x_2), \quad v_2 = 0, \quad v_3 = 0,$$

where f is a known function of x_1.

6.23 For viscous compressible flows (in spatial description), show that

$$\tilde{\sigma} - p = \left(\lambda + \tfrac{2}{3}\mu\right)\frac{1}{\rho}\frac{D\rho}{Dt},$$

where $\tilde{\sigma} = -\sigma_{ii}/3$ is the mean stress and p is the thermodynamic pressure.

6.24 *The Navier–Stokes equations.* Show that for a compressible fluid, the Cauchy equations of motion (5.3.10) can be expressed as

$$\rho\frac{D\mathbf{v}}{Dt} = \rho\mathbf{f} - \boldsymbol{\nabla}p + (\lambda + \mu)\,\boldsymbol{\nabla}(\boldsymbol{\nabla}\cdot\mathbf{v}) + \mu\nabla^2\mathbf{v}.$$

Simplify the equation for (a) an incompressible fluid and (b) hydrostatic state of stress.

6.25 Show that for an incompressible fluid the equation of motion simplifies to

$$\frac{D}{Dt}(\rho\mathbf{v}) = \rho\mathbf{f} - \boldsymbol{\nabla}p + \mu\nabla^2\mathbf{v}.$$

6.26 Show that for the two-dimensional flow of an incompressible Newtonian fluid with $\boldsymbol{\nabla}\times\mathbf{f} = 0$, where $\mathbf{f}$ is the body force vector (measured per unit volume), the vorticity $\mathbf{w}$ [see Eq. (3.6.5)] satisfies the diffusion equation

$$\rho\frac{D\mathbf{w}}{Dt} = \mu\nabla^2\mathbf{w}.$$

6.27 *Stokesian fluid.* A Stokesian fluid is one in which (a) the stress tensor σ is a continuous function of the rate of deformation tensor $\mathbf{D}$ and the local thermodynamic state (i.e., may depend on temperature), but independent of other kinematic variables; (b) σ is not an explicit function of position $\mathbf{x}$; (c) constitutive behavior is isotropic; and (d) the stress is hydrostatic when the rate of deformation is zero, $\mathbf{D} = \mathbf{0}$. Consider the following constitutive equation for a Stokesian fluid:

$$\sigma = -p\mathbf{I} + \mu\mathbf{D} + \beta\,\mathbf{D}\cdot\mathbf{D} \quad (\sigma_{ij} = -p\,\delta_{ij} + \mu D_{ij} + \beta\,D_{ik}D_{kj}).$$

Write the equations of motion (5.3.10) in terms of p and $\mathbf{D}$ for a Stokesian fluid. Note that a linear Stokesian fluid is a Newtonian fluid.

6.28 *Irrotational motion.* The velocity field $\mathbf{v}$ is said to be irrotational when the vorticity is zero, $\mathbf{w} = \mathbf{0}$. Then there exists a velocity potential $\phi(\mathbf{x}, t)$ such that $\mathbf{v} = \boldsymbol{\nabla}\phi$. Show that the Navier–Stokes equations of Problem **6.24** can be expressed in the form

$$\rho\boldsymbol{\nabla}\left[\frac{\partial\phi}{\partial t} + \tfrac{1}{2}(\boldsymbol{\nabla}\phi)^2\right] = \rho\mathbf{f} - \boldsymbol{\nabla}p + (\lambda + 2\mu)\boldsymbol{\nabla}(\nabla^2\phi).$$

6.29 Show that in the case of irrotational body force $\mathbf{f} = -\boldsymbol{\nabla}V$ and when p is a function only of ρ

$$\frac{\partial\phi}{\partial t} + \tfrac{1}{2}(\boldsymbol{\nabla}\phi)^2 + V + P(\rho) - \frac{1}{\rho}(\lambda + 2\mu)\nabla^2\phi = g(t),$$

where $P(\rho) = \int_{p_0}^{p} dp/\rho$, p_0 is a constant, and $g(t)$ is a function of time only.

Heat Transfer

6.30 Show that for an isotropic Newtonian fluid the energy equation can be expressed in the form

$$\rho \frac{De}{Dt} = \mathbf{\nabla} \cdot (k\mathbf{\nabla}\theta) - p\,J_1 + (\lambda + 2\mu)J_1^2 - 4\mu J_2 + \rho\,r,$$

where J_1 and J_2 are the principal invariants of $\mathbf{D}$ [see Eq. (3.4.36)], and k is the conductivity.

6.31 Show that for an isotropic Newtonian fluid the energy equation can be expressed in the form

$$\rho\theta \frac{D\eta}{Dt} = \mathbf{\nabla} \cdot (k\mathbf{\nabla}\theta) + (\lambda + 2\mu)J_1^2 - 4\mu J_2 + \rho\,r,$$

where θ is the absolute temperature, η is the entropy, J_1 and J_2 are the principal invariants of $\mathbf{D}$, and k is the conductivity. *Hint:* $\theta\,d\eta = de + p\,d(1/\rho)$ and $d/dt = D/Dt$.

6.32 The thermal stress coefficients, β_{ij}, measure the increases in the stress components per unit decrease in temperature with no change in the strain, that is,

$$\beta_{ij} = -\frac{\partial\sigma_{ij}}{\partial\theta}\bigg|_{\varepsilon=\text{const}}.$$

Deduce from the above equation the result

$$\beta_{ij} = \rho_0 \frac{\partial\eta}{\partial\varepsilon_{ij}}.$$

6.33 The specific heat at constant strain is defined by

$$c_v = \frac{\partial e}{\partial\theta}\bigg|_{\varepsilon=\text{const}}.$$

Deduce from the above equation the result

$$c_v = -\theta \frac{\partial^2\Psi}{\partial\theta^2}.$$

6.34 Consider a reference state at zero strain and temperature θ_0, and expand $\Psi(\theta, \varepsilon)$ in Taylor's series about this state up to quadratic terms in θ and ε_{ij} to derive the constitutive equations, Eq. (6.8.20), for linear thermoelasticity. Specialize the relations to the isotropic case.

LINEARIZED ELASTICITY

You cannot depend on your eyes when your imagination is out of focus.
— Mark Twain (1835–1910)

Research is to see what everybody else has seen, and to think what nobody else has thought.
— Albert Szent-Gyoergi (1893-1986)

7.1 Introduction

This chapter is dedicated to the study of deformation and stress in solid bodies under a prescribed set of forces and kinematic constraints. In a majority of problems, we assume that stresses and strains are small so that linear strain-displacement relations and Hooke's law are valid, and we use appropriate governing equations derived using the Lagrangian description in the previous chapters to solve them for stresses and displacements. In the linearized elasticity we assume that the geometric changes are so small that we neglect squares of the displacement gradients, that is, $|\nabla \mathbf{u}|^2 \approx 0$, and do not make a distinction between the deformed and undeformed geometries, between the second Piola–Kirchhoff stress tensor $\mathbf{S}$ and the Cauchy stress tensor $\boldsymbol{\sigma}$, and between the current coordinates $\mathbf{x}$ and the material coordinates $\mathbf{X}$ (and use $\boldsymbol{\sigma}$ and $\mathbf{x}$). Mathematically, we seek solutions to coupled partial differential equations over an elastic domain occupied by the reference (or undeformed) configuration of the body, subject to specified boundary conditions on displacements or forces. Such problems are called boundary value problems of elasticity.

Most practical problems of even linearized elasticity involve geometries that are complicated, and analytical solutions to such problems cannot be obtained. Therefore, the objective here is to familiarize the reader with certain solution methods as applied to simple boundary value problems. Boundary value problems discussed in most elasticity books are about the same, and they illustrate the methodologies used in the analytical solution of problems of elasticity. Although is a book on a first course in continuum mechanics, typical solid mechanics problems discussed in most elasticity books, for example, Timoshenko and Goodier (1970), Slaughter (2002), and Sadd (2004) are covered. The methods discussed here may not be directly useful in solving practical engineering problems, but the discussion provides certain insights into the formulation and solution of boundary value problems. These insights are useful irrespective of the specific problems or methods of solution presented here.

7.2 Governing Equations

7.2.1 Preliminary Comments

It is useful to summarize the equations of linearized elasticity for use in the later sections of this chapter. The governing equations of a three-dimensional elastic body involve (1) 6 strain-displacement relations among 9 variables, namely 6 components of strain tensor $\boldsymbol{\varepsilon}$ and 3 components of displacement vector $\mathbf{u}$; (2) 3 equations of motion among 6 components of stress tensor $\boldsymbol{\sigma}$, assuming symmetry of the stress tensor; and (3) 6 stress–strain equations among 6 stress and 6 strain components that are already counted. Thus, there are a total of 15 coupled equations among 15 scalar field variables. These equations are listed here in vector form and Cartesian, cylindrical, and spherical component forms for an isotropic body occupying a domain Ω with closed boundary Γ in the reference configuration. Figures 7.2.1(a)–(c) show the normal stress components in the three coordinate systems; shear stress components should be obvious (as well as all of the strain components).

7.2.2 Summary of Equations

All of the equations derived in Chapters 3, 4, 5, and 6 in material description are presented here. Throughout this chapter, we use the following notations: $\mathbf{x} = \mathbf{X}$, $\boldsymbol{\varepsilon} = \mathbf{E}$, and $\boldsymbol{\sigma} = \mathbf{S}$.

7.2.2.1 Strain-displacement equations

The linearized strain-displacement relations are summarized here:

Vector form:

$$\boldsymbol{\varepsilon} = \frac{1}{2}\left[\boldsymbol{\nabla}\mathbf{u} + (\boldsymbol{\nabla}\mathbf{u})^{\mathsf{T}}\right] \tag{7.2.1}$$

Rectangular Cartesian component form: (u_x, u_y, u_z)

$$\varepsilon_{xx} = \frac{\partial u_x}{\partial x}, \qquad\qquad \varepsilon_{xy} = \frac{1}{2}\left(\frac{\partial u_x}{\partial y} + \frac{\partial u_y}{\partial x}\right)$$

$$\varepsilon_{xz} = \frac{1}{2}\left(\frac{\partial u_x}{\partial z} + \frac{\partial u_z}{\partial x}\right), \qquad \varepsilon_{yy} = \frac{\partial u_y}{\partial y} \tag{7.2.2}$$

$$\varepsilon_{yz} = \frac{1}{2}\left(\frac{\partial u_y}{\partial z} + \frac{\partial u_z}{\partial y}\right), \qquad \varepsilon_{zz} = \frac{\partial u_z}{\partial z}$$

Component form in cylindrical coordinates: (u_r, u_θ, u_z)

$$\varepsilon_{rr} = \frac{\partial u_r}{\partial r}, \qquad\qquad \varepsilon_{r\theta} = \frac{1}{2}\left(\frac{1}{r}\frac{\partial u_r}{\partial \theta} + \frac{\partial u_\theta}{\partial r} - \frac{u_\theta}{r}\right)$$

$$\varepsilon_{rz} = \frac{1}{2}\left(\frac{\partial u_r}{\partial z} + \frac{\partial u_z}{\partial r}\right), \qquad \varepsilon_{\theta\theta} = \frac{u_r}{r} + \frac{1}{r}\frac{\partial u_\theta}{\partial \theta} \tag{7.2.3}$$

$$\varepsilon_{z\theta} = \frac{1}{2}\left(\frac{\partial u_\theta}{\partial z} + \frac{1}{r}\frac{\partial u_z}{\partial \theta}\right), \qquad \varepsilon_{zz} = \frac{\partial u_z}{\partial z}$$

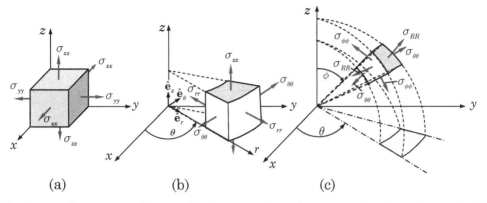

Fig. 7.2.1: Components of a second-order tensor (stress) on a typical volume element in (a) Cartesian, (b) cylindrical, and (c) spherical coordinate systems.

Component form in spherical coordinates: (u_R, u_ϕ, u_θ)

$$\varepsilon_{RR} = \frac{\partial u_R}{\partial R}, \quad \varepsilon_{\phi\phi} = \frac{1}{R}\left(\frac{\partial u_\phi}{\partial \phi} + u_R\right)$$

$$\varepsilon_{R\phi} = \tfrac{1}{2}\left(\frac{1}{R}\frac{\partial u_R}{\partial \phi} + \frac{\partial u_\phi}{\partial R} - \frac{u_\phi}{R}\right)$$

$$\varepsilon_{R\theta} = \tfrac{1}{2}\left(\frac{1}{R\sin\phi}\frac{\partial u_R}{\partial \theta} + \frac{\partial u_\theta}{\partial R} - \frac{u_\theta}{R}\right) \qquad (7.2.4)$$

$$\varepsilon_{\phi\theta} = \tfrac{1}{2}\frac{1}{R}\left(\frac{1}{\sin\phi}\frac{\partial u_\phi}{\partial \theta} + \frac{\partial u_\theta}{\partial \phi} - u_\theta\cot\phi\right)$$

$$\varepsilon_{\theta\theta} = \frac{1}{R\sin\phi}\left(\frac{\partial u_\theta}{\partial \theta} + u_R\sin\phi + u_\phi\cos\phi\right)$$

7.2.2.2 Equations of motion

The linearized equations of motion, under the assumption that $\boldsymbol{\sigma}^{\mathrm{T}} = \boldsymbol{\sigma}$, are summarized. The equilibrium equations are obtained by setting the acceleration terms to zero. Here $\mathbf{f}$ is the body force vector measured per unit mass.

Vector form

$$\boldsymbol{\nabla}\cdot\boldsymbol{\sigma} + \rho_0\mathbf{f} = \rho_0\frac{\partial^2\mathbf{u}}{\partial t^2} \qquad (7.2.5)$$

Rectangular Cartesian component form: $(\sigma_{xx}, \sigma_{yy}, \sigma_{xy}, \cdots)$

$$\frac{\partial \sigma_{xx}}{\partial x} + \frac{\partial \sigma_{xy}}{\partial y} + \frac{\partial \sigma_{xz}}{\partial z} + \rho_0 f_x = \rho_0\frac{\partial^2 u_x}{\partial t^2}$$

$$\frac{\partial \sigma_{yx}}{\partial x} + \frac{\partial \sigma_{yy}}{\partial y} + \frac{\partial \sigma_{yz}}{\partial z} + \rho_0 f_y = \rho_0\frac{\partial^2 u_y}{\partial t^2} \qquad (7.2.6)$$

$$\frac{\partial \sigma_{zx}}{\partial x} + \frac{\partial \sigma_{zy}}{\partial y} + \frac{\partial \sigma_{zz}}{\partial z} + \rho_0 f_z = \rho_0\frac{\partial^2 u_z}{\partial t^2}$$

Component form in cylindrical coordinates: $(\sigma_{rr}, \sigma_{\theta\theta}, \sigma_{r\theta}, \cdots)$

$$\frac{\partial \sigma_{rr}}{\partial r} + \frac{1}{r}\frac{\partial \sigma_{r\theta}}{\partial \theta} + \frac{\partial \sigma_{rz}}{\partial z} + \frac{1}{r}(\sigma_{rr} - \sigma_{\theta\theta}) + \rho_0 f_r = \rho_0 \frac{\partial^2 u_r}{\partial t^2}$$

$$\frac{\partial \sigma_{\theta r}}{\partial r} + \frac{1}{r}\frac{\partial \sigma_{\theta\theta}}{\partial \theta} + \frac{\partial \sigma_{\theta z}}{\partial z} + \frac{\sigma_{\theta r} + \sigma_{r\theta}}{r} + \rho_0 f_\theta = \rho_0 \frac{\partial^2 u_\theta}{\partial t^2} \qquad (7.2.7)$$

$$\frac{\partial \sigma_{zr}}{\partial r} + \frac{1}{r}\frac{\partial \sigma_{z\theta}}{\partial \theta} + \frac{\partial \sigma_{zz}}{\partial z} + \frac{\sigma_{zr}}{r} + \rho_0 f_z = \rho_0 \frac{\partial^2 u_z}{\partial t^2}$$

Component form in spherical coordinates: $(\sigma_{RR}, \sigma_{\phi\phi}, \sigma_{R\phi}, \cdots)$

$$\frac{\partial \sigma_{RR}}{\partial R} + \frac{1}{R}\frac{\partial \sigma_{R\phi}}{\partial \phi} + \frac{1}{R\sin\phi}\frac{\partial \sigma_{R\theta}}{\partial \theta} + \frac{1}{R}(2\sigma_{RR} - \sigma_{\phi\phi} - \sigma_{\theta\theta} + \sigma_{R\phi}\cot\phi)$$

$$+ \rho_0 f_R = \rho_0 \frac{\partial^2 u_R}{\partial t^2}$$

$$\frac{\partial \sigma_{\phi R}}{\partial R} + \frac{1}{R}\frac{\partial \sigma_{\phi\phi}}{\partial \phi} + \frac{1}{R\sin\phi}\frac{\partial \sigma_{\phi\theta}}{\partial \theta} + \frac{1}{R}[(\sigma_{\phi\phi} - \sigma_{\theta\theta})\cot\phi + \sigma_{R\phi} + 2\sigma_{\phi R}]$$

$$+ \rho_0 f_\phi = \rho_0 \frac{\partial^2 u_\phi}{\partial t^2} \qquad (7.2.8)$$

$$\frac{\partial \sigma_{\theta R}}{\partial R} + \frac{1}{R}\frac{\partial \sigma_{\theta\phi}}{\partial \phi} + \frac{1}{R\sin\phi}\frac{\partial \sigma_{\theta\theta}}{\partial \theta} + \frac{1}{R}[(\sigma_{\phi\theta} + \sigma_{\theta\phi})\cot\phi + \sigma_{R\theta}]$$

$$+ \rho_0 f_\theta = \rho_0 \frac{\partial^2 u_\theta}{\partial t^2}$$

7.2.2.3 Constitutive equations

The stress–strain equations of a linear, isotropic, elastic body are presented here.

Vector form

$$\boldsymbol{\sigma} = 2\mu\boldsymbol{\varepsilon} + \lambda\,(\mathrm{tr}\,\boldsymbol{\varepsilon})\,\mathbf{I} \qquad (7.2.9)$$

Rectangular Cartesian, cylindrical, and spherical component forms

$$\begin{Bmatrix} \sigma_{11} \\ \sigma_{22} \\ \sigma_{33} \\ \sigma_{23} \\ \sigma_{13} \\ \sigma_{12} \end{Bmatrix} = \frac{E}{(1+\nu)(1-2\nu)} \begin{bmatrix} 1-\nu & \nu & \nu & 0 & 0 & 0 \\ \nu & 1-\nu & \nu & 0 & 0 & 0 \\ \nu & \nu & 1-\nu & 0 & 0 & 0 \\ 0 & 0 & 0 & \frac{1-2\nu}{2} & 0 & 0 \\ 0 & 0 & 0 & 0 & \frac{1-2\nu}{2} & 0 \\ 0 & 0 & 0 & 0 & 0 & \frac{1-2\nu}{2} \end{bmatrix} \begin{Bmatrix} \varepsilon_{11} \\ \varepsilon_{22} \\ \varepsilon_{33} \\ 2\varepsilon_{23} \\ 2\varepsilon_{13} \\ 2\varepsilon_{12} \end{Bmatrix}$$

$$(7.2.10)$$

Here the subscripts 1, 2, and 3 take x, y, and z for rectangular Cartesian coordinates; r, θ, and z for cylindrical coordinates; and R, ϕ, and θ for spherical coordinates. The Lamé constants μ and λ are related to Young's modulus E and Poisson's ratio ν by

$$\mu = G = \frac{E}{2(1+\nu)}, \qquad \lambda = \frac{\nu E}{(1+\nu)(1-2\nu)} \,. \qquad (7.2.11)$$

Equations (7.2.1)–(7.2.10) are valid for all problems of linearized elasticity; various problems differ from each other only in (a) geometry of the domain, (b) boundary conditions, and (c) values of the material parameters E and ν. The general form of the boundary condition is presented next.

7.2.2.4 Boundary conditions

Vector form

$$\mathbf{u} = \hat{\mathbf{u}} \ \text{ on } \ \Gamma_u, \quad \hat{\mathbf{n}} \cdot \boldsymbol{\sigma} = \hat{\mathbf{t}} \ \text{ on } \ \Gamma_\sigma. \tag{7.2.12}$$

Component form

$$u_i = \hat{u}_i \ \text{ on } \ \Gamma_u, \quad \sigma_{ij} n_j = \hat{t}_i \ \text{ on } \ \Gamma_\sigma, \tag{7.2.13}$$

where Γ_σ and Γ_u are disjoint portions (except for a point) of the boundary whose union is equal to the total boundary Γ, and quantities with a hat are specified values. Note that only one element of the pair (t_i, u_i), for any $i = 1, 2, 3$, may be specified at a point on the boundary. The indices (1, 2, 3) may take the values of (x, y, z), (r, θ, z), and (R, ϕ, θ).

7.2.2.5 Compatibility conditions

In addition to the 15 equations listed in (7.2.1), (7.2.5), and (7.2.9), there are 6 *compatibility conditions* among 6 components of strain:

$$\nabla \times (\nabla \times \boldsymbol{\varepsilon})^{\mathrm{T}} = \mathbf{0}, \quad e_{ikr} e_{jls} \varepsilon_{ij,k\ell} = 0. \tag{7.2.14}$$

Recall that the compatibility equations are necessary and sufficient conditions on the strain field to ensure the existence of a corresponding displacement field. Associated with each displacement field, there is a unique strain field as given by Eq. (7.2.1) and there is no need to use the compatibility conditions. The compatibility conditions are required only when the strain or stress field is given and the displacement field is to be determined.

In most formulations of boundary value problems of elasticity, one does not use the 15 equations in 15 unknowns. Most often, the 15 equations are reduced to either 3 equations in terms of displacement field or 6 equations in terms of stress field. The two sets of equations are presented next.

7.2.3 The Navier Equations

The 15 equations can be combined into 3 equations by substituting strain-displacement equations into the stress–strain relations and the result into the equations of equilibrium. We shall carry out this process using the Cartesian component form (in index notation) and then express the final result in vector as well as Cartesian component forms.

The Cartesian component form of Eq. (7.2.9) is

$$\sigma_{ij} = \mu \left(u_{i,j} + u_{j,i} \right) + \lambda u_{k,k} \delta_{ij}. \tag{7.2.15}$$

Substituting into Eq. (7.2.5), we arrive at the equations

$$\rho_0 \frac{\partial^2 u_i}{\partial t^2} = \sigma_{ji,j} + \rho_0\, f_i$$
$$= \mu\,(u_{i,jj} + u_{j,ij}) + \lambda u_{k,ki} + \rho_0\, f_i$$
$$= \mu u_{i,jj} + (\mu + \lambda) u_{j,ji} + \rho_0\, f_i. \tag{7.2.16}$$

Thus, we have

$$\mu \nabla^2 \mathbf{u} + (\mu + \lambda) \boldsymbol{\nabla}\left(\boldsymbol{\nabla} \cdot \mathbf{u}\right) + \rho_0 \mathbf{f} = \rho_0 \frac{\partial^2 \mathbf{u}}{\partial t^2},$$
$$\mu u_{i,jj} + (\mu + \lambda) u_{j,ji} + \rho_0\, f_i = \rho_0 \frac{\partial^2 u_i}{\partial t^2}. \tag{7.2.17}$$

These are called *Lamé–Navier equations* of elasticity, and they represent the equilibrium equations expressed in terms of the displacement field. The boundary conditions (7.2.13) can be expressed in terms of the displacement field as

$$[n_j \mu\,(u_{i,j} + u_{j,i}) + n_i \lambda u_{k,k}] = \hat{t}_i \quad \text{on} \ \ \Gamma_\sigma, \qquad u_i = \hat{u}_i \ \ \text{on} \ \ \Gamma_u. \tag{7.2.18}$$

Equations (7.2.17) and (7.2.18) together describe the boundary value problem of linearized elasticity.

7.2.4 The Beltrami–Michell Equations

Alternative to the formulation of Section 7.2.3, the 12 equations from (7.2.5) and (7.2.9) and 6 equations from (7.2.14) can be combined into 6 equations in terms of the stress field. Substitution of the constitutive (strain-stress) equations

$$\varepsilon_{ij} = \frac{1}{E}\left[(1+\nu)\sigma_{ij} - \nu\sigma_{kk}\delta_{ij}\right] \tag{7.2.19}$$

into the compatibility equations (7.2.14) yields

$$0 = e_{ikr} e_{jls}\, \varepsilon_{ij,k\ell}$$
$$= e_{ikr} e_{jls}\left[(1+\nu)\sigma_{ij,k\ell} - \nu\sigma_{mm,k\ell}\delta_{ij}\right]$$
$$= (1+\nu) e_{ikr} e_{jls}\sigma_{ij,k\ell} - \nu e_{ikr} e_{ils}\sigma_{mm,k\ell}$$
$$= (1+\nu) e_{ikr} e_{jls}\sigma_{ij,k\ell} - \nu\left(\delta_{k\ell}\delta_{rs} - \delta_{ks}\delta_{\ell r}\right)\sigma_{mm,k\ell}$$
$$= (1+\nu) e_{ikr} e_{jls}\sigma_{ij,k\ell} - \nu\left(\delta_{rs}\sigma_{mm,kk} - \sigma_{mm,rs}\right). \tag{7.2.20}$$

In view of the identity

$$e_{ikr} e_{jls} = \begin{vmatrix} \delta_{ij} & \delta_{i\ell} & \delta_{is} \\ \delta_{kj} & \delta_{k\ell} & \delta_{ks} \\ \delta_{rj} & \delta_{r\ell} & \delta_{rs} \end{vmatrix} = \delta_{ij}\delta_{k\ell}\delta_{rs} - \delta_{ij}\delta_{ks}\delta_{r\ell} - \delta_{kj}\delta_{i\ell}\delta_{rs} + \delta_{kj}\delta_{r\ell}\delta_{is}$$
$$+ \delta_{rj}\delta_{i\ell}\delta_{ks} - \delta_{rj}\delta_{k\ell}\delta_{is}, \tag{7.2.21}$$

Eq. (7.2.20) simplifies to

$$\delta_{rs}\sigma_{ii,jj} - \sigma_{ii,rs} - (1+\nu)\left(\delta_{rs}\sigma_{ij,ij} + \sigma_{rs,ii} - \sigma_{is,ir} - \sigma_{ir,is}\right) = 0. \tag{7.2.22}$$

Contracting the indices r and s $(s \to r)$ gives

$$2\sigma_{ii,jj} - (1 + \nu)(\sigma_{ij,ij} + \sigma_{jj,ii}) = 0.$$

Simplifying the above result, we obtain

$$\sigma_{ii,jj} = \frac{(1+\nu)}{(1-\nu)}\sigma_{ij,ij}. \qquad (7.2.23)$$

Substituting this result back into Eq. (7.2.22) leads to

$$\sigma_{ij,kk} + \frac{1}{1+\nu}\sigma_{kk,ij} = \frac{\nu}{1-\nu}\sigma_{rs,rs}\delta_{ij} + \sigma_{kj,ki} + \sigma_{ki,kj}. \qquad (7.2.24)$$

Next, we use the equilibrium equations to compute the second derivative of the stress components, $\sigma_{rs,rk} = -\rho_0\, f_{s,k}$. We have

$$\sigma_{ij,kk} + \frac{1}{1+\nu}\sigma_{kk,ij} = -\frac{\nu\rho_0}{1-\nu}f_{k,k}\delta_{ij} - \rho_0\left(f_{j,i} + f_{i,j}\right), \qquad (7.2.25)$$

or in vector form

$$\nabla^2\boldsymbol{\sigma} + \frac{1}{1+\nu}\nabla[\nabla(\operatorname{tr}\boldsymbol{\sigma})] = -\frac{\nu\rho_0}{1-\nu}(\nabla \cdot \mathbf{f})\mathbf{I} - \rho_0\left[\nabla\mathbf{f} + (\nabla\mathbf{f})^{\mathsf{T}}\right]. \qquad (7.2.26)$$

The 6 equations in (7.2.25) or (7.2.26), called *Michell's equations*, provide the necessary and sufficient conditions for an equilibrated stress field to be compatible with the displacement field in the body. The traction boundary conditions in Eq. (7.2.13) are valid for this formulation.

When the body force is uniform, we have $\nabla \cdot \mathbf{f} = 0$ and $\nabla\mathbf{f} = \mathbf{0}$, and Michell's equations (7.2.26) reduce to *Beltrami's equations*

$$\nabla^2\boldsymbol{\sigma} + \frac{1}{1+\nu}\nabla[\nabla(\operatorname{tr}\boldsymbol{\sigma})] = \mathbf{0}, \quad \sigma_{ij,kk} + \frac{1}{1+\nu}\sigma_{kk,ij} = 0. \qquad (7.2.27)$$

7.3 Solution Methods

7.3.1 Types of Problems

The equilibrium problems, also called boundary value problems, of elasticity can be classified into three types on the basis of the nature of specified boundary conditions. They are outlined next.

Type I. Boundary value problems in which all specified boundary conditions are of the displacement type

$$\mathbf{u} = \hat{\mathbf{u}} \text{ on } \Gamma, \qquad (7.3.1)$$

are called boundary value problems of type I or *displacement boundary value problems*.

Type II. Boundary value problems in which all specified boundary conditions are of the traction type,

$$\mathbf{t} = \hat{\mathbf{t}} \text{ on } \Gamma, \qquad (7.3.2)$$

are called boundary value problems of type II or *stress boundary value problems*. Such boundary value problems are rare because most practical problems involve specifying displacements that eliminate rigid-body motion.

Type III. Boundary value problems in which all specified boundary conditions are of the mixed type,

$$\mathbf{u} = \hat{\mathbf{u}} \ \text{ on } \ \Gamma_u \ \text{ and } \ \mathbf{t} = \hat{\mathbf{t}} \ \text{ on } \ \Gamma_\sigma, \tag{7.3.3}$$

are called boundary value problems of type III or *mixed boundary value problems*. Most practical problems, including contact problems, fall into this category.

7.3.2 Types of Solution Methods

An *exact solution* of a problem is one that satisfies the governing differential equation(s) at every point of the domain as well as the boundary conditions exactly. In general, finding exact solutions of elasticity problems is not simple owing to complicated geometries and boundary conditions. An *approximate solution* is one that satisfies governing differential equations as well as the boundary conditions approximately. *Numerical solutions* are approximate solutions that are developed using a numerical method, such as the finite difference method, the finite element method, the boundary element method, and other methods. Often one seeks approximate solutions of practical problems using numerical methods. The phrase *analytical solution* is used to indicate that the solution, exact or approximate, is obtained using analytical means rather than by numerical methods. Also, one may obtain exact solution to an idealized (or approximate) mathematical model of the actual problem. Most of the exact solutions found in textbooks fall into this category.

There are several types of solution methods for finding analytical solutions [see Slaughter (2002)]. The most common methods are described here.

1. The *inverse method* is one in which one finds the solution for displacement, strain, and stress fields by solving the governing equations of elasticity, and then tries to find a problem with geometry and boundary conditions to which the fields correspond. This approach is more common with mathematicians than with engineers.

2. The *semi-inverse method* is one in which the solution form in terms of unknown functions is arrived at with the help of a qualitative understanding of the problem characteristics. The unknown functions are determined to satisfy the governing equations. In identifying a solution form, often assumptions are made about the displacement or stress field (in addition to the constitutive behavior) to reduce a three-dimensional problem to a two-dimensional or even one-dimensional problem. Very few problems of elasticity have exact solutions, and the assumed fields in most cases are approximate. The semi-inverse method is the most commonly used approach in solid mechanics.

3. The *method of potentials* is one in which potential functions (with unknowns) are introduced to trivially satisfy some or all of the governing equations, and the functions are then determined using the remaining governing equations as well as boundary conditions of the problem. The potential functions are then used to determine stresses, strains, and displacements.

4. *Variational methods* are those that make use of extremum (i.e., minimum or maximum) and stationary principles, which are equivalent to the governing equations and some of the boundary conditions of the problem. The principles are cast in terms of strain energy, work done by loads, and kinetic energy of the system. The variational methods have the added advantage of being approximate methods. Variational methods form the basis of certain numerical methods such as the finite element method.

Other analytical methods include complex variable methods, integral transform methods, perturbation methods, method of multiple scales, and so on. In the remainder of this chapter, we consider mostly the semi-inverse method and the method of potentials to formulate and solve certain problems of linearized elasticity.

7.3.3 Examples of the Semi-inverse Method

In the first problem (spherical pressure vessel) considered in this section, the displacement field is assumed in terms of an unknown function, and then the equations of elasticity or their equivalents are used to determine differential equations governing the unknown function. In the second problem (deformation of a prismatic bar under its own weight), the state of stress is assumed in terms of an unknown function and the equations of elasticity are used to determine the unknown function, strains, and displacements. In the first problem, even though the semi-inverse method is used, the assumed form of the solution happens to be exact. This is not the case in most problems of elasticity. These two examples illustrate the general methodology of solving elasticity problems by the semi-inverse method. The key element of the approach is to gain sufficient qualitative understanding of solution (displacements and stresses) before identifying the solution form.

Example 7.3.1 ───

Consider an isotropic, hollow spherical pressure vessel of internal radius a and outside radius b. The vessel is pressurized at $r = a$ as well as at $r = b$ with pressures p_a and p_b, respectively, as shown in Fig. 7.3.1(a). Determine the displacements, strains, and stresses in the pressure vessel.

Solution: We use the spherical coordinate system to formulate the problem. Based on the spherical symmetry of the geometry, boundary conditions, and material properties, we note that the solution also exhibits spherical symmetry, that is, the solution does not depend on ϕ and θ coordinates [see Fig. 7.3.1(b)]. In fact, the only nonzero displacement is u_R, and it is only a function of the radial distance R. Thus, this three-dimensional elasticity problem can be formulated as a two-dimensional one without any approximation.

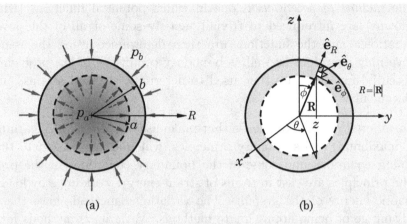

Fig. 7.3.1: A spherical pressure vessel.

For this problem, only the following stress boundary conditions (BVP type II) are known:

$$\text{At } R = a: \ \hat{\mathbf{n}} = -\hat{\mathbf{e}}_R, \ \mathbf{t} = p_a\,\hat{\mathbf{e}}_R \ \text{or} \ \sigma_{RR} = -p_a, \ \sigma_{R\phi} = \sigma_{R\theta} = 0,$$
$$\text{At } R = b: \ \hat{\mathbf{n}} = \hat{\mathbf{e}}_R, \ \mathbf{t} = -p_b\,\hat{\mathbf{e}}_R \ \text{or} \ \sigma_{RR} = -p_b, \ \sigma_{R\phi} = \sigma_{R\theta} = 0. \tag{7.3.4}$$

Based on our qualitative understanding of the solution to the problem, we begin with the assumed displacement field

$$u_R = U(R), \ \ u_\phi = u_\theta = 0, \tag{7.3.5}$$

where $U(R)$ is an unknown function to be determined such that the equations of elasticity and boundary conditions of the problem are satisfied. If we cannot find $U(R)$ that satisfies the governing equations, then we must abandon the assumption in Eq. (7.3.5).

The only nonzero strains associated with the displacement field (7.3.5) are [see Eq. (7.2.4)]

$$\varepsilon_{RR} = \frac{dU}{dR}, \ \ \varepsilon_{\phi\phi} = \varepsilon_{\theta\theta} = \frac{1}{R}U(R). \tag{7.3.6}$$

The nonzero stresses are

$$\sigma_{RR} = 2\mu\varepsilon_{RR} + \lambda\left(\varepsilon_{RR} + \varepsilon_{\phi\phi} + \varepsilon_{\theta\theta}\right) = (2\mu + \lambda)\frac{dU}{dR} + 2\lambda\frac{U}{R},$$

$$\sigma_{\phi\phi} = 2\mu\varepsilon_{\phi\phi} + \lambda\left(\varepsilon_{RR} + \varepsilon_{\phi\phi} + \varepsilon_{\theta\theta}\right) = 2(\mu + \lambda)\frac{U}{R} + \lambda\frac{dU}{dR}, \tag{7.3.7}$$

$$\sigma_{\theta\theta} = \sigma_{\phi\phi}.$$

The last two equations of equilibrium, Eq. (7.2.8) without the body force and acceleration terms, are trivially satisfied, and the first equation reduces to

$$\frac{d\sigma_{RR}}{dR} + \frac{1}{R}\left(2\sigma_{RR} - \sigma_{\phi\phi} - \sigma_{\theta\theta}\right) = 0, \tag{7.3.8}$$

which can be expressed in terms of the displacement function $U(R)$ using Eq. (7.3.7)

$$(2\mu + \lambda)\frac{d^2U}{dR^2} + \frac{2\lambda}{R}\frac{dU}{dR} - 2\lambda\frac{U}{R^2}$$

$$+ \frac{1}{R}\left[2(2\mu + \lambda)\frac{dU}{dR} + 4\lambda\frac{U}{R} - 2\lambda\frac{dU}{dR} - 4(\mu + \lambda)\frac{U}{R}\right] = 0. \tag{7.3.9}$$

Simplifying the expression, we obtain

$$R^2\frac{d^2U}{dR^2} + 2R\frac{dU}{dR} - 2U = 0. \tag{7.3.10}$$

The linear differential equation (7.3.10) can be transformed to one with constant coefficients by a change of independent variable, $R = e^\xi$ (or $\xi = \ln R$). Using the chain rule of differentiation, we obtain

$$\frac{dU}{dR} = \frac{dU}{d\xi}\frac{d\xi}{dR} = \frac{1}{R}\frac{dU}{d\xi}, \quad \frac{d^2U}{dR^2} = \frac{d}{dR}\left(\frac{1}{R}\frac{dU}{d\xi}\right) = \frac{1}{R^2}\left(-\frac{dU}{d\xi} + \frac{d^2U}{d\xi^2}\right).$$

Substituting the above expressions into (7.3.10), we obtain

$$\frac{d^2U}{d\xi^2} + \frac{dU}{d\xi} - 2U = 0. \tag{7.3.11}$$

Seeking solution in the form $U(\xi) = e^{m\xi}$ and substituting it into Eq. (7.3.11), we obtain $(m-1)(m+2) = 0$. Hence, the general solution to the problem is

$$U(\xi) = c_1 e^\xi + c_2 e^{-2\xi}. \tag{7.3.12}$$

Changing back to the original independent variable R, the radial displacement is

$$u_R(R) = U(R) = c_1 R + \frac{c_2}{R^2}, \tag{7.3.13}$$

where the constants c_1 and c_2 are to be determined using the boundary conditions in Eq. (7.3.4). Hence, we must compute σ_{RR},

$$\sigma_{RR} = (2\mu + \lambda)\left(c_1 - c_2\frac{2}{R^3}\right) + 2\lambda\left(c_1 + c_2\frac{1}{R^3}\right)$$
$$= (2\mu + 3\lambda)c_1 - 4\mu c_2\frac{1}{R^3}. \tag{1}$$

Applying the stress boundary conditions in (7.3.5) and (7.3.6), we obtain

$$(2\mu + 3\lambda)c_1 - \frac{4\mu c_2}{a^3} = -p_a,$$
$$(2\mu + 3\lambda)c_1 - \frac{4\mu c_2}{b^3} = -p_b. \tag{2}$$

Solving for the constants c_1 and c_2, we obtain

$$c_1 = \frac{1}{(2\mu + 3\lambda)}\left(\frac{p_a a^3 - p_b b^3}{b^3 - a^3}\right), \quad c_2 = \frac{a^3 b^3}{4\mu}\left(\frac{p_a - p_b}{b^3 - a^3}\right). \tag{3}$$

Finally, the displacement u_R and stresses σ_{RR}, $\sigma_{\phi\phi}$, and $\sigma_{\theta\theta}$ in the sphere are given by

$$u_R(R) = \frac{R}{(2\mu + 3\lambda)}\left(\frac{p_a a^3 - p_b b^3}{b^3 - a^3}\right) + \frac{a^3 b^3}{4\mu R^2}\left(\frac{p_a - p_b}{b^3 - a^3}\right), \tag{7.3.14}$$

$$\sigma_{RR} = \left(\frac{p_a a^3 - p_b b^3}{b^3 - a^3}\right) - \frac{a^3 b^3}{R^3}\left(\frac{p_a - p_b}{b^3 - a^3}\right),$$
$$\sigma_{\phi\phi} = \sigma_{\theta\theta} = \left(\frac{p_a a^3 - p_b b^3}{b^3 - a^3}\right) + \frac{a^3 b^3}{2R^3}\left(\frac{p_a - p_b}{b^3 - a^3}\right). \tag{7.3.15}$$

Since the off-diagonal elements of the stress tensor are zero, that is, $\sigma_{R\phi} = \sigma_{R\theta} = \sigma_{\phi\theta} = 0$, σ_{RR}, $\sigma_{\phi\phi}$, and $\sigma_{\theta\theta}$ are the principal stresses, with $\hat{\mathbf{e}}_R$, $\hat{\mathbf{e}}_\phi$, and $\hat{\mathbf{e}}_\theta$ being the principal directions, respectively.

Example 7.3.2

Consider a prismatic bar with dimensions $2a \times 2b \times L$ and mass density ρ in a gravitational field $\mathbf{g} = -g\,\hat{\mathbf{e}}_3$. The top surface of the bar is attached to a rigid support in such a way that $u = v = w = 0$ at $x = y = 0, z = L$, as shown in Fig. 7.3.2. Use the semi-inverse method to determine the displacements i the body.

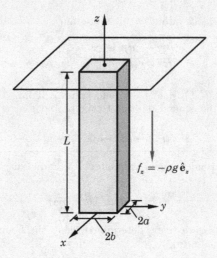

Fig. 7.3.2: Deformation of a prismatic bar under its own weight.

Solution: First we summarize the boundary conditions. We have

$$\mathbf{u}(0,0,L) = \mathbf{0}, \quad \mathbf{t}(x,y,0) = \mathbf{0}, \quad \mathbf{t}(x,\pm b,z) = \mathbf{0}, \quad \mathbf{t}(\pm a, y, z) = \mathbf{0}. \tag{7.3.16}$$

Thus, we find that

$$\sigma_{xx} = \sigma_{yy} = \sigma_{xy} = \sigma_{xz} = \sigma_{yz} = 0,$$

on the boundary, except at the point $x = y = 0$ and $z = L$. Since there are no other geometric constraints (that is, the body is free to change its geometry), it does not develop the stresses $\sigma_{xx}, \sigma_{yy}, \sigma_{xy}, \sigma_{xz}$, and σ_{yz}. Thus, we use the semi-inverse method, where we assume that

$$\sigma_{zz} = S(z), \quad \sigma_{xx} = \sigma_{yy} = \sigma_{xy} = \sigma_{xz} = \sigma_{yz} = 0. \tag{7.3.17}$$

The boundary conditions require that $S(0) = 0$. The first two equations of equilibrium are satisfied trivially and the third equation reduces to

$$\frac{dS}{dz} = \rho g \quad \rightarrow \quad S(z) = \gamma\,z + c \quad (\gamma = \rho g). \tag{1}$$

The constant of integration, c, is zero in order to satisfy the boundary condition $S(0) = 0$. Thus, the stress field is

$$\sigma_{zz} = \gamma\,z, \quad \sigma_{xx} = \sigma_{yy} = \sigma_{xy} = \sigma_{xz} = \sigma_{yz} = 0. \tag{2}$$

The stress compatibility conditions in Eq. (7.2.26) are trivially satisfied.

The strains are given by

$$\varepsilon_{xx} = \varepsilon_{yy} = -\frac{\nu}{E}\gamma\,z, \quad \varepsilon_{zz} = \frac{1}{E}\gamma\,z, \quad \varepsilon_{xy} = \varepsilon_{xz} = \varepsilon_{yz} = 0. \tag{3}$$

The corresponding displacement field is determined from the strain-displacement boundary conditions:

$$\varepsilon_{zz} = \frac{1}{E}\gamma\,z, \quad \rightarrow \quad u_z = \frac{1}{2E}\gamma\,z^2 + h(x,y),$$

$$2\varepsilon_{xz} = \frac{\partial u_x}{\partial z} + \frac{\partial u_z}{\partial x} = 0, \quad \rightarrow \quad \frac{\partial u_x}{\partial z} = -\frac{\partial h}{\partial x}, \tag{4}$$

where h is a function to be determined. Integrating Eq. (4), we obtain

$$u_x = -\frac{\partial h}{\partial x} z + g(x, y),$$ (5)

where g is a function to be determined. Similarly,

$$2\varepsilon_{yz} = \frac{\partial u_y}{\partial z} + \frac{\partial u_z}{\partial y} = 0, \quad \frac{\partial u_y}{\partial z} = -\frac{\partial h}{\partial y}, \quad u_y = -\frac{\partial h}{\partial y} z + f(x, y),$$ (6)

where f is a function to be determined. Now comparing ε_{xx} from Eq. (3) with that computed from Eq. (5), we obtain

$$-\frac{\nu}{E} \gamma z = -\frac{\partial^2 h}{\partial x^2} z + \frac{\partial g}{\partial x}.$$

We see that, because it must hold for any z,

$$\frac{\partial^2 h}{\partial x^2} = \frac{\nu}{E} \gamma, \quad \frac{\partial g}{\partial x} = 0 \quad \rightarrow \quad g = G(y).$$ (7)

Similarly, comparing ε_{yy} from Eq. (3) with that computed from Eq. (6), we obtain

$$-\frac{\nu}{E} \gamma z = -\frac{\partial^2 h}{\partial y^2} z + \frac{\partial f}{\partial y},$$

we see that, since it must hold for any z,

$$\frac{\partial^2 h}{\partial y^2} = \frac{\nu}{E} \gamma, \quad \frac{\partial f}{\partial y} = 0 \quad \rightarrow \quad f = F(x).$$ (8)

From $\varepsilon_{xy} = 0$, we see that

$$-2\frac{\partial^2 h}{\partial x \partial y} z + \frac{\partial g}{\partial y} + \frac{\partial f}{\partial x} = 0.$$ (9)

This gives the result

$$\frac{\partial^2 h}{\partial x \partial y} = 0, \quad \frac{dG}{dy} + \frac{dF}{dx} = 0 \quad \rightarrow G(y) = c_1 y + c_2, \quad F(x) = -c_1 x + c_3.$$ (10)

Conditions in Eqs. (6)–(8) imply that h is of the form

$$h(x, y) = \frac{\nu}{2E} \gamma (x^2 + y^2) + c_4 x + c_5 y + c_6,$$ (7.3.18)

where c_i are constants. In summary, we have

$$u_x = -\frac{\partial h}{\partial x} z + g(x, y) = -\frac{\nu}{E} \gamma x z - c_4 z + c_1 y + c_2,$$

$$u_y = -\frac{\partial h}{\partial y} z + f(x, y) = -\frac{\nu}{E} \gamma y z - c_5 z - c_1 x + c_3,$$ (7.3.19)

$$u_z = \frac{\gamma}{2E} [z^2 + \nu(x^2 + y^2)] + c_4 x + c_5 y + c_6.$$

The displacement boundary conditions in Eq. (7.3.16) give $c_2 = c_3 = 0$, and $c_6 = -\gamma L^2/2E$, which correspond to the translational rigid-body motions. To remove the six rigid-body rotations, we may require

$$\frac{\partial u_x}{\partial y} = \frac{\partial u_y}{\partial x} = \frac{\partial u_y}{\partial z} = \frac{\partial u_z}{\partial y} = \frac{\partial u_y}{\partial z} = \frac{\partial u_x}{\partial z} = \frac{\partial u_z}{\partial x} = 0 \quad \text{at } x = y = 0, \text{ and } z = L,$$

which yield all other constants to be zero, giving the final displacement field

$$u_x = -\frac{\partial h}{\partial x} z + g(x, y) = -\frac{\nu}{E} \gamma x z,$$

$$u_y = -\frac{\partial h}{\partial y} z + f(x, y) = -\frac{\nu}{E} \gamma y z,$$ (7.3.20)

$$u_z = \frac{\gamma}{2E} [z^2 + \nu(x^2 + y^2)].$$

7.3.4 Stretching and Bending of Beams

In this section, we use the semi-inverse method to formulate equations governing stretching and bending of prismatic members. Using a set of assumptions concerning the kinematics of deformation of the members, the form of the displacement field is identified. We consider the prismatic bar shown in Fig. 7.3.3. The bar has a length L and has rectangular cross section of dimensions $b \times h$, b being the width and h being the height, such that $b < h << L$. We set up a coordinate system such that the x-axis is along the length of the beam through its geometric centroid, y-axis is transverse to the length of the beam, and the z-axis is out of the plane of the page, as shown in Fig. 7.3.3. A distributed load $q(x)$ (measured per unit length) acts along the length of the beam in the xy-plane in the positive y-direction, a distributed load $f(x)$ (measured per unit length) acts along the center line of the beam in the x-direction, and a point load F_0 acts at a distance $x = a$ from the left end. The bar is geometrically constrained at the right end in such a way that all three displacements are zero there. Thus, the boundary conditions are

$$\mathbf{u}(L, y, z) = \mathbf{0}, \quad \sigma_{zz} = \sigma_{xz} = \sigma_{yz} = 0 \text{ on faces } z = \pm b/2 \text{ for all } x, y,$$
$$\sigma_{yy}(x, h/2, 0) = q(x), \quad \sigma_{yy}(x, -h/2, 0) = 0, \quad \sigma_{xy}(x, \pm h/2, 0) = 0, \quad (7.3.21)$$
$$\sigma_{yz}(x, \pm h/2, z) = 0, \quad \sigma_{xx}(0, y, z) = 0, \quad \sigma_{xy}(0, y, z) = 0, \quad \sigma_{xz}(0, y, z) = 0.$$

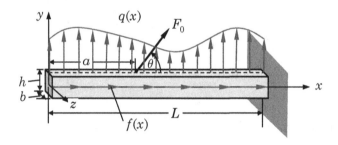

Fig. 7.3.3: A prismatic bar under various loads.

Solving the problem for exact displacements, strains, and stresses that satisfy the boundary conditions in Eq. (7.3.21) and the equilibrium equations of elasticity is an impossible task. We can formulate it as an equivalent problem of finding the solution that satisfies statically equivalent[1] stress boundary conditions and through-thickness-integrated equations of elasticity. Such formulation reduces the three-dimensional elasticity problem to a one-dimensional elasticity problem, known as the *beam bending* problem. Once again, we use the semi-inverse method, and assume a form the displacement field.

We seek a solution $(u_x, u_y, 0)$ based on the following assumptions: the transverse normal lines, such as AB shown in Fig. 7.3.4(a), (1) remain straight, (2)

---

[1]The phrase "statically equivalent" means that the two distributions of forces have the same resultant force and resultant moment.

are inextensible, and (3) rotate such that they remain normal to the middle surface after deformation. These assumptions are known as the *Euler–Bernoulli hypothesis* of beam bending. The first two assumptions together amount to neglecting Poisson's effect and the transverse normal strain (i.e., $\varepsilon_{yy} = 0$). The third assumption is to neglect the transverse shear strain $\varepsilon_{xy} = 0$. We assume that the deformation is only two-dimensional (in the plane of the page). This requires that the applied loads be in the xy plane so that stretching and bending are in the xy plane, and there is no rotation about the x axis.

The Euler–Bernoulli hypothesis is satisfied by the following form of the displacement field:

$$\mathbf{u} = \left[u(x) - y\frac{\partial u_y}{\partial x} \right] \hat{\mathbf{e}}_x + u_y\, \hat{\mathbf{e}}_y,$$

$$u_x = u(x) - y\frac{dv}{dx}, \quad u_y = v(x), \quad u_z = 0,$$

(7.3.22)

where $u(x)$ and $v(x)$ are functions to be determined by requiring that the equilibrium equations of elasticity are satisfied in an integral sense, as explained shortly. From the assumed form of the displacement field, we see that the displacement component u_x consists of two parts: stretching displacement $u(x)$ of all lines parallel to the x-axis and the displacement $-y(dv/dx)$ due to bending action, which is proportional to the distance y measured from the middle plane. The transverse displacement $u_y = v(x)$ is independent of the y-coordinate, a consequence of the inextensibility assumption.

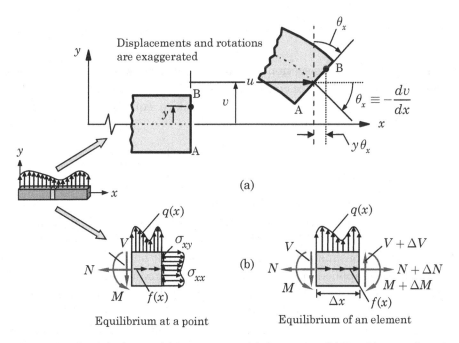

(a)

(b)

Equilibrium at a point   Equilibrium of an element

Fig. 7.3.4: Bending of a beam. (a) Kinematics of deformation. (b) Equilibrium of an element of the beam.

The only nonzero strain and corresponding stress components corresponding to the assumed displacement field are [$\nu = 0$ in writing the stress–strain relation $\sigma_{xx} = (2\mu + \lambda)\varepsilon_{xx} = E\varepsilon_{xx}$ but not in the relation $2\mu = 2G = E/(1 + \nu)$]

$$\varepsilon_{xx} = \frac{du}{dx} - y\frac{d^2v}{dx^2}, \tag{7.3.23}$$

$$\sigma_{xx} = E\left(\frac{du}{dx} - y\frac{d^2v}{dx^2}\right), \tag{7.3.24}$$

where E is Young's modulus of the material.

Since we cannot satisfy the equations of equilibrium, Eq. (7.2.6), without the inertia terms, at every point of the beam, we derive equations of equilibrium by considering a typical element of the beam, as shown in Fig. 7.3.4(b). Summing the forces and moments on the element, we obtain

sum of the forces in the x-direction: $\dfrac{dN}{dx} + f(x) = 0,$ (7.3.25)

sum of the forces in the z-direction: $\dfrac{dV}{dx} + q(x) = 0,$ (7.3.26)

sum of the moments about the y-axis: $V - \dfrac{dM}{dx} = 0,$ (7.3.27)

where $N(x)$ is the axial force, $M(x)$ is the bending moment, and $V(x)$ is the shear force. These quantities are known as the *stress resultants*, and they can be defined in terms of the stresses σ_{xx} and σ_{xy} as

$$N(x) = \int_A \sigma_{xx}\, dA, \quad M(x) = \int_A y\,\sigma_{xx}\, dA, \quad V(x) = \int_A \sigma_{xy}\, dA, \tag{7.3.28}$$

where $A = bh$ is the cross-sectional area. One can show that the equilibrium equations (7.3.25)–(7.3.27) are equivalent to the following two stress equilibrium equations ($\sigma_{xz} = \sigma_{zz} = \sigma_{yz} = 0$; hence, the third equation of equilibrium is trivially satisfied):

$$\frac{\partial \sigma_{xx}}{\partial x} + \frac{\partial \sigma_{xy}}{\partial y} = 0, \quad \frac{\partial \sigma_{xy}}{\partial x} + \frac{\partial \sigma_{yy}}{\partial y} = 0.$$

This is left as an exercise for the reader (see Problem **7.13**).

From the constitutive relation $\sigma_{xy} = 2G\varepsilon_{xy}$, we have $\sigma_{xy} = 0$ and, therefore, $V = 0$ from Eq. (7.3.28). Although the transverse shear force V is zero from the kinematic assumptions made here, in reality it cannot be zero as it is responsible for supporting the applied vertical loads on the beam, as can be seen from Eq. (7.3.27). This is the flaw in the Euler–Bernoulli beam theory, which can be overcome by ignoring the definition of V in Eq. (7.3.28) and calculating it using Eq. (7.3.27). That is, substitute for V from Eq. (7.3.27) into Eq. (7.3.26) and obtain only two equations of equilibrium:

$$-\frac{dN}{dx} = f(x), \quad -\frac{d^2M}{dx^2} = q(x). \tag{7.3.29}$$

The stress resultants (N, M) can be related back to the unknown functions (u, v) as [because the x-axis is taken through the geometric centroid of the cross section, we have $\int_A y \, dA = 0$]:

$$N(x) = \int_A \sigma_{xx} \, dA = EA\frac{du}{dx}, \quad M(x) = \int_A y \, \sigma_{xx} \, dA = -EI\frac{d^2v}{dx^2}, \qquad (7.3.30)$$

where I is the moment of inertia about the axis of bending (i.e., z-axis). From Eqs. (7.3.24) and (7.3.30), we can express σ_{xx} in terms of the stress resultants N and M as

$$\sigma_{xx} = E\varepsilon_{xx} = \frac{N(x)}{A} + \frac{M(x)y}{I}. \qquad (7.3.31)$$

Finally, we have two equations of equilibrium governing u and v

$$-\frac{d}{dx}\left(EA\frac{du}{dx}\right) = f(x), \quad \frac{d^2}{dx^2}\left(EI\frac{d^2v}{dx^2}\right) = q(x). \qquad (7.3.32)$$

Note that the two equations are not coupled, that is, each equation can be solved independent of the other. Indeed, when no axial loads are applied on the beam, we have $u = 0$ everywhere in the beam. Conversely, when no bending loads are applied on the beam, we have $v = 0$ everywhere. The former case is known as the beam bending problem and the latter as the bar problem. The two equations in (7.3.32) are subjected to boundary conditions of the type

$$u = \hat{u}, \quad N = \hat{N}; \quad v = \hat{v}, \quad -\frac{dv}{dx} = \hat{\theta}, \quad M = \hat{M}, \quad V = \hat{V}, \qquad (7.3.33)$$

Only one element of each of the following three pairs should be specified at a boundary point:

$$(u, N), \quad (v, V), \quad (\theta, M). \qquad (7.3.34)$$

This completes the formulation of the Euler–Bernoulli beam theory. Next, we consider an example.

Example 7.3.3

Consider the cable-supported beam shown in Fig. 7.3.5(a). The beam as well as the cable are made of homogeneous, linear elastic, isotropic materials, with constant geometric properties. Determine the displacements (u, v) and the force F_c in the cable.

Solution: Figure 7.3.5(b) contains the effect of the cable force on the beam. We begin with the first equation in (7.3.32) and integrate it twice with respect to x and obtain

$$E_b A_b \frac{du}{dx} = c_1, \quad E_b A_b u(x) = c_1 x + c_2, \qquad (1)$$

where the constants of integration, c_1 and c_2, are determined using the boundary conditions

$$u(L_b) = 0, \quad \left[E_b A_b \frac{du}{dx}\right]_{x=0} = -F_c \cos\alpha. \qquad (2)$$

We obtain $c_1 = -F_c \cos\alpha$ and $c_2 = F_c L_b \cos\alpha$, and the solution becomes

$$u(x) = \frac{F_c L_b}{E_b A_b}\left(1 - \frac{x}{L_b}\right)\cos\alpha. \qquad (3)$$

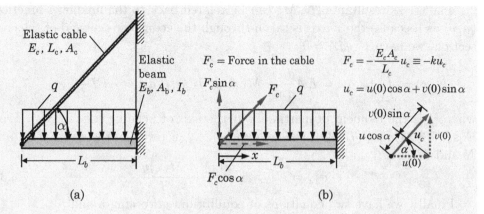

Fig. 7.3.5: A cable-supported beam.

Next, we consider the second equation in (7.3.32) and integrate it four times with respect to x and obtain

$$\frac{d}{dx}\left(E_b I_b \frac{d^2 v}{dx^2}\right) = -qx + c_3,$$

$$E_b I_b \frac{d^2 v}{dx^2} = -q\frac{x^2}{2} + c_3 x + c_4,$$

$$E_b I_b \frac{dv}{dx} = -q\frac{x^3}{6} + c_3 \frac{x^2}{2} + c_4 x + c_5,$$

$$E_b I_b\, v(x) = -q\frac{x^4}{24} + c_3 \frac{x^3}{6} + c_4 \frac{x^2}{2} + c_5 x + c_6,$$

(4)

where the constants of integration, c_3, c_4, c_5, and c_6 are obtained with the help of the boundary conditions

$$V(0) = -F_c \sin\alpha, \quad M(0) = 0, \quad \left[\frac{dv}{dx}\right]_{x=L_b} = 0, \quad v(L_b) = 0.$$

(5)

We obtain

$$c_3 = F_c \sin\alpha, \quad c_4 = 0, \quad c_5 = \frac{qL_b^3}{6} - \frac{F_c L_b^2}{2}\sin\alpha, \quad c_6 = -\frac{qL_b^4}{8} + \frac{F_c L_b^3}{3}\sin\alpha.$$

The solution is given by

$$v(x) = -\frac{qL_b^4}{24E_b I_b}\left[3 - 4\frac{x}{L_b} + \left(\frac{x}{L_b}\right)^4\right] + \frac{F_c L_b^3}{6E_b I_b}\left[2 - 3\frac{x}{L_b} + \left(\frac{x}{L_b}\right)^3\right]\sin\alpha.$$

(6)

The displacements at $x = 0$ are

$$u(0) = \frac{F_c L_b}{E_b A_b}\cos\alpha, \quad v(0) = -\frac{qL_b^4}{8E_b I_b} + \frac{F_c L_b^3}{3E_b I_b}\sin\alpha.$$

(7)

To determine the cable force, F_c, first we note that

$$u_c = u(0)\,\cos\alpha + v(0)\,\sin\alpha,$$

(8)

and calculate F_c from (u_c is in the opposite direction to F_c)

$$F_c = -\frac{E_c A_c}{L_c}u_c = -\frac{E_c A_c}{L_c}\left(\frac{F_c L_b}{E_b A_b}\cos^2\alpha + \frac{F_c L_b^3}{3E_b I_b}\sin^2\alpha - \frac{qL_b^4}{8E_b I_b}\sin\alpha\right),$$

(9)

or

$$F_c = \frac{qL_b^4}{8E_b I_b}\sin\alpha\left[\frac{L_c}{E_c A_c} + \frac{L_b}{E_b A_b}\cos^2\alpha + \frac{L_b^3}{3E_b I_b}\sin^2\alpha\right]^{-1}.$$

(10)

7.3.5 Superposition Principle

An advantage of linear boundary value problems is that the principle of super-position holds. The principle of superposition is said to hold for a solid body if the displacements obtained under two sets of boundary conditions and forces are equal to the sum of the displacements that would be obtained by applying each set of boundary conditions and forces separately.

To be more specific, consider the following two sets of boundary conditions and forces

$$\text{Set 1:} \quad \mathbf{u} = \mathbf{u}^{(1)} \text{ on } \Gamma_u; \quad \mathbf{t} = \mathbf{t}^{(1)} \text{ on } \Gamma_\sigma; \quad \mathbf{f} = \mathbf{f}^{(1)} \text{ in } \Omega, \quad (7.3.35)$$
$$\text{Set 2:} \quad \mathbf{u} = \mathbf{u}^{(2)} \text{ on } \Gamma_u; \quad \mathbf{t} = \mathbf{t}^{(2)} \text{ on } \Gamma_\sigma; \quad \mathbf{f} = \mathbf{f}^{(2)} \text{ in } \Omega, \quad (7.3.36)$$

where the specified data $(\mathbf{u}^{(1)}, \mathbf{t}^{(1)}, \mathbf{f}^{(1)})$ and $(\mathbf{u}^{(2)}, \mathbf{t}^{(2)}, \mathbf{f}^{(2)})$ are independent of the deformation. Suppose that the solution to the two problems be $\mathbf{u}^{(1)}(\mathbf{x})$ and $\mathbf{u}^{(2)}(\mathbf{x})$, respectively. The superposition of the two sets of boundary conditions is

$$\mathbf{u} = \mathbf{u}^{(1)} + \mathbf{u}^{(2)} \text{ on } \Gamma_u; \quad \mathbf{t} = \mathbf{t}^{(1)} + \mathbf{t}^{(2)} \text{ on } \Gamma_\sigma; \quad \mathbf{f} = \mathbf{f}^{(1)} + \mathbf{f}^{(2)} \text{ in } \Omega. \quad (7.3.37)$$

Because of the linearity of the elasticity equations, the solution of the boundary value problem with the superposed data is $\mathbf{u}(\mathbf{x}) = \mathbf{u}^{(1)}(\mathbf{x}) + \mathbf{u}^{(2)}(\mathbf{x})$ in Ω. This is known as the *superposition principle*.

The principle of superposition can be used to represent a linear problem with complicated boundary conditions and/or loads as a combination of linear problems that are equivalent to the original problem. Example 7.3.4 illustrates this point.

Example 7.3.4

Consider the indeterminate beam shown in Fig. 7.3.6(a). Determine the deflection of point A using the principle of superposition.

Solution: The problem can be viewed as one equivalent to the two beam problems shown in Fig. 7.3.6(b). The sum of the deflections from each problem is the solution of the original problem. Within the restrictions of the linear Euler–Bernoulli beam theory, the deflections are linear functions of the loads. Therefore, the principle of superposition is valid. Thus, the transverse displacement of the original beam can be determined as the sum of the displacements of the individual beams shown in Fig. 7.3.6(b):

$$v(x) = \frac{q_0 L^4}{24EI} \left[3 - 4\frac{x}{L} + \left(\frac{x}{L}\right)^4 \right] - \frac{F_s L^3}{6EI} \left[2 - 3\frac{x}{L} + \left(\frac{x}{L}\right)^3 \right]. \tag{1}$$

In particular, the deflection v_A at point A is equal to the sum of v_A^q and v_A^s due to the distributed load q_0 and spring force F_s, respectively, at point A:

$$v_A = v_A^q + v_A^s = \frac{q_0 L^4}{8EI} - \frac{F_s L^3}{3EI}. \tag{2}$$

Because the spring force F_s is equal to $k v_A$, we can calculate v_A from

$$v_A = \frac{q_0 L^4}{8EI \left(1 + \frac{kL^3}{3EI}\right)}. \tag{3}$$

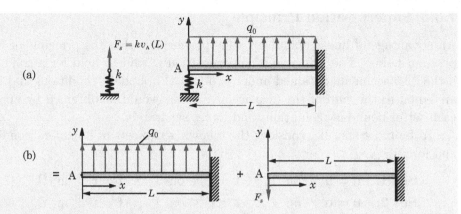

Fig. 7.3.6: Representation of an indeterminate beam as a superposition of two determinate beams.

7.3.6 Uniqueness of Solutions

Although the existence of solutions is a difficult question to answer, the uniqueness of solutions is rather easy to prove for linear boundary value problems of elasticity. Consider the problem of finding the solution to the Navier equations (7.2.17) of linearized elasticity, for a given body force $\mathbf{f}$ and boundary conditions

$$\mathbf{u} = \hat{\mathbf{u}} \text{ on } \Gamma_u, \tag{7.3.38}$$

$$\mathbf{t} = \hat{\mathbf{t}} \text{ on } \Gamma_\sigma. \tag{7.3.39}$$

Now suppose that for this set of loads and boundary conditions, there exist two distinct solutions, $\mathbf{u}^{(1)}(\mathbf{x}, t)$ and $\mathbf{u}^{(2)}(\mathbf{x}, t)$. Associated with the two displacement fields, we can compute the strains and stress fields $(\boldsymbol{\varepsilon}^{(1)}, \boldsymbol{\sigma}^{(1)})$ and $(\boldsymbol{\varepsilon}^{(2)}, \boldsymbol{\sigma}^{(2)})$. Then the difference $\mathbf{u}^d(\mathbf{x}, t) \equiv \mathbf{u}^{(1)}(\mathbf{x}, t) - \mathbf{u}^{(2)}(\mathbf{x}, t)$ satisfies the homogeneous form of the Navier equation (with $\mathbf{f}^d = \mathbf{f}^{(1)} - \mathbf{f}^{(2)} = \mathbf{0}$, because the applied forces and boundary values are the same for both solutions)

$$\mu\nabla^2\mathbf{u}^d + (\mu + \lambda)\nabla(\nabla \cdot \mathbf{u}^d) = \mathbf{0} \text{ in } \Omega, \tag{7.3.40}$$

as well as the homogeneous forms of the boundary conditions

$$\mathbf{u}^d = \mathbf{0} \text{ on } \Gamma_u, \tag{7.3.41}$$

$$\mathbf{t}^d = \mathbf{0} \text{ on } \Gamma_\sigma. \tag{7.3.42}$$

Because no work is done on the body by external forces (because $\mathbf{f}^d$ and $\mathbf{t}^d$ are zero), the strain energy density U_0 stored in the body is zero. Noting that the strain energy density U_0 (measured per unit volume)

$$U_0(\boldsymbol{\varepsilon}) = \frac{\lambda}{2}(\text{tr}\,\boldsymbol{\varepsilon})^2 + \mu\,\text{tr}(\boldsymbol{\varepsilon} \cdot \boldsymbol{\varepsilon}), \quad U_0(\varepsilon_{ij}) = \mu\varepsilon_{ij}\varepsilon_{ij} + \tfrac{1}{2}\lambda(\varepsilon_{kk})^2, \tag{7.3.43}$$

is a positive-definite function of the strains [see Eqs. (6.3.36) and (6.3.37)],

$$U_0(\boldsymbol{\varepsilon}) > 0 \text{ whenever } \boldsymbol{\varepsilon} \neq \mathbf{0}, \text{ and } U_0(\boldsymbol{\varepsilon}) = 0 \text{ only when } \boldsymbol{\varepsilon} = \mathbf{0}, \tag{7.3.44}$$

we conclude that the strain field ε^d is zero and hence the stress field σ^d is also zero:

$$\varepsilon^d = \varepsilon^{(1)} - \varepsilon^{(2)} = \mathbf{0}, \quad \sigma^d = \sigma^{(1)} - \sigma^{(2)} = \mathbf{0}, \tag{7.3.45}$$

implying that the strain and stress fields associated with the two distinct displacements $\mathbf{u}^{(1)}$ and $\mathbf{u}^{(2)}$ are the same, that is, they are unique. Also, $\varepsilon^d = \mathbf{0}$ implies that $\nabla \mathbf{u}^d = \mathbf{0}$, which corresponds to a rigid-body motion. For type I and type III problems, the displacement boundary conditions eliminate the rigid-body motion and, therefore, the displacements are unique for type I and type III problems. For boundary value problems of type II, the displacements are determined within the quantities representing rigid-body motions.

7.4 Clapeyron's, Betti's, and Maxwell's Theorems

7.4.1 Clapeyron's Theorem

The principle of superposition is *not* valid for energies because they are quadratic functions of displacements or forces. In other words, when a linear elastic body $\mathcal{B}$ is subjected to more than one external force, the total work done due to external forces is *not* equal to the sum of the works that are obtained by applying the single forces separately. However, there exist theorems that relate the work done in linear elastic solids by two different forces applied in different orders. We will consider them in this section.

Recall from Chapter 6 that the strain energy density due to linear elastic deformation is given by[2]

$$\begin{aligned} U_0 &= \tfrac{1}{2} \varepsilon : \mathbf{C} : \varepsilon = \tfrac{1}{2} \varepsilon : \sigma \\ &= \tfrac{1}{2} C_{ijkl} \varepsilon_{kl} \varepsilon_{ij} = \tfrac{1}{2} \sigma_{ij} \varepsilon_{ij}. \end{aligned} \tag{7.4.1}$$

The total strain energy stored in the body $\mathcal{B}$ occupying the region Ω with surface Γ is equal to

$$U = \int_\Omega U_0 \, d\mathbf{x} = \tfrac{1}{2} \int_\Omega \sigma : \varepsilon \, d\mathbf{x} = \tfrac{1}{2} \int_\Omega \sigma_{ij} \varepsilon_{ij} \, d\mathbf{x}. \tag{7.4.2}$$

The total work done by the body force $\mathbf{f}$ (measured per unit volume) and surface traction $\mathbf{t}$ (measured per unit area) in moving through their respective displacements $\mathbf{u}$ is given by

$$W_E = \int_\Omega \mathbf{f} \cdot \mathbf{u} \, d\mathbf{x} + \oint_\Gamma \mathbf{t} \cdot \mathbf{u} \, ds. \tag{7.4.3}$$

When $\mathbf{u} = \mathbf{0}$ on a portion Γ_u of the boundary Γ, the surface integral in Eq. (7.4.3) becomes

$$\int_{\Gamma_\sigma} \mathbf{t} \cdot \mathbf{u} \, ds, \quad \text{where} \quad \Gamma_\sigma = \Gamma - \Gamma_u.$$

---

[2]In this chapter U_0 is measured per unit volume as opposed to per unit mass.

Owing to the symmetry of the stress tensor, $\sigma_{ij} = \sigma_{ji}$, we can write $\sigma_{ij}\varepsilon_{ij} = \sigma_{ij}u_{i,j}$. Consequently, the strain energy U can be expressed as

$$U = \frac{1}{2}\int_\Omega \sigma_{ij}\,\varepsilon_{ij}\,d\mathbf{x} = \frac{1}{4}\int_\Omega \sigma_{ij}\,(u_{i,j} + u_{j,i})\,d\mathbf{x} = \frac{1}{2}\int_\Omega \sigma_{ij}\,u_{i,j}\,d\mathbf{x}$$

$$= -\frac{1}{2}\int_\Omega \sigma_{ij,j}\,u_i\,d\mathbf{x} + \frac{1}{2}\oint_\Gamma n_j\sigma_{ij}u_i\,ds$$

$$= \frac{1}{2}\int_\Omega f_i\,u_i\,d\mathbf{x} + \frac{1}{2}\oint_\Gamma t_i u_i\,ds = \frac{1}{2}\int_\Omega \mathbf{f}\cdot\mathbf{u}\,d\mathbf{x} + \frac{1}{2}\oint_\Gamma \mathbf{t}\cdot\mathbf{u}\,ds,$$

where, in arriving at the last line, we have used the stress equilibrium equation $\sigma_{ij,j} + f_i = 0$, Cauchy's formula $t_i = \sigma_{ij}n_j$, and the divergence theorem. Thus, the total strain energy stored in a body undergoing linear elastic deformation is also equal to the one-half of the work done by applied forces

$$\frac{1}{2}\int_\Omega \boldsymbol{\sigma} : \boldsymbol{\varepsilon}\,d\mathbf{x} = \frac{1}{2}\int_\Omega \mathbf{f}\cdot\mathbf{u}\,d\mathbf{x} + \frac{1}{2}\oint_\Gamma \mathbf{t}\cdot\mathbf{u}\,ds. \tag{7.4.4}$$

The first term on the right-hand side represents the work done by body force $\mathbf{f}$ in moving through the displacement $\mathbf{u}$ while the second term represents the work done by surface force $\mathbf{t}$ in moving through the displacements $\mathbf{u}$ during the deformation. Equation (7.4.4) is known as *Clapeyron's theorem*. The next three examples illustrate the usefulness of the theorem.

Example 7.4.1

Consider a linear elastic spring with spring constant k. Let F be the external force applied on the spring to elongate it and u be the resulting elongation of the spring (see Fig. 7.4.1). Verify Clapeyron's theorem.

Solution: The internal force developed in the spring is $F_s = ku$. The work done by F_s in moving through an increment of displacement du is $F_s\,du$. The total strain energy stored in the spring is

$$U = \int_0^u F_s\,du = \int_0^u ku\,du = \frac{1}{2}ku^2. \tag{7.4.5}$$

The work done by external force F is equal to Fu. But by equilibrium, $F = F_s = ku$. Hence,

$$U = \frac{1}{2}ku^2 = \frac{1}{2}Fu,$$

which proves Clapeyron's theorem.

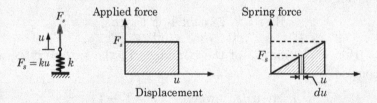

Fig. 7.4.1: Strain energy stored in a linear elastic spring.

Example 7.4.2

Consider a uniform elastic bar of length L, cross-sectional area A, and modulus of elasticity E. The bar is fixed at $x = 0$ and subjected to a tensile force of P at $x = L$, as shown in Fig. 7.4.2. Determine the axial displacement $u(L)$ using Clapeyron's theorem.

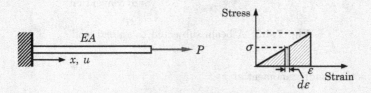

Fig. 7.4.2: A bar subjected to an end load.

Solution: If the axial displacement in the bar is equal to $u(x)$, then the work done by external point force P is equal to $W = Pu(L)$. The strain energy in the bar is given by

$$U = \frac{1}{2} \int_A \int_0^L \sigma_{xx}\varepsilon_{xx}\, dx\, dA = \frac{EA}{2}\int_0^L \varepsilon_{xx}^2\, dx = \frac{EA}{2}\int_0^L \left(\frac{du}{dx}\right)^2 dx = \frac{1}{2}\int_0^L \frac{N^2}{EA}\, dx. \quad (7.4.6)$$

Hence, by Clapeyran's theorem we have

$$\frac{Pu(L)}{2} = \frac{EA}{2}\int_0^L \left(\frac{du}{dx}\right)^2 dx.$$

To make use of the above equation to determine $u(x)$, let us assume that $u(x) = u(L)x/L$, which certainly satisfies the geometric boundary condition, $u(0) = 0$. Then we have

$$u(L) = \frac{EA}{P}\int_0^L \left(\frac{du}{dx}\right)^2 dx = \frac{EA}{PL}[u(L)]^2,$$

or $u(L) = PL/AE$ and the solution is $u(x) = Px/AE$, which happens to coincide with the exact solution to the problem.

Example 7.4.3

Consider a cantilever beam of length L and flexural rigidity EI and bent by a point load F at the free end (see Fig. 7.4.3). Determine $v(0)$ using Clapeyron's theorem.

Solution: By Clapeyron's theorem we have

$$\frac{1}{2}Fv(0) = \frac{1}{2}\int_A \int_0^L \sigma_{xx}\varepsilon_{xx}\, dx\, dA.$$

But according to the Euler–Bernoulli beam theory, the strain and stress in the beam are given by

$$\varepsilon_{xx} = -y\frac{d^2v}{dx^2}, \quad \sigma_{xx} = E\varepsilon_{xx} = -Ey\frac{d^2v}{dx^2}, \quad (7.4.7)$$

where v is the transverse deflection. Then we have

$$\frac{1}{2}Fv(0) = \frac{1}{2}\int_A \int_0^L E\varepsilon_{xx}^2\, dx\, dA = \frac{1}{2}\int_A \int_0^L Ey^2\left(\frac{d^2v}{dx^2}\right)^2 dA\, dx$$

$$= \frac{1}{2}\int_0^L EI\left(\frac{d^2v}{dx^2}\right)^2 dx = \frac{1}{2}\int_0^L \frac{M^2}{EI}\, dx, \quad (7.4.8)$$

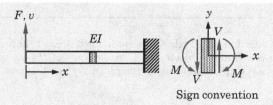

Fig. 7.4.3: A beam subjected to an end load.

where $M(x)$ is the bending moment at x

$$M(x) = \int_A y\sigma_{xx}\, dA = -E\int_A y^2 \frac{d^2v}{dx^2}\, dA = -EI\frac{d^2v}{dx^2}. \tag{7.4.9}$$

Equation (7.4.8) can be used to determine the deflection $v(0)$. The bending moment at any point x is $M(x) = -Fx$. Hence, we have

$$Fv(0) = \frac{1}{EI}\int_0^L F^2 x^2\, dx = \frac{F^2 L^3}{3EI} \quad \text{or} \quad v(0) = \frac{FL^3}{3EI}. \tag{7.4.10}$$

7.4.2 Betti's Reciprocity Theorem

Consider the equilibrium state of a linear elastic solid under the action of two different external forces, $\mathbf{F}_1$ and $\mathbf{F}_2$, as shown in Fig. 7.4.4. Since the order of application of the forces is arbitrary for linearized elasticity, we suppose that force $\mathbf{F}_1$ is applied first. Let W_1 be the work done by $\mathbf{F}_1$. Then, we apply force $\mathbf{F}_2$ at some other point of the body, which does work W_2. This work is the same as that produced by force $\mathbf{F}_2$, if it alone were acting on the body. However, when force $\mathbf{F}_2$ is applied, force $\mathbf{F}_1$ (which is already acting on the body) does additional work because its point of application is displaced due to the deformation caused by force $\mathbf{F}_2$. Let us denote this work by W_{12}, which is the work done by force F_1 due to the application of force F_2. Thus the total work done by the application of forces $\mathbf{F}_1$ and $\mathbf{F}_2$, $\mathbf{F}_1$ first and $\mathbf{F}_2$ next, is

$$W = W_1 + W_2 + W_{12}. \tag{7.4.11}$$

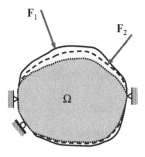

Fig. 7.4.4: Configurations of an elastic body due to the application of loads $\mathbf{F}_1$ and $\mathbf{F}_2$. —— Undeformed configuration. - - - - Deformed configuration after the application of $\mathbf{F}_1$. ...... Deformed configuration after the application of $\mathbf{F}_2$.

Work W_{12}, which can be positive or negative, is zero if and only if the displacement of the point of application of force $\mathbf{F}_1$ produced by force $\mathbf{F}_2$ is zero or perpendicular to the direction of $\mathbf{F}_1$. Now suppose that we change the order of application of the forces, that is, force $\mathbf{F}_2$ is applied first and force $\mathbf{F}_1$ is applied next. Then the total work done is equal to

$$\bar{W} = W_1 + W_2 + W_{21}, \tag{7.4.12}$$

where W_{21} (note the order of the subscripts) is the work done by force F_2 due to the application of force F_1. The work done in both cases should be the same because, at the end, the body is loaded by the same pair of external forces. Thus, we have $W = \bar{W}$, or

$$W_{12} = W_{21}. \tag{7.4.13}$$

Equation (7.4.13) is a mathematical statement of Betti's (1823–1892) reciprocity theorem: *If a linear elastic body is subjected to two different sets of forces, the work done by the first system of forces in moving through the displacements produced by the second system of forces is equal to the work done by the second system of forces in moving through the displacements produced by the first system of forces.* Applied to a three-dimensional elastic body Ω with closed surface s, Eq. (7.4.13) takes the form

$$\int_\Omega \mathbf{f}^{(1)} \cdot \mathbf{u}^{(2)} \, d\mathbf{x} + \oint_s \mathbf{t}^{(1)} \cdot \mathbf{u}^{(2)} \, ds = \int_\Omega \mathbf{f}^{(2)} \cdot \mathbf{u}^{(1)} \, d\mathbf{x} + \oint_s \mathbf{t}^{(2)} \cdot \mathbf{u}^{(1)} \, ds, \tag{7.4.14}$$

where $\mathbf{u}^{(i)}$ are the displacements produced by body forces $\mathbf{f}^{(i)}$ and surface forces $\mathbf{t}^{(i)}$. The usefulness of Betti's (also Maxwell's) reciprocity theorem is that it allows us to compute the the displacements or forces at points other than where the forces are applied; that is, the theorem does not allow us to determine the displacement of a point where the force is applied.

The proof of Betti's reciprocity theorem is straightforward. Let W_{12} denote the work done by forces $(\mathbf{f}^{(1)}, \mathbf{t}^{(1)})$ acting through the displacement $\mathbf{u}^{(2)}$ produced by the forces $(\mathbf{f}^{(2)}, \mathbf{t}^{(2)})$. Then

$$
\begin{aligned}
W_{12} &= \int_\Omega \mathbf{f}^{(1)} \cdot \mathbf{u}^{(2)} \, d\mathbf{x} + \oint_s \mathbf{t}^{(1)} \cdot \mathbf{u}^{(2)} \, ds \\
&= \int_\Omega f_i^{(1)} u_i^{(2)} \, d\mathbf{x} + \oint_s t_i^{(1)} u_i^{(2)} \, ds \\
&= \int_\Omega f_i^{(1)} u_i^{(2)} \, d\mathbf{x} + \oint_s n_j \sigma_{ji}^{(1)} u_i^{(2)} \, ds \\
&= \int_\Omega f_i^{(1)} u_i^{(2)} \, d\mathbf{x} + \int_\Omega \left(\sigma_{ji}^{(1)} u_i^{(2)} \right)_{,j} \, d\mathbf{x} \\
&= \int_\Omega \left(\sigma_{ij,j}^{(1)} + f_i^{(1)} \right) u_i^{(2)} \, d\mathbf{x} + \int_\Omega \sigma_{ij}^{(1)} u_{i,j}^{(2)} \, d\mathbf{x} \\
&= \int_\Omega \sigma_{ij}^{(1)} u_{i,j}^{(2)} \, d\mathbf{x} = \int_\Omega \sigma_{ij}^{(1)} \varepsilon_{ij}^{(2)} \, d\mathbf{x}. \tag{7.4.15}
\end{aligned}
$$

Using Hooke's law $\sigma_{ij}^{(1)} = C_{ijk\ell}\,\varepsilon_{k\ell}^{(1)}$, we obtain

$$W_{12} = \int_{\Omega} C_{ijk\ell}\,\varepsilon_{k\ell}^{(1)}\,\varepsilon_{ij}^{(2)}\ dx. \tag{7.4.16}$$

Since $C_{ijk\ell} = C_{k\ell ij}$, it follows that

$$W_{12} = \int_{\Omega} C_{ijk\ell}\,\varepsilon_{k\ell}^{(1)}\,\varepsilon_{ij}^{(2)}\ dx = \int_{\Omega} C_{k\ell ij}\,\varepsilon_{ij}^{(2)}\,\varepsilon_{k\ell}^{(1)}\ dx = \int_{\Omega} \sigma_{k\ell}^{(2)}\,\varepsilon_{k\ell}^{(1)}\ dx = W_{21}. \tag{7.4.17}$$

Thus, we have established the equality in Eq. (7.4.14). From Eq. (7.4.17), we also have

$$\int_{\Omega} \sigma_{ij}^{(1)}\,\varepsilon_{ij}^{(2)}\ dx = \int_{\Omega} \sigma_{ij}^{(2)}\,\varepsilon_{ij}^{(1)}\ dx,$$
$$\int_{\Omega} \boldsymbol{\sigma}^{(1)} : \boldsymbol{\varepsilon}^{(2)}\ dx = \int_{\Omega} \boldsymbol{\sigma}^{(2)} : \boldsymbol{\varepsilon}^{(1)}\ dx. \tag{7.4.18}$$

Example 7.4.4

(a) Consider a cantilever beam of length L subjected to two different types of loads: a concentrated load F at the free end and a uniformly distributed load of intensity q throughout the span (see Fig. 7.4.5). Verify that the work done by the point load F in moving through the displacement v^q produced by q is equal to the work done by the distributed force q in moving through the displacement v^F produced by the point load F, $W_{12} = W_{21}$.

(b) A load $P = 4000$ lb acting at a point A of a beam produces 0.25 in. at point B and 0.75 in. at point C of the beam. Find the deflection of point A produced by loads 4500 lb and 2000 lb acting at points B and C, respectively.

Solution: (a) The deflection $v^F(x)$ due to the concentrated load alone is

$$v^F(x) = \frac{FL^3}{6EI}\left[2 - 3\frac{x}{L} + \left(\frac{x}{L}\right)^3\right],$$

and the deflection equation due to the distributed load is

$$v^q(x) = \frac{qL^4}{24EI}\left[3 - 4\frac{x}{L} + \left(\frac{x}{L}\right)^4\right].$$

The work done by load F in moving through the displacement due to the application of the uniformly distributed load q is

$$W_{12} = Fv^q(0) = \frac{FqL^4}{8EI}.$$

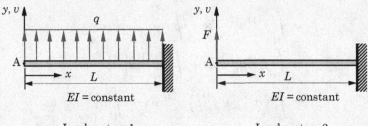

Fig. 7.4.5: A cantilever beam subjected to two different types of loads.

The work done by the uniformly distributed q in moving through the displacement field due to the application of point load F is

$$W_{21} = \int_0^L q\,v^F(x)\,dx = \int_0^L q\,\frac{F}{6EI}\left(x^3 - 3L^2 x + 2L^3\right)dx = \frac{FqL^4}{8EI},$$

which is in agreement with W_{12}.

(b) From Betti's reciprocity theorem and the principle of superposition, we have

$$F_B \cdot v_{BA} + F_C \cdot v_{CA} = F_A \cdot v_{AB} + F_A \cdot v_{AC} = F_A \cdot v_A$$

where $v_A = v_{AB} + v_{AC}$, $\ F_A = 4{,}000$ lb, $\ F_B = 4{,}500$ lb, $\ F_C = 2{,}000$ lb. We obtain

$$v_A = \frac{F_B \cdot v_{BA} + F_C \cdot v_{CA}}{F_A} = \frac{4500 \times 0.25 + 2000 \times 0.75}{4000} = 0.65625 \text{ in}.$$

7.4.3 Maxwell's Reciprocity Theorem

An important special case of Betti's reciprocity theorem is given by Maxwell's (1831–1879) reciprocity theorem. Maxwell's theorem was given in 1864, whereas Betti's theorem was given in 1872. Therefore, it may be considered that Betti generalized the work of Maxwell.

Consider a linear elastic solid subjected to force $\mathbf{F}^1$ of unit magnitude acting at point 1, and force $\mathbf{F}^2$ of unit magnitude acting at a different point 2 of the body. Let $\mathbf{u}_{12}$ be the displacement of point 1 in the direction of force $\mathbf{F}^1$ produced by unit force $\mathbf{F}^2$, and $\mathbf{u}_{21}$ be the displacement of point 2 in the direction of force $\mathbf{F}^2$ produced by unit force $\mathbf{F}^1$ (see Fig. 7.4.6). From Betti's theorem it follows that

$$\mathbf{F}^1 \cdot \mathbf{u}_{12} = \mathbf{F}^2 \cdot \mathbf{u}_{21} \quad \text{or} \tag{7.4.19}$$

$$\mathbf{u}_{12} = \mathbf{u}_{21}. \tag{7.4.20}$$

Equation (7.4.19) is a statement of Maxwell's theorem. If $\hat{\mathbf{e}}_1$ and $\hat{\mathbf{e}}_2$ denote the unit vectors along forces $\mathbf{F}^1$ and $\mathbf{F}^2$, respectively, Maxwell's theorem states that the displacement of point 1 in the $\hat{\mathbf{e}}_1$ direction produced by unit force acting at point 2 in the $\hat{\mathbf{e}}_2$ direction is equal to the displacement of point 2 in the $\hat{\mathbf{e}}_2$ direction produced by unit force acting at point 1 in the $\hat{\mathbf{e}}_1$ direction. We close this section with several examples of the use of Maxwell's theorem.

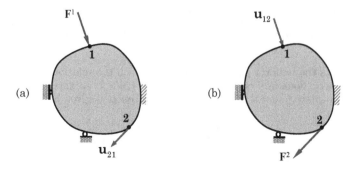

Fig. 7.4.6: Configurations of the body discussed in Maxwell's theorem.

Example 7.4.5

Consider a cantilever beam ($E = 24 \times 10^6$ psi, $I = 120$ in.^{4}) of length 12 ft. subjected to a point load 4000 lb at the free end, as shown in Fig. 7.4.7(a). Use Maxwell's theorem to find the deflection at a point 3 ft. from the free end.

Solution: By Maxwell's theorem, the displacement v_{BC} at point B ($x = 3$ ft.) produced by the 4000-lb load at point C ($x = 0$) is equal to the deflection v_{CB} at point C produced by applying the 4000-lb load at point B. Let v_B and θ_B denote the deflection and slope, respectively, at point B owing to load $F = 4000$ lb applied at point B, as shown in Fig. 7.4.7(b). The deflection at point B ($x = b = 3$ ft.) caused by load $F = 4000$ lb at point C ($x = 0$) is ($v_B = Fa^3/3EI$ and $\theta_B = Fa^2/2EI$)

$$
\begin{aligned}
v_{BC} = v_{CB} &= v_B + (3 \times 12)\theta_B \\
&= \frac{4000(9 \times 12)^3}{3EI} + \frac{(3 \times 12)4000(9 \times 12)^2}{2EI} \\
&= \frac{243 \times 6000 \times (12)^3}{24 \times 10^6 \times 120} = 0.8748 \text{ in.}
\end{aligned}
$$

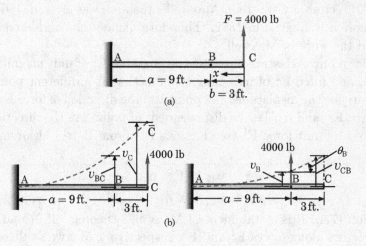

Fig. 7.4.7: The cantilever beam of Example 7.4.5.

Example 7.4.6

Consider a circular plate of radius a with an axisymmetric boundary condition, and subjected to an asymmetric loading of the type (see Fig. 7.4.8)

$$
q(r, \theta) = q_0 + q_1 \frac{r}{a} \cos\theta, \tag{7.4.20}
$$

where q_0 represents the uniform part of the load for which the solution can be determined for various axisymmetric boundary conditions [see Reddy (2007)]. In particular, the deflection of a clamped circular plate under a point load F_0 at the center is given by

$$
v(r) = \frac{F_0 a^2}{16\pi D}\left[1 - \frac{r^2}{a^2} + 2\frac{r^2}{a^2}\ln\left(\frac{r}{a}\right)\right]. \tag{7.4.21}
$$

Use the Betti/Maxwell reciprocity theorem to determine the center deflection of a clamped plate under an asymmetric distributed load.

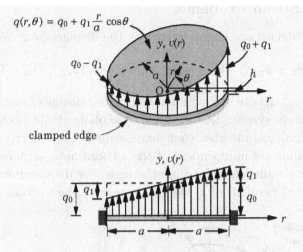

Fig. 7.4.8: A circular plate subjected to an asymmetric loading.

Solution: By Maxwell's theorem, the work done by a point load F_0 at the center of the plate due to the deflection (at the center) v_c caused by the distributed load $q(r,\theta)$ is equal to the work done by the distributed load $q(r,\theta)$ in moving through the displacement $v_0(r)$ caused by the point load F_0 at the center (it is not necessary to make $F_0 = 1$ because it will cancel out from both sides). The center deflection of a clamped circular plate under asymmetric load given in Eq. (7.4.20) is $v_c = v(0)$:

$$F_0\,v_c = \frac{F_0 a^2}{16\pi D} \int_0^{2\pi} \int_0^a \left(q_0 + \frac{q_1}{a}r\cos\theta\right)\left[1 - \frac{r^2}{a^2}\left(1 - 2\ln\frac{r}{a}\right)\right] r\,dr\,d\theta$$

$$v_c = \frac{q_0 a^2}{16\pi D} \int_0^a \left(r - \frac{r^3}{a^2} - \frac{2}{a^2}r^3\ln\frac{r}{a}\right) dr = \frac{q_0 a^4}{64 D}, \tag{7.4.22}$$

where the following integral identity is used in arriving at the result:

$$\int r^n \ln(\alpha r)\,dr = \frac{r^{n+1}}{n+1}\ln(\alpha r) - \frac{r^{n+1}}{(n+1)^2}, \qquad \alpha = \text{constant}. \tag{7.4.23}$$

7.5 Solution of Two-Dimensional Problems

7.5.1 Introduction

In a class of problems in elasticity, due to geometry, material properties, boundary conditions and external applied loads, the solutions (that is, displacements and stresses) are not dependent on one of the coordinates. Such problems are called *plane elasticity* problems. The plane elasticity problems considered here are grouped into *plane strain* and *plane stress* problems. Both classes of problems are described by a set of two *coupled* partial differential equations expressed in terms of two dependent variables that represent the two components of the displacement vector. The governing equations of plane strain problems differ from those of the plane stress problems only in the coefficients of the differential equations, as shown shortly. The discussion here is limited to isotropic materials.

7.5.2 Plane Strain Problems

Plane strain problems are characterized by the displacement field

$$\mathbf{u} = u_x\,\hat{\mathbf{e}}_x + u_y\,\hat{\mathbf{e}}_y \quad [u_x = u_x(x,y),\ u_y = u_y(x,y),\ u_z = 0], \qquad (7.5.1)$$

where (u_x, u_y, u_z) denote the components of the displacement vector $\mathbf{u}$ in the (x, y, z) coordinate system. An example of a plane strain problem is provided by the long cylindrical member (not necessarily of circular cross section) under external loads that are independent of the z-coordinate, as shown in Fig. 7.5.1. For cross sections sufficiently far from the ends, the displacement u_z is zero and u_x and u_y are independent of z, that is, a state of plane strain exists.

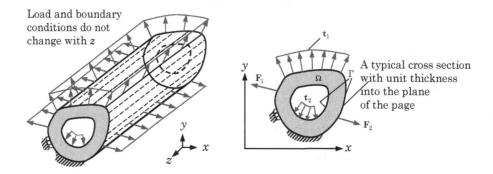

Fig. 7.5.1: An example of a plane strain problem.

The displacement field (7.5.1) results in the following strain field:

$$\varepsilon_{xx} = \frac{\partial u_x}{\partial x}, \quad 2\varepsilon_{xy} = \frac{\partial u_x}{\partial y} + \frac{\partial u_y}{\partial x}, \quad \varepsilon_{yy} = \frac{\partial u_y}{\partial y},$$

$$\varepsilon_{xz} = \varepsilon_{yz} = \varepsilon_{zz} = 0, \qquad (7.5.2)$$

The stress components are calculated using the stress–strain relations [see Eq. (7.2.9); also note $\lambda/(\mu + \lambda) = 2\nu$]

$$\sigma_{xx} = (2\mu + \lambda)\varepsilon_{xx} + \lambda\,\varepsilon_{yy}, \quad \sigma_{yy} = (2\mu + \lambda)\varepsilon_{yy} + \lambda\,\varepsilon_{xx}, \quad \sigma_{xy} = 2\mu\,\varepsilon_{xy},$$

$$\sigma_{zz} = \lambda\,(\varepsilon_{xx} + \varepsilon_{yy}) = \nu\,(\sigma_{xx} + \sigma_{yy}), \quad \sigma_{xz} = 0, \quad \sigma_{yz} = 0. \qquad (7.5.3)$$

Writing in terms of E and ν directly from Eq. (7.2.10), we have

$$\begin{Bmatrix} \sigma_{xx} \\ \sigma_{yy} \\ \sigma_{xy} \end{Bmatrix} = \frac{E}{(1+\nu)(1-2\nu)} \begin{bmatrix} 1-\nu & \nu & 0 \\ \nu & 1-\nu & 0 \\ 0 & 0 & \frac{(1-2\nu)}{2} \end{bmatrix} \begin{Bmatrix} \varepsilon_{xx} \\ \varepsilon_{yy} \\ 2\varepsilon_{xy} \end{Bmatrix}. \qquad (7.5.4)$$

The equations of equilibrium of three-dimensional linear elasticity, with the body force components

$$f_x = f_x(x,y), \quad f_y = f_y(x,y), \quad f_z = 0, \qquad (7.5.5)$$

reduce to the following two equations:

$$\frac{\partial \sigma_{xx}}{\partial x} + \frac{\partial \sigma_{xy}}{\partial y} + f_x = 0, \tag{7.5.6}$$

$$\frac{\partial \sigma_{xy}}{\partial x} + \frac{\partial \sigma_{yy}}{\partial y} + f_y = 0. \tag{7.5.7}$$

The boundary conditions are either the stress type

$$\left.\begin{array}{c} t_x \equiv \sigma_{xx} n_x + \sigma_{xy} n_y = \hat{t}_x \\ t_y \equiv \sigma_{xy} n_x + \sigma_{yy} n_y = \hat{t}_y \end{array}\right\} \quad \text{on } \Gamma_\sigma, \tag{7.5.8}$$

or the displacement type

$$u_x = \hat{u}_x, \quad u_y = \hat{u}_y \quad \text{on } \Gamma_u. \tag{7.5.9}$$

Here (n_x, n_y) denote the components (or direction cosines) of the unit normal vector on the boundary Γ, Γ_σ and Γ_u are disjoint (i.e., nonoverlapping) portions of the boundary Γ such that their sum is equal to the total boundary

$$\Gamma = \Gamma_\sigma + \Gamma_u, \quad \Gamma_\sigma \cap \Gamma_u = \text{empty}, \tag{7.5.10}$$

$\hat{t}_x$ and $\hat{t}_y$ are the components of the specified traction vector, and $\hat{u}_x$ and $\hat{u}_y$ are the components of the specified displacement vector. Only one element of each pair, (u_x, t_x) and (u_y, t_y), should be specified at a boundary point.

The preceding discussion can be extended to plane strain problems in cylindrical coordinates. We now consider an example of a plane strain problem.

Example 7.5.1 ———————————————————————————————

Consider an isotropic, hollow circular cylinder of internal radius a and outside radius b. The cylinder is held between rigid supports such that $u_z = 0$ at $z = \pm L/2$, pressurized at $r = a$ as well as at $r = b$, and is rotating with a uniform speed of ω about its axis (i.e., the z-axis), as shown in Fig. 7.5.2. Determine the displacements, strains, and stresses in the cylinder.

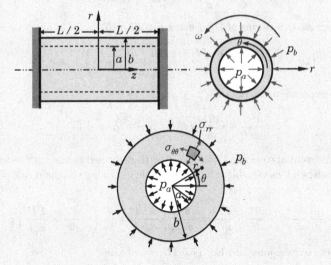

Fig. 7.5.2: Rotating cylindrical pressure vessel.

Solution: Because of the geometry and boundary conditions (material is isotropic), the cylindrical coordinate system (r, θ, z) is most convenient to formulate the problem. The rotation of the cylinder about its own axis generates a radial (centrifugal) force of magnitude $\rho_0 \omega^2 r$ at a distance r. Thus, the body force vector is $\mathbf{f} = \rho_0 \omega^2 r \, \hat{\mathbf{e}}_r$. Also, we find that the problem has symmetry about $z = 0$, and the plane $z = 0$ has exactly the same boundary conditions as the plane $z = L/2$. Therefore, we find that the problem has symmetry about $z = L/4$. This way, it is clear that we can consider any section of unit length of the cylinder to determine the displacements, strains, and stresses. In other words, it is a plane strain problem, and the solution is independent of θ (due to the axisymmetric geometry, forces, and material properties) and z. In fact, the only nonzero displacement is u_r, and it is only a function of the radial coordinate r. The problem has only stress boundary conditions (BVP type II),

$$\text{At } r = a: \quad \hat{\mathbf{n}} = -\hat{\mathbf{e}}_r, \quad \mathbf{t} = p_a \, \hat{\mathbf{e}}_r \quad \text{or} \quad \sigma_{rr} = -p_a, \quad \sigma_{r\theta} = 0, \tag{7.5.11}$$

$$\text{At } r = b: \quad \hat{\mathbf{n}} = \hat{\mathbf{e}}_r, \quad \mathbf{t} = -p_b \, \hat{\mathbf{e}}_r \quad \text{or} \quad \sigma_{rr} = -p_b, \quad \sigma_{r\theta} = 0. \tag{7.5.12}$$

We begin with the assumed displacement field

$$u_r = U(r), \quad u_\theta = u_z = 0, \tag{7.5.13}$$

where $U(r)$ is an unknown function to be determined such that the equations of elasticity and boundary conditions of the problem are satisfied. The strains are [see Eq. (7.2.3)]

$$\varepsilon_{rr} = \frac{dU}{dr}, \quad \varepsilon_{\theta\theta} = \frac{U}{r}, \quad \varepsilon_{zz} = \varepsilon_{r\theta} = \varepsilon_{rz} = \varepsilon_{\theta z} = 0. \tag{7.5.14}$$

The stresses are determined using the stress–strain relations in Eq. (7.2.9)

$$\sigma_{rr} = 2\mu\varepsilon_{rr} + \lambda\left(\varepsilon_{rr} + \varepsilon_{\theta\theta}\right) = (2\mu + \lambda)\frac{dU}{dr} + \lambda\frac{U}{r},$$

$$\sigma_{\theta\theta} = 2\mu\varepsilon_{\theta\theta} + \lambda\left(\varepsilon_{rr} + \varepsilon_{\theta\theta}\right) = (2\mu + \lambda)\frac{U}{r} + \lambda\frac{dU}{dr}, \tag{7.5.15}$$

$$\sigma_{zz} = \lambda\left(\varepsilon_{rr} + \varepsilon_{\theta\theta}\right) = \lambda\left(\frac{dU}{dr} + \frac{U}{r}\right).$$

All other stresses, $\sigma_{r\theta}$, σ_{rz}, and $\sigma_{\theta z}$, are zero.

The last two equations of equilibrium, Eq. (7.2.7) without the acceleration terms, are trivially satisfied, and the first equation reduces to

$$\frac{d\sigma_{rr}}{dr} + \frac{1}{r}\left(\sigma_{rr} - \sigma_{\theta\theta}\right) + \rho_0\omega^2 r = 0,$$

which can be expressed in terms of $U(r)$ using Eq. (7.5.15)

$$(2\mu + \lambda)\frac{d^2U}{dr^2} + \lambda\frac{d}{dr}\left(\frac{U}{r}\right) + \frac{2\mu}{r}\left(\frac{dU}{dr} - \frac{U}{r}\right) + \rho_0\omega^2 r = 0. \tag{7.5.16}$$

Simplifying the expression, we obtain

$$r^2\frac{d^2U}{dr^2} + r\frac{dU}{dr} - U = -\alpha r^3, \quad \alpha = \frac{\rho_0\omega^2}{2\mu + \lambda}. \tag{7.5.17}$$

The linear differential equation (7.5.17) can be transformed to one with constant coefficients by a change of independent variable, $r = e^\xi$ (or $\xi = \ln r$). Using the chain rule of differentiation, we obtain

$$\frac{dU}{dr} = \frac{dU}{d\xi}\frac{d\xi}{dr} = \frac{1}{r}\frac{dU}{d\xi}, \quad \frac{d^2U}{dr^2} = \frac{d}{dr}\left(\frac{1}{r}\frac{dU}{d\xi}\right) = \frac{1}{r^2}\left(-\frac{dU}{d\xi} + \frac{d^2U}{d\xi^2}\right). \tag{7.5.18}$$

Substituting these expressions into Eq. (7.5.17), we obtain

$$\frac{d^2U}{d\xi^2} - U = -\alpha e^{3\xi}. \tag{7.5.19}$$

Seeking a solution in the form, $U(\xi) = e^{m\xi}$, we obtain $m = \pm 1$, and the total solution to the problem is

$$U(\xi) = c_1 e^{\xi} + c_2 e^{-\xi} - \frac{\alpha}{8} e^{3\xi}. \tag{7.5.20}$$

Changing back to the original independent variable r, the radial displacement is

$$u_r(r) = U(r) = c_1 r + \frac{c_2}{r} - \frac{\alpha}{8} r^3, \tag{7.5.21}$$

where the constants c_1 and c_2 are to be determined using the boundary conditions in Eqs. (7.5.11) and (7.5.12). Hence, we must compute σ_{rr},

$$\sigma_{rr} = (2\mu + \lambda)\left(c_1 - \frac{c_2}{r^2} - \frac{3\alpha}{8} r^2\right) + \lambda\left(c_1 + \frac{c_2}{r^2} - \frac{\alpha}{8} r^2\right)$$

$$= 2(\mu + \lambda)c_1 - 2\mu\frac{c_2}{r^2} - \frac{(3\mu + 2\lambda)\alpha}{4} r^2. \tag{7.5.22}$$

Applying the stress boundary conditions in (7.5.11) and (7.5.12), we obtain

$$2(\mu + \lambda)c_1 - 2\mu\frac{c_2}{a^2} - \frac{(3\mu + 2\lambda)\alpha}{4} a^2 = -p_a,$$

$$2(\mu + \lambda)c_1 - 2\mu\frac{c_2}{b^2} - \frac{(3\mu + 2\lambda)\alpha}{4} b^2 = -p_b. \tag{7.5.23}$$

Solving for the constants c_1 and c_2,

$$c_1 = \frac{1}{2(\mu + \lambda)}\left[\left(\frac{p_a a^2 - p_b b^2}{b^2 - a^2}\right) + (b^2 + a^2)\frac{(3\mu + 2\lambda)}{(2\mu + \lambda)}\frac{\rho_0\omega^2}{4}\right],$$

$$c_2 = \frac{a^2 b^2}{2\mu}\left[\left(\frac{p_a - p_b}{b^2 - a^2}\right) + \frac{(3\mu + 2\lambda)}{(2\mu + \lambda)}\frac{\rho_0\omega^2}{4}\right]. \tag{7.5.24}$$

Finally, the displacement u_r and stress σ_{rr} in the cylinder are given by

$$u_r = \frac{1}{2(\mu + \lambda)}\left[\left(\frac{p_a a^2 - p_b b^2}{b^2 - a^2}\right) + (b^2 + a^2)\frac{(3\mu + 2\lambda)}{(2\mu + \lambda)}\frac{\rho_0\omega^2}{4}\right] r$$

$$+ \frac{a^2 b^2}{2\mu}\left[\left(\frac{p_a - p_b}{b^2 - a^2}\right) + \frac{(3\mu + 2\lambda)}{(2\mu + \lambda)}\frac{\rho_0\omega^2}{4}\right]\frac{1}{r} - \frac{\rho_0\omega^2}{8(2\mu + \lambda)} r^3, \tag{7.5.25}$$

$$\sigma_{rr} = \left[\left(\frac{p_a a^2 - p_b b^2}{b^2 - a^2}\right) + (b^2 + a^2)\frac{(3\mu + 2\lambda)}{(2\mu + \lambda)}\frac{\rho_0\omega^2}{4}\right]$$

$$- \frac{a^2 b^2}{r^2}\left[\left(\frac{p_a - p_b}{b^2 - a^2}\right) + \frac{(3\mu + 2\lambda)}{(2\mu + \lambda)}\frac{\rho_0\omega^2}{4}\right] - \frac{(3\mu + 2\lambda)\alpha}{4} r^2. \tag{7.5.26}$$

Similarly, stresses $\sigma_{\theta\theta}$ and σ_{zz} can be computed.

7.5.3 Plane Stress Problems

A state of *plane stress* is one in which the stresses associated with one of the coordinates (z) are zero and the other stresses are functions of the remaining two coordinates $(x$ and $y)$:

$$\sigma_{xz} = \sigma_{yz} = \sigma_{zz} = 0,$$

$$\sigma_{xx} = \sigma_{xx}(x, y), \quad \sigma_{xy} = \sigma_{xy}(x, y), \quad \sigma_{yy} = \sigma_{yy}(x, y). \tag{7.5.27}$$

An example of a plane stress problem is provided by a thin plate subjected to loads in the xy plane that are independent of z, as shown in Fig. 7.5.3. The top and bottom surfaces of the plate are assumed to be traction-free, and $f_z = 0$ and $u_z = 0$.

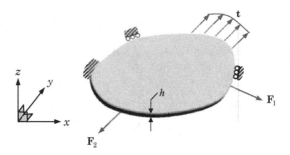

Fig. 7.5.3: A thin plate in a state of plane stress.

The stress–strain relations of a plane stress state for an isotropic material are obtained by inverting the strain–stress relations in Eq. (6.3.23):

$$\left\{ \begin{array}{c} \sigma_{xx} \\ \sigma_{yy} \\ \sigma_{xy} \end{array} \right\} = \frac{E}{1 - \nu^2} \begin{bmatrix} 1 & \nu & 0 \\ \nu & 1 & 0 \\ 0 & 0 & \frac{(1-\nu)}{2} \end{bmatrix} \left\{ \begin{array}{c} \varepsilon_{xx} \\ \varepsilon_{yy} \\ 2\varepsilon_{xy} \end{array} \right\}. \tag{7.5.28}$$

The equations of equilibrium as well as boundary conditions of a plane stress problem are the same as those listed in Eqs. (7.5.6)–(7.5.9). Note that the governing equations of plane stress and plane strain differ from each other only on account of the difference in the constitutive equations for the two cases.

Example 7.5.2 ————————————————————————

Consider a thin, uniform, solid circular disk of radius a, spinning at a constant angular velocity of ω, as shown in Fig. 7.5.4. Use the semi-inverse method to determine the displacements, strains, and stresses in the disk.

Solution: This problem is almost the same as the problem of the rotating cylinder considered in Example 7.5.1. The difference is that the cylinder problem was one of plane strain and the present thin disk problem is one of plane stress. First, we set up the polar cylindrical coordinate system (r, θ), with the origin at the center of the disk, r being the radial coordinate

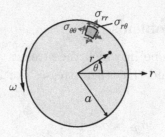

Fig. 7.5.4: Thin, uniform, spinning solid disk.

and θ the circumferential coordinate. The boundary conditions are

$$u_r(0, \theta) = \text{finite}, \quad \sigma_{rr}(a, \theta) = \sigma_{r\theta}(a, \theta) = 0. \tag{1}$$

Because of the axisymmetry of the geometry, boundary conditions, and material, the disk experiences only a radial displacement field that varies only with r. Using the semi-inverse method, we assume

$$u_r = U(r), \quad u_\theta = 0, \tag{2}$$

where $U(r)$ is an unknown function to be determined such that the equations of elasticity and boundary conditions of the problem are satisfied. The strains associated with the displacement field (2) are

$$\varepsilon_{rr} = \frac{dU}{dr}, \quad \varepsilon_{\theta\theta} = \frac{U}{r}, \quad \varepsilon_{r\theta} = 0. \tag{3}$$

The stresses are determined using the stress–strain relations for *plane stress*, Eq. (7.5.28). We obtain

$$\sigma_{rr} = \frac{E}{1 - \nu^2}(\varepsilon_{rr} + \nu\varepsilon_{\theta\theta}) = \frac{E}{1 - \nu^2}\frac{dU}{dr} + \frac{E\nu}{1 - \nu^2}\frac{U}{r},$$

$$\sigma_{\theta\theta} = \frac{E}{1 - \nu^2}(\nu\varepsilon_{rr} + \varepsilon_{\theta\theta}) = \frac{E\nu}{1 - \nu^2}\frac{dU}{dr} + \frac{E}{1 - \nu^2}\frac{U}{r}. \tag{4}$$

The shear stress $\sigma_{r\theta}$ is zero.

The first two equations of equilibrium, Eq. (7.2.7) without the acceleration terms, are trivially satisfied, and the first equation reduces to

$$\frac{d\sigma_{rr}}{dr} + \frac{1}{r}(\sigma_{rr} - \sigma_{\theta\theta}) + \rho_0\omega^2 r = 0,$$

where $\rho_0 f_r = \rho_0\omega^2 r$. The above equation can be expressed in terms of $U(r)$ using Eq. (4)

$$\frac{E}{1 - \nu^2}\left[\frac{d^2 U}{dr^2} + \nu\frac{d}{dr}\left(\frac{U}{r}\right) + \frac{(1 - \nu)}{r}\left(\frac{dU}{dr} - \frac{U}{r}\right)\right] + \rho_0\omega^2 r = 0. \tag{5}$$

Simplifying the expression, we obtain

$$\frac{d}{dr}\left[\frac{1}{r}\frac{d}{dr}(rU)\right] = -\alpha r, \quad \alpha = \left(\frac{1 - \nu^2}{E}\right)\rho_0\omega^2, \tag{6}$$

where we have used the identities

$$\frac{1}{r}\left(\frac{dU}{dr} - \frac{U}{r}\right) = \frac{d}{dr}\left(\frac{U}{r}\right), \quad \frac{d}{dr}\left(\frac{dU}{dr} + \frac{U}{r}\right) = \frac{d}{dr}\left[\frac{1}{r}\frac{d}{dr}(rU)\right]. \tag{7}$$

The solution to Eq. (6) is given by

$$u_r(r) = U(r) = \frac{c_1}{2}r + \frac{c_2}{r} - \frac{\alpha}{8}r^3, \tag{8}$$

where the constants c_1 and c_2 are to be determined using the boundary conditions in Eq. (1). The fact that u_r is finite (i.e., bounded) at $r = 0$ requires $c_2 = 0$. Then we have

$$U(r) = \frac{c_1}{2}r - \frac{\alpha}{8}r^3, \quad \frac{dU}{dr} = -\frac{3}{8}\alpha r^2 + \frac{c_1}{2}, \quad \frac{U}{r} = -\frac{1}{8}\alpha r^2 + \frac{c_1}{2}. \tag{9}$$

Computing σ_{rr} using Eq. (4), we obtain

$$\sigma_{rr} = \frac{E}{1 - \nu^2}\left(\frac{dU}{dr} + \nu\frac{U}{r}\right) = \frac{E\alpha}{(1 - \nu^2)}\left[-\frac{3 + \nu}{8}\alpha r^2 + \frac{1 + \nu}{2}c_1\right]. \tag{10}$$

Then $\sigma_{rr}(a, \theta) = 0$ gives

$$c_1 = \frac{1}{4}\left(\frac{3 + \nu}{1 + \nu}\right)\alpha a^2. \tag{11}$$

Thus, the solution in Eq. (8) becomes

$$u(r) = \frac{1}{4}\left(\frac{3+\nu}{1+\nu}\right)\alpha a^2 r - \frac{\alpha}{8}r^3 = \frac{(1-\nu)}{8E}\left[2(3+\nu)a^2 - (1+\nu)r^2\right]\rho_0\omega^2 r. \tag{12}$$

The stresses σ_{rr} and $\sigma_{\theta\theta}$ are

$$\sigma_{rr}(r) = \frac{(3+\nu)}{8}\left(a^2 - r^2\right)\rho_0\omega^2$$

$$\sigma_{\theta\theta}(r) = \frac{1}{8}\left[(3+\nu)a^2 - (1+3\nu)r^2\right]\rho_0\omega^2 \tag{13}$$

The values of the maximum displacement and maximum stresses are

$$u_{\max} = u_r(a) = \frac{(1-\nu)(5+\nu)}{8E}\rho_0\omega^2 a^3,$$

$$\sigma_{\max} = \sigma_{rr}(0) = \sigma_{\theta\theta}(0) = \frac{(3+\nu)}{8}\rho_0\omega^2 a^2. \tag{14}$$

7.5.4 Unification of Plane Strain and Plane Stress Problems

The equilibrium equations (7.5.6) and (7.5.7), which are valid for both plane stress and plane strain, can be expressed in index notation as

$$\sigma_{\beta\alpha,\beta} + f_\alpha = 0, \tag{7.5.29}$$

To unify the formulation for plane strain and plane stress, we introduce the parameter s:

$$s = \begin{cases} \frac{1}{1-\nu}, & \text{for plane strain} \\ 1+\nu, & \text{for plane stress.} \end{cases} \tag{7.5.30}$$

Then the constitutive equations of plane stress as well as plane strain can be expressed as

$$\sigma_{\alpha\beta} = 2\mu\left[\varepsilon_{\alpha\beta} + \left(\frac{s-1}{2-s}\right)\varepsilon_{\gamma\gamma}\delta_{\alpha\beta}\right],$$

$$\varepsilon_{\alpha\beta} = \frac{1}{2\mu}\left[\sigma_{\alpha\beta} - \left(\frac{s-1}{s}\right)\sigma_{\gamma\gamma}\delta_{\alpha\beta}\right], \tag{7.5.31}$$

where α, β, and γ take values of 1 and 2 (or x and y). The compatibility equation (3.7.4) for plane stress and plane strain problems takes the form

$$\varepsilon_{\alpha\alpha,\beta\beta} - \varepsilon_{\alpha\beta,\alpha\beta} = 0 \quad (\alpha, \beta = 1, 2), \tag{7.5.32}$$

or, in terms of stress components,

$$\nabla^2\sigma_{\alpha\alpha} = -s\, f_{\alpha,\alpha}. \tag{7.5.33}$$

Comparing the constitutive equations of plane strain and plane stress, Eqs. (7.5.4) and (7.5.28), it is clear that the plane strain equations can be transformed to corresponding plane stress equations, and vice versa, by a simple change in material parameters, as follows:

Plane stress to plane strain: $E \rightarrow \frac{E}{1-\nu^2}$ and $\nu \rightarrow \frac{\nu}{1-\nu}$,

Plane strain to plane stress: $E \rightarrow \frac{(1+2\nu)E}{(1+\nu)^2}$ and $\nu \rightarrow \frac{\nu}{1+\nu}$. $\tag{7.5.34}$

7.5.5 Airy Stress Function

Airy stress function is a potential function introduced to identically satisfy the equations of equilibrium, Eqs. (7.5.6) and (7.5.7). First, we assume that the body force vector $\mathbf{f}$ is derivable from a scalar potential V_f such that

$$\mathbf{f} = -\boldsymbol{\nabla} V_f \quad \text{or} \quad f_x = -\frac{\partial V_f}{\partial x}, \quad f_y = -\frac{\partial V_f}{\partial y}. \tag{7.5.35}$$

When body forces are derivable from a potential V_f, they are said to be *conservative*. Next, we introduce the Airy stress function $\Phi(x, y)$ such that

$$\sigma_{xx} = \frac{\partial^2 \Phi}{\partial y^2} + V_f, \quad \sigma_{yy} = \frac{\partial^2 \Phi}{\partial x^2} + V_f, \quad \sigma_{xy} = -\frac{\partial^2 \Phi}{\partial x \partial y}. \tag{7.5.36}$$

This definition of $\Phi(x, y)$ automatically satisfies the equations of equilibrium (7.5.6) and (7.5.7).

The stresses derived from Eq. (7.5.36) are subject to the compatibility conditions (7.5.33). Substituting for $\sigma_{\alpha\beta}$ in terms of Φ from Eq. (7.5.36) into Eq. (7.5.33), we obtain

$$\nabla^4 \Phi + (2 - s)\nabla^2 V_f = 0, \tag{7.5.37}$$

where $\nabla^4 = \nabla^2\nabla^2$ is the *biharmonic operator*, which, in two dimensions, has the form

$$\nabla^4 = \frac{\partial^4}{\partial x^4} + 2\frac{\partial^4}{\partial x^2 \partial y^2} + \frac{\partial^4}{\partial y^4}.$$

If the body forces are zero, we have $V_f = 0$ and Eq. (7.5.37) reduces to the *biharmonic equation*

$$\nabla^4 \Phi = 0. \tag{7.5.38}$$

In cylindrical coordinate system, Eqs. (7.5.35) and (7.5.36) take the form

$$f_r = -\frac{\partial V_f}{\partial r}, \quad f_\theta = -\frac{1}{r}\frac{\partial V_f}{\partial \theta}, \tag{7.5.39}$$

$$\sigma_{rr} = \frac{1}{r}\frac{\partial \Phi}{\partial r} + \frac{1}{r^2}\frac{\partial^2 \Phi}{\partial \theta^2} + V_f, \quad \sigma_{\theta\theta} = \frac{\partial^2 \Phi}{\partial r^2} + V_f, \quad \sigma_{r\theta} = -\frac{\partial}{\partial r}\left(\frac{1}{r}\frac{\partial \Phi}{\partial \theta}\right). \tag{7.5.40}$$

The biharmonic operator $\nabla^4 = \nabla^2\nabla^2$ can be expressed using the definition of ∇^2 in a cylindrical coordinate system

$$\nabla^2 = \frac{\partial^2}{\partial r^2} + \frac{1}{r}\frac{\partial}{\partial r} + \frac{1}{r^2}\frac{\partial^2}{\partial \theta^2}. \tag{7.5.41}$$

In summary, the solution to a plane elastic problem using the Airy stress function involves finding the solution to Eq. (7.5.37) and satisfying the boundary conditions of the problem. The most difficult part is finding a solution to the fourth-order equation (7.5.37) over a given domain. Often the form of the Airy stress function is obtained by either the inverse method or semi-inverse method. Next we consider several examples of the Airy stress function approach. Additional examples can be found in the books by Timoshenko and Goodier (1970) and Slaughter (2002).

Example 7.5.3

Suppose that the Airy stress function is a second-order polynomial (the lowest order that gives a nonzero stress field) of the form

$$\Phi(x, y) = c_1 xy + c_2 x^2 + c_3 y^2. \tag{7.5.42}$$

Assuming that the body force field is zero, determine if the constants c_1, c_2, and c_3 correspond to a possible state of stress for some boundary value problem (the inverse method).

Solution: Clearly, the biharmonic equation is trivially satisfied by Φ in Eq. (7.5.42). The corresponding stress field is

$$\sigma_{xx} = \frac{\partial^2 \Phi}{\partial y^2} = 2c_3, \quad \sigma_{yy} = \frac{\partial^2 \Phi}{\partial x^2} = 2c_2, \quad \sigma_{xy} = -\frac{\partial^2 \Phi}{\partial x \partial y} = -c_1. \tag{7.5.43}$$

Thus, the constants represent a uniform stress state throughout the body, and it is independent of the geometry. Thus, there are infinite number of problems for which the stress field is a solution. In particular, the rectangular domain with the boundary stresses shown in Fig. 7.5.5 is one such problem.

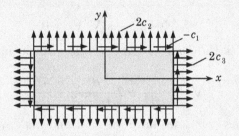

Fig. 7.5.5: A plane problem with uniform stress field.

Example 7.5.4

Take the Airy stress function to be a third-order polynomial of the form

$$\Phi(x, y) = c_1 xy + c_2 x^2 + c_3 y^2 + c_4 x^2 y + c_5 xy^2 + c_6 x^3 + c_7 y^3. \tag{7.5.44}$$

Assuming that the body force field is zero, determine the stress field and identify a possible boundary value problem.

Solution: We note that $\nabla^4 \Phi = 0$ for any c_i. The corresponding stress field is

$$\sigma_{xx} = 2c_3 + 2c_5 x + 6c_7 y, \quad \sigma_{yy} = 2c_2 + 2c_4 y + 6c_6 y, \quad \sigma_{xy} = -c_1 - 2c_4 x - 2c_5 y. \tag{7.5.45}$$

Again, there are an infinite number of problems for which the stress field is a solution. In particular, for $c_1 = c_2 = c_3 = c_4 = c_5 = c_6 = 0$, the solution corresponds to a thin beam in pure bending (see Fig. 7.5.6).

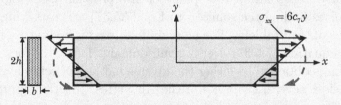

Fig. 7.5.6: A thin beam in pure bending.

Example 7.5.5

Take the Airy stress function to be a fourth-order polynomial of the form (omit terms that were already considered in the last two cases)

$$\Phi(x,y) = c_8 x^2 y^2 + c_9 x^3 y + c_{10} x y^3 + c_{11} x^4 + c_{12} y^4, \qquad (7.5.46)$$

and determine the stress field and associated boundary value problems.

Solution: Computing $\nabla^4 \Phi$ and equating it to zero (body force field is zero) we find that

$$c_8 + 3(c_{11} + c_{12}) = 0.$$

Thus out of five constants only four of them are independent. The corresponding stress field is

$$\sigma_{xx} = 2c_8 x^2 + 6c_{10} xy + 12c_{12} y^2 = -6c_{11} x^2 + 6c_{10} xy + 6c_{12}(2y^2 - x^2)$$

$$\sigma_{yy} = 2c_8 y^2 + 6c_9 xy + 12c_{11} x^2 = 6c_9 xy + 6c_{11}(2x^2 - y^2) - 6c_{12} y^2 \qquad (7.5.47)$$

$$\sigma_{xy} = -4c_8 xy - 3c_9 x^2 - 3c_{10} y^2 = 12c_{11} xy + 12c_{12} xy - 3c_9 x^2 - 3c_{10} y^2.$$

By suitable adjustment of the constants, we can obtain various loads on rectangular plates. For instance, taking all coefficients except c_{10} equal to zero, we obtain

$$\sigma_{xx} = 6c_{10} xy, \quad \sigma_{yy} = 0, \quad \sigma_{xy} = -3c_{10} y^2.$$

7.5.6 Saint-Venant's Principle

A boundary value problem of elasticity requires the boundary conditions to be known in the form of stresses or displacements [see Eqs. (7.5.8) and (7.5.9)] at *every point* of the boundary. As shown in Example 7.5.3, the boundary forces are distributed as a function of the distance along the boundary. If the boundary forces are distributed in any other form (other than per unit surface area), the boundary conditions cannot be expressed as point wise quantities.

For example, consider the cantilever beam with an end load, as shown in Fig. 7.5.7. At $x = 0$, where $\hat{n} = -\hat{e}_x$, we are required to specify $t_x = -\sigma_{xx}$ and $t_y = -\sigma_{xy}$ (because u_x and u_y are clearly not zero there). There is no problem in stating that $\sigma_{xx}(0, y) = 0$, but σ_{xy} is not known point wise. We can possibly say that the integral of $t_y = -\sigma_{xy}$ over the beam cross section must be equal to P:

$$\int_A t_y(0, y)\, dA = -\int_A \sigma_{xy}(0, y)\, dA = P,$$

which is not equal to specifying σ_{xy} point wise. If we state that $\sigma_{xy}(0, y) = -P/A$, where A is the cross-sectional area of the beam, then we have a inconsistency that σ_{xy} is nonzero from the left face and zero from the bottom and top faces of the beam. Thus, there is a stress singularity at points $(x, y) = (0, \pm h)$. We also have a different type of singularity at points $(x, y) = (L, \pm h)$. Strictly speaking, such problems do not admit exact elasticity solutions. We must overcome such singularities by reformulating the problem as one that admits an engineering solution.

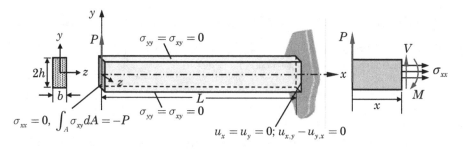

Fig. 7.5.7: A cantilever beam under an end load.

Analytical (approximate) solutions for such problems, when they exist, show that a change in the distribution of the load on the end, without change of the resultant, alters the stress significantly only near the end. *Saint-Venant's principle* says that the effect of the change in the boundary condition from point wise specification to a *statically equivalent* condition (that is, the same net force and moment due to the distributed forces and stresses) is local; that is, the solutions obtained with the two sets of boundary conditions are approximately the same at points sufficiently far from the points where the elasticity boundary conditions are replaced with statically equivalent boundary conditions. Of course, the phrase "sufficiently far" is rather ambiguous. The distance is often taken to be equal to or greater than the length scale of the portion of the boundary where the boundary conditions are replaced. In the case of the beam shown in Fig. 7.5.7, the distance is $2h$ (height of the beam). In the next example, we discuss an engineering solution to the problem shown in Fig. 7.5.7.

Example 7.5.6

Here we consider the problem of a cantilever beam with an end load, as shown in Fig. 7.5.7. The problem can be treated as a plane stress if the beam is of small thickness b compared to the height, $b \ll h$ (of course, $h \ll L$ to call it a beam). If the beam is a portion of a very long slab, in the thickness direction, it can be treated as a plane strain problem. Write the boundary conditions and determine the Airy stress function, stresses, and displacements of the problem.

Solution: The boundary conditions are of mixed type (see Fig. 7.5.7): The tractions are specified on the boundaries $x = 0$ and $y = \pm h$, while the displacements are specified on the boundary $x = L$. However, boundary conditions of plane elasticity can be written only on $x = L$ and $y = \pm h$. On $x = 0$, we know only the total force in the y-direction and not the associated stress. Hence, it must be written as an integral condition on stress $\sigma_{xy}(0, y)$. Thus, we have

$$\sigma_{xx}(0, y) = 0, \quad \sigma_{xy}(x, -h) = 0, \quad \sigma_{yy}(x, -h) = 0,$$
$$\sigma_{xy}(x, h) = 0, \quad \sigma_{yy}(x, h) = 0, \tag{7.5.48}$$

$$u_x(L, y) = 0, \quad u_y(L, y) = 0, \tag{7.5.49}$$

$$b \int_{-h}^{h} \sigma_{xy}(0, y) \, dy = -P. \tag{7.5.50}$$

Due to the boundary condition in Eq. (7.5.50), the resulting boundary value problem is not an exact elasticity problem in the sense that boundary values are not specified point wise. If P is replaced with a shear stress condition $\sigma_{xy}(0, y) = \tau_0$, it is a proper elasticity boundary condition, but even in this case there is a singularity at $x = L$ and $y = h$.

This problem is discussed in most elasticity and continuum mechanics books, despite the fact that it is not a well-posed problem owing to point singularities at the corners of the domain. Therefore, the solution being sought is an approximate solution, which is a reasonable one, by Saint-Venant's principle, away from the isolated points of singularity.

The semi-inverse method allows us to identify the form of the Airy stress function. The knowledge of the stress distributions from the elementary theory of beams provides the needed clue to identify the terms in the Airy stress function. Recall the following stress field from the Euler–Bernoulli beam theory [see Section 7.3.4, Eq. (7.3.31)]:

$$\sigma_{xx} = \frac{M(x)y}{I}, \quad \sigma_{yy} = 0, \quad \sigma_{xy} = \frac{V(x)Q(y)}{Ib}, \tag{7.5.51}$$

where M is the bending moment and V is the shear force [see Eq. (7.5.24)]:

$$M = \int_A y\sigma_{xx}\, dA, \quad V = \int_A \sigma_{xy}\, dA, \tag{7.5.52}$$

I is the moment of inertia about the z-axis, and Q is the first moment of area

$$I = \int_A y^2\, dA = 2bh^3/3, \quad Q(y) = \int_{\bar{A}} y\, dA = b\int_y^h y\, dy. \tag{7.5.53}$$

Here $\bar{A}$ denotes the cross-sectional area between line y and the top of the beam. Clearly, Q is a quadratic function of y. We also note that $M(x)$ is a linear function of x while V is a constant for the problem at hand. Therefore, σ_{xx} is linear in x, σ_{xy} is a quadratic in y, and $\sigma_{yy} = 0$ at $y = \pm h$ for any x (except possibly at $x = L$). Using this qualitative information and definitions (7.5.35), in the absence of body forces (i.e., $V_f = 0$), we take the Airy stress function to be

$$\Phi(x,y) = x(c_1 y + c_2 y^2 + c_3 y^3). \tag{7.5.54}$$

Note that only the first and third terms are dictated by the stress field in a beam. The second term is added to make it a complete quadratic polynomial in y. Also, Φ cannot have terms higher than x because of the boundary condition $\sigma_{yy}(x, \pm h) = 0$. The nonzero stresses are

$$\sigma_{xx} = \frac{\partial^2 \Phi}{\partial y^2} = x\,(2c_2 + 6c_3 y), \quad \sigma_{yy} = \frac{\partial^2 \Phi}{\partial x^2} = 0, \quad \sigma_{xy} = -\frac{\partial^2 \Phi}{\partial x \partial y} = -(c_1 + 2c_2 y + 3c_3 y^2). \tag{7.5.55}$$

The choice in (7.5.54) satisfies the biharmonic equation for any values of c_1, c_2, and c_3. We determine the constants c_i using the stress boundary conditions in Eqs. (7.5.48) and (7.5.50). The stress boundary conditions $\sigma_{xx}(0, y) = 0$ and $\sigma_{yy}(x, \pm h) = 0$ are trivially satisfied. We have

$$\sigma_{xy}(x, \pm h) = 0 \rightarrow c_1 - 2c_2 h + 3c_3 h^2 = 0 \quad \text{and} \quad c_1 + 2c_2 h + 3c_3 h^2 = 0,$$

which yield $c_2 = 0$ and $c_1 = -3h^2 c_3$. Lastly, we have

$$b\int_{-h}^h \sigma_{xy}(0, y)\, dy = -P \rightarrow -2hb(c_1 + h^2 c_3) = -P. \tag{7.5.56}$$

Thus, the constants c_i are

$$c_1 = \frac{3P}{4bh}, \quad c_2 = 0, \quad c_3 = -\frac{P}{4bh^3}, \tag{7.5.57}$$

and the Airy stress function becomes

$$\Phi(x,y) = -\frac{Pxy}{6I}\left(y^2 - 3h^2\right). \tag{7.5.58}$$

The stresses from Eq. (7.5.55) are [$I = 2bh^3/3 = Ah^2/3$, where $A = 2bh$ is the area of cross section of the beam]

$$\sigma_{xx} = -\frac{6Pxy}{4bh^3} = -\frac{Pxy}{I}, \quad \sigma_{yy} = 0,$$

$$\sigma_{xy} = -\frac{3P}{4bh} + \frac{3Py^2}{4bh^3} = -\frac{P}{2I}\left(h^2 - y^2\right). \tag{7.5.59}$$

The stresses in Eq. (7.5.59) are exactly those predicted by the classical (i.e., Euler–Bernoulli) beam theory, where $M(x) = -Px$ and $V = -P$. This is not surprising because our choice of terms in the Airy stress function was dictated by the form of the stress field from the classical beam theory. This also indicates that we cannot obtain any better stress field than the elementary beam theory for the boundary conditions (7.5.48)–(7.5.50).

The strain field associated with the stress field in Eq. (7.5.59) is computed using the strain–stress relations in Eq. (6.3.32):

$$
\begin{aligned}
\varepsilon_{xx} &= \tfrac{1}{E}\sigma_{xx} = -\tfrac{1}{EI}Pxy, \\
\varepsilon_{yy} &= -\tfrac{\nu}{E}\sigma_{xx} = \tfrac{\nu}{EI}Pxy, \\
\varepsilon_{xy} &= \tfrac{(1+\nu)}{E}\sigma_{xy} = -\tfrac{(1+\nu)}{2EI}P(h^2 - y^2),
\end{aligned}
\tag{7.5.60}
$$

where ν is the Poisson ratio and E is Young's modulus. The strain field in Eq. (7.5.60) is the same as that in Eq. (3.7.12), with $x_1 = x$ and $x_2 = y$. Therefore, the displacements are the same as those determined in Example 3.7.2, namely in Eq. (3.7.22), with $u_1 = u_x$ and $u_2 = u_y$:

$$
u_x(x,y) = \frac{PL^3}{6EI}\frac{y}{L}\left\{3\left[1 - \left(\frac{x}{L}\right)^2\right] + (2+\nu)\left(\frac{y}{h}\right)^2\left(\frac{h}{L}\right)^2 - 3(1+\nu)\left(\frac{h}{L}\right)^2\right\},
$$

$$
u_y(x,y) = \frac{PL^3}{6EI}\left\{2 - 3\frac{x}{L}\left[1 - \nu\left(\frac{y}{h}\right)^2\left(\frac{h}{L}\right)^2\right] + \frac{x^3}{L^3} + 3(1+\nu)\left(\frac{h}{L}\right)^2\left(1 - \frac{x}{L}\right)\right\}.
$$

As $(h/L)^2 \to 0$, we recover the Euler–Bernoulli beam solution.

Example 7.5.7

Consider a thin rectangular plate of length $2a$, width $2b$, and thickness h that has a circular hole of radius R at the center of the plate. A uniform traction of magnitude σ_0 is applied to the ends of the plate, as shown in Fig. 7.5.8. Determine the stress field in the plate under the assumption that $R \ll b$.

Solution: The boundary conditions of the problem are

$$
\sigma_{xx}(\pm a, y) = \sigma_0, \quad \sigma_{xy}(\pm a, y) = 0, \quad \sigma_{yy}(x, \pm b) = 0, \quad \sigma_{xy}(x, \pm b) = 0, \tag{1}
$$

$$
\sigma_{rr}(R, \theta) = 0, \quad \sigma_{r\theta}(R, \theta) = 0. \tag{2}
$$

Since the hole is assumed to be very small compared to the height of the plate (i.e., $R \ll b$), we can solve the problem for a stress field inside a circular region of radius $b > c > R$, as shown in Fig. 7.5.8. The stresses at radius c are essentially the same as in the plate without the hole (a consequence of Saint-Venant's principle). We use the cylindrical coordinate system to determine the stress field inside the circle of radius c.

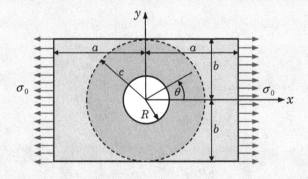

Fig. 7.5.8: A thin rectangular plate with a central hole.

Recall from Eq. (4.3.7) the transformation equations between $(\sigma_{xx}, \sigma_{yy}, \sigma_{xy})$ and $(\sigma_{11} = \sigma_{rr}, \sigma_{22} = \sigma_{\theta\theta}, \sigma_{12} = \sigma_{r\theta})$:

$$
\sigma_{rr} = \sigma_{xx} \cos^2\theta + \sigma_{yy} \sin^2\theta + \sigma_{xy} \sin 2\theta,
$$
$$
\sigma_{\theta\theta} = \sigma_{xx} \sin^2\theta + \sigma_{yy} \cos^2\theta - \sigma_{xy} \sin 2\theta, \tag{3}
$$
$$
\sigma_{r\theta} = -\frac{1}{2}(\sigma_{xx} - \sigma_{yy})\sin 2\theta + \sigma_{xy} \cos 2\theta.
$$

Using the transformation equations in Eq. (3), we can write the boundary conditions at $r = c$ for any θ as ($\sigma_{yy} = 0$ and $\sigma_{xy} = 0$):

$$
\sigma_{rr}(c, \theta) = \sigma_0 \cos^2\theta = \frac{\sigma_0}{2}(1 + \cos 2\theta),
$$
$$
\sigma_{\theta\theta}(c, \theta) = \sigma_0 \sin^2\theta = \frac{\sigma_0}{2}(1 - \cos 2\theta), \tag{4}
$$
$$
\sigma_{r\theta}(c, \theta) = -\frac{\sigma_0}{2}\sin 2\theta.
$$

The form of the boundary conditions in Eq. (4) suggests that the Airy stress function Φ should be of the form

$$
\Phi(r, \theta) = G(r) + F(r) \cos 2\theta, \tag{5}
$$

with $G(r)$ and $F(r)$ satisfying [because $\nabla^2\nabla^2\Phi = \nabla^2\nabla^2 G(r) + \nabla^2\nabla^2(F\cos 2\theta) = 0$ implies that $\tilde{\nabla}^2\tilde{\nabla}^2 G(r) = 0$ and $\hat{\nabla}^2\hat{\nabla}^2 F = 0$]

$$
\tilde{\nabla}^2\tilde{\nabla}^2 G = \left(\frac{d^2}{dr^2} + \frac{1}{r}\frac{d}{dr}\right)^2 G(r) = 0, \quad \hat{\nabla}^2\hat{\nabla}^2 F = \left(\frac{d^2}{dr^2} + \frac{1}{r}\frac{d}{dr} - \frac{4}{r^2}\right)^2 F(r) = 0. \tag{6}
$$

The general solutions to the equations in (6) are of the form

$$
F(r) = \frac{c_1}{r^2} + c_2 + c_3 r^2 + c_4 r^4, \quad G(r) = c_5 + c_6 \ln r + c_7 r^2 + c_8 r^2 \ln r, \tag{7}
$$

and we have

$$
\frac{dF}{dr} = -\frac{2c_1}{r^3} + 2c_3 r + 4c_4 r^3, \qquad \frac{d^2F}{dr^2} = \frac{6c_1}{r^4} + 2c_3 + 12c_4 r^2,
$$
$$
\frac{dG}{dr} = \frac{c_6}{r} + 2c_7 r + rc_8(1 + 2\ln r), \qquad \frac{d^2G}{dr^2} = -\frac{c_6}{r^2} + 2c_7 + c_8(3 + 2\ln r),
$$
$$
\frac{\partial\Phi}{\partial r} = \left[\frac{c_6}{r} + 2c_7 r + rc_8(1 + 2\ln r)\right] + \left(-\frac{2c_1}{r^3} + 2c_3 r + 4c_4 r^3\right)\cos 2\theta,
$$
$$
\frac{\partial^2\Phi}{\partial r^2} = \left[-\frac{c_6}{r^2} + 2c_7 + c_8(3 + 2\ln r)\right] + \left(\frac{6c_1}{r^4} + 2c_3 + 12c_4 r^2\right)\cos 2\theta, \tag{8}
$$
$$
\frac{\partial\Phi}{\partial\theta} = -2\left(\frac{c_1}{r^2} + c_2 + c_3 r^2 + c_4 r^4\right)\sin 2\theta,
$$
$$
\frac{\partial^2\Phi}{\partial\theta^2} = -4\left(\frac{c_1}{r^2} + c_2 + c_3 r^2 + c_4 r^4\right)\cos 2\theta,
$$
$$
\frac{\partial^2\Phi}{\partial\theta\partial r} = -2\left(-\frac{2c_1}{r^3} + 2c_3 r + 4c_4 r^3\right)\sin 2\theta.
$$

Substituting the expressions from Eq. (7) into Eq. (5) and using the definition of the stress components, we obtain

$$
\sigma_{rr} = \frac{1}{r}\frac{\partial\Phi}{\partial r} + \frac{1}{r^2}\frac{\partial^2\Phi}{\partial\theta^2} = \frac{c_6}{r^2} + 2c_7 + c_8(1 + 2\ln r) - \left(\frac{6c_1}{r^4} + \frac{4c_2}{r^2} + 2c_3\right)\cos 2\theta,
$$
$$
\sigma_{\theta\theta} = \frac{\partial^2\Phi}{\partial r^2} = -\frac{c_6}{r^2} + 2c_7 + c_8(3 + 2\ln r) + \left(\frac{6c_1}{r^4} + 2c_3 + 12c_4 r^2\right)\cos 2\theta, \tag{9}
$$
$$
\sigma_{r\theta} = -\frac{\partial}{\partial r}\left(\frac{1}{r}\frac{\partial\Phi}{\partial\theta}\right) = \frac{1}{r^2}\frac{\partial\Phi}{\partial\theta} - \frac{1}{r}\frac{\partial^2\Phi}{\partial\theta\partial r} = \left(-\frac{6c_1}{r^4} - \frac{2c_2}{r^2} + 2c_3 + 6c_4 r^2\right)\sin 2\theta.
$$

Note that c_5 does not enter the calculation of stresses. The boundary conditions in Eqs. (2) and (4) are used to determine the remaining constants. As $r \to \infty$, the expressions for stresses in (8) must approach those in Eq. (4). For this to happen, c_4 and c_8 must be zero and $2c_7 = \sigma_0/2$ and $2c_3 = -\sigma_0/2$. The boundary conditions in Eq. (2) yield the following relations among the remaining constants:

$$\frac{c_6}{R^2} + 2c_7 = 0, \quad \frac{6c_1}{R^4} + \frac{4c_2}{R^2} + 2c_3 = 0, \quad -\frac{6c_1}{R^4} - \frac{2c_2}{R^2} + 2c_3 = 0. \tag{10}$$

Solving these equations, we obtain

$$c_1 = -\frac{\sigma_0 R^4}{4}, \quad c_2 = \frac{\sigma_0 R^2}{2}, \quad c_3 = -\frac{\sigma_0}{4}, \quad c_4 = 0, \quad c_6 = -\frac{\sigma_0 R^2}{2}, \quad c_7 = \frac{\sigma_0}{4}, \quad c_8 = 0. \tag{11}$$

Substituting these values into Eq. (8), we obtain

$$\sigma_{rr} = \frac{\sigma_0}{2}\left[\left(1 - \frac{R^2}{r^2}\right) + \left(1 + \frac{3R^4}{r^4} - \frac{4R^2}{r^2}\right)\cos 2\theta\right],$$

$$\sigma_{\theta\theta} = \frac{\sigma_0}{2}\left[\left(1 + \frac{R^2}{r^2}\right) - \left(1 + \frac{3R^4}{r^4}\right)\cos 2\theta\right], \tag{11}$$

$$\sigma_{r\theta} = -\frac{\sigma_0}{2}\left(1 - \frac{3R^4}{r^4} + \frac{2R^2}{r^2}\right)\sin 2\theta.$$

The maximum normal stress occurs at $(r, \theta) = (R, \pm 90°)$ and shear stress at $(r, \theta) = (\sqrt{3}R, -45°)$:

$$\sigma_{max} = \sigma_{\theta\theta}(R, \pm 90°) = 3\sigma_0, \quad \sigma_{r\theta}(\sqrt{3}R, -45°) = \frac{2}{3}\sigma_0. \tag{12}$$

7.5.7 Torsion of Cylindrical Members

The stress function approach used to study a number of plane elasticity problems in the previous sections is also useful in studying torsion of noncircular cylindrical members. However, we cannot use the Airy stress function here because the present problem does not fall into the category of plane elasticity problems. The governing equations for this problem must be developed from basic principles. The problem was first studied by Saint-Venant using the semi-inverse method.

Consider a cylindrical member of noncircular cross-section and length L and subjected to an end torque $\mathbf{T} = T\hat{\mathbf{e}}_z$, as shown in Fig. 7.5.9(a). The lateral surface of the cylinder is free of tractions. Saint-Venant studied the problem under the following assumptions:

1. The projection of each cross section onto the xy-plane rotates about the z-axis (taken through the geometric centroid of the cross section) with no in-plane distortion.

2. The amount of rotation of each cross section is proportional to its distance from the end of the cylinder, $\Theta = \theta z$, where Θ is the *twist* and θ is the *twist per unit length*.

3. Each cross section's out-of-plane distortion is the same and its magnitude is proportional to θ.

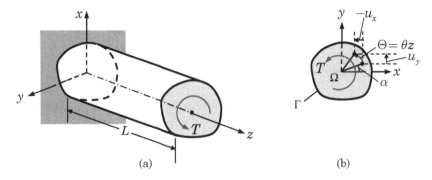

Fig. 7.5.9: (a) Torsion of a cylindrical member. (b) A typical cross section.

In view of the aforementioned assumptions, our attention is focused on a typical cross section of the cylinder; the plane of the cross section is denoted by Ω and its boundary by Γ, as shown in Fig. 7.5.9(b). Our interest is to determine the shear stresses, σ_{xz} and σ_{yz}, produced by the torque, because they are needed in the design of shafts used, for example, in power transmission. There are two different formulations to study the problem. One is based on the *warping function* and the other on *Prandtl stress function*. The details of these two formulations are presented next.

7.5.7.1 Warping function

The displacements of a typical point (r, α) in Ω can be computed as follows [refer to Fig. 7.5.9(b)]:

$$
\begin{aligned}
u_x &= r \cos(\Theta + \alpha) - r \cos \alpha = x(\cos \Theta - 1) - y \sin \Theta, \\
u_y &= r \sin(\Theta + \alpha) - r \sin \alpha = x \sin \Theta + y(\cos \Theta - 1).
\end{aligned}
\tag{7.5.61}
$$

The third assumption implies that

$$
u_z = \theta \psi(x, y),
\tag{7.5.62}
$$

where ψ denotes the warping function. If $\Theta = \theta z$ is very small compared to unity, $\Theta \ll 1$, the displacement field becomes (because $\cos \Theta \approx 0$ and $\sin \Theta \approx \Theta$)

$$
u_x = -\theta yz, \quad u_y = \theta xz, \quad u_z = \theta \psi(x, y).
\tag{7.5.63}
$$

Since we started with an assumed displacement field (a semi-inverse method), we only make sure that the equations of equilibrium are satisfied (and the compatibility equations are automatically met). Toward this end, we compute strains first and then stresses. The linear strain-displacement relations $(\varepsilon_{ij} = \frac{1}{2}(u_{i,j} + u_{j,i})$ give $\varepsilon_{xx} = 0$, $\varepsilon_{yy} = 0$, $\varepsilon_{zz} = 0$, $\varepsilon_{xy} = 0$, and

$$
\varepsilon_{xz} = \frac{\theta}{2}\left(\frac{\partial \psi}{\partial x} - y\right), \quad \varepsilon_{yz} = \frac{\theta}{2}\left(\frac{\partial \psi}{\partial y} + x\right).
\tag{7.5.64}
$$

The stresses are computed using the constitutive equations of an isotropic material, $\sigma_{ij} = 2\mu\varepsilon_{ij} + \lambda\varepsilon_{kk}\delta_{ij}$. We find that $\sigma_{xx} = \sigma_{yy} = \sigma_{zz} = \sigma_{xy} = 0$, and

$$\sigma_{xz} = \mu\theta\left(\frac{\partial\psi}{\partial x} - y\right), \quad \sigma_{yz} = \mu\theta\left(\frac{\partial\psi}{\partial y} + x\right). \tag{7.5.65}$$

Thus, a cross section of the cylinder experiences only the shear stresses σ_{xz} and σ_{yz}; the projected shear traction vector at a point (x, y) of a cross section is $\mathbf{t}(\hat{\mathbf{e}}_z) = \sigma_{xz}\,\hat{\mathbf{e}}_x + \sigma_{yz}\,\hat{\mathbf{e}}_y$.

Assuming that the body forces are negligible, the first two equilibrium equations $(\sigma_{ij,j} = 0)$ are trivially satisfied. The third equilibrium equation reduces to

$$\frac{\partial\sigma_{xz}}{\partial x} + \frac{\partial\sigma_{yz}}{\partial y} = 0 \quad\Rightarrow\quad \mu\theta\left(\frac{\partial^2\psi}{\partial x^2} + \frac{\partial^2\psi}{\partial y^2}\right) = 0 \text{ in } \Omega. \tag{7.5.66}$$

The boundary conditions on the lateral surface of the cylinder, that is, on the boundary Γ, where the unit normal is given by $\hat{\mathbf{n}} = n_x\hat{\mathbf{e}}_x + n_y\hat{\mathbf{e}}_y$, are $t_x = t_y = t_z = 0$. Because all but σ_{xz} and σ_{yz} are zero and $n_z = 0$, the boundary conditions $t_x = t_y = 0$ are trivially satisfied. The remaining boundary conditions $t_z = 0$ yield

$$t_z = \sigma_{xz}n_x + \sigma_{yz}n_y = \mu\theta\left(\frac{\partial\psi}{\partial x} - y\right)n_x + \mu\theta\left(\frac{\partial\psi}{\partial y} + x\right)n_y = 0. \tag{7.5.67}$$

From Fig. 7.5.10, we note that n_x and n_y can be calculated as

$$n_x = \frac{dy}{ds}, \quad n_y = -\frac{dx}{ds}, \quad \hat{\mathbf{n}} = \frac{dy}{ds}\hat{\mathbf{e}}_x - \frac{dx}{ds}\hat{\mathbf{e}}_y. \tag{7.5.68}$$

Then the boundary condition in Eq. (7.5.67) becomes

$$\left(\frac{\partial\psi}{\partial x} - y\right)\frac{dy}{ds} - \left(\frac{\partial\psi}{\partial y} + x\right)\frac{dx}{ds} = 0 \text{ on } \Gamma. \tag{7.5.69}$$

Thus, the boundary value problem becomes one of finding ψ such that

$$\nabla^2\psi = 0 \text{ in } \Omega, \quad \left(\frac{\partial\psi}{\partial x} - y\right)\frac{dy}{ds} - \left(\frac{\partial\psi}{\partial y} + x\right)\frac{dx}{ds} = 0 \text{ on } \Gamma. \tag{7.5.70}$$

Once $\psi(x, y)$ is known, the shear stresses can be computed from Eq. (7.5.65).

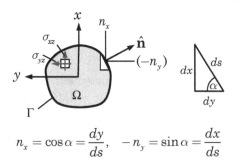

$$n_x = \cos\alpha = \frac{dy}{ds}, \quad -n_y = \sin\alpha = \frac{dx}{ds}$$

Fig. 7.5.10: Calculation of the direction cosines.

Example 7.5.8

Consider the case in which $\psi = 0$ and use the inverse method to determine the problem (i.e., cross section of the cylinder) for which it corresponds. Also, determine the stresses σ_{xz} and σ_{yz} as well as the projected shear stress magnitude in terms of the shear modulus $\mu = G$ and the applied torque T.

Solution: For $\psi = 0$, $\nabla^2 \psi = 0$ is trivially satisfied. The boundary condition in Eq. (7.5.69) becomes

$$y\frac{dy}{ds} + x\frac{dx}{ds} = 0 \quad \rightarrow \quad \frac{d}{ds}(x^2 + y^2) = 0 \text{ or } x^2 + y^2 = \text{constant}, c^2 \tag{1}$$

on the boundary Γ. This equation corresponds to that of a circle with Γ being the boundary of a circle of radius c and Ω being the interior of the circle; that is, the cross section of the cylinder is a circle of radius c.

The stresses are

$$\sigma_{xz} = -\mu\theta y, \quad \sigma_{yz} = \mu\theta x. \tag{2}$$

To express the stresses in terms of the torque, we write the equilibrium of moments about the z-axis. We obtain

$$T = \int_\Omega (x\sigma_{yz} - y\sigma_{xz})\, dx\, dy = \mu\theta \int_\Omega (x^2 + y^2)\, dx\, dy = \mu\theta c^2 \frac{\pi c^2}{2} = \mu\theta\frac{\pi c^4}{2}, \tag{3}$$

or $\mu\theta = 2T/\pi c^4$. Note that

$$\int_\Omega (x^2 + y^2)\, dx\, dy \equiv J$$

is the polar moment of inertia. Then the stresses in Eq. (2) can be expressed in terms of T as

$$\sigma_{xz} = -\frac{2T}{\pi c^4}y, \quad \sigma_{yz} = \frac{2T}{\pi c^4}x. \tag{4}$$

The projected shear stress magnitude at any point on the cross section is

$$\tau = |\mathbf{t}(\hat{\mathbf{e}}_z)| = \sqrt{\sigma_{xz}^2 + \sigma_{yz}^2} = \frac{2T}{\pi c^4}\sqrt{(x^2 + y^2)} = \frac{2Tr}{\pi c^4}. \tag{5}$$

Clearly, the maximum shear stress is $\tau_{\max} = \frac{2T}{\pi c^3}$.

The exact solutions of the torsion problem (7.5.70) are possible for elliptical and rectangular cross sections. For geometrically complicated cross sections, one must use numerical methods.

7.5.7.2 Prandtl's stress function

Here we begin with an assumed stress field. We note that the following stresses are identically zero for the problem:

$$\sigma_{xx} = \sigma_{yy} = \sigma_{zz} = \sigma_{xy} = 0. \tag{7.5.71}$$

Therefore, only stress equilibrium equation left to be satisfied is

$$\frac{\partial\sigma_{xz}}{\partial x} + \frac{\partial\sigma_{yz}}{\partial y} = 0. \tag{7.5.72}$$

We choose to satisfy this equation identically by introducing a stress function $\Psi(x, y)$, called the *Prandtl stress function*, such that

$$\sigma_{xz} = \frac{\partial \Psi}{\partial y}, \quad \sigma_{yz} = -\frac{\partial \Psi}{\partial x}. \tag{7.5.73}$$

Since we started with the stress field, the stress function Ψ is subject to satisfying the strain compatibility conditions in Eqs. (3.7.7) and (3.7.8), which can be expressed in terms of the stresses, as given in Eq. (7.2.27). For the case at hand, Eq. (7.2.27) takes the form $\sigma_{3\alpha,\beta\beta} = 0$, for $\alpha, \beta = 1, 2$:

$$\frac{\partial^2 \sigma_{xz}}{\partial x^2} + \frac{\partial^2 \sigma_{xz}}{\partial y^2} = 0 \;\Rightarrow\; \frac{\partial}{\partial y}\left(\frac{\partial^2 \Psi}{\partial x^2} + \frac{\partial^2 \Psi}{\partial y^2}\right) = 0,$$

$$\frac{\partial^2 \sigma_{yz}}{\partial x^2} + \frac{\partial^2 \sigma_{yz}}{\partial y^2} = 0 \;\Rightarrow\; \frac{\partial}{\partial x}\left(\frac{\partial^2 \Psi}{\partial x^2} + \frac{\partial^2 \Psi}{\partial y^2}\right) = 0. \tag{7.5.74}$$

From these two equations it follows that Ψ is governed by the equation

$$\nabla^2 \Psi = c, \tag{7.5.75}$$

where c is a constant. Equation (7.5.75) must be solved subject to the traction-free boundary condition on the lateral surface Γ:

$$\sigma_{xz} n_x + \sigma_{yz} n_y = \frac{\partial \Psi}{\partial y}\frac{dy}{ds} + \frac{\partial \Psi}{\partial x}\frac{dx}{ds} \equiv \frac{d\Psi}{ds} = 0 \;\text{ on } \Gamma, \tag{7.5.76}$$

That is, Ψ is a constant, say K, on Γ. For multiply connected cross sections, the constant K on different boundaries, in general, has different values. For simply connected cross sections, we can arbitrarily set the constant to zero, $K = $ because the constant does not contribute to the stress field. In summary, the Prandtl stress function is determined from solving the boundary value problem

$$\nabla^2 \Psi = c \;\text{ in } \Omega, \quad \Psi = 0 \;\text{ on } \Gamma. \tag{7.5.77}$$

The warping function $\psi(x, y)$ is related to the Prandtl stress function by

$$\frac{\partial \psi}{\partial x} = \frac{1}{\mu\theta}\frac{\partial \Psi}{\partial y} + y, \quad \frac{\partial \psi}{\partial y} = -\frac{1}{\mu\theta}\frac{\partial \Psi}{\partial x} - x. \tag{7.5.78}$$

The two equations in (7.5.78) can be combined by differentiating the first one with respect to y and the second one with respect to x and eliminating ψ to obtain

$$-\nabla^2 \Psi = 2\mu\theta \;\text{ in } \Omega, \quad \Psi = 0 \;\text{ on } \Gamma. \tag{7.5.79}$$

Once Ψ is known, the stresses can be determined from Eq. (7.5.73).

As in the case of the warping function, exact solutions of the torsion problem (7.5.79) are possible for a few simple cross sections. For geometrically complicated cross sections, one must use numerical methods. In general, solving Eq. (7.5.79) is simpler than solving Eq. (7.5.70) because of the complicated boundary condition in Eq. (7.5.69). To solve Eq. (7.5.79), one assumes Ψ to be in the form $\Psi = Af(x, y)$, where A is a constant and $f(x, y)$ is a sufficiently differentiable (i.e., $\nabla^2 f \neq 0$) function that is identically zero on the boundary. If $-\nabla^2 f$ is a nonzero constant c (so that Ac can be equated to $2\mu\theta$), we solve for the constant A and obtain the complete solution. If $\nabla^2 f$ is not a constant, an exact solution is not possible, although an approximate solution can be obtained. Next we consider an example.

Example 7.5.9

Consider a cylindrical shaft of elliptical cross section, Ω. The boundary Γ is the ellipse with semi-axes a and b:

$$\Gamma = \left\{(x,y) : \frac{x^2}{a^2} + \frac{y^2}{b^2} = 1\right\}. \tag{1}$$

Determine the Prandtl stress function and the shear stresses.

Solution: We select Ψ to be

$$\Psi(x,y) = A\left(\frac{x^2}{a^2} + \frac{y^2}{b^2} - 1\right), \tag{2}$$

where A is a constant to be determined such that Eq. (7.5.79)$_1$ is satisfied. Since the boundary condition $\Psi = 0$ on Γ is satisfied, we substitute Ψ from Eq. (2) into $-\nabla^2\Psi = 2\mu\theta$ and obtain

$$-2A\left(\frac{1}{a^2} + \frac{1}{b^2}\right) = 2\mu\theta \Rightarrow A = -\frac{\mu\theta a^2 b^2}{a^2 + b^2}. \tag{3}$$

The Prandtl stress function is then given by

$$\Psi(x,y) = \frac{\mu\theta a^2 b^2}{a^2 + b^2}\left(1 - \frac{x^2}{a^2} - \frac{y^2}{b^2}\right). \tag{4}$$

For solid cylinders of elliptic cross section, the twist per unit length θ can be related to the applied torque T by

$$T = \int_\Omega (x\sigma_{yz} - y\sigma_{xz})\, dx\, dy = -\int_\Omega \left(x\frac{\partial\Psi}{\partial x} + y\frac{\partial\Psi}{\partial y}\right) dx\, dy$$

$$= -\int_\Omega \left[\frac{\partial(x\Psi)}{\partial x} + \frac{\partial(y\Psi)}{\partial y}\right] dx\, dy + 2\int_\Omega \Psi(x,y)\, dx\, dy$$

$$= -\oint_\Gamma (x\Psi\, dy + y\Psi\, dx) + 2\int_\Omega \Psi(x,y)\, dx\, dy = 2\int_\Omega \Psi(x,y)\, dx\, dy, \tag{5}$$

where we used the fact that $\Psi = 0$ on Γ of a solid cylinder. For the problem at hand we obtain

$$T = 2\int_\Omega \Psi(x,y)\, dx\, dy = \frac{\pi\mu\theta a^3 b^3}{(a^2 + b^2)}. \tag{6}$$

Then the stresses σ_{xz} and σ_{yz} are calculated using Eq. (7.5.73) as

$$\sigma_{xz} = -\frac{2\mu\theta a^2}{a^2 + b^2}y = -\frac{2T}{\pi ab^3}y, \quad \sigma_{yz} = -\frac{2\mu\theta b^2}{a^2 + b^2}x = \frac{2T}{\pi a^3 b}x. \tag{7}$$

For $b < a$, the maximum shear stress occurs at $(x,y) = (0, \pm b)$, and the shear stress magnitude is

$$\tau_{max} = \frac{2\mu\theta a^2 b}{a^2 + b^2} = \frac{2T}{\pi ab^4}. \tag{8}$$

For solid circular cylinders, $b = a$, Eqs. (7) and (8) yield the same results as in Example 7.5.7 with $c = a = b$.

The warping function can be determined from Eq. (7.5.78)$_1$ by integrating with respect to x and setting the integration constant to zero:

$$\psi(x,y) = \frac{b^2 - a^2}{a^2 + b^2}xy \rightarrow u_z = \theta\psi(x,y) = \frac{b^2 - a^2}{a^2 + b^2}\theta xy = -\frac{(a^2 - b^2)T}{\mu\pi a^3 b^3}xy. \tag{9}$$

7.6 Methods Based on Total Potential Energy

7.6.1 Introduction

In Chapter 5 of this book, laws of physics (or conservation principles) and vector mechanics are used to derive the equations governing continua. These equations, as applied to solid bodies, can also be formulated by means of variational principles. Variational principles have played an important role in solid mechanics. The principle of minimum total potential energy, for example, can be regarded as a substitute for the equations of equilibrium of elastic bodies. Similarly, Hamilton's principle can be used in lieu of the equations governing dynamical systems, and the variational forms presented by Biot replace certain equations in linear continuum thermodynamics.

The use of variational principles makes it possible to concentrate in a single functional all of the intrinsic features of the problem at hand: the governing equations, the boundary conditions, initial conditions, constraint conditions, and even jump conditions. Variational principles can serve to derive not only the governing equations but they also suggest nature of the boundary conditions. Finally, and perhaps most importantly, variational principles provide a natural means for seeking approximate solutions; they are at the heart of the most powerful approximate methods in use in mechanics (e.g., the traditional Ritz and Galerkin methods, and the finite element method). In many cases they can also be used to establish upper and/or lower bounds on approximate solutions.

This section is devoted to the study of the principle of minimum total potential energy and its applications. To keep the scope of the chapter within reasonable limits, only key elements of the principle are presented here. Additional information can be found in the textbook by Reddy (2002).

7.6.2 The Variational Operator

Mathematically speaking, an integral of the form

$$I(u) = \int_\Omega F(\mathbf{x}, u, \nabla u) \, d\mathbf{x}$$

whose value is a real number, that is, I is a mapping that transforms functions u from a function space into a real number field, is called a *functional*. Note that $F(\mathbf{x}, u, \nabla u)$ does not qualify as a functional because it is a function and not a real number. An example of a functional is provided by the strain energy U of an elastic body. In particular,

$$U = \frac{EA}{2} \int_0^L \left(\frac{du}{dx} \right)^2 dx$$

is a functional.

As in the case of the minimum of an ordinary function $f(x)$, the minimum of a functional $I(u)$ involves differentiation with respect to the dependent variable(s). The derivative with respect to a dependent variable is known as the *Gâuteax derivative*, which is defined as

$$\delta F(u) \equiv \frac{d}{d\epsilon} F(u + \epsilon v)\Big|_{\epsilon=0} \tag{7.6.1}$$

and we say that $\delta F(u)$ is the first variation of the function $F(u)$ in the direction of v. The quantity ϵv is denoted as δu, and it is called the *first variation* of u. The operator δ itself is known as the *variational operator*.

The variational operator δ acts much like a total differential operator d, except that it operates with respect to the dependent variable(s) rather than the independent variables, like the coordinate x and time t. Indeed, the laws of *variation* of sums, products, ratios, and powers of functions of a dependent variable u are completely analogous to the corresponding laws of differentiation; that is, the variational calculus (i.e., calculus with δ) resembles the differential calculus. For example, if $F_1 = F_1(u)$ and $F_2 = F_2(u)$ are functions of a dependent variable u, we have

$$\delta(F_1 \pm F_2) = \delta F_1 \pm \delta F_2.$$
$$\delta(F_1\, F_2) = \delta F_1\, F_2 + F_1 \delta F_2.$$
$$\delta\left(\frac{F_1}{F_2}\right) = \frac{\delta F_1\, F_2 - F_1\, \delta F_2}{F_2^2}. \tag{7.6.2}$$
$$\delta(F_1)^n = n(F_1)^{n-1}\delta F_1.$$

If $G = G(u, v, w)$ is a function of several dependent variables u, v, and w, and possibly their derivatives, the total variation is the sum of partial variations:

$$\delta G = \delta_u G + \delta_v G + \delta_w G, \tag{7.6.3}$$

where, for example, δ_u denotes the partial variation with respect to u. The variational operator can be interchanged with differential and integral operators:

$$\delta(\nabla u) = \nabla(\delta u). \tag{7.6.4}$$

$$\delta\left(\int_\Omega u\, d\mathbf{x}\right) = \int_\Omega \delta u\, d\mathbf{x}. \tag{7.6.5}$$

Equations (7.6.2)–(7.6.5) are valid in multiple dimensions and for functions that depend on more than one dependent variable.

Similar to the necessary and sufficient conditions from the calculus of variations for the minimum of a functional, the conditions for the minimum of a functional are

$$\delta I = 0 \text{ (necessary condition)}, \tag{7.6.7}$$
$$\delta^2 I > 0 \text{ (sufficient condition)}. \tag{7.6.8}$$

When I denotes a certain energy functional in solid mechanics, the necessary condition (7.6.7) yields some associated governing equations, which are equivalent to those derived from the conservation principles of mechanics. However, Eq. (7.6.7) also gives the form of boundary conditions. The equations obtained in Ω from the necessary condition (7.6.7) for equilibrium problems are known as the *Euler equations* (or the *Euler–Lagrange equations* for dynamical systems) and those obtained on Γ (or on a portion of Γ) are known as the *natural boundary conditions*.

The variational principles of solid mechanics can be classified into three categories [see Oden and Reddy (1982) and Reddy (2002)]: (1) variational principles involving (energy) functionals that involve the primary variables such as displacements and temperature are called *primal principles*; (2) variational principles that are based on functionals containing the secondary variables such as stresses and heat flux are called *dual principles*; and (3) variational principles based on functionals that include both primary and secondary variables (e.g., both stresses and displacements, or stresses, strains, and displacements) are called *mixed principles*. In this section we consider the variational principle based on the *total potential energy functional* for linear elastic bodies that contains the displacement field as the dependent variables.

7.6.3 The Principle of the Minimum Total Potential Energy

7.6.3.1 Construction of the total potential energy functional

Recall from Sections 6.2 and 7.5 that for elastic bodies (in the absence of temperature variations) there exists a strain energy density function U_0 (measured per unit volume) such that [see Eq. (6.2.15)]

$$\boldsymbol{\sigma} = \frac{\partial U_0}{\partial \boldsymbol{\varepsilon}} \qquad \left(\sigma_{ij} = \frac{\partial U_0}{\partial \varepsilon_{ij}} \right). \tag{7.6.9}$$

The strain energy density U_0 is a function of strains at a point and is assumed to be positive definite. For linear elastic bodies (that is, obeying the generalized Hooke's law), the strain energy density is given by [see Eq. (6.3.1) or (7.6.9)]

$$U_0 = \frac{1}{2}\boldsymbol{\sigma} : \boldsymbol{\varepsilon} = \frac{1}{2}\sigma_{ij}\varepsilon_{ij} = \frac{1}{2}c_{ijkl}\varepsilon_{ij}\varepsilon_{kl}. \tag{7.6.10}$$

Hence, the total strain energy of the body $\mathcal{B}$ occupying volume Ω is given by

$$U = \int_\Omega U_0(\varepsilon_{ij}) \, d\mathbf{x} = \frac{1}{2}\int_\Omega \boldsymbol{\sigma} : \boldsymbol{\varepsilon} \, d\mathbf{x} = \frac{1}{2}\int_\Omega \sigma_{ij}\varepsilon_{ij} \, d\mathbf{x}. \tag{7.6.11}$$

The total work done by applied body force $\mathbf{f}$ and surface force $\mathbf{t}$ is given by [see Eq. (7.6.11)]

$$V = -\left[\int_\Omega \mathbf{f} \cdot \mathbf{u} \, d\mathbf{x} + \oint_\Gamma \mathbf{t} \cdot \mathbf{u} \, ds \right], \tag{7.6.12}$$

where the minus sign in the expression for V indicates that the work is expended, whereas U in Eq. (7.6.11) is the available strain energy stored in body $\mathcal{B}$. The total potential energy (functional) of body $\mathcal{B}$ is the sum of the strain energy stored in the body and the work done by external forces

$$\begin{aligned}
\Pi = U + V &= \frac{1}{2}\int_\Omega \boldsymbol{\sigma} : \boldsymbol{\varepsilon} \, d\mathbf{x} - \left[\int_\Omega \mathbf{f} \cdot \mathbf{u} \, d\mathbf{x} + \oint_\Gamma \mathbf{t} \cdot \mathbf{u} \, ds \right] \\
&= \frac{1}{2}\int_\Omega \sigma_{ij}\varepsilon_{ij} \, d\mathbf{x} - \left[\int_\Omega f_i u_i \, d\mathbf{x} + \oint_\Gamma t_i u_i \, ds \right].
\end{aligned} \tag{7.6.13}$$

The *principle of minimum total potential energy* can be stated as follows: If a body is in equilibrium, of all *admissible* displacement fields $\mathbf{u}$ the one $\mathbf{u}_0$ that makes the total potential energy a minimum corresponds to the equilibrium solution:

$$\Pi(\mathbf{u}_0) \leq \Pi(\mathbf{u}). \tag{7.6.14}$$

An admissible displacement is the one that satisfies the geometric constraints of the problem.

7.6.3.2 Euler's equations and natural boundary conditions

Here, we illustrate how the Navier equations of elasticity, Eq. (7.2.17) and the traction boundary conditions in Eq. (7.2.18), can be derived as the Euler equations using the principle of minimum total potential energy. Consider a linear elastic body $\mathcal{B}$ occupying volume Ω with boundary Γ and subjected to body force $\mathbf{f}$ (measured per unit volume) and surface traction $\hat{\mathbf{t}}$ on portion Γ_σ of the surface. We assume that the displacement vector $\mathbf{u}$ is specified to be $\hat{\mathbf{u}}$ on the remaining portion, Γ_u, of the boundary $(\Gamma = \Gamma_u \cup \Gamma_\sigma)$. Therefore, $\delta\mathbf{u} = \mathbf{0}$ on Γ_u.

The total potential energy functional is given by (summation on repeated indices is implied throughout this discussion)

$$\Pi(\mathbf{u}) = \int_\Omega \left(\frac{1}{2}\sigma_{ij}\,\varepsilon_{ij} - f_i\,u_i \right) dx - \int_{\Gamma_\sigma} \hat{t}_i\,u_i\,ds, \tag{7.6.15}$$

The first term under the volume integral represents the strain energy density of the elastic body, the second term represents the work done by the body force $\mathbf{f}$, and the third term represents the work done by the specified traction $\hat{\mathbf{t}}$.

The strain-displacement relations and stress–strain relations for an isotropic elastic body are given by Eqs. (7.2.1) and (7.2.9), respectively. Substituting Eqs. (7.2.1) and (7.2.9) into Eq. (7.6.15), we obtain

$$\Pi(\mathbf{u}) = \int_\Omega \left[\frac{\mu}{4}(u_{i,j} + u_{j,i})(u_{i,j} + u_{j,i}) + \frac{\lambda}{2}u_{i,i}\,u_{k,k} - f_i\,u_i \right] dx$$
$$- \int_{\Gamma_\sigma} \hat{t}_i u_i\,ds. \tag{7.6.16}$$

Setting the first variation of Π to zero (that is, using the principle of minimum total potential energy), we obtain

$$0 = \int_\Omega \left[\frac{\mu}{2}(\delta u_{i,j} + \delta u_{j,i})(u_{i,j} + u_{j,i}) + \lambda\delta u_{i,i}u_{k,k} - f_i\delta u_i \right] dx$$
$$- \int_{\Gamma_\sigma} \hat{t}_i\,\delta u_i\,ds, \tag{7.6.17}$$

wherein the product rule of variation is used and similar terms are combined. Next, we use the component form of the gradient theorem to relieve δu_i of any derivative so that we can use the fundamental lemma of variational calculus to set the coefficients of δu_i to zero in Ω and on the portion of Γ where δu_i is arbitrary. Using the gradient theorem, we can write

$$\int_\Omega \delta u_{i,j} \left(u_{i,j} + u_{j,i}\right) d\mathbf{x} = -\int_\Omega \delta u_i \left(u_{i,j} + u_{j,i}\right)_{,j} d\mathbf{x} + \oint_\Gamma \delta u_i \left(u_{i,j} + u_{j,i}\right) n_j \, ds,$$

where n_j denotes the jth direction cosine of the unit normal vector to the surface $\hat{n}$. Using this result in Eq. (7.6.17) we arrive at

$$0 = \int_\Omega \left[-\frac{\mu}{2} \left(u_{i,j} + u_{j,i}\right)_{,j} \delta u_i - \frac{\mu}{2} \left(u_{i,j} + u_{j,i}\right)_{,i} \delta u_j - \lambda u_{k,ki} \delta u_i - f_i \delta u_i\right] d\mathbf{x}$$

$$+ \oint_\Gamma \left[\frac{\mu}{2} \left(u_{i,j} + u_{j,i}\right) \left(n_j \delta u_i + n_i \delta u_j\right) + \lambda u_{k,k} n_i \, \delta u_i\right] ds - \int_{\Gamma_\sigma} \delta u_i \hat{t}_i \, ds$$

$$= \int_\Omega \left[-\mu \left(u_{i,j} + u_{j,i}\right)_{,j} - \lambda u_{k,ki} - f_i\right] \delta u_i \, d\mathbf{x}$$

$$+ \oint_\Gamma \left[\mu \left(u_{i,j} + u_{j,i}\right) + \lambda u_{k,k} \, \delta_{ij}\right] n_j \, \delta u_i \, ds - \int_{\Gamma_\sigma} \delta u_i \, \hat{t}_i \, ds. \qquad (7.6.18)$$

In arriving at the last step, a change of dummy indices is made to combine terms.

Recognizing that the expression inside the square brackets of the closed surface integral is nothing but σ_{ij} and $\sigma_{ij} n_j = t_i$ by Cauchy's formula, we can write

$$\oint_\Gamma \left[\mu \left(u_{i,j} + u_{j,i}\right) + \lambda u_{k,k} \delta_{ij}\right] n_j \delta u_i \, ds = \oint_\Gamma t_i \, \delta u_i \, ds.$$

This boundary expression resulting from the "integration-by-parts" to relieve $\delta\mathbf{u}$ of any derivatives is used to classify the variables of the problem. The coefficient of δu_i is called the *secondary variable*, and the varied quantity itself (without the variational symbol) is called the *primary variable*. Thus, u_i is the primary variable and t_i is the corresponding secondary variable. They always appear in pairs, and only one element of the pair may be specified at any boundary point. Specification of a primary variable is called the *essential boundary condition* and specification of a secondary variable is termed the *natural boundary condition*. They are also known as the geometric and force boundary conditions, respectively. In applied mathematics, they are known as the *Dirichlet boundary condition* and the *Neumann boundary condition*, respectively.

Returning to the boundary integral, it can be expressed as the sum of integrals on Γ_u and Γ_σ:

$$\oint_\Gamma t_i \delta u_i \, ds = \int_{\Gamma_u} t_i \delta u_i \, ds + \int_{\Gamma_\sigma} t_i \delta u_i \, ds = \int_{\Gamma_\sigma} t_i \delta u_i \, ds.$$

The integral over Γ_u is set to zero because $\mathbf{u}$ is specified there, that is, $\delta\mathbf{u} = \mathbf{0}$. Hence, Eq. (7.6.18) becomes

$$0 = \int_\Omega \left[-\mu \left(u_{i,j} + u_{j,i}\right)_{,j} - \lambda u_{k,ki} - f_i\right] \delta u_i \, d\mathbf{x} + \int_{\Gamma_\sigma} \delta u_i \left(t_i - \hat{t}_i\right) ds. \qquad (7.6.19)$$

Using the fundamental lemma of calculus of variations, we set the coefficients of δu_i in Ω and δu_i on Γ_σ from Eq. (7.6.19) to zero separately and obtain

$$\mu\, u_{i,jj} + (\mu + \lambda)u_{k,ki} + f_i = 0 \text{ in } \Omega, \tag{7.6.20}$$

$$n_j \sigma_{ij} - \hat{t}_i = 0 \text{ on } \Gamma_\sigma, \tag{7.6.21}$$

for $i = 1, 2, 3$. Equation (7.6.20) represents the Navier equations of elasticity (7.2.17), and the natural boundary conditions (7.6.21) are the same as the traction boundary conditions listed in Eq.(7.2.18).

7.6.3.3 Minimum property of the total potential energy functional

To show that the total potential energy of a linear elasticity body is indeed the minimum at its equilibrium configuration, we consider the total potential energy functional [more general than the one considered in Eq. (7.6.14)]:

$$\Pi(\mathbf{u}) = \int_\Omega \left(\tfrac{1}{2} C_{ijk\ell}\, \varepsilon_{k\ell}\, \varepsilon_{ij} - f_i\, u_i \right) d\mathbf{x} - \int_{\Gamma_\sigma} \hat{t}_i\, u_i\, ds, \tag{7.6.22}$$

where $C_{ijk\ell}$ are the components of the fourth-order elasticity tensor.

Let $\mathbf{u}$ be the true displacement field and $\bar{\mathbf{u}}$ be an arbitrary but admissible displacement field. Then $\bar{\mathbf{u}}$ is of the form

$$\bar{\mathbf{u}} = \mathbf{u} + \alpha \mathbf{v},$$

where α is a real number and $\mathbf{v}$ is a sufficiently differentiable function that satisfies the homogeneous form of the essential boundary condition $\mathbf{v} = \mathbf{0}$ on Γ_u. Then $\Pi(\bar{\mathbf{u}})$ is given by

$$\Pi(\mathbf{u} + \alpha \mathbf{v}) = \int_\Omega \left[\tfrac{1}{2} C_{ijk\ell} \left(\varepsilon_{k\ell} + \alpha g_{k\ell} \right)\left(\varepsilon_{ij} + \alpha g_{ij} \right) - f_i \left(u_i + \alpha v_i \right) \right] d\mathbf{x}$$

$$- \int_{\Gamma_\sigma} \hat{t}_i (u_i + \alpha v_i)\, ds,$$

where

$$g_{ij} = \tfrac{1}{2}(v_{i,j} + v_{j,i}).$$

Collecting the terms, we obtain (because $C_{ijk\ell} = C_{k\ell ij}$)

$$\Pi(\bar{\mathbf{u}}) = \Pi(\mathbf{u}) + \alpha \left[\int_\Omega \left(-f_i v_i + C_{ijk\ell}\varepsilon_{k\ell}g_{ij} + \tfrac{1}{2}\alpha\, C_{ijk\ell}g_{ij}g_{k\ell} \right) d\mathbf{x} - \int_{\Gamma_\sigma} \hat{t}_i v_i\, ds \right], \tag{7.6.23}$$

Using the equilibrium equations (7.2.5) and the generalized Hooke's law $\sigma_{ij} = C_{ijk\ell}\varepsilon_{k\ell}$ we can write

$$-\int_\Omega f_i v_i\, d\mathbf{x} = \int_\Omega \sigma_{ij,j} v_i\, d\mathbf{x} = \int_\Omega C_{ijk\ell}\, \varepsilon_{k\ell,j}\, v_i\, d\mathbf{x}$$

$$= -\int_\Omega C_{ijk\ell}\varepsilon_{k\ell}v_{i,j}\, d\mathbf{x} + \int_{\Gamma_\sigma} C_{ijk\ell}\, \varepsilon_{k\ell}\, v_i\, n_j\, ds$$

$$= -\int_\Omega C_{ijk\ell}\, \varepsilon_{k\ell}\, g_{ij}\, d\mathbf{x} + \int_{\Gamma_\sigma} \hat{t}_i v_i\, ds, \tag{7.6.24}$$

where the condition $v_i = 0$ on Γ_u is used in arriving at the last step. Substituting Eq. (7.6.24) into Eq. (7.6.23), we arrive at

$$\Pi(\bar{\mathbf{u}}) = \Pi(\mathbf{u}) + \frac{\alpha^2}{2} \int_\Omega C_{ijk\ell}\, g_{ij}\, g_{k\ell}\, d\mathbf{x}. \qquad (7.6.25)$$

In view of the nonnegative nature of the second term on the right-hand side of Eq. (7.6.23), it follows that

$$\Pi(\bar{\mathbf{u}}) \geq \Pi(\mathbf{u}), \qquad (7.6.26)$$

and $\Pi(\bar{\mathbf{u}}) = \Pi(\mathbf{u})$ only if the quadratic expression $\frac{1}{2}C_{ijk\ell}g_{ij}g_{k\ell}$ is zero. Owing to the positive-definiteness of the strain energy density, the quadratic expression is zero only if $v_i = 0$, which in turn implies $\bar{u}_i = u_i$. Thus, Eq. (7.6.26) implies that of all admissible displacement fields the body can assume, the true one is that which makes the total potential energy a minimum. Next, we consider an example to illustrate the use of the principle of minimum total potential energy.

Example 7.6.1

Consider the bending of a beam according to the Euler–Bernoulli beam theory (see Section 7.3.4). Construct the total potential energy functional and then determine the governing equation and boundary conditions of the problem.

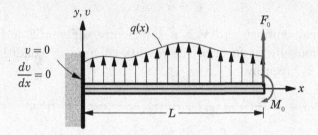

Fig. 7.6.1: A beam with applied loads.

Solution: The total potential energy of a cantilever beam under pure bending by distributed transverse force $q(x)$ and point load F_0 (see Fig. 7.6.1) with the assumption of small strains and displacements for the linear elastic case (i.e., obeys Hooke's law) is given by

$$\Pi(v) = \frac{1}{2}\int_0^L \left[EI\left(\frac{d^2v}{dx^2}\right)^2 \right] dx - \left[\int_0^L q(x)v(x)\, dx + F_0 v(L) + M_0\left(-\frac{dv}{dx}\right)_{x=L} \right], \qquad (7.6.27)$$

where L is the length, A the cross-sectional area, I moment of inertia about the axis (y) of bending, and E is Young's modulus of the beam. The first term represents the strain energy U (see Example 7.4.3); the second term represents the work done by the applied distributed load $q(x)$ in moving through the deflection $v(x)$; the third terms represents the work done by the point load F_0 in moving through the displacement $v(L)$; and the last term represents the work done by moment M_0 in moving through the rotation $\theta_x(L) = (-\frac{dv}{dx})_{x=L}$.

Applying the principle of minimum total potential energy, $\delta\Pi = 0$, we obtain

$$0 = \delta\Pi = \int_0^L EI\frac{d^2v}{dx^2}\frac{d^2\delta v}{dx^2}\, dx - \left[\int_0^L q\delta v\, dx + F_0\delta v(L) + M_0\left(-\frac{d\delta v}{dx}\right)_{x=L} \right]. \qquad (1)$$

Next, we carry out integration-by-parts on the first term to relieve δv of any derivative so that we can use the fundamental lemma of variational calculus to obtain the Euler equation:

$$0 = \int_0^L \frac{d^2}{dx^2}\left(EI\frac{d^2 v}{dx^2}\right)\delta v\, dx + \left[EI\frac{d^2 v}{dx^2}\frac{d\delta v}{dx} - \frac{d}{dx}\left(EI\frac{d^2 v}{dx^2}\right)\delta v\right]_0^L$$
$$- \left[\int_0^L q\delta v\, dx + F_0\delta v(L) + M_0\left(-\frac{d\delta v}{dx}\right)_{x=L}\right]. \tag{2}$$

The boundary terms resulting from integration-by-parts allows us to classify the boundary conditions of the problem. The quantities with δ, δv, and $\delta(dv/dx)$ in the boundary terms, indicate that v and (dv/dx) (removing the variational operator from the quantities) are the quantities whose specification constrains the beam geometrically. These variables are called the *primary variables*:

$$v;\qquad \frac{dv}{dx}. \tag{7.6.28}$$

Thus, the deflection v and slope (or rotation) dv/dx are the primary variables of the problem. The expressions that are coefficients of δv and $\delta(dv/dx)$ in the boundary terms are called the *secondary variables*:

$$\delta w:\quad \frac{d}{dx}\left(EI\frac{d^2 v}{dx^2}\right);\quad \delta\left(\frac{dv}{dx}\right):\quad EI\frac{d^2 v}{dx^2}. \tag{7.6.29}$$

It is clear that the secondary variables are nothing but the shear force $V(x) = dM/dx$ and bending moment $M(x)$

$$V(x) = -\frac{d}{dx}\left(EI\frac{d^2 v}{dx^2}\right);\quad M(x) = -EI\frac{d^2 v}{dx^2}. \tag{7.6.30}$$

Only one element of each of the pairs (v, V) and $(dv/dx, M)$ may be specified at a point. Note that the identification of the primary and secondary variables is unique. Specifying a primary variable constitutes a *geometric* or *essential* boundary condition, and specification of a secondary variable constitutes a *force* or *natural* boundary condition.

Returning to the expression in Eq. (2), first we collect the coefficients of δv in $(0, L)$ together and set them to zero, because δv is arbitrary in $(0, L)$. We obtain the Euler equation

$$\frac{d^2}{dx^2}\left(EI\frac{d^2 v}{dx^2}\right) - q(x) = 0,\quad 0 < x < L. \tag{7.6.31}$$

Equation (7.6.31) can also be derived from vector mechanics by considering an element of the beam and summing the forces and moments, and then relating the bending moment M to the deflection v, as discussed in Section 7.3.4.

Now considering all boundary terms in Eq. (2), we conclude that

$$\left[\frac{d}{dx}\left(EI\frac{d^2 v}{dx^2}\right)\right]_{x=0}\delta v(0) = 0,\quad \left[-\frac{d}{dx}\left(EI\frac{d^2 v}{dx^2}\right) - F_0\right]_{x=L}\delta v(L) = 0, \tag{3}$$

$$\left(EI\frac{d^2 v}{dx^2}\right)_{x=0}\left(\frac{d\delta v}{dx}\right)_{x=0} = 0,\quad \left(EI\frac{d^2 v}{dx^2} + M_0\right)_{x=L}\left(\frac{d\delta v}{dx}\right)_{x=L} = 0. \tag{4}$$

If either of the quantities δw and $(d\delta v/dx)$ is zero at $x = 0$ or $x = L$, if v or (dv/dx) is specified, the corresponding variations vanish because a specified quantity cannot be varied; the vanishing of the coefficients of δv and $(d\delta v/dx)$ at points where the geometric boundary conditions are not specified provides the natural boundary conditions. Various combinations of one variable from each of the pairs (v, V) and (θ_x, M), where $\theta_x = -(dv/dx)$, define beams with different boundary conditions.

As an example, suppose that the beam is clamped (i.e., fixed or built-in) at $x = 0$ and free at $x = L$ (a cantilever beam), as shown in Fig. 7.6.1. Then, $\delta v(0) = 0$ and $(d\delta v(0)/dx) = 0$, and the corresponding secondary variables, namely the shear force and bending moment, are unknown there. Since the free end, $x = L$, is subjected to an upward transverse force F_0 and clockwise bending moment M_0, the force or natural boundary conditions become

$$\left[-\frac{d}{dx}\left(EI\frac{d^2v}{dx^2} \right) - F_0 \right]_{x=L} = 0, \quad \left(EI\frac{d^2v}{dx^2} + M_0 \right)_{x=L} = 0. \tag{5}$$

Since the secondary variables are known at $x = L$, we will not know the corresponding primary variables until the problem is solved. Another example is provided by a simply supported (or hinged at both ends) beam without any applied moments at the supports. Then we have the following boundary conditions:

$$v(0) = 0, \quad \left(EI\frac{d^2v}{dx^2} \right)_{x=0} = 0; \quad v(L) = 0, \quad \left(EI\frac{d^2v}{dx^2} \right)_{x=L} = 0. \tag{6}$$

7.6.4 Castigliano's theorem I

Suppose that the displacement field of a solid body can be expressed in terms of the displacements of a finite number of points $\mathbf{x}_i$ $(i = 1, 2, \cdots N)$ as

$$\mathbf{u}(\mathbf{x}) = \sum_{i=1}^{N} \mathbf{u}_i\, \phi_i(\mathbf{x}), \tag{7.6.35}$$

where $\mathbf{u}_i$ are unknown displacement parameters, called *generalized displacements*, and ϕ_i are known functions of position, called *interpolation functions* with the property that ϕ_i is unity at the ith point (i.e., $\mathbf{x} = \mathbf{x}_i$) and zero at all other points ($\mathbf{x}_j$, $j \neq i$). Then it is possible to represent the strain energy U and potential energy V due to applied loads in terms of the generalized displacements $\mathbf{u}_i$. Then the principle of minimum total potential energy can be written as

$$\delta\Pi = \delta U + \delta V = 0 \;\Rightarrow\; \delta U = -\delta V \quad \text{or} \quad \frac{\partial U}{\partial \mathbf{u}_i}\cdot\delta\mathbf{u}_i = -\frac{\partial V}{\partial \mathbf{u}_i}\cdot\delta\mathbf{u}_i, \tag{7.6.36}$$

where sum on repeated indices is implied. Since

$$\frac{\partial V}{\partial \mathbf{u}_i} = -\mathbf{F}_i$$

and $\delta\mathbf{u}_i$ are arbitrary, it follows that

$$\left(\frac{\partial U}{\partial \mathbf{u}_i} - \mathbf{F}_i \right)\cdot\delta\mathbf{u}_i = 0 \quad \text{or} \quad \frac{\partial U}{\partial \mathbf{u}_i} = \mathbf{F}_i. \tag{7.6.37}$$

Equation (7.6.37) is known as Castigliano's theorem I.

When applied to a structure loaded by generalized point loads $\mathbf{F}_i$ with associated generalized displacements $\mathbf{u}_i$, both having the same sense, Castigliano's theorem I gives

$$\frac{\partial U}{\partial \mathbf{u}_i} = \mathbf{F}_i. \tag{7.6.38}$$

It is clear from the derivation that Castigliano's theorem I is a special case of the principle of minimum total potential energy.

Application of Castigliano's theorem I to structural members (trusses and frames) can be found in many books [see Reddy (2002) and references therein]. In Example 7.6.2, an application of Castigliano's theorem I to beams is illustrated.

Example 7.6.2 ──────────────

Consider a straight beam of length L and constant bending stiffness EI (modulus E and moment of inertia I about the axis of bending y). If Δ_i are the generalized displacements and Q_i are the generalized point loads at the ends of the beam segment, as shown in Fig. 7.6.2, use Castigliano's theorem I to establish a relationship between the generalized displacements and generalized forces.

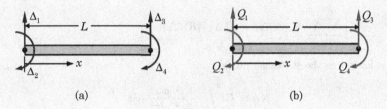

(a)   (b)

Fig. 7.6.2: (a) Generalized displacements. (b) Generalized forces.

Solution: The equilibrium equation of a beam segment according to the Euler–Bernoulli beam theory (see Example 7.6.1) is

$$EI\frac{d^4v}{dx^4} = 0. \tag{7.6.39}$$

The exact solution to this fourth-order equation is

$$v(x) = c_1 + c_2 x + c_3 x^2 + c_4 x^3, \tag{7.6.40}$$

where c_i $(i = 1, 2, 3, 4)$ are constants of integration, which we express in terms of the deflections and rotations at the two ends of an element beam of length L. Let

$$\Delta_1 \equiv v(0) = c_1, \quad \Delta_2 \equiv \left(-\frac{dv}{dx}\right)_{x=0} = -c_2,$$

$$\Delta_3 \equiv v(L) = c_1 + c_2 L + c_3 L^2 + c_4 L^3, \tag{7.6.41}$$

$$\Delta_4 \equiv \left(-\frac{dv}{dx}\right)_{x=L} = -c_2 - 2c_3 L - 3c_4 L^2.$$

Clearly, Δ_1 and Δ_3 are the values of the transverse deflection v at $x = 0$ and $x = L$, respectively, and Δ_2 and Δ_4 are the rotations $-dv/dx$, measured positive clockwise, at $x = 0$ and $x = L$, respectively; see Fig. 7.6.2(b).

The reason for picking two deflection values and two rotations, as opposed to four deflections at four points of the beam, needs to be understood. From Example 7.6.1, it is clear that both v and dv/dx are the primary (kinematic) variables, which must be continuous at every point of the beam. If we were to join two such beam segments (possibly made of different bending stiffness EI), the kinematic variables can be made continuous by equating the like degrees of freedom at the point common to the two segments.

The four equations in Eq. (7.6.41) can be solved for c_i in terms of Δ_i, called *generalized displacements*, which will serve as the generalized coordinates for the application of Castigliano's theorem I. Substituting the result into Eq. (7.6.40) yields

$$v(x) = \phi_1(x)\Delta_1 + \phi_2(x)\Delta_2 + \phi_3(x)\Delta_3 + \phi_4(x)\Delta_4 = \sum_{i=1}^{4} \phi_i(x)\Delta_i, \tag{7.6.42}$$

where $\phi_i(x)$ $(i = 1, 2, 3, 4)$ are known as the *Hermite cubic polynomials*

$$\phi_1(x) = 1 - 3\left(\frac{x}{L}\right)^2 + 2\left(\frac{x}{L}\right)^3, \qquad \phi_3(x) = \left(\frac{x}{L}\right)^2\left(3 - 2\frac{x}{L}\right),$$

$$\phi_2(x) = -x\left[1 - 2\left(\frac{x}{L}\right) + \left(\frac{x}{L}\right)^2\right], \quad \phi_4(x) = x\frac{x}{L}\left(1 - \frac{x}{L}\right). \tag{7.6.43}$$

We note that Eq. (7.6.42) has the same form as Eq. (7.6.35).

The strain energy of the beam now can be expressed in terms of the generalized coordinates Δ_i $(i = 1, 2, 3, 4)$ as

$$U = \frac{EI}{2}\int_0^L \left(\frac{d^2v}{dx^2}\right)^2 dx = \frac{EI}{2}\int_0^L \left(\sum_{i=1}^{4}\Delta_i\frac{d^2\phi_i}{dx^2}\right)\left(\sum_{j=1}^{4}\Delta_j\frac{d^2\phi_j}{dx^2}\right)dx$$

$$= \frac{1}{2}\sum_{i=1}^{4}\sum_{j=1}^{4}K_{ij}\Delta_i\Delta_j = \tfrac{1}{2}\{\Delta\}^{\mathrm{T}}[K]\{\Delta\} \tag{7.6.44}$$

where $[K]$ is known as the stiffness matrix

$$K_{ij} = EI\int_0^L \frac{d^2\phi_i}{dx^2}\frac{d^2\phi_j}{dx^2}\,dx. \tag{7.6.45}$$

Note that $[K]$ is symmetric ($K_{ij} = K_{ji}$). By carrying out the indicated integration, K_{ij} can be evaluated, as will be shown shortly.

Although we assumed that there is no distributed transverse load on the beam, as per Eq. (7.6.39), if there were a distributed load $q(x)$, acting upward, it can be converted to statically equivalent generalized point loads at the end points of the beam segment by

$$q_i = \int_0^L q(x)\phi_i(x)\,dx, \quad i = 1, 2, 3, 4. \tag{7.6.46}$$

The transverse point loads q_1 and q_3 and bending moments q_2 and q_4 together are statically equivalent (that is, satisfy the force and moment equilibrium conditions of the beam) to the distributed load $q(x)$ on the beam, as shown in Fig. 7.6.3(a). We distinguish between q_i and Q_i, because the latter are generalized point loads that are not due to the distributed load, $q(x)$; Q_i are the reactions at the ends of the beam, as shown in Fig. 7.6.3(b). The work done by external loads is

$$V = -\sum_{i=1}^{4}(q_i + Q_i)\Delta_i. \tag{7.6.47}$$

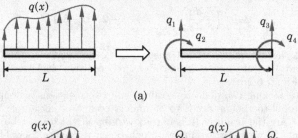

(a)

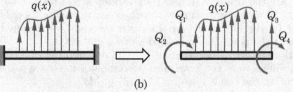

(b)

Fig. 7.6.3: (a) Statically equivalent generalized loads q_i due to distributed load $q(x)$ and (b) Generalized reaction forces.

Using Castigliano's theorem I, we obtain the required relations between the generalized displacements $\{\Delta\}$ and generalized forces $\{Q\}$

$$\frac{\partial U}{\partial \Delta_i} = -\frac{\partial V}{\partial \Delta_i} \quad \Rightarrow \quad \sum_{j=1}^{4} K_{ij}\Delta_j = Q_i + q_i \quad \text{or} \quad [K]\{\Delta\} = \{q\} + \{Q\},$$

or, in explicit matrix form

$$\frac{2EI}{L^3} \begin{bmatrix} 6 & -3L & -6 & -3L \\ -3L & 2L^2 & 3L & L^2 \\ -6 & 3L & 6 & 3L \\ -3L & L^2 & 3L & 2L^2 \end{bmatrix} \begin{Bmatrix} \Delta_1 \\ \Delta_2 \\ \Delta_3 \\ \Delta_4 \end{Bmatrix} = \begin{Bmatrix} q_1 \\ q_2 \\ q_3 \\ q_4 \end{Bmatrix} + \begin{Bmatrix} Q_1 \\ Q_2 \\ Q_3 \\ Q_4 \end{Bmatrix}. \tag{7.6.48}$$

It can be verified that the stiffness matrix $[K]$ is singular because the rigid-body motions (that is, rigid-body translation and rotation) of the beam segment are not eliminated.

Example 7.6.3

Consider a beam fixed at $x = 0$ (this geometric condition eliminates the rigid-body motion), supported at $x = L$ by a linear elastic spring with spring constant k, subjected to uniformly distributed load of intensity q_0, and clockwise bending moment M_0 at $x = L$, as shown in Fig. 7.6.4. Determine the elongation $w(L)$ in the spring.

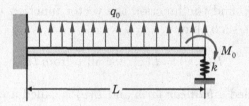

Fig. 7.6.4: A beam fixed at $x = 0$ and supported by a spring at $x = L$.

Solution: The geometric boundary conditions at $x = 0$ require that $\Delta_1 = \Delta_2 = 0$. These conditions remove the rigid-body modes of vertical translation and rotation about the y-axis. The force boundary conditions at $x = L$ require $Q_3 = -F_s = -kv(L) = -k\Delta_3$ and $Q_4 = M_0$. For uniformly distributed load acting upward, $q(x) = q_0$, the load vector $\{q\}$ is given by

$$\begin{Bmatrix} q_1 \\ q_2 \\ q_3 \\ q_4 \end{Bmatrix} = \frac{q_0 L}{12} \begin{Bmatrix} 6 \\ -L \\ 6 \\ L \end{Bmatrix}. \tag{7.6.49}$$

Then we have

$$\frac{2EI}{L^3} \begin{bmatrix} 6 & -3L & -6 & -3L \\ -3L & 2L^2 & 3L & L^2 \\ -6 & 3L & 6 & 3L \\ -3L & L^2 & 3L & 2L^2 \end{bmatrix} \begin{Bmatrix} 0 \\ 0 \\ \Delta_3 \\ \Delta_4 \end{Bmatrix} = \frac{q_0 L}{12} \begin{Bmatrix} 6 \\ -L \\ 6 \\ L \end{Bmatrix} + \begin{Bmatrix} Q_1 \\ Q_2 \\ -k\Delta_3 \\ M_0 \end{Bmatrix}. \tag{7.6.50}$$

Thus, there are four equations in four unknowns, Q_1, Q_2, Δ_3, and Δ_4. Since the last two equations contain Δ_3 and Δ_4 as the only unknowns, we can write

$$\begin{bmatrix} \frac{12EI}{L^3} + k & \frac{6EI}{L^2} \\ \frac{6EI}{L^2} & \frac{4EI}{L} \end{bmatrix} \begin{Bmatrix} \Delta_3 \\ \Delta_4 \end{Bmatrix} = \frac{q_0 L}{12} \begin{Bmatrix} 6 \\ L \end{Bmatrix} + \begin{Bmatrix} 0 \\ M_0 \end{Bmatrix}. \tag{7.6.51}$$

Solving for $\Delta_3 = v(L)$ and $\Delta_4 = -(dv/dx)(L)$, we obtain

$$\Delta_3 = \left(q_0 L^2 - 4M_0\right) \frac{3L^2}{8EI\,(3+\alpha)}, \quad \alpha = \frac{kL^3}{EI},$$

$$\Delta_4 = -\frac{q_0 L^3}{48EI} \frac{(24-\alpha)}{(3+\alpha)} + \frac{M_0 L}{4EI} \frac{(12+\alpha)}{(3+\alpha)}. \tag{7.6.52}$$

The reactions at the fixed end can be determined using the first two equations in Eq. (7.6.50):

$$\left\{ \begin{array}{c} Q_1 \\ Q_2 \end{array} \right\} = \frac{2EI}{L^3} \begin{bmatrix} -6 & -3L \\ 3L & L^2 \end{bmatrix} \left\{ \begin{array}{c} \Delta_3 \\ \Delta_4 \end{array} \right\} - \frac{q_0 L}{12} \left\{ \begin{array}{c} 6 \\ -L \end{array} \right\}. \tag{7.6.53}$$

The solution obtained in Eqs. (7.6.52) and (7.6.53) is exact because the representation in Eq. (7.6.42) is the exact solution of Eq. (7.6.31) when EI is a constant and the distributed load $q(x)$ is replaced by statically equivalent point forces and moments.

7.6.5 The Ritz Method

7.6.5.1 The variational problem

The Ritz method, named after German engineer W. Ritz (1878–1909), is a numerical method of solving problems posed in terms of solving the *variational problem* of the type: find the function the vector function $u(x)$ from a suitable space $\mathcal{U}$ of functions such that

$$B(u, v) = L(v) \quad \text{for all } v \text{ from } \mathcal{U}. \tag{7.6.54}$$

where $B(u, v)$ is called a *bilinear form* and $L(v)$ is called a *linear form*, with the properties

$$B(\alpha u_1 + \beta u_2, v) = \alpha B(u_1, v) + \beta B(u_2, v) \text{ (linearity in the first argument)}$$
$$B(u, \alpha v_1 + \beta v_2) = \alpha B(u, v_1) + \beta B(u, v_2) \text{ (linearity in the second argument)}$$
$$L(\alpha v_1 + \beta v_2) = \alpha L(v_1) + \beta L(v_2), \tag{7.6.55}$$

for any real numbers α and β and dependent variables u, u_1, u_2, v, v_1, and v_2. The bilinear form is said to be symmetric if $B(u, v) = B(v, u)$ (that is, u and v can be interchanged without changing the value of B).

Some space and mathematical concepts from functional analysis are required to formally introduce the properties of the space $\mathcal{U}$, though these would distract the reader from the focus of the book. However, it suffices to say that the space $\mathcal{U}$ possesses properties of an inner product space, that is, functions u from $\mathcal{U}$ are sufficiently differentiable to a certain order as dictated by the functional $I(u)$, u and its various derivatives are square-integrable in the sense that

$$\int_\Omega |u(x)|^2 \, dx < \infty, \quad \int_\Omega |\nabla u(x)|^2 \, dx < \infty, \quad \text{and so on,}$$

and an inner product can be defined between any two elements u and v of the space $\mathcal{U}$.

Whenever $B(\cdot, \cdot)$ is bilinear and symmetric and $L(\cdot)$ is linear, a quadratic functional can be defined [see Reddy (2002)]:

$$I(u) = \tfrac{1}{2}B(u, u) - L(u), \qquad (7.6.56)$$

such that $\delta I = 0$ gives the variational problem in Eq. (7.6.54):

$$0 = \delta I(u) = \tfrac{1}{2}\left[B(\delta u, u) + B(u, \delta u)\right] - L(\delta u) = B(\delta u, u) - L(\delta u),$$

which is the same as Eq. (7.6.54) with $\delta u = v$.

As an example of $I(u)$ in linearized elasticity, we consider the axial deformation of a uniform bar with an end spring. The governing equation is

$$-\frac{d}{dx}\left(EA\frac{du}{dx}\right) = f(x), \quad 0 < x < L, \qquad (7.6.57)$$

and the boundary conditions are

$$u(0) = 0, \quad \left[\left(EA\frac{du}{dx}\right) + ku(x)\right]_{x=L} = P, \qquad (7.6.58)$$

where $E = E(x)$ is Young's modulus, $A = A(x)$ is the cross-sectional area, L is the length, k is the spring constant, $f(x)$ is the distributed axial load, and P is the axial load at $x = L$, as shown in Fig. 7.6.5. Equations (7.6.57) and (7.6.58) are equivalent to minimizing the total potential energy functional $I(u) = \Pi(u)$:

$$\Pi(u) = \int_0^L \frac{EA}{2}\left(\frac{du}{dx}\right)^2 dx + \frac{k}{2}[u(L)]^2 - \left[\int_0^L fu\,dx + Pu(L)\right]$$
$$= \tfrac{1}{2}B(u, u) - L(u), \qquad (7.6.59)$$

subjected to the geometric boundary condition, $u(0) = 0$. The bilinear and linear forms in this case are

$$B(u, v) = \int_0^L EA\frac{du}{dx}\frac{dv}{dx}\,dx + k\,u(L)v(L),$$
$$\qquad (7.6.60)$$
$$L(v) = \int_0^L fv\,dx + Pv(L).$$

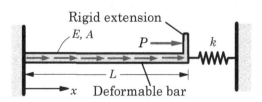

Fig. 7.6.5: Axial deformation of a uniform bar with an end spring.

Another example of the functional I is provided by the total potential energy functional Π in Eq. (7.6.27) associated with the bending of straight beams according to the Euler–Bernoulli beam theory. The bilinear and linear forms in this case are

$$B(v, w) = \int_0^L EI \frac{d^2v}{dx^2} \frac{d^2w}{dx^2} \, dx,$$

$$L(w) = \int_0^L qw \, dx + F_0 w(L) + M_0 \alpha(L),$$

(7.6.61)

where $\alpha = -(dw/dx)$.

7.6.5.2 Description of the method

In the Ritz method, we seek an approximation $U_N(x)$ of $u(x)$, for a fixed and preselected N, in the form

$$u(x) \approx U_N(x) = c_i \phi_i(x) + \phi_0(x),$$

(7.6.62)

where summation on repeated index i is implied (over the range of 1 to N), $\phi_i(x)$ are appropriately selected approximation functions, and c_i are unknown parameters. In view of the fact that the natural boundary conditions of the problem are included in the functional $I(u)$, we require the approximate solution U_N to satisfy only the geometric boundary conditions. In order that U_N satisfies the geometric boundary conditions *for any* c_i, it is convenient to choose the approximation in the form (7.6.62) and require $\phi_0(x)$ to satisfy the actual specified geometric boundary conditions. For instance, if $u(x)$ is specified to be $\hat{u}$ at $x = 0$, we require $\phi_0(x)$ be such that $\phi_0(0) = \hat{u}$, while ϕ_i are required to satisfy the homogeneous form of the geometric boundary condition, $\phi_i(0) = 0$. This follows from

$$0 = U_N(0) = \sum_{i=1}^N c_i \phi_i(0) + \phi_0(0) = \sum_{i=1}^N c_i \phi_i(0) + \hat{u}.$$

Because $U_N(0) = \hat{u}$, it follows that

$$\sum_{i=1}^N c_i \phi_i(0) = 0 \;\rightarrow\; \phi_i(0) = 0 \text{ for all } i = 1, 2, \ldots, N.$$

Thus, $\phi_i(x)$ must satisfy the *homogeneous form* of specified essential boundary conditions, and they must be sufficiently differentiable as required by the functional $I(U)$. The parameters c_i are determined by the condition that $I(U_N)$ is the minimum, that is, $\delta I(U) = 0$.

The *approximation functions* ϕ_0 and ϕ_i should be such that the substitution of Eq. (7.6.54) into $\delta \Pi$ results in N linearly independent equations for the parameters c_j ($j = 1, 2, \ldots, N$) so that the system has a solution. To ensure that the algebraic equations resulting from the Ritz approximation have a solution, and the approximate solution $U_N(x)$ converges to the true solution $u(x)$ of the problem as the value of N is increased, ϕ_i ($i = 1, 2, \ldots, N$) and ϕ_0 must satisfy certain requirements, as outlined next.

(1) ϕ_0 must satisfy the *specified* geometric boundary conditions. It is identically zero if all of the specified essential boundary conditions are homogeneous, that is, $\phi_0(x) = 0$. $(7.6.63)_1$

(2) ϕ_i $(i = 1, 2, \ldots, N)$ must satisfy the following three conditions: (a) be continuous, as required by the quadratic functional $I(u)$; (b) satisfy the *homogeneous form* of the specified essential boundary conditions; and (c) the set $\{\phi_i\}$ be linearly independent and complete; completeness refers to the property that all lower order terms up to the highest desired term must be included. $(7.6.63)_2$

Substituting $U_N(c_1, c_2, \cdots, c_N)$ into the total potential energy functional Π in Eq. (7.6.56), we obtain Π as a function of the parameters c_1, c_2, $\cdots$, c_N (after carrying out the indicated integration with respect to x):

$$\Pi = \tfrac{1}{2}B(c_j\phi_j + \phi_0, c_k\phi_k + \phi_0) - L(c_k\phi_k + \phi_0)$$
$$= \tfrac{1}{2}\left[c_j c_k B(\phi_j, \phi_k) + 2c_j B(\phi_j, \phi_0) + B(\phi_0, \phi_0)\right] - c_k L(\phi_k) - L(\phi_0).$$

Then c_i are determined (or adjusted) such that $\delta\Pi = 0$; in other words, we minimize Π with respect to c_i, $i = 1, 2, \cdots, N$:

$$0 = \delta\Pi = \frac{\partial\Pi}{\partial c_1}\delta c_1 + \frac{\partial\Pi}{\partial c_2}\delta c_2 + \ldots + \frac{\partial\Pi}{\partial c_N}\delta c_N = \sum_{j=1}^{N}\frac{\partial\Pi}{\partial c_j}\delta c_j.$$

Because the set $\{c_i\}$ is linearly independent, it follows that

$$0 = \frac{\partial\Pi}{\partial c_i} = \tfrac{1}{2}\left[c_k B(\phi_i, \phi_k) + c_j B(\phi_j, \phi_i) + 2B(\phi_i, \phi_0)\right] - L(\phi_i)$$
$$= c_j B(\phi_i, \phi_j) + B(\phi_i, \phi_0) - L(\phi_i) \quad \text{for} \quad i = 1, 2, \cdots, N.$$

or

$$\mathbf{Bc} = \mathbf{R}, \tag{7.6.64}$$

where

$$B_{ij} = B(\phi_i, \phi_j), \quad R_i = L(\phi_i) - B(\phi_i, \phi_0). \tag{7.6.65}$$

Equation (7.6.64) consists of N linear algebraic equations among N parameters, c_1, c_2, $\cdots$, c_N. Once the parameters are determined from Eq. (7.6.64), the solution U_N to the problem is given by Eq. (7.6.62). We consider couple of examples next.

Example 7.6.4 ───

Formulate the N-parameter Ritz solution $U_N(x)$ of the bar problem described by Eqs. (7.6.57) and (7.6.58) for $AE(x) = a_0(2 - \frac{x}{L})$, $k = 0$, and $f(x) = f_0$, and determine the Ritz solutions for $N = 1$ and $N = 2$ [see Reddy (2002)].

Solution: First, we must select the approximation functions ϕ_0 and ϕ_i. Apart from the guidelines given in (7.6.63), the selection of the coordinate functions is largely arbitrary. As a general rule, the coordinate functions ϕ_i should be selected from an admissible set (that is, those meeting the two conditions listed earlier), from the lowest order to a desirable order, without missing any intermediate terms (i.e., the completeness property). Also, ϕ_0 should be any lowest order (including zero) that satisfied the specified essential boundary conditions of the problem; $\phi_0(x)$ has no continuity (differentiability) requirement.

For the problem at hand, $\phi_0 = 0$ because the specified essential boundary condition is homogeneous, $u(0) = 0$; this homogeneous essential boundary condition requires us to find $\phi_1(x)$ such that $\phi_1(0) = 0$ and it is differentiable at least once with respect to x because $\Pi(u)$ involves the first derivatives of $u \approx U_N$. If an algebraic polynomial is to be selected, the lowest order polynomial that has a nonzero first derivative is

$$\phi_1(x) = a + bx,$$

where a and b are constants. The condition $\phi_1(0) = 0$ gives $a = 0$. Since b is arbitrary, we take it to be unity (any nonzero constant will be absorbed into c_1). When $N > 1$, property 2(c) in Eq. $(7.6.63)_2$ requires that ϕ_i, $i > 1$, should be selected such that the set $\{\phi_i\}_{i=1}^{N}$ is linearly independent and makes the set complete. In the present case, this is done by choosing ϕ_2 to be x^2. Clearly, $\phi_2(x) = x^2$ meets the conditions $\phi_2(0) = 0$, linearly independent of $\phi_1(x) = x$ (i.e., ϕ_2 is not a constant multiple of ϕ_1), and the set $\{x, x^2\}$ is complete (i.e., no other admissible term up to quadratic is omitted). In other words, in selecting coordinate functions of a given degree, one should not omit any lower-order terms that are admissible. Otherwise, the approximate solution will never converge to the exact solution, no matter how many terms are used in the Ritz approximation, as the exact solution may have those lower order terms that were omitted in the approximate solution. We choose

$$U_2(x) = c_1\phi_1 + c_2\phi_2 + \cdots + c_N\phi_N(x) = c_i\,\phi_i(x). \tag{1}$$

For the choice of algebraic polynomials, the N-parameter Ritz coefficients B_{ij} are [see Eq. $(7.6.60)$]

$$B_{ij} = B(\phi_i, \phi_j) = \int_0^L a_0 \left(1 - \frac{x}{L}\right) \frac{d\phi_i}{dx}\frac{d\phi_j}{dx}\,dx + k\,\phi_i(L)\phi_j(L)$$

$$= a_0\,ij \int_0^L \left(1 - \frac{x}{L}\right) x^{i+j-2}\,dx + k(L)^{i+j}$$

$$= a_0 \frac{ij(1+i+j)}{(i+j-1)(i+j)}(L)^{i+j-1} + k(L)^{i+j}, \tag{2}$$

$$R_i = \int_0^L f\phi_i\,dx + P\phi_i(L) = \frac{f_0}{i+1}(L)^{i+1} + P(L)^i. \tag{3}$$

Note that $k = 0$ for the problem at hand.

For one-term approximation $(N = 1)$, we have

$$a_{11} = \frac{3}{2}a_0 L, \quad b_1 = \frac{1}{2}f_0 L^2 + PL,$$

$$c_1 = \frac{b_1}{a_{11}} = \frac{6}{9a_0 L}\left(\frac{3}{6}f_0 L^2 + PL\right) = \frac{f_0 L + 2P}{3a_0},$$

and the one-parameter Ritz solution is

$$U_1(x) = \frac{f_0 L + 2P}{3a_0}x. \tag{4}$$

For $N = 2$, we have

$$a_{11} = \frac{3}{2}a_0 L, \quad a_{12} = a_{21} = \frac{4}{3}a_0 L^2, \quad a_{22} = \frac{5}{3}a_0 L^3,$$

$$b_1 = \frac{1}{2}f_0 L^2 + P_0 L, \quad b_2 = \frac{1}{3}f_0 L^3 + P_0 L^2.$$

The Ritz equations can be written in matrix form as

$$\frac{a_0 L}{6}\begin{bmatrix} 9 & 8L \\ 8L & 10L^2 \end{bmatrix}\begin{Bmatrix} c_1 \\ c_2 \end{Bmatrix} = \frac{f_0 L^2}{6}\begin{Bmatrix} 3 \\ 2L \end{Bmatrix} + PL\begin{Bmatrix} 1 \\ L \end{Bmatrix},$$

whose solution by is

$$c_1 = \frac{1}{a_0}\left(\frac{7}{13}f_0 + \frac{6}{13}P\right), \quad c_2 = \frac{3}{13a_0L}(-f_0L + P).$$

Hence, the two-parameter Ritz solution is

$$U_2(x) = \frac{7f_0L + 6P}{13a_0}x + \frac{3(P - f_0L)}{13a_0L}x^2. \tag{5}$$

The exact solution of Eqs. (7.6.57) and (7.6.58) with $u(0) = 0$, $k = 0$, $EA = a_0[2 - (x/L)]$, and $f = f_0$ is

$$u(x) = \frac{f_0L}{a_0}x + \frac{(f_0L - P)L}{a_0}\log\left(1 - \frac{x}{2L}\right) \tag{6}$$

$$\approx \frac{f_0L + P}{2a_0}x + \frac{P - f_0L}{8a_0L}x^2 + \frac{P - f_0L}{24a_0L^2}x^3 + \dots \tag{7}$$

Table 7.6.1 contains a comparison of the Ritz coefficients c_i for $N = 1, 2, \dots, 8$ with the exact coefficients in Eq. (7) for $L = 10$ ft., $a_0 = 180 \times 10^6$ lb, $f_0 = 0$, and $P = 10 \times 10^6$ lb. Clearly the Ritz coefficients c_i converge to the exact ones as N goes from 1 to 8.

Table 7.6.1: The Ritz coefficients* for the axial deformation of an isotropic elastic bar subjected to axial force.

n	$\bar{c}_1$	$\bar{c}_2$	$\bar{c}_3$	$\bar{c}_4$	$\bar{c}_5$	$\bar{c}_6$	$\bar{c}_7$	$\bar{c}_8$
1	37.037							
2	25.641	12.821						
3	28.219	4.409	4.879					
4	27.691	7.788	0.000	3.029				
5	27.794	6.701	3.389	−1.040	1.664			
6	27.775	7.009	1.904	2.012	−1.142	0.952		
7	27.778	6.929	2.453	0.320	1.447	−0.980	0.560	
8	27.778	6.948	2.272	1.094	−0.287	1.136	−0.769	0.336
Exact	27.778	6.944	2.315	0.868	0.347	0.145	0.062	0.027

* $\bar{c}_i = c_i \times 10^{5+i}$.

Example 7.6.5

Consider a simply supported beam of length L and constant bending stiffness EI, subjected to uniformly distributed transverse load q_0. Determine the transverse displacement $v(x)$ of the beam using the Ritz method with $N = 2$ and $N = 3$.

Solution: The Ritz equations are obtained from Eq. (7.6.64) and (7.6.65), where the bilinear and linear forms defined by Eq. (7.6.61) with $F_0 = M_0 = 0$. We choose an N-parameter Ritz approximation of the form

$$v(x) \approx V_2(x) = c_1\phi_1(x) + c_2\phi_2(x) + \dots + c_N\phi_N(x) + \phi_0(x).$$

Noting that the essential boundary conditions are $v(0) = v(L) = 0$, we select ϕ_0 to be zero. As far as ϕ_i are concerned, we choose them to vanish at $x = 0$ and $x = L$. Thus, we can choose

$$\phi_1 = x(L - x), \quad \phi_2 = x^2(L - x), \quad \phi_i(x) = x^i(L - x), \quad \cdots, \quad \phi_N(x) = x^N(L - x), \tag{1}$$

and

$$\frac{d\phi_i}{dx} = iLx^{i-1} - (i+1)x^i, \quad \frac{d^2\phi_i}{dx^2} = i(i-1)Lx^{i-2} - (i+1)ix^i. \tag{2}$$

Substituting these expressions into B_{ij} and R_i, we obtain

$$
\begin{aligned}
B_{ij} &= \int_0^L EI \frac{d^2\phi_i}{dx^2} \frac{d^2\phi_j}{dx^2} \, dx \\
&= \int_0^L EI \left[i(i-1)Lx^{i-2} - (i+1)ix^i \right] \left[j(j-1)Lx^{j-2} - (j+1)jx^j \right] dx \quad (2)
\end{aligned}
$$

$$
R_i = \int_0^L q_0 \phi_i \, dx = \int_0^L q_0(x^i L - x^{i+1}) \, dx.
$$

For $N = 2$, we have

$$
EIL \begin{bmatrix} 4 & 2L \\ 2L & 4L^2 \end{bmatrix} \begin{Bmatrix} c_1 \\ c_2 \end{Bmatrix} = \frac{q_0 L^3}{12} \begin{Bmatrix} 2 \\ L \end{Bmatrix}. \quad (3)
$$

The solution of these equations yields the result $c_1 = q_0 L^2 / 24EI$ and $c_2 = 0$, and the two-parameter Ritz solution becomes

$$
V_2(x) = \frac{q_0 L^4}{24EI} \left(\frac{x}{L} - \frac{x^2}{L^2} \right). \quad (4)
$$

For $N = 3$ we obtain

$$
EIL \begin{bmatrix} 4 & 2L & 2L^2 \\ 2L & 4L^2 & 4L^3 \\ 2L^2 & 4L^3 & 4.8L^4 \end{bmatrix} \begin{Bmatrix} c_1 \\ c_2 \\ c_3 \end{Bmatrix} = \frac{q_0 L^3}{12} \begin{Bmatrix} 2 \\ L \\ 0.6L^2 \end{Bmatrix}, \quad (5)
$$

and we obtain $c_1 = c_2 L = -c_3 L^2 = q_0 L^2 / 24EI$. Hence, the three-parameter Ritz solution is

$$
V_3(x) = \frac{q_0 L^4}{24EI} \left(\frac{x}{L} - 2\frac{x^3}{L^3} + \frac{x^4}{L^4} \right), \quad (6)
$$

which coincides with the exact solution of the beam problem.

Example 7.6.6

Consider the Poisson equation governing the Prandtl stress function Ψ, Eq. (7.5.79):

$$
-\nabla^2 \Psi = 2\mu\theta \quad \text{in } \Omega, \quad \Psi = 0 \quad \text{on } \Gamma. \quad (1)
$$

If the cross section of the cylinder is a square, $\Omega = \{-a \le x, \, y \le a\}$, as shown in Fig. 7.6.6, determine the stress function and compute the shear stresses σ_{xz} and σ_{yz} using a one-parameter Ritz approximation.

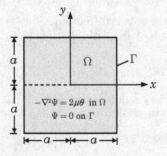

Fig. 7.6.6: Torsion of a cylinder of square cross section.

Solution: The functional associated with Eq. (1) is

$$\Pi(\Psi) = \frac{1}{2} \int_{-a}^{a} \int_{-a}^{a} \left[\left(\frac{\partial \Psi}{\partial x} \right)^2 + \left(\frac{\partial \Psi}{\partial y} \right)^2 \right] dx\, dy - 2\mu\theta \int_{-a}^{a} \int_{-a}^{a} \Psi \, dx\, dy. \tag{2}$$

and the bilinear and linear forms are

$$B(\Psi, \Phi) = \int_{-a}^{a} \int_{-a}^{a} \left(\frac{\partial \Psi}{\partial x} \frac{\partial \Phi}{\partial x} + \frac{\partial \Psi}{\partial y} \frac{\partial \Phi}{\partial y} \right) dx\, dy, \quad L(\Phi) = 2\mu\theta \int_{-a}^{a} \int_{-a}^{a} \Phi \, dx\, dy. \tag{3}$$

For $N = 1$ we choose the function

$$\phi_1 = (a^2 - x^2)(a^2 - y^2), \tag{4}$$

and obtain

$$B_{ij} = \int_{-a}^{a} \int_{-a}^{a} \left(\frac{\partial \phi_i}{\partial x} \frac{\partial \phi_j}{\partial x} + \frac{\partial \phi_i}{\partial y} \frac{\partial \phi_j}{\partial y} \right) dx\, dy = \frac{256}{45},$$

$$R_i = 2\mu\theta \int_{-a}^{a} \int_{-a}^{a} \phi_i \, dx\, dy = 2\mu\theta \frac{16}{9a^2}, \tag{5}$$

and the one-parameter solution is given by

$$\Psi_1(x,y) = \frac{5\mu\theta a^2}{8} \left(1 - \frac{x^2}{a^2} \right) \left(1 - \frac{y^2}{a^2} \right). \tag{6}$$

$$\sigma_{xz} = -\frac{5\mu\theta a}{4} \frac{x}{a} \left(1 - \frac{y^2}{a^2} \right), \quad \sigma_{yz} = \frac{5\mu\theta a}{4} \frac{y}{a} \left(1 - \frac{y^2}{a^2} \right).$$

For $N = 2$ with

$$\phi_1 = (a^2 - x^2)(a^2 - y^2), \quad \phi_2 = (x^2 + y^2)\phi_1, \cdots. \tag{7}$$

we obtain

$$a^8 \begin{bmatrix} \frac{256}{45} & \frac{1024}{525} a^2 \\ \frac{1024}{525} a^2 & \frac{11264}{4725} a^4 \end{bmatrix} \begin{Bmatrix} c_1 \\ c_2 \end{Bmatrix} = 2\mu\theta a^6 \begin{Bmatrix} \frac{16}{9} \\ \frac{32}{45} a^2 \end{Bmatrix}, \tag{8}$$

whose solution yields

$$c_1 = \frac{1295}{2216a^2} \mu\theta, \quad c_2 = \frac{525}{4432a^4} \mu\theta. \tag{9}$$

The two-parameter Ritz solution is given by

$$\Psi_2(x,y) = \frac{\mu\theta a^2}{4432} [2590 + 525(\bar{x}^2 + \bar{y}^2)](1 - \bar{x}^2)(1 - \bar{y}^2), \tag{10}$$

where $\bar{x} = x/a$ and $\bar{y} = y/a$.

The exact solution to Eq. (1) can be obtained using the separation of variables method, and it is given by

$$\Psi(x,y) = \frac{32\mu\theta a^2}{\pi^3} \sum_{n=1,3,5\cdots}^{\infty} \frac{1}{n^3} (-1)^{(n-1)/2} \left[1 - \frac{\cosh(n\pi y/2a)}{\cosh(n\pi/2)} \right] \cos \frac{n\pi x}{2a}. \tag{11}$$

The exact value of Ψ at the center of the region is

$$\Psi(0,0) = 0.5884\mu\theta a^2,$$

whereas the two-parameter Ritz solution is $0.5844\mu\theta a^2$, which has an error of only 0.68%.

Although the problem is presented here as one governing the Prandtl stress function for the torsion of a cylindrical member, the equation arises, among others, in connection with the transverse deflection of a membrane fixed on all sides and subjected to uniform pressure f_0 (in place of $2\mu\theta$) and in conduction heat transfer in a square region with internal heat generation of f_0 unit area. The function u denotes the deflection u in the case of a membrane and the temperature T in the case of conduction heat transfer. Thus the results obtained can also be interpreted for these two problems.

7.7 Hamilton's Principle

7.7.1 Introduction

The principle of total potential energy discussed in the previous section can be generalized to initial value problems, that is, problems involving time, and the principle is known as *Hamilton's principle*. In Hamilton's principle the system under consideration is assumed to be characterized by two energy functions: the *kinetic energy* K and the total *potential energy* Π. For *discrete* systems (i.e., systems with a finite number of degrees of freedom), these energies can be described in terms of a finite number of generalized coordinates and their derivatives with respect to time t. For *continuous* systems (that is, systems that are described by an infinite number of generalized coordinates), the energies can be expressed in terms of the dependent variables of the problem that are functions of position and time.

7.7.2 Hamilton's Principle for a Rigid Body

To gain a simple understanding of Hamilton's principle, consider a single particle or a rigid-body (which is a collection of particles, the distance between which is unaltered at all times) of mass m moving under the influence of a force (see Reddy, 2002) $\mathbf{F} = \mathbf{F}(\mathbf{r}, t)$. The path $\mathbf{r}(t)$ followed by the particle is related to the force $\mathbf{F}$ and mass m by the principle of balance of linear momentum (i.e., Newton's second law of motion):

$$\mathbf{F}(\mathbf{r}, t) = \frac{d}{dt}\left(m\frac{d\mathbf{r}}{dt}\right). \tag{7.7.1}$$

A path that differs from the actual path is expressed as $\mathbf{r} + \delta\mathbf{r}$, where $\delta\mathbf{r}$ is the variation of the path for any arbitrarily *fixed* time t. We suppose that the actual path $\mathbf{r}$ and the *varied* path differ except at two distinct times t_1 and t_2, that is, $\delta\mathbf{r}(t_1) = \delta\mathbf{r}(t_2) = \mathbf{0}$. Taking the scalar product of Eq. (7.7.1) with the variation $\delta\mathbf{r}$, and integrating with respect to time between t_1 and t_2, we obtain

$$\int_{t_1}^{t_2}\left[\frac{d}{dt}\left(m\frac{d\mathbf{r}}{dt}\right) - \mathbf{F}(\mathbf{r}, t)\right] \cdot \delta\mathbf{r}\, dt = 0. \tag{7.7.2}$$

Integration-by-parts of the first term in Eq. (7.7.2) yields

$$-\int_{t_1}^{t_2}\left[m\frac{d\mathbf{r}}{dt} \cdot \frac{d\delta\mathbf{r}}{dt} + \mathbf{F}(\mathbf{r}, t) \cdot \delta\mathbf{r}\right] dt + \left(m\frac{d\mathbf{r}}{dt} \cdot \delta\mathbf{r}\right)\Big|_{t_1}^{t_2} = 0. \tag{7.7.3}$$

The last term in Eq. (7.7.3) vanishes because $\delta\mathbf{r}(t_1) = \delta\mathbf{r}(t_2) = \mathbf{0}$. Also, note that

$$m\frac{d\mathbf{r}}{dt} \cdot \frac{d\delta\mathbf{r}}{dt} = \delta\left[\frac{m}{2}\frac{d\mathbf{r}}{dt} \cdot \frac{d\mathbf{r}}{dt}\right] \equiv \delta K, \tag{7.7.4}$$

where K is the kinetic energy of the particle or a rigid-body

$$K = \frac{1}{2}m\frac{d\mathbf{r}}{dt} \cdot \frac{d\mathbf{r}}{dt} = \frac{1}{2}m\mathbf{v} \cdot \mathbf{v}, \tag{7.7.5}$$

and δK is called the *virtual kinetic energy*. The expression $\mathbf{F}(\mathbf{r}, t) \cdot \delta \mathbf{r}$ is called the *virtual work done by external forces* and denoted by

$$\delta W_E = -\mathbf{F}(\mathbf{r}, t) \cdot \delta \mathbf{r}. \tag{7.7.6}$$

The minus sign indicates that the work is done by external force $\mathbf{F}$ *on* the body in moving through the displacement $\delta \mathbf{r}$. Equation (7.7.3) now takes the form

$$\int_{t_1}^{t_2} (\delta K - \delta W_E)dt = 0, \tag{7.7.7}$$

which is known as the *general form of Hamilton's principle* for a single particle or rigid body. Note that a particle or a rigid-body has no strain energy Π because the distance between the particles is unaltered.

Suppose that the force $\mathbf{F}$ is conservative (i.e., the sum of the potential and kinetic energies is conserved) such that it can be replaced by the gradient of a potential

$$\mathbf{F} = -\text{grad } V, \tag{7.7.8}$$

where $V = V(\mathbf{r}, t)$ is the *potential energy due to the loads* on the body. Then Eq. (7.7.7) can be expressed in the form

$$\delta \int_{t_1}^{t_2} (K - V) \, dt = 0, \tag{7.7.9}$$

because $(\mathbf{r} = x_i \hat{\mathbf{e}}_i)$

$$\text{grad } V \cdot \delta \mathbf{r} = \frac{\partial V}{\partial x_i} \delta x_i = \delta V(\mathbf{x}).$$

The difference between the kinetic and potential energies is called the *Lagrangian function*

$$L \equiv K - V. \tag{7.7.10}$$

Equation (7.7.9) is known as Hamilton's principle for the conservative motion of a particle (or a rigid body). The principle can be stated as follows: *The motion of a particle acted on by conservative forces between two arbitrary instants of time t_1 and t_2 is such that the line integral over the Lagrangian function is an extremum for the path motion.* Stated in other words, of all possible paths that the particle could travel from its position at time t_1 to its position at time t_2, its actual path will be one for which the integral

$$I \equiv \int_{t_1}^{t_2} L \, dt \tag{7.7.11}$$

is an extremum (i.e., a minimum, maximum, or an inflection).

If the path $\mathbf{r}$ can be expressed in terms of the generalized coordinates $q_i (i = 1, 2, 3)$, the Lagrangian function can be written in terms of q_i and their time derivatives

$$L = L(q_1, q_2, q_3, \dot{q}_1, \dot{q}_2, \dot{q}_3). \tag{7.7.12}$$

Then the condition for the extremum of I in (7.7.11) results in the equation (note that $\delta q_i = 0$ at t_1 and t_2)

$$\delta I = \delta \int_{t_1}^{t_2} L(q_1, q_2, q_3, \dot{q}_1, \dot{q}_2, \dot{q}_3) dt = 0$$

$$= \int_{t_1}^{t_2} \sum_{i=1}^{3} \left[\frac{\partial L}{\partial q_i} - \frac{d}{dt}\left(\frac{\partial L}{\partial \dot{q}_i}\right) \right] \delta q_i \, dt. \tag{7.7.13}$$

When all q_i are linearly independent (i.e., no constraints among q_i), the variations δq_i are independent of each other for all t, except that all $\delta q_i = 0$ at t_1 and t_2. Therefore, the coefficients of $\delta q_1, \delta q_2$, and δq_3 vanish separately:

$$\frac{\partial L}{\partial q_i} - \frac{d}{dt}\left(\frac{\partial L}{\partial \dot{q}_i}\right) = 0, \quad i = 1, 2, 3. \tag{7.7.14}$$

These equations are called the *Lagrange equations of motion*. Recall that in Section 7.6 (for a static case) these equations were also called the Euler equations. For the dynamic case involving deformable solids, these equations will be called the *Euler–Lagrange equations*.

When the forces are not conservative, we must deal with the general form of Hamilton's principle in Eq. (7.7.7). In this case, there exists no functional I that must be an extremum. If the virtual work can be expressed in terms of the generalized coordinates q_i by

$$\delta W_E = -\left(F_1 \, \delta q_1 + F_2 \, \delta q_2 + F_3 \, \delta q_3\right), \tag{7.7.15}$$

where F_i are the *generalized forces*, then we can write Eq. (7.7.14) as

$$\int_{t_1}^{t_2} \sum_{i=1}^{3} \left[\frac{\partial K}{\partial q_i} - \frac{d}{dt}\left(\frac{\partial K}{\partial \dot{q}_i}\right) + F_i \right] \delta q_i \, dt = 0, \tag{7.7.16}$$

and the Euler-Lagrange equations for the nonconservative forces are given by

$$\delta q_i : \quad \frac{\partial K}{\partial q_i} - \frac{d}{dt}\left(\frac{\partial K}{\partial \dot{q}_i}\right) + F_i = 0, \quad i = 1, 2, 3. \tag{7.7.17}$$

Example 7.7.1 _____

Consider the planar motion of a pendulum that consists of a mass m attached at the end of a rigid massless rod of length L that pivots about a fixed point O, as shown in Fig. 7.7.1. Determine the equation of motion.

Solution: The position of the mass can be expressed in terms of the generalized coordinates $q_1 = l$ and $q_2 = \theta$, measured from the vertical position. Because l is a constant, we have $\dot{q}_1 = 0$ and θ is the only independent generalized coordinate. The force $\mathbf{F}$ acting on the mass m is the component of the gravitational force,

$$\mathbf{F} = mg\left(\cos\theta \, \hat{\mathbf{e}}_r - \sin\theta \, \hat{\mathbf{e}}_\theta\right) \equiv F_r \, \hat{\mathbf{e}}_r + F_\theta \, \hat{\mathbf{e}}_\theta. \tag{7.7.18}$$

The component along $\hat{\mathbf{e}}_r$ does no work because $q_1 = l$ is a constant. The second component, F_θ, is derivable from the potential $(\nabla V = -F_\theta \, \hat{\mathbf{e}}_\theta)$:

$$V(\theta) = -\left[-mgl(1 - \cos\theta)\right] = mgl(1 - \cos\theta), \tag{7.7.19}$$

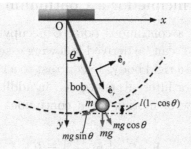

Fig. 7.7.1: Planar motion of a pendulum.

where V represents the potential energy of the mass m at any instant of time with respect to the static equilibrium position $\theta = 0$, and ∇ is the gradient operator in the polar coordinate system:

$$\nabla = \hat{\mathbf{e}}_r \frac{\partial}{\partial r} + \frac{\hat{\mathbf{e}}_\theta}{r} \frac{\partial}{\partial \theta}. \tag{7.7.20}$$

Thus the kinetic energy and the potential energy due to external load are given by

$$K = \frac{m}{2}(\ell\dot{\theta})^2, \quad V = mgl(1 - \cos\theta),$$
$$\delta K = ml^2 \dot{\theta}\,\delta\dot{\theta}, \quad \delta V = mgl\sin\theta\,\delta\theta = -F_\theta(l\,\delta\theta). \tag{7.7.21}$$

Therefore, the Lagrangian function L is a function of θ and $\dot{\theta}$. The Euler–Lagrange equation is given by

$$\delta q_2 = \delta\theta : \quad \frac{\partial L}{\partial \theta} - \frac{d}{dt}\left(\frac{\partial L}{\partial \dot{\theta}}\right) = 0,$$

which yields

$$-mgl\sin\theta - \frac{d}{dt}(ml^2\,\dot{\theta}) = 0 \quad \text{or} \quad \ddot{\theta} + \frac{g}{l}\sin\theta = 0 \quad (F_\theta = ml\,\ddot{\theta}). \tag{7.7.22}$$

Equation (7.7.22) represents a second-order nonlinear differential equation governing θ. For small angular motions, Eq. (7.7.22) can be linearized by replacing $\sin\theta \approx \theta$:

$$\ddot{\theta} + \frac{g}{\ell}\theta = 0. \tag{7.7.23}$$

Now suppose that the mass experiences a resistance force $\mathbf{F}^*$ proportional to its speed (e.g., the mass m is suspended in a medium with viscosity μ). According to Stokes's law,

$$\mathbf{F}^* = -6\pi\mu a l\dot{\theta}\,\hat{\mathbf{e}}_\theta, \tag{7.7.24}$$

where μ is the viscosity of the surrounding medium, a is the radius of the bob, and $\hat{\mathbf{e}}_\theta$ is the unit vector tangential to the circular path. The resistance of the massless rod supporting the bob is neglected. The force $\mathbf{F}^*$ is not derivable from a potential function (i.e., nonconservative). Thus, we have one part of the force (i.e., gravitational force) conservative and the other (i.e., viscous force) nonconservative. Hence, we use Hamilton's principle expressed by Eq. (7.7.14) or Eq. (7.7.17) with

$$\delta W_E = \delta V - \mathbf{F}^* \cdot (l\delta\theta\,\hat{\mathbf{e}}_\theta) = \left(mgl\sin\theta + 6\pi\mu a l^2\,\dot{\theta}\right)\delta\theta \equiv -F_\theta l\,\delta\theta.$$

Then the equation of motion is given by $[K = K(\dot{\theta})]$:

$$-\frac{d}{dt}\left(\frac{\partial K}{\partial \dot{\theta}}\right) + F_\theta l = 0 \quad \text{or} \quad \ddot{\theta} + \frac{g}{l}\sin\theta + \frac{6\pi a\mu}{m}\dot{\theta} = 0. \tag{7.7.25}$$

The coefficient $c = 6\pi a\mu/m$ is called the *damping* coefficient.

7.7.3 Hamilton's Principle for a Continuum

Hamilton's principle for a continuous body $\mathcal{B}$ occupying configuration κ with volume Ω and boundary Γ can be derived following essentially the same ideas as discussed for a particle or a rigid body. In contrast to a rigid body, a continuum is characterized by strain (or internal) energy U, in addition to the kinetic energy K. Newton's second law of motion for a continuous body can be written in general terms as

$$\mathbf{F} - \frac{\partial}{\partial t}\left(m \frac{\partial \mathbf{v}}{\partial t}\right) = \mathbf{0}, \tag{7.7.26}$$

where m is the mass, $\mathbf{v}(\mathbf{x}, t) = \partial \mathbf{u}/\partial t$ is the velocity vector, $\mathbf{u}(\mathbf{x}, t)$ is the displacement vector, and $\mathbf{F}$ is the resultant of *all* forces acting on the body $\mathcal{B}$. The actual path $\mathbf{u}$ followed by a material particle in position $\mathbf{x}$ in the body is varied, consistent with kinematic (essential) boundary conditions on Γ, to $\mathbf{u} + \delta\mathbf{u}$, where $\delta\mathbf{u}$ is the admissible variation (or virtual displacement) of the path. We assume that the varied path differs from the actual path except at initial and final times, t_1 and t_2, respectively. Thus, an admissible variation $\delta\mathbf{u}$ satisfies the conditions,

$$\delta\mathbf{u}(\mathbf{x}, t) = \mathbf{0} \text{ on } \Gamma_u \text{ for all } t,$$
$$\delta\mathbf{u}(\mathbf{x}, t_1) = \delta\mathbf{u}(\mathbf{x}, t_2) = \mathbf{0} \text{ for all } \mathbf{x}, \tag{7.7.27}$$

where Γ_u denotes the portion of the boundary Γ of the body where the displacement vector $\mathbf{u}$ is specified.

The work done on body $\mathcal{B}$ at time t by the resultant force $\mathbf{F}$, which consists of body force $\mathbf{f}$ and specified surface traction $\hat{\mathbf{t}}$ in moving through respective virtual displacements $\delta\mathbf{u}$, is given by

$$\int_\Omega \mathbf{f} \cdot \delta\mathbf{u} \, d\mathbf{x} + \int_{\Gamma_\sigma} \hat{\mathbf{t}} \cdot \delta\mathbf{u} \, ds - \int_\Omega \boldsymbol{\sigma} : \delta\boldsymbol{\varepsilon} \, d\mathbf{x}, \tag{7.7.28}$$

where $\boldsymbol{\sigma}$ and $\boldsymbol{\varepsilon}$ are the stress and strain tensors, and Γ_σ is the portion of the boundary Γ on which tractions are specified ($\Gamma = \Gamma_u \cup \Gamma_\sigma$). The last term in Eq. (7.7.28) is known as the *virtual work stored in the body* due to deformation. The strains $\delta\boldsymbol{\varepsilon}$ are assumed to be compatible in the sense that the strain-displacement relations (7.2.1) are satisfied. The work done by the inertia force $\partial(m\mathbf{v})/\partial t$ in moving through the virtual displacement $\delta\mathbf{u}$ is given by

$$\int_\Omega \frac{\partial}{\partial t}\left(\rho_0 \frac{\partial \mathbf{u}}{\partial t}\right) \cdot \delta\mathbf{u} \, d\mathbf{x}, \tag{7.7.29}$$

where ρ_0 is the mass density of the medium ($m = \rho_0 \, d\mathbf{x}$). We have, analogous to Eq. (7.7.2) for a rigid body, the result

$$\int_{t_1}^{t_2}\left\{\int_\Omega \frac{\partial}{\partial t}\left(\rho_0 \frac{\partial \mathbf{u}}{\partial t}\right) \cdot \delta\mathbf{u} \, d\mathbf{x} - \left[\int_\Omega (\mathbf{f} \cdot \delta\mathbf{u} - \boldsymbol{\sigma} : \delta\boldsymbol{\varepsilon}) \, d\mathbf{x} + \int_{\Gamma_\sigma} \hat{\mathbf{t}} \cdot \delta\mathbf{u} \, ds\right]\right\} dt = 0,$$

or

$$-\int_{t_1}^{t_2}\left[\int_\Omega \rho_0 \frac{\partial \mathbf{u}}{\partial t} \cdot \frac{\partial \delta\mathbf{u}}{\partial t} \, d\mathbf{x} + \int_\Omega (\mathbf{f} \cdot \delta\mathbf{u} - \boldsymbol{\sigma} : \delta\boldsymbol{\varepsilon}) \, d\mathbf{x} + \int_{\Gamma_\sigma} \hat{\mathbf{t}} \cdot \delta\mathbf{u} \, ds\right] dt = 0. \tag{7.7.30}$$

In arriving at the expression in Eq. (7.7.30), integration-by-parts is used on the first term; the integrated terms vanish because of the initial and final conditions in Eq. (7.7.27). Equation (7.7.30) is known as the general form of Hamilton's principle for a continuous medium – conservative or not, and elastic or not.

For an elastic body, we recall from the previous sections that the forces $\mathbf{f}$ and $\mathbf{t}$ are conservative,

$$\delta V = - \left(\int_\Omega \mathbf{f} \cdot \delta \mathbf{u} \, dx + \int_{\Gamma_\sigma} \hat{\mathbf{t}} \cdot \delta \mathbf{u} \, ds \right), \tag{7.7.31}$$

and that there exists a strain energy density function $U_0 = U_0(\varepsilon)$ such that

$$\sigma = \frac{\partial U_0}{\partial \varepsilon}. \tag{7.7.32}$$

Substituting Eqs. (7.7.31) and (7.7.32) into Eq. (7.7.30), we obtain

$$\delta \int_{t_1}^{t_2} [K - (V + U)] dt = 0, \tag{7.7.33}$$

where K and U are the kinetic and strain energies:

$$K = \int_\Omega \frac{\rho_0}{2} \frac{\partial \mathbf{u}}{\partial t} \cdot \frac{\partial \mathbf{u}}{\partial t} \, dx, \quad U = \int_\Omega U_0 \, dx. \tag{7.7.34}$$

Equation(7.7.33) represents Hamilton's principle for an elastic body. Recall that the sum of the strain energy U and potential energy V of external forces, $U + V$, is called the total potential energy, Π, of the body. For bodies involving no motion (that is, forces are applied sufficiently slowly such that the motion is independent of time, and the inertia forces are negligible), Hamilton's principle (7.7.33) reduces to the principle of virtual displacements. Equation (7.7.33) may be viewed as the dynamics version of the principle of virtual displacements.

The Euler–Lagrange equations associated with the Lagrangian, $L = K - \Pi$, can be obtained from Eq. (7.7.33):

$$0 = \delta \int_{t_1}^{t_2} L(\mathbf{u}, \nabla \mathbf{u}, \dot{\mathbf{u}}) \, dt$$

$$= \int_{t_1}^{t_2} \left[\int_\Omega \left(\rho_0 \frac{\partial^2 \mathbf{u}}{\partial t^2} - \nabla \cdot \sigma - \mathbf{f} \right) \cdot \delta \mathbf{u} \, dx + \int_{\Gamma_\sigma} (\mathbf{t} - \hat{\mathbf{t}}) \cdot \delta \mathbf{u} \, ds \right] dt, \tag{7.7.35}$$

where $\mathbf{t} = \sigma \cdot \hat{\mathbf{n}}$. In arriving at Eq. (7.7.35) from Eq. (7.7.33), we have used integration-by-parts, gradient theorems, and Eqs. (7.7.27)$_1$. Since $\delta \mathbf{u}$ is arbitrary for t, $t_1 < t < t_2$, and for $\mathbf{x}$ in Ω and also on Γ_σ, it follows that

$$\rho_0 \frac{\partial^2 \mathbf{u}}{\partial t^2} - \nabla \cdot \sigma - \mathbf{f} = 0 \quad \text{in } \Omega \text{ for } t > 0, \tag{7.7.36}$$

$$\mathbf{t}(s, t) - \hat{\mathbf{t}}(s, t) = 0 \quad \text{on } \Gamma_\sigma \text{ for } t > 0. \tag{7.7.37}$$

Equations (7.7.36) are the Euler–Lagrange equations for an elastic body. Equations (7.7.36) are also subject to initial conditions of the form

$$\mathbf{u}(\mathbf{x}, 0) = \mathbf{u}_0(\mathbf{x}), \quad \dot{\mathbf{u}}(\mathbf{x}, 0) = \mathbf{v}_0(\mathbf{x}), \tag{7.7.38}$$

where $\mathbf{u}_0$ and $\dot{\mathbf{v}}_0$ are the initial displacement and initial velocity vectors, respectively.

Example 7.7.2 ——————————————————————

The displacement field for pure bending (i.e., omit the axial displacement u) of a beam according to the Euler–Bernoulli beam theory is (see Section 7.3.4)

$$u_1(x, z, t) = -y\frac{\partial v}{\partial x}, \quad u_2 = 0, \quad u_3(x, t) = v(x, t), \tag{1}$$

where v is the transverse displacement. Determine the equations of motion of the Euler–Bernoulli beam theory.

Solution: The Lagrange function associated with the dynamics of the Euler–Bernoulli beam is given by $L = K - (U + V)$, where

$$K = \int_0^L \int_A \left[\frac{\rho_0}{2}\left(-z\frac{\partial^2 v}{\partial x \partial t}\right)^2 + \frac{\rho_0}{2}\left(\frac{\partial w}{\partial t}\right)^2 \right] dA\, dx$$

$$= \int_0^L \left[\frac{\rho_0 I}{2}\left(\frac{\partial^2 v}{\partial x \partial t}\right)^2 + \frac{\rho_0 A}{2}\left(\frac{\partial v}{\partial t}\right)^2 \right] dx, \tag{2}$$

$$U = \int_0^L \int_A \frac{E}{2}\left(-y\frac{\partial^2 v}{\partial x^2}\right)^2 dA\, dx$$

$$= \int_0^L \frac{EI}{2}\left(\frac{\partial^2 v}{\partial x^2}\right)^2 dx, \tag{3}$$

$$V = -\int_0^L q v\, dx, \tag{4}$$

where q is the transverse distributed load. In arriving at the expressions for K and U, we have used the fact that the x-axis coincides with the geometric centroidal axis, $\int_A y\, dA = 0$.

The Hamilton principle gives

$$0 = \delta \int_0^T (K - U - V)\, dt$$

$$= \int_0^T \int_0^L \left[\rho_0 I \frac{\partial \dot{v}}{\partial x}\frac{\partial \delta \dot{v}}{\partial x} + \rho_0 A \dot{v}\, \delta \dot{v} - EI\frac{\partial^2 v}{\partial x^2}\frac{\partial^2 \delta v}{\partial x^2} + q\, \delta v \right] dx\, dt. \tag{5}$$

The Euler–Lagrange equation obtained from Eq. (5) is the equation of motion governing the Euler–Bernoulli beam theory

$$\frac{\partial^2}{\partial x \partial t}\left(\rho_0 I \frac{\partial^2 v}{\partial x \partial t}\right) - \frac{\partial}{\partial t}\left(\rho_0 A \frac{\partial v}{\partial t}\right) - \frac{\partial^2}{\partial x^2}\left(EI\frac{\partial^2 v}{\partial x^2}\right) + q = 0, \tag{7.7.39}$$

for $0 < x < L$ and $t > 0$. The first term is the contribution due to rotary inertia. The boundary and initial conditions associated with Eq. (7.7.39) are

$$\text{Boundary conditions:} \quad \text{specify: } v \text{ or } \frac{\partial}{\partial x}\left(EI\frac{\partial^2 v}{\partial x^2}\right) + \rho_0 I \frac{\partial^3 v}{\partial t^2 \partial x},$$

$$\text{specify: } \frac{\partial v}{\partial x} \text{ or } EI\frac{\partial^2 v}{\partial x^2}, \tag{7.7.40}$$

$$\text{Initial conditions:} \quad \text{specify: } v(x, 0) \text{ and } \dot{v}(x, 0).$$

Example 7.7.3

Suppose that the Euler–Bernoulli beam of Example 7.7.2 experiences two types of viscous (velocity-dependent) damping: (1) viscous resistance to transverse displacement of the beam and (2) a viscous resistance to straining of the beam material. If the resistance to transverse velocity is denoted by $c(x)$, the corresponding damping force is given by $q_D(x,t) = c(x)\dot{v}$. If the resistance to strain velocity is c_s, the damping stress is $\sigma^D_{xx} = c_s \dot{\varepsilon}_{xx}$. Derive the equations of motion of the beam with both types of damping.

Solution: We must add the following terms due to damping to the expression in Eq. (5) of Example 7.7.2:

$$
-\int_0^T \left[\int_\Omega \sigma_D \delta\varepsilon \, d\mathbf{x} + \int_0^L q_D \delta v \, dx \right] dt
$$

$$
= -\int_0^T \left[\int_0^L \int_A c_s \left(-z\frac{\partial^3 v}{\partial x^2 \partial t} \right) \left(-z\frac{\partial^2 \delta v}{\partial x^2} \right) dA \, dx + \int_0^L q_D \delta v \, dx \right] dt
$$

$$
= -\int_0^T \int_0^L \left(I c_s \frac{\partial^3 v}{\partial x^2 \partial t} \frac{\partial^2 \delta v}{\partial x^2} + c\frac{\partial v}{\partial t} \delta v \right) dx \, dt. \tag{1}
$$

Then the expression in Eq. (5) of Example 7.7.2 becomes

$$
0 = \int_0^T \int_0^L \left[\rho_0 I \frac{\partial \dot{v}}{\partial x} \frac{\partial \delta v}{\partial x} + \rho_0 A \dot{v} \, \delta \dot{v} - EI\frac{\partial^2 v}{\partial x^2}\frac{\partial^2 \delta v}{\partial x^2} + q\, \delta v \right] dx \, dt
$$

$$
- \int_0^T \int_0^L \left(I c_s \frac{\partial^3 v}{\partial x^2 \partial t} \frac{\partial^2 \delta v}{\partial x^2} + c\frac{\partial w}{\partial t}\delta v \right) dx \, dt \tag{2}
$$

and the Euler–Lagrange equation is

$$
\frac{\partial^2}{\partial x \partial t}\left(\rho_0 I \frac{\partial^2 v}{\partial x \partial t} \right) - \frac{\partial}{\partial t}\left(\rho_0 A \frac{\partial v}{\partial t} \right) - \frac{\partial^2}{\partial x^2}\left(EI\frac{\partial^2 v}{\partial x^2} \right)
$$

$$
- \frac{\partial^2}{\partial x^2}\left(I c_s \frac{\partial^3 v}{\partial x^2 \partial t} \right) - c\frac{\partial v}{\partial t} + q = 0. \tag{3}
$$

The boundary and initial conditions for this case are

Boundary conditions: specify: v or $\dfrac{\partial}{\partial x}\left(EI\dfrac{\partial^2 v}{\partial x^2} + I c_s \dfrac{\partial^3 v}{\partial x^2 \partial t} \right) + \rho_0 I \dfrac{\partial^3 v}{\partial t^2 \partial x}$,

specify: $\dfrac{\partial v}{\partial x}$ or $EI\dfrac{\partial^2 v}{\partial x^2} + I c_s \dfrac{\partial^3 v}{\partial x^2 \partial t}$, $\qquad$ (4)

Initial conditions: specify: $v(x,0)$ and $\dot{v}(x,0)$.

7.8 Summary

This is a very comprehensive chapter on linearized elasticity. Beginning with a summary of the linearized elasticity equations that include the Navier equations and the Beltrami–Michell equations of elasticity, the three types of boundary value problems and the principle of superposition were discussed. The Clapeyron theorem and Betti and Maxwell reciprocity theorems and their applications were also presented. Various methods of solutions, namely, the inverse method, the semi-inverse method, the method of potentials, and variational methods are discussed. The two-dimensional elasticity problems, plane strain and plane stress, are formulated, and their solutions by the inverse method and the Airy stress function method are presented. Analytical solutions of a number of standard boundary value problems of elasticity using the Airy stress function are discussed. Torsion of cylindrical members is also presented. The principle of minimum total potential energy and its special case, the Castigliano theorem I, are discussed. The Ritz method is introduced as a general method of solving problems formulated as variational problems of finding u such that $B(u, v) = L(v)$ holds for all v. Lastly, Hamilton's principle for problems of dynamics is presented. A number of examples are included throughout the chapter.

Solution of elasticity problems discussed in this chapter requires an understanding of the problem from the aspect of suitable boundary conditions; existence of solution symmetries, if any; and the qualitative nature of the solution. Only then one may choose a solution method that suits its solution strategy. An insight into the problem is necessary for the use of the semi-inverse method. If one makes assumptions on the basis of a qualitative understanding of the problem and solves the boundary value problem, then the assumptions are likely to be correct. If not, the assumptions need to be modified. Also, most real-world problems do not admit exact or analytical solutions, and approximate solutions are the only alternative. The theoretical formulation of a problem based on the principles of mechanics is a necessary first and most important step even when one considers its solution by a numerical method. Therefore, a course on continuum mechanics or elasticity helps in correctly formulating the governing equations of boundary value problems of mechanics.

Problems

STRAINS, STRESSES, AND STRAIN ENERGY

7.1 Define the *deviatoric* components of stress and strain as follows:

$$s_{ij} \equiv \sigma_{ij} - \tfrac{1}{3}\sigma_{kk}\delta_{ij}, \quad e_{ij} \equiv \varepsilon_{ij} - \tfrac{1}{3}\varepsilon_{kk}\delta_{ij}.$$

Determine the constitutive relation between s_{ij} and e_{ij} for an isotropic material.

7.2 For each of the displacement fields given below, sketch the displaced positions in the x_1x_2-plane of the points initially on the sides of the square shown in Fig. P7.2.

    (a) $\mathbf{u} = \tfrac{\alpha}{2}(x_2\,\hat{\mathbf{e}}_1 + x_1\,\hat{\mathbf{e}}_2)$. (b) $\mathbf{u} = \tfrac{\alpha}{2}(-x_2\,\hat{\mathbf{e}}_1 + x_1\,\hat{\mathbf{e}}_2)$. (c) $\mathbf{u} = \alpha\, x_1\,\hat{\mathbf{e}}_2$.

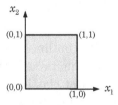

Fig. P7.2

7.3 For each of the displacement fields in Problem **7.2**, determine the components of (a) the Green–Lagrange strain tensor **E**, (b) the infinitesimal strain tensor ε, (c) the infinitesimal rotation tensor $\boldsymbol{\Omega}$, and (d) the infinitesimal rotation vector $\boldsymbol{\omega}$ (see Sections 3.4 and 3.5 for the definitions).

7.4 Similar to Cauchy's formula for a stress tensor, one can think of a similar formula for the strain tensor,

$$\varepsilon_n = \varepsilon \cdot \hat{\mathbf{n}},$$

where ε_n represents the strain vector in the direction of the unit normal vector, $\hat{\mathbf{n}}$. Determine the longitudinal strain corresponding to the displacement field $\mathbf{u} = \frac{\alpha}{2}\left(x_2\,\hat{\mathbf{e}}_1 + x_1\,\hat{\mathbf{e}}_2\right)$ in the direction of the vector $\hat{\mathbf{e}}_1 + \hat{\mathbf{e}}_2$.

7.5 For the displacement vector given in the cylindrical coordinate system

$$\mathbf{u} = Ar\,\hat{\mathbf{e}}_r + Brz\,\hat{\mathbf{e}}_\theta + C\sin\theta\,\hat{\mathbf{e}}_z,$$

where A, B, and C are constants, determine the infinitesimal strain components in the cylindrical coordinate system.

7.6 The displacement vector at a point referred to the basis $(\hat{\mathbf{e}}_1, \hat{\mathbf{e}}_2, \hat{\mathbf{e}}_3)$ is $\mathbf{u} = 2\hat{\mathbf{e}}_1 + 2\hat{\mathbf{e}}_2 - 4\hat{\mathbf{e}}_3$. Determine $\bar{u}_i$ with respect to the basis $(\hat{\bar{\mathbf{e}}}_1, \hat{\bar{\mathbf{e}}}_2, \hat{\bar{\mathbf{e}}}_3)$, where $\hat{\bar{\mathbf{e}}}_1 = (2\hat{\mathbf{e}}_1 + 2\hat{\mathbf{e}}_2 + \hat{\mathbf{e}}_3)/3$ and $\hat{\bar{\mathbf{e}}}_2 = (\hat{\mathbf{e}}_1 - \hat{\mathbf{e}}_2)/\sqrt{2}$.

7.7 Express Navier's equations of elasticity (7.2.17) in the cylindrical coordinate system.

7.8 An isotropic body ($E = 210$ GPa and $\nu = 0.3$) with two-dimensional state of stress experiences the following displacement field (in mm):

$$u_1 = 3x_1^2 - x_1^3 x_2 + 2x_2^3, \quad u_2 = x_1^3 + 2x_1 x_2,$$

where x_i are in meters. Determine the stresses and rotation of the body at point $(x_1, x_2) = (0.05, 0.02)$ m.

7.9 A two-dimensional state of stress exists in a body with the following components of stress:

$$\sigma_{11} = c_1 x_2^3 + c_2 x_1^2 x_2 - c_3 x_1, \quad \sigma_{22} = c_4 x_2^3 - c_5, \quad \sigma_{12} = c_6 x_1 x_2^2 + c_7 x_1^2 x_2 - c_8,$$

where c_i are constants. Assuming that the body forces are zero, determine the conditions on the constants so that the stress field is in equilibrium and satisfies the compatibility equations.

7.10 Express the strain energy for a linear isotropic body in terms of the (a) strain components and (b) stress components.

7.11 A rigid uniform member ABC of length L, pinned at A and supported by linear elastic springs, each of stiffness k, at B and C, is shown in Fig. P7.11. Find the total strain energy of the system when the point C is displaced vertically down by the amount u_C.

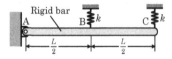

Fig. P7.11

7.12 Repeat Problem **7.11** when the springs are nonlinearly elastic, with the force deflection relationship, $F = ku^2$, where k is a constant.

7.13 Consider the equations of motion of 2-D elasticity (in the x- and z-coordinates) in the absence of body forces:

$$\frac{\partial\sigma_{xx}}{\partial x} + \frac{\partial\sigma_{xy}}{\partial y} = \rho_0\frac{\partial^2 u_x}{\partial t^2}$$

$$\frac{\partial\sigma_{xy}}{\partial x} + \frac{\partial\sigma_{yy}}{\partial y} = \rho_0\frac{\partial^2 u_y}{\partial t^2}$$

For a beam of uniform height h and width b, integrate the preceding equations with respect to y from $-h/2$ to $h/2$, and express the results in terms of the stress resultants N and V defined in Eq. (7.3.28). Use the following boundary conditions:

$$\sigma_{xy}(x, h/2) - \sigma_{xy}(x, -h/2) = f(x)/b, \quad \sigma_{xy}(x, h/2) + \sigma_{xy}(x, -h/2) = 0,$$

$$\sigma_{yy}(x, -h/2) = 0, \quad b\sigma_{yy}(x, h/2) = q$$

Next, multiply the first equation of motion with y and integrate it with respect to y from $-h/2$ to $h/2$, and express the results in terms of the stress resultants M and V defined in Eq. (7.3.28).

7.14 For the plane elasticity problems shown in Figs. P7.14(a)-(d), write the boundary conditions and classify them into type I, type II, or type III.

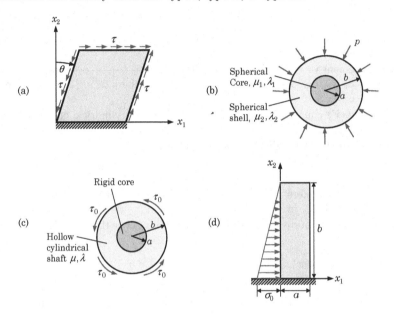

Fig. P7.14

CLAPEYRON'S, BETTI'S, AND MAXWELL'S THEOREMS

7.15 Consider a cantilever beam of length L, constant bending stiffness EI, and with right end $(x = L)$ fixed, as shown in Fig. P7.15. If the left end $(x = 0)$ is subjected to a moment M_0, use Clapeyron's theorem to determine the rotation (in the direction of the moment) at $x = 0$.

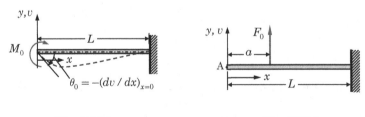

Fig. P7.15   Fig. P7.16

7.16 Consider a cantilever beam of length L, constant bending stiffness EI, and with the right end fixed, as shown in Fig. P7.16. If a point load F_0 is applied at a distance a from the free end, determine the deflection $v(a)$ using Clapeyron's theorem.

7.17 Determine the deflection at the midspan of a cantilever beam subjected to a uniformly distributed load q_0 throughout the span and a point load F_0 at the free end, as shown in Fig. P7.17. Use Maxwell's theorem and superposition.

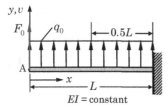
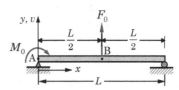

Fig. P7.17   **Fig. P7.18**

7.18 Consider a simply supported beam of length L subjected to a concentrated load F_0 at the midspan and a bending moment M_0 at the left end, as shown in Fig. P7.18. Verify that Betti's theorem holds.

7.19 Use the reciprocity theorem to determine the deflection $v_c = v(0)$ at the center of a simply supported circular plate under asymmetric loading (see Fig. P7.19):

$$q(r, \theta) = q_0 + q_1 \frac{r}{a} \cos \theta.$$

The deflection $v(r)$ due to a point load F_0 at the center of a simply supported circular plate is

$$v(r) = \frac{F_0 a^2}{16 \pi D} \left[\left(\frac{3+\nu}{1+\nu} \right) \left(1 - \frac{r^2}{a^2} \right) + 2 \left(\frac{r}{a} \right)^2 \log \left(\frac{r}{a} \right) \right],$$

where $D = Eh^3 / [12(1 - \nu^2)]$ and h is the plate thickness.

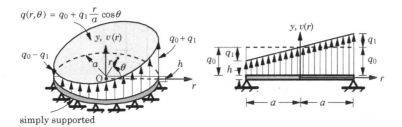

Fig. P7.19

7.20 Use the reciprocity theorem to determine the center deflection $v_c = v(0)$ of a simply supported circular plate under loading $q(r) = q_0(1 - r/a)$.

7.21 Use the reciprocity theorem to determine the center deflection $v_c = v(0)$ of a clamped circular plate under loading $q(r) = q_0(1 - r/a)$. The deflection due to a point load F_0 at the center of a clamped circular plate is given in Eq. (7.4.21).

7.22 Determine the center deflection $v_c = v(0)$ of a clamped circular plate subjected to a point load F_0 at a distance b from the center (and for some θ) using the reciprocity theorem.

7.23 Rewrite Eq. (7.3.10) in a form suitable for direct integration and obtain the solution given in Eq. (7.3.13). *Hint:* Note that $\frac{dU}{dR} + \frac{2U}{R} = \frac{1}{R^2} \frac{d}{dR} \left(R^2 U \right)$.

SOLUTION OF ELASTICITY PROBLEMS

7.24 Verify that the compatibility equation (3.7.4) takes the form

$$\varepsilon_{\alpha\alpha,\beta\beta} - \varepsilon_{\alpha\beta,\alpha\beta} = 0 \quad (\alpha, \beta = 1, 2), \tag{1}$$

or, in terms of stress components for the plane stress case,

$$\nabla^2 \sigma_{\alpha\alpha} = -(1+\nu) f_{\alpha,\alpha}. \tag{2}$$

7.25 Rewrite Eq. (7.5.17) in a form suitable for direct integration and obtain the solution given in Eq. (7.5.21). *Hint:* Note that $\frac{1}{r}\frac{dU}{dr} - \frac{U}{r^2} = \frac{d}{dr}\left(\frac{U}{r}\right)$.

7.26 Show that the solution to the differential equation for $G(r)$ in Eq. (6) is indeed given by the first equation in Eq. (7) of Example 7.5.7. *Hint:* Note that (verify to yourself)

$$\frac{d^2 G}{dr^2} + \frac{1}{r}\frac{dG}{dr} = \frac{1}{r}\frac{d}{dr}\left(r\frac{dG}{dr}\right); \quad \int r\ln r\,dr = \frac{r^2}{2}\left(\ln r - \frac{1}{2}\right).$$

7.27 Show that the solution to the differential equation for $F(r)$ in Eq. (6) is indeed given by the second equation in Eq. (7) of Example 7.5.7. *Hint:* Note that (verify to yourself)

$$\frac{d^2 F}{dr^2} + \frac{1}{r}\frac{dF}{dr} - \frac{4F}{r^2} = \frac{1}{r^3}\frac{d}{dr}\left(r^3\frac{dF}{dr} - 2r^2 F\right).$$

7.28 The only nonzero stress in a prismatic bar of length L, made of an isotropic material (E and ν), is $\sigma_{11} = -M_0 x_3/I$, where M_0 is the bending moment and I is the moment inertia about the x_2-axis, respectively. Determine the three-dimensional displacement field. Eliminate the rigid-body translations and rotations requiring that $\mathbf{u} = \mathbf{0}$ and $\mathbf{\Omega} = \mathbf{0}$ at $\mathbf{x} = \mathbf{0}$.

7.29 A solid circular cylindrical body of radius a and height h is placed between two rigid plates, as shown in Fig. P7.29. The plate at B is held stationary and the plate at A is subjected to a downward displacement of δ. Using a suitable coordinate system, write the boundary conditions for the two cases: (a) When the cylindrical object is bonded to the plates at A and B. (b) When the plates at A and B are frictionless.

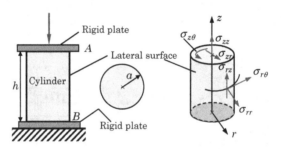

Fig. P7.29

7.30 The lateral surface of a homogeneous, isotropic, solid circular cylinder of radius a, length L, and mass density ρ is bonded to a rigid surface. Assuming that the ends of the cylinder at $z = 0$ and $z = L$ are traction-free (see Fig. P7.30), determine the displacement and stress fields in the cylinder due to its own weight.

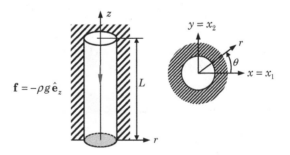

Fig. P7.30

7.31 An external hydrostatic pressure of magnitude p is applied to the surface of a spherical body of radius b with a concentric *rigid* spherical inclusion of radius a, as shown in Fig. P7.31. Determine the displacement and stress fields in the spherical body. Using the stress field obtained, determine the stresses at the surface of a rigid inclusion in an infinite elastic medium.

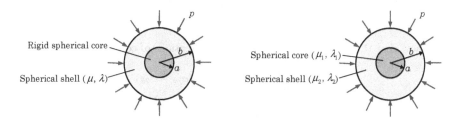

Fig. P7.31   Fig. P7.32

7.32 Consider the concentric spheres shown in Fig. P7.32. Suppose that the core is elastic and the outer shell is subjected to external pressure p (both are linearly elastic). Assuming Lamé constants of μ_1 and λ_1 for the core and μ_2 and λ_2 for the outer shell, and that the interface is perfectly bonded at $r = a$, determine the displacements of the core as well as for the shell.

7.33 Consider a long hollow circular shaft with a *rigid* internal core (a cross section of the shaft is shown in Fig. P7.33). Assuming that the inner surface of the shaft at $r = a$ is perfectly bonded to the rigid core and the outer boundary at $r = b$ is subjected to a uniform *shearing* traction of magnitude τ_0, find the displacement and stress fields in the problem.

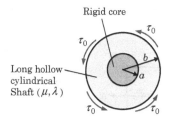

Fig. P7.33

AIRY STRESS FUNCTION

7.34 For the plane stress field

$$\sigma_{xx} = cxy, \quad \sigma_{xy} = 0.5c(h^2 - y^2), \quad \sigma_{yy} = 0,$$

where c and h are constants, (a) show that it is in equilibrium under a zero body force, and (b) find an Airy stress function $\Phi(x, y)$ corresponding to it.

7.35 In cylindrical coordinates, we assume that the body force vector $\mathbf{f}$ is derivable from the scalar potential $V_f(r, \theta)$:

$$\mathbf{f} = -\nabla V_f \quad \left(f_r = -\frac{\partial V_f}{\partial r}, \quad f_\theta = -\frac{1}{r}\frac{\partial V_f}{\partial \theta} \right), \tag{1}$$

and define the Airy stress function $\Phi(r, \theta)$ such that

$$\sigma_{rr} = \frac{1}{r}\frac{\partial \Phi}{\partial r} + \frac{1}{r^2}\frac{\partial^2 \Phi}{\partial \theta^2} + V_f,$$

$$\sigma_{\theta\theta} = \frac{\partial^2 \Phi}{\partial r^2} + V_f, \tag{2}$$

$$\sigma_{r\theta} = -\frac{\partial}{\partial r}\left(\frac{1}{r}\frac{\partial \Phi}{\partial \theta}\right).$$

Show that this choice trivially satisfies the equations of equilibrium

$$\frac{\partial \sigma_{rr}}{\partial r} + \frac{1}{r}\frac{\partial \sigma_{r\theta}}{\partial \theta} + \frac{1}{r}(\sigma_{rr} - \sigma_{\theta\theta}) + f_r = 0,$$

$$\frac{\partial \sigma_{\theta r}}{\partial r} + \frac{1}{r}\frac{\partial \sigma_{\theta\theta}}{\partial \theta} + \frac{2\sigma_{r\theta}}{r} + f_\theta = 0. \tag{3}$$

The tensor form of the compatibility condition in Eq. (7.5.33) is invariant.

7.36 Interpret the stress field obtained with the Airy stress function in Eq. (7.5.42) when all constants except c_3 are zero. Use the domain shown in Fig. 7.5.6 to sketch the stress field.

7.37 Interpret the following stress field obtained in Example 7.5.5 using the domain shown in Fig. 7.5.6:

$$\sigma_{xx} = 6c_{10}xy, \quad \sigma_{yy} = 0, \quad \sigma_{xy} = -3c_{10}y^2.$$

Assume that c_{10} is a positive constant.

7.38 Compute the stress field associated with the Airy stress function

$$\Phi(x, y) = Ax^5 + Bx^4y + Cx^3y^2 + Dx^2y^3 + Exy^4 + Fy^5.$$

Interpret the stress field for the case in which constants A, B, and C are zero. Use the rectangular domain shown in Fig. P7.38 to sketch the stress field on its boundaries.

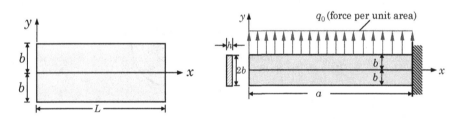

Fig. P7.38   Fig. P7.39

7.39 Determine the Airy stress function for the stress field of the beam shown in Fig. P7.39 and evaluate the stress field.

7.40 Investigate what problem is solved by the Airy stress function

$$\Phi = \frac{3A}{4b}\left(xy - \frac{xy^3}{3b^2}\right) + \frac{B}{4b}y^2,$$

where A and B are constants. Use the domain in Fig. P7.38 to sketch the stress field.

7.41 Show that the Airy stress function

$$\Phi(x, y) = \frac{q_0}{8b^3}\left[x^2\left(y^3 - 3b^2y + 2b^3\right) - \frac{1}{5}y^3\left(y^2 - 2b^2\right)\right]$$

satisfies the compatibility condition. Determine the stress field and find what problem it corresponds to when applied to the region $-b \leq y \leq b$ and $x = 0, L$ (see Fig. P7.38).

7.42 The thin cantilever beam shown in Fig. P7.42 is subjected to a uniform shearing traction of magnitude τ_0 along its upper surface. Determine if the Airy stress function

$$\Phi(x, y) = \frac{\tau_0}{4} \left(xy - \frac{xy^2}{b} - \frac{xy^3}{b^2} + \frac{ay^2}{b} + \frac{ay^3}{b^2} \right)$$

satisfies the compatibility condition and stress boundary conditions of the problem.

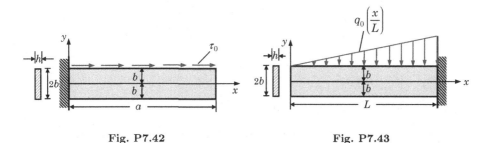

| Fig. P7.42 | Fig. P7.43 |

7.43 Consider the problem of a cantilever beam carrying a uniformly varying distributed transverse load, as shown in Fig. P7.43. The following Airy stress function is suggested (explain the terms to yourself):

$$\Phi(x, y) = Axy + Bx^3 + Cx^3y + Dxy^3 + Ex^3y^3 + Fxy^5.$$

Determine each of the constants and find the stress field.

7.44 The curved beam shown in Fig. P7.44 is curved along a circular arc. The beam is fixed at the upper end and it is subjected at the lower end to a distribution of tractions statically equivalent to a force per unit thickness $\mathbf{P} = -P\hat{\mathbf{e}}_1$. Assume that the beam is in a state of plane strain/stress. Show that an Airy stress function of the form

$$\Phi(r) = \left(Ar^3 + \frac{B}{r} + Cr \log r \right) \sin \theta$$

provides an approximate solution to this problem and solve for the values of the constants A, B, and C.

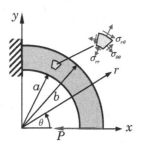

Fig. P7.44

7.45 Determine the stress field in a semi-infinite plate due to a normal load, f_0 force/unit length, acting on its edge, as shown in Fig. P7.45. Use the following Airy stress function (that satisfies the compatibility condition $\nabla^4 \Phi = 0$):

$$\Phi(r, \theta) = A\theta + Br^2\theta + Cr\theta \sin \theta + Dr\theta \cos \theta,$$

where A, B, C, and D are constants [see Eq. (7.5.40) for the definition of stress components in terms of the Airy stress function Φ]. Neglect the body forces (i.e., $V_f = 0$). *Hint:* Stresses must be single-valued. Determine the constants using the boundary conditions of the problem.

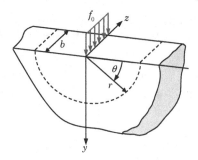

Fig. P7.45

TORSION OF CYLINDRICAL MEMBERS

7.46 Show that the resultant forces in the three coordinate directions on the end surface (i.e., $z = L$ face) are zero. Also show that the resultant moments about the x- and y-axes on the end surface are also zero.

7.47 Use the warping function $\psi(x, y) = kxy$, where k is a constant, to determine the cross section for which it is the solution. Determine the value of k in terms of the geometric parameters of the cross section and evaluate stresses in terms of these parameters and μ.

7.48 Consider a cylindrical member with the equilateral triangular cross section shown in Fig. P7.48. Show that the exact solution for the problem can be obtained and that the twist per unit length θ and stresses σ_{xz} and σ_{yz} are given by

$$\theta = \frac{5\sqrt{3}T}{27\mu a^4}, \quad \sigma_{xz} = \frac{\mu\theta}{a}(x - a)y, \quad \sigma_{yz} = \frac{\mu\theta}{2a}(x^2 + 2ax - y^2).$$

Hint: First write the equations for the three sides of the triangle (that is, $y = mx + c$, where m denotes the slope and c denotes the intercept), with the coordinate system shown in the figure, and then take the product of the three equations to construct the stress function. Also note that

$$\int_\Omega F(x, y)\, dx\, dy = \int_{-2a}^{a} \int_{-\frac{x+2a}{\sqrt{3}}}^{\frac{x+2a}{\sqrt{3}}} F(x, y)\, dy\, dx.$$

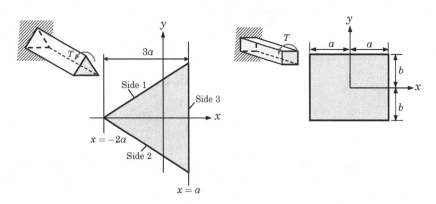

Fig. P7.48   **Fig. P7.49**

7.49 Consider torsion of a cylindrical member with the rectangular cross section shown in Fig. P7.46. Determine if a function of the form

$$\Psi = A\left(\frac{x^2}{a^2} - 1\right)\left(\frac{y^2}{b^2} - 1\right),$$

where A is a constant, can be used as a Prandtl stress function.

7.50 From Example 7.5.8, we know that for circular cylindrical members we have $\psi = 0$. Use the cylindrical coordinate system to show that $\sigma_{zr} = 0$ and $\sigma_{z\alpha} = Tr/J$, where J is the polar moment of inertia.

ENERGY AND VARIATIONAL METHODS

7.51 *Timoshenko beam theory.* Consider the displacement field

$$u_1(x, y) = y\phi(x), \quad u_2(x, y) = v(x), \quad u_3 = 0, \tag{1}$$

where $v(x)$ is the transverse deflection and ϕ is the rotation about the z-axis. Follow the developments of Section 7.3.4 and Example 7.6.1 (see Fig. 7.6.1) to develop the total potential energy functional

$$\Pi(u, w, \phi) = \frac{1}{2} \int_0^L \left[EI \left(\frac{d\phi}{dx} \right)^2 + GA \left(\frac{dv}{dx} + \phi \right)^2 - qv \right] dx - F_0 v(L) - M_0 \phi(L),$$

where EI is the bending stiffness and GA is the shear stiffness (E and G are Young's modulus and shear modulus, respectively, A is the cross-sectional area, and I is the moment of inertia). Then derive the Euler equations and the natural boundary conditions of the Timoshenko beam theory.

7.52 Identify the bilinear and linear forms associated with the quadratic functional of the Timoshenko beam theory in Problem **7.51**.

7.53 The total potential energy functional for a membrane stretched over domain $\Omega \in \Re^2$ is given by

$$\Pi(u) = \int_\Omega \left\{ \frac{T}{2} \left[\left(\frac{\partial u}{\partial x_1} \right)^2 + \left(\frac{\partial u}{\partial x_2} \right)^2 \right] - fu \right\} d\mathbf{x},$$

where $u = u(x_1, x_2)$ denotes the transverse deflection of the membrane, T is the tension in the membrane, and $f = f(x_1, x_2)$ is the transversely distributed load on the membrane. Determine the governing differential equation and the permissible boundary conditions for the problem (that is, identify the essential and natural boundary conditions of the problem) using the principle of minimum total potential energy.

7.54 Use the results of Example 7.6.2 to obtain the deflection at the center of a clamped-clamped beam (length $2L$ and $EI = $ constant) under uniform load of intensity q_0 and supported at the center by a linear elastic spring (k).

7.55 Use the results of Example 7.6.2 to obtain the deflection $v(L)$ and slopes $(-dv/dx)(L)$ and $(-dv/dx)(2L)$ under a point load F_0 for the beam shown in Fig. P7.55. It is sufficient to set up the three equations for the three unknowns.

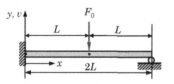
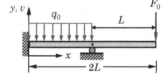

Fig. P7.55   Fig. P7.56

7.56 Use the results of Example 7.6.2 to obtain the deflection $v(2L)$ and slopes at $x = L$ and $x = 2L$ for the beam shown in Fig. P7.56. It is sufficient to set up the three equations for the three unknowns.

7.57 Consider an arbitrary triangular, plane elastic domain Ω of thickness h and made of orthotropic material. Suppose that the body is free of body forces but subjected to tractions on its sides, as shown in Fig. P7.57. Use Catigliano's theorem I and derive a relationship between the point displacements and the corresponding forces at the vertices of the triangle.

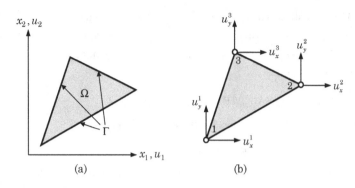

7.58 Find a two-parameter Ritz approximation of the transverse deflection of a simply supported beam (constant EI) on an elastic foundation (modulus k) that is subjected to a uniformly distributed load, q_0. Use (a) algebraic and (b) trigonometric polynomials.

7.59 Establish the total potential energy functional in Eq. (2) of Example 7.6.6.

7.60 Determine a one-parameter Ritz approximation $U_1(x)$ of $u(x)$, which is governed by the equation (like the equation governing the Prandtl stress function over square cross section of 2 units)

$$-\left(\frac{\partial^2 u}{\partial x^2} + \frac{\partial^2 u}{\partial y^2}\right) = f_0 \text{ in a unit square,}$$

subjected to the boundary conditions

$$u(1, y) = u(x, 1) = 0, \quad \frac{\partial u}{\partial x}\bigg|_{(0,y)} = \frac{\partial u}{\partial y}\bigg|_{(x,0)} = 0.$$

Take the origin of the coordinate system at the lower left corner of the unit square.

HAMILTON'S PRINCIPLE

7.61 Find Beltrami–Michell equations for dynamic elasticity.

7.62 Extend Clapeyron's Theorem to the dynamic case by starting with the expression

$$\int_0^T (U - K)\, d\mathbf{x},$$

where K is the kinetic energy.

7.63 Consider a pendulum of mass m_1 with a flexible suspension, as shown in Fig. P7.63. The hinge of the pendulum is in a block of mass m_2, which can move up and down between the frictionless guides. The block is connected by a linear spring (of spring constant k) to an immovable support. The coordinate x is measured from the position of the block in which the system remains stationary. Derive the Euler–Lagrange equations of motion for the system.

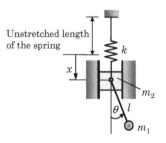

Fig. P7.63

7.64 A chain of total length L and mass m per unit length slides down from the edge of a smooth table. Assuming that the chain is rigid, find the equation of motion governing the chain (see Example 5.3.3).

7.65 Consider a cantilever beam supporting a lumped mass M at its end (J is the mass moment of inertia), as shown in Fig. P7.65. Derive the equations of motion and natural boundary conditions for the problem using the Euler–Bernoulli beam theory.

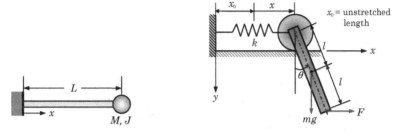

Fig. P7.65   Fig. P7.66

7.66 Derive the equations of motion of the system shown in Fig. P7.66. Assume that the mass moment of inertia of the link about its mass center is $J = m\Omega^2$, where Ω is the radius of gyration.

7.67 Derive the equations of motion of the *Timoshenko beam theory*, starting with the displacement field (including the axial displacement, u):

$$u_1(x, y, t) = u(x, t) + y\phi(x, t), \quad u_2 = v(x, t), \quad u_3 = 0.$$

Assume that the beam is subjected to distributed axial load $f(x, t)$ and transverse load $q(x, t)$, and that the x-axis coincides with the geometric centroid.

7.68 Derive the equations of motion of the third-order *Reddy beam theory* based on the displacement field

$$u_1(x, y, t) = u(x, t) + y\phi(x, t) - c_1 y^3 \left(\phi + \frac{\partial v}{\partial x}\right) \tag{1}$$

$$u_2(x, y, t) = v(x, t), \quad u_3 = 0,$$

where $c_1 = 4/(3h^2)$. Assume that the beam is subjected to distributed axial load $f(x, t)$ and transverse load $q(x, t)$, and that the x-axis coincides with the geometric centroid.

7.69 Consider a uniform cross-sectional bar of length L, with the left end fixed and the right end connected to a rigid support via a linear elastic spring (with spring constant k), as shown in Fig. P7.69. Determine the first two natural axial frequencies of the bar using the Ritz method. *Hint:* The kinetic energy K and the strain energy U associated with the axial motion of the member are given by

$$K = \int_0^L \frac{\rho A}{2} \left(\frac{\partial u}{\partial t}\right)^2 dx, \quad U = \int_0^L \frac{EA}{2} \left(\frac{\partial u}{\partial x}\right)^2 dx + \frac{k}{2}[u(L, t)]^2. \tag{1}$$

Use Hamilton's principle to obtain the variational equation, and for periodic motion assume

$$u(x, t) = u_0(x)e^{i\omega t}, \quad i = \sqrt{-1}, \tag{2}$$

where ω is the frequency of natural vibration, and $u_0(x)$ is the amplitude, to reduce the variational statement to

$$0 = \int_0^L \left(\rho A\omega^2 u_0 \delta u_0 - EA\frac{du_0}{dx}\frac{d\delta u_0}{dx}\right) dx - ku_0(L)\delta u_0(L). \tag{3}$$

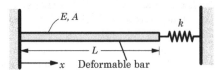

Fig. P7.69

7.70 Consider the equation

$$-\nabla^2 u = \lambda u, \quad \text{in } \Omega, \quad u = 0 \quad \text{on } \Gamma, \tag{1}$$

where Ω is the triangular domain shown in Fig. P7.48 and Γ is its boundary. Equation (1) describes a nondimensional form of the equation governing the natural vibration of a triangular membrane of side a; mass density ρ; and tension T ($\lambda = \rho a^2 \omega^2 / T$, ω being the natural frequency of vibration). Determine the fundamental frequency (that is, determine λ) of vibration by using a one-parameter Ritz approximation of the problem.

FLUID MECHANICS AND HEAT TRANSFER

It is not uncommon for engineers to accept the reality of phenomena that are not yet under-
stood, as it is very common for physicists to disbelieve the reality of phenomena that seem to
contradict contemporary beliefs of physics.   —— Henry H. Bauer (1931–)

8.1 Governing Equations

8.1.1 Preliminary Comments

Matter exists, in a majority of cases, only in two states: solid and fluid.[1] The
difference between the two is that a solid can resist shear force in static deformation whereas a fluid cannot. Shear force acting on a fluid causes it to deform
continuously. Thus, a fluid at rest can take only hydrostatic pressure and no
shear stress. Therefore, the stress vector at a point in a fluid at rest can be
expressed as

$$\mathbf{t}^{(\hat{\mathbf{n}})} = \hat{\mathbf{n}} \cdot \boldsymbol{\sigma} = -p\,\hat{\mathbf{n}} \ \text{ or } \ \boldsymbol{\sigma} = -p\,\mathbf{I}, \tag{8.1.1}$$

where $\hat{\mathbf{n}}$ is the unit vector normal to the surface and p is called the *hydrostatic*
pressure. It is clear from Eq. (8.1.1) that hydrostatic pressure is equal to the
negative of the mean stress

$$p = -\frac{1}{3}\sigma_{ii} = -\tilde{\sigma}. \tag{8.1.2}$$

In general, for a compressible fluid, p is related to temperature θ and density
ρ by an equation of the form

$$F(p, \rho, \theta) = 0. \tag{8.1.3}$$

This equation is called the *equation of state*. Recall from Section 6.5.3 that the
hydrostatic pressure p is not equal, in general, to the thermodynamic pressure
p appearing in the constitutive equation of a fluid in motion [see Eq. (6.5.1)]:

$$\boldsymbol{\sigma} = \mathbf{F}(\mathbf{D}) - p\mathbf{I} = \boldsymbol{\tau} - p\mathbf{I}, \tag{8.1.4}$$

where $\boldsymbol{\tau}$ is the viscous stress tensor, which is a function of the motion, namely,
the rate of deformation tensor $\mathbf{D}$; $\boldsymbol{\tau}$ vanishes when a fluid is at rest.

---

[1]In mathematical representation of matter, we may even entertain states such as "fluid-like"
or "solid-like" materials, which we do not deal with here.

Fluid mechanics is a branch of mechanics that deals with the effects of fluids at rest (statics) or in motion (dynamics) on surfaces where they come in contact. Fluids do not have the so-called natural state to which they return on removal of forces causing deformation. Therefore, we use a spatial (or Eulerian) description to write the governing equations. Pertinent equations are summarized next for an isotropic, Newtonian fluid. *Heat transfer* is a branch of engineering that deals with the transfer of thermal energy within a medium or from one medium to another due to a temperature difference. In this chapter, we study some typical problems of fluid mechanics and heat transfer.

8.1.2 Summary of Equations

The basic equations of viscous fluids are listed here. The number of equations, N_{eq}, and the number of new dependent variables, N_{var}, for three-dimensional problems are listed in parentheses. The stress tensor is assumed to be symmetric. All variables used here are the same as those introduced in the previous chapters, except that we shall use T in place of θ for the absolute temperature (because θ is used as a coordinate in the cylindrical and spherical coordinate systems).

Continuity equation ($N_{eq} = 1$, $N_{var} = 4$)

$$\frac{\partial \rho}{\partial t} + \text{div}(\rho \mathbf{v}) = 0; \quad \frac{D\rho}{Dt} + \rho \frac{\partial v_i}{\partial x_i} = 0 \tag{8.1.5}$$

Equations of motion ($N_{eq} = 3$, $N_{var} = 6$)

$$\boldsymbol{\nabla} \cdot \boldsymbol{\sigma} + \rho \mathbf{f} = \rho \left(\frac{\partial \mathbf{v}}{\partial t} + \mathbf{v} \cdot \boldsymbol{\nabla} \mathbf{v} \right); \quad \frac{\partial \sigma_{ji}}{\partial x_j} + \rho f_i = \rho \frac{Dv_i}{Dt} \tag{8.1.6}$$

Energy equation ($N_{eq} = 1$, $N_{var} = 4$)

$$\rho \frac{De}{Dt} = \boldsymbol{\sigma} : \mathbf{D} - \boldsymbol{\nabla} \cdot \mathbf{q} + \rho r_h; \quad \rho \frac{De}{Dt} = \sigma_{ij} D_{ij} - \frac{\partial q_i}{\partial x_i} + \rho r_h \tag{8.1.7}$$

Constitutive equation ($N_{eq} = 6$, $N_{var} = 7$)

$$\boldsymbol{\sigma} = 2\mu \mathbf{D} + \lambda (\text{tr} \, \mathbf{D}) \mathbf{I} - p \mathbf{I}; \quad \sigma_{ij} = 2\mu D_{ij} + \lambda D_{kk} \delta_{ij} - p \delta_{ij} \tag{8.1.8}$$

Heat conduction equation ($N_{eq} = 3$, $N_{var} = 1$)

$$\mathbf{q} = -k \boldsymbol{\nabla} T; \quad q_i = -k \frac{\partial T}{\partial x_i}. \tag{8.1.9}$$

Kinetic equation of state ($N_{eq} = 1$, $N_{var} = 0$)

$$p = p(\rho, T) \tag{8.1.10}$$

Caloric equation of state ($N_{eq} = 1$, $N_{var} = 0$)

$$e = e(\rho, T) \tag{8.1.11}$$

Rate of deformation-velocity equations ($N_{\text{eq}} = 6$, $N_{\text{var}} = 0$)

$$\mathbf{D} = \frac{1}{2}\left[\boldsymbol{\nabla}\mathbf{v} + (\boldsymbol{\nabla}\mathbf{v})^{\mathrm{T}}\right]; \quad D_{ij} = \frac{1}{2}\left(\frac{\partial v_i}{\partial x_j} + \frac{\partial v_j}{\partial x_i}\right) \tag{8.1.12}$$

Material time derivative

$$\frac{D}{Dt} \equiv \frac{\partial}{\partial t} + \mathbf{v}\cdot\boldsymbol{\nabla}; \quad \frac{D}{Dt} \equiv \frac{\partial}{\partial t} + v_i\frac{\partial}{\partial x_i} \tag{8.1.13}$$

Thus, there are 22 equations and 22 variables.

8.2 Fluid Mechanics Problems

8.2.1 Governing Equations of Viscous Fluids

Here we summarize the governing equations of fluid flows for the isothermal case. As in elasticity, the number of equations of fluid flow can be combined to obtain a smaller number of equations in as many unknowns. For instance, Eqs. (8.1.5), (8.1.6), (8.1.8), and (8.1.12) can be combined to yield the following equations:

$$\frac{\partial \rho}{\partial t} + \boldsymbol{\nabla}\cdot(\rho\mathbf{v}) = 0; \quad \frac{\partial \rho}{\partial t} + \frac{\partial(\rho v_i)}{\partial x_i} = 0, \tag{8.2.1}$$

$$\mu\nabla^2\mathbf{v} + (\mu+\lambda)\boldsymbol{\nabla}(\boldsymbol{\nabla}\cdot\mathbf{v}) - \boldsymbol{\nabla}p + \rho\mathbf{f} = \rho\left(\frac{\partial\mathbf{v}}{\partial t} + \mathbf{v}\cdot\boldsymbol{\nabla}\mathbf{v}\right);$$
$$\mu v_{i,jj} + (\mu+\lambda)v_{j,ji} - \frac{\partial p}{\partial x_i} + \rho f_i = \rho\left(\frac{\partial v_i}{\partial t} + v_j\frac{\partial v_i}{\partial x_j}\right). \tag{8.2.2}$$

Equations in (8.2.1) and (8.2.2) are known as the *Navier–Stokes equations.*

Equations (8.2.1) and (8.2.2) together contain four equations in five unknowns (v_1, v_2, v_3, ρ, p). For compressible fluids, Eqs. (8.2.1) and (8.2.2) are appended with Eqs. (8.1.7) and (8.1.9)–(8.1.11). For the isothermal case, Eqs. (8.2.1) and (8.2.2) are appended with Eq. (8.1.10), where $p = p(\rho)$.

For incompressible fluids, ρ is a known function of position, and thus we have four equations in four unknowns,

$$\boldsymbol{\nabla}\cdot\mathbf{v} = 0; \quad \frac{\partial v_i}{\partial x_i} = 0, \tag{8.2.3}$$

$$\mu\nabla^2\mathbf{v} - \boldsymbol{\nabla}p + \rho\mathbf{f} = \rho\left(\frac{\partial\mathbf{v}}{\partial t} + \mathbf{v}\cdot\boldsymbol{\nabla}\mathbf{v}\right);$$
$$\mu v_{i,jj} - \frac{\partial p}{\partial x_i} + \rho f_i = \rho\left(\frac{\partial v_i}{\partial t} + v_j\frac{\partial v_i}{\partial x_j}\right). \tag{8.2.4}$$

The expanded forms of these four equations in rectangular Cartesian and orthogonal curvilinear (i.e., cylindrical and spherical) coordinate systems are presented next.

Cartesian coordinate system (x, y, z); $v_1 = v_x$, $v_2 = v_y$, and $v_3 = v_z$:

$$\frac{\partial v_x}{\partial x} + \frac{\partial v_y}{\partial y} + \frac{\partial v_z}{\partial z} = 0 \tag{8.2.5}$$

$$\mu \left(\frac{\partial^2 v_x}{\partial x^2} + \frac{\partial^2 v_x}{\partial y^2} + \frac{\partial^2 v_x}{\partial z^2} \right) - \frac{\partial p}{\partial x} + \rho f_x = \rho \left(\frac{\partial v_x}{\partial t} + v_x \frac{\partial v_x}{\partial x} + v_y \frac{\partial v_x}{\partial y} + v_z \frac{\partial v_x}{\partial z} \right),$$

$$(8.2.6)$$

$$\mu \left(\frac{\partial^2 v_y}{\partial x^2} + \frac{\partial^2 v_y}{\partial y^2} + \frac{\partial^2 v_y}{\partial z^2} \right) - \frac{\partial p}{\partial y} + \rho f_y = \rho \left(\frac{\partial v_y}{\partial t} + v_x \frac{\partial v_y}{\partial x} + v_y \frac{\partial v_y}{\partial y} + v_z \frac{\partial v_y}{\partial z} \right),$$

$$(8.2.7)$$

$$\mu \left(\frac{\partial^2 v_z}{\partial x^2} + \frac{\partial^2 v_z}{\partial y^2} + \frac{\partial^2 v_z}{\partial z^2} \right) - \frac{\partial p}{\partial z} + \rho f_z = \rho \left(\frac{\partial v_z}{\partial t} + v_x \frac{\partial v_z}{\partial x} + v_y \frac{\partial v_z}{\partial y} + v_z \frac{\partial v_z}{\partial z} \right).$$

$$(8.2.8)$$

Cylindrical coordinate system (r, θ, z); $v_1 = v_r$, $v_2 = v_\theta$, and $v_3 = v_z$

$$\frac{1}{r} \frac{\partial (r v_r)}{\partial r} + \frac{1}{r} \frac{\partial v_\theta}{\partial \theta} + \frac{\partial v_z}{\partial z} = 0 \tag{8.2.9}$$

$$\mu \left[\frac{\partial}{\partial r} \left(\frac{1}{r} \frac{\partial}{\partial r} (r v_r) \right) + \frac{1}{r^2} \left(\frac{\partial^2 v_r}{\partial \theta^2} - 2 \frac{\partial v_\theta}{\partial \theta} \right) + \frac{\partial^2 v_r}{\partial z^2} \right] - \frac{\partial p}{\partial r} + \rho f_r$$

$$= \rho \left(\frac{\partial v_r}{\partial t} + v_r \frac{\partial v_r}{\partial r} + \frac{v_\theta}{r} \frac{\partial v_r}{\partial \theta} - \frac{v_\theta^2}{r} + v_z \frac{\partial v_z}{\partial z} \right) \tag{8.2.10}$$

$$\mu \left[\frac{\partial}{\partial r} \left(\frac{1}{r} \frac{\partial}{\partial r} (r v_\theta) \right) + \frac{1}{r^2} \left(\frac{\partial^2 v_\theta}{\partial \theta^2} + 2 \frac{\partial v_r}{\partial \theta} \right) + \frac{\partial^2 v_\theta}{\partial z^2} \right] - \frac{\partial p}{\partial \theta} + \rho f_\theta$$

$$= \rho \left(\frac{\partial v_\theta}{\partial t} + v_r \frac{\partial v_\theta}{\partial r} + \frac{v_\theta}{r} \frac{\partial v_\theta}{\partial \theta} + \frac{v_r v_\theta}{r} + v_z \frac{\partial v_\theta}{\partial z} \right) \tag{8.2.11}$$

$$\mu \left[\frac{1}{r} \frac{\partial}{\partial r} \left(r \frac{\partial v_z}{\partial r} \right) + \frac{1}{r^2} \frac{\partial^2 v_z}{\partial \theta^2} + \frac{\partial^2 v_z}{\partial z^2} \right] - \frac{\partial p}{\partial z} + \rho f_z$$

$$= \rho \left(\frac{\partial v_z}{\partial t} + v_r \frac{\partial v_z}{\partial r} + \frac{v_\theta}{r} \frac{\partial v_z}{\partial \theta} + v_z \frac{\partial v_z}{\partial z} \right) \tag{8.2.12}$$

Spherical coordinate system (R, ϕ, θ); $v_1 = v_R$, $v_2 = v_\phi$, and $v_3 = v_\theta$

$$2 \frac{v_R}{R} + \frac{\partial v_R}{\partial R} + \frac{1}{R \sin \phi} \frac{\partial (v_\phi \sin \phi)}{\partial \phi} + \frac{1}{R \sin \phi} \frac{\partial v_\theta}{\partial \theta} = 0 \tag{8.2.13}$$

$$\mu \left[\frac{1}{R^2} \frac{\partial^2}{\partial R^2} (R^2 v_R) + \frac{1}{R^2 \sin \phi} \frac{\partial}{\partial \phi} \left(\sin \phi \frac{\partial v_R}{\partial \phi} \right) + \frac{1}{R^2 \sin^2 \phi} \frac{\partial^2 v_R}{\partial \theta^2} \right] - \frac{\partial p}{\partial R} + \rho f_R$$

$$= \rho \left[\frac{\partial v_R}{\partial t} + v_R \frac{\partial v_R}{\partial R} + \frac{v_\phi}{R} \frac{\partial v_R}{\partial \phi} + \frac{v_\theta}{R \sin \phi} \frac{\partial v_R}{\partial \theta} - \left(\frac{v_\phi^2 + v_\theta^2}{R} \right) \right] \tag{8.2.14}$$

$$\mu\left[\frac{1}{R^2}\frac{\partial}{\partial R}\left(R^2\frac{\partial v_\phi}{\partial R}\right)+\frac{1}{R^2}\frac{\partial}{\partial\phi}\left(\frac{1}{\sin\phi}\frac{\partial}{\partial\phi}(v_\phi\sin\phi)\right)+\frac{1}{R^2\sin^2\phi}\frac{\partial^2 v_\phi}{\partial\theta^2}\right.$$

$$\left.+\frac{2}{R^2}\left(\frac{\partial v_R}{\partial\phi}-\frac{\cos\phi}{\sin^2\phi}\frac{\partial v_\theta}{\partial\theta}\right)\right]-\frac{1}{R}\frac{\partial p}{\partial\phi}+\rho f_\phi$$

$$=\rho\left(\frac{\partial v_\phi}{\partial t}+v_R\frac{\partial v_\phi}{\partial R}+\frac{v_\phi}{R}\frac{\partial v_\phi}{\partial\phi}+\frac{v_\theta}{R\sin\phi}\frac{\partial v_\phi}{\partial\theta}+\frac{v_R v_\phi}{R}-\frac{v_\theta^2\cot\phi}{R}\right) \qquad (8.2.15)$$

$$\mu\left[\frac{1}{R^2}\frac{\partial}{\partial R}\left(R^2\frac{\partial v_\theta}{\partial R}\right)+\frac{1}{R^2}\frac{\partial}{\partial\phi}\left(\frac{1}{\sin\phi}\frac{\partial}{\partial\phi}(v_\theta\sin\phi)\right)+\frac{1}{R^2\sin^2\phi}\frac{\partial^2 v_\theta}{\partial\theta^2}\right.$$

$$\left.+\frac{2}{R^2\sin\phi}\left(\frac{\partial v_R}{\partial\theta}+\cot\phi\frac{\partial v_\phi}{\partial\theta}\right)\right]-\frac{1}{R\sin\phi}\frac{\partial p}{\partial\theta}+\rho f_\theta$$

$$=\rho\left(\frac{\partial v_\theta}{\partial t}+v_R\frac{\partial v_\theta}{\partial R}+\frac{v_\phi}{R}\frac{\partial v_\theta}{\partial\phi}+\frac{v_\theta}{R\sin\phi}\frac{\partial v_\theta}{\partial\theta}+\frac{v_\theta v_R}{R}+\frac{v_\theta v_\phi}{R}\cot\phi\right) \qquad (8.2.16)$$

In general, finding exact solutions of the Navier–Stokes equations is an impossible task. The principal reason is the nonlinearity of the equations, and consequently, the principle of superposition is not valid. In the following sections, we shall find exact solutions of Eqs. (8.2.3) and (8.2.4) for certain flow problems for which the convective terms (i.e., $\mathbf{v}\cdot\nabla\mathbf{v}$) vanish and problems become linear. Of course, even for linear problems, flow geometry must be simple to be able to determine the exact solution. The books by Bird, et al. (1960) and Schlichting (1979) contain a number of such problems, and we discuss a few of them here. Like in linearized elasticity, often the semi-inverse method is used to obtain the solutions.

For several classes of flows with constant density and viscosity, the differential equations are expressed in terms of a potential function, called *stream function*, ψ. For two-dimensional planar problems (where $v_z = 0$ and data as well as the solution do not depend on z), the stream function is defined by

$$v_x = -\frac{\partial\psi}{\partial y}, \qquad v_y = \frac{\partial\psi}{\partial x}. \qquad (8.2.17)$$

This definition of ψ automatically satisfies the continuity equation (8.2.5):

$$\frac{\partial v_x}{\partial x}+\frac{\partial v_y}{\partial y}=-\frac{\partial^2\psi}{\partial x\partial y}+\frac{\partial^2\psi}{\partial x\partial y}=0.$$

Next, we determine the equation governing the stream function ψ. Recall the definition of the vorticity $\mathbf{w}=\frac{1}{2}\nabla\times\mathbf{v}$. In two dimensions, the only nonzero component of the vorticity vector is ζ ($\mathbf{w}=w_z\,\hat{\mathbf{e}}_z$)

$$\mathbf{w}=\frac{1}{2}\nabla\times\mathbf{v}, \qquad w_z=\frac{1}{2}\left(\frac{\partial v_y}{\partial x}-\frac{\partial v_x}{\partial y}\right). \qquad (8.2.18)$$

Substituting the definition (8.2.17) into Eq. (8.2.18), we obtain

$$\mathbf{w}=w_z\,\hat{\mathbf{e}}_z=\frac{1}{2}\left(\frac{\partial^2\psi}{\partial x^2}+\frac{\partial^2\psi}{\partial y^2}\right)\hat{\mathbf{e}}_z=\frac{1}{2}\nabla^2\psi\,\hat{\mathbf{e}}_z. \qquad (8.2.19)$$

Next, recall the vorticity equation (see Problems **5.17** and **6.26**):

$$\frac{\partial \mathbf{w}}{\partial t} + (\mathbf{v} \cdot \boldsymbol{\nabla})\mathbf{w} = (\mathbf{w} \cdot \boldsymbol{\nabla})\mathbf{v} + \nu \nabla^2 \mathbf{w}, \quad \nu = \frac{\mu}{\rho}. \tag{8.2.20}$$

For two-dimensional flows the vorticity vector $\mathbf{w}$ is perpendicular to the plane of the flow and, therefore, $(\mathbf{w} \cdot \boldsymbol{\nabla})\mathbf{v}$ is zero. Then

$$\frac{\partial \mathbf{w}}{\partial t} + \mathbf{v} \cdot \boldsymbol{\nabla}\mathbf{w} = \nu \nabla^2 \mathbf{w}. \tag{8.2.21}$$

Substituting Eq. (8.2.19) into the vorticity equation (8.2.21), we obtain

$$\frac{\partial \nabla^2 \psi}{\partial t} + (\mathbf{v} \cdot \boldsymbol{\nabla})(\nabla^2 \psi) = \nu \nabla^4 \psi. \tag{8.2.22}$$

In the rectangular Cartesian coordinate system, Eq. (8.2.22) has the form

$$\frac{\partial \nabla^2 \psi}{\partial t} + \left(-\frac{\partial \psi}{\partial y}\frac{\partial \nabla^2 \psi}{\partial x} + \frac{\partial \psi}{\partial x}\frac{\partial \nabla^2 \psi}{\partial y} \right) = \nu \nabla^4 \psi. \tag{8.2.23}$$

In the cylindrical coordinate system, the stream function ψ is related to the velocities

$$v_r = -\frac{1}{r}\frac{\partial \psi}{\partial \theta}, \quad v_\theta = \frac{\partial \psi}{\partial r}, \tag{8.2.24}$$

and the governing equation (8.2.23) takes the form

$$\frac{\partial \nabla^2 \psi}{\partial t} + \frac{1}{r}\left(-\frac{\partial \psi}{\partial \theta}\frac{\partial \nabla^2 \psi}{\partial r} + \frac{\partial \psi}{\partial r}\frac{\partial \nabla^2 \psi}{\partial \theta} \right) = \nu \nabla^4 \psi, \tag{8.2.25}$$

where ∇^2 is given in Table 2.4.2 for the cylindrical coordinate system.

In the spherical coordinate system, the stream function ψ is defined by

$$v_R = -\frac{1}{R^2 \sin\phi}\frac{\partial \psi}{\partial \phi}, \quad v_\phi = \frac{1}{R \sin\phi}\frac{\partial \psi}{\partial R}, \tag{8.2.26}$$

and Eq. (8.2.23) has the form

$$\frac{\partial \tilde{\nabla}^2 \psi}{\partial t} + \frac{1}{R^2 \sin\phi}\left(-\frac{\partial \psi}{\partial \phi}\frac{\partial \tilde{\nabla}^2 \psi}{\partial R} + \frac{\partial \psi}{\partial R}\frac{\partial \tilde{\nabla}^2 \psi}{\partial \phi} \right) = \nu \tilde{\nabla}^4 \psi, \tag{8.2.27}$$

$$\tilde{\nabla}^2 = \frac{\partial^2}{\partial R^2} + \frac{\sin\phi}{R^2}\frac{\partial}{\partial \phi}\left(\frac{1}{\sin\phi}\frac{\partial}{\partial \phi} \right).$$

8.2.2 Inviscid Fluid Statics

For incompressible inviscid fluids (i.e., fluids with zero viscosity), the constitutive equation for stress is [see Eq. (6.5.16)]

$$\boldsymbol{\sigma} = -p\mathbf{I} \quad (\sigma_{ij} = -p\delta_{ij}),$$

where p is the hydrostatic pressure, the equations of motion (8.1.6) reduce to

$$-\boldsymbol{\nabla}p + \rho\mathbf{f} = \rho\frac{D\mathbf{v}}{Dt}. \tag{8.2.28}$$

The body force in hydrostatics problem often represents the gravitational force, $\rho \mathbf{f} = -\rho g\,\hat{\mathbf{e}}_3$, where the positive x_3-axis is taken positive upward. Consequently, the equations of motion reduce to

$$-\frac{\partial p}{\partial x_1} = \rho a_1, \quad -\frac{\partial p}{\partial x_2} = \rho a_2, \quad -\frac{\partial p}{\partial x_3} = \rho g + \rho a_3, \tag{8.2.29}$$

where $a_i = \dot{v}_i$ is the ith component of acceleration.

For steady flows with constant velocity field, the equations in (8.2.29) simplify to

$$-\frac{\partial p}{\partial x_1} = 0, \quad -\frac{\partial p}{\partial x_2} = 0, \quad -\frac{\partial p}{\partial x_3} = \rho g. \tag{8.2.30}$$

The first two equations in (8.2.30) imply that $p = p(x_3)$. Integrating the third equation with respect to x_3, we obtain

$$p(x_3) = -\rho g x_3 + c_1,$$

where c_1 is the constant of integration, which can be evaluated using the pressure boundary condition at $x_3 = H$, where H is the height of the column of liquid; see Fig. 8.2.1(a). On the free surface we have $p = p_0$, where p_0 is the atmospheric pressure. Then the constant of integration is $c_1 = p_0 + \rho g H$ and we have

$$p(x_3) = \rho g(H - x_3) + p_0. \tag{8.2.31}$$

For the unsteady case in which the fluid in a rectangular container moves at a constant acceleration a_1 in the x_1-direction, the equations of motion in Eq. (8.2.29) become

$$-\frac{\partial p}{\partial x_1} = \rho a_1, \quad -\frac{\partial p}{\partial x_2} = 0, \quad -\frac{\partial p}{\partial x_3} = \rho g, \tag{8.2.32}$$

From the second equation it follows that $p = p(x_1, x_3)$. Integrating the first equation with respect to x_1, we obtain

$$p(x_1, x_3) = -\rho a_1 x_1 + f(x_3),$$

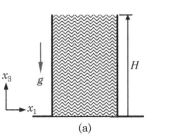
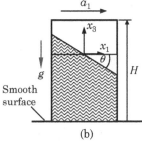

(a)   (b)

Fig. 8.2.1: (a) Column of liquid of height H. (b) A container of fluid moving with a constant acceleration, $\mathbf{a} = a_1\hat{\mathbf{e}}_1$.

where $f(x_3)$ is a function of x_3 alone. Substituting the preceding equation into the third equation in (8.2.32), and integrating with respect to x_3, we arrive at

$$f(x_3) = \rho g x_3 + c_2, \quad p(x_1, x_3) = -\rho a_1 x_1 + \rho g x_3 + c_2,$$

where c_2 is a constant of integration. If $x_3 = 0$ is taken on the free surface of the fluid in the container, then $p = p_0$ at $x_1 = x_3 = 0$, giving $c_2 = p_0$. Thus,

$$p(x_1, x_3) = p_0 - \rho a_1 x_1 + \rho g x_3. \tag{8.2.33}$$

Equation (8.2.33) suggests that the free surface (which is a plane), where $p = p_0$, is given by the equation $a_1 x_1 = g x_3$. The orientation of the plane is given by the angle θ, as shown in Fig. 8.2.1(b), where

$$\tan \theta = \frac{dx_3}{dx_1} = \frac{a_1}{g}. \tag{8.2.34}$$

When the fluid is a perfect gas, the constitutive equation for pressure is the equation of state

$$p = \rho RT, \tag{8.2.35}$$

where T is the absolute temperature (in degrees Kelvin) and R is the gas constant $(m \cdot N/kg \cdot K)$. If the perfect gas is at rest at a constant temperature, then we have

$$\frac{p}{p_0} = \frac{\rho}{\rho_0}, \tag{8.2.36}$$

where ρ_0 is the density at pressure p_0. From the third equation in (8.2.30), we have

$$dx_3 = -\frac{1}{\rho g} dp = -\frac{p_0}{\rho_0 g} \frac{dp}{p}.$$

Integrating from $x_3 = x_3^0$ to x_3, we obtain

$$x_3 - x_3^0 = -\frac{p_0}{\rho_0 g} \ln\left(\frac{p}{p_0}\right) \quad \text{or} \quad p = p_0 \exp\left(-\frac{x_3 - x_3^0}{p_0/\rho_0 g}\right). \tag{8.2.37}$$

8.2.3 Parallel Flow (Navier–Stokes Equations)

A flow is called *parallel* if only one velocity component is nonzero (i.e., all fluid particles moving in the same direction). Suppose that the x_1-axis is taken along the flow direction, and let $v_2 = v_3 = 0$ as well as the body forces f_1, f_2, and f_3 be negligible. Then it follows from Eq. (8.2.3) that

$$\frac{\partial v_1}{\partial x_1} = 0 \quad \rightarrow \quad v_1 = v_1(x_2, x_3, t). \tag{8.2.38}$$

Thus, for a parallel flow, we have

$$v_1 = v_1(x_2, x_3, t), \quad v_2 = v_3 = 0. \tag{8.2.39}$$

Consequently, the three equations of motion in (8.2.4) simplify to the following linear differential equations:

$$-\frac{\partial p}{\partial x_1} + \mu \left(\frac{\partial^2 v_1}{\partial x_2^2} + \frac{\partial^2 v_1}{\partial x_3^2} \right) = \rho \frac{\partial v_1}{\partial t}, \quad \frac{\partial p}{\partial x_2} = 0, \quad \frac{\partial p}{\partial x_3} = 0. \tag{8.2.40}$$

The last two equations in (8.2.40) imply that p is only a function of x_1. Thus, given the pressure gradient dp/dx_1, the first equation in (8.2.40) can be used to determine v_1. Next we consider some specific examples.

Example 8.2.1

Steady flow of viscous incompressible fluid between parallel plates. Consider a steady flow (i.e., $\partial v_1/\partial t = 0$) in a channel with two parallel flat walls (see Fig. 8.2.2). Let the distance between the two walls be b. Using the alternative notation, $x_1 = x$, $x_2 = y$, and $v_1 = v_x$, Eq. (8.2.40) can be reduced to the following boundary value problem:

$$\mu \frac{d^2 v_x}{dy^2} = \frac{dp}{dx}, \quad 0 < y < b \tag{1}$$

$$v_x(0) = 0, \quad v_x(b) = U.$$

When $U = 0$, the flow is known as the *Poiseuille flow*, and when $U \neq 0$, the flow is termed as the *Couette flow*. Determine the velocity distributions.

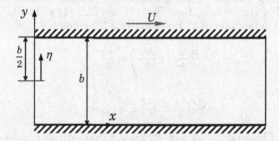

Fig. 8.2.2: Parallel flow through a straight channel.

Solution: The solution to the Couette flow problem described by Eq. (1) is

$$v_x(y) = \frac{y}{b} U - \frac{b^2}{2\mu} \frac{dp}{dx} \frac{y}{b} \left(1 - \frac{y}{b} \right), \quad 0 < y < b, \tag{2}$$

$$\bar{v}_x(\bar{y}) = \bar{y} + f \bar{y} (1 - \bar{y}), \quad \bar{v}_x = \frac{v_x}{U}, \quad \bar{y} = \frac{y}{b}, \quad f = -\frac{b^2}{2\mu U} \frac{dp}{dx}. \tag{3}$$

In the case of Poiseuille flow, the solution in Eqs. (2) and (3) reduces to

$$v_x(y) = -\frac{b^2}{2\mu} \frac{dp}{dx} \frac{y}{b} \left(1 - \frac{y}{b} \right), \quad 0 < y < b, \tag{4}$$

$$v_x(\eta) = -\frac{1}{2\mu} \frac{dp}{dx} \left(\frac{b^2}{4} - \eta^2 \right), \quad \eta = y - \frac{b}{2}, \quad -\frac{b}{2} < \eta < \frac{b}{2}. \tag{5}$$

Figures 8.2.3(a) and 8.2.3(b) show the velocity distributions for the two cases, $U = 0$ (Poiseuille flow) and $U \neq 0$ (Couette flow).

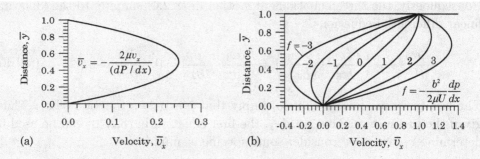

Fig. 8.2.3: Velocity distribution $\bar{v}_x(y)$ for (a) the Poiseuille flow and (b) the Couette flow.

Example 8.2.2

Steady flow of a viscous incompressible fluid through a circular pipe. The steady flow through a long, straight, horizontal circular pipe is another problem that admits exact solution to the Navier–Stokes equations. Use the cylindrical coordinate system with r being the radial coordinate, and the z-coordinate is taken along the axis of the pipe of radius R_0. Assume that the velocity components v_r and v_θ in the radial and tangential directions, respectively, are zero. Determine the (a) velocity $v_z = v_z(r)$, (b) volume rate of flow, and (c) shear stress τ_{rz} at the wall.

Solution: (a) The continuity equation (8.2.9) for the axisymmetric flow (i.e., the flow field is independent of θ) implies that the velocity component parallel to the axis of the pipe, v_z, is only a function of r. For steady flow, Eqs. (8.2.10) and (8.2.11) yield $(\partial p/\partial r) = (\partial p/\partial \theta) = 0$, implying that p is only a function of z. Equation (8.2.12) simplifies to

$$\frac{\mu}{r}\frac{d}{dr}\left(r\frac{dv_z}{dr}\right) = \frac{dp}{dz},\tag{1}$$

whose solution is given by

$$v_z(r) = \frac{1}{4\mu}\frac{dp}{dz}r^2 + A\log r + B,\tag{2}$$

where the constants of integration, A and B, are determined using the boundary conditions (of vanishing shear stress at the center of the pipe and zero velocity at the wall),

$$\text{at } r = 0: \quad r\tau_{rz} \equiv r\mu\left(\frac{\partial v_r}{\partial z} + \frac{\partial v_z}{\partial r}\right) = 0; \quad \text{and} \quad \text{at } r = R_0: \quad v_z = 0.\tag{3}$$

We find that

$$A = 0, \quad B = -\frac{R_0^2}{4\mu}\frac{dp}{dz},\tag{4}$$

and the solution becomes

$$v_z(r) = -\frac{1}{4\mu}\frac{dp}{dr}\left(R_0^2 - r^2\right) = -\frac{R_0^2}{4\mu}\frac{dp}{dr}\left(1 - \frac{r^2}{R_0^2}\right).\tag{5}$$

Thus, the velocity over the cross section of the pipe varies as a paraboloid of revolution. The maximum velocity occurs along the axis of the pipe and it is equal to

$$(v_z)_{\max} = v_z(0) = -\frac{R_0^2}{4\mu}\frac{dp}{dz}.\tag{6}$$

(b) The volume rate of flow through the pipe is

$$Q = \int_0^{2\pi}\int_0^{R_0} v_z(r)\, r\, dr\, d\theta = \frac{\pi R_0^4}{8\mu}\left(-\frac{dp}{dz}\right).\tag{7}$$

(c) The wall shear stress is (the minus sign implies that the direction is opposite to the flow)

$$\tau_w = -\mu \left(\frac{dv_z}{dr} \right)_{r=R_0} = \frac{R_0}{2} \frac{dp}{dz}. \tag{8}$$

Example 8.2.3

Unsteady flow of a viscous incompressible fluid through a circular pipe. Consider unsteady flow of a viscous fluid of constant ρ and μ through a long horizontal circular pipe of length L and radius R_0. Assume that the fluid is initially at rest. At $t = 0$ a pressure gradient dp/dz, assumed to be independent of t, is applied to the system. Determine the velocity profile as a function of time for $t > 0$.

Solution: Let $v_z(r,t) = V(r,t)$. Then Eq. (8.2.12) takes the form

$$\rho \frac{\partial V}{\partial t} = \frac{\mu}{r} \frac{\partial}{\partial r} \left(r \frac{\partial V}{\partial r} \right) - \frac{dp}{dz}, \quad 0 < r < R_0, \quad t > 0. \tag{1}$$

The boundary conditions in Eq. (3) of Example 8.2.2 are still valid for this problem for all $t > 0$. The initial condition is

$$V(r,0) = 0, \quad 0 < r < R_0. \tag{2}$$

To solve the problem, we introduce the following dimensionless variables:

$$\bar{V} = -\frac{4\mu L}{(dp/dz)R_0^2} V; \quad \xi = \frac{r}{R_0}; \quad \tau = \frac{\mu}{\rho R_0^2} t. \tag{3}$$

Then the governing equation in Eq. (1), boundary conditions in Eq. (3) of Example 8.2.2, and the initial condition in Eq. (3) become, respectively,

$$\frac{\partial \bar{V}}{\partial \tau} = \frac{1}{\xi} \frac{\partial}{\partial \xi} \left(\xi \frac{\partial \bar{V}}{\partial \xi} \right) + 4, \quad 0 < \xi < 1, \quad \tau > 0, \tag{4}$$

$$\text{B.C.: } \bar{V}(1,\tau) = 0, \ \bar{V}(0,\tau) \text{ is finite}; \quad \text{I.C.: } \bar{V}(\xi,0) = 0, \tag{5}$$

where the condition $d\bar{V}/d\xi = 0$ at $\xi = 0$ is replaced with $\bar{V}(0,\tau) = $ finite, and they are equivalent.

Next, we seek the solution $\bar{V}(\xi,\tau)$ as the sum of steady state solution $\bar{V}(\xi,\tau) \rightarrow \bar{V}_\infty(\xi)$ as $\tau \rightarrow \infty$ and transient solution $\bar{V}_\tau(\xi,\tau)$ such that

$$-4 = \frac{1}{\xi} \frac{d}{d\xi} \left(\xi \frac{d\bar{V}_\infty}{d\xi} \right), \tag{6}$$

$$\frac{\partial \bar{V}_\tau}{\partial \tau} = \frac{1}{\xi} \frac{\partial}{\partial \xi} \left(\xi \frac{\partial \bar{V}_\tau}{\partial \xi} \right). \tag{7}$$

Equation (6) is subjected to the boundary conditions

$$\bar{V}_\infty(1) = 0, \quad \bar{V}_\infty(0) \text{ is finite}; \tag{8}$$

Equation (7) is to be solved with the boundary and initial conditions:

$$\text{B.C.: } \bar{V}_\tau(1,\tau) = 0, \ \bar{V}_\tau(0,\tau) \text{ is finite}; \quad \text{I.C.: } \bar{V}_\tau(\xi,0) = -\bar{V}_\infty. \tag{9}$$

The solution of Eqs. (6) and (8) is

$$\bar{V}_\infty(\xi) = 1 - \xi^2. \tag{10}$$

The solution to Eqs. (7) and (9) can be obtained using the separation of variables technique. We assume a solution in the form

$$\bar{V}_\tau(\xi, \tau) = X(\xi)T(\tau), \tag{11}$$

and substitute into Eq. (7) to obtain

$$\frac{1}{T}\frac{dT}{d\tau} = \frac{1}{X}\frac{1}{\xi}\frac{d}{d\xi}\left(\xi\frac{dX}{d\xi}\right). \tag{12}$$

Because the left-hand side is a function of τ alone and the right-hand side is a function of ξ alone, it follows that both sides must be equal to a constant, which we choose to designate as $-\alpha^2$ (because the solution must be a decay type in τ and periodic in ξ). Thus, we have

$$\frac{dT}{d\tau} + \alpha^2 T = 0 \quad \rightarrow \quad T(\tau) = Ae^{-\alpha^2\tau}, \tag{13}$$

and

$$\frac{1}{\xi}\frac{d}{d\xi}\left(\xi\frac{dX}{d\xi}\right) + \alpha^2 X = 0 \quad \rightarrow \quad X(\xi) = C_1 J_0(\alpha\xi) + C_2 Y_0(\alpha\xi), \tag{14}$$

where J_0 and Y_0 are the zero-order Bessel functions of the first and second kind, respectively. The constants A, C_1, and C_2 must be determined such that the initial conditions in Eq. (9) are satisfied. The condition that $\bar{V}_\tau(0, \tau)$ be finite requires $X(0)$ to be finite. Since $Y_0(0) = -\infty$, it follows that $C_2 = 0$. The boundary condition $\bar{V}_\tau(1, \tau) = 0$ requires $X(1) = J_0(\alpha) = 0$. The function $J_0(\alpha)$ is an oscillating function, and thus has the following zeros [i.e., the roots of $J_0(\alpha_n) = 0$]:

$$\alpha_1 = 2.4048, \quad \alpha_2 = 5.5201, \quad \alpha_3 = 8.6537, \quad \alpha_4 = 11.7915, \quad \alpha_5 = 14.9309, \cdots . \tag{15}$$

Thus the total solution can be written as

$$\bar{V}_\tau(\xi, \tau) = \sum_{n=1}^{\infty} C_n e^{-\alpha_n^2 \tau} J_0(\alpha_n \xi). \tag{16}$$

The constants $C_n = AC_{1n}$ are determined using the initial condition in Eq. (9). We have

$$\bar{V}_\tau(\xi, 0) = -\bar{V}_\infty = -(1 - \xi^2) = \sum_{n=1}^{\infty} C_n J_0(\alpha_n \xi). \tag{17}$$

The functions $J_0(\alpha_n)$ satisfy the following orthogonality condition:

$$\int_0^1 J_0(\alpha_n \xi) J_0(\alpha_m \xi)\, \xi\, d\xi = \begin{cases} 0, & m \neq n \\ \beta_n, & m = n, \end{cases} \tag{18}$$

where β_n is given by

$$\beta_n = \int_0^1 [J_0(\alpha_n \xi)]^2\, \xi\, d\xi = \frac{1}{2}[J_1(\alpha_m)]^2,$$
$$\int_0^1 J_0(\alpha_n \xi)(1 - \xi^2)\xi\, d\xi = \frac{4J_1(\alpha_n)}{\alpha_n^3}. \tag{19}$$

The above integrals are evaluated using some standard relations for the Bessel functions. Thus, we obtain

$$C_n = -\frac{8}{\alpha_n^3 J_1(\alpha_n)}. \tag{20}$$

The final expression for the velocity $\bar{V}(\xi)$ is

$$\bar{V}(\xi, \tau) = (1 - \xi^2) - 8\sum_{n=1}^{\infty} \frac{J_0(\alpha_n \xi)}{\alpha_n^3 J_1(\alpha_n)} e^{-\alpha_n^2 \tau}. \tag{21}$$

8.2.4 Problems with Negligible Convective Terms

The exact solution of the Navier–Stokes equations is made difficult by the presence of the convective (nonlinear) terms, $\mathbf{v} \cdot \nabla \mathbf{v}$. When the motion is assumed to be very slow, the convective terms are very small compared to the viscous terms $\mu \nabla^2 \mathbf{v}$ and can be neglected, resulting in linear equations of motion. Such flows are called *creeping flows*, and the Navier–Stokes equations without the convective terms are often called the *Stokes equations*. For creeping flows of viscous incompressible fluids the governing equations (8.2.1) and (8.2.2) reduce to

$$\nabla \cdot \mathbf{v} = 0, \quad \frac{\partial v_i}{\partial x_i} = 0, \tag{8.2.41}$$

$$\rho \frac{\partial \mathbf{v}}{\partial t} = \mu \nabla^2 \mathbf{v} - \nabla p + \rho \mathbf{f}, \quad \rho \frac{\partial v_i}{\partial t} = \mu v_{i,jj} - \frac{\partial p}{\partial x_i} + \rho f_i. \tag{8.2.42}$$

These simplified equations can be solved to determine the flow field in some cases, which are discussed in Examples 8.2.4 and 8.2.5.

Example 8.2.4 ────────────────────────────

Flow of a viscous incompressible fluid around a sphere. Consider the steady slow flow of a viscous incompressible fluid around a sphere of radius R_0. The fluid approaches the sphere in the z direction at a velocity V_∞, as shown in Fig. 8.2.4. Neglecting the convective terms in Eq. (8.2.27), the governing equation (with no dependence on θ, and omitting v_θ terms) in terms of the stream function $\psi = \psi(R, \phi)$ is $\tilde{\nabla}^4 \psi = 0$ or

$$\left[\frac{\partial^2}{\partial R^2} + \frac{\sin \phi}{R^2} \frac{\partial}{\partial \phi} \left(\frac{1}{\sin \phi} \frac{\partial}{\partial \phi} \right) \right]^2 \psi(R, \phi) = 0. \tag{1}$$

This equation must be solved subjected to the boundary conditions

$$v_R(R_0, \phi) = -\frac{1}{R_0^2 \sin \phi} \frac{\partial \psi}{\partial \phi} \bigg|_{R=R_0} = 0,$$

$$v_\phi(R_0, \phi) = \frac{1}{R_0 \sin \phi} \frac{\partial \psi}{\partial R} \bigg|_{R=R_0} = 0, \tag{2}$$

$$v_R(R, \phi) = V_\infty \cos \phi \quad \text{and} \quad v_\phi = -V_\infty \sin \phi \quad \text{at} \quad R = \infty.$$

The first two conditions reflect the attachment of the viscous fluid to the surface of the sphere.

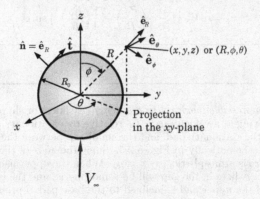

Fig. 8.2.4: Creeping flow around a sphere.

The third condition implies that $v_R = V_\infty$ far from the sphere,

$$\psi \to -\frac{1}{2} V_\infty R^2 \sin^2 \phi \quad \text{as} \quad R \to \infty. \tag{3}$$

Determine the velocity components v_R and v_ϕ by solving Eq. (1) for ψ subject to the boundary conditions in Eqs. (2) and (3).

Solution: The condition in Eq. (3) suggests the following form of the solution:

$$\psi(R, \phi) = f(R) \sin^2 \phi. \tag{4}$$

Substituting Eq. (4) into Eq. (1) gives

$$\left(\frac{d^2}{dR^2} - \frac{2}{R^2} \right) \left(\frac{d^2}{dR^2} - \frac{2}{R^2} \right) f(R) = 0, \tag{5}$$

or

$$\frac{d}{dR} \left\{ \frac{1}{R^2} \frac{d}{dR} \left[R^2 \frac{d}{dR} \left(\frac{1}{R^2} \frac{d}{dR}(Rf) \right) \right] \right\} = 0. \tag{6}$$

Successive integrations gives the result

$$f(R) = \frac{c_1}{R} + c_2 R + c_3 R^2 + c_4 R^4. \tag{7}$$

Satisfaction of the boundary condition in Eq. (3) requires that $f(R)$ can contain terms up to R^2 and $f(R) = -V_\infty/2$ at infinity, thus giving $c_4 = 0$ and $c_3 = -V_\infty/2$. The solution is

$$\psi(R, \phi) = f(R) \sin^2 \phi = \left(\frac{c_1}{R} + c_2 R - \frac{V_\infty}{2} R^2 \right) \sin^2 \phi. \tag{8}$$

The velocity components are [see Eq. (8.2.26)]

$$v_R = -\frac{1}{R^2 \sin \phi} \frac{\partial \psi}{\partial \phi} = \left(V_\infty - 2\frac{c_1}{R^3} - 2\frac{c_2}{R} \right) \cos \phi,$$
$$v_\phi = \frac{1}{R \sin \phi} \frac{\partial \psi}{\partial R} = \left(-V_\infty - \frac{c_1}{R^3} + \frac{c_2}{R} \right) \sin \phi. \tag{9}$$

Applying the boundary conditions in Eq. (2) gives $c_1 = -V_\infty R_0^3/4$ and $c_2 = 3V_\infty R_0/4$ so that the velocity distributions are (see Problem **8.15** for the shear stress and pressure distributions)

$$v_R = V_\infty \left[1 - \frac{3}{2}\left(\frac{R_0}{R} \right) + \frac{1}{2}\left(\frac{R_0}{R} \right)^3 \right] \cos \phi,$$
$$v_\phi = -V_\infty \left[1 - \frac{3}{4}\left(\frac{R_0}{R} \right) - \frac{1}{4}\left(\frac{R_0}{R} \right)^3 \right] \sin \phi. \tag{10}$$

Example 8.2.5

Flow of a viscous incompressible lubricant in a bearing. A slider (or slipper) bearing consists of a short sliding pad moving at a velocity $v_x = U_0$ relative to a stationary pad inclined at a small angle with respect to the stationary pad, and the small gap between the two pads is filled with a lubricant, as shown schematically in Fig. 8.2.5. Since the ends of the bearing are generally open, the pressure there is atmospheric, say $p = p_0$. When the upper pad is parallel to the base plate, the pressure everywhere in the gap will be atmospheric, and the bearing cannot support any transverse load. If the upper pad is inclined to the base pad, a pressure distribution is set up in the gap. For large values of U_0, the pressure generated can be of sufficient magnitude to support heavy loads normal to the base pad.

When the width of the gap and the angle of inclination are small, one may assume that $v_y = 0$ and $v_z = 0$ and the pressure is only a function of x. Assuming a two-dimensional state of flow in the xy-plane and a small angle of inclination of the stationary pad, and neglecting the normal stress gradient (in comparison with the shear stress gradient), the equations governing the flow of the lubricant between the pads can be reduced to [see Schlichting (1979)]

$$\mu \frac{\partial^2 v_x}{\partial y^2} = \frac{dp}{dx}, \quad \frac{dp}{dx} = \frac{6\mu U_0}{h^2}\left(1 - \frac{H}{h}\right), \quad 0 < x < L, \tag{1}$$

where

$$h(x) = h_1 + \frac{h_2 - h_1}{L}x, \quad H = \frac{2h_1 h_2}{h_1 + h_2}. \tag{2}$$

Determine the velocity and pressure distributions.

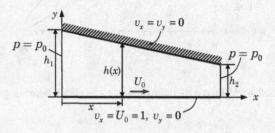

Fig. 8.2.5: Schematic of a slider bearing.

Solution: The solution of Eq. (1), subject to the boundary conditions $v_x(0,0) = U_0$ and $v_x(x,h) = 0$ is

$$v_x(x,y) = \left(U_0 - \frac{h^2}{2\mu}\frac{dp}{dx}\frac{y}{h}\right)\left(1 - \frac{y}{h}\right), \tag{3}$$

$$p(x) = \frac{6\mu U_0 L(h_1 - h)(h - h_2)}{h^2(h_1^2 - h_2^2)}, \tag{4}$$

$$\sigma_{xy}(x,y) = \mu\frac{\partial v_x}{\partial y} = \frac{dp}{dx}\left(y - \frac{h}{2}\right) - \mu\frac{U_0}{h}. \tag{5}$$

Numerical results are obtained using the following parameters:

$$h_1 = 2h_2 = 8 \times 10^{-4} \text{ ft}, \quad L = 0.36 \text{ ft}, \quad \mu = 8 \times 10^{-4} \text{ lb/ft}^2, \quad U_0 = 30 \text{ ft}. \tag{6}$$

Table 8.2.1 contains numerical values of the velocity, pressure, and shear stress as a function of position. Figure 8.2.6 contains plots of the velocity v_x at $x = 0$, 0.18, and 0.36 ft, while Fig. 8.2.7 contains plots of pressure and shear stress as a function of x at $y = 0$.

Table 8.2.1: Comparison of finite element solutions velocities with the analytical solutions for viscous fluid in a slider bearing.

$\bar{y}$	$v_x(0,y)$	$\bar{y}$	$v_x(0.18,y)$	$\bar{y}$	$v_x(0.36,y)$	x	$\bar{p}(x,0)$	$-\sigma_{xy}(x,0)$
0.0	30.000	0.00	30.000	0.00	30.000	0.01	7.50	59.99
1.0	22.969	0.75	25.156	0.50	29.531	0.03	22.46	59.89
2.0	16.875	1.50	20.625	1.00	28.125	0.05	37.29	59.67
3.0	11.719	2.25	16.406	1.50	25.781	0.07	51.89	59.30
4.0	7.500	3.00	12.500	2.00	22.500	0.09	66.12	58.77
5.0	4.219	3.75	8.906	2.50	18.281	0.27	129.60	38.40
6.0	1.875	4.50	5.625	3.00	13.125	0.29	118.57	32.71
7.0	0.469	5.25	2.656	3.50	7.031	0.31	99.58	25.70
8.0	0.000	6.00	0.000	4.00	0.000	0.33	70.30	17.04

$\bar{x} = 10x, \quad \bar{y} = y \times 10^4, \quad \bar{p} = p \times 10^{-2}.$

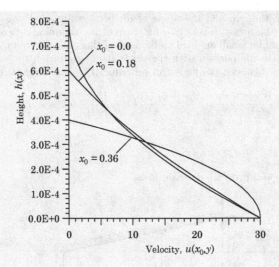

Fig. 8.2.6: Velocity distributions for the slider bearing problem.

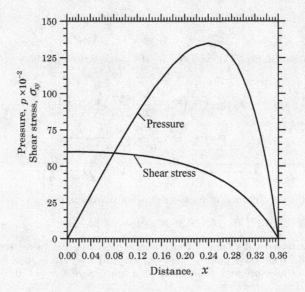

Fig. 8.2.7: Pressure and shear stress distributions for the slider bearing problem.

8.2.5 Energy Equation for One-Dimensional Flows

Various forms of the energy equation derived in the preceding sections are valid for any continuum. For simple, one-dimensional flow problems (i.e., problems with one stream of fluid particles), the equations derived are too complicated to be of use. In this section a simple form of the energy equation is derived for use with one-dimensional fluid flow problems.

The first law of thermodynamics for a system occupying the domain (control volume) Ω can be written as

$$\frac{D}{Dt} \int_\Omega \rho\epsilon \, dV = W_{\text{net}} + H_{\text{net}}, \qquad (8.2.43)$$

where ϵ is the total energy stored per unit mass, W_{net} is the net rate of work transferred into the system, and H_{net} is the net rate of heat transfer into the system. The total stored energy per unit mass ϵ consists of the internal energy per unit mass e, the kinetic energy per unit mass $v^2/2$, and the potential energy per unit mass gz (g is the gravitational acceleration and z is the vertical distance above a reference value):

$$\epsilon = e + \frac{v^2}{2} + gz. \qquad (8.2.44)$$

The rate of work done in the absence of body forces is given by ($\boldsymbol{\sigma} = \boldsymbol{\tau} - p\mathbf{I}$)

$$W_{\text{net}} = W_{\text{shaft}} - \oint_\Gamma p\mathbf{v} \cdot \hat{\mathbf{n}} \, ds, \qquad (8.2.45)$$

where p is the pressure (normal stress) and W_{shaft} is the rate of work done by the tangential force due to shear stress (e.g., in rotary devices such as fans, propellers, and turbines).

Using the Reynolds transport theorem (5.2.40) and Eqs. (8.2.44) and (8.2.45), we can write (8.2.43) as

$$\frac{\partial}{\partial t} \int_\Omega \rho\epsilon \, dV + \oint_\Gamma \left(e + \frac{p}{\rho} + \frac{v^2}{2} + gz \right) \rho\mathbf{v} \cdot \hat{\mathbf{n}} \, ds = W_{\text{shaft}} + H_{\text{net}}. \qquad (8.2.46)$$

If only one stream of fluid (compressible or incompressible) enters the control volume, the integral over the control surface in Eq. (8.2.46) can be written as

$$\left(e + \frac{p}{\rho} + \frac{v^2}{2} + gz \right)_{\text{out}} (\rho Q)_{\text{out}} - \left(e + \frac{P}{\rho} + \frac{v^2}{2} + gz \right)_{\text{in}} (\rho Q)_{\text{in}}, \qquad (8.2.47)$$

where ρQ denotes the mass flow rate. Finally, if the flow is steady, Eq. (8.2.46) can be written as

$$\left(e + \frac{p}{\rho} + \frac{v^2}{2} + gz \right)_{\text{out}} (\rho Q)_{\text{out}} - \left(e + \frac{p}{\rho} + \frac{v^2}{2} + gz \right)_{\text{in}} (\rho Q)_{\text{in}} = W_{\text{shaft}} + H_{\text{net}}. \qquad (8.2.48)$$

In writing Eq. (8.2.48), it is assumed that the flow is one-dimensional and the velocity field is uniform. If the velocity profile at sections crossing the control surface is not uniform, a correction must be made to Eq. (8.2.48). In particular, when the velocity profile is not uniform, the integral

$$\oint_\Gamma \frac{v^2}{2} \rho\mathbf{v} \cdot \hat{\mathbf{n}} \, ds$$

cannot be replaced with $(v^2/2)(\rho Q) = \rho A v^3/2$, where A is the cross-sectional area of the flow because integral of v^3 is different when v is uniform or varies across the section. If we define the ratio, called the *kinetic energy coefficient*

$$\alpha = \frac{\oint_\Gamma \frac{v^2}{2} \rho\mathbf{v} \cdot \hat{\mathbf{n}} \, ds}{(\rho Q v^2/2)}, \qquad (8.2.49)$$

Eq. (8.2.48) can be expressed as

$$\left(e + \frac{p}{\rho} + \frac{\alpha v^2}{2} + gz\right)_{\text{out}} (\rho Q)_{\text{out}} - \left(e + \frac{p}{\rho} + \frac{\alpha v^2}{2} + gz\right)_{\text{in}} (\rho Q)_{\text{in}} = W_{\text{shaft}} + H_{\text{net}}.$$

(8.2.50)

An example of the application of energy equation (8.2.50) is presented next.

Example 8.2.6

A pump delivers water at a steady rate of Q_0 (gal/min.), as shown in Fig. 8.2.8. If the left side pipe is of diameter d_1 (in.) and the right side pipe is of diameter d_2 (in.), and the pressures in the two pipes are p_1 and p_2 (psi), respectively, determine the horsepower (hp) required by the pump if the rise in the internal energy across the pump is e. Assume that there is no change of elevation in water level across the pump, and the pumping process is adiabatic (i.e., the heat transfer rate is zero). Use the following data ($\alpha = 1$):

$$\rho = 1.94 \text{ slugs/ft}^3, \quad d_1 = 4 \text{ in.}, \quad d_2 = 1 \text{ in.},$$

$$p_1 = 20 \text{ psi}, \quad p_2 = 50 \text{ psi}, \quad Q_0 = 350 \text{ gal/min.}, \quad e = 3300 \text{ lb-ft/slug}.$$

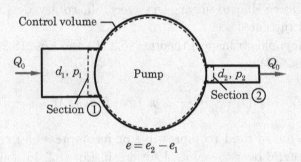

$$e = e_2 - e_1$$

Fig. 8.2.8: The pump considered in Example 5.4.1.

Solution: We take the control volume between the entrance and exit sections of the pump, as shown in dotted lines in Fig. 8.2.8. The mass flow rate entering and exiting the pump is the same (conservation of mass) and equal to

$$\rho Q_0 = \frac{1.94 \times 350}{7.48 \times 60} = 1.513 \text{ slugs/s}.$$

The velocities at Sections 1 and 2 are (converting all quantities to proper units) are

$$v_1 = \frac{Q_0}{A_1} = \frac{350}{7.48 \times 60} \frac{4 \times 144}{16\pi} = 8.94 \text{ ft/s},$$

$$v_2 = \frac{Q_0}{A_2} = \frac{350}{7.48 \times 60} \frac{4 \times 144}{\pi} = 143 \text{ ft/s}.$$

For adiabatic flow $H_{\text{net}} = 0$, the potential energy term is zero on account of no elevation difference between the entrance and exits, and $e = e_2 - e_1 = 3300$ ft-lb/slug. Thus, we have

$$W_{\text{shaft}} = \rho Q_0 \left[\left(e + \frac{p}{\rho} + \frac{v^2}{2}\right)_2 - \left(e + \frac{p}{\rho} + \frac{v^2}{2}\right)_1\right]$$

$$= (1.513) \left[3300 + \frac{(50 - 20) \times 144}{1.94} + \frac{(143)^2 - (8.94)^2}{2}\right] \frac{1}{550} = 43.22 \text{ hp}.$$

8.3 Heat Transfer Problems

8.3.1 Governing Equations

For heat transfer in incompressible fluids, neglecting mechanical stresses, the internal energy e is only a function of the absolute temperature T, $e = e(T)$ and ρ independent of T. For heat transfer in a solid medium, all of the velocity components and the dissipation $\Phi = \boldsymbol{\tau} : \mathbf{D}$ should be set to zero. Then we have

$$\rho\frac{De}{Dt} = \rho\frac{de}{dT}\frac{DT}{Dt}. \tag{8.3.1}$$

The quantity de/dT is the specific heat at constant volume, c_v:

$$c_v = \left(\frac{de}{dT}\right)\Big|_{\text{constant volume}}. \tag{8.3.2}$$

Then the energy equation [see Eqs. (8.1.7) and (8.1.9)] takes the form

$$\rho c_v\frac{DT}{Dt} = \boldsymbol{\nabla}\cdot(\mathbf{k}\cdot\boldsymbol{\nabla}T) + \rho r_h, \tag{8.3.3}$$

or, in index notation

$$\rho c_v\frac{DT}{Dt} = \frac{\partial}{\partial x_i}\left(k_{ij}\frac{\partial T}{\partial x_j}\right) + \rho r_h.$$

For heat transfer in fluids at constant pressure or heat transfer in solids, c_v is replaced with c_p, the specific heat at constant pressure. The expanded forms of Eq. (8.3.3) in rectangular Cartesian coordinates (x, y, z), cylindrical coordinates (r, θ, z), and spherical coordinates (R, ϕ, θ) are presented here for the isotropic case in which k and μ are constants (and $c_v \to c_p$).

Cartesian coordinate system (x, y, z)

$$\rho c_p\left(\frac{\partial T}{\partial t} + v_x\frac{\partial T}{\partial x} + v_y\frac{\partial T}{\partial y} + v_z\frac{\partial T}{\partial z}\right) = k\left(\frac{\partial^2 T}{\partial x^2} + \frac{\partial^2 T}{\partial y^2} + \frac{\partial^2 T}{\partial z^2}\right)$$

$$+ 2\mu\left[\left(\frac{\partial v_x}{\partial x}\right)^2 + \left(\frac{\partial v_y}{\partial y}\right)^2 + \left(\frac{\partial v_z}{\partial z}\right)^2\right] + \mu\left[\left(\frac{\partial v_x}{\partial y} + \frac{\partial v_y}{\partial x}\right)^2\right.$$

$$\left. + \left(\frac{\partial v_x}{\partial z} + \frac{\partial v_z}{\partial x}\right)^2 + \left(\frac{\partial v_y}{\partial z} + \frac{\partial v_z}{\partial y}\right)^2\right] + \rho r_h \tag{8.3.4}$$

Cylindrical coordinate system (r, θ, z)

$$\rho c_p\left(\frac{\partial T}{\partial t} + v_r\frac{\partial T}{\partial r} + \frac{v_\theta}{r}\frac{\partial T}{\partial \theta} + v_z\frac{\partial T}{\partial z}\right) = k\left[\frac{1}{r}\frac{\partial}{\partial r}\left(r\frac{\partial T}{\partial r}\right) + \frac{1}{r^2}\frac{\partial^2 T}{\partial \theta^2} + \frac{\partial^2 T}{\partial z^2}\right]$$

$$+ 2\mu\left\{\left(\frac{\partial v_r}{\partial r}\right)^2 + \left[\frac{1}{r}\left(\frac{\partial v_\theta}{\partial \theta} + v_r\right)\right]^2 + \left(\frac{\partial v_z}{\partial z}\right)^2\right\}$$

$$+ \mu\left\{\left(\frac{\partial v_\theta}{\partial z} + \frac{1}{r}\frac{\partial v_z}{\partial \theta}\right)^2 + \left(\frac{\partial v_z}{\partial r} + \frac{\partial v_r}{\partial z}\right)^2 + \left[\frac{1}{r}\frac{\partial v_r}{\partial \theta} + r\frac{\partial}{\partial r}\left(\frac{v_\theta}{r}\right)\right]^2\right\} + \rho r_h$$

$$\tag{8.3.5}$$

Spherical coordinate system (R, ϕ, θ)

$$\rho c_p \left(\frac{\partial T}{\partial t} + v_R \frac{\partial T}{\partial R} + \frac{v_\phi}{R} \frac{\partial T}{\partial \phi} + \frac{v_\theta}{R \sin \phi} \frac{\partial T}{\partial \theta} \right)$$

$$= k \left[\frac{1}{R^2} \frac{\partial}{\partial R} \left(R^2 \frac{\partial T}{\partial R} \right) + \frac{1}{R^2 \sin \phi} \frac{\partial}{\partial \phi} \left(\sin \phi \frac{\partial T}{\partial \phi} \right) + \frac{1}{R^2 \sin^2 \phi} \frac{\partial^2 T}{\partial \theta^2} \right]$$

$$+ 2\mu \left[\left(\frac{\partial v_R}{\partial R} \right)^2 + \left(\frac{1}{R} \frac{\partial v_\phi}{\partial \phi} + \frac{v_R}{R} \right)^2 + \left(\frac{1}{r \sin \phi} \frac{\partial v_\theta}{\partial \theta} + \frac{v_R}{R} + \frac{v_\phi \cot \phi}{R} \right)^2 \right]$$

$$+ \mu \left\{ \left[R \frac{\partial}{\partial R} \left(\frac{v_\phi}{r} \right) + \frac{1}{R} \frac{\partial v_R}{\partial \phi} \right]^2 + \left[\frac{1}{R \sin \phi} \frac{\partial v_R}{\partial \theta} + R \frac{\partial}{\partial R} \left(\frac{v_\theta}{R} \right) \right]^2 \right.$$

$$\left. + \left[\frac{\sin \phi}{R} \frac{\partial}{\partial \phi} \left(\frac{v_\theta}{\sin \phi} \right) + \frac{1}{R \sin \phi} \frac{\partial v_\phi}{\partial \theta} \right]^2 \right\} + \rho r_h \tag{8.3.6}$$

8.3.2 Heat Conduction in a Cooling Fin

Heat transfer from a surface to the surrounding fluid medium can be increased by attaching thin strips, called *fins*, of conducting material to the surface, as shown in Fig. 8.3.1(a). We assume that the fins are very long in the y-direction, and heat conducts only along the x-direction and convects through the lateral surface, i.e., $T = T(x, t)$. This assumption reduces the three-dimensional problem to a one-dimensional problem. By setting the velocity components to zero in Eq. (8.3.3) and noting that $T = T(x, t)$, we obtain

$$\rho c_p \frac{\partial T}{\partial t} = k \frac{\partial^2 T}{\partial x^2} + \rho r_h. \tag{8.3.7}$$

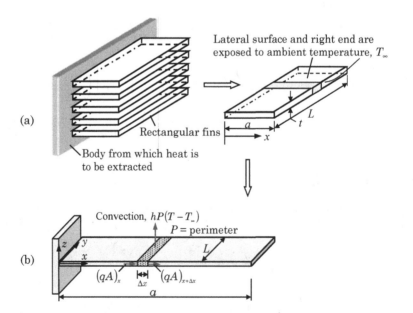

Fig. 8.3.1: Heat transfer in a cooling fin.

Equation (8.3.7) does not account for the cross-sectional area of the fin and convective heat transfer through the surface. Therefore, we derive the governing equation from the first principles, accounting for the cross-sectional change with x and convection from the surface. We assume steady heat conduction.

Consider an element of length Δx at a distance x in the fin, as shown in Fig. 8.3.1(b). The balance of energy in the element requires that

$$(qA)_x - (qA)_{x+\Delta x} - hP\Delta x(T - T_\infty) + \rho r_h(\frac{A_x + A_{x+\Delta x}}{2})\Delta x = 0, \qquad (8.3.8)$$

where q is the heat flux, A is the cross-sectional area (which can be a function of x), P is the perimeter, h is the film conductance, and Q is internal heat generation per unit mass (which is zero in the case of fins). Dividing throughout by Δx and taking the limit $\Delta x \to 0$, we obtain

$$-\frac{d}{dx}(qA) - hP(T - T_\infty) + \rho r_h A = 0. \qquad (8.3.9)$$

Using Fourier's law, $q = -k(dT/dx)$, where k is thermal conductivity of the fin, we obtain

$$\frac{d}{dx}\left(kA\frac{dT}{dx}\right) - hP(T - T_\infty) + \rho r_h A = 0, \qquad (8.3.10)$$

which governs one-dimensional heat transfer in a solid whose cross section A may be a function of x.

Example 8.3.1

Convection heat transfer in a fin. Determine the temperature distribution in a fin of length a, cross-sectional area A, and conductivity k (assume that k and A are constant). Suppose that the left end is maintained at temperature T_0 and the surface as well as the right end are exposed to a surrounding medium with temperature T_∞ and heat transfer coefficient h (relative to the fin). Assume that there is no internal generation (i.e., $r_h = 0$).

Solution: Equation (8.3.4) must be solved subject to the boundary conditions

$$T(0) = T_0, \qquad \left[kA\frac{dT}{dx} + hA(T - T_\infty)\right]_{x=a} = 0. \qquad (1)$$

The second boundary condition is a statement of the balance of energy (conductive and convective) at $x = a$.

We introduce the following nondimensional quantities for convenience of solving the problem analytically:

$$\theta = \frac{T - T_\infty}{T_0 - T_\infty}, \quad \xi = \frac{x}{a}, \quad m^2 = \frac{hP}{kA}a^2, \quad N = \frac{ha}{k}. \qquad (2)$$

Then Eqs. (8.3.10) and (1) take the form

$$\frac{d^2\theta}{d\xi^2} - m^2\theta = 0 \text{ for } 0 < \xi < 1; \quad \theta(0) = 1, \quad \left[\frac{d\theta}{d\xi} + N\theta\right]_{\xi=1} = 0. \qquad (3)$$

The general solution to the differential equation in (3) is

$$\theta(\xi) = c_1 \cosh m\xi + c_2 \sinh m\xi, \quad 0 < \xi < 1,$$

where the constants c_1 and c_2 are to be determined using the boundary conditions in (3). We obtain

$$\theta(0) = 1 \Rightarrow c_1 = 1; \qquad \left[\frac{d\theta}{d\xi} + N\theta\right]_{\xi=1} = 0 \Rightarrow c_2 = -\frac{m\sinh m + N\cosh m}{m\cosh m + N\sinh m}, \qquad (4)$$

and the solution becomes

$$\theta(\xi) = \frac{\cosh m\xi\,(m\cosh m + N\sinh m) - (m\sinh m + N\cosh m)\sinh m\xi}{m\cosh m + N\sinh m}$$

$$= \frac{m\cosh m(1-\xi) + N\sinh m(1-\xi)}{m\cosh m + N\sinh m}, \qquad 0 < \xi < 1. \qquad (5)$$

The *effectiveness of a fin* is defined by (omitting the end effects)

$$H_e = \frac{\text{Actual heat convected by the fin surface}}{\text{Heat that would be convected if the fin surface were held at } T_0}$$

$$= \frac{\int_0^L \int_0^a h(T - T_\infty)\,dx\,dy}{\int_0^L \int_0^a h(T_0 - T_\infty)\,dx\,dy} = \int_0^1 \theta(\xi)\,d\xi$$

$$= \int_0^1 \frac{m\cosh m(1-\xi) + N\sinh m(1-\xi)}{m\cosh m + N\sinh m}\,d\xi$$

$$= \frac{1}{m}\frac{m\sinh m + N(\cosh m - 1)}{m\cosh m + N\sinh m}. \qquad (6)$$

8.3.3 Axisymmetric Heat Conduction in a Circular Cylinder

Here we consider heat transfer in a long circular cylinder (see Fig. 8.3.2). If the boundary conditions and material of the cylinder are axisymmetric, that is, independent of the circumferential coordinate θ, it is sufficient to consider a typical rz-plane, where r is the radial coordinate and z is the axial coordinate. Further, if the cylinder is very long, say 10 diameters in length, then heat transfer along a typical radial line is all we need to determine; thus, the problem is reduced to a one-dimensional one.

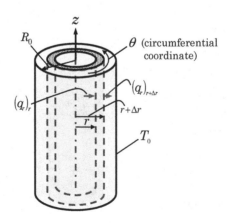

Fig. 8.3.2: Heat conduction in a circular cylinder.

The governing equation for this one-dimensional heat transfer in solids ($\mathbf{v} = 0$) can be obtained from Eq. (8.3.5) as

$$\rho c_p \frac{\partial T}{\partial t} = \frac{1}{r} \frac{\partial}{\partial r} \left(kr \frac{\partial T}{\partial r} \right) + \rho r_h(r), \quad 0 < r < R_0, \tag{8.3.11}$$

where ρr_h is internal heat generation per unit volume and R_0 is the radius of the cylinder. For example, in the case of an electric wire of circular cross section and electrical conductivity k_e (1/Ohm/m) heat is produced at the rate of

$$\rho r_h = \frac{I^2}{k_e}, \tag{8.3.12}$$

where I is electric current density (amps/m^2) passing through the wire. Equation (8.3.11) is to be solved subject to the appropriate initial condition and boundary conditions at $r = 0$ and $r = R_0$.

Example 8.3.2

Steady-state heat transfer in a long cylinder. Consider steady-state heat transfer in an isotropic cylinder of radius R_0, when there is a uniform internal heat generation of $\rho r_h = g$ and the surface of the cylinder is subjected to a temperature $T(R_0) = T_0$. Assuming that the cylinder is very long (so that there is no conduction of heat along the z-direction), determine the temperature distribution in the cylinder.

Solution: Owing to the axisymmetry (from geometry, material, and boundary conditions view points), the problem becomes one of solving the boundary value problem [the governing equation is deduced from Eq. (8.3.5)]:

$$k \frac{1}{r} \frac{d}{dr} \left(r \frac{dT}{dr} \right) + g = 0, \quad 0 < r < R_0; \quad (rq_r)_{r=0} = \left[-kr \frac{dT}{dr} \right]_{r=0} = 0, \quad T(R_0) = T_0. \tag{1}$$

The general solution is given by

$$T(r) = -\frac{gr^2}{4k} + c_1 \log r + c_2. \tag{2}$$

The constants c_1 and c_2 are determined using the boundary conditions:

$$(rq_r)(0) = 0 \Rightarrow c_1 = 0; \quad T(R_0) = T_0 \Rightarrow c_2 = T_0 + \frac{gR_0^2}{4k}.$$

Hence the solution (1) becomes

$$T(r) = T_0 + \frac{gR_0^2}{4k} \left(1 - \frac{r^2}{R_0^2} \right), \tag{3}$$

which is a quadratic function of the radial distance r. The heat flux is given by

$$q(r) = -k \frac{dT}{dr} = \frac{gr}{2}, \tag{4}$$

and the total heat flow at the surface of the cylinder is

$$Q = 2\pi R_0 L \, q(R_0) = \pi R_0^2 L g.$$

Example 8.3.3

Unsteady heat transfer in a long cylinder. Determine the transient (i.e., time-dependent) temperature profile in the cylinder of Example 8.3.2 when it is subjected to the following initial and boundary conditions:

$$\text{Initial condition:} \qquad\qquad T(r,0) = 0$$

$$\text{Boundary conditions:} \quad T(R_0, t) = 0, \quad (rq_r)_{r=0} = \left[-kr\frac{\partial T}{\partial r}\right]_{r=0} = 0. \tag{1}$$

The temperature rise is solely due to the internal heat generation g.

Solution: The governing differential equation is

$$k\frac{1}{r}\frac{\partial}{\partial r}\left(r\frac{\partial T}{\partial r}\right) + g = \rho c_p \frac{\partial T}{\partial t}, \quad 0 < r < R_0. \tag{2}$$

Let us introduce the following variables:

$$\theta = \frac{4k}{gR_0^2}T; \quad \xi = \frac{r}{R_0}; \quad \tau = \frac{k}{\rho c_p R_0^2}t. \tag{3}$$

Then Eq. (2) becomes

$$\frac{\partial T}{\partial \tau} = \frac{1}{\xi}\frac{\partial}{\partial \xi}\left(\xi\frac{\partial \theta}{\partial \xi}\right) + 4, \quad 0 < \xi < 1, \tag{4}$$

and the initial and boundary conditions become

$$\theta(\xi,0) = 0, \quad T(1,\tau) = 0, \quad \left[-\xi\frac{\partial \theta}{\partial \xi}\right]_{\xi=0} = 0. \tag{5}$$

The problem described by Eqs. (4) and (5) is equivalent to solving the problem described by Eqs. (6) and (7) of Example 8.2.3, where the finite-valuedness of θ at $\xi = 0$ is replaced by its derivative being zero there, both giving the same result [i.e., $C_2 = 0$ in Eq. (14) of Example 8.2.3]. Therefore, the transient solution of Eqs. (4) and (5) is given by Eq. (21) of Example 8.2.3:

$$\theta(\xi,\tau) = (1 - \xi^2) - 8\sum_{n=1}^{\infty}\frac{J_0(\alpha_n\xi)}{\alpha_n^3 J_1(\alpha_n)}e^{-\alpha_n^2\tau}, \tag{6}$$

where α_n are the roots of the zero-order Bessel function of the first kind, $J_0(\alpha_n) = 0$, and $J_1(\alpha_n)$ is the first-order Bessel function of the first kind. The first five roots of $J_0(\alpha_n) = 0$ are given in Eq. (15) of Example 8.2.3.

Several representative examples of one-dimensional heat transfer were discussed in the foregoing examples; numerous other problems that differ only in terms of the boundary conditions can be solved using the approaches presented there. Problems involving multiple materials can also be solved by imposing the continuity of temperature and balance of heats at the dissimilar material interfaces. The most important step in simplifying a problem to a one-dimensional one is to identify *solution* symmetry that may reduce the dimensionality of the problem. For example, for a hollow sphere with uniform temperatures at inner and outer surfaces, the temperature distribution with uniform internal heat generation can be solved as a one-dimensional problem.

8.3.4 Two-Dimensional Heat Transfer

Many problems of heat transfer require two- or three-dimensional analysis. Analytical solutions of these problems are limited to simple geometries and boundary conditions. Here we present an example of a steady-state, two-dimensional, heat transfer problem with the help of the separation-of-variables technique.

Example 8.3.4

Steady-state heat transfer in a rectangular plate. Consider steady-state heat conduction in a rectangular plate made of an isotropic material with conductivity k and sinusoidal temperature distribution on one edge, as shown in Fig. 8.3.3. Assume that there is no internal heat generation ($Q = 0$). Determine the temperature distribution in the plate.

Solution: The governing equation for this problem is a special case of Eq. (8.3.4), where all velocity terms as well as the time-derivative terms are set to zero:

$$k \left(\frac{\partial^2 T}{\partial x^2} + \frac{\partial^2 T}{\partial y^2} \right) = 0. \tag{1}$$

The boundary conditions are

$$T(x,0) = 0, \quad T(0,y) = 0, \quad T(a,y) = 0, \quad T(x,b) = T_0 \sin \frac{\pi x}{a}. \tag{2}$$

The classical approach to an analytical solution of the Laplace or Poisson equation over a regular (i.e., rectangular or circular) domain is the separation-of-variables technique. In this technique, we assume the temperature $T(x,y)$ to be of the form

$$T(x,y) = X(x)Y(y), \tag{3}$$

where X is a function of x alone and Y is a function of y alone. Substituting Eq. (3) into Eq. (1) and rearranging the terms, we obtain

$$\frac{1}{X} \frac{d^2 X}{dx^2} = -\frac{1}{Y} \frac{d^2 Y}{dy^2}. \tag{4}$$

Since the left-hand side is a function of x alone and the right-hand side is a function of y alone, it follows that both sides must be equal to a constant, which we choose to be $-\lambda^2$ (because the solution must be periodic in x so as to satisfy the boundary condition on the edge $y = b$). Thus, we have

$$\frac{d^2 X}{dx^2} + \lambda^2 X = 0, \quad \frac{d^2 Y}{dy^2} - \lambda^2 Y = 0, \tag{5}$$

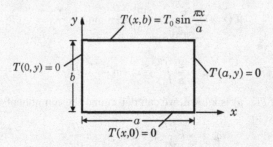

Fig. 8.3.3: Heat conduction in a rectangular plate.

whose general solutions are

$$X(x) = c_1 \cos \lambda x + c_2 \sin \lambda x, \quad Y(y) = c_3 e^{-\lambda y} + c_4 e^{\lambda y}. \tag{6}$$

The solution $T(x, y)$ is given by

$$T(x, y) = (c_1 \cos \lambda x + c_2 \sin \lambda x) \left(c_3 e^{-\lambda y} + c_4 e^{\lambda y} \right). \tag{7}$$

The constants c_i $(i = 1, 2, 3, 4)$ are determined using the boundary conditions in Eq. (2). We obtain

$$T(x, 0) = 0 \;\rightarrow\; (c_1 \cos \lambda x + c_2 \sin \lambda x)(c_3 + c_4) = 0 \;\rightarrow\; c_3 = -c_4,$$

$$T(0, y) = 0 \;\rightarrow\; c_1 \left(c_3 e^{-\lambda y} + c_4 e^{\lambda y} \right) = 0 \;\rightarrow\; c_1 = 0,$$

$$T(a, y) = 0 \;\rightarrow\; c_2 \sin \lambda a \left(c_3 e^{-\lambda y} + c_4 e^{\lambda y} \right) = 0 \;\rightarrow\; \sin \lambda a = 0.$$

The last conclusion is reached because $c_2 = 0$ will make the whole solution trivial. We have

$$\sin \lambda a = 0 \;\rightarrow\; \lambda a = n\pi \;\text{ or }\; \lambda_n = \frac{n\pi}{a}. \tag{8}$$

The solution in Eq. (7) now can be expressed as

$$T(x, y) = \sum_{n=1}^{\infty} A_n \sin \frac{n\pi x}{a} \sinh \frac{n\pi y}{a}. \tag{9}$$

The constants A_n, $n = 1, 2, \cdots$ are determined using the remaining boundary condition. We have

$$T(x, b) = T_0 \sin \frac{\pi x}{a} = \sum_{n=1}^{\infty} A_n \sin \frac{n\pi x}{a} \sinh \frac{n\pi b}{a}.$$

Multiplying both sides with $\sin(m\pi x/a)$ and integrating from 0 to a and using the orthogonality of the sine functions

$$\int_0^a \sin \frac{n\pi x}{a} \sin \frac{m\pi x}{a} \, dx = \begin{cases} 0, & m \neq n \\ \frac{a}{2}, & m = n \end{cases}$$

we obtain

$$A_1 = \frac{T_0}{\sinh \frac{n\pi b}{a}}, \quad A_n = 0 \text{ for } n \neq 1.$$

Hence, the final solution is

$$T(x, y) = T_0 \frac{\sinh \frac{\pi y}{a}}{\sinh \frac{\pi b}{a}} \sin \left(\frac{\pi x}{a} \right). \tag{10}$$

When the boundary condition at $y = b$ is replaced with $T(x, b) = f(x)$, then the solution is given by

$$T(x, y) = \sum_{n=1}^{\infty} A_n \frac{\sinh \frac{n\pi y}{a}}{\sinh \frac{n\pi b}{a}} \sin \left(\frac{n\pi x}{a} \right), \tag{11}$$

with A_n given by

$$A_n = \frac{2}{a} \int_0^a f(x) \sin \frac{n\pi x}{a} \, dx. \tag{12}$$

Once the temperature $T(x, y)$ is known, we can determine the components of heat flux, q_x and q_y, from Fourier's law:

$$q_x = -k \frac{\partial T}{\partial x}, \quad q_y = -k \frac{\partial T}{\partial y}. \tag{13}$$

8.3.5 Coupled Fluid Flow and Heat Transfer

Many real-world systems involve coupled heat transfer and fluid flows. Internal combustion engines and solar collectors are good examples (they also involve stress and deformation due to temperature differences). Although the study such complex problems is outside the scope of this first book on continuum mechanics, an example in which the fluid flow is coupled to heat transfer is presented next.

Example 8.3.5 ——————————————————————————————

Coupled heat transfer and fluid flow. Consider the fully developed, incompressible, steady-state Couette flow between parallel plates with zero pressure gradient (see Section 8.2). Suppose that the top plate moving with a velocity U and maintained at a temperature T_1 and the bottom plate is stationary and maintained at temperature T_0 (see Fig. 8.3.4). Assuming fully developed temperature profile and zero internal heat generation, determine the temperature field.

Solution: For fully developed temperature field, we can assume that $T = T(y)$. Then the energy equation (8.3.4) reduces to

$$k\frac{d^2T}{dy^2} + \mu\left(\frac{dv_x}{dy}\right)^2 = 0 \rightarrow \frac{d^2T}{dy^2} = -\frac{\mu}{kb^2}U^2. \tag{1}$$

The solution of this equation is

$$T(y) = -\frac{\mu U^2}{kb^2}\frac{y^2}{2} + Ay + B,$$

where the constants A and B are determined using the boundary conditions $T(0) = T_0$ and $T(b) = T_1$. We obtain

$$T(0) = T_0 \Rightarrow B = T_0; \quad T(b) = T_1 \Rightarrow A = \frac{T_1 - T_0}{b} + \frac{\mu U^2}{2kb}. \tag{2}$$

Thus the temperature field in the channel is given by

$$\frac{T(y) - T_0}{T_1 - T_0} = \frac{y}{b} + \frac{\mu U^2}{2k(T_1 - T_0)}\frac{y}{b}\left(1 - \frac{y}{b}\right). \tag{3}$$

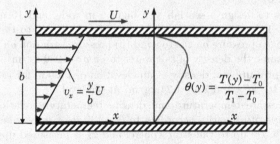

Fig. 8.3.4: Velocity and temperature distributions for Couette flow.

8.4 Summary

In this chapter, analytical solutions of selective problems in fluid mechanics and heat transfer are presented. First, a summary of the equations as applied to viscous incompressible fluids are presented in Cartesian, cylindrical, and spherical coordinate systems. The following example problems involving viscous incompressible fluid were formulated and their analytical solutions were determined: (1) a problem of inviscid fluid statics (using Cartesian coordinates); (2) steady flow of a viscous fluid between parallel plates (using Cartesian coordinates); (3) steady flow of a viscous fluid through a circular pipe (using cylindrical coordinates); (4) unsteady flow of a viscous fluid through a circular pipe (using cylindrical coordinates); (5) steady flow of a viscous fluid around a sphere (using spherical coordinates); and (6) steady flow of a viscous fluid through a slider bearing (using Cartesian coordinates). The energy equation for one-dimensional uniform, steady flow of incompressible fluid is also developed and an example of its application is presented.

In the heat transfer section, the energy equation is specialized to heat transfer in fluids or solids, and the equation is expressed in Cartesian, cylindrical, and spherical coordinate systems. The example problems discussed include (1) one-dimensional, convection heat transfer in a fin; (2) steady-state heat transfer in a long cylinder; (3) unsteady in a long cylinder; (4) steady-state heat transfer in a rectangular plate; and (5) coupled steady flow of a viscous incompressible fluid between differentially heated plates.

For all problems discussed in this chapter, emphasis is placed on the formulation of the problem and identifying and simplifying the pertinent equations for the problem under consideration.

Problems

Note: Problems **6.20–6.34** on fluid flow and heat transfer are suitable for this chapter.

FLUID MECHANICS

8.1 Assume that the velocity components in an incompressible flow are independent of the x coordinate and $v_z = 0$ to simplify the continuity equation (8.2.5) and the equations of motion (8.2.6)–(8.2.8).

8.2 An engineer is to design a sea lab 4 m high, 5 m wide, and 10 m long to withstand submersion to 120 m, measured from the surface of the sea to the top of the sea lab. Determine the (a) pressure on the top and (b) pressure variation on the side of the cubic structure. Assume the density of salt water to be $\rho = 1020$ kg/m^3.

8.3 Compute the pressure and density at an elevation of 1600 m for isothermal conditions. Assume $P_0 = 10^2$ kPa and $\rho_0 = 1.24$ kg/m^3 at sea level.

8.4 Derive the pressure–temperature and density–temperature relations for an ideal gas when the temperature varies according to $\theta(x_3) = \theta_0 + mx_3$, where m is taken to be $m = -0.0065°$C/m up to the stratosphere, and x_3 is measured upward from sea level. *Hint:* Use Eq. (8.2.35) and the third equation in Eq. (8.2.32).

8.5 Consider the steady flow of a viscous incompressible Newtonian fluid down an inclined surface of slope α under the action of gravity (see Fig. P8.5). The thickness of the fluid perpendicular to the plane is h and the pressure on the free surface is p_0, a constant. Use the semi-inverse method (i.e., assume the form of the velocity field) to determine the pressure and velocity field.

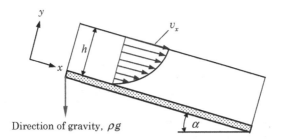

Fig. P8.5

8.6 Two immiscible fluids are flowing in the x-direction in a horizontal channel of length L and width $2b$ under the influence of a fixed pressure gradient. The fluid rates are adjusted such that the channel is half filled with fluid I (denser phase) and half filled with fluid II (less dense phase). Assuming that the gravity of the fluids is negligible, determine the velocity field. Use the geometry and coordinate system shown in Fig. P8.6.

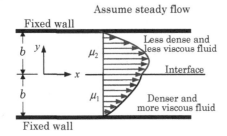

Fig. P8.6

8.7 Consider the steady flow of a viscous, incompressible fluid in the annular region between two coaxial circular cylinders of radii R_0 and αR_0, $\alpha < 1$, as shown in Fig. P8.7. Take $\bar{p} = p + \rho g z$. Determine the velocity and shear stress distributions in annulus.

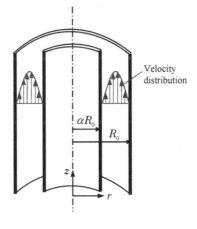

Fig. P8.7

8.8 Consider a steady, isothermal, incompressible fluid flowing between two vertical concentric long circular cylinders with radii R_1 and R_2. If the outer one rotating with an angular velocity Ω, show that the Navier–Stokes equations reduce to the following equations governing the circumferential velocity v_θ and pressure p:

$$\rho\frac{v_\theta^2}{r} = \frac{\partial p}{\partial r}, \quad \mu\frac{d}{dr}\left(\frac{1}{r}\frac{d}{dr}(rv_\theta)\right) = 0, \quad 0 = -\frac{\partial p}{\partial z} + \rho r_h.$$

Determine the velocity v_θ and shear stress $\tau_{r\theta}$ distributions.

8.9 Consider an isothermal, incompressible fluid flowing radially between two concentric porous spherical shells. Assume steady flow with v_R. Simplify the continuity and momentum equations for the problem.

8.10 A fluid of constant density and viscosity is in a cylindrical container of radius R_0 and the container is rotated about its axis with an angular velocity of Ω. Use the cylindrical coordinate system with the z-coordinate along the cylinder axis. Let the body force vector to be equal to $\rho\mathbf{f} = -g\hat{\mathbf{e}}_z$. Assume that $v_r = u = 0$ and $v_z = w = 0$, and $v_\theta = v_\theta(r)$ and simplify the governing equations. Determine $v_\theta(r)$ from the second momentum equation subject to the boundary condition $v_\theta(R_0) = \Omega r$. Then evaluate the pressure p from the remaining equations.

8.11 Consider the *unsteady* parallel flow on a flat plate (or plane wall). Assume that the motion is started impulsively from rest. Take the x-coordinate along the plate and the y-coordinate perpendicular to the wall. Assume that only nonzero velocity component is $v_x = v_x(y, t)$ and that the pressure p is a constant. Show that the Navier–Stokes equations for this case are simplified to

$$\rho\frac{\partial v_x}{\partial t} = \mu\frac{\partial^2 v_x}{\partial y^2}, \quad 0 < y < \infty. \tag{1}$$

Solve Eq. (1) for $v_x(y)$ using the following initial and boundary conditions:

$$\begin{aligned}
\text{Initial condition} \quad & v_x(y, 0) = 0, \\
\text{Boundary conditions} \quad & v_x(0, t) = U_0, \quad v_x(\infty, t) = 0.
\end{aligned} \tag{2}$$

Hint: Introduce a new coordinate η by assuming $\eta = y/(2\sqrt{\nu t})$, where ν is the kinematic viscosity $\nu = \mu/\rho$, and seek a solution in the form $v_x(\eta) = U_0 f(\eta)$. The solution is obtained in terms of the *complementary error function*:

$$\text{erfc}\,\eta = \frac{2}{\sqrt{\pi}}\int_\eta^\infty e^{-\eta^2}\,d\eta = 1 - \text{erf}\,\eta = 1 - \frac{2}{\sqrt{\pi}}\int_0^\eta e^{-\eta^2}\,d\eta, \tag{3}$$

where erf η is the *error function*.

8.12 Solve Eq. (1) of Problem **8.11** for the following boundary conditions (i.e., flow near an oscillating flat plate)

$$\begin{aligned}
\text{Initial condition} \quad & v_x(y, 0) = 0, \\
\text{Boundary conditions} \quad & v_x(0, t) = U_0\cos nt, \quad u(\infty, t) = 0.
\end{aligned} \tag{1}$$

In particular, obtain the solution

$$v_x(y, t) = U_0 e^{\lambda y}\cos(nt - \lambda y), \quad \lambda = \sqrt{\frac{\rho n}{2\mu}}. \tag{2}$$

8.13 Show that the components of the viscous stress tensor τ [see Eq. (6.5.7)] for an isotropic, viscous, Newtonian fluid in cylindrical coordinates are related to the velocity gradients by

$$\tau_{rr} = 2\mu\frac{\partial v_r}{\partial r} + \lambda\nabla\cdot\mathbf{v}, \qquad \tau_{\theta\theta} = 2\mu\left(\frac{1}{r}\frac{\partial v_\theta}{\partial\theta} + \frac{v_r}{r}\right) + \lambda\nabla\cdot\mathbf{v},$$

$$\tau_{zz} = 2\mu\frac{\partial v_z}{\partial z} + \lambda\nabla\cdot\mathbf{v}, \qquad \tau_{r\theta} = \mu\left[r\frac{\partial}{\partial r}\left(\frac{v_\theta}{r}\right) + \frac{1}{r}\frac{\partial v_r}{\partial\theta}\right],$$

$$\tau_{z\theta} = \mu\left(\frac{\partial v_\theta}{\partial z} + \frac{1}{r}\frac{\partial v_z}{\partial\theta}\right), \qquad \tau_{zr} = \mu\left(\frac{\partial v_z}{\partial r} + \frac{\partial v_r}{\partial z}\right),$$

$$\nabla\cdot\mathbf{v} = \frac{1}{r}\frac{\partial(rv_r)}{\partial r} + \frac{1}{r}\frac{\partial v_\theta}{\partial\theta} + \frac{\partial v_z}{\partial z}.$$

8.14 Show that the components of the viscous stress tensor τ [see Eq. (6.5.7)] for an isotropic, viscous, Newtonian fluid in spherical coordinates are related to the velocity gradients by

$$\tau_{RR} = 2\mu \frac{\partial v_R}{\partial R} + \lambda \boldsymbol{\nabla} \cdot \mathbf{v}, \qquad \tau_{\phi\phi} = 2\mu \left(\frac{1}{R} \frac{\partial v_\theta}{\partial \theta} + \frac{v_R}{R} \right) + \lambda \boldsymbol{\nabla} \cdot \mathbf{v}$$

$$\tau_{\theta\theta} = 2\mu \left(\frac{1}{R \sin \phi} \frac{\partial v_\phi}{\partial \phi} + \frac{v_R}{R} + \frac{v_\phi \cot \phi}{R} \right) + \lambda \boldsymbol{\nabla} \cdot \mathbf{v}$$

$$\tau_{R\phi} = \mu \left[R \frac{\partial}{\partial R} \left(\frac{v_\phi}{R} \right) + \frac{1}{R} \frac{\partial v_R}{\partial \phi} \right], \qquad \tau_{R\theta} = \mu \left[\frac{1}{r \sin \phi} \frac{\partial v_R}{\partial \theta} + R \frac{\partial}{\partial R} \left(\frac{v_\theta}{R} \right) \right]$$

$$\tau_{\phi\theta} = \mu \left[\frac{\sin \phi}{R} \frac{\partial}{\partial \phi} \left(\frac{v_\theta}{\sin \phi} \right) + \frac{1}{R \sin \phi} \frac{\partial v_\phi}{\partial \theta} \right]$$

$$\boldsymbol{\nabla} \cdot \mathbf{v} = \frac{1}{R^2} \frac{\partial (R^2 v_R)}{\partial R} + \frac{1}{R \sin \phi} \frac{\partial}{\partial \phi} (v_\phi \sin \phi) + \frac{1}{R \sin \phi} \frac{\partial v_\theta}{\partial \theta}.$$

8.15 Use the velocity field in Eq. (10) of Example 8.2.4 to show that the shear stress component $\tau_{r\phi}$ and pressure p are

$$\tau_{R\phi} = \frac{3\mu V_\infty}{2R_0} \left(\frac{R_0}{R} \right)^4 \sin \phi, \quad p = p_0 - \rho g z - \frac{3\mu V_\infty}{2R_0} \left(\frac{R_0}{R} \right)^2 \cos \phi,$$

where p_0 is the pressure in the plane $z = 0$ far away from the sphere and $-\rho g z$ is the contribution of the fluid weight (hydrostatic effect).

8.16 Derive the following vorticity equation for a fluid of constant density and viscosity:

$$\frac{\partial \mathbf{w}}{\partial t} + (\mathbf{v} \cdot \boldsymbol{\nabla}) \mathbf{w} = (\mathbf{w} \cdot \boldsymbol{\nabla}) \mathbf{v} + \nu \nabla^2 \mathbf{w},$$

where $\mathbf{w} = \boldsymbol{\nabla} \times \mathbf{v}$ and $\nu = \mu/\rho$.

8.17 *Bernoulli's equations.* Consider a flow with hydrostatic pressure, $\sigma = -p\mathbf{I}$ and conservative body force $\mathbf{f} = -\boldsymbol{\nabla}\phi$.

    (a) For steady flow, show that

$$\mathbf{v} \cdot \boldsymbol{\nabla} \left(\frac{v^2}{2} + \phi \right) + \frac{1}{\rho} \mathbf{v} \cdot \boldsymbol{\nabla} p = 0.$$

    (b) For steady and irrotational (i.e., $\boldsymbol{\nabla} \times \mathbf{v} = 0$) flow, show that

$$\boldsymbol{\nabla} \left(\frac{v^2}{2} + \phi \right) + \frac{1}{\rho} \boldsymbol{\nabla} p = 0.$$

8.18 Use Bernoulli's equation (which is valid for *steady, frictionless, incompressible flow*) derived in Problem **8.17** to determine the velocity and discharge of the fluid at the exit of the nozzle in the wall of the reservoir shown in Fig. P8.18.

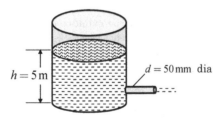

Fig. P8.18

8.19 The fan shown in Fig. P8.19 moves air ($\rho = 1.23$ kg/m^3) at a mass flow rate of 0.1 kg/min. The upstream side of the fan is connected to a pipe of diameter $d_1 = 50$ mm, the flow is laminar, the velocity distribution is parabolic, and the kinetic energy coefficient is $\alpha = 2$. The downstream of the fan is connected to a pipe of diameter $d_2 = 25$ mm, the flow is turbulent, the velocity profile is uniform, and the kinetic energy coefficient is $\alpha = 1$. If the rise in static pressure between upstream and downstream is 100 Pa and the fan motor draws 0.15 W, determine the loss $(-H_{\text{net}})$.

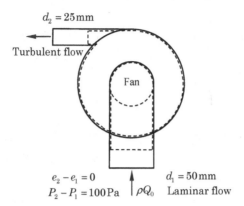

Fig. P8.19

8.20 Show that for a steady creeping flow of a viscous incompressible fluid in the absence of body forces, the Navier–Stokes equations become

$$\nabla p = \mu \nabla^2 \mathbf{v}.$$

HEAT TRANSFER

8.21 In heat transfer, one often neglects the strain energy part of the internal energy e and assumes that e depends only on the temperature θ, $e = e(\theta)$. Show that the energy equation (5.4.11) reduces to

$$\rho c \frac{D\theta}{Dt} = \nabla \cdot (\mathbf{k} \cdot \nabla \theta) + \rho r_h,$$

where $c = \partial e / \partial \theta$. State the assumptions under which the equation is derived.

8.22 Consider a long electric wire of length L and cross section with radius R_0 and electrical conductivity k_e [1/(Ohm·m)]. An electric current with current density I (amps/m²) is passing through the wire. The transmission of an electric current is an irreversible process in which some electrical energy is converted into thermal energy (heat). The rate of heat production per unit volume is given by

$$\rho r_h = \frac{I^2}{k_e}.$$

Assuming that the temperature rise in the cylinder is small enough not to affect the thermal or electrical conductivities and heat transfer is one-dimensional along the radius of the cylinder, derive the governing equation using balance of energy.

8.23 Solve the equation derived in Problem **8.22** using the boundary conditions

$$q(0) = \text{finite}, \quad T(R_0) = T_0.$$

8.24 A slab of length L is initially at temperature $f(x)$. For times $t > 0$ the boundaries at $x = 0$ and $x = L$ are kept at temperatures T_0 and T_L, respectively. Obtain the temperature distribution in the slab as a function of position x and time t.

8.25 A slab of unit height, $0 \leq x \leq 1$, is initially kept at temperature $T(x,0) = T_0(1 - x^2) = f(x)$. For times $t > 0$, the boundary at $x = 0$ is kept insulated and the boundary at $x = 1$ is kept at zero temperature, $T_1 = 0$. Determine the temperature distribution $T(x,t)$.

8.26 Determine the steady-state temperature distribution in an isotropic hollow sphere (internal and external radii are a and b, respectively) with uniform temperatures at the inner (T_i) and outer (T_0) surfaces but without internal heat generation.

8.27 Obtain the steady-state temperature distribution $T(x, y)$ in a rectangular region, $0 \leq x \leq a$, $0 \leq y \leq b$ for the boundary conditions

$$q_x(0, y) = 0, \quad q_y(x, b) = 0, \quad q_x(a, y) + hT(a, y) = 0, \quad T(x, 0) = f(x).$$

8.28 Consider the steady flow through a long, straight, horizontal circular pipe of radius R_0. The velocity field is given by (see Example 8.2.2)

$$v_r = 0, \quad v_\theta = 0, \quad v_z(r) = -\frac{R_0^2}{4\mu} \frac{dp}{dr} \left(1 - \frac{r^2}{R_0^2}\right). \tag{1}$$

If the pipe is maintained at a temperature T_0 on the surface, determine the steady-state temperature distribution in the fluid.

8.29 Consider the free convection problem of flow between two parallel plates of different temperature. A fluid density with density ρ and viscosity μ is placed between two *vertical* plates a distance $2a$ apart, as shown in Fig. P8.29. Suppose that the plate at $x = a$ is maintained at a temperature T_1 and the plate at $x = -a$ is maintained at a temperature T_2, with $T_2 > T_1$. Assuming that the plates are very long in the y-direction and hence that the temperature and velocity fields are only a function of x, determine the temperature $T(x)$ and velocity $v_y(x)$. Assume that the volume rate of flow in the upward moving stream is the same as that in the downward moving stream and the pressure gradient is solely due to the weight of the fluid.

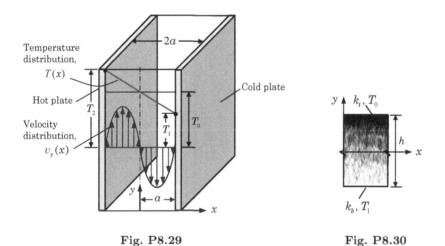

Fig. P8.29   **Fig. P8.30**

8.30 Determine the steady-state temperature distribution through an infinite slab of height h, thickness b (see Fig. P8.30), and made of isotropic material whose conductivity changes with the height according to the equation

$$k(y) = (k_t - k_b)f(y) + k_b, \quad f(y) = \left(\frac{1}{2} + \frac{y}{h}\right)^n, \tag{1}$$

where k_t and k_b are the values of the conductivities of the top and bottom surfaces, and n is a constant. Assume that the top is maintained at temperature T_0 and the bottom is maintained at temperature T_1.

LINEARIZED VISCOELASTICITY

In questions of science, the authority of a thousand is not worth the humble reasoning of a single individual. — Galileo Galilei (1564-1642)

All truths are easy to understand once they are discovered; the point is to discover them. — Galileo Galilei (1564-1642)

9.1 Introduction

9.1.1 Preliminary Comments

The simplest class of deformable solids are thermoelastic solids, which are elastic, nondissipative, and have no memory. When deformable solids also have the mechanism of dissipation, they are termed *thermoviscelastic solids,* which may or may not have memory. In this chapter we consider thermoviscoelastic solids with memory. When we restrict the case of infinitesimal deformations, then we have linear viscoelastic solids with or without memory.

Constitutive relations for linearized viscoelastic solids can be derived using one of two approaches. In the first approach, the entropy inequality is used to provide guidance. Some of the elements of this approach were discussed in Section 6.6.3, and they are helpful also for viscoelastic solids with memory. The alternative is to use a phenomenological approach, in which the observed physics is incorporated into a mathematical model that does not violate laws of physics, although the entropy inequality does not play a direct role. In this chapter, we consider the phenomenological approach of developing constitutive models for linear thermoviscoelasticity. There are many examples of viscoelastic materials with memory. Metals at elevated temperatures, concrete, and polymers are examples of viscoelastic behavior. As stated in Chapter 6, constitutive behavior is determined through experiments, and mathematical models of the constitutive behavior are developed to simulate the response, that is, experimental results are used to determine the parameters of the model and validate it. The mathematical models of the viscoelastic constitutive behavior are needed, just like Hooke's law, to analytically or numerically determine the system response. In this section we study mathematical models of linear viscoelastic behavior. The characteristics of a viscoelastic material are that they (a) have a time-dependent stress response and (b) even may have permanent deformation (i.e., do not return to original geometry after the removal of forces causing the deformation).

The viscoelastic response characteristics of a material are determined often using (1) creep tests, (2) stress relaxation tests, or (3) dynamic response to loads varying sinusoidally with time. The *creep* test involves determining the strain response under a constant stress, and it is done under uniaxial tensile stress owing to its simplicity. Application of a constant stress σ_0 produces a strain that, in general, contains three components: (1) an instantaneous, (2) a plastic, and (3) a delayed reversible component:

$$\varepsilon(t) = \left[J_\infty + \frac{t}{\eta_0} + \psi(t) \right] \sigma_0,$$

where $J_\infty \sigma_0$ is the instantaneous component of strain, η_0 is the Newtonian viscosity coefficient, and $\psi(t)$ is the creep function such that $\psi(0) = 0$. The *relaxation* test involves determination of stress under constant strain. Application of a constant strain ε_0 produces a stress that contains two components

$$\sigma(t) = [E_0 + \phi(t)] \varepsilon_0,$$

where E_0 is the static elastic modulus and $\phi(t)$ is the relaxation function such that $\phi(0) = 0$.

A qualitative understanding of actual viscoelastic behavior of materials can be gained through spring-and-dashpot models. For a linear response, combinations of linear elastic springs and linear viscous dashpots are used. Two simple spring-and-dashpot models are the *Maxwell model* and the *Kelvin–Voigt model*. The Maxwell model characterizes a viscoelastic fluid while the Kelvin–Voigt model represents a viscoelastic solid. Other combinations of these models are also used. The mathematical models to be discussed here provide some insight into the creep and relaxation characteristics of viscoelastic responses, but they may not represent a satisfactory quantitative behavior of any real material. A combination of the Maxwell and Kelvin–Voigt models may represent the creep and/or relaxation responses of some materials.

9.1.2 Initial Value Problem, the Unit Impulse, and the Unit Step Function

The governing equations of the mathematical models involving springs and dashpots are ordinary differential equations in time t. These equations relate stress σ to strain ε and they have the general form

$$P(\sigma) = Q(\varepsilon), \tag{9.1.1}$$

where P and Q are differential operators of order m and n, respectively,

$$P = \sum_{m=0}^{M} p_m \frac{d^m}{dt^m}, \quad Q = \sum_{n=0}^{N} q_n \frac{d^n}{dt^n}. \tag{9.1.2}$$

The coefficients p_m and q_n are known in terms of the spring constants k_i and dashpot constants η_i of the model. Equation (9.1.1) is solved either for $\varepsilon(t)$ for a specified $\sigma(t)$ (creep response) or for $\sigma(t)$ for a given $\varepsilon(t)$ (relaxation response). Since Eq. (9.1.1) is an mth-order differential equation for the relaxation response or nth-order equation for the creep response, we must know as many initial values, that is, values at time $t = 0$, of σ or ε

$$
\sigma(0) = \sigma_0, \quad \dot{\sigma}(0) = \dot{\sigma}_0, \quad \ldots, \quad \left(\frac{d^{M-1}\sigma}{dt^{M-1}}\right)_{t=0} = \sigma_0^{(M-1)}
$$

$$
\left[\varepsilon(0) = \varepsilon_0, \quad \dot{\varepsilon}(0) = \dot{\varepsilon}_0, \quad \ldots, \quad \left(\frac{d^{N-1}\varepsilon}{dt^{N-1}}\right)_{t=0} = \varepsilon_0^{(N-1)}\right], \tag{9.1.3}
$$

where $\sigma_0^{(i)}$, for example, denotes the value of the ith time derivative of $\sigma(t)$ at time $t = 0$. Equation (9.1.1) together with (9.1.3) defines an initial value problem.

In the forthcoming sections, we will study the creep and relaxation responses of the mathematical models of viscoelasticity under applied inputs. The applied stress or strain can be in the form of a *unit impulse* or a *unit step function*. The unit impulse, also known as the *Dirac delta function*, is defined as

$$
\delta(t - t_0) = 0, \quad \text{for } t \neq t_0,
$$

$$
\int_{-\infty}^{\infty} \delta(t - t_0)dt = 1. \tag{9.1.4}
$$

The units of the Dirac delta function are $1/s = s^{-1}$. A plot of the Dirac delta function is shown in Fig. 9.1.1(a). The time interval in which the Dirac delta function is nonzero is defined to be infinitely small, say ϵ. The Dirac delta function can be used to represent an arbitrary point value F_0 at $t = t_0$ as a function of time:

$$
f(t) = F_0\delta(t - t_0); \quad \int_{-\infty}^{\infty} f(t)\,dt = \int_{-\infty}^{\infty} F_0\delta(t - t_0)\,dt = F_0, \tag{9.1.5}
$$

where $f(t)$ has units of F_0 per second. The unit step function is defined as [see Fig. 9.1.1(b)]

$$
H(t - t_0) = \begin{cases} 0, & \text{for } t < t_0, \\ 1, & \text{for } t > t_0. \end{cases} \tag{9.1.6}
$$

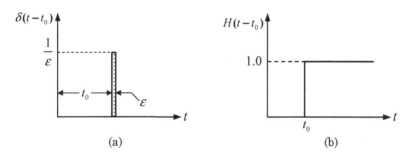

(a)   (b)

Fig. 9.1.1: (a) The Dirac delta function. (b) Unit step function.

Clearly, the function $H(t)$ is discontinuous at $t = t_0$, where its value jumps from 0 to 1. The unit step function is dimensionless. The unit step function $H(t)$, when it multiplies an arbitrary function $f(t)$, sets the portion of $f(t)$ corresponding to $t < 0$ to zero while leaving the portion corresponding to $t > 0$ unchanged. The Dirac delta function is viewed as the derivative of the unit step function; conversely, the unit step function is the integral of the Dirac delta function

$$\delta(t) = \frac{dH(t)}{dt}; \quad H(t) = \int_{-\infty}^{t} \delta(\xi)\, d\xi. \tag{9.1.7}$$

9.1.3 The Laplace Transform Method

Solving the ordinary differential equations in time, arising in the study of the creep or relaxation response, is not easy. The Laplace transform method is widely used to solve linear differential equations, especially those governing initial-value problems. The significant feature of the method is that it allows in a natural way the use of singularity functions such as the Dirac delta function and the unit step function in the data of the problem. Here we review the method in the context of solving initial value problems.

The (one-sided) *Laplace transformation* of a function $f(t)$, denoted $\bar{f}(s)$, is defined as

$$\bar{f}(s) \equiv \mathcal{L}[f(t)] = \int_{0}^{\infty} e^{-st} f(t)\, dt, \tag{9.1.8}$$

where s is, in general, a complex quantity referred as a *subsidiary variable*, and the function e^{-st} is known as the *kernel* of the transformation. The Laplace transforms of some functions are given in Table 9.1.1. The table can also be used for inverse transforms. Examples 9.1.1. and 9.1.2 illustrate the use of the Laplace transform method in the solution of differential equations.

Table 9.1.1: The Laplace transforms of some standard functions.

$f(t)$	$\bar{f}(s)$
$f(t)$	$\int_{0}^{\infty} e^{-st} f(t)\, dt$
$\dot{f} \equiv \frac{df}{dt}$	$s\bar{f}(s) - f(0)$
$\ddot{f} \equiv \frac{d^2 f}{dt^2}$	$s^2 \bar{f}(s) - sf(0) - \dot{f}(0)$
$f^{(n)}(t) \equiv \frac{d^n f}{dt^n}$	$s^n \bar{f}(s) - s^{n-1} f(0) - s^{n-2} \dot{f}(0) - \cdots - f^{(n-1)}(0)$
$\int_{0}^{t} f(\xi)\, d\xi$	$\frac{1}{s} \bar{f}(s)$
$\int_{0}^{t} \int_{0}^{\tau} f(\xi)\, d\xi\, d\tau$	$\frac{1}{s^2} \bar{f}(s)$
$\int_{0}^{t} f_1(\xi) f_2(t - \xi)\, d\xi$	$\bar{f}_1(s) \bar{f}_2(s)$
$H(t)$	$\frac{1}{s}$
$\delta(t) = \dot{H}(t)$	1
$\dot{\delta}(t) = \ddot{H}(t)$	s
$\delta^{(n)}(t)$	s^n
t	$\frac{1}{s^2}$
t^n	$\frac{n!}{s^{n+1}}$

Table 9.1.1: The Laplace transforms (continued).

$f(t)$	$\bar{f}(s)$
$tf(t)$	$-\bar{f}'(s)$
$t^n f(t)$	$(-1)^n \bar{f}^{(n)}(s)$
$\frac{1}{t} f(t)$	$\int_s^\infty f(\xi)\, d\xi$
$e^{at} f(t)$	$\bar{f}(s-a)$
e^{at}	$\frac{1}{s-a}$
te^{at}	$\frac{1}{(s-a)^2}$
$t^n e^{at}$	$\frac{n!}{(s-a)^{n+1}}, \quad n = 0, 1, 2, \cdots$
$e^{at} - e^{bt}$	$\frac{a-b}{(s-a)(s-b)}$
$\left(ae^{at} - be^{bt}\right)$	$\frac{s(a-b)}{(s-a)(s-b)}$
$\sin at$	$\frac{a}{s^2+a^2}$
$\cos at$	$\frac{s}{s^2+a^2}$
$\sinh at$	$\frac{a}{s^2-a^2}$
$\cosh at$	$\frac{s}{s^2-a^2}$
$t \sin at$	$\frac{2as}{(s^2+a^2)^2}$
$t \cos at$	$\frac{s^2-a^2}{(s^2+a^2)^2}$
$e^{bt} \sin at$	$\frac{a}{(s-b)^2+a^2}$
$e^{bt} \cos at$	$\frac{s-b}{(s-b)^2+a^2}$
$1 - \cos at$	$\frac{a^2}{s(s^2+a^2)}$
$at - \sin at$	$\frac{a^3}{s^2(s^2+a^2)}$
$\sin at - at \cos at$	$\frac{2a^3}{(s^2+a^2)^2}$
$\sin at + at \cos at$	$\frac{2as^2}{(s^2+a^2)^2}$
$\cos at - \cos bt$	$\frac{(b^2-a^2)s}{(s^2+a^2)^2(s^2+b^2)}, \quad b^2 \neq a^2$
$\sin at \cosh at - \cos at \sinh at$	$\frac{4a^3 s}{s^4+4a^4}$
$\sin at \sinh at$	$\frac{2a^2 s}{s^4+4a^4}$
$\sinh at - \sin at$	$\frac{2a^3}{(s^4-a^4)}$
$\cosh at - \cos at$	$\frac{2a^2 s}{s^4-a^4}$
$\sqrt{t}$	$\frac{\sqrt{\pi}}{2} s^{-3/2}$
$\frac{1}{\sqrt{\pi t}}$	$\frac{1}{\sqrt{s}}$
$J_0(at)$	$\frac{1}{\sqrt{s^2+a^2}}$
$\frac{e^{bt} - e^{at}}{t}$	$\log \frac{s-a}{s-b}$
$\frac{1}{t}(1 - \cos at)$	$\frac{1}{2} \log \frac{s^2+a^2}{s^2}$
$\frac{1}{t}(1 - \cosh at)$	$\frac{1}{2} \log \frac{s^2-a^2}{s^2}$
$\frac{1}{t} \sin kt$	$\arctan \frac{k}{s}$

$J_0(at)$ is the Bessel function of the first kind.

Example 9.1.1

Consider the first-order differential equation

$$b\frac{du}{dt} + cu = f_0, \tag{1}$$

where b, c, and f_0 are constants. Equation (1) is subjected to zero initial condition, $u(0) = 0$. Determine the solution using the Laplace transform method.

Solution: The Laplace transform of the equation gives

$$(bs + c)\,\bar{u} = \frac{f_0}{s} \quad \text{or} \quad \bar{u}(s) = \frac{f_0}{bs\left(s + \frac{c}{b}\right)}. \tag{2}$$

To invert Eq. (2) to determine $u(t)$, we rewrite the above expression as (i.e., split into partial fractions; see Problem **9.1** for an explanation of the *method of partial fractions*):

$$\bar{u}(s) = \frac{f_0}{c}\left(\frac{1}{s} - \frac{1}{s + \alpha}\right), \quad \alpha = \frac{c}{b}, \tag{3}$$

The inverse transform is given by (see Table 9.1.1)

$$u(t) = \frac{f_0}{c}\left(1 - e^{-\alpha t}\right). \tag{4}$$

When b and c are positive real numbers, $u(t)$ approaches f_0/c as $t \to \infty$.

Example 9.1.2

Consider the second-order differential equation

$$a\frac{d^2 u}{dt^2} + b\frac{du}{dt} + cu = f_0, \tag{1}$$

where a, b, c, and f_0 are constants. The equation is to be solved subjected to zero initial conditions, $u(0) = \dot{u}(0) = 0$. Determine the solution using the Laplace transform method.

Solution: The Laplace transform of the equation gives

$$\left(as^2 + bs + c\right)\bar{u} = \frac{f_0}{s}, \tag{2}$$

or

$$\bar{u}(s) = \frac{f_0}{s(as^2 + bs + c)}.$$

To invert Eq. (2) to determine $u(t)$, first we write $as^2 + bs + c$ as $a(s + \alpha)(s + \beta)$, where α and β are the roots of the equation $as^2 + bs + c = 0$:

$$\alpha = \frac{1}{2a}\left(b - \sqrt{b^2 - 4ac}\right), \quad \beta = \frac{1}{2a}\left(b + \sqrt{b^2 - 4ac}\right), \tag{3}$$

so that

$$\bar{u}(s) = \frac{f_0}{as(s + \alpha)(s + \beta)}. \tag{4}$$

The actual nature of the solution $u(t)$ depends on the nature of the roots α and β in Eq. (3). There are three possible cases, depending on whether $b^2 - 4ac > 0$, $b^2 - 4ac = 0$, or $b^2 - 4ac < 0$. We discuss them under the assumption that a, b, and c are positive real numbers.

Case 1. When $b^2 - 4ac > 0$ the roots are real, positive, and unequal. Then, we can rewrite Eq. (4) as

$$\bar{u} = \frac{f_0}{a}\left[\frac{A}{s} + \frac{B}{s+\alpha} + \frac{C}{s+\beta}\right],$$

and then we can use the inverse Laplace transform to obtain $u(t)$. The constants A, B, and C satisfy the relations

$$A + B + C = 0, \quad (\alpha + \beta)A + \beta B + \alpha C = 0, \quad \alpha\beta A = 1$$

The solution of these equations is

$$A = \frac{1}{\alpha\beta}, \quad B = \frac{1}{\alpha(\beta - \alpha)}, \quad C = \frac{1}{\beta(\beta - \alpha)}.$$

Thus, we have

$$\bar{u}(s) = \frac{f_0}{a}\left[\frac{1}{\alpha\beta s} - \frac{1}{\alpha(\beta - \alpha)(s + \alpha)} + \frac{1}{\beta(\beta - \alpha)(s + \beta)}\right]. \tag{5}$$

The inverse transform gives

$$u(t) = \frac{f_0}{a\alpha\beta}\left[1 - \frac{\beta}{\beta - \alpha}e^{-\alpha t} + \frac{\alpha}{\beta - \alpha}e^{-\beta t}\right]$$

$$= \frac{f_0}{a\alpha\beta(\beta - \alpha)}\left[\beta\left(1 - e^{-\alpha t}\right) - \alpha\left(1 - e^{-\beta t}\right)\right]. \tag{6}$$

Hence, $u(t)$ approaches $f_0/a\alpha(\beta - \alpha)$ as $t \to \infty$.

Case 2. When $b^2 - 4ac = 0$ the roots are real, positive, and equal, $\alpha = \beta = b/2a$. Then Eq. (4) takes the form

$$\bar{u}(s) = \frac{f_0}{as(s + \alpha)^2} = \frac{f_0}{a\alpha}\left[\frac{1}{\alpha}\left(\frac{1}{s} - \frac{1}{s + \alpha}\right) - \frac{1}{(s + \alpha)^2}\right]. \tag{7}$$

The inverse Laplace transform gives

$$u(t) = \frac{f_0}{a\alpha^2}\left[1 - (1 + \alpha t)e^{-\alpha t}\right]. \tag{8}$$

Hence, $u(t)$ approaches $4f_0a/b^2$ as $t \to \infty$.

Case 3. When $b^2 - 4ac < 0$ the roots are complex and they appear in complex conjugate pairs

$$\alpha = \alpha_1 - i\alpha_2, \quad \beta = \alpha_1 + i\alpha_2; \qquad \alpha_1 = \frac{b}{2a}, \quad \alpha_2 = \sqrt{4ac - b^2}. \tag{9}$$

From Eq. (6), we obtain

$$u(t) = \frac{f_0}{a\alpha\beta(\beta - \alpha)}e^{-\alpha_1 t}\left[\beta\left(1 - e^{i\alpha_2 t}\right) - \alpha\left(1 - e^{-i\alpha_2 t}\right)\right]$$

$$= \frac{f_0}{a(\alpha_1^2 + \alpha_2^2)}e^{-\alpha_1 t}\left(1 - \cos\alpha_2 t - \frac{\alpha_1}{\alpha_2}\sin\alpha_2 t\right). \tag{10}$$

Hence, $u(t)$ approaches zero as $t \to \infty$.

9.2 Spring and Dashpot Models

9.2.1 Creep Compliance and Relaxation Modulus

The equations relating stress σ and strain ε in spring-dashpot models are ordinary differential equations and they have the general form given in Eq. (9.1.1). The solution of Eq. (9.1.1) to determine $\sigma(t)$ for a given $\varepsilon(t)$ (relaxation response) or to determine $\varepsilon(t)$ for given $\sigma(t)$ (creep response) is made easy by the Laplace transform method. In this section, we shall study several standard spring-dashpot models for their constitutive models and creep and relaxation responses.

First we note certain features of the general constitutive equation (9.1.1). In general, the creep response and relaxation response are of the form

$$\varepsilon(t) = J(t)\sigma_0, \tag{9.2.1}$$
$$\sigma(t) = Y(t)\varepsilon_0, \tag{9.2.2}$$

where $J(t)$ is called the *creep compliance* and $Y(t)$ the *relaxation modulus* associated with (9.1.1). The function $J(t)$ is the strain per unit of applied stress, and $Y(t)$ is the stress per unit of applied strain. By definition, both $J(t)$ and $Y(t)$ are zero for all $t < 0$.

The Laplace transform of Eq. (9.1.1) for creep response and relaxation response have the forms

$$\text{Creep response} \quad \bar{Q}_s\bar{\varepsilon}(s) = \bar{P}_s\bar{\sigma}(s) = \frac{1}{s}\bar{P}_s\sigma_0, \tag{9.2.3}$$

$$\text{Relaxation response} \quad \bar{P}_s\bar{\sigma}(s) = \bar{Q}_s\bar{\varepsilon}(s) = \frac{1}{s}\bar{Q}_s\varepsilon_0, \tag{9.2.4}$$

where

$$\bar{P}_s = \sum_{m=0}^{M} p_m s^m, \quad \bar{Q}_s = \sum_{n=0}^{N} q_n s^n. \tag{9.2.5}$$

The Laplace transforms of Eqs. (9.2.1) and (9.2.2) are

$$\bar{\varepsilon}(s) = \bar{J}(s)\sigma_0, \tag{9.2.6}$$
$$\bar{\sigma}(s) = \bar{Y}(s)\varepsilon_0, \tag{9.2.7}$$

Comparing Eq. (9.2.3) with (9.2.6) and Eq. (9.2.4) with (9.2.7), we obtain

$$\bar{J}(s) = \frac{1}{s}\frac{\bar{P}_s}{\bar{Q}_s}, \quad \bar{Y}(s) = \frac{1}{s}\frac{\bar{Q}_s}{\bar{P}_s}. \tag{9.2.8}$$

It also follows that the Laplace transforms of the creep compliance and relaxation modulus are related by

$$\bar{J}(s)\,\bar{Y}(s) = \frac{1}{s^2} \quad \text{or} \quad t = \int_0^t Y(t-t')\,J(t')\,dt'. \tag{9.2.9}$$

Thus, once we know creep compliance $J(t)$, we can determine the relaxation modulus $Y(t)$ and vice versa:

$$Y(t) = \mathcal{L}^{-1}\left[\frac{1}{s^2\bar{J}(s)}\right], \quad J(t) = \mathcal{L}^{-1}\left[\frac{1}{s^2\bar{Y}(s)}\right]. \tag{9.2.10}$$

Although creep and relaxation tests have the advantage of simplicity, there are also shortcomings. The first shortcoming is that uniaxial creep and relaxation test procedures assume the stress to be uniformly distributed through the specimen, with the lateral surfaces being free to expand and contract. This condition cannot be satisfied at the ends of a specimen that is attached to a test machine. The second shortcoming involves the dynamic effects that are encountered in obtaining data at short times. The relaxation and creep functions that are determined through Eqs. (9.2.1) and (9.2.2) are based on the assumption that all transients excited through the dynamic response of specimen and testing machine are neglected.

9.2.2 Maxwell Element

The Maxwell element of Fig. 9.2.1 consists of a linear elastic spring element in series with a dashpot element. The stress–strain relation for the model is developed using the following stress–strain relationships of individual elements:

$$\sigma = k\varepsilon, \quad \sigma = \eta\dot{\varepsilon}, \tag{9.2.11}$$

where k is the spring elastic constant, η is the dashpot viscous constant, and the superposed dot indicates time derivative. It is understood that the spring element responds instantly to a stress, while the dashpot cannot respond instantly (because its response is rate dependent). Let ε_1 be the strain in the spring and ε_2 be the strain in the dashpot. Note that when elements are connected in series, each element carries the same amount of stress while the strains are different in each element. We have

$$\dot{\varepsilon} = \dot{\varepsilon}_1 + \dot{\varepsilon}_2 = \frac{\dot{\sigma}}{k} + \frac{\sigma}{\eta}$$

or

$$\sigma + \frac{\eta}{k}\frac{d\sigma}{dt} = \eta\frac{d\varepsilon}{dt} \quad [P(\sigma) = Q(\varepsilon)]. \tag{9.2.12}$$

Thus, we have $M = N = 1$ [see Eqs. (9.1.1) and (9.1.2)] and $p_0 = 1$, $p_1 = \eta/k$, $q_0 = 0$ and $q_1 = \eta$.

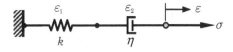

Fig. 9.2.1: The Maxwell element.

9.2.2.1 Creep response

Let $\sigma = \sigma_0 H(t)$. Then the differential equation in Eq. (9.2.12) simplifies to

$$q_1\frac{d\varepsilon}{dt} = p_1\sigma_0\delta(t) + p_0\sigma_0 H(t). \tag{9.2.13}$$

The Laplace transform of Eq. (9.2.13) is

$$q_1 \left[s\bar{\varepsilon}(s) - \varepsilon(0) \right] = \sigma_0 \left(p_1 + \frac{p_0}{s} \right).$$

Assuming that $\varepsilon(0) = 0$, we obtain

$$\bar{\varepsilon}(s) = \sigma_0 \left(\frac{p_1}{q_1 s} + \frac{p_0}{q_1 s^2} \right).$$

The inverse transform gives the creep response

$$\varepsilon(t) = \frac{\sigma_0}{q_1} (p_1 + p_0 t) = \frac{\sigma_0}{k} \left(1 + \frac{t}{\tau} \right) \quad \text{for } t > 0, \qquad (9.2.14)$$

where τ is the *retardation time* or *relaxation time*

$$\tau = \frac{p_1}{p_0} = \frac{\eta}{k}. \qquad (9.2.15)$$

Note that $\varepsilon(0^+) = \sigma_0/k$. The coefficient of σ_0 in Eq. (9.2.14) is called the *creep compliance*, denoted by $J(t)$

$$J(t) = \frac{1}{k} \left(1 + \frac{t}{\tau} \right). \qquad (9.2.16)$$

The creep response of the Maxwell model is shown in Fig. 9.2.2(a).

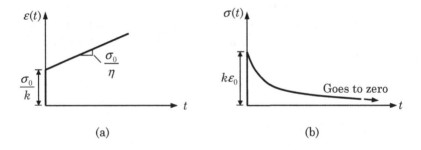

Fig. 9.2.2: (a) Creep and (b) relaxation responses of the Maxwell element.

9.2.2.2 Relaxation response

Let $\varepsilon = \varepsilon_0 H(t)$. Then Eq. (9.2.12) reduces to

$$p_1 \frac{d\sigma}{dt} + p_0 \sigma = q_1 \varepsilon_0 \delta(t). \qquad (9.2.17)$$

The Laplace transform of the above equation is

$$p_1 \left(s\bar{\sigma} - \sigma(0) \right) + p_0 \bar{\sigma} = q_1 \varepsilon_0.$$

Using the initial condition $\sigma(0) = 0$, we write

$$\bar{\sigma}(s) = \varepsilon_0 \left(\frac{q_1}{p_0 + p_1 s} \right) = \frac{q_1}{p_1} \varepsilon_0 \left(\frac{1}{\frac{p_0}{p_1} + s} \right),$$

whose inverse transform is

$$\sigma(t) = \frac{q_1}{p_1}\varepsilon_0 e^{-p_0 t/p_1} = k\varepsilon_0 e^{-t/\tau} \equiv \varepsilon_0\, Y(t), \quad \text{for } t > 0. \qquad (9.2.18)$$

The coefficient of ε_0 in Eq. (9.2.18), $Y(t)$, is called the *relaxation modulus*

$$Y(t) = ke^{-t/\tau}, \quad \tau = p_1/p_0 = \eta/k. \qquad (9.2.19)$$

The relaxation response of the Maxwell model is shown in Fig. 9.2.2(b).

The relaxation modulus $Y(t)$ can also be obtained from Eq. (9.2.10). First note from Eq. (9.2.16) that

$$\bar{J}(s) = \frac{1}{ks^2}\left(s + \frac{1}{\tau}\right).$$

Then, using Eq. (9.2.10), we obtain

$$Y(t) = \mathcal{L}^{-1}\left[\frac{1}{s^2\,\bar{J}(s)}\right] = \mathcal{L}^{-1}\left[\frac{k}{(s + \frac{1}{\tau})}\right] = ke^{-t/\tau},$$

which is the same as that in Eq. (9.2.19).

Figure 9.2.3 shows the creep and relaxation responses of the Maxwell model in a standard test in which the stress and strain are monitored to see the creep and relaxation during the same test.

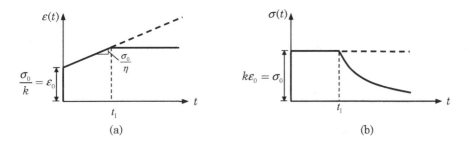

Fig. 9.2.3: (a) Creep and (b) relaxation responses of the Maxwell model in a standard test.

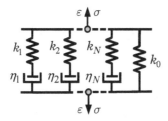

Fig. 9.2.4: The generalized Maxwell model.

A *generalized Maxwell model* consists of N Maxwell elements in parallel and a spring (k_0) in series, as shown in Fig. 9.2.4. The relaxation response of the generalized Maxwell model is of the form [see Eq. (9.2.18)]

$$\sigma(t) = \varepsilon_0 \left[k_0 + \sum_{n=1}^{N} k_n\, e^{-\frac{t}{\tau_n}} \right], \qquad \tau_n = \frac{\eta_n}{k_n}. \tag{9.2.20}$$

The relaxation modulus of the generalized Maxwell model is

$$Y(t) = k_0 + \sum_{n=1}^{N} k_n\, e^{-\frac{t}{\tau_n}}. \tag{9.2.21}$$

9.2.3 Kelvin–Voigt Element

The Kelvin–Voigt element of Fig. 9.2.5 consists of a linear elastic spring element in parallel with a dashpot element. The stress–strain relation for the model is derived as follows. Let σ_1 be the stress in the spring and σ_2 be the stress in the dashpot. Note that each element carries the same amount of strain. Then

$$\sigma = \sigma_1 + \sigma_2 = k\varepsilon + \eta\frac{d\varepsilon}{dt}. \tag{9.2.22}$$

We have $p_0 = 1$, $q_0 = k$, and $q_1 = \eta$.

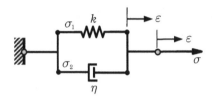

Fig. 9.2.5: The Kelvin–Voigt solid element.

9.2.3.1 Creep response

Let $\sigma = \sigma_0 H(t)$. Then the differential equation in Eq. (9.2.22) becomes

$$q_1\frac{d\varepsilon}{dt} + q_0\varepsilon = p_0\sigma_0 H(t). \tag{9.2.23}$$

The Laplace transform of the equation yields (with zero initial condition)

$$\bar{\varepsilon}(s) = \frac{p_0\sigma_0}{q_1}\frac{1}{s\left(s + \frac{q_0}{q_1}\right)} = \frac{p_0\sigma_0}{q_0}\left[\frac{1}{s} - \frac{1}{\left(s + \frac{q_0}{q_1}\right)}\right].$$

The inverse is

$$\varepsilon(t) = \frac{p_0\sigma_0}{q_0}\left(1 - e^{-\frac{q_0}{q_1}t}\right) = \frac{\sigma_0}{k}\left(1 - e^{-\frac{t}{\tau}}\right) \equiv \sigma_0\, J(t). \tag{9.2.24}$$

Thus, the creep compliance of the Kelvin–Voigt model is

$$J(t) = \frac{1}{k}\left(1 - e^{-\frac{t}{\tau}}\right), \qquad \tau = \frac{q_1}{q_0} = \frac{\eta}{k}. \tag{9.2.25}$$

The creep response of the Kelvin–Voigt model is shown in Fig. 9.2.6(a). Note that in the limit $t \to \infty$, the strain attains the value $\varepsilon_\infty = \sigma_0/k$.

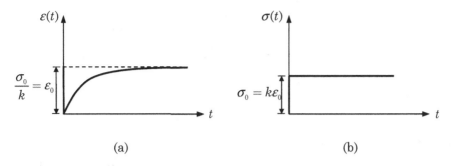

Fig. 9.2.6: (a) Creep and (b) relaxation responses of the Kelvin–Voigt element.

9.2.3.2 Relaxation response

Let $\varepsilon(t) = \varepsilon_0 H(t)$ in Eq. (9.2.22). We obtain

$$\sigma(t) = \varepsilon_0 \left[q_0 H(t) + q_1 \delta(t) \right] \equiv Y(t)\varepsilon_0, \quad Y(t) = \left[kH(t) + \eta\delta(t) \right]. \quad (9.2.26)$$

Alternatively, we can also determine $Y(t)$ using Eq. (9.2.9)

$$\bar{Y}(s) = \frac{1}{s^2 \bar{J}(s)}, \quad s^2 \bar{J}(s) = \frac{s}{\eta} \frac{1}{s + \frac{k}{\eta}} \quad \rightarrow \quad \bar{Y}(s) = \frac{k}{s} + \eta,$$

from which we obtain $Y(t)$ as given in Eq. (9.2.26). The relaxation response of the Kelvin–Voigt model is shown in Fig. 9.2.6(b). The creep and relaxation responses in the standard test of the Kelvin–Voigt model are shown in Figs. 9.2.7(a) and 9.2.7(b), respectively.

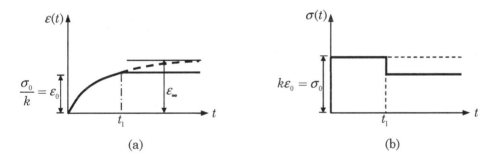

Fig. 9.2.7: A standard test of a Kelvin–Voigt solid.

A *generalized Kelvin–Voigt model* consists of N Kelvin–Voigt elements in series along with a spring element, as shown in Fig. 9.2.8; it can be used to fit creep data to a high degree. The creep compliance of the generalized Kelvin–Voigt model is [see Eq. (9.2.25)]

$$J(t) = \frac{1}{k_0} + \sum_{n=1}^{N} \frac{1}{k_n} \left(1 - e^{-\frac{t}{\tau_n}} \right), \quad \tau_n = \frac{\eta_n}{k_n}. \quad (9.2.27)$$

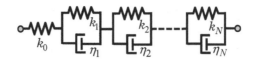

Fig. 9.2.8: Generalized Kelvin-Voigt model.

9.2.4 Three-Element Models

There are two three-element models, as shown in Figs. 9.2.9(a) and 9.2.9(b). In the first one an extra spring element is added in series to the Kelvin–Voigt element, and in the second one a spring element is added in parallel to the Maxwell element. The constitutive equations for the these models are discussed here.

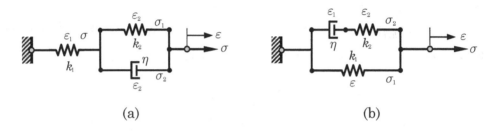

(a)   (b)

Fig. 9.2.9: Three-element models.

For the three-element model in Fig. 9.2.9(a), we have

$$\sigma = \sigma_1 + \sigma_2, \quad \varepsilon = \varepsilon_1 + \varepsilon_2, \quad \sigma_1 = k_2\varepsilon_2, \quad \sigma_2 = \eta\dot{\varepsilon}_2, \quad \varepsilon_1 = \frac{\sigma}{k_1}, \qquad (9.2.28)$$

Using the relations in (9.2.28) we obtain

$$\frac{\eta}{k_1}\frac{d\sigma}{dt} + \left(1 + \frac{k_2}{k_1}\right)\sigma = k_2\varepsilon + \eta\frac{d\varepsilon}{dt}. \qquad (9.2.29)$$

Equation (9.2.29) is of the form $P(\sigma) = Q(\varepsilon)$

$$p_0\sigma + p_1\frac{d\sigma}{dt} = q_0\varepsilon + q_1\frac{d\varepsilon}{dt},$$

$$p_0 = 1 + \frac{k_2}{k_1}, \quad p_1 = \frac{\eta}{k_1}, \quad q_0 = k_2, \quad q_1 = \eta. \qquad (9.2.30)$$

For the three-element model shown in Fig. 9.2.9(b), we have

$$\sigma = \sigma_1 + \sigma_2, \quad \varepsilon = \varepsilon_1 + \varepsilon_2, \quad \varepsilon_1 = \frac{\sigma_2}{k_2}, \quad \dot{\varepsilon}_2 = \frac{\sigma_2}{\eta}, \quad \varepsilon = \frac{\sigma_1}{k_1}, \qquad (9.2.31)$$

Combining the above relations, we arrive at

$$\frac{1}{\eta}\sigma + \frac{1}{k_2}\frac{d\sigma}{dt} = \frac{k_1}{\eta}\varepsilon + \left(1 + \frac{k_1}{k_2}\right)\frac{d\varepsilon}{dt},$$

or

$$p_0\sigma + p_1\frac{d\sigma}{dt} = q_0\varepsilon + q_1\frac{d\varepsilon}{dt},$$

$$p_0 = \frac{1}{\eta}, \quad p_1 = \frac{1}{k_2}, \quad q_0 = \frac{k_1}{\eta}, \quad q_1 = 1 + \frac{k_1}{k_2}. \tag{9.2.32}$$

Apparently, the three-element model represents the constitutive behavior of an ideal cross-linked polymer.

The creep and relaxation response of the three-element model shown in Fig. 9.2.9(a) are studied next. Substituting $\sigma(t) = \sigma_0 H(t)$ into Eq. (9.2.30), we obtain

$$p_0\sigma_0 H(t) + p_1\sigma_0\delta(t) = q_0\varepsilon + q_1\frac{d\varepsilon}{dt}. \tag{9.2.33}$$

The Laplace transform of Eq. (9.2.33) yields

$$(q_0 + q_1 s)\,\bar{\varepsilon}(s) = \sigma_0\left(\frac{p_0}{s} + p_1\right) \quad \text{or} \quad \bar{\varepsilon}(s) = \sigma_0\frac{(p_0 + p_1 s)}{s(q_0 + q_1 s)}, \tag{9.2.34}$$

where zero initial conditions are used. We rewrite the above expression in a form suitable for inversion back to the time domain:

$$\bar{\varepsilon}(s) = \sigma_0\left[\frac{p_0}{q_0}\left(\frac{1}{s} - \frac{1}{\frac{q_0}{q_1} + s}\right) + \frac{p_1}{q_1}\frac{1}{\left(\frac{q_0}{q_1} + s\right)}\right]. \tag{9.2.35}$$

Using the inverse Laplace transform, we obtain

$$\varepsilon(t) = \sigma_0\left[\frac{p_0}{q_0}\left(1 - e^{-\frac{t}{\tau}}\right) + \frac{p_1}{q_1}e^{-\frac{t}{\tau}}\right], \quad \tau = \frac{q_1}{q_0},$$

$$= \sigma_0\left[\frac{k_1 + k_2}{k_1 k_2}\left(1 - e^{-\frac{t}{\tau}}\right) + \frac{1}{k_1}e^{-\frac{t}{\tau}}\right], \quad \tau = \frac{\eta}{k_2}. \tag{9.2.36}$$

Thus, the creep compliance is given by

$$J(t) = \left[\frac{k_1 + k_2}{k_1 k_2}\left(1 - e^{-\frac{t}{\tau}}\right) + \frac{1}{k_1}e^{-\frac{t}{\tau}}\right] = \frac{1}{k_1} + \frac{1}{k_2}\left(1 - e^{-\frac{t}{\tau}}\right). \tag{9.2.37}$$

For the relaxation response, let $\varepsilon(t) = \varepsilon_0 H(t)$ in Eq. (9.2.30) and obtain

$$p_0\sigma + p_1\frac{d\sigma}{dt} = q_0\varepsilon_0 H(t) + q_1\varepsilon_0\delta(t). \tag{9.2.38}$$

The Laplace transform of the equation is

$$(p_0 + p_1 s)\,\bar{\sigma}(s) = \varepsilon_0\left(\frac{q_0}{s} + q_1\right) \quad \text{or} \quad \bar{\sigma}(s) = \varepsilon_0\frac{(q_0 + q_1 s)}{s(p_0 + p_1 s)}, \tag{9.2.39}$$

where zero initial conditions are used. We rewrite the above expression in the form

$$\bar{\sigma}(s) = \varepsilon_0\left[\frac{q_0}{p_0}\left(\frac{1}{s} - \frac{1}{\frac{p_0}{p_1} + s}\right) + \frac{q_1}{p_1}\frac{1}{\left(\frac{p_0}{p_1} + s\right)}\right]. \tag{9.2.40}$$

Using the inverse Laplace transform, we obtain

$$\sigma(t) = \varepsilon_0 \left[\frac{q_0}{p_0} \left(1 - e^{-\frac{t}{\tau}}\right) + \frac{q_1}{p_1} e^{-\frac{t}{\tau}} \right], \quad \tau = \frac{p_1}{p_0},$$

$$= \varepsilon_0 \left[\frac{k_1 k_2}{k_1 + k_2} \left(1 - e^{-\frac{t}{\tau}}\right) + k_1 e^{-\frac{t}{\tau}} \right], \quad \tau = \frac{\eta}{k_1 + k_2}.$$

$$(9.2.41)$$

Thus, the relaxation modulus is given by

$$Y(t) = \left[\frac{k_1 k_2}{k_1 + k_2} \left(1 - e^{-\frac{t}{\tau}}\right) + k_1 e^{-\frac{t}{\tau}} \right], \quad \tau = \frac{\eta}{k_1 + k_2}. \qquad (9.2.42)$$

Determination of the creep and relaxation responses of the three-element model in Fig. 9.2.9(b), also known as the *standard linear solid*, is considered in Example 9.2.1 (also, see Problem **9.3**).

9.2.5 Four-Element Models

The four-element models, such as the ones shown in Figs. 9.2.10(a) and 9.2.10(b), have constitutive relations that involve second-order derivatives of stress and/or strain. Here we discuss the creep response of such models in general terms. The determination of relaxation response follows along similar lines to what is discussed for creep response.

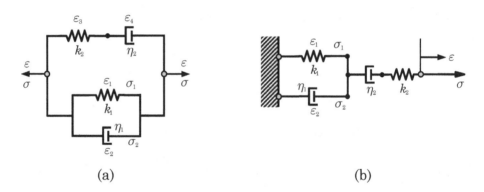

(a)   (b)

Fig. 9.2.10: Four-element models.

Consider the second-order differential equation in time

$$p_0 \sigma + p_1 \dot{\sigma} + p_2 \ddot{\sigma} = q_0 \varepsilon + q_1 \dot{\varepsilon} + q_2 \ddot{\varepsilon}. \qquad (9.2.43)$$

Let $\sigma(t) = \sigma_0 H(t)$. We have

$$p_0 \sigma_0 H(t) + p_1 \sigma_0 \delta(t) + p_2 \sigma_0 \dot{\delta}(t) = q_0 \varepsilon + q_1 \dot{\varepsilon} + q_2 \ddot{\varepsilon}. \qquad (9.2.44)$$

Taking the Laplace transform and assuming homogeneous initial conditions, we obtain

$$\sigma_0 \left(\frac{p_0}{s} + p_1 + p_2 s \right) = \left(q_0 + q_1 s + q_2 s^2 \right) \bar{\varepsilon}(s), \qquad (9.2.45)$$

or

$$\bar{\varepsilon}(s) = \sigma_0 \frac{p_0 + p_1 s + p_2 s^2}{s(q_0 + q_1 s + q_2 s^2)}. \tag{9.2.46}$$

To invert Eq. (9.2.46) to determine $\varepsilon(t)$, first we write $q_2 s^2 + q_1 s + q_0$ as $q_2(s + \alpha)(s + \beta)$, where α and β are the roots of the equation $q_2 s^2 + q_1 s + q_0 = 0$:

$$\alpha = \frac{1}{2q_2}\left(q_1 - \sqrt{q_1^2 - 4q_2 q_0}\right), \quad \beta = \frac{1}{2q_2}\left(q_1 + \sqrt{q_1^2 - 4q_2 q_0}\right), \tag{9.2.47}$$

so that

$$\bar{\varepsilon}(s) = \sigma_0 \frac{p_0 + p_1 s + p_2 s^2}{q_2 s(s + \alpha)(s + \beta)}. \tag{9.2.48}$$

We write the solution in three parts for the case of real and unequal roots with $q_0 \neq 0$, $q_1 \neq 0$, and $q_2 \neq 0$:

$$\bar{\varepsilon}_1(s) = \sigma_0 \frac{p_0}{q_2}\left[\frac{1}{\alpha\beta s} - \frac{1}{\alpha(\beta - \alpha)(s + \alpha)} + \frac{1}{\beta(\beta - \alpha)(s + \beta)}\right], \tag{9.2.49}$$

$$\bar{\varepsilon}_2(s) = \sigma_0 \frac{p_1}{q_2}\left[\frac{1}{(\beta - \alpha)(s + \alpha)} - \frac{1}{(\beta - \alpha)(s + \beta)}\right], \tag{9.2.50}$$

$$\bar{\varepsilon}_3(s) = \sigma_0 \frac{p_2}{q_2}\left[-\frac{\alpha}{(\beta - \alpha)(s + \alpha)} + \frac{\beta}{(\beta - \alpha)(s + \beta)}\right]. \tag{9.2.51}$$

The solution is obtained by taking inverse Laplace transform

$$\varepsilon(t) = \frac{\sigma_0}{q_2}\left\{p_0\left[\frac{1}{\alpha\beta} - \frac{e^{-\alpha t}}{\alpha(\beta - \alpha)} + \frac{e^{-\beta t}}{\beta(\beta - \alpha)}\right]\right.$$
$$\left. + p_1\left[\frac{e^{-\alpha t}}{(\beta - \alpha)} - \frac{e^{-\beta t}}{(\beta - \alpha)}\right] + p_2\left[-\frac{\alpha e^{-\alpha t}}{(\beta - \alpha)} + \frac{\beta e^{-\beta t}}{(\beta - \alpha)}\right]\right\}. \tag{9.2.52}$$

When $q_2 = 0$, $q_1 \neq 0$, and $q_0 \neq 0$, Eq. (9.2.46) takes the form (with $\alpha = q_0/q_1$)

$$\bar{\varepsilon}(s) = \frac{\sigma_0}{q_1}\left[\frac{p_0}{\alpha}\left(\frac{1}{s} - \frac{1}{s + \alpha}\right) + \frac{p_1}{s + \alpha} + p_2\left(1 - \frac{\alpha}{s + \alpha}\right)\right], \tag{9.2.53}$$

and the solution is given by

$$\varepsilon(t) = \frac{\sigma_0}{q_1}\left[\frac{p_0}{\alpha}\left(1 - e^{-\alpha t}\right) + p_1 e^{-\alpha t} + p_2\left(\delta(t) - \alpha e^{-\alpha t}\right)\right]. \tag{9.2.54}$$

The Dirac delta function indicates that the model lacks impact response. That is, if a Dirac delta function appears in a relaxation function $Y(t)$, a finite stress is not sufficient to produce at once a finite strain, and an infinite one is needed.

When $q_0 = 0$, $q_1 \neq 0$, and $q_2 \neq 0$, Eq. (9.2.46) takes the form (with $\alpha = q_1/q_2$)

$$\bar{\varepsilon}(s) = \frac{\sigma_0}{q_2}\left[\frac{p_0}{\alpha^2}\left(\frac{\alpha}{s^2} - \frac{1}{s} + \frac{1}{s + \alpha}\right) + \frac{p_1}{\alpha}\left(\frac{1}{s} - \frac{1}{s + \alpha}\right) + \frac{p_2}{s + \alpha}\right], \tag{9.2.55}$$

and the solution is given by

$$\varepsilon(t) = \frac{\sigma_0}{q_2} \left[\frac{p_0 t}{\alpha} + \frac{1}{\alpha} \left(p_1 - \frac{p_0}{\alpha} \right) \left(1 - e^{-\alpha t} \right) + p_2 e^{-\alpha t} \right]. \tag{9.2.56}$$

This completes the general discussion of the creep response of four-element models. For the relaxation response the role of p's and q's is exchanged. Alternatively, we can use Eq. (9.2.10) to determine $Y(t)$.

Example 9.2.1

Consider the differential equation in Eq. (9.2.32),

$$p_0 \sigma + p_1 \dot{\sigma} = q_0 \varepsilon + q_1 \dot{\varepsilon}, \tag{1}$$

with

$$p_0 = \frac{1}{\eta}, \quad p_1 = \frac{1}{k_2}, \quad p_2 = 0, \quad q_0 = \frac{k_1}{\eta}, \quad q_1 = \frac{k_1 + k_2}{k_2}, \quad q_2 = 0. \tag{2}$$

Determine the creep and relaxation response.

Solution: From Eq. (9.2.54), we have the creep response $(\alpha = q_0/q_1)$:

$$\varepsilon(t) = \sigma_0 \frac{k_2}{k_1 + k_2} \left[\frac{1}{\alpha \eta} \left(1 - e^{-\alpha t} \right) + \frac{1}{k_2} e^{-\alpha t} \right]$$

$$= \sigma_0 \left[\frac{1}{k_1} \left(1 - e^{-\alpha t} \right) + \frac{1}{k_1 + k_2} e^{-\alpha t} \right], \quad \alpha = \frac{k_1 k_2}{\eta(k_1 + k_2)}. \tag{3}$$

Thus, the creep compliance of the three-element model in Fig. 9.2.9(b) is given by

$$J(t) = \frac{1}{k_1} \left(1 - e^{-\alpha t} \right) + \frac{1}{k_1 + k_2} e^{-\alpha t}. \tag{4}$$

The relaxation response is $\sigma(t) = Y(t)\varepsilon_0$ with $Y(t)$ computed as follows. We have

$$\bar{Y}(s) = \frac{1}{s^2 \bar{J}(s)}, \quad \bar{J}(s) = \frac{1}{k_1} \left(\frac{1}{s} - \frac{1}{s+\alpha} \right) + \frac{1}{k_1 + k_2} \frac{1}{s+\alpha},$$

and

$$s^2 \bar{J}(s) = \frac{s \left(s + \frac{k_2}{\eta} \right)}{(k_1 + k_2)(s + \alpha)}, \quad \frac{1}{s^2 \bar{J}(s)} = \frac{k_2}{s + \frac{k_2}{\eta}} + \frac{k_1}{s}. \tag{5}$$

Thus, the relaxation modulus is

$$Y(t) = k_1 + k_2 e^{-t/\tau}, \quad \tau = \frac{\eta}{k_2}. \tag{6}$$

Example 9.2.2

Consider the differential equation

$$\ddot{\varepsilon} + \frac{k_2}{\eta_2} \dot{\varepsilon} = \frac{1}{k_1} \ddot{\sigma} + \left(\frac{1}{\eta_1} + \frac{1}{\eta_2} + \frac{k_2}{k_1 \eta_2} \right) \dot{\sigma} + \frac{k_2}{\eta_1 \eta_2} \sigma. \tag{1}$$

Thus, we have

$$q_0 = 0, \quad q_1 = \frac{k_2}{\eta_2}, \quad q_2 = 1, \quad p_0 = \frac{k_2}{\eta_1 \eta_2}, \quad p_1 = \frac{1}{\eta_1} + \frac{1}{\eta_2} + \frac{k_2}{k_1 \eta_2}, \quad p_2 = \frac{1}{k_1}. \tag{2}$$

Determine the creep and relaxation response of the model.

Solution: The creep response is given by Eq. (9.2.56)

$$\varepsilon(t) = \frac{\sigma_0}{q_2}\left[\frac{p_0 t}{\alpha} + \frac{1}{\alpha}\left(p_1 - \frac{p_0}{\alpha}\right)\left(1 - e^{-\alpha t}\right) + p_2 e^{-\alpha t}\right]$$

$$= \sigma_0\left[\frac{1}{k_1} + \frac{t}{\eta_1} + \frac{1}{k_2}\left(1 - e^{-t/\tau}\right)\right], \quad \tau = \frac{1}{\alpha} = \frac{\eta_2}{k_2}. \tag{3}$$

Thus, the creep compliance is

$$J(t) = \frac{1}{k_1} + \frac{t}{\eta_1} + \frac{1}{k_2}\left(1 - e^{-t/\tau}\right). \tag{4}$$

To compute the relaxation modulus, we compute

$$\bar{J}(s) = \frac{1}{k_1 s} + \frac{1}{\eta_1 s^2} + \frac{1}{k_2}\left(\frac{1}{s} - \frac{1}{s + \frac{1}{\tau}}\right), \tag{5}$$

$$s^2\bar{J}(s) = \frac{s}{k_1} + \frac{1}{\eta_1} + \frac{1}{\eta_2}\left(\frac{s}{s + \frac{1}{\tau}}\right) = \frac{as^2 + bs + c}{d\left(s + \frac{1}{\tau}\right)}, \tag{6}$$

where

$$a = \eta_1\eta_2, \quad b = (k_1 + k_2)\eta_1 + k_1\eta_2, \quad c = k_1 k_2, \quad d = k_1\eta_1\eta_2. \tag{6}$$

Then

$$\bar{Y}(s) = \frac{1}{s^2\bar{J}} = \frac{d\left(s + \frac{1}{\tau}\right)}{as^2 + bs + c} = \frac{d}{a}\left(\frac{A}{s + \alpha} + \frac{B}{s + \beta}\right), \tag{7}$$

where

$$\alpha = \frac{b}{2a} + \frac{1}{2a}\sqrt{b^2 - 4ac}, \quad \beta = \frac{b}{2a} - \frac{1}{2a}\sqrt{b^2 - 4ac} \tag{8}$$

$$A = -\frac{k_2 - \eta_2\alpha}{\eta_2(\alpha - \beta)}, \quad B = \frac{k_2 - \eta_2\beta}{\eta_2(\alpha - \beta)}.$$

It can be shown that $b^2 > 4ac$ and $\alpha > \beta > 0$ for $k_i > 0$ and $\eta_i > 0$. Hence, we have

$$Y(t) = \frac{k_1\eta_1}{\sqrt{b^2 - 4ac}}\left[-(k_2 - \eta_2\alpha)e^{-\alpha t} + (k_2 - \eta_2\beta)e^{-\beta t}\right]$$

$$= \frac{k_1\eta_1}{\sqrt{b^2 - 4ac}}\left[k_2\left(e^{-\beta t} - e^{-\alpha t}\right) + \eta_2\left(\alpha e^{-\alpha t} - \beta e^{-\beta t}\right)\right]. \tag{9}$$

9.3 Integral Constitutive Equations

9.3.1 Hereditary Integrals

The spring-and-dashpot elements are discrete models and are governed by differential equations. At $t = 0$ a stress σ_0 applied suddenly produces a strain $\varepsilon(t) = J(t)\sigma_0$ (see Fig. 9.3.1). If the stress σ_0 is maintained unchanged, then $\varepsilon(t) = J(t)\sigma_0$ describes the strain for all $t > 0$. If we treat the material as linear, we can use the principle of linear superposition to calculate the strain produced in a given direction by the action of several loads of different magnitudes. If, at $t = t_1$, some more stress $\Delta\sigma_1$ is applied, then additional strain is produced which is proportional to $\Delta\sigma_1$ and depends on the same creep compliance. This additional strain is measured for $t > t'$. Hence, the total strain for $t > t_1$ is the sum of strain due to σ_0 and that due to $\Delta\sigma_1$:

$$\varepsilon(t) = J(t)\sigma_0 + J(t - t_1)\Delta\sigma_1. \tag{9.3.1}$$

Similarly, if additional stress $\Delta\sigma_2$ is applied at time $t = t_2$, then the total strain for $t > t_2$ is

$$\varepsilon(t) = J(t)\sigma_0 + J(t - t_1)\Delta\sigma_1 + J(t - t_2)\Delta\sigma_2$$

$$= J(t)\sigma_0 + \sum_{i=1}^{2} J(t - t_i)\Delta\sigma_i. \tag{9.3.2}$$

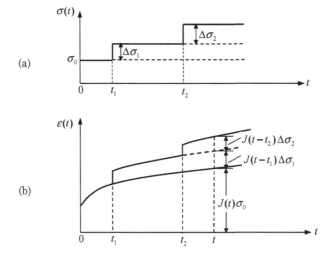

(a)

(b)

Fig. 9.3.1: Strain response due to σ_0 and $\Delta\sigma_i$.

If the stress applied is an arbitrary function of t, it can be divided into the first part $\sigma_0 H(t)$ and a sequence of infinitesimal stress increments $d\sigma(t') H(t-t')$ (see Fig. 9.3.2). The corresponding strain at time t can be written (using the

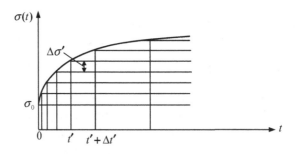

Fig. 9.3.2: Linear superposition to derive hereditary integral.

Boltzman's superposition principle) as

$$\varepsilon(t) = J(t)\sigma_0 + \int_0^t J(t - t') d\sigma(t') = J(t)\sigma_0 + \int_0^t J(t - t') \frac{d\sigma(t')}{dt'} dt'. \quad (9.3.3)$$

Equation (9.3.3) indicates that the strain at any given time depends on all that has happened before, that is, on the entire stress history $\sigma(t')$ for $t' < t$. This is in contrast to the elastic material whose strain depends only on the stress acting at that time only. Equation (9.3.3) is called a *hereditary integral*.

Equation (9.3.3) can be written in alternate form:

$$\varepsilon(t) = J(t)\sigma(0) + \left[J(t - t')\, \sigma(t') \right]_0^t - \int_0^t \frac{dJ(t - t')}{dt'} \sigma(t')\, dt'$$

$$= J(0)\, \sigma(t) + \int_0^t \frac{dJ(t - t')}{d(t - t')} \sigma(t')\, dt' \quad (9.3.4)$$

$$= J(0)\, \sigma(t) + \int_0^t \frac{dJ(\tau)}{d\tau} \sigma(t - \tau)\, d\tau. \quad (9.3.5)$$

Note that Eq. (9.3.3) separates the strain caused by initial stress $\sigma(0)$ and that caused by stress increments. On the other hand, Eq. (9.3.5) separates the strain into the part that would occur if the total stress $\sigma(t)$ were applied at time t and an additional strain was produced due to creep.

It is possible to include the initial part due to σ_0 into the integral. For example, Eq. (9.3.3) can be written as

$$\varepsilon(t) = \int_{-\infty}^t J(t - t') \frac{d\sigma(t')}{dt'} dt'. \quad (9.3.6)$$

The fact that $J(t) = 0$ for $t < 0$ is used in writing the above integral, which is known as *Stieljes integral*.

Arguments similar to those presented for the creep compliance can be used to derive the hereditary integrals for the relaxation modulus $Y(t)$. If the strain history is known as a function of time, $\varepsilon(t)$, the stress is given by

$$\sigma(t) = Y(t)\varepsilon(0) + \int_0^t Y(t - t') \frac{d\varepsilon(t')}{dt'} dt' \quad (9.3.7)$$

$$= Y(0)\, \varepsilon(t) + \int_0^t \frac{dY(t')}{dt'} \varepsilon(t - t')\, dt' \quad (9.3.8)$$

$$= \int_{-\infty}^t Y(t - t') \frac{d\varepsilon(t')}{dt'} dt'. \quad (9.3.9)$$

The thermodynamic restriction on $Y(t)$ is that it be nonnegative and be a monotonically decreasing function with finite limit for $t \to \infty$.

Example 9.3.1

Consider the stress history shown in Fig. 9.3.3. Write the hereditary integral in Eq. (9.3.4) for the Maxwell model and Kelvin–Voigt model.

Solution: The creep compliance of the Maxwell model is given in Eq. (9.2.16) as $J(t) = (1/k + t/\eta)$ with $J(0) = 1/k$. Then the strain response according to the hereditary integral in Eq. (9.3.4) is given by

$$\text{For } t < t_1 : \quad \varepsilon(t) = \sigma_1 \frac{t}{t_1}\frac{1}{k} + \frac{\sigma_1}{t_1}\int_0^t t'\frac{1}{\eta}\,dt' = \frac{\sigma_1 t}{\eta t_1}\left(\frac{\eta}{k} + \frac{t}{2}\right). \tag{1}$$

$$\text{For } t > t_1 : \quad \varepsilon(t) = \sigma_1 \frac{1}{k} + \frac{\sigma_1}{t_1}\int_0^{t_1} t'\frac{1}{\eta}\,dt' + \sigma_1 \int_{t_1}^t 1\cdot\frac{1}{\eta}dt'$$

$$= \frac{\sigma_1}{\eta}\left(\frac{\eta}{k} + \frac{t_1}{2} + t\right). \tag{2}$$

Note that by setting $t_1 = 0$, we obtain the same result as in Eq. (9.2.14).

The creep compliance of the Kelvin–Voigt model is given in Eq. (9.2.25). Then the strain response according to the hereditary integral in Eq. (9.3.4) is given by

$$\text{For } t < t_1 : \quad \varepsilon(t) = \sigma_1 \frac{t}{t_1}\cdot 0 + \frac{\sigma_1}{\eta t_1}\int_0^t t' e^{-(t-t')/\tau}\,dt'$$

$$= \frac{\sigma_1}{kt_1}\left[t - \frac{\eta}{k}\left(1 - e^{-t/\tau}\right)\right]. \tag{3}$$

$$\text{For } t > t_1 : \quad \varepsilon(t) = \frac{\sigma_1}{\eta t_1}\int_0^{t_1} t' e^{-(t-t')/\tau}\,dt' + \frac{\sigma_1}{\eta}\int_{t_1}^t e^{-(t-t')/\tau}\,dt'$$

$$= \frac{\sigma_1}{k}\left[1 + \frac{\eta}{kt_1}\left(1 - e^{t_1/\tau}\right)e^{-t/\tau}\right]. \tag{4}$$

By setting $t_1 = 0$ in Eq. (4), we obtain (use the L'Hospital rule to deal with a zero divided by zero condition) the same strain response as in Eq. (9.2.24). Note that for $t \to \infty$, the strain goes to $\varepsilon = \sigma_1/k$, the same limit as if σ_1 were applied suddenly at $t = 0$ or $t = t_1$. This implies that the stress history is wiped out if sufficient time has elapsed. Thus, the Kelvin–Voigt model represents the behavior of an elastic solid.

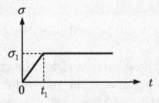

Fig. 9.3.3: Stress history.

9.3.2 Hereditary Integrals for Deviatoric Components

The one-dimensional linear viscoelastic stress–strain relations developed in the previous sections can be extended in a straightforward manner to those relating the deviatoric stress components to the deviatoric strain components. Recall that the deviatoric stress and strain tensors are defined as

$$\text{Deviatoric stress} \quad \boldsymbol{\sigma}' \equiv \boldsymbol{\sigma} - \tfrac{1}{3}\tilde{\sigma}\mathbf{I}, \quad (\sigma'_{ij} = \sigma_{ij} - \tfrac{1}{3}\sigma_{kk}\delta_{ij}), \tag{9.3.10}$$

$$\text{Deviatoric strain} \quad \boldsymbol{\varepsilon}' \equiv \boldsymbol{\varepsilon} - \tfrac{1}{3}\mathrm{tr}(\varepsilon), \quad (\varepsilon'_{ij} = \varepsilon_{ij} - \tfrac{1}{3}\varepsilon_{kk}\delta_{ij}), \tag{9.3.11}$$

where $\tilde{\sigma}$ is the mean stress and e is the dilatation

$$\text{trace of stress tensor:} \quad \tilde{\sigma} \equiv \sigma_{ii}, \qquad \text{dilatation:} \quad \tilde{e} \equiv \varepsilon_{ii}. \qquad (9.3.12)$$

The constitutive equations between the deviatoric components of a linear elastic isotropic material are

$$\tilde{\sigma} = 3K\tilde{e}, \quad \boldsymbol{\sigma}' = 2\mu\boldsymbol{\varepsilon}' \ (\sigma'_{ij} = 2\mu\,\varepsilon'_{ij}). \qquad (9.3.13)$$

Here K denotes the bulk modulus and μ is the Lamé constant (the same as the shear modulus), which are related to Young's modulus E and Poisson's ratio ν with the thermodynamic restrictions

$$K = \frac{E}{3(1 - 2\nu)}, \quad \mu = G = \frac{E}{2(1 + \nu)}, \qquad (9.3.14)$$

$$E > 0, \quad G > 0, \quad -1 < \nu < 0.5 \text{ (in most cases } 0 < \nu < 0.5).$$

The linear viscoelastic strain–stress and stress–strain relations for the deviatoric components in Cartesian coordinates are

$$\varepsilon'_{ij}(t) = \int_{-\infty}^{t} J_s(t - t') \frac{d\sigma'_{ij}}{dt'} \, dt', \qquad (9.3.15)$$

$$\varepsilon_{kk}(t) = \int_{-\infty}^{t} J_d(t - t') \frac{d\sigma_{kk}}{dt'} \, dt', \qquad (9.3.16)$$

$$\sigma'_{ij}(t) = 2 \int_{-\infty}^{t} G(t - t') \frac{d\varepsilon'_{ij}}{dt'} \, dt', \qquad (9.3.17)$$

$$\sigma_{kk}(t) = 3 \int_{-\infty}^{t} K(t - t') \frac{d\varepsilon_{kk}}{dt'} \, dt', \qquad (9.3.18)$$

where $J_s(t)$ is the creep compliance in shear and J_d is the creep compliance in dilation. The thermodynamic restriction on J, J_d, $G(t)$, and $K(t)$ are that they be positive with finite values as $t \to \infty$. The general stress–strain relations may be written as

$$\sigma_{ij}(t) = 2 \int_{-\infty}^{t} G(t - t') \frac{d\varepsilon_{ij}(t')}{dt'} \, dt'$$
$$+ \delta_{ij} \int_{-\infty}^{t} \left[K(t - t') - \frac{2}{3}G(t - t') \right] \frac{d\varepsilon_{kk}(t')}{dt'} \, dt', \qquad (9.3.19)$$

$$\varepsilon_{ij}(t) = \int_{-\infty}^{t} J_s(t - t') \frac{d\sigma_{ij}(t')}{dt'} \, dt'$$
$$+ \frac{1}{3}\delta_{ij} \int_{-\infty}^{t} \left[J_d(t - t') - J_s(t - t') \right] \frac{d\sigma_{kk}(t')}{dt'} \, dt'. \qquad (9.3.20)$$

The Laplace transforms of Eqs. (9.3.15)–(9.3.18) are

$$\bar{\varepsilon}'_{ij}(s) = s\,\bar{J}_s(s)\,\bar{\sigma}'_{ij}(s), \quad \bar{\sigma}'_{ij}(s) = 2s\,\bar{G}(s)\,\bar{\varepsilon}'_{ij}(s), \qquad (9.3.21)$$

$$\bar{\varepsilon}_{kk}(s) = s\,\bar{J}_d(s)\,\bar{\sigma}_{kk}(s), \quad \bar{\sigma}_{kk}(s) = 3s\,\bar{K}(s)\,\bar{\varepsilon}_{kk}(s). \qquad (9.3.22)$$

from which it follows that

$$2\bar{G}(s) = \frac{1}{s^2 \, \bar{J}_s(s)}, \tag{9.3.23}$$

$$3\bar{K}(s) = \frac{1}{s^2 \, \bar{J}_d(s)}. \tag{9.3.24}$$

9.3.3 The Correspondence Principle

There exists a certain correspondence between the elastic and viscoelastic solutions of a boundary value problem. The correspondence allows us to obtain solutions of a viscoelastic problem from that of the corresponding elastic problem.

Consider a one-dimensional elastic problem, such as a bar or beam, carrying certain applied loads F_i^0, $i = 1, 2, \cdots$. Suppose that the stress induced is σ^e. The strain is

$$\varepsilon^e = \sigma^e / E. \tag{9.3.25}$$

Then consider the same structure, but made of a viscoelastic material. Assume that the same loads are applied at time $t = 0$ and then held constant

$$F_i(t) = F_i^0 H(t).$$

The stress in the viscoelastic beam is $\sigma(t) = \sigma^e H(t)$. The strain in the viscoelastic structure is

$$\varepsilon(t) = J(t)\sigma^e. \tag{9.3.26}$$

For any time t the strain in the viscoelastic structure is like the strain in an elastic beam of modulus $E = 1/J(t)$. Thus, we have the following *correspondence principle* (Part 1): If a viscoelastic structure is subjected to loads that are all applied simultaneously at $t = 0$ and then held constant, its stresses are the same as those in an elastic structure under the same loads, and its time-dependent strains and displacements are obtained from those of the elastic structure by replacing E by $1/J(t)$.

Next, consider an elastic structure in which the displacements are prescribed and held constant. Suppose that the displacement in the structure is u^e. The strain ε^e can be computed from the displacement u^e using an appropriate kinematic relation, and stress σ using the constitutive equation

$$\sigma = E\varepsilon^e. \tag{9.3.27}$$

Then consider the same structure but made of a viscoelastic material. If we prescribe deflection $u(t) = u^e H(t)$, the strains produced are $\varepsilon(t) = \varepsilon^e H(t)$. The strain will produce a stress

$$\sigma(t) = Y(t)\varepsilon^e. \tag{9.3.28}$$

For any time t the stress in the viscoelastic structure is like the stress in an elastic beam of modulus $E = Y(t)$. Thus, we have the second part of the *correspondence principle*: If a viscoelastic structure is subjected to displacements that are all imposed at $t = 0$ and then held constant, its displacements and strains are the same as those in the elastic structure under the same displacements, and its time-dependent stresses are obtained from those of the elastic structure by replacing E by $Y(t)$.

The foregoing ideas for step loads or step displacements can be generalized to loads and displacements that are arbitrary functions of time. Let $v^e(\mathbf{x})$ be the deflection of a structure made of elastic material and subjected to a load $f_0(\mathbf{x})$. Then by the correspondence principle, the deflection of the same structure but made of viscoelastic material with creep compliance $J(t)$ and subjected to the step load $f(\mathbf{x}, t) = f_0(\mathbf{x})H(t)$ is

$$v(\mathbf{x}, t) = J(t)v^e(\mathbf{x}). \tag{9.3.29}$$

If the load history is of general type, $f(\mathbf{x}, t) = f_0(\mathbf{x})g(t)$, we can break the load history into a sequence of infinitesimal steps $dg(t')$, as shown in Fig. 9.3.4. Then we can write the solution in the form of a hereditary integral

$$v(\mathbf{x}, t) = v^e(\mathbf{x}) \left[g(0)J(t) + \int_0^t J(t - t') \frac{dg(t')}{dt'} \, dt' \right]. \tag{9.3.30}$$

Next we consider a number of examples to illustrate how to determine the viscoelastic response.

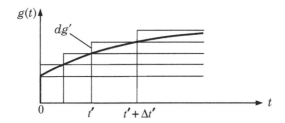

Fig. 9.3.4: Load history as a sequence of infinitesimal load steps.

Example 9.3.2 ———————————————————————————

Consider a simply supported beam, as shown in Fig. 9.3.5. At time $t = 0$ a point load F_0 is placed at the center of the beam. Determine the viscoelastic center deflections using Maxwell's and Kelvin's models.

Solution: The deflection at the center of the elastic beam is

$$v_c^e = \frac{F_0 L^3}{48EI}. \tag{1}$$

For a viscoelastic beam, we replace $1/E$ with creep compliance $J(t)$ of a chosen viscoelastic material (e.g., Maxwell model, Kelvin model, etc.)

$$v_c^v(t) = J(t) \frac{F_0 L^3}{48I}. \tag{2}$$

Using the Maxwell model, we can write [see Eq. (9.2.16)]

$$v_c^v(t) = \frac{1}{k}\left(1 + \frac{t}{\tau}\right)\frac{F_0 L^3}{48I}, \quad \tau = \frac{\eta}{k}. \tag{3}$$

For the Kelvin model, we obtain [see Eq. (9.2.25)]

$$v_c^v(t) = \frac{1}{k}\left(1 - e^{-t/\tau}\right)\frac{F_0 L^3}{48I}, \quad \tau = \frac{\eta}{k}. \tag{4}$$

Clearly, the response is quite different for the two materials.

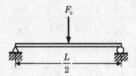

Fig. 9.3.5: A simply supported beam with a central point load.

Example 9.3.3

Consider the simply supported beam of Fig. 9.3.5 but with specified deflection v_c at the center of the beam. Determine the load for the viscoelastic response of the Maxwell and Kelvin models.

Solution: The force required to deflect the elastic beam at the center by v_c is

$$F^e = \frac{48EIv_c}{L^3}. \tag{1}$$

To obtain the load for a viscoelastic beam, we replace E with relaxation modulus $Y(t)$ of the viscoelastic material used

$$F^e(t) = Y(t)\frac{48Iv_c}{L^3}. \tag{2}$$

For the Maxwell model, we have the result [see Eq. (9.2.19)]

$$F_c^v(t) = ke^{-t/\tau}\frac{48Iv_c}{L^3}, \quad \tau = \frac{\eta}{k}, \tag{3}$$

and for the Kelvin model, we obtain [see Eq. (9.2.26)]

$$F_c^v(t) = k\left[H(t) + \tau\,\delta(t)\right]\frac{48Iv_c}{L^3}, \quad \tau = \frac{\eta}{k}. \tag{4}$$

Example 9.3.4

Consider a simply supported beam with a uniformly distributed load of intensity q_0 as shown in Fig. 9.3.6(a). Determine the viscoelastic deflection at the center.

Solution: The elastic deflection of the beam is given by

$$v^e(x) = \frac{q_0 L^4}{24EI}\left[\left(\frac{x}{L}\right) - 2\left(\frac{x}{L}\right)^3 + \left(\frac{x}{L}\right)^4\right]. \tag{1}$$

The midspan deflection is

$$v_c^e(L/2) = \frac{5q_0 L^4}{384EI}.$$ (2)

For the load history shown in Fig. 9.3.6(b), the midspan deflection of the viscoelastic beam is

$$v_c^v(L/2, t) = \frac{5q_0 L^4}{384I} \frac{1}{t_1} \int_0^t J(t - t')\, dt', \quad 0 < t < t_1,$$ (3)

$$v_c^v(L/2, t) = \frac{5q_0 L^4}{384I} \frac{1}{t_1} \int_0^{t_1} J(t - t')\, dt', \quad t > t_1.$$ (4)

For example, if we use the Kelvin–Voigt model, we obtain $(\tau = \eta/k)$

$$v_c^v(L/2, t) = \frac{5q_0 L^4}{384I} \frac{1}{kt_1} \left[t - \frac{\eta}{k} \left(1 - e^{-t/\tau} \right) \right], \quad 0 < t < t_1,$$ (5)

$$v_c^v(L/2, t) = \frac{5q_0 L^4}{384I} \frac{1}{k} \left[1 + \frac{\eta}{kt_1} \left(1 - e^{t_1/\tau} \right) e^{-t/\tau} \right], \quad t > t_1.$$ (6)

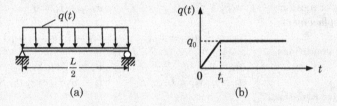

Fig. 9.3.6: A simply supported beam with a uniform load.

9.3.4 Elastic and Viscoelastic Analogies

In this section we examine the analogies between the field equations of elastic and viscoelastic bodies. These analogies help us to solve viscoelastic problems when solutions to the corresponding elastic problem are known. The field equations are summarized in Table 9.3.1 for the two cases. The Laplace transformed equations of elastic and viscoelastic bodies are summarized in Table 9.3.2. The correspondence is more apparent. A comparison of the Laplace transformed elastic and viscoelastic equations reveals the following correspondence

$$\sigma_{ij}^e(\mathbf{x}) \sim \bar{\sigma}_{ij}^v(\mathbf{x}, s), \quad \varepsilon_{ij}^e(\mathbf{x}) \sim \bar{\varepsilon}_{ij}^v(\mathbf{x}, s),$$ (9.3.31)

$$G^e(\mathbf{x}) \sim \bar{G}^*(\mathbf{x}, s) = s\bar{G}(\mathbf{x}, s) \quad K^e(\mathbf{x}) \sim \bar{K}^*(\mathbf{x}, s) = s\bar{K}(\mathbf{x}, s).$$ (9.3.32)

This correspondence allows us to use the solution of an elastic boundary value problem to obtain the transformed solution of the associated viscoelastic boundary value problem by simply replacing the elastic material properties G and K with G^* and K^*. One needs only to invert the solution to obtain the time-dependent viscoelastic solution. This analogy does not apply to problems for which the boundary conditions are time dependent.

Table 9.3.1: Field equations of elastic and viscoelastic bodies.

Type of equation	Elasticity	Viscoelasticity
Equations of motion	$\sigma_{ij,j} + f_i = \rho \ddot{u}_i$	$\sigma_{ij,j} + f_i = \rho \ddot{u}_i$
Strain–displacement equations	$\varepsilon_{ij} = \frac{1}{2}(u_{i,j} + u_{j,i})$	$\varepsilon_{ij} = \frac{1}{2}(u_{i,j} + u_{j,i})$
Boundary conditions	$u_i = \hat{u}_i$ on S_1	$u_i = \hat{u}_i$ on S_1
	$t_i \equiv n_j \sigma_{ji} = \hat{t}_i$ on S_2	$t_i \equiv n_j \sigma_{ji} = \hat{t}_i$ on S_2
Constitutive equations	$\sigma'_{ij} = 2G\varepsilon'_{ij}$	$\sigma'_{ij} = 2\int_{-\infty}^{t} G(t-t') \frac{d\varepsilon'_{ij}}{dt'}\, dt'$
	$\sigma_{kk} = 3K\varepsilon_{kk}$	$\sigma_{kk} = 3\int_{-\infty}^{t} K(t-t') \frac{d\varepsilon_{kk}}{dt'}\, dt'$

Table 9.3.2: Field equations of elastic and Laplace transformed viscoelastic bodies for the quasi-static case.

Type of equation	Elasticity	Viscoelasticity
Equations of motion	$\sigma_{ij,j} + f_i = 0$	$\bar{\sigma}_{ij,j} + \bar{f}_i = 0$
Strain–displacement equations	$\varepsilon_{ij} = \frac{1}{2}(u_{i,j} + u_{j,i})$	$\bar{\varepsilon}_{ij} = \frac{1}{2}(\bar{u}_{i,j} + \bar{u}_{j,i})$
Boundary conditions	$u_i = \hat{u}_i$ on S_1	$\bar{u}_i = \hat{\bar{u}}_i$ on S_1
	$t_i \equiv n_j \sigma_{ji} = \hat{t}_i$ on S_2	$\bar{t}_i \equiv n_j \bar{\sigma}_{ji} = \hat{\bar{t}}_i$ on S_2
Constitutive equations**	$\sigma'_{ij} = 2G\varepsilon'_{ij}$	$\bar{\sigma}'_{ij} = 2s\bar{G}(s)\bar{\varepsilon}'_{ij} = 2G^*(s)\bar{\varepsilon}'_{ij}$
	$\sigma_{kk} = 3K\varepsilon_{kk}$	$\bar{\sigma}_{kk} = 3s\bar{K}(s)\,\bar{\varepsilon}_{kk} = 3K^*(s)\,\bar{\varepsilon}_{kk}$

** $G^*(s) = s\bar{G}(s), \quad K^*(s) = s\bar{K}(s)$.

The analogy also holds for the dynamic case, but it is between the Laplace transformed elastic variables and viscoelastic variables:

$$\bar{\sigma}^e_{ij}(\mathbf{x}, s) \sim \bar{\sigma}^v_{ij}(\mathbf{x}, s), \quad \bar{\varepsilon}^e_{ij}(\mathbf{x}, s) \sim \bar{\varepsilon}^v_{ij}(\mathbf{x}, s), \tag{9.3.33}$$

$$\bar{G}^e(\mathbf{x}, s) \sim \bar{G}^*(\mathbf{x}, s) = s\bar{G}(\mathbf{x}, s) \quad \bar{K}^e(\mathbf{x}, s) \sim \bar{K}^*(\mathbf{x}, s) = s\bar{K}(\mathbf{x}, s). \tag{9.3.34}$$

Next we consider two examples of application of the elastic–viscoelastic analogy.

Example 9.3.5

The structure shown in Fig. 9.3.7 consists of a viscoelastic rod and elastic rod connected in parallel to a rigid bar. The cross-sectional areas of the rods are the same. The modulus of the material of the rods are

$$\begin{array}{ll} \text{Viscoelastic rod:} & E(t) = 2\mu H(t) + 2\eta\delta(t), \\ \text{Elastic rod:} & E = \text{Young's modulus} = \text{constant.} \end{array} \tag{1}$$

If a load of $F(t) = F_0 H(t)$ acts on the rigid bar and the rigid bar is maintained horizontal, determine the resulting displacement of the rigid bar.

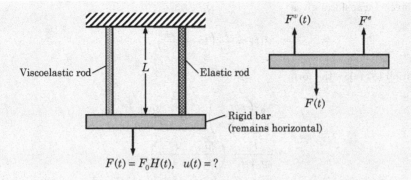

Fig. 9.3.7: Elastic-viscoelastic bar system.

Solution: Let u^e and $u^v(t)$ be the axial displacements in elastic and viscoelastic rods, respectively. Then the axial strains in elastic and viscoelastic rods are given by

$$\varepsilon^e = \frac{u^e}{L}, \quad \varepsilon^v(t) = \frac{u^v(t)}{L}. \tag{2}$$

The strain-stress relations for the two rods are

$$\varepsilon^e = \frac{\sigma^e}{E^e}, \quad \varepsilon^v(t) = \int_{-\infty}^{t} J(t-\tau) \frac{d\sigma^v}{d\tau}. \tag{3}$$

The axial stresses in elastic and viscoelastic rods are given by

$$\sigma^e = \frac{F^e}{A}, \quad \sigma^v(t) = \frac{F^v(t)}{A}. \tag{4}$$

From Eqs. (9.3.53)–(9.3.55) we have

$$u^e = \frac{F^e L}{E^e A}, \quad u^v(t) = \frac{L}{A} \int_{-\infty}^{t} J(t-\tau) \frac{dF^v}{d\tau} \, d\tau, \tag{5}$$

where F^e and F^v are the axial forces in the elastic and viscoelastic rods, respectively. The geometric compatibility requires $u^e = u^v$, giving

$$\frac{F^e L}{A E^e} = \frac{L}{A} \int_{-\infty}^{t} J(t-\tau) \frac{dF^v}{d\tau} \, d\tau$$

or

$$F^e = E^e \int_{-\infty}^{t} J(t-\tau) \frac{dF^v}{d\tau} \, d\tau. \tag{6}$$

The force equilibrium requires

$$F(t) = F^v(t) + F^e = F^v(t) + E^e \int_{-\infty}^{t} J(t-\tau) \frac{dF^v}{d\tau} \, d\tau, \tag{7}$$

which is an integro-differential equation for $F^v(t)$.

Using the Laplace transform, we obtain

$$\frac{F_0}{s} = \left(1 + E^e s \bar{J}\right) \bar{F}^v. \tag{8}$$

Since $s\bar{J} = \frac{1}{sE}$, we can write

$$\bar{J}(s) = \frac{1}{s^2 \bar{E}} = \frac{1}{s(2\eta s + 2\mu)} = \frac{1}{2\mu} \left(\frac{1}{s} - \frac{1}{s + \frac{\mu}{\eta}}\right), \tag{9}$$

and the inverse transform gives

$$J(t) = \frac{1}{2\mu}\left(1 - e^{-\frac{\mu t}{\eta}}\right).$$ (10)

Equation (8) takes the form

$$\bar{F}^v = \frac{F_0}{s}\left(\frac{s + \frac{\mu}{\eta}}{s + \alpha}\right), \quad \alpha = \frac{2\mu + E^e}{2\eta},$$

$$= \frac{F_0}{2\mu + E^e}\left(\frac{2\mu}{s} - \frac{E^e}{s + \alpha}\right).$$ (11)

The inverse transform gives the force in the viscoelastic rod

$$F^v(t) = \frac{F_0}{2\mu + E^e}\left(2\mu - E^e e^{-\alpha t}\right).$$ (12)

Then from Eq. (5) we have

$$\bar{u}^v(s) = \frac{L}{A}s\bar{J}\bar{F}^v = \frac{F_0 L}{As(s + \alpha)} = \frac{F_0 L}{A(2\mu + E^e)}\left[\frac{1}{s} - \frac{1}{s + \alpha}\right].$$ (13)

The inverse transform yields the displacement

$$u^v(t) = \frac{F_0 L}{A(2\mu + E^e)}\left(1 - e^{-\alpha t}\right).$$ (14)

Example 9.3.6

Consider an isotropic, hollow, thick-walled, long circular cylinder of internal radius a and outside radius b (see Example 7.5.1 for the elastic solution for a more general problem). The cylinder is held between rigid supports such that $u_z = 0$ at $z = \pm L/2$ (i.e., plane strain state) and pressurized at $r = a$, as shown in Fig. 9.3.8. The material is assumed to be elastic in dilatation and viscoelastic in shear with response defined by a standard linear solid, whose constitutive equation is

$$\frac{1}{\eta}\sigma + \frac{1}{k_2}\frac{d\sigma}{dt} = \frac{k_1}{\eta}\varepsilon + \left(1 + \frac{k_1}{k_2}\right)\frac{d\varepsilon}{dt}.$$ (1)

Determine the displacements, strains, and stresses in the cylinder. Assume zero initial conditions.

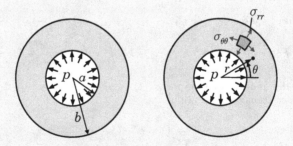

Fig. 9.3.8: Elastic–viscoelastic bar system.

Solution: First we identify the constitutive relations for a material that is elastic in dilatation and viscoelastic in shear, defined by a standard linear solid (see Problem **9.2**):

$$\text{elastic, dilatation:}\quad \tilde{\sigma} = 3K\tilde{e}, \tag{2}$$

$$\text{viscoelastic, shear:}\quad p_0\,s_{ij} + p_1\frac{ds_{ij}}{dt} = q_0\,e_{ij} + q_1\frac{de_{ij}}{dt}, \tag{3}$$

where $s_{ij} = \sigma'_{ij}$ and $e_{ij} = \varepsilon'_{ij}$ [see Eqs. (9.3.10)–(9.3.13)], and $(k_i \to 2G_i)$

$$p_0 = 1, \quad p_1 = \frac{\eta}{2G_2}, \quad q_0 = 2G_1, \quad q_1 = \eta\left(\frac{G_1+G_2}{G_2}\right). \tag{4}$$

Taking the Laplace transfor of Eq. (3) and using Eq. (9.3.21), we obtain $(p_0 = 1)$

$$2s\bar{G}(s) = \frac{q_1 + q_2 s}{p_1 + p_2 s} \quad \to \quad 2\bar{G}(s) = \frac{A}{s} + \frac{B}{p_1(p_1^{-1}+s)}, \tag{5}$$

with $A = q_0$ and $B = q_1 - q_0 p_1$. Hence, we have

$$2G(t) = q_0 + \left(\frac{q_1}{p_1} - q_0\right)e^{-t/p_1} = 2G_1 + 2G_2\,e^{-(2G_2 t/\eta)}. \tag{6}$$

The elastic solution (in dilatation) is known from Example 7.5.1 as

$$u_r(r) = \frac{1}{2(\mu+\lambda)}\left(\frac{p}{b^2-a^2}\right)r + \frac{a^2 b^2}{2\mu}\left(\frac{p}{b^2-a^2}\right)\frac{1}{r} = \frac{(1+\nu)pa^2 b}{E(b^2-a^2)}\left[(1-\nu)\frac{r}{b}+\frac{b}{r}\right], \tag{7}$$

$$\sigma_{rr}(r) = \left(\frac{pa^2}{b^2-a^2}\right)\left(1-\frac{b^2}{r^2}\right), \quad \sigma_{\theta\theta}(r) = \left(\frac{pa^2}{b^2-a^2}\right)\left(1+\frac{b^2}{r^2}\right). \tag{8}$$

Using the relations between (E, ν) and (G, K),

$$E = \frac{9KG}{3K+G}, \quad \nu = \frac{3K-2G}{6K+2G}, \tag{9}$$

the displacement u_r can be expressed in terms of G and K as

$$u_r = \frac{(1+\nu)pa^2 b}{2G(b^2-a^2)}\left[\frac{3G}{3K+G}\frac{r}{b}+\frac{b}{r}\right]. \tag{10}$$

Since the stresses do not depend on the material parameters, they are valid for the viscoelastic case. As for the displacement, we use the viscoelastic analogy and write

$$\bar{u}_r(r,s) = \frac{(1+\nu)pa^2 b}{2\bar{G}(s)(b^2-a^2)}\left[\frac{3\bar{G}(s)}{3\bar{K}(s)+\bar{G}(s)}\frac{r}{b}+\frac{b}{r}\right], \tag{12}$$

where $\bar{K} = K$, and $\bar{G}(s)$ is given by Eq. (5)

$$\bar{G}(s) = \frac{1}{2}\left[\frac{q_0}{s} + \frac{q_1 - q_0 p_1}{p_1(p_1^{-1}+s)}\right]. \tag{13}$$

The inversion of Eq. (12) to the time domain is algebraically complicated, but can be done with the help of the inverse transforms given Table 9.1.1. The viscoelastic solution for the displacement is

$$u_r(r,t) = \frac{pa^2 b}{(b^2-a^2)}\left\{3\frac{r}{b}\left[\frac{1}{6K+q_0}+\left(\frac{p_1}{6Kp_1+q_1}-\frac{1}{6K+q_0}\right)\exp\left(-\frac{6K+q_0}{6Kp_1+q_1}t\right)\right]\right.$$
$$\left.+\frac{1}{q_0}\frac{b}{r}\left[1+\left(\frac{q_0 p_1}{q_1}-1\right)\exp\left(-\frac{q_0}{q_1}t\right)\right]\right\}. \tag{14}$$

9.4 Summary

This chapter is dedicated to an introduction to linearized viscoelasticity. Beginning with simple spring-dashpot models of Maxwell and Kelvin–Voigt, three- and four-element models and integral constitutive models are discussed, and their creep and relaxation responses are derived. The discussion is then generalized to derive integral constitutive equations of linearized viscoelastic materials. Analogies between elastic and viscoelastic solutions are discussed. Applications of the analogies to the solutions of some typical problems from mechanics of materials are presented. This chapter constitutes a good introduction to a more complete course on theory of viscoelasticity.

Problems

9.1 *Method of partial fractions.* Suppose that we have a ratio of polynomials of the type

$$\frac{\bar{F}(s)}{\bar{G}(s)},$$

where $\bar{F}(s)$ is a polynomial of degree m and $\bar{G}(s)$ is a polynomial of degree n, with $n > m$. We wish to write in the form

$$\frac{\bar{F}(s)}{\bar{G}(s)} = \frac{c_1}{s + \alpha_1} + \frac{c_2}{s + \alpha_2} + \frac{c_3}{s + \alpha_3} + \cdots + \frac{c_n}{s + \alpha_n},$$

where c_i and α_i are constants to be determined using

$$c_i = \lim_{s \to -\alpha_i} \frac{(s + \alpha_i)\bar{F}(s)}{\bar{G}(s)}, \quad n = 1, 2, \cdots, n.$$

It is understood that $\bar{G}(s)$ is equal to the product $\bar{G}(s) = (s + \alpha_1)(s + \alpha_2)\ldots(s + \alpha_n)$. If

$$\bar{F}(s) = s^2 - 6, \quad \bar{G}(s) = s^3 + 4s^2 + 3s,$$

determine c_i.

9.2 Given the following transformed function

$$\bar{u}_r(r, s) = \frac{(1 + \nu)pa^2 b}{2\bar{G}(s)(b^2 - a^2)} \left[\frac{3\bar{G}(s)}{3\bar{K}(s) + \bar{G}(s)} \frac{r}{b} + \frac{b}{r} \right], \tag{1}$$

where $\bar{K} = K$, p, a, b, p_1, q_0, and q_1 are constants, and $\bar{G}(s)$ is given by

$$\bar{G}(s) = \frac{1}{2} \left[\frac{q_0}{s} + \frac{q_1 - q_0 p_1}{p_1(p_1^{-1} + s)} \right], \tag{2}$$

determine its Laplace inverse, $u_r(r, t)$.

9.3 Determine the creep and relaxation responses of the three-element model (i.e., standard linear solid) of Fig. 9.2.9(b) following the procedure used in Eqs. (9.2.33)–(9.2.42) for the three-element model shown Fig. 9.2.9(a). In particular show that the creep compliance function is the creep compliance function is

$$J(t) = \frac{1}{k_1} - \frac{k_2}{k_1(k_1 + k_2)} e^{-t/\tau},$$

and the relaxation function is

$$Y(t) = k_1 + k_2 e^{-t/\tau}.$$

9.4 Derive the governing differential equation for the spring-dashpot model shown in Fig. P9.4. Determine the creep compliance $J(t)$ and relaxation modulus $Y(t)$ associated with the model.

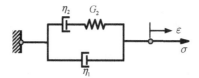

Fig. P9.4

9.5 Determine the relaxation modulus $Y(t)$ of the three-element model of Fig. 9.2.9(a) using Eq. (9.2.10) and the creep compliance in Eq. (9.2.37) [i.e., verify the result in Eq. (9.2.42)].

9.6 Derive the governing differential equation for the mathematical model obtained by connecting the Maxwell element in *series* with the Kelvin–Voigt element (see Fig. P9.6).

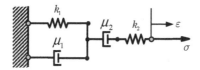

Fig. P9.6

9.7 Determine the creep compliance $J(t)$ and relaxation modulus $Y(t)$ of the four-element model of Problem **9.6**.

9.8 Derive the governing differential equation for the mathematical model obtained by connecting the Maxwell element in *parallel* with the Kelvin-Voigt element (see Fig. P9.8).

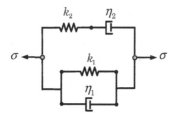

Fig. P9.8

9.9 Derive the governing differential equation of the four-parameter solid shown in Fig. P9.9. Show that it degenerates into the Kelvin–Voigt solid when its components parts are made equal.

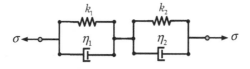

Fig. P9.9

9.10 Determine the creep compliance $J(t)$ and relaxation modulus $Y(t)$ of the four-element model of Problem **9.8**.

9.11 If a strain of $\varepsilon(t) = \varepsilon_0 t$ is applied to the four-element model of Problem **9.8**, determine the stress $\sigma(t)$ using a suitable hereditary integral [use $Y(t)$ from Problem **9.10**].

9.12 For the three-element model of Fig. 9.2.9(b), determine the stress $\sigma(t)$ when the applied strain is $\varepsilon(t) = \varepsilon_0 + \varepsilon_1 t$, where ε_0 and ε_1 are constants.

9.13 Determine expressions for the (Laplace) transformed modulus $\bar{E}(s)$ and Poisson's ratio $\bar{\nu}$ in terms of the transformed bulk modulus $\bar{K}(s)$ and transformed shear modulus $\bar{G}(s)$.

9.14 Evaluate the hereditary integral in Eq. (9.3.4) for the three-element model of Fig. 9.2.9(a) and stress history shown in Fig. 9.3.3.

9.15 Given that the shear creep compliance of a Kelvin–Voigt viscoelastic material is

$$J(t) = \frac{1}{2G_0}\left(1 - e^{-t/\tau}\right),$$

where G_0 and τ are material constants, determine the following properties for this material:

(a) shear relaxation modulus, $2G(t)$,

(b) the differential operators P and Q of Eq. (9.1.1),

(c) integral form of the stress–strain relation, and

(d) integral form of the strain–stress relation.

9.16 The strain in a uniaxial viscoelastic bar whose viscoelastic modulus is $E(t) = E_0/(1 + t/C)$ is $\varepsilon(t) = At$, where E_0, C, and A are constants. Determine the stress $\sigma(t)$ in the bar.

9.17 Determine the free end deflection $w^v(t)$ of a cantilever beam of length L, second moment of inertia I, and subjected to a point load $F(t)$ at the free end, for the cases (a) $F(t) = F_0 H(t)$ and (b) $F(t) = F_0 e^{-\alpha t}$. The material of the beam has the relaxation modulus of $E(t) = Y(t) = A + Be^{-\alpha t}$.

9.18 A cantilever beam of length L is made of a viscoelastic material that can be represented by the three-parameter solid shown in Fig. 9.2.9(a). The beam carries a load of $F(t) = F_0 H(t)$ at its free end. Assuming that the second moment of area of the beam is I, determine the tip deflection.

9.19 A simply supported beam of length L, second moment of area I is made from the Kelvin–Voigt type viscoelastic material whose compliance constitutive response is

$$J(t) = \frac{1}{E_0}\left(1 - e^{-t/\tau}\right),$$

where E_0 and τ are material constants. The beam is loaded by a transverse distributed load

$$q(x,t) = q_0\left(1 - \frac{x}{L}\right)t^2 = f(x)\,g(t),$$

where q_0 is the intensity of the distributed load at $x = 0$ and $g(t) = t^2$. Determine the deflection and stress in the viscoelastic beam using the Euler–Bernoulli beam theory.

9.20 The pin-connected structure shown in Fig. P9.20 is made from an incompressible viscoelastic material whose shear response can be expressed as

$$P = 1 + \frac{\eta}{\mu}\frac{d}{dt}, \quad Q = \eta\frac{d}{dt},$$

where η and μ are material constants. The structure is subjected to a time-dependent vertical force $F(t)$, as shown in Fig. P9.20. Determine the vertical load $F(t)$ required to produce this deflection history. Assume that member AB has a cross-sectional area $A_1 = 9/16$ in.^{2} and member BC has a cross-sectional area $A_2 = 125/48$ in.^{2}.

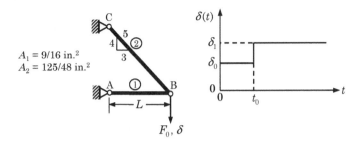

Fig. P9.20

9.21 Consider a hallow thick-walled spherical pressure vessel composed of two different viscoelastic materials, as shown in Fig. P9.21. *Formulate* (you need not obtain complete solution to) the boundary value problem from which the stress and displacement fields may be determined.

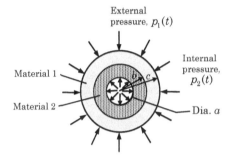

Fig. P9.21

9.22 The linear elastic solution for axial stress $\sigma_{xx}(x,y)$ and transverse displacement $v(y)$, based on the Euler–Bernoulli beam theory, of a cantilever beam of length L, flexural stiffness EI, and loaded at the free end with F_0, as shown in Fig. P9.22, is

$$\sigma_{xx} = -\frac{F_0 xy}{EI}, \qquad v(x) = \frac{F_0 L^3}{6EI}\left(2 - 3\frac{x}{L} + \frac{x^2}{L^2}\right).$$

Determine the viscoelastic counterparts of the stress and displacement using the viscoelastic analogy and Kelvin–Voigt model.

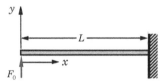

Fig. P9.22

References
for Additional Reading

Although the book is largely self-contained, readers may find the following references useful for further study, additional examples and exercises, and alternative explanations of various concepts discussed in this book.

1. R. Aris, *Vectors, Tensors, and the Basic Equations in Fluid Mechanics*, Prentice-Hall, Englewood Cliffs, NJ (1962).

2. E. Betti, "Teoria della' Elasticita," *Nuovo Cimento*, Serie 2, Tom VII and VIII (1872).

3. R. B. Bird, W. E. Stewart, and E. N. Lightfoot, *Transport Phenomena*, John Wiley & Sons, New York (1960).

4. R. B. Bird, R. C. Armstrong, and O. Hassager, *Dynamics of Polymeric Liquids, Vol. 1: Fluid Mechanics*, 2nd ed., John Wiley & Sons, New York (1971).

5. J. Bonet and R. D. Wood, *Nonlinear Continuum Mechanics for Finite Element Analysis*, 2nd ed., Cambridge University Press, New York (2008).

6. A. C. Eringen and G. W. Hanson, *Nonlocal Continuum Field Theories*, Springer-Verlag, New York (2002).

7. R. T. Fenner and J. N. Reddy, *Mechanics of Solids ad Structures*, 2nd ed., CRC Press, Boca Raton, FL (2012).

8. W. Flügge, *Viscoelasticity*, 2nd ed., Springer-Verlag, New York (1975).

9. M. E. Gurtin, *An Introduction to Continuum Mechanics*, Elsevier Science & Technology, San Diego, CA (1981).

10. M. E. Gurtin, E. Fried, and L. Anand, *The Mechanics and Thermodynamics of Continua*, Cambridge University Press, New York (2010).

11. K. D. Hjelmstad, *Fundamentals of Structural Mechanics*, Prentice–Hall, Englewood Cliffs, NJ (1997).

12. G. A. Holzapfel, *Nonlinear Solid Mechanics*, John Wiley & Sons, New York (2001).

13. H. Hochstadt, *Special Functions of Mathematical Physics*, Holt, Reinhart and Winston, New York (1961).

14. J. D. Jackson, *Classical Electrodynamics*, 2nd ed., John Wiley & Sons, New York (1975).

15. W. Jaunzemis, *Continuum Mechanics*, Macmillan, New York (1967).

16. E. Kreyszig, *Introduction Functional Analysis with Applications*, John Wiley & Sons, New York (1978).

17. E. H. Lee, "Viscoelasticity," in *Handbook of Engineering Mechanics*, McGraw-Hill, New York (1962).

18. L. E. Malvern, *Introduction to the Mechanics of a Continuous Medium*, Prentice Hall, Englewood Cliffs, NJ (1997).

19. J. C. Maxwell, "On the calculation of the equilibrium and the stiffness of frames," *Philosophical Magazine, Ser. 4*, **27**, 294 (1864).

20. N. I. Mushkelishvili, *Some Basic Problems of the Mathematical Theory of Elasticity*, Noordhoff, Gröningen, The Netherlands (1963).

21. Naghdi, P. M., *P. M. Naghdi's Notes on Continuum Mechanics*, Department of Mechanical Engineering, University of California, Berkeley, 2001. These class notes were developed (continuously since 1960) by Professor P. M. Naghdi (1924–1994) for a first course on continuum mechanics. The handwritten notes (by Professor Naghdi) were typed and later digitized (presumably, under the direction of Professor J. Casey).

22. A. W. Naylor and G. R. Sell, *Linear Operator Theory in Engineering and Science*, Holt, Reinhart and Winston, New York (1971).

23. W. Noll, *The Non-Linear Field Theories of Mechanics*, 3rd ed., Springer-Verlag, New York (2004).

24. R. W. Ogden, *Non-Linear Elastic Deformations*, Halsted (John Wiley & Sons), New York (1984).

25. J. T. Oden and J. N. Reddy, *Variational Principles in Theoretical Mechanics*, 2nd ed., Springer-Verlag, Berlin (1983).

26. J. N. Reddy, *Applied Functional Analysis and Variational Methods in Engineering*, McGraw-Hill, New York (1986); reprinted by Krieger, Malabar, FL (1991).

27. J. N. Reddy, *Energy Principles and Variational Methods in Applied Mechanics*, 2nd ed., John Wiley & Sons, New York (2002).

28. J. N. Reddy, *Mechanics of Laminated Composite Plates and Shells. Theory and Analysis*, 2nd ed., CRC Press, Boca Raton, FL (2004).

29. J. N. Reddy, *An Introduction to the Finite Element Method*, 3rd ed., McGraw-Hill, New York (2006).

30. J. N. Reddy and D. K. Gartling, *The Finite Element Method in Heat Transfer and Fluid Dynamics*, 2nd ed., CRC Press, Boca Raton, FL (2001).

31. J. N. Reddy and M. L. Rasmussen, *Advanced Engineering Analysis*, John Wiley & Sons, New York (1982); reprinted by Krieger, Malabar, FL (1991).

32. H. Schlichting, *Boundary Layer Theory*, (translated from German by J. Kestin), 7th ed., McGraw-Hill, New York (1979).

33. L. A. Segel, *Mathematics Applied to Continuum Mechanics (with additional material on elasticity by G. H. Handelman)*, Dover Publications, New York (1987).

34. W. S. Slaughter, *The Linearized Theory of Elasticity*, Birkhäser, Boston (2002).

35. I. S. Sokolnikoff, *Mathematical Theory of Elasticity*, 2nd ed., McGraw-Hill, New York; reprinted by Krieger, Melbourne, FL (1956).

36. M. H. Sadd, *Elasticity: Theory, Applications, and Numerics*, Academic Press, New York (2004).

37. S. P. Timoshenko and J. N. Goodier, *Theory of Elasticity*, 3rd ed., McGraw-Hill, New York (1970).

38. C. A. Truesdell, *Elements of Continuum Mechanics*, 2nd printing, Springer-Verlag, New York (1984).

39. C. Truesdell and R. A. Toupin, "The Classical Field Theories," in *Encyclopedia of Physics, III/1*, S. Flügge (ed.), Springer-Verlag, Berlin (1965).

40. C. Truesdell and W. Noll, "The Non-Linear Field Theories of Mechanics," in *Encyclopedia of Physics, III/3*, S. Flügge (ed.), Springer-Verlag, Berlin (1965).

Answers to Selected Problems

If a man is in too big a hurry to give up an error he is liable to give up some truth with it.

—— Wilbur Wright (1867–1912)

Chapter 1

1.1 The equation of motion is

$$\frac{dv}{dt} + \alpha v^2 = g, \quad \alpha = \frac{c}{m}.$$

1.2 The energy balance gives

$$-\frac{d}{dx}(Aq) + \beta P(T_\infty - T) + Ag = 0.$$

1.3 The equation governing the transverse deflection v is

$$\frac{d^2}{dx^2}\left(EI\frac{d^2v}{dx^2}\right) = q(x).$$

1.4 The conservation of mass gives

$$\frac{d(Ah)}{dt} = q_i - q_0,$$

where A is the area of cross section of the tank $(A = \pi D^2/4)$ and ρ is the mass density of the liquid.

1.5 $p = 2t_s/R$.

Chapter 2

2.1 The equation of the required line is $\mathbf{r} = \mathbf{A} + \alpha\mathbf{B}$.

2.3 The equation for the required plane is $(\mathbf{C} - \mathbf{A}) \times (\mathbf{B} - \mathbf{A}) \cdot (\mathbf{r} - \mathbf{A}) = 0$, where $\mathbf{r}$ is the position vector.

2.8 The vector sum of the areas is $\mathbf{A} \times \mathbf{B} + \mathbf{B} \times \mathbf{C} + \mathbf{C} \times \mathbf{A} + (\mathbf{C} - \mathbf{A}) \times (\mathbf{B} - \mathbf{A})$.

2.11 The vectors are linearly dependent.

2.12 (c) $A_1 = -13$, $A_2 = 21$, and $A_3 = 19$.

2.13 (a) $\hat{\mathbf{e}}_1 = \frac{1}{\sqrt{2}}\left(\hat{\mathbf{i}}_1 + \hat{\mathbf{i}}_3\right)$, $\hat{\mathbf{e}}_2 = \frac{1}{3\sqrt{2}}\left(-\hat{\mathbf{i}}_1 + 4\hat{\mathbf{i}}_2 + \hat{\mathbf{i}}_3\right)$, and $\hat{\mathbf{e}}_3 = \frac{1}{3}\left(2\hat{\mathbf{i}}_1 + \hat{\mathbf{i}}_2 - 2\hat{\mathbf{i}}_3\right)$.

2.17 $\frac{2}{\rho}$.

2.19 Use the two properties of determinants: (1) $\det([S][T]) = \det[S] \cdot \det[T]$ and (2) $\det [S]^{\mathrm{T}} = \det[S]$.

2.20 (a) Let $\hat{\mathbf{e}}_i = \delta_{ip}\hat{\mathbf{e}}_p$, $\hat{\mathbf{e}}_j = \delta_{jq}\hat{\mathbf{e}}_q$, and $\hat{\mathbf{e}}_k = \delta_{kr}\hat{\mathbf{e}}_r$.

2.22 (a) The transformation matrix is

$$\begin{bmatrix} \frac{1}{\sqrt{3}} & \frac{-1}{\sqrt{3}} & \frac{1}{\sqrt{3}} \\ \frac{2}{\sqrt{14}} & \frac{3}{\sqrt{14}} & \frac{1}{\sqrt{14}} \\ \frac{-4}{\sqrt{42}} & \frac{1}{\sqrt{42}} & \frac{5}{\sqrt{42}} \end{bmatrix}.$$

2.24 Follows from the definition

$$[L] = \begin{bmatrix} \frac{1}{\sqrt{2}} & 0 & \frac{1}{\sqrt{2}} \\ \frac{1}{2} & \frac{1}{\sqrt{2}} & -\frac{1}{2} \\ -\frac{1}{2} & \frac{1}{\sqrt{2}} & \frac{1}{2} \end{bmatrix}.$$

2.26 The determinants are (a) -8. (b) -5.

2.27 (b) The inverse is

$$[A]^{-1} = -\frac{1}{5} \begin{bmatrix} 1 & -2 & -1 \\ -4 & 3 & -1 \\ 3 & -1 & -3 \end{bmatrix}.$$

2.28 (b) Positive. (c) Not positive.

2.29 The positive matrix associated with $[Q]$ is

$$\begin{bmatrix} 2 & 1 & 1 \\ 1 & 2 & 3 \\ 1 & 3 & 5 \end{bmatrix}.$$

2.35 Use the gradient theorem with $\phi = 1$.

2.36 Use the divergence theorem.

2.41 (d) Note that

$$\mathbf{S}_i \cdot (\mathbf{S}_j \times \mathbf{S}_k) = \begin{vmatrix} S_{i1} & S_{i2} & S_{i3} \\ S_{j1} & S_{j2} & S_{j3} \\ S_{k1} & S_{k2} & S_{k3} \end{vmatrix}.$$

2.42 (a) $S_{ii} = 12$. (c) $S_{ij}S_{ij} = 281$. (e) $\{C\} = \{18 \ 15 \ 34\}^{\mathrm{T}}$.

2.44 The components $\bar{A}_i$ are given by

$$\left\{ \begin{array}{c} \bar{A}_1 \\ \bar{A}_2 \\ \bar{A}_3 \end{array} \right\} = \left\{ \begin{array}{c} 1 - \frac{3\sqrt{3}}{2} \\ 1 \\ \frac{5\sqrt{3}}{2} \end{array} \right\}.$$

2.48 Obtain Part (c) of Problem 2.39, which is the required result.

2.51 Use the del operator from Table 2.4.2 to compute the divergence of the tensor $\mathbf{S}$.

2.55 (a) $\lambda_1 = 3.0$, $\lambda_2 = 2(1 + \sqrt{5}) = 6.472$, $\lambda_3 = 2(1 - \sqrt{5}) = -2.472$. The eigenvector components A_i associated with λ_3 are $\hat{\mathbf{A}}^{(3)} = \pm(0.5257, 0.8507, 0)$.

(c) The characteristic polynomial is $[-\lambda^2 + 6\lambda - 8](\lambda - 1) = 0$. The eigenvalues are $\lambda_1 = 4$, $\lambda_2 = 2$, $\lambda_3 = 1$.

(d) The characteristic polynomial is $[-\lambda^2 + 5\lambda - 6](\lambda + 1) = 0$. The eigenvalues are $\lambda_1 = 3$, $\lambda_2 = 2$, $\lambda_3 = -1$. The eigenvector associated with $\lambda_1 = 3$ is $\hat{\mathbf{A}}^{(1)} = \pm\frac{1}{\sqrt{2}}(1, 0, 1)$.

2.56 (a) The characteristic polynomial is $-\lambda^3 + 6\lambda^2 + 78\lambda - 108 = 0$. The eigenvalues are $\lambda_1 = 11.8242$, $\lambda_2 = 1.2848$, $\lambda_3 = -7.1090$.

(b) The characteristic polynomial is $-\lambda^3 + 5\lambda^2 - 6\lambda + 1 = 0$. The eigenvalues are $\lambda_1 = 3.24698$, $\lambda_2 = 1.55496$, $\lambda_3 = 0.19806$. The eigenvectors are $\hat{\mathbf{A}}^{(1)} = \pm(0.328, -0.737, 0.591)$; $\hat{\mathbf{A}}^{(2)} = \pm(0.591, -0.328, -0.737)$; $\hat{\mathbf{A}}^{(3)} = \pm(0.737, 0.591, 0.328)$.

(c) The characteristic polynomial is $-\lambda^3 + \lambda^2 + \lambda - 1 = 0$, and the eigenvalues are $\lambda_1 = -1$, $\lambda_2 = 1$, $\lambda_3 = 1$. The eigenvectors are (not normalized) $\hat{\mathbf{A}}^{(1)} = (1, -1, 1)$; $\hat{\mathbf{A}}^{(2)} = \pm(1, 0, -1)$; $\hat{\mathbf{A}}^{(3)} = \pm(-1, 0, 1)$.

(d) The characteristic polynomial is $-\lambda^3 + 5\lambda^2 - 2\lambda - 8 = 0$, and the eigenvalues are $\lambda_1 = -1$, $\lambda_2 = 2$, $\lambda_3 = 4$. The eigenvectors are (not normalized) $\hat{\mathbf{A}}^{(1)} = (-0.5, 1, 0.16667)$; $\hat{\mathbf{A}}^{(2)} = \pm(0, 0, 1)$; $\hat{\mathbf{A}}^{(3)} = \pm(1, 0.5, 0.5)$.

2.57 $\lambda_1 = 9$, $\lambda_2 = -9$, $\lambda_3 = -18$.

2.58 $p([A]) = \begin{bmatrix} 1 & 0 \\ 0 & 1 \end{bmatrix}.$

2.60 The inverse is

$$[S]^{-1} = \frac{1}{12} \begin{bmatrix} 7 & -2 & 1 \\ -2 & 4 & -2 \\ 1 & -2 & 7 \end{bmatrix}.$$

Chapter 3

3.1 $\mathbf{v} = \frac{\mathbf{x}}{1+t}$, $\mathbf{a} = \mathbf{0}$.

3.2 $\boldsymbol{\chi}(\mathbf{x}) = (4X_1 + X_2 - 2X_1X_2)\,\hat{\mathbf{e}}_1 + (-X_1 + 3X_2 + 2X_1X_2)\,\hat{\mathbf{e}}_2 + X_3\hat{\mathbf{e}}_3$.

3.5 (c) $a_1 = a^2 \frac{e^{at}}{1+e^{at}} x_1$, $\quad a_2 = 4a^2 \frac{e^{-2at}}{1+e^{-2at}} x_2$, $\quad a_3 = 0$.

3.6 $u_1 = 3X_1 + X_2 - 2X_1X_2$, $\quad u_2 = -X_1 + 2X_2 + 2X_1X_2$, $\quad u_3 = 0$.

3.7 (a) $[F] = \begin{bmatrix} 1 & t^2 & 0 \\ t^2 & 1 & 0 \\ 0 & 0 & 1 \end{bmatrix}$. (c) $\left\{ \begin{matrix} 1 \\ 2 \\ 1 \end{matrix} \right\}$.

3.8 (b) $[C] = [B] = \begin{bmatrix} k_1^2 & 0 & 0 \\ 0 & k_2^2 & 0 \\ 0 & 0 & k_3^2 \end{bmatrix}$.

3.9 (a) $[F] = \begin{bmatrix} k_1 & e_0k_2 & 0 \\ 0 & k_2 & 0 \\ 0 & 0 & k_3 \end{bmatrix}$. (b) $[B] = \begin{bmatrix} k_1^2 + e_0^2k_2^2 & e_0k_2^2 & 0 \\ e_0k_2^2 & k_2^2 & 0 \\ 0 & 0 & k_3^2 \end{bmatrix}$.

3.10 (b) The displacements in the spatial description are

$$u_1(\mathbf{x}) = x_1\,(1 - \cos At) + x_2 \sin At,$$
$$u_2(\mathbf{x}) = -x_1 \sin At + x_2\,(1 - \cos At),$$
$$u_3(\mathbf{x}) = \frac{Bt}{1 + Bt}\, x_3.$$

3.11 (a) $u_1(\mathbf{X}) = AX_2$, $\quad u_2(\mathbf{X}) = BX_1$, $\quad u_3(\mathbf{X}) = 0$. (c) $2[E] = \begin{bmatrix} B^2 & A+B & 0 \\ A+B & A^2 & 0 \\ 0 & 0 & 0 \end{bmatrix}$.

3.13 $[F] = \begin{bmatrix} 1 & t^2 & 0 \\ t^2 & 1 & 0 \\ 0 & 0 & 1 \end{bmatrix}$.

3.15 (c) $[F] = \begin{bmatrix} \cosh t & \sinh t & 0 \\ \sinh t & \cosh t & 0 \\ 0 & 0 & 1 \end{bmatrix}$.

3.17 (b) The angle ABC after deformation is $90 - \beta$, where $\cos \beta = \frac{\mu}{\sqrt{1+\mu^2}}$, and $\mu = 2/(2-\gamma)$.

3.18 (a) $[E] = \begin{bmatrix} 6 & 7 & 0 \\ 7 & 8 & 0 \\ 0 & 0 & 0 \end{bmatrix}$, $\gamma = 1$.

3.19 $u_1 = \frac{e_0}{b} X_2$, $\quad u_2 = 0$, $\quad u_3 = 0$.

3.20 $u_1 = \left(\frac{e_0}{b^2}\right)X_2^2$, $\quad u_2 = 0$, $\quad u_3 = 0$, and $E_{11} = 0$, $\quad E_{12} = \frac{e_0}{b^2} X_2$, $\quad E_{22} = \frac{1}{2}\left(2X_2\frac{e_0}{b^2}\right)^2$.

3.21 $u_1 = e_1 \frac{X_1}{a} \frac{X_2}{b}$, $\quad u_2 = e_2 \frac{X_1}{a} \frac{X_2}{b}$.

3.22 $u_1 = -0.2X_1 + 0.5X_2$, $\quad u_2 = 0.2X_1 - 0.1X_2 + 0.1X_1X_2$.

3.23 $\varepsilon_{rr} = A$, $\quad \varepsilon_{\theta\theta} = A$, $\quad \varepsilon_{z\theta} = \frac{1}{2}\left(Br + \frac{C}{r}\cos\theta\right)$.

3.25 The linear components are given by $\varepsilon_{11} = 3X_1^2X_2 + c_1(2c_2^3 + 3c_2^2X_2 - X_2^3)$, $\quad \varepsilon_{22} = -(2c_2^3 + 3c_2^2X_2 - X_2^3 + 3c_1X_1^2X_2)$, $2\varepsilon_{12} = X_1\left[X_1^2 + c_1(3c_2^2 - 3X_2^2)\right] - 3c_1X_1X_2^2$.

3.26 (b) $E'_{11}(= E_{nn}) \approx \frac{ae_0}{a^2+b^2}$, $\quad E'_{12}(= E_{ns}) \approx \frac{e_0}{2b}\left(\frac{a^2-b^2}{a^2+b^2}\right)$.

3.27 The principal strains are $\varepsilon_1 = 0$ and $\varepsilon_2 = 10^{-3}$ in./in. The principal direction associated with $\varepsilon_1 = 0$ is $\mathbf{A}_1 = \hat{\mathbf{e}}_1 - 2\hat{\mathbf{e}}_2$ and that associated with $\varepsilon = 10^{-3}$ is $\mathbf{A}_2 = 2\hat{\mathbf{e}}_1 + \hat{\mathbf{e}}_2$.

3.29 The only nonzero linear strains are: $\varepsilon_{rr} = \frac{dU}{dr}$, $\quad \varepsilon_{\theta\theta} = \frac{U}{r}$.

3.30 The only nonzero linear strains are: $E_{RR} = \frac{dU}{dR} + \frac{1}{2}\left(\frac{dU}{dR}\right)^2$, $\quad E_{\phi\phi} = E_{\theta\theta} = \frac{U}{R} + \frac{1}{2}\left(\frac{U}{R}\right)^2$.

3.35 Use the definition of J

$$\begin{vmatrix} \frac{\partial x_1}{\partial X_1} & \frac{\partial x_1}{\partial X_2} & \frac{\partial x_1}{\partial X_3} \\ \frac{\partial x_2}{\partial X_1} & \frac{\partial x_2}{\partial X_2} & \frac{\partial x_2}{\partial X_3} \\ \frac{\partial x_3}{\partial X_1} & \frac{\partial x_3}{\partial X_2} & \frac{\partial x_3}{\partial X_3} \end{vmatrix}$$

and the hints to arrive at the result.

3.37 Use the definition of $\mathbf{C}$ in terms of $\mathbf{E}$ and Eq. (3.6.14).

3.38 Use the definition of $\mathbf{e}$ and follow the procedure of Section 3.6.2.

3.40 Use the expression in index notation from Eq. (3.7.11) to establish the symmetry of $\mathbf{S}$.

3.42 $u_1 = cX_1X_2^2$ and $u_2 = cX_1^2X_2$.

3.43 (b) The strain field is *not* compatible.

3.45 The function $f(X_2, X_3)$ is of the form $f(X_2, X_3) = A + BX_2 + CX_3$, where A, B, and C are arbitrary constants.

3.48 Take the time derivative of the identity $\mathbf{Q} \cdot \mathbf{Q}^{\mathrm{T}} = \mathbf{I}$.

3.51 Not objective.

3.55 Use the definition (3.6.3) and Eqs. (3.6.14) and (3.9.1) as well as the symmetry of $\mathbf{U}$ to establish the result.

3.56 $[C] == \frac{1}{2} \begin{bmatrix} \sqrt{5}+1 & \sqrt{5}-1 & 0 \\ \sqrt{5}-1 & \sqrt{5}+1 & 0 \\ 0 & 0 & 6 \end{bmatrix}$.

3.57 $[U] = \begin{bmatrix} 2.2313 & 0.1455 \\ 0.1455 & 1.0671 \end{bmatrix}, \ [V] = \begin{bmatrix} 1.0671 & 0.1455 \\ 0.1455 & 2.2313 \end{bmatrix}$.

3.58 $[U] = \frac{1}{2\sqrt{2}} \begin{bmatrix} 3+\sqrt{3} & 3-\sqrt{3} & 0 \\ 3-\sqrt{3} & 1+3\sqrt{3} & 0 \\ 0 & 0 & 2\sqrt{2} \end{bmatrix}, \ [V] = \frac{1}{\sqrt{2}} \begin{bmatrix} \sqrt{3}+1 & \sqrt{3}-1 & 0 \\ \sqrt{3}-1 & \sqrt{3}+1 & 0 \\ 0 & 0 & \sqrt{2} \end{bmatrix}$.

3.59 $[U] = \begin{bmatrix} 2.414 & 0 & 0 \\ 0 & 0.414 & 0 \\ 0 & 0 & 1.0 \end{bmatrix}, \ [V] = \begin{bmatrix} 2.121 & 0.707 & 0 \\ 0.707 & 0.707 & 0 \\ 0.000 & 0.000 & 1 \end{bmatrix}$.

3.60 $[\sqrt{C}] = \frac{1}{2} \begin{bmatrix} \sqrt{5}+1 & \sqrt{5}-1 & 0 \\ \sqrt{5}-1 & \sqrt{5}+1 & 0 \\ 0 & 0 & 6 \end{bmatrix}$.

3.61 $[U] = \frac{1}{2\sqrt{2}} \begin{bmatrix} 3+\sqrt{3} & 3-\sqrt{3} & 0 \\ 3-\sqrt{3} & 1+3\sqrt{3} & 0 \\ 0 & 0 & 2\sqrt{2} \end{bmatrix}, \ [R] = \frac{1}{2\sqrt{2}} \begin{bmatrix} \sqrt{3}+1 & \sqrt{3}-1 & 0 \\ 1-\sqrt{3} & \sqrt{3}+1 & 0 \\ 0 & 0 & 2\sqrt{2} \end{bmatrix},$

and $[V] = \frac{1}{\sqrt{2}} \begin{bmatrix} \sqrt{3}+1 & \sqrt{3}-1 & 0 \\ \sqrt{3}-1 & \sqrt{3}+1 & 0 \\ 0 & 0 & \sqrt{2} \end{bmatrix}$.

3.63 With respect to the basis $\hat{\mathbf{e}}_i$ we have

$$U = \begin{bmatrix} 0.707 & 0.707 & 0 \\ 0.707 & 2.121 & 0 \\ 0 & 0 & 1.0 \end{bmatrix} \text{ and } [V] = \begin{bmatrix} 2.121 & 0.707 & 0 \\ 0.707 & 0.707 & 0 \\ 0 & 0 & 1.0 \end{bmatrix}.$$

Chapter 4

4.2 Partial answer:
on FG: $\mathbf{t} = 2\hat{\mathbf{j}}$, on EF: $\mathbf{t} = -3\hat{\mathbf{j}}$,
on HA: $\mathbf{t} = \mathbf{0}$, on AB: unknown.

4.3 (a(i)) $\mathbf{t}^{\hat{n}} = 2(\hat{\mathbf{e}}_1 + 7\hat{\mathbf{e}}_2 + \hat{\mathbf{e}}_3)$. (c) $\sigma_n = -\frac{22}{3} = -7.33$ MPa, $\sigma_s = 12.26$ MPa.

4.4 (a) $\mathbf{t}^{\hat{n}} = \frac{1}{\sqrt{3}}(5\hat{\mathbf{e}}_1 + 5\hat{\mathbf{e}}_2 + 9\hat{\mathbf{e}}_3)$ MPa. (b) $\sigma_n = 6.3333$ MPa, $\sigma_s = 1.8856$ MPa.

4.5 $\sigma_n = -\frac{17}{6} = -2.833$ MPa, $\sigma_s = 8.67$ MPa.

4.6 (b) $t_n = -16.67$ MPa, $t_s = 52.7$ MPa.

4.7 $\sigma_n = 0.3478$ MPa, $\sigma_s = 4.2955$ MPa.

4.8 (b) $|\mathbf{t}| = 3.86$ MPa, and $t_{nn} = 3.357$ MPa.

(c) $[\bar{\sigma}] = \frac{1}{18}\begin{bmatrix} 32 & -42 & -28\sqrt{2} \\ -42 & 72 & -24\sqrt{2} \\ -28\sqrt{2} & -24\sqrt{2} & -32 \end{bmatrix}$.

4.9 (b) $|\mathbf{t}| = 3.1396$ MPa. (c) $t_{nn} = 0.6429$ MPa, and $t_{ns} = 3.073$ MPa.

4.10 (a) $\mathbf{t}_{nn} = P(\hat{\mathbf{e}}_1 + \hat{\mathbf{e}}_2 - \hat{\mathbf{e}}_3)$ and $\mathbf{t}_{ns} = P(\hat{\mathbf{e}}_2 + \hat{\mathbf{e}}_3)$.

4.11 $\sigma_0 = 3$; $\hat{n} = \pm\frac{1}{\sqrt{14}}(3\hat{\mathbf{e}}_1 - 2\hat{\mathbf{e}}_2 + \hat{\mathbf{e}}_3)$.

4.13 $\sigma_n = -40.8$ MPa, $\sigma_s = -20.67$ MPa.

4.14 $\sigma_n = 3.84$ MPa, $\sigma_s = -17.99$ MPa.

4.15 $\sigma_n = 95$ MPa, $\sigma_s = -15$ MPa.

4.16 $\sigma_n = -76.6$ MPa, $\sigma_s = 32.68$ MPa.

4.17 $\sigma_0 = 140$ MPa, $\sigma_s = -90$ MPa.

4.18 $\sigma_{p1} = 6.6568$ MPa, $\sigma_{p2} = 1$ MPa, $\sigma_{p3} = -4.6568$ MPa.

4.19 $\sigma_{p1} = 97.2$ MPa, $\sigma_{p2} = -7.2$ MPa.

4.20 $\sigma_{p1} = 121.98$ MPa, $\sigma_{p2} = -81.98$ MPa.

4.21 (a) $\sigma_1 = -15$ MPa, $\sigma_2 = 6$ MPa, $\sigma_3 = 15$ MPa. (b) $\sigma_1 = 11.824$ MPa, $\mathbf{n}^{(1)} = \pm(1, 0.462, 0.814)$.

4.22 $\tilde{\sigma} = 2$ and $[\sigma'] = \begin{bmatrix} 1 & 5 & 8 \\ 5 & -1 & 0 \\ 8 & 0 & 0 \end{bmatrix}$.

4.23 $\lambda_1' = \frac{2}{3}$, $\lambda_2' = \frac{5}{3}$, $\lambda_3' = -\frac{7}{3}$; $\hat{n}^{(1)} = -0.577\hat{\mathbf{e}}_1 + 0.577\hat{\mathbf{e}}_1 + 0.577\hat{\mathbf{e}}_3$.

4.24 $\lambda_1 = -13.5416$ MPa, $\lambda_2 = 8.0$ MPa, $\lambda_3 = 16.5416$ MPa.

4.25 $\lambda_1 = 6.856$, $\lambda_2 = -10.533$, $\lambda_3 = -3.323$.

4.26 $\sigma_1 = 25$ MPa, $\sigma_2 = 50$ MPa, $\sigma_3 = 75$ MPa;
$\hat{n}^{(1)} = \pm\left(\frac{3}{5}\hat{\mathbf{e}}_1 - \frac{4}{5}\hat{\mathbf{e}}_3\right)$, $\hat{n}^{(2)} = \pm\hat{\mathbf{e}}_2$, $\hat{n}^{(3)} = \pm\left(\frac{4}{5}\hat{\mathbf{e}}_1 + \frac{3}{5}\hat{\mathbf{e}}_3\right)$.

4.27 (a) $\rho\mathbf{f} = B\hat{\mathbf{e}}_2$. (c) $\sigma_{p1} = -\sqrt{2}AB$, $\sigma_{p2} = 0$, $\sigma_{p3} = \sqrt{2}AB$. (d) $(\sigma_{ns})_{\max} = \sqrt{2}AB$.

4.29 $\rho\mathbf{f} = 0$.

4.30 $\rho\mathbf{f} = -6x_2\,\hat{\mathbf{e}}_1 \times 10^6$ N/m^3.

4.31 Not satisfied.

4.32 $\rho\mathbf{f} = -4Bx_3\,\hat{\mathbf{e}}_3 \times 10^6$ N/m^3.

4.33 $\rho\mathbf{f} = -4\hat{\mathbf{e}}_3$.

4.34 $\rho\mathbf{f} = -4cx_3\,\hat{\mathbf{e}}_3$.

4.35 (c) $\mathbf{T} = AB\sigma_0\,\hat{\mathbf{E}}_3$ and $\tilde{\mathbf{T}} = \frac{AB}{C}\sigma_0\,\hat{\mathbf{E}}_2$.

4.36 (c) $\mathbf{T} = -AC\sigma_0\,\hat{\mathbf{e}}_2$ MPa, $\tilde{\mathbf{T}} = -\frac{AC}{B}\sigma_0\,\hat{\mathbf{e}}_2$MPa.

Chapter 5

5.1 First show that $\nabla\left(\frac{v^2}{2}\right) - \mathbf{v} \times \nabla \times \mathbf{v} = \mathbf{v} \cdot \nabla\mathbf{v}$.

5.5 Use Table 2.4.2 and replace $\mathbf{A}$ with $\rho\mathbf{v}$.

5.6 Use Table 2.4.2 and replace $\mathbf{A}$ with $\rho\mathbf{v}$, and note that $2\frac{A_R}{R} + \frac{\partial A_R}{\partial R} = \frac{1}{R^2}\frac{\partial(R^2 A_R)}{\partial R}$.

5.7 (a) Satisfies. (b) Satisfies.

5.8 $Q = \frac{b}{6}\left(3v_0 - c\right)$ m^3/(s.m).

5.9 $F_n = 5.118\,$N.

5.10 $F_n = 88.15\,$N, $Q_L = 0.01273$ m^3/s, and $Q_R = 0.00424$ m^3/s.

5.11 (a) $F = 24.12$ N. (b) $F = 12.06$ N.

5.12 $v(t) = \sqrt{\frac{g}{L}(x^2 - x_0^2)}$, $a(t) = \frac{g}{L}x(t)$ and $v(t_0) = \sqrt{\frac{g}{L}(L^2 - x_0^2)} \approx \sqrt{gL}$ when $L \gg x_0$.

5.17 Proving this identity requires the proof of the following identities: $\mathbf{v} \cdot \nabla\mathbf{v} = \nabla\left(\frac{v^2}{2}\right) - \mathbf{v} \times \nabla \times \mathbf{v}$ and $\nabla \times (\mathbf{A} \times \mathbf{B}) = \mathbf{B} \cdot \nabla\mathbf{A} - \mathbf{A} \cdot \nabla\mathbf{B} + \mathbf{A}\,\nabla \cdot \mathbf{B} - \mathbf{B}\,\nabla \cdot \mathbf{A}$.

5.18 $\rho f_1 = 0$, $\rho f_2 = 0$, and $\rho f_3 = -4abx_3$.

5.19 $\rho\mathbf{f} = -x_2\,\hat{\mathbf{e}}_1 + 2x_1\,\hat{\mathbf{e}}_2$.

5.20 Certain conditions have to be met on c_i.

5.21 Satisfied.

5.22 Satisfied only if $B + 2C = 0$.

5.23 $\sigma_{12} = -P\left(h^2 - x_2^2\right)/2I_3$ and $\sigma_{22} = 0$, where $I_3 = 2bh^3/3$.

5.24 $\sigma_{12} = -q_0x_1\left(h^2 - x_2^2\right)/2I_3$ and $\sigma_{22} = \frac{q_0x_2}{6I_3}\left(3h^2 - x_2^2\right) - \frac{q_0}{2b}$.

5.25 $c_3 = 0$, $c_2 + c_6 = 0$, $c_7 = 0$ (when $C_{10} \neq 0$); $c6 + 3c_4 = 0$; all other constants are arbitrary.

5.26 (a) No restrictions on c_i. (b) S_{rr} does not satisfy the equilibrium equations.

5.27 (a) $T = 0.15$ N-m. (b) When $T = 0$, $\omega_0 = 477.5$ rpm.

5.28 $\omega = 16.21$ rad/s $= 154.8$ rpm.

5.32 $\sigma : \mathbf{W} = 0$ because of the skew symmetry of $\mathbf{W}$.

Chapter 6

6.1 Note that
$$[R^\theta]^{-1} = [R^{-\theta}], \quad [T^\theta]^{-1} = [T^{-\theta}], \quad [T^\theta] = [R^{-\theta}]^{\mathrm{T}}, \quad [T^{-\theta}] = [R^\theta]^{\mathrm{T}}.$$

6.2 Carrying out the matrix multiplications indicated, we obtain (only selective coefficients are given here)

$$\bar{S}_{11} = S_{11}\cos^4\theta - 2S_{16}\cos^3\theta\sin\theta + (2S_{12} + S_{66})\cos^2\theta\sin^2\theta$$
$$- 2S_{26}\cos\theta\sin^3\theta + S_{22}\sin^4\theta$$
$$\bar{S}_{12} = S_{12}\cos^4\theta + (S_{16} - S_{26})\cos^3\theta\sin\theta + (S_{11} + S_{22} - S_{66})\cos^2\theta\sin^2\theta$$
$$+ (S_{26} - S_{16})\cos\theta\sin^3\theta + S_{12}\sin^4\theta$$
$$\bar{S}_{22} = S_{22}\cos^4\theta + 2S_{26}\cos^3\theta\sin\theta + (2S_{12} + S_{66})\cos^2\theta\sin^2\theta$$
$$+ 2S_{16}\cos\theta\sin^3\theta + S_{11}\sin^4\theta$$
$$\bar{S}_{33} = S_{33}$$
$$\bar{S}_{66} = S_{66}(\cos^2\theta - \sin^2\theta)^2 + 4(S_{16} - S_{26})(\cos^2\theta - \sin^2\theta)\cos\theta\sin\theta$$
$$+ 4(S_{11} + S_{22} - 2S_{12})\cos^2\theta\sin^2\theta$$

$$\bar{S}_{44} = S_{44} \cos^2 \theta + 2S_{45} \cos \theta \sin \theta + S_{55} \sin^2 \theta$$
$$\bar{S}_{55} = S_{55} \cos^2 \theta + S_{44} \sin^2 \theta - 2S_{45} \cos \theta \sin \theta. \tag{1}$$

$$\bar{C}_{11} = C_{11} \cos^4 \theta - 4C_{16} \cos^3 \theta \sin \theta + 2(C_{12} + 2C_{66}) \cos^2 \theta \sin^2 \theta$$
$$\qquad - 4C_{26} \cos \theta \sin^3 \theta + C_{22} \sin^4 \theta$$
$$\bar{C}_{12} = C_{12} \cos^4 \theta + 2(C_{16} - C_{26}) \cos^3 \theta \sin \theta + (C_{11} + C_{22} - 4C_{66}) \cos^2 \theta \sin^2 \theta$$
$$\qquad + 2(C_{26} - C_{16}) \cos \theta \sin^3 \theta + C_{12} \sin^4 \theta$$
$$\bar{C}_{22} = C_{22} \cos^4 \theta + 4C_{26} \cos^3 \theta \sin \theta + 2(C_{12} + 2C_{66}) \cos^2 \theta \sin^2 \theta$$
$$\qquad + 4C_{16} \cos \theta \sin^3 \theta + C_{11} \sin^4 \theta$$
$$\bar{C}_{33} = C_{33}$$
$$\bar{C}_{66} = 2(C_{16} - C_{26}) \cos^3 \theta \sin \theta + (C_{11} + C_{22} - 2C_{12} - 2C_{66}) \cos^2 \theta \sin^2 \theta$$
$$\qquad + 2(C_{26} - C_{16}) \cos \theta \sin^3 \theta + C_{66}(\cos^4 \theta + \sin^4 \theta)$$
$$\bar{C}_{44} = C_{44} \cos^2 \theta + C_{55} \sin^2 \theta + 2C_{45} \cos \theta \sin \theta$$
$$\bar{C}_{55} = C_{55} \cos^2 \theta + C_{44} \sin^2 \theta - 2C_{45} \cos \theta \sin \theta. \tag{2}$$

6.6 $\begin{Bmatrix} \sigma_{11} \\ \sigma_{22} \\ \sigma_{23} \\ \sigma_{13} \\ \sigma_{12} \end{Bmatrix} = 10^6 \begin{Bmatrix} 37.8 \\ 43.2 \\ 27.0 \\ 21.6 \\ 0.0 \\ 5.4 \end{Bmatrix}$ Pa.

6.7 $I_1 = 108$ MPa, $I_2 = 2,507.76$ MPa^2, $I_3 = 25,666.67$ MPa^3; $J_1 = 500 \times 10^{-6}$, $J_2 = 235 \times 10^{-9}$, $J_3 = -32 \times 10^{-12}$.

6.8 $I_1 = 78.8$ MPa, $I_2 = 1,062.89$ MPa^2, $I_3 = 17,368.75$ MPa^3.

6.9 $J_1 = 66.65 \times 10^{-6}$, $J_2 = 63,883.2 \times 10^{-12}$, $J_3 = 244,236 \times 10^{-18}$.

6.10 $\begin{Bmatrix} \sigma_1 \\ \sigma_2 \\ \sigma_6 \end{Bmatrix} = \frac{1}{1 - \nu_{12}\nu_{21}} \begin{bmatrix} E_1 & \nu_{12}E_2 & 0 \\ \nu_{21}E_1 & E_2 & 0 \\ 0 & 0 & (1 - \nu_{12}\nu_{21})G_{12} \end{bmatrix} \begin{Bmatrix} \varepsilon_1 \\ \varepsilon_2 \\ \varepsilon_6 \end{Bmatrix}.$

6.11 $\mathbf{S} = \frac{\partial \Psi}{\partial \mathbf{E}} = \lambda \, (\text{tr } \mathbf{E})\mathbf{I} + 2\mu\mathbf{E}.$

6.12 $\Psi(\boldsymbol{\sigma}) = \frac{1}{2E} \left[(1 + \nu)\text{tr}(\boldsymbol{\sigma}^2) - \nu(\text{tr } \boldsymbol{\sigma})^2 \right].$

6.13 $E > 0, \quad G > 0, \quad K > 0, \quad -1 < \nu < 0.5.$

6.17 $\sigma_{rr} = (2\mu + \lambda)\frac{dU}{dr} + \lambda\frac{U}{r}, \quad \sigma_{\theta\theta} = (2\mu + \lambda)\frac{U}{r} + \lambda\frac{dU}{dr}, \quad \sigma_{zz} = \lambda\left(\frac{dU}{dr} + \frac{U}{r}\right).$

6.18 $\sigma_{RR} = (2\mu + \lambda)\frac{dU}{dR} + 2\lambda\frac{U}{R}, \quad \sigma_{\phi\phi} = 2(\mu + \lambda)\frac{U}{R} + \lambda\frac{dU}{dR}, \quad \sigma_{\theta\theta} = \sigma_{\phi\phi}.$

6.20 $\tau_{11} = 0, \quad \tau_{22} = \frac{2\mu k}{1 + kt}, \quad \tau_{12} = \mu\left(\frac{4tk}{(1+kt)^2}x_2\right).$

6.23 Note that $\boldsymbol{\nabla} \cdot \mathbf{v} = \frac{1}{\rho}\frac{D\rho}{Dt}.$

6.25 First show that $\frac{D}{Dt}(\rho\mathbf{v}) = \rho\frac{D\mathbf{v}}{Dt}.$

6.29 A direct result from the solution to Problem **6.28**.

6.32 Begin with $\Psi = e - \eta\theta = \Psi(\theta, \boldsymbol{\varepsilon})$ and obtain $\dot{\Psi} = -\eta\dot{\theta} + \frac{1}{\rho}\sigma_{ij}\dot{\varepsilon}_{ij}.$

Chapter 7

7.1 $s_{ij} = 2\mu \, e_{ij}.$

7.3 For $\mathbf{u} = \frac{\alpha}{2}(x_2\,\hat{\mathbf{e}}_1 + x_1\,\hat{\mathbf{e}}_2)$, (a) $E_{11} = \frac{1}{2}\left(\frac{\alpha}{2}\right)^2$, $E_{22} = \frac{1}{2}\left(\frac{\alpha}{2}\right)^2$, $2E_{12} = \alpha.$

7.4 $\varepsilon_n = \boldsymbol{\varepsilon} \cdot \hat{\mathbf{n}} = \frac{\alpha}{2\sqrt{2}}(\hat{\mathbf{e}}_1 + \hat{\mathbf{e}}_2).$

7.5 $\varepsilon_{rr} = A, \; \varepsilon_{\theta\theta} = A, \; \varepsilon_{z\theta} = \frac{1}{2}\left(Br + C\frac{1}{r}\cos\theta\right).$

7.6 $\{\bar{u}\} = \left\{ \begin{array}{c} \frac{4}{3} \\ 0 \\ \frac{20}{3\sqrt{2}} \end{array} \right\}.$

7.7 We obtain

$$\mu \left[\frac{1}{r} \frac{\partial}{\partial r} \left(r \frac{\partial u_r}{\partial r} \right) + \frac{1}{r^2} \frac{\partial^2 u_r}{\partial \theta^2} - \frac{u_r}{r^2} - \frac{2}{r^2} \frac{\partial u_\theta}{\partial \theta} + \frac{\partial^2 u_r}{\partial z^2} \right]$$
$$+ (\lambda + \mu) \frac{\partial}{\partial r} \left\{ \frac{1}{r} \left[\frac{\partial(r u_r)}{\partial r} + \frac{\partial u_\theta}{\partial \theta} + r \frac{\partial u_z}{\partial z} \right] \right\} + \rho_0 f_r = \rho_0 \frac{\partial^2 u_r}{\partial t^2},$$

$$\mu \left[\frac{1}{r} \frac{\partial}{\partial r} \left(r \frac{\partial u_\theta}{\partial r} \right) + \frac{1}{r^2} \frac{\partial^2 u_\theta}{\partial \theta^2} + \frac{2}{r^2} \frac{\partial u_r}{\partial \theta} - \frac{1}{r^2} u_\theta + \frac{\partial^2 u_\theta}{\partial z^2} \right]$$
$$+ (\lambda + \mu) \frac{1}{r} \frac{\partial}{\partial \theta} \left\{ \frac{1}{r} \left[\frac{\partial(r u_r)}{\partial r} + \frac{\partial u_\theta}{\partial \theta} + r \frac{\partial u_z}{\partial z} \right] \right\} + \rho_0 f_\theta = \rho_0 \frac{\partial^2 u_\theta}{\partial t^2},$$

$$\mu \left[\frac{1}{r} \frac{\partial}{\partial r} \left(r \frac{\partial u_z}{\partial r} \right) + \frac{1}{r^2} \frac{\partial^2 u_z}{\partial \theta^2} + \frac{\partial^2 u_z}{\partial z^2} \right]$$
$$+ (\lambda + \mu) \frac{\partial}{\partial z} \left\{ \frac{1}{r} \left[\frac{\partial(r u_r)}{\partial r} + \frac{\partial u_\theta}{\partial \theta} + r \frac{\partial u_z}{\partial z} \right] \right\} + \rho_0 f_z = \rho_0 \frac{\partial^2 u_z}{\partial t^2}.$$

7.8 $\sigma_{11} = 96.88$ MPa, $\sigma_{22} = 64.597$ MPa, $\sigma_{33} = 48.443$ MPa, $\sigma_{12} = 4.02$ MPa.

7.9 $c_6 = -c_2 = -3c_4$, and c_1 and c_5 are arbitrary.

7.10 $U(\varepsilon) = \frac{1}{2} \int_\Omega (2\mu \, \varepsilon_{ij}\varepsilon_{ij} + \lambda \varepsilon_{ii}\varepsilon_{jj}) \, d\mathbf{x}$, $U\sigma = \frac{1}{2} \int_\Omega \frac{1}{E} [(1+\nu)\sigma_{ij}\sigma_{ij} - \nu\sigma_{ii}\sigma_{jj}] \, d\mathbf{x}.$

7.11 $U = \frac{5}{8} k u_c^2.$

7.12 $U = \frac{3}{8} k u_c^3.$

7.14 (a) The boundary conditions are $u_1 = u_2 = 0$ on $x_2 = 0$; $t_n = 0$, $t_s = \tau$ on plane with normal $\hat{\mathbf{n}} = (\cos\theta \, \hat{\mathbf{e}}_1 - \sin\theta \, \hat{\mathbf{e}}_2)$, $t_n = 0$, $t_s = \tau$ on plane with normal $\hat{\mathbf{n}} = \hat{\mathbf{e}}_2$, $t_n = 0$, $t_s = \tau$ on plane with normal $\hat{\mathbf{n}} = -(\sin\theta \, \hat{\mathbf{e}}_1 + \cos\theta \, \hat{\mathbf{e}}_2)$. The boundary value problem is of type III.

7.15 $\theta(0) = \frac{M_0 L}{EI}.$

7.16 $w(a) = \frac{F_0}{3EI}(L-a)^3.$

7.17 $v(\frac{L}{2}) = -\left(\frac{5F_0 L^3}{48EI} + \frac{17q_0 L^4}{384EI} \right).$

7.19 $v_c = \frac{q_0 a^4}{64D} \left(\frac{5+\nu}{1+\nu} \right).$

7.20 $v_c = \frac{q_0 a^4}{(1+\nu)D} \left(\frac{5+\nu}{64} - \frac{6+\nu}{150} \right).$

7.21 $v_c = \frac{43}{4800} \frac{q_0 a^4}{D}.$

7.22 $v_{cb} \frac{Q_0 b^2}{16\pi D} \left(2\log\frac{b}{a} + \frac{a^2}{b^2} - 1 \right).$

7.25 Note that $-\alpha r = \frac{d^2 U}{dr^2} + \frac{1}{r} \frac{dU}{dr} - \frac{U}{r^2} = \frac{d}{dr}\left(\frac{dU}{dr} + \frac{U}{r} \right) = \frac{d}{dr}\left[\frac{1}{r} \frac{d}{dr}(rU) \right].$

7.28 The displacement field is

$$u_1 = -\frac{M_0}{EI} x_1 x_3 - c_4 x_3 + c_1 x_2 + c_2,$$

$$u_2 = \frac{\nu M_0}{EI} x_2 x_3 + c_5 x_3 - c_1 x_1 + c_3,$$

$$u_3 = \frac{M_0}{2EI} [x_1^2 + \nu(x_3^2 - x_2^2)] + c_4 x_1 + c_5 x_2 + c_6.$$

The constants c_i are to be determined using the displacement boundary conditions.

7.29 (a) At $z = 0$: $u_r = u_\theta = u_z = 0$;   At $z = h$: $u_r = u_\theta = 0$, $u_z = -\delta$;
      (b) At $z = 0$: $\sigma_{zr} = \sigma_{z\theta} = 0$, $u_z = 0$;   At $z = h$: $\sigma_{zr} = \sigma_{z\theta} = 0$, $u_z = -\delta$.

7.30 $u_z(r) = -\frac{\rho g a^2}{4\mu} \left(1 - \frac{r^2}{a^2} \right)$, $\sigma_{\theta z} = 0$, $\sigma_{zr} = \frac{\rho g}{2} r.$

7.31 $u_R(R) = -\frac{b^3 pR}{3Kb^3 + 4\mu a^3}\left(1 - \frac{a^3}{R^3}\right),$ $\sigma_{RR}(R) = -\left[\frac{1 + 2\alpha(a^3/R^3)}{1+\beta}\right]p,$

$\alpha = \frac{2\mu}{3K},$ $\beta = 2\alpha\frac{a^3}{b^3},$ $\sigma_{\theta\theta}(R) = \sigma_{\phi\phi}(R) = -\left[\frac{1 - \alpha(a^3/R^3)}{1+\beta}\right]p.$

7.32 We have

$$u_R^{(1)}(R) = A_1 R + \frac{B_1}{R^2}, \quad \sigma_{RR}^{(1)} = 3K_1 A_1 - \frac{4\mu_1}{R^3}B_1 \quad \text{(core)}$$

$$u_R^{(2)}(R) = A_2 R + \frac{B_2}{R^2}, \quad \sigma_{RR}^{(2)} = 3K_2 A_2 - \frac{4\mu_2}{R^3}B_2 \quad \text{(shell)}$$

where $3K_i = 2\mu_i + 3\lambda_i$ $(i = 1, 2)$. The four constants can be determined using the following four conditions: $\sigma_{RR}^{(2)}(b) = -p;$ $\sigma_{RR}^{(1)}(a) = \sigma_{RR}^{(2)}(a);$ $u_R^{(1)}(a) = u_R^{(2)}(a);$ $B_1 = 0$ by symmetry.

7.33 $u_\theta(r) = \frac{\tau_0 b^2}{2\mu a}\left(\frac{r}{a} - \frac{a}{r}\right),$ $\sigma_{r\theta} = \frac{b^2 \tau_0}{r^2}.$

7.34 $\Phi(x, y) = \frac{1}{6}c(-3h^2 xy + xy^3).$

7.36 It is nothing but a centroidally loaded uniaxial member.

7.38 $\sigma_{xx} = 2D\left(3x^2 y - 2y^3\right),$ $\sigma_{yy} = 2Dy^3,$ $\sigma_{xy} = -6Dxy^2.$

7.39 The stress field is

$$\sigma_{xx} = \frac{3q_0}{10}\left(\frac{y}{b} + \frac{5a^2}{2b^2}\frac{x^2}{a^2}\frac{y}{b} - \frac{5}{3}\frac{y^3}{b^3}\right), \quad \sigma_{yy} = \frac{q_0}{4}\left(-2 - 3\frac{y}{b} + \frac{y^3}{b^3}\right), \quad \sigma_{xy} = \frac{3q_0 a}{4b}\frac{x}{a}\left(1 - \frac{y^2}{b^2}\right).$$

7.42 The stresses are given by

$$\sigma_{xx} = \frac{\tau_0}{4}\left(-\frac{2x}{b} - \frac{6xy}{b^2} + \frac{2a}{b} + \frac{6ay}{b^2}\right), \quad \sigma_{yy} = 0,$$

$$\sigma_{xy} = -\frac{\tau_0}{4}\left(1 - \frac{2y}{b} - \frac{3y^2}{b^2}\right).$$

7.43 The stresses are given by

$$\sigma_{xx} = \frac{q_0}{20}\frac{x}{L}\frac{y}{b}\left(-6 - 5\frac{x^2}{L^2}\frac{L^2}{b^2} + 10\frac{y^2}{b^2}\right),$$

$$\sigma_{yy} = \frac{q_0}{4}\frac{x}{L}\left(2 + 3\frac{y}{b} + \frac{y^3}{b^3}\right)$$

$$\sigma_{xy} = -\frac{q_0}{40}\left(\frac{b}{L} + 15\frac{L}{b}\frac{x^2}{L^2} - 6\frac{b}{L}\frac{y^2}{b^2} - 15\frac{L}{b}\frac{x^2}{L^2}\frac{y^2}{b^2} + 5\frac{b}{L}\frac{y^4}{b^4}\right).$$

7.44 The stresses are

$$\sigma_{rr} = 2A\left(r + \frac{a^2 b^2}{r^3} - \frac{a^2 + b^2}{r}\right)\sin\theta,$$

$$\sigma_{\theta\theta} = 2A\left(3r - \frac{a^2 b^2}{r^3} - \frac{a^2 + b^2}{r}\right)\sin\theta,$$

$$\sigma_{r\theta} = -2A\left(r + \frac{a^2 b^2}{r^3} - \frac{a^2 + b^2}{r}\right)\cos\theta.$$

7.45 $\sigma_{rr} = -\frac{2f_0}{\pi r}\sin\theta,$ $\sigma_{\theta\theta} = 0,$ $\sigma_{r\theta} = 0.$

7.46 $F_x = \int_\Omega \sigma_{xz}\, dx\, dy = \mu\theta \int_\Omega \left(\frac{\partial\psi}{\partial x} - y\right) dx\, dy.$

7.47 The stresses are $\sigma_{xz} = -\frac{2\mu\theta a^2}{a^2 + b^2}y,$ $\sigma_{yz} = \frac{2\mu\theta b^2}{a^2 + b^2}x.$

7.48 The stresses are

$$\sigma_{31} = \frac{\mu\theta}{a}x_2(x_1 - a), \quad \sigma_{32} = \frac{\mu\theta}{2a}\left(x_1^2 + 2ax_1 - x_2^2\right).$$

The angle of twist is $\theta = \frac{5\sqrt{3}T}{27\mu a^4}.$

7.49 No exact solution can be found (without going through the separation of variables technique).

7.50 $\sigma_{z\alpha} = \mu\theta\left(\frac{1}{r}\frac{\partial\psi}{\partial\alpha} + r\right)$, $\sigma_{zr} = \mu\theta\frac{\partial\psi}{\partial r}$.

7.51 The Euler equations are

$$\delta v: \qquad -\frac{d}{dx}\left[GA\left(\phi + \frac{dv}{dx}\right)\right] - q = 0$$

$$\delta\phi: \qquad -\frac{d}{dx}\left(EI\frac{d\phi}{dx}\right) + GA\left(\phi + \frac{dv}{dx}\right) = 0.$$

7.52 Take the first variation of the functional given in Problem **7.51** to identify the bilinear and linear forms.

7.53 $-T\left(\frac{\partial^2 u}{\partial x_1^2} + \frac{\partial^2 u}{\partial x_2^2}\right) - f = 0$ in Ω

7.54 $v(0) = -\frac{q_0 L^4}{24EI + kL^3}$.

7.55 $\frac{2EI}{L^3}\begin{bmatrix} 12 & 0 & -3L \\ 0 & 4L^2 & L^2 \\ -3L & L^2 & 2L^2 \end{bmatrix}\begin{Bmatrix} \Delta_3^{(1)} \\ \Delta_4^{(1)} \\ \Delta_4^{(2)} \end{Bmatrix} = \begin{Bmatrix} -F_0 \\ 0 \\ 0 \end{Bmatrix}$.

7.56 $\frac{2EI}{L^3}\begin{bmatrix} 4L^2 & 3L & L^2 \\ 3L & 6 & 3L \\ L^2 & 3L & 2L^2 \end{bmatrix}\begin{Bmatrix} \Delta_4^{(1)} \\ \Delta_3^{(2)} \\ \Delta_4^{(2)} \end{Bmatrix} = -\frac{q_0 L}{12}\begin{Bmatrix} L \\ 0 \\ 0 \end{Bmatrix} + \begin{Bmatrix} 0 \\ -F_0 \\ 0 \end{Bmatrix}$.

7.58 (a) $c_1 = \frac{q_0 L^3}{12\Delta}\left(6EIL^3 + \frac{kL^7}{420}\right)$ and $c_2 = 0$.

7.59 $\Pi(\Psi) = \frac{1}{2G}\int_\Omega\left[\left(\frac{\partial\Psi}{\partial y}\right)^2 + \left(-\frac{\partial\Psi}{\partial y}\right)^2\right]dx\,dy - 2\theta\int_\Omega\Psi\,dx\,dy$.

7.60 $U_1 = \frac{3}{8}f_0(x-1)(y-1)$ or $U_1 = \frac{5f_0}{16}(x^2-1)(y^2-1)$.

7.62 $\int_0^T(U-K)dt = \frac{1}{2}\int_0^T\left(\int_\Omega f_i u_i\,dx + \oint_\Gamma t_i u_i\,ds\right)dt$.

7.63 $L = \frac{1}{2}m_1\left[l^2\dot\theta^2 + \dot x^2 - 2l\dot x\dot\theta\sin\theta\right] + \frac{1}{2}m_2\dot x^2 + m_1 g(x - l\cos\theta) + m_2 gx + \frac{1}{2}k(x+h)^2$, where h is the elongation in the spring due to the masses $h = \frac{g}{k}(m_1 + m_2)$.

7.64 $\ddot x - \frac{g}{l}x = 0$.

7.65 $\frac{\partial}{\partial t}\left(\rho A\frac{\partial v}{\partial t}\right) + \frac{\partial^2}{\partial x^2}\left(EI\frac{\partial^2 v}{\partial x^2}\right) - \frac{\partial^2}{\partial x\partial t}\left(\rho I\frac{\partial^2 v}{\partial x\partial t}\right) = q$.

7.66 The Euler–Lagrange equations are

$$m(\ddot x + l\ddot\theta\cos\theta - l\dot\theta^2\sin\theta) + kx = F,$$

$$m\left[l\ddot x\cos\theta + (l^2 + \Omega^2)\ddot\theta\right] + mgl\sin\theta = 2aF\cos\theta.$$

7.67 The Euler–Lagrange equations are

$$\delta u: \qquad -\frac{\partial N_{xx}}{\partial x} - f + \frac{\partial}{\partial t}\left(m_0\frac{\partial u}{\partial t}\right) = 0,$$

$$\delta v: \qquad -\frac{\partial Q_x}{\partial x} - q + \frac{\partial}{\partial t}\left(m_0\frac{\partial v}{\partial t}\right) = 0,$$

$$\delta\phi: \qquad -\frac{\partial M_{xx}}{\partial x} + Q_x + \frac{\partial}{\partial t}\left(m_2\frac{\partial\phi}{\partial t}\right) = 0.$$

7.68 The Euler–Lagrange equations are

$$\delta u: \qquad \frac{\partial N_{xx}}{\partial x} = I_0\frac{\partial^2 u}{\partial t^2}, \tag{8}$$

$$\delta v: \qquad \frac{\partial\bar Q_x}{\partial x} + c_1\frac{\partial^2 P_{xx}}{\partial x^2} + q = I_0\frac{\partial^2 v}{\partial t^2} + c_1\left(J_4\frac{\partial^3\phi}{\partial x\partial t^2} - c_1 I_6\frac{\partial^4 v}{\partial x^2\partial t^2}\right), \tag{9}$$

$$\delta\phi: \qquad \frac{\partial\bar M_{xx}}{\partial x} - \bar Q_x = K_2\frac{\partial^2\phi}{\partial t^2} - c_1 J_4\frac{\partial^3 v}{\partial x\partial t^2}. \tag{10}$$

7.69 $\omega_1 = \frac{2.038}{L}\sqrt{\frac{E}{\rho}}, \quad \omega_2 = \frac{6.206}{L}\sqrt{\frac{E}{\rho}}.$

7.70 Use $\phi_1(x,y) = xy(x+y-1).$

Chapter 8

8.2 (a) The pressure at the top of the sea lab is $p = 1.2\text{MN/m}^2.$

8.3 $\rho = 1.02 \text{ kg/m}^3.$

8.4 $p = p_0\left(1 + \frac{mx_3}{\theta_0}\right)^{-g/mR}, \quad \rho = \rho_0\left(1 + \frac{mx_3}{\theta_0}\right)^{-g/mR}.$

8.5 $p(y) = p_0 + \rho g h \cos\alpha\left(1 - \frac{y}{h}\right), \quad U(y) = \frac{\rho g h^2 \sin\alpha}{2\mu}\left(2\frac{y}{h} - \frac{y^2}{h^2}\right).$

8.7 The shear stress is given by

$$\tau_{rz} = -\left(\frac{d\bar{p}}{dz}\frac{r}{2} + \frac{1}{r}c_1\right) = -\frac{d\bar{p}}{dz}\frac{R_0}{4}\left[2\left(\frac{r}{R_0}\right) + (1-\alpha^2)\frac{1}{\log\alpha}\left(\frac{R_0}{r}\right)\right].$$

8.8 The velocity field is

$$v_\theta(r) = \frac{\Omega R_1^2}{R_1^2 - R_2^2}\left(r - \frac{R_2^2}{r}\right).$$

If $R_1 = R_0$ and $R_2 = \alpha R_0$ with $0 < \alpha < 1$, we have

$$v_\theta(r) = \frac{\Omega R_0}{1 - \alpha^2}\left(\frac{r}{R_0} - \alpha^2\frac{R_0}{r}\right).$$

The shear stress distribution is given by $\tau_{r\theta} = -2\mu\Omega\frac{\alpha^2}{1-\alpha^2}\left(\frac{R_0}{r}\right)^2.$

8.9 The continuity equation simplifies to $R^2 v_R = c_1$, a constant.

8.10 $p = -\rho g z + \frac{1}{2}\rho\Omega^2 r^2 + c$, where $c = p_0 + \rho g z_0.$

8.11 $f(\eta) = 1 - \frac{2}{\sqrt{\pi}}\int_0^\eta e^{-\xi^2}\,d\xi.$

8.12 $v_x(y,t) = U_0 e^{-\eta}\cos(nt - \eta).$

8.15 Use the definition of the shear stress $\tau_{R\phi} = \mu\left[r\frac{\partial}{\partial R}\left(\frac{v_\phi}{R}\right) + \frac{1}{R}\frac{\partial v_R}{\partial\phi}\right]$ and the ϕ-momentum equation to obtain the required result.

8.16 Take the curl of the equation of motion and then use the result of Problem **5.17**.

8.17 For Part (a), make use of the result of Problem **5.16**; for Part (b) use the fact that the curl of the gradient of a function is zero.

8.18 $v_2 = 9.9 \text{ m/s}$ and $Q = 0.01945 \text{ m}^3/\text{s} = 19.45 \text{ liters/s}.$

8.19 $-H_{\text{net}} = 5.3665 \text{ N}\cdot\text{m/kg}.$

8.21 $\rho c\frac{D\theta}{Dt} = \nabla\cdot(\mathbf{k}\cdot\nabla\theta) + \rho r_h.$

8.22 $-\frac{d}{dr}(rq_r) + r\rho r_h = 0.$

8.23 $T(r) = T_0 + \frac{\rho r_h R_0^2}{4k}\left[1 - \left(\frac{r}{R_0}\right)^2\right].$

8.24 $\theta(x,t) = \sum_{n=1}^\infty B_n \sin\lambda_n x\, e^{-\alpha\lambda_n^2 t}$, with $B_n = \frac{2}{L}\int_0^L f(x)\sin\lambda_n x\,dx.$

8.25 $T(x,t) = 4T_0\sum_{n=1}^\infty \frac{\sin\lambda_n}{\lambda_n^3}\sin\lambda_n x\, e^{-\alpha\lambda_n^2 t}.$

8.26 $\frac{T - T_i}{T_0 - T_i} = \frac{b}{b-a}\left(1 - \frac{a}{R}\right).$

8.27 $T(x,y) = \sum_{n=1}^\infty A_n \frac{\cosh\lambda_n(b-y)}{\cosh\lambda_n b}\cos\lambda_n x$, with $A_n = \frac{2}{a}\int_0^a f(x)\cos\lambda_n x\,dx.$

8.28 $T(r) = T_0 - \frac{\mu\alpha^2 R_0^3}{9k}\left[1 - \left(\frac{r}{R_0}\right)^3\right].$

8.29 $v_y(x) = \frac{\rho_r\beta_r g a^2(T_2 - T_1)}{12\mu}\left[\left(\frac{x}{a}\right)^3 - \left(\frac{x}{a}\right)\right].$

8.30 $T(y) = A\int_{-h/2}^{h/2}\frac{1}{k(y)}dy + B$, where the constants A and B are to be determined using the boundary conditions at $y = h/2$ and $y = -h/2.$

Chapter 9

9.1 $-2H(t) + 2.5e^{-t} + 0.5e^{-3t}$.

9.2 The result is given in Eq. (14) of Example 9.4.6.

9.3 $J(t) = \frac{1}{k_1} - \frac{k_2}{k_1(k_1+k_2)}e^{-t/\tau}$, $Y(t) = k_1 + k_2 e^{-t/\tau}$.

9.4 $J(t) = \left[\frac{t}{\eta_1+\eta_2} + \frac{1}{G_2}\left(\frac{\eta_2}{\eta_1+\eta_2}\right)^2\left(1 - e^{-\alpha_2 t}\right)\right]$, $Y(t) = \eta_1\delta(t) + G_2 e^{-t/\tau_2}$, $\tau_2 = \frac{1}{\alpha_1} = \frac{\eta_2}{G_2}$.

9.5 $Y(t) = \frac{k_1 k_2}{k_1+k_2}\left(1 - e^{-\lambda t}\right) + k_1 e^{-\lambda t}$, $\lambda = \frac{k_1+k_2}{\eta}$.

9.6 $q_1\dot{\varepsilon} + q_2\ddot{\varepsilon} = p_0\sigma + p_1\dot{\sigma} + p_2\ddot{\sigma}$, where $p_0 = \frac{k_1}{\mu_1\mu_2}$, $p_1 = \frac{k_1}{k_2\mu_1} + \frac{1}{\mu_1} + \frac{1}{\mu_2}$, $p_2 = \frac{1}{k_2}$, $q_1 = \frac{k_1}{\mu_1}$ $q_2 = 1$.

9.7 $Y(t) = \frac{k_1 k_2 \mu_2}{\lambda_1 - \lambda_2}\left[(\lambda_1 - \alpha)e^{-\lambda_1 t} - (\lambda_2 - \alpha)e^{-\lambda_2 t}\right]$.

9.8 $q_0\varepsilon + q_1\dot{\varepsilon} + q_2\ddot{\varepsilon} = p_0\sigma + p_1\dot{\sigma}$, where $p_0 = \frac{1}{\eta_2}$, $p_1 = \frac{1}{k_2}$, $q_0 = \frac{k_1}{\eta_2}$, $q_1 = 1 + \frac{k_1}{k_2} + \frac{\eta_1}{\eta_2}$, $q_2 = \frac{\eta_1}{k_2}$.

9.9 $p_0\sigma + p_1\dot{\sigma} = q_0\varepsilon + q_1\dot{\varepsilon} + q_2\ddot{\varepsilon}$, where $p_0 = k_1 + k_2$, $p_1 = \eta_1 + \eta_2$, $q_0 = k_1 k_2$, $q_1 = k_1\eta_2 + k_2\eta_1$, $q_2 = \eta_1\eta_2$.

9.10 The creep compliance is

$$J(t) = \frac{1}{q_2}\left\{p_0\left[\frac{1}{\alpha\beta} - \frac{e^{-\alpha t}}{\alpha(\beta - \alpha)} + \frac{e^{-\beta t}}{\beta(\beta - \alpha)}\right]\right.$$

$$\left. + p_1\left[\frac{e^{-\alpha t}}{(\beta - \alpha)} - \frac{e^{-\beta t}}{(\beta - \alpha)}\right] + p_2\left[-\frac{\alpha e^{-\alpha t}}{(\beta - \alpha)} + \frac{\beta e^{-\beta t}}{(\beta - \alpha)}\right]\right\}.$$

The relaxation modulus is $Y(t) = k_1 + k_2 e^{-\alpha t} + \eta_1\delta(t)$.

9.11 $\sigma(t) = \left[k_1 + k_2 e^{-\alpha t} + \eta_1\delta(t)\right]\varepsilon_0 + \varepsilon_0\left[tk_1 + \frac{k_2}{\alpha}\left(1 - e^{-\alpha t}\right) + \eta_1 H(t)\right]$.

9.12 $Y(t) = k_1 + k_2 e^{-t/\tau}$, $\tau = \frac{\eta}{k_2}$.

9.13 $\bar{E}(s) = \frac{9\bar{K}(s)\bar{G}(s)}{3\bar{K}(s)+\bar{G}(s)}$, $s\bar{\nu}(s) = \frac{3\bar{K}(s)-2\bar{G}(s)}{2[3\bar{K}(s)+\bar{G}(s)]}$.

9.14 $\varepsilon(t) = \sigma_1\left(\frac{t}{k_1} + \frac{1}{k_2}e^{-t/\tau}\right)$, for $t > t_0$.

9.15 (a) $2G(t) = 2G_0\left[H(t) + \tau\delta(t)\right]$. (c) $\sigma'_{ij}(t) = 2G(t)\varepsilon'_{ij}(0) + 2\int_0^t G(t-t')\frac{d\varepsilon'_{ij}(t')}{dt'}\,dt'$.

9.16 $\sigma(t) = \ln(1 + t/C)$.

9.17 (a) $w^v(L,t) = \frac{P_0 L^3}{3E_0 I}\left[-\frac{B}{A}e^{-\frac{A\alpha}{E_0}t} + \frac{E_0}{A}H(t)\right]$. (b) $w^v(L,t) = \frac{P_0 L^3}{3E_0 I}e^{-\frac{A\alpha}{E_0}t}$.

9.18 $w^v(L,t) = \frac{P_0 L^3}{3I}\left[\frac{p_0}{q_0}H(t) + \left(\frac{q_0 p_1 - q_0 p_1}{q_1 q_0}\right)e^{-(q_1/p_1)t}\right]$.

9.19 $w^v(x,t) = \frac{q_0 L^4}{360I}\left(1 - \frac{x}{L}\right)\left[7 - 10\left(1 - \frac{x}{L}\right)^2 + 3\left(1 - \frac{x}{L}\right)^4\right]h(t)$, where

$h(t) = \frac{2\tau^2}{E_0}\left(1 - e^{-t/\tau}\right) + \frac{\tau^2}{E_0}\left(\frac{t}{\tau}\right)\left[\left(\frac{t}{\tau}\right) - 2\right]$. $\sigma(x,t) = -Ez\frac{\partial^2 w^v}{\partial x^2} = \frac{q_0 L^2 z}{60I}\left(1 - \frac{x}{L}\right)\frac{x}{L}h(t)$.

9.20 $P(t) = \frac{1}{2L}\left[\delta_0 E(t) + (\delta_1 - \delta_0)E(t - t_0)\right]$.

9.21 The Laplace transformed viscoelastic solutions for the displacements and stresses are obtained from

$$\bar{u}_r(r,s) = \bar{A}_i(s)r + \frac{\bar{B}_i(s)}{r^2},$$

$$\sigma_{rr}(r,s) = (2\mu + 3\lambda)\bar{A}_i(s) - \frac{4\mu}{r^3}\bar{B}_i(s),$$

$$\sigma_{\theta\theta}(r,s) = \sigma_{\phi\phi}(r,s) = [2s\bar{\mu}(s) + 3s\bar{\lambda}(s)]\bar{A}_i(s) + \frac{4s\bar{\mu}(s)}{r^3}\bar{B}_i(s),$$

where $\bar{A}_i(s)$ and $\bar{B}_i(s)$ are the same as A_i and B_i with ν and E replaced by $s\bar{\nu}(s)$ and $s\bar{E}(s)$, respectively.

9.22 $\bar{v}(x,t) = \frac{F_0 L^3}{\eta 6I}\left(2 - 3\frac{x}{L} + \frac{x^2}{L^2}\right)e^{-kt/\eta}$.

Index

Printed in the United States
By Bookmasters